CROSS-CULTURAL FAMILY RESEARCH AND PRACTICE

CROSS-CULTURAL FAMILY RESEARCH AND PRACTICE

Edited by

W. KIM HALFORD
School of Psychology, University of Queensland, St. Lucia, QLD, Australia

FONS VAN DE VIJVER†
Professor, School of Social and Behavioural Sciences, Tilburg University, Tilburg, the Netherlands

ELSEVIER

ACADEMIC PRESS
An imprint of Elsevier

† Deceased

Academic Press is an imprint of Elsevier
125 London Wall, London EC2Y 5AS, United Kingdom
525 B Street, Suite 1650, San Diego, CA 92101, United States
50 Hampshire Street, 5th Floor, Cambridge, MA 02139, United States
The Boulevard, Langford Lane, Kidlington, Oxford OX5 1GB, United Kingdom

Library of Congress Cataloging-in-Publication Data
A catalog record for this book is available from the Library of Congress

British Library Cataloguing-in-Publication Data
A catalogue record for this book is available from the British Library

ISBN 978-0-12-815493-9

For information on all Academic Press publications
visit our website at https://www.elsevier.com/books-and-journals

Publisher: Nikki Levy
Acquisitions Editor: Joslyn Chaiprasert-Paguio
Editorial Project Manager: Sara Pianavilla
Production Project Manager: Niranjan Bhaskaran
Cover Designer: Greg Harris

Typeset by SPi Global, India

Dedication

The topic of family spans my professional interests and also is deeply personal for me. My decades of practice as a couple and family therapist, my research career focused on families, and my personal life with my wife Barbara and our family, all convince me of the centrality of family life to humans. I dedicate this book to my dear, late friend and colleague Fons van de Vijver, with whom I began this book; and my family of wife Barbara, sons James and Chris, daughters-in-law Ai Ito Halford and Rocio del Carmen Ponce Reyes, and my grandchildren, Vera Halford-Ponce, Aya Halford-Ito, Nao Halford-Ito, and Kaz Halford-Ito.

W. Kim Halford

Contents

Part 3 Influences on family functioning

Part 5 Conclusion

Contributors

Haleama Al Sabbah
College of Natural and Health Sciences, Zayed University, Dubai, United Arab Emirates

Olivia Altamirano
Department of Psychology, University of Miami, Coral Gables, FL, United States

Ariel Bartolo-Kira
Department of Psychology, Florida Atlantic University, Boca Raton, FL, United States

David F. Bjorklund
Department of Psychology, Florida Atlantic University, Boca Raton, FL, United States

Marc H. Bornstein
Eunice Kennedy Shriver National Institute of Child Health and Human Development, Bethesda, MD; UNICEF, New York, NY, United States; Institute for Fiscal Studies, London, United Kingdom

Adam Bourne
Australian Research Centre in Sex Health and Society, La Trobe University, Melbourne, VIC, Australia

Pierre-André Chiappori
Department of Economics, Columbia University, New York, NY, United States

Elysia Choi
Department of Applied Psychology, New York University, New York, NY, United States

Brendan Churchill
School of Social and Political Sciences, The University of Melbourne, Parkville, VIC, Australia

Clare Clingain
Department of Applied Psychology, New York University, New York, NY, United States

Frank M. Dattilio
Department of Psychiatry, Harvard Medical School, Boston, MA, United States

Fons van de Vijver
Department of Culture Studies, Tilburg University, Tilburg, Netherlands

Norman B. Epstein
Department of Family Science, University of Maryland, College Park, MD, United States

Mariana K. Falconier
Department of Family Science, University of Maryland, College Park, MD, United States

Arlette Gautier
Department of Sociology, University of Brest, LABERS, Brest, France

W. Kim Halford
School of Psychology, University of Queensland, St. Lucia, QLD, Australia

Belinda Hewitt
School of Social and Political Sciences, The University of Melbourne, Parkville, VIC, Australia

Richard E. Heyman
Family Translational Research Group, New York University, New York, NY, United States

E. Jeffrey Hill
School of Family Life, Brigham Young University, Provo, UT, United States

Erin Kramer Holmes
School of Family Life, Brigham Young University, Provo, UT, United States

Joop T.V.M. de Jong
Amsterdam Institute for Social Science Research, University of Amsterdam; Amsterdam University Medical Center, Amsterdam, Netherlands; Boston University School of Medicine, Boston, MA, USA

Susan Kline
School of Communication, The Ohio State University, Columbus, OH, United States

Jennifer E. Lansford
Sanford School of Public Policy, Duke University, Durham, NC, United States

Ron J. Lesthaeghe
Sociology Research Center, Vrije Universiteit of Brussels, Brussels, Belgium

Daisy Lopez
Department of Psychology, University of Miami, Coral Gables, FL, United States

Anthony Lyons
Australian Research Centre in Sex Health and Society, La Trobe University, Melbourne, VIC, Australia

Cari McIlduff
Parenting and Family Support Centre, The University of Queensland, Brisbane, QLD, Australia

Merranda McLaughlin
Department of Psychology, University of Miami, Coral Gables, FL, United States

Danielle M. Mitnick
Family Translational Research Group, New York University, New York, NY, United States

José Alberto Molina
Department of Economics, University of Zaragoza, Zaragoza, Spain; IZA Institute of Labor Economics, Bonn, Germany

Alyson J. Myers
Department of Psychology, Florida Atlantic University, Boca Raton, FL, United States

Aleja M. Parsons
Family Translational Research Group, New York University, New York, NY, United States

Christopher A. Pepping
School of Psychology & Public Health, La Trobe University, Melbourne, VIC, Australia

Richard J. Petts
Department of Sociology, Ball State University, Muncie, IN, United States

Jennifer Power
Australian Research Centre in Sex Health and Society, La Trobe University, Melbourne, VIC, Australia

David L. Sam
Department of Psychosocial Science and Center for International Health, University of Bergen, Bergen, Norway

Matthew R. Sanders
Parenting and Family Support Centre, The University of Queensland, Brisbane, QLD, Australia

Mary B. Setrana
Centre for Migration Studies, University of Ghana, Legon-Accra, Ghana; Department of Sociology, University of Johannesburg, Johannesburg, South Africa

Esther J. Sin
Department of Applied Psychology, New York University, New York, NY, United States

Saumya Singh
Parenting and Family Support Centre, The University of Queensland, Brisbane, QLD, Australia

Meghna Singhal
Parenting and Family Support Centre, The University of Queensland, Brisbane, QLD, Australia

Selcuk R. Sirin
Department of Applied Psychology, New York University, New York, NY, United States

Amy M. Smith Slep
Family Translational Research Group, New York University, New York, NY, United States

Janet Stanley
Sustainable Society Institute, School of Design, University of Melbourne, Melbourne, VIC, Australia

Clare R. Thomas
School of Family Life, Brigham Young University, Provo, UT, United States

Karen M.T. Turner
Parenting and Family Support Centre, The University of Queensland, Brisbane, QLD, Australia

Amy Weisman de Mamani
Department of Psychology, University of Miami, Coral Gables, FL, United States

Shuangyue Zhang
Department of Communication Studies, Sam Houston State University, Huntsville, TX, United States

About the editors

W. Kim Halford is emeritus professor of clinical psychology at the University of Queensland in Brisbane, Australia. Kim has (co) authored more than 200 scientific publications, and his work has been quoted about 8000 times. He has researched predominantly on couples and family psychology, with a strong emphasis on the development and dissemination of innovative couple therapy programs, and cross-cultural perspectives. He is a fellow of the Association of Behavioral and Cognitive Therapy, the Australian Association of Cognitive and Behavior Therapy, and the Australian Psychological Society; and was awarded the President's Award for distinguished contribution to psychology from the Australian Psychological Society, and the Distinguished Career Award from the Australian Association of Cognitive and Behavior Therapy. Kim has served on the editorial boards of leading international journals including the *Journal of Consulting and Clinical Psychology*, *Journal of Family Psychology*, *Behavior Therapy*, and *Family Process*. He has published nine books, including one edited handbook. Kim is the author of the Couple CARE relationship education and therapy program.

The late **Fons van de Vijver** was a professor of cross-cultural psychology at Tilburg University in the Netherlands. He has (co) authored about 500 publications, mainly in the domain of cross-cultural psychology. His work has been quoted around 26,000 times. He was the 2013 recipient of the 2013 International Award of the American Psychological Association (for his contributions to international cooperation and to the advancement of knowledge of psychology), the 2014 recipient of the IAAP Fellows Award (of the International Association of Applied Psychology for contributions to applied psychology), and the 2014 recipient of the Sindbad Award of the Dutch Psychological Association (for his contributions to intercultural psychology). He is a former editor of the *Journal of Cross-Cultural Psychology* (2001–2007), and serves on the board of various journals. He is a former president of Division 2 (Assessment and Evaluation) of the International Association of Applied Psychology, the European Association of Psychological Assessment, and President of the International Association for Cross-Cultural Psychology.

A Personal Tribute to Fons van de Vijver (1952–2019)

Alphonsius Josephus Rachel (Fons) van de Vijver was born on October 4, 1952 in Koewacht, the Netherlands. He died suddenly on June 1, 2019, he was 66 years old. The village in which he grew up was close to the Netherlands-Belgian border and he commented to me that he observed differences in how the Belgians and Dutch lived, and felt a sense of otherness from the Belgians. He was fascinated by the cultural differences he observed, and was curious about how they came about. He retained that curiosity about culture all his life, and became a world leading cross-cultural psychologist.

Fons van de Vijver and I met in 2010, when he was visiting the University of Queensland (UQ) where I was a professor of clinical psychology. A PhD student of mine at the time, Danika Hiew, and I were working on a project studying cultural influences on the relationships of Chinese-Western intercultural couples. Danika somehow learned that Fons was visiting UQ, and that he had published widely in cross-cultural psychology, and had written a leading text on cross-cultural psychology methods. So we asked him to meet to discuss about Danika's PhD research. In that first meeting, I learned that this man was not just a fabulous scholar with an incredibly sharp mind, he also was warm, kind, and patient. A meeting or two later Fons agreed to become a co-supervisor of Danika's PhD.

Across the next decade Fons and I grew to be research collaborators and close friends. Fons and his wife Lona had a long-term plan to retire to Australia, and had bought a house in Maidenwell in rural Queensland. They visited Australia twice per year to engage with Maidenwell and Australia. Fons and Lona and my wife Barbara and I caught up regularly during their visits. Fons and I began collaborating on series of studies examining culture and couple relationships. In 2016, I spent most of a sabbatical leave with Fons at Tilburg University, where we wrote up papers from our shared work, played with ideas, went on bicycle rides together, and ate splendidly.

In mid-2018 Fons retired, as planned, from Tilburg University, and he and Lona moved to Maidenwell. At the end of 2018, I also retired from the University of Queensland. For both Fons and me, retirement meant we shed some onerous responsibilities, which gave each of us more free time to pursue research ideas and passions. Around that time Fons and I discussed working on the current book together. We developed a proposal, invited the chapter authors to contribute, negotiated publication with Elsevier Press, and thoroughly enjoyed shaping the ideas together.

I anticipated an enduring friendship in retirement with Fons. In addition to our shared scholarly interest, he and I were both keen on physical activity, and often did active things together. The last time we bicycled together was up near Lake Cootharbar in Queensland, where my wife and I have a cottage. I had to ask him to go slower up the hills as, despite my view that I was pretty fit, I struggled to keep up. When we swam out the front of our cottage in the lake, Fons would plough off into the distance, while I slowly stroked along in his wake. In the morning of May 31, 2019 we exchanged emails about some of the chapters we had read for the book, and both of us felt pleased about the progress. It came as a profound shock when Lona rang on the morning of June 1, 2019 to say that Fons had died of a cerebral hemorrhage, which had occurred while he was out riding near Maidenwell the previous afternoon.

Fons was an incredibly humble man, he never spoke of his many achievements. Over time, I gradually learned from others that he was not just a professor of cross-cultural psychology at Tilburg University in the Netherlands, but also held professorial appointments at North-West University in South Africa, the National Research University in Russia, and the University of Queensland in Australia. He (co)authored about 500 publications, primarily in the area of cross-cultural psychology, and was a world leader in the refinement of research methods in cross-cultural psychology. He was a superb mentor, and supervised more than 40 PhD students and 5 postdoctoral fellows, as well as provided guidance to countless colleagues. He had served as president of Division 2 (Assessment and Evaluation) of the International Association of Applied Psychology, president of the European Association of Psychological Assessment, president of the International Association for Cross-Cultural Psychology, and was long-term editor of the Journal of Cross-Cultural Psychology. In 2013 he received the International Award of the American Psychological Association for his contributions to international cooperation and to the advancement of knowledge of psychology.

He was also the 2014 recipient of the Fellows Award of the International Association of Applied Psychology.

All those formal recognitions and achievements capture only some aspects of Fons. He was curious and interested in lots of things, and fascinated by why people do the things they do. Fons was a great scholar of the highest integrity. He was the most superb collaborator, got things done efficiently, and seemed always to add an important idea to any conversation he was part of. He was patient and kind. He was incredibly productive, and even in retirement he worked every day. Yet he was interested in so many other things, the environment, world politics, rural Queensland, culture, food, and wine. Fons fully engaged with people around him, was a proud, active member of the Maidenwell Rural Volunteer Fire Brigade, and he and Lona served as an informal IT help desk for the local community.

Completing this book is bitter sweet. I am pleased to complete the work Fons and I began together, and cherish that I worked on this project with him. Like so many of his colleagues, friends, and former students, I miss him deeply. The outpouring of grief from so many people around the world on his passing was deeply moving; you can read some of this in the online tribute pages to him at his former university of Tilburg, and on several international and national organization's web pages. I wrote this tribute to celebrate the life of a very special human being.

Fons is survived by Lona Goudswaard, Fons' life partner and wife.

W. Kim Halford

Preface

There are various scientific disciplines that have studied families. Interestingly, many disciplines have also conducted culture-comparative studies. Almost all studies in the area are monodisciplinary. Consequently, there is a host of studies (and data) that address similar issues (such as the link between cultural factors and family functioning) from different scientific perspectives. As editors we firmly believe that there is value in integrating these perspectives. The current book is intended to achieve two key objectives. First, the book is aimed at presenting the state of the art in cross-cultural family perspectives, and as editors we have encouraged authors to employ an integrative, multidisciplinary perspective. Second, the book is intended to provide cultural theory and data to guide the practice of family therapies, and describe how family therapies can be delivered effectively across diverse cultures.

Cultural factors are relevant for the family in four ways. Firstly, a comparison of nation states provides information about how country differences in background factors (e.g., economic and religious differences) are functionally related to how families are structured and function. Secondly, modern societies are multicultural. In Australia around 30% of all new marriages have at least one partner who was not born in Australia; the number for the United States is 15% (with markedly higher percentages for some ethnic groups, such as Japanese Americans with over 40% of intercultural marriages); the number for Germany is 13%; the number for Japan is 3%. These acculturation issues, characteristic of people in the diaspora, have been studied extensively. The authors address the implications for families in the book. Thirdly, there is a recent interest in comparing acculturation issues across nation states so as to understand why immigrant groups do better in some countries than in others and why some receiving countries are more (or less) successful in dealing with immigrants. Finally, many different professionals work with families to try to help them have healthy, positive family lives. Family therapy originally developed in Western high-income countries, and the practice of family therapies needs to be culturally appropriate.

Two overarching question are pursued in the current book. First, what is universal to the form and functioning of all human families, and what is more culturally specific? Second, how can family therapy be provided in ways that are culturally safe and appropriate in diverse cultural contexts? The final chapter aims to integrate the perspectives developed in the various chapters, and describe the universalities and cultural specificities in human families, and the implications for family therapy practice.

Acknowledgments

I express heartfelt thanks to Don Baucom and Chris Pepping who read earlier drafts of Chapter 17, and Bill von Hippel who read an earlier draft of Chapter 21. As always, their scholarship and keen intellect provided very helpful guidance. For the friendship; the inspiring conversations about families, research, and practice; and making this academic life meaningful and fun; my special thanks to Don Baucom, Guy Bodenman, Dean Busby, Andy Christensen, Brian Doss, Kurt Hahlweg, Rick Heyman, Danika Hiew, Tom Holman, Somboon Jarukasemthawee, Jeff Larson, Patrick Leung, Howard Markman, Dan O'Leary, Chris Pepping, Jemima Petch, Kullaya Pisitungkagarn, Galena Rhoades, Matt Sanders, Tamara Sher, Amy Slep-Smith, Doug Snyder, and Scott Stanley. A special thanks to Bob Weiss, who helped me get a start, always encouraged me to savor outrageous fun, and introduced me to many of the people in the preceding list.

Special thanks to the Elsevier team, particularly Dennis McGonagle for the support in developing the book concept and getting the book under way, Sara Pianavilla for the editorial support during the writing process, and to Niranja Bhaskaran for editorial work in the production process.

Thank you Fons, for the friendship and for being a wonderful collaborator on this book. Your death still stuns me, leaves me aching, but I am so grateful for the time we had together, including working on the start of this book together.

Lastly, and most importantly, I thank my very special family: Barbara my wife of 40+ years with whom I built this life and who indulges my academic passions (obsessions?); our sons James and Chris, their wives Rocio and Ai-chan, and our grandchildren Aya, Nao, Kaz, and Vera. In the good times and the hard, we are there for each other, and that makes life much richer and more wondrous.

W. Kim Halford

PART 1
Introduction

CHAPTER 1

An introduction to families and culture: Research and practice

W. Kim Halford[a] and Fons van de Vijver[b]
[a]School of Psychology, University of Queensland, St. Lucia, QLD, Australia
[b]Department of Culture Studies, Tilburg University, Tilburg, Netherlands

Family is central to humans' lives. In the World Values Survey (2016) conducted across 60 countries, more than 85,000 adults rated the importance of a range of aspects of life. On a 4-point scale (1 = *very important* to 4 = *not at all important*), 92% of respondents rated family as very important, whereas work (63%), religion (50%), friends (47%), leisure (36%), and politics (15%) were much less frequently rated as very important. In the 60 countries surveyed, the modal response in every single country was to rate family life as very important. Across diverse cultures drawn from all regions of the world, the mean overall rated importance of family was 1.10 ($SD = 0.35$), which is very close to the maximum possible importance rating of 1.00. (NB: Lower scores correspond to higher rated importance.) This rating contrasts with the lower mean importance ratings for other significant aspects of life: work ($M = 1.51$, $SD = 0.79$), friends ($M = 1.67$, $SD = 0.73$), religion ($M = 1.89$, $SD = 1.05$), and leisure ($M = 1.89$, $SD = 0.83$).

The global pattern of attaching very high importance in life to family might suggest that there are cross-cultural human universals in the forms and functions of family life. Consistent with this possible universality, evolution has shaped the way humans reproduce and care for offspring, and family plays a central role in these functions (Buss, 2016). At the same time, there are large bodies of evidence documenting important cultural differences in some of the forms and functions of families. For example, in most Western cultures (e.g., United States, Australia, Germany), a committed couple relationship is seen by most adults as distinct from the families of origin of the partners, and there are certain matters only discussed within the boundaries of the couple relationship (Epstein, Chen, & Beyder-Kamjou, 2005). The satisfaction of the adult partners with that relationship is central to, and defines, the success of the couple relationship (Coontz, 2005; Halford, 2011).

Cross-Cultural Family Research and Practice
https://doi.org/10.1016/B978-0-12-815493-9.00001-6

In contrast, in traditional Chinese culture, the couple relationship is considered as an extension of the parent's family (Shi & Wang, 2009), and filial piety—obeying and showing respect for one's parents—is considered a virtue in Chinese culture (Chan, Ng, & Hui, 2010). In these contexts, the extent of family approval of how the couple interacts with the extended family is central to the view of the relationship as successful (Coontz, 2005). Thus, family form can vary (the couple-based nuclear family vs extended family group), as can family function (meeting the needs of the adult partners and offspring, vs meeting the needs of the whole extended family).

The aim of this book is to provide a multidisciplinary synthesis of research on what is common to all human families, what varies cross-culturally, and the implications for practice in promoting positive family functioning. The book draws on insights from diverse disciplines including anthropology, communication studies, demography, economics, evolutionary biology, law, political science, psychiatry, psychology, and sociology. The book content falls conceptually into five broad areas, this introduction plus four more sections. The second (next) section is an analysis of the forms and functions of families. The third section analyzes influences on family functioning, and the role of culture in moderating the effects of these influences. The fourth section is a review of family interventions, with a particular focus on how to deliver family interventions in a culturally informed and appropriate manner. The fifth and final section consists of a conclusions chapter.

Family forms and functions

This section analyzes the universals, and the cross-cultural differences, in families' form and function. It begins with Chapter 2 by Bjorklund, Myers, and Bartolo-Kira, which analyzes evolutionary influences on human families. As the chapter describes, humans are a distinctive species in that we humans have extended to an unprecedented level the primate trend toward a large brain, complex social groups, and an extended period of immaturity. The extended period of immaturity not only allows the human brain to continue to develop after birth to give a complex behavioral repertoire, but also makes human offspring depend on caregivers for survival. Caregivers are predominantly family members, and successful raising of offspring requires family members to cooperate. Thus, humans are cooperative family breeders, and family serves the evolutionally critical function of promoting the survival of young humans. Bjorklund and colleagues describe how parents, offspring, and other family members have evolved particular patterns of within-family behavior that enhances survival.

Bjorklund and colleagues also note that evolution operates relatively slowly, and humans evolved behavioral tendencies that influence family functioning, rather than fixed behaviors, as this allows flexibility in response to changing environments. Our human characteristics evolved in a time referred to as the environment of evolutionary adaptation, when humans were living a hunter-gatherer existence. Cultural changes have modified the forms, and to some extent the functions, of families, and these cultural influences are explored in the subsequent chapters.

Chapter 3 by Hewitt and Churchill and Chapter 4 by Lesthaeghe use a demographic lens to examine cross-cultural similarities and differences in a family life. As Hewitt and Churchill documented, family life has undergone dramatic transformations in the last few hundred years, with global declines in the rates of marriage, increasing age at marriage, increasing rates of divorce, changes in gender roles within the family, and shifts in who is allowed to marry. They concluded that marriage and family formation are continuing to change, and while there are some consistent global trends (e.g., a decline in marriage rates), there seems not to be convergence toward a particular family form but rather diversification of family forms within countries.

Lesthaeghe critically analyzes the rise of one family form, unwedded cohabitation. He documents how the trends toward cohabitation vary greatly across countries, in terms of rises in rates of cohabitation, differences in overall rates of cohabitation, and who within particular countries cohabits. Importantly, Lesthaeghe documents how cultural contexts moderate the association of cohabitation with the functions of family, cultural contexts include variables like the level of income inequality, welfare state provisions, and religiosity.

Gautier in Chapter 5 examines family law across 189 countries. She describes the large variations cross-nationally in laws on who can marry, such as the minimum age for marriage, whether the law permits polygamous relationships, and the legality of same sex marriage. Gautier also describes the large between-country variations in laws about the relationship within couples. She analyzes how many such laws disadvantage women. For example, numerous countries have laws requiring wives to be obedient to their husbands, and also that women can only undertake certain activities with the husband's permission (e.g., open a bank account, obtain a passport). Moreover, many countries have laws that specify property settlement arrangements after divorce that disadvantage women economically.

Chiappori and Molina in Chapter 6 use an economic framework to analyze couple relationships cross-culturally. Their analysis starts from the economic argument that people live in couple and family relationships because that arrangement allows more efficient use of economic resources than living as individuals. They then analyze what influences spousal bargaining power to decide on the expenditure of family resources, drawing on studies in high-, middle-, and low-income countries. They concluded that spousal bargaining power strongly depends on the cultural context, and that the spousal bargaining power moderates the effects of social policy designed to reduce family disadvantage.

Zhang and Kline in Chapter 7 critically review the research on the association between couple communication and couple relationship satisfaction. They noted that, cross-culturally, there is an association between relationship satisfaction and high rates of positive communication (e.g., positive affect, validation, and warmth) and low rates of negative communication (e.g., hostility, dominance, and invalidation). However, the normative rates of particular behaviors, and the strength of the association between satisfaction and communication behaviors, vary considerably cross-culturally. They conclude by considering the implications of cross-cultural variation in communication for couple therapies, which often focus on promoting positive couple communication.

Parsons, Heyman, Mitnick, and Slep in Chapter 8 review an important but dark side of family functioning, the occurrence of violence and abuse in families. In their chapter, they explore the definition of what constitutes partner abuse and neglect, and what constitutes child abuse and neglect. They consider whether such definitions are culturally determined; to what extent it is possible to establish worldwide, cross-culturally meaningful, standards that define unacceptable violence and abuse; and how to understand how systems of power, privilege, and oppression operate with particular cultural contexts.

Chapter 9 by Pepping, Power, Bourne, and Lyons reviews the form and functions of lesbian, gay, bisexual, and transgender (LGBT) families in a cross-cultural context. The chapter describes the changing legal recognition of LGBT families in many parts of the world, and the implications of the different laws for LGBT family functioning. The authors present evidence that most people in LGBT communities desire to have a satisfying couple relationship, and describe their aspirations for those relationships. The chapter analyzes the distinctive challenges for LGBT couples who wish to be parents, and the laws surrounding issues like surrogacy, artificial insemination, and adoption as they affect LGBT couples. Finally, the chapter looks at the negative effects of stigma and discrimination against LGBT families, and suggests how to attenuate these negative effects.

Influences on family

This section analyzes how culture moderates the effects of many influences on family functioning. The influences considered range from those that have an impact on all families, like work-family balance, through to influences that can have a major impact but affect only some families, such as migration and war.

The first chapter (Chapter 10) is by Holmes, Thomas, Petts, and Hill, who address the interface between work and family. They describe how work and family domains can have positive and negative effects on each other. The authors explained a conceptual model in which work-family system interaction are determined, at least in part, by the cultural context within which the family and work systems exist. The authors then describe the large variations that exist cross-culturally in gender roles, maternity and parental leave provisions, availability of flexible work arrangements, and child care practices. They analyzed how these cultural factors moderate the work-family interface.

The next chapter (Chapter 11) of this section by Al Sabbah examines how families and physical health influence each other, and how that association is moderated by culture. The author describes how ill health affects the whole family, and the way in which family interaction can have effects on health. There is an analysis of a family influences on a range of health-related behaviors, including diet, exercise, and alcohol and other drug use, and how these family effects are moderated by broader cultural values.

Chapter 12 by Stanley analyzes how government social policy impacts families, with particular attention to how poor policy might inadvertently drive social exclusion and disadvantage. The chapter uses the city of Melbourne, Australia as a case example of how social policy impacts family functioning. The analysis highlights that, even within a high-income country like Australia, issues like poor urban planning, shortfalls in infrastructure spending, and victim-blaming social welfare policies can negatively impact family life.

In Chapter 13 by Sam and Setrana, and Chapter 14 by Sirin, Sin, Clingain, and Choi, the authors present an overview of the vast numbers of families migrating from one country to another globally, explore how migration impacts families, and analyze cultural moderators of those impacts. Sam and Setrana emphasized that migration is an enabler of human well-being by allowing people to move to where they are safe and/or where their skills are needed and valued. At the same time, migration has profound effects on families, both positive and negative. They analyzed in detail some of the challenges facing children who migrate, and the factors that enhance adjustment outcomes for families and children who migrate. The authors pay particular attention to the changing profile of migrants, and the gendered nature of the experiences of family migration.

Sirin and colleagues in Chapter 14 analyze the effects of antiimmigrant sentiment in destination countries, focusing particularly on the United States and Europe. They explore an integrative risk and resilience model of adaptation to a new cultural context. They then describe the content of the current antiimmigrant policies pursued by a number of national governments, and the impact of those policies on immigrant family well-being. The chapter also contains consideration of how the family can be a source of resilience in individuals acculturating to a new country of residence.

In Chapter 15 de Jong describes the challenges for families coming to terms with armed conflict. He documents the horrifically large number of individuals exposed to war, and explains how culture moderates the effects of that exposure on individuals and families. He analyzes the extent to which posttraumatic stress is a universal vs culturally specific phenomenon, and the varying approaches taken in different cultures to address the consequences of war-related trauma. He then reviews the role of family therapy in addressing war-related trauma, and how culturally informed approaches can be developed and applied.

Family interventions

This section shifts the emphasis in content from research on culture and families, to practice of family interventions in a culturally informed manner. Of course, research and practice should inform each other, and there is only a partial separation of the topics of research and practice between the "Family Interventions" section and other sections of the book. Several of the chapters in "Family Forms and Functions" and "Influences on Family" sections have contents on practice of family interventions. For example, Chapter 9 by Pepping and colleagues in the "Family Forms and Functions" section has numerous suggestions on how to ensure that couple interventions are appropriate to work with LGBTQ families. Similarly, in the "Influences on Family" section, Chapter 15 by de Jong reviews the content and effectiveness of family therapy for addressing the effects of war-related trauma. Conversely, in the "Family Interventions" section, some chapters include research evidence on family forms, functions, and influence on family functioning. For example, Chapter 17 by Halford and van de Vijver analyzes research on cultural differences in the forms and functions of couple relationships, before focusing on the culturally informed practice of couple interventions. Similarly, Chapter 18 by Lansford and Bornstein and Chapter 19 by Turner, Singhal, McIlduff, Singh, and Sanders examine research on cultural differences in parenting practices before analyzing culturally informed parenting interventions.

In the first chapter of the "Family Interventions" section, which is Chapter 16 by Epstein, Falconier, and Dattilio, the authors describe the development of Couple and Family Therapy (CFT). They argue that the development of CFT historically was largely an endeavor conducted in high-income Western countries, most importantly in the United States. The chapter describes how models of family therapy that were developed in the United States migrated to many other countries, often with little attention being paid to cultural differences. The authors explain how models of training, professional accreditation, and licensure have been developed for CFT practitioners around the globe. Then they analyze a range of cultural factors that potentially moderate the effects of CFT, including cultural views about the causes and treatment of psychological disorders; cultural values around individualism-collectivism, family boundaries, gender roles, and family roles; the role of professional helpers in different cultures; and style of communication. The chapter concludes with recommendations including the need for much more research on the effects of CFT in diverse cultural contexts.

Chapter 17 on Couple Relationships was authored by the editors of the current volume, Halford and van de Vijver. We begin by considering the varying forms and function of couple relationships across time, and across cultures, and how form and function influence each other. We use research on relationship standards (beliefs as about what is needed in a great couple relationship) to explore cross-cultural consistencies and variations in couple relationships. That is followed by a critical review of research on the effectiveness of couple interventions, with a particular focus on the cross-cultural applicability of such interventions.

Lansford and Bornstein are the authors of Chapter 18, and they focus on parenting interventions and culture. They begin with an analysis of the key responsibilities of parenting, and then describe how parenting differs across cultural contexts. They then discuss the United Nations Convention on the Rights of the Child, and relate this convention to national laws and policies related to parenting. There is a critical review of a range of parenting programs designed to enhance child outcomes across low-, middle-, and high-income countries, with a careful consideration of how the specific targets for enhanced child outcomes are selected in particular cultural contexts. Finally, the authors offer some guidelines for adapting parenting programs for different cultural contexts.

In Chapter 19, Turner, Singhal, McIlduff, Singh, and Sanders describe the development and cross-cultural dissemination of Triple P—The Positive Parenting Program. Triple P is an exemplar of a Western-developed intervention that has been continually refined by research, and adapted for use in a very diverse range of cultures. Turner and colleagues present evidence of Triple P effectiveness across a diverse array of cultural contexts. They then describe a series of principles, drawn largely from their experience of disseminating Triple P, that promote cost-effective adaptation of evidence-based parenting programs.

In Chapter 20, Weisman de Mamani, Altamirano, McLaughlin, and Lopez review the use of family interventions in the treatment of individuals suffering from psychosis. The authors begin outlining the rationale for a family intervention focusing on the treatment of psychoses, and then outlining the research on the effectiveness of that approach. There is a consideration of how culture moderates family responses to a relative with a psychosis, and analysis of the implications for family interventions. The authors then describe a culturally informed approach to family interventions for psychosis.

Chapter 21 is a conclusions chapter written by the editors, which seeks to integrate the major themes developed in the preceding chapters. It analyzes the consistencies and the variations in family forms and functions across cultures. The chapter also considers how to address issues of culture in the development and effective implementation of family-based interventions.

References

Buss, D. M. (2016). *The evolution of desire: Strategies of human mating* (4th ed.). New York: Basic Books.

Chan, D.K.-S., Ng, T.T.-T., & Hui, C.-M. (2010). Interpersonal relationships in rapidly changing Chinese societies. In M. H. Bond (Ed.), *The Oxford handbook of Chinese psychology*. (pp. 515–532). Oxford: Oxford University Press. https://doi.org/10.1093/oxfordhb/9780199541850.001.0001.

Coontz, S. (2005). *Marriage, a history: From obedience to intimacy, or how love conquered marriage.* New York: Viking Press.

Epstein, N. B., Chen, F., & Beyder-Kamjou, I. (2005). Relationship standards and marital satisfaction in Chinese and American couples. *Journal of Marital and Family Therapy, 31*, 59–74. https://doi.org/10.1111/j.1752-0606.2005.tb01543.x.

Halford, W. K. (2011). *Marriage and relationship education: What works and how to provide it.* New York: Guilford.

Shi, L., & Wang, L. (2009). A multilevel contextual model for couples from Mainland China. In M. Rastogi & V. Thomas (Eds.), *Multicultural couple therapy* (pp. 297–316). Thousand Oaks, CA: Sage Publications.

World Values Survey. (2016). http://www.worldvaluessurvey.org/wvs.jsp.

PART 2

Essential forms and functions of families

CHAPTER 2

Human child-rearing and family from an evolutionary perspective

David F. Bjorklund, Alyson J. Myers, and Ariel Bartolo-Kira
Department of Psychology, Florida Atlantic University, Boca Raton, FL, United States

Homo sapiens are unusual mammals, but mammals nonetheless. As mammals, they possess certain features that facilitate the rearing of offspring on the one hand and constrain it on the other. Mothers are highly invested in their children, with conception and prenatal development occurring within their bodies and nursing being the sole source of nutrition available to their young. As such, mothers *are* the environment of infant mammals, serving as a filter by which their young experience the world, exposing them to immune factors, hormones, and other chemicals, initially through the placenta and later through milk. Mothers and their offspring also appear to have evolved psychological mechanisms designed to foster attachment and thus survival. Fathers, in contrast, are less invested by obligation, in part, because there is always some degree of *paternity uncertainty* ("mother's baby, father's maybe"), so that their evolutionary fitness may be better served by seeking additional mating opportunities rather than investing time and effort in their mates' offspring, which may not be theirs. In fact, for most mammals, a father's postnatal contribution is not required for offspring success (Trivers, 1972).

Of course, in other ways, humans are atypical mammals. They are among the 5% of mammals in which fathers regularly devote some postnatal parenting effort (Clutton-Brock, 1991). Human males are not the most invested fathers in the mammal world—that distinction goes to the monogamous South American titi (*Callicebus cupreus*) and owl (*Aotus azarae*) monkeys in which fathers carry infants on their backs most of the day, spending more time with their babies than the mothers. However, men across the world and throughout history have provided—if not a lot of direct childcare—substantial indirect economic support to their mates and children. This increased paternal investment relative to most mammals was likely dictated by humans' unique natural history.

Cross-Cultural Family Research and Practice
https://doi.org/10.1016/B978-0-12-815493-9.00002-8

In this chapter, the complexities that comprise life as a human, and more particularly as human parents and children, are discussed from an evolutionary perspective. After defining some central concepts of evolutionary biology and psychology, we examine the evolution of human parenting and evolved mechanisms for parents' investment in children and children's solicitation of investment from parents. While parents and children have similar goals—survival to reproductive age—they have different evolved psychological mechanisms for investment. Parents have evolved mechanisms to prevent overinvestment in any one child and children have evolved mechanisms for soliciting as much investment as possible from parents. We also examine the role of the *environment of evolutionary adaptedness* for humans' unique psychological features, discussing the role of our ancestral hunter/gatherer lifestyle. Most current psychological research is conducted with WEIRD (Western, Educated, Industrialized, Rich, Democratic) societies, ignoring that conditions in which our species evolved are quite different than conditions today. We explore this topic by discussing the flexibility children have to adapt to a broad variety of child-rearing practices and living conditions, including different physical environments, cultural practices, and family structures. Lastly, we present *life history theory* as a framework for explaining how developmental plasticity in interaction with an animal's ecology produces adaptive individual differences, and we examine these differences in a broad range of rearing environments.

Before beginning our examination of human child-rearing and family from an evolutionary perspective, we believe that defining a few concepts central to evolutionary theory may be helpful to some readers. Modern evolutionary theory stems from the publication of Charles Darwin's seminal work, *The Origin of Species*, published in 1859. Darwin, a naturalist and biologist, was not the first to suggest that species evolve (or in Darwin's terms, "descent with modification"); his major contribution was to propose *natural selection* as the mechanism for how evolutionary changes occurred. There are four components to natural selection: (1) more individuals in a generation are born than will survive to reproduce; (2) there is variation in important characteristics among individuals; (3) these variations are inheritable; and (4) features (variations) that result in individuals surviving and reproducing tend to be passed on to future generations (or "selected for") as a result of interactions between individuals and the local environment, whereas features associated with early death or failure to reproduce tend not to be "selected for" (or are "selected against") and are reduced in frequency or eliminated. Evolutionary theory became integrated with genetic theory in the early decades of the 20th century, so that evolution came to be defined as changes in gene distribution within a population over time. A major theoretical advancement occurred in 1964 when William Hamilton published his theory of *inclusive fitness* (Hamilton, 1964). The theory proposes that organisms behave in ways that cause their genes to be passed onto future generations, regardless of whether or not those genes are a result of direct inheritance through reproduction (i.e., having children) or through the reproduction of relatives, who share genes with an organism. Thus, an animal can increase its inclusive fitness directly by reproducing or indirectly by fostering the success of individuals who possess copies of its genes (e.g., siblings, grandoffspring, cousins).

Another concept central to evolutionary theory is that of *adaptation*. According to Buss, Haselton, Shackelford, Bleske, and Wakefield (1998), "An adaptation may be defined as an inherited and reliably developing characteristic that came into existence as a feature of a species through natural selection because it helped to directly or indirectly facilitate reproduction during the period of its evolution" (p. 535). Adaptations can be for physical attributes—the opposable thumb, for example, which is important in making and using tools—or for behavior or cognition, for instance, feeling jealous when someone threatens an important relationship, which can translate into action being taken to protect that relationship. An adaptation can be inferred by its reliability (it develops in all members of a species, or all members of one sex, in all species-typical environments), its efficiency (it solves a problem effectively), and its improbable usefulness (relatively guaranteed functionality with low cost and high gain).

Evolutionary psychology applies the principles of natural selection to explain human behavior. Perhaps the most important concept in evolutionary psychology is that of *evolved psychological mechanisms.* That is, what evolved over the course of humans' ancestral past were information-processing systems within organisms, shaped by natural selection to solve recurrent adaptive problems faced by our ancestors, such as determining what is good to eat, finding and keeping a mate, and forming attachments between mothers and infants to keep babies alive. These mechanisms are implicit in nature, operating below the level of consciousness, although for humans, they can potentially become available to self-awareness. Evolutionary psychological mechanisms and adaptations are not necessarily the most ideal solutions to adaptive problems, nor might they continue to be adaptive as the surrounding environment changes; however, they have allowed our ancestors to survive, which has allowed for the continuation of our species.

One frequent criticism of evolutionary approaches to human behavior is that they necessarily imply that if a certain pattern is "natural" or has "evolved" it is also "normal," or justifiable in some other way. The idea that something is inherently good because it is natural, the so-called *naturalistic fallacy*, is precisely that, a fallacy. It is one thing to *understand* the past and the evolutionary influences on contemporary behavior, but it is quite another thing to *justify* that behavior or to propose its inevitability or social desirability.

We would like to make one final point about evolutionary explanations of human behavior. An evolutionary perspective does *not* imply genetic determinism, that is, "genes determine behavior." Rather, ever since Darwin, evolutionists have recognized that changes occur as a result of the interaction of an organism with its environment. We, in particular, take an evolutionary *developmental* perspective, noting that it is variation of development that produces the stuff upon which natural selection works and that *plasticity*—the ability to modify morphology, behavior, or cognition—is an evolved characteristic of animals, especially young animals, and particularly of *Homo sapiens*.

The evolution of human parenting

Human parenting practices across the globe are highly diverse. (We discuss more about this later in this chapter.) Yet, there is a common evolved foundation upon which all human child-rearing practices are built and that differentiates humans from most other mammals. At the core of human parenting is the necessity to care for an extremely helpless infant, whose helplessness extends well past infancy.

For reasons that are still hotly debated, *Homo sapiens* exacerbated a general primate trend, evolving a large brain, complex social groups, and an extended period of immaturity. According to the *social brain hypothesis* (Alexander, 1989; Bjorklund & Rosenberg, 2005; Dunbar, 2003), the large brain—nearly three times the relative size of those of our chimpanzee and bonobo cousins—was needed to deal with the complexity of human social groups and social relationships. To develop such a brain, however, much postnatal growth is required; brains and the skulls that hold them can only get so large before they are unable to pass through the birth canal of a bipedal woman. As a result, much development that would have occurred prenatally in other primates had to occur postnatally in humans, extending the period of dependency for infants and juveniles. Perhaps due to similar selection pressures, humans added a new life stage—childhood—between infancy and the juvenile period in which children are no longer nursing though are still highly dependent on adults for care. Humans also extended the length of the juvenile period (referred to as middle childhood by psychologists) and the preadult stage, inventing the stage of adolescence (Bogin, 2001). Speculatively, these extensions allowed for the expansion of consciousness within humans' ancestors and provided a framework for unbridled innovation.

As for other mammals, the task of caring for young offspring went primarily to our foremothers, but because of the demands brought about by prolonged immaturity, mothers alone were rarely able to successfully raise a child to adulthood. Humans became *cooperative breeders* (Hrdy, 2009, 2016), with people other than the mother (alloparents) helping to care for infants and young children, which some have speculated led to enhanced prosocial and cognitive abilities (e.g., van Schaik & Burkart, 2010).

Although most alloparents were females, fathers' contributions became important. Commitment to his mate and her offspring increased the chances of his children's survival and success. For example, Geary (2000) noted a relationship between infant and child mortality rates and marital status across a wide range of cultures. For instance, Indonesian children of divorced parents were found to have a 12% higher mortality rate than those children of monogamously married couples, and a similar pattern was observed in 11 of the 14 developing nations studied. To promote commitment from fathers, men and women developed emotional bonds (they "fell in love" and also experienced sexual jealousy)—which typically lasted long enough for a child to be marginally independent. They also developed *biparental families* whose primary function was to raise children (Buss, 2011; Fisher, 2004). By committing to a mate, a man also gained increased paternity certainty—knowing with reasonable certainty that any children born to his mate were fathered by him—permitting him to invest time and resources into his mate's offspring, increasing the chances that these children do in fact carry copies of his genes. Increased paternity certainty promotes paternal investment, which in turn, promotes increased likelihood of child survival.

Humans are not the only animal that raises offspring in families. Many bird species form monogamous relationships, often for life, with both males and females contributing to their offsprings' nurturance. Among primates, a number of monkey species similarly forms monogamous partnerships, with both parents involved in caring for the young (recall the owl and titi monkeys described earlier). Among the great apes, polygamous gorilla fathers protect and occasionally play with their offspring. In general, however, birds and nonhuman primates show little variability in family structure.

Although humans' natural history played a critical role in shaping male-female relations, the formation of families, and the extended period of dependency needed to rear a child to adulthood, natural selection also shaped the psychologies of individual parents and children. Although parents and children share the same long-term goal—survival of the child to adulthood—the interests of parents and children are not identical. As we discuss in the next section, both parents and children have evolved psychological mechanisms designed to maximize their fitness and these are sometimes at odds.

Evolved mechanisms for investing in children (and getting investment from parents)

Parental investment refers to the amount of resources any given parent provides to his or her children. Investment can come in many forms, including nutrition, caregiving, instruction, or even attention. Although today we may think of differential investment in terms of how much money parents put away for children in a college fund or whether children receive piano, dance, or karate lessons, in many parts of the world, and certainly for our ancestors, investment from parents could mean the difference between life and death. Ancestral parents had limited resources (including time) to devote to their children and so evolved mechanisms to evaluate whether and how much to invest in any given child based on a variety of factors, such as a child's health, their own health, age, or social support (see Bjorklund & Myers, 2019). Children, however, are not passive participants; rather, they have evolved psychological mechanisms to solicit resources from their parents, increasing their chances for survival. While children will try to solicit as many resources as possible from their parents, parents are reluctant to give all of their resources to just one child. If parents have multiple children, they will distribute resources among their current children and may consider having more children in the future. In other words, from a parent's perspective, there is a trade-off between investment in parenting and investment in reproduction.

Robert Trivers (1972) developed *parental investment theory* to describe this trade-off between investment in parenting and in reproduction. There is a conflict between how much effort and investment parents devote to raising a child and how much effort and investment they devote to reproduction (or mating). In mammals, including humans, maternal investment is obligatory. Throughout the duration of a mammal's pregnancy, the father is not required to be in any way present to ensure his offspring's survival. The mother is a necessary presence in ensuring the survival of the offspring, and, after birth, she is typically the sole source of nutrition throughout nursing. Even after nursing, there is a substantial amount of investment involved, as humans have a prolonged juvenile period requiring dependence on parents for resources. According to parental investment theory, compared to males, females are less inclined to engage in risky behavior because they are of greater evolutionary value (the death of a mother almost guarantees the death of her children; not so the death of a father). Males conversely are more inclined to engage in risky behavior (to increase their status and mating opportunities) and are less integral to the survival of their offspring (Campbell, 1999, 2013). While fathers have no postcopulatory obligatory investment, they frequently provide resources in terms of shelter, food, or protection in both traditional and contemporary societies, and their investment is related to their children's success. For instance, fathers' emotional and economic investment in their children has been shown to be positively related with their children's eventual social status, educational achievement, social competence, psychological well-being, and ultimate survival (see Geary, 2000).

Following parental investment theory, males and females evolved different psychologies with respect to mating and parenting. Before parenting can begin, of course, one must first find a mate. Certain "universals" of mate selection have been proposed and supported by evolutionary theory. Women, because they are the more investing sex with respect to parenting, have evolved to be more hesitant to consent to sex than men (a single act of intercourse can result in 9 months gestation and several years of nursing for a woman) and put greater value in a potential mate's ability and willingness to invest resources (time, food, money) in them and their offspring than do men (e.g., Buss, 2016; Buss et al., 1990). This is especially the case when environments require substantial investments in children or biparental care (Gangestad & Simpson, 2000). Both men and women value good health in a partner, which is often reflected in physical attractiveness, although this preference for an attractive mate is substantially greater for men than for women, presumably because of its association with fertility. For example, one feature associated with facial attractiveness is symmetry (i.e., the left and right side of the face are similar), which is an indication of lack of developmental perturbations and thus good health and, in women, potential fertility (e.g., Jasienska, Lipson, Ellison, Thune, & Ziomkiewicz, 2006). Men also show a decided preference for women with a waist-to-hip ratio of about .7 (the bottom two-thirds of an hourglass shape). Researchers have reported that this value is associated with a woman's health and fertility (e.g., Singh, 1993, 2002; for reviews see Sugiyama, 2005, and Weeden & Sabini, 2005, although recent research has questioned this interpretation, Lassek & Gaulin, 2018). In addition, men place a high value on youth in a mate, which again is an indication of potential fertility (Buss, 2016).

Once men and women become partners and parents, they must decide how much time, effort, and resources to invest in their children. Although parents want to distribute their resources to their children judiciously, children want to maximize their own fitness and will attempt to solicit as many resources as they can from their parents. Trivers's (1974) theory of *parent-offspring conflict* best illustrates this concept. Trivers suggested that parents and children will have direct conflict over three major issues:

(1) the amount of time parents should invest in any given offspring (children want parents to invest for a longer period of time than parents are willing to invest);

(2) the amount of investment that a parent should give to an individual offspring (each child seeks to solicit more investment from parents than received by his or her siblings);

(3) the amount of altruism that an individual's offspring should show toward other relatives, including siblings (parents want their children to show greater altruism toward their siblings than the children want to show).

Both parents and children have evolved psychological mechanisms to deal with parent-child conflict. Parents have evolved mechanisms to invest wisely in their children, and children have evolved mechanisms to solicit as many resources as possible from their parents.

Parents' evolved psychological mechanisms

Ancestral parents needed to evaluate an infant or child's worth to determine how much investment to give to any individual offspring. While all parents want their children to succeed and reach reproductive age (and this was surely true of our foreparents), the reality is that some children have a better chance of success than others. Investing in children who display cues of health and rigor is beneficial to parents, so that resources are not "wasted" on a child who is unlikely to survive to reproductive age regardless of investment. As parents must be concerned with investment in all current and potential children, they must protect themselves from overinvesting. Thus, natural selection has selected for parents who are adept at recognizing cues of potential reproductive success in their offspring and who, when necessary, will appropriately reduce investment in children (Hrdy, 1999, 2009). There are different cues that can lead to reduced investment from parents, including a child's health and age, mother's social support, mother's reproductive status, and the mother's social and economic conditions.

From the perspective of well-resourced, contemporary society, it may seem counterintuitive for a parent to invest less in a child who is sickly; however, when examined through the lens of evolutionary theory, it is a reasonable alternative. When resources are scarce, it behooves parents to invest more in a child who is more likely to reach reproductive age and pass on his or her genes than to waste resources on a child displaying cues indicative of poor health. It is not necessarily that parents are making these harsh decisions consciously, but rather, these decisions may be occurring at an unconscious level.

In modern society, if a sickly infant becomes overly burdensome to parents, the state or other societal institutions will step in to assist with caregiving. However, in our ancestral past, there was no intervention by a state or religious organization to assist parents in the caretaking of a sickly child. Instead, the burden of caring for the infant would fall primarily on the mother, and taking care of a sickly child could jeopardize not only the health of her other current or potential offspring but also her own health. Therefore, ancestral women evolved mechanisms to recognize cues and evaluate the health of their children, allowing them to invest less (or none at all) if needed. Parents in Western societies have been shown to invest more in a healthier child than a child born prematurely or with some sort of ailment, as the healthier child is more likely to survive (e.g., Bugental, Beaulieu, & Silbert-Geiger, 2010). Modern women still have the same evolved mechanisms as their ancestors as demonstrated in a variety of studies (e.g., Daly & Wilson, 1981; Mann, 1992). For example, Daly and Wilson (1981) found that children with intellectual or congenital defects (e.g., Down syndrome, cystic fibrosis, spina bifida) experienced physical abuse 2–10 times more frequently than nonaffected children. A study of less drastic differential investment examined maternal interactions with premature and low-birthweight twins (Mann, 1992). Mann (1992) reported that mothers showed a higher frequency of positive maternal behaviors (kissing, gazing, holding, etc.) toward the healthier of the two 8-month-old infants. Despite the fact that the sicklier baby was sometimes more vocal, the healthier twin still received more positive maternal behaviors than the sickly twin. To reiterate, these behaviors are often enacted unconsciously as a means of enhancing inclusive fitness.

In addition to a child's health, a child's age may influence parental investment decisions. The reproductive value of a child increases with age, meaning that parents are more likely to invest in older rather than younger children in times when resources are limited (Daly & Wilson, 1988). In a cross-cultural study using the Human Relations Area Files, Daly and Wilson (1988) reported that children born into environments where resources were scarce had a higher likelihood of being abandoned or killed at birth than children born into more resource-rich environments. Under such conditions, when a mother did not have the resources to invest in two children, if there were an older sibling, the younger one would be abandoned. The death rates for infants were relatively high for our ancestors (Volk & Atkinson, 2013). As such, a child who had already survived his or her first year of life would be more likely to reach adulthood, increasing his or her reproductive value relative to a newborn.

While there are child-focused cues that lead to differential investment, there are also cues related to the mother, including social support and the mother's reproductive status. Human infants and children have an extended period of immaturity, and social support is an important factor in raising children. Social support can be provided by both relatives and nonrelatives, and importantly, the mother's mate. If social support is plentiful, the mother is able to invest more in her children. However, if social support is low, resource availability is restricted. A marriage partner can indicate greater access to resources and social support. In an extreme example, unmarried mothers are more likely to commit infanticide (the killing of an infant) than married women (Daly & Wilson, 1988). Within the first year of life, a child is most at risk for being killed, and following this year, the probability of filicide (killing a child after 1 year of age) steadily decreases (Daly & Wilson, 1988). From an evolutionary perspective, a lack of a committed partner may be indicative of an inability to raise a child to reproductive age, and a woman's best option would be to cease investment in an offspring through abandonment or infanticide. The mother can then attempt to reproduce later when social and economic resources are more abundant. This would especially be the case if the mother was young. A younger woman has more reproductive years ahead of her, and ceasing investment in an infant is not as worrisome as she has many more childbearing years remaining to reproduce again when resources and opportunities are more optimal (Bjorklund & Pellegrini, 2002). In contrast, an older woman would have fewer reproductive years left, and divesting in an offspring may not be a viable option, as this could potentially be her last (or only) child. Consistent with this idea, younger mothers in the United States have a higher likelihood of neglecting or abusing their children than do older mothers (Lee & George, 1999). In Canada, teenage mothers were more likely to commit infanticide than mothers in their 20s (Daly & Wilson, 1983). This pattern also holds true cross-culturally. For example, among the Ayoreo, a group of nomadic foragers spanning Paraguay and Bolivia, when a newborn is sickly or deformed or if economic prospects are lacking, mothers in their teens are more likely to commit infanticide than mothers in their 20s (Bugos & McCarthy, 1984). If a child is killed on the day of his or her birth, this is referred to as neonaticide. Based on a sample of 110 Italian cases of mothers who killed 123 of their offspring from 1976 to 2010, Camperio Ciani and Fontanesi (2012) reported that neonaticide usually occurs when the mother is not in a committed relationship, financially struggling, and young. Camperio Ciani and Fontanesi proposed that neonaticide is adaptive and is the only child-killing type that can be considered reproductive disinvestment in comparison to infanticide and filicide, which involve more psychopathological patterns of killing.

Because of their greater obligatory investment, natural selection has seemingly been more potent in shaping female psychologies with respect to child-rearing than that of males'. However, because men's investment in their offspring is related to their children's survival and success, they too have evolved mechanisms to promote their fitness with respect to parenting. There is evidence, for example, that men are able to use olfactory cues to identify children who are biologically related to them. In one study, Dubas, Heijkoop, and Aken (2009) examined 78 Dutch families (66 mothers, 39 fathers, and 99 children) in which children wore long-sleeved T-shirts over the course of three consecutive nights. Parents were then blindfolded and asked to smell a control T-shirt as well as those of their own child and other children. They were then asked to report the degree of the pleasantness of the T-shirt's odor in addition to identifying whether or not the shirt had been initially worn by their child. The researchers reported that fathers' olfactory recognition was significantly related to their parental investment. Fathers who had been able to identify their children's odors as opposed to those unidentifiable odors of control children reported greater affection and feelings of attachment toward their children and displayed less ignoring behaviors than fathers who could not identify their children's odors. This effect was not found for mothers. The ability to identify genetically related children by smell can serve to prevent a man from allocating and potentially depleting his resources when it is not in his best inclusive interest to do so. (Because of internal fertilization and gestation, a mother always knows a baby is hers.) In a similar T-shirt-smelling study, fathers demonstrated aversions to the smell of T-shirts worn by their daughters (and vice versa), but not their sons, and brothers and sisters showed aversions to T-shirts worn by their opposite-sex siblings though not their same-sex siblings. These aversions were reported whether people could correctly identify who had worn the T-shirt or not. Mothers did not show any consistent aversions. These aversions (father-daughter; brother-sister) represent the greatest danger for incest and suggest that olfactory cues evolved as one mechanism for incest avoidance (Weisfeld, Czilli, Phillips, Gal, & Lichtman, 2003).

That fathers discriminate how much they invest in offspring dependent on paternity certainty can be seen in how much men invest in their stepchildren. A man's stepchildren, of course, are not genetically related to him, so that men invest in these children at all reflects the flexibility of humans' mating and parenting strategies. In fact, rather than being viewed as an investment in parenting, stepfather investment has been viewed as an investment in mating: Men invest in their mate's existing children in order to gain sexual access to their mother. This is reflected in research examining stepfather investment in their stepchildren (e.g., Anderson, Kaplan, & Lancaster, 1999; Flinn, 1988; Marlowe, 1999). For example, Anderson et al. (1999) identified four classes of relationships that exist between men and the children they invest in: genetic offspring of current mates, genetic offspring of previous mates, step offspring of current mates, and step offspring of previous mates. Men from Albuquerque, New Mexico were interviewed regarding their reproductive behavior and their marital, employment, parenting, and reproductive histories. In order to determine the paternal allocation of resources, measures were developed that recorded the probability that a child attends college, the probability that a child who attends college will receive money for it, current financial expenditures on children, and how often each week men spend time with their children (aged 5–12 years). In accordance with parental investment theory, Anderson et al. reported that there was significant variance across the four classes of relationship and that men invested more in the children of their current mates, who were shown to receive the highest amounts of investment in comparison to the stepchildren of previous mates who received the least. This pattern is consistent with the position that males invest more in children of their current mates as a means of securing their current mating opportunity.

It is evident that various factors affect whether or not parents will invest in their children and if they do, how much exactly that winds up being. It is important to keep in mind the interactionist nature of evolutionary theory as it relates to psychological mechanisms; the parent is sensitive to the child as the child is sensitive to the parent. Cues are exchanged because the environment interacts with the individuals who in turn, interact with the environment.

Children's evolved psychological mechanisms

As we have discussed, parents, especially mothers, have evolved psychological mechanisms to ensure investment is distributed appropriately among children. However, infants and children are not passive recipients of their parents' attention. Rather, infants have also evolved psychological mechanisms to solicit as many resources as they can from their parents as aforementioned. According to Trivers (1974), children have evolved psychological "weapons" to combat parents' efforts to thwart overinvestment.

Perhaps the most important set of adaptations infants and children have evolved are those related to infant-mother attachment. *Attachment* is a biologically based motivational system that evolved to protect children and to motivate adults to provide care (Bowlby, 1969; Del Giudice, 2009). Children evolved a set of mechanisms to achieve this attachment, and these mechanisms interact with parental mechanisms to promote close emotional bonds between infants and their parents and for parents to provide care. One example of this is *neonatal imitation* in which newborns match adults' facial gestures (Meltzoff & Moore, 1977). Although such matching behavior was initially thought to reflect early social learning, subsequent research has shown that neonates do not actually match the facial gestures they are shown, although they do increase the expression of other facial gestures (Oostenbroek et al., 2016; see Bjorklund, 2018). A more likely interpretation of newborns' responses to viewing adults' facial gestures is that this matching behavior evolved as a reflex-like mechanism that promotes infant and mother interactions during a time when infants are unable to intentionally control their own behavior (Bjorklund, 1987). Neonatal imitation disappears around 2 months of age when higher-cortical brain areas are able to influence infants' intentional actions (see Periss & Bjorklund, 2011).

Infants are also more likely to attend to the biological movement (Bardi, Regolin, & Simion, 2011, 2014) and to face-like stimuli (Easterbrook, Kisilevsky, Hains, & Muir, 1999; Mondloch et al., 1999). These are mechanisms that influence and promote attachment between mothers and infants, strengthening their bonds and thus promoting caretaking and investment behaviors by the mother.

Another weapon that enhances the survival of an infant or child is jealousy, reflected by an infant's protesting when a caretaker's attention is diverted to another baby. By at least 9 months of age, infants will exhibit *jealousy protest* when their mothers are paying attention to another infant (Hart, 2018). In a series of experiments, infants watched as their mothers talked and paid attention to either a life-like doll or a book. The infants' negative affect, reduced exploration, requests for the mothers' attention, and heightened arousal were recorded. Infants demonstrated greater signs of distress when their mother was attending to the baby doll rather than the book, indicating that the infants were not merely disturbed by the lack of attention, but especially by the fact that their mother was attending to another baby. The other baby was considered a potential social rival for the mother's attention (Hart & Carrington, 2002; Hart, Carrington, Tronick, & Carroll, 2004; Mize, Pineda, Blau, Marsh, & Jones, 2014); this jealousy protest was exhibited reliably by 9 months of age. Hart (2018) suggested that jealousy protest is an evolved adaptation that is related to attachment and tends to be exhibited most reliably when a competitor infant could have been born. In ancestral times, infant mortality was high (Volk & Atkinson, 2013), and losing exclusive access to a mother for nursing and nurturing due to the birth of a sibling would have likely jeopardized the health and survival of the existing infant, who would still be highly dependent on his or her mother. Although the jealousy-like behaviors observed in infancy may have different underlying cognitive and neurological bases than those observed late in the second year of life, they serve a similar adaptive function—protecting an important social relationship—and thus qualify as an evolved adaptation to secure their mothers' attention and caretaking behaviors (see Myers & Bjorklund, 2018).

Other infant psychological weapons serve to activate evolved mechanisms in adults. These evolved mechanisms promote social relations with their parents and can be behavioral, vocal, or physical (Bowlby, 1969; Goetz, Keltner, & Simon-Thomas, 2010). Behavioral cues include smiling or clumsy movements (Bowlby, 1969). Vocal cues include crying, which can be a reliable sign of health (DeVries, 1984; Soltis, 2004). Lastly, physical cues include a baby-schema for faces, including flat noses, chubby cheeks, rounded and large head relative to body size, large eyes, and short and broad extremities (Lorenz, 1943). These features tend to elicit greater caretaking behaviors from adults. For example, compared to less-cute infants, adults view cuter infants (i.e., those whose features more closely match the prototypical baby-schema) more positively on attributes such as warmth and honesty (e.g., Leibenluft, Gobbini, Harrison, & Haxby, 2004; Senese et al., 2013). Adults also exhibit more affectionate interactions (Langlois, Ritter, Casey, & Sawin, 1995), express greater motivation for caregiving (e.g., Glocker et al., 2009), make hypothetical adoption decisions about (Waller, Volk, & Quinsey, 2004), and express greater empathy for (Machluf & Bjorklund, 2016) cuter than less-cute infants (see Franklin & Volk, 2018; Kringelbach, Stark, Alexander, Bornstein, & Stein, 2016; Lucion et al., 2017 for reviews). For example, in one set of studies, photos of infants' faces were manipulated to create faces with high (round face and high forehead) and low (narrow face and low forehead) baby-schema features. Adults were asked to rate the cuteness of the infant in the photo as well as their motivation for caregiving for each depicted infant (Glocker et al., 2009). High baby-schema faces were rated as cuter and elicited greater motivation for caregiving than the low baby-schema faces. In other research, Sprengelmeyer et al. (2009) manipulated the baby-schema characteristics of infants' faces and reported that sensitivity to cuteness cues was greater for women than men and especially greater for premenopausal women. (Older and menopausal women's sensitivity to cuteness cues was similar to those of men.) The authors suggested that this pattern reflects a hormonal link between infant cues for cuteness and motivation to provide care, with younger women being especially sensitive to the baby-schema cues. Consistent with these findings, the preference for immature faces is typically not seen until adolescence, at approximately 15 years of age, suggesting that this preference or bias may be related to the onset of possible parenthood (Borgi, Cogliati-Dezza, Brelsford, Meints, & Cirulli, 2014; Fullard & Reiling, 1976; Gross, 1997). Other research has shown that cuter, more attractive faces continue to elicit positive responses from adults until approximately 4.5 years of age, after which ratings of likeability and attractiveness of children's faces are similar to ratings of adult faces (Luo, Li, & Lee, 2011).

Although baby faces lose some of their potency to positively influence caregiving motivation toward the end of the preschool years, adults now tend to respond positively to other cues of immaturity (and thus the need for care), including young children's voices (rating immature voices more positively than mature voices, Hernández Blasi, Bjorklund, Agut, Lozano, & Martínez, 2018) and to some aspects of children's verbal expressions of cognitive immaturity (e.g., "The sun's not out because it's mad"). In a series of experiments, adults rated hypothetical children expressing such cognitively immature statements highly on items of positive affect (e.g., likeable, friendly, feel more nurturant toward) and helplessness, and lower on items of negative affect (e.g., sneaky, feel more irritated with, feel more aggravated with) than children expressing more mature cognition (e.g., "The sun's not out because a cloud moved in front of it") (Bjorklund, Hernández Blasi, & Periss, 2010; Hernández-Blasi, Bjorklund, & Ruiz-Soler, 2015). Thus, although facial cuteness seems to lose it effectiveness in eliciting caregiving as children age, it is replaced by other physical (vocal) and cognitive cues, which may be more indicative of older children's need for caregiving than physical appearance. Similar to research using infant faces, this positive bias toward some types of cognitive immaturity is not observed until middle adolescence, suggesting that the bias may be related to the possible onset of parenthood (Hernández Blasi & Bjorklund, 2018; Hernández Blasi, Bjorklund, & Ruiz-Soler, 2017; Periss, Hernández Blasi, & Bjorklund, 2012).

Childhood and parenting in the environment of evolutionary adaptedness

An animal's adaptations evolved over different periods of a species' phylogeny. The *environment of evolutionary adaptedness* (EEA) refers to the time and ecology when an adaptation arose or became established. Roughly speaking, it refers to the environment to which a species is adapted. Evolutionary psychologists argue that the EEA for most of humans' unique psychological features extends over the last 2 million years or so when our ancestors lived as hunter/gatherers. Almost 60,000 years ago and perhaps as early as 300,000 years ago, anatomically modern humans had evolved brains and psychological mechanisms adapted to a hunter/gatherer lifestyle. Although human culture has changed much in the last 60,000 years, fossil evidence suggests that human anatomy has not. Modern *Homo sapiens* have similar brains and psychological adaptations as our ancient relatives. Contemporary peoples' adaptations evolved for life in these ancient ecologies, and thus can sometimes result in a mismatch between psychological mechanisms evolved to deal with a foraging lifestyle and current environments.

What was the environment and lifestyle of our ancestors like? We can never know for certain, of course, but using contemporary hunter/gatherer groups as models for how our ancestors lived, we can get a good idea of what the childhoods of our ancestors might have been like (see Bjorklund & Pellegrini, 2002; Bowlby, 1969; Gray, 2013; Kaplan, Hill, Lancaster, & Hurtado, 2000; Konner, 2010; Lancy, 2015). Moreover, several theorists have argued that the childhoods of contemporary hunter/gatherers—the closest model we have for how our ancestors lived—should be the standard on which we evaluate the child-rearing practices of different cultures today (Gray, 2013; Konner, 2010).

Based on how modern hunter/gatherers live, our nomadic ancestors likely lived in small groups, probably fewer than 100 individuals, although they sometimes had contact with other groups who shared the same area. They made their living by gathering, mostly the work of women, by hunting, mostly the work of men, and by scavenging. As for most mammals, childcare was primarily performed by females, although rarely only mothers. In addition to mothers, infants and young children were cared for by alloparents, consisting mainly of female relatives (grandmothers, sisters, and older offspring), also by unrelated females, and by fathers, who likely provided protection, nurturance, and support in the form of food and other tangible resources for their children and their children's mother.

Of particular interest when examining the care provided by alloparents is the care provided by grandparents. When grandparents invest in grandchildren it benefits their inclusive fitness because such care will increase the likelihood of their genetic kin's survival, and thus, their own genetic line's survival. In order for investment in a grandchild to benefit a grandparent's inclusive fitness, the grandparent must evaluate the genetic relatedness of a grandchild. Investment in an unrelated child is costly. In terms of paternity certainty, a paternal grandparent cannot be any more certain of the relatedness of the child than a father can. However, maternal grandparents are certain of their relationship to the mother (their daughter) and her child (their grandchild). From an inclusive fitness perspective, it is reasonable to predict that maternal grandparents, especially maternal grandmothers, will invest more in terms of resources and care to their grandchildren than will paternal grandparents (Bjorklund & Pellegrini, 2002).

This pattern of grandparental investment has been found in multiple studies across a variety of cultures (e.g., Danielsbacka & Tanskanen, 2012; Danielsbacka, Tanskanen, Jokela, & Rotkirch, 2011; Huber & Breedlove, 2007; Laham, Gonsalkorale, & von Hippel, 2005; see Euler & Michalski, 2007 for a review). In one of the first studies to investigate grandparental investment from an evolutionary perspective, Euler and Weitzel (1996) asked

603 adults who had all 4 grandparents alive for the duration of their youth about the amount of care and solicitude they received from each grandparent up until the age of 7. The researchers found that the adults' maternal grandmothers provided the most amount of care to their grandchildren, followed by maternal grandfathers. Paternal grandfathers provided the least amount of care. This pattern is consistent with the predictions based on paternity certainty and inclusive fitness theory. Genetic certainty is highest in maternal grandmothers, which serves as a cue for investment. In contrast, paternity certainty is low for paternal grandparents, especially paternal grandfathers. In terms of inclusive fitness, it is costly to invest care and resources into children whose genetic relatedness to you is uncertain. Similar patterns of investment have been found for aunts and uncles, with maternal aunts investing the most in their nieces and nephews and paternal uncles investing the least (e.g., Gaulin, McBurney, & Brakeman-Wartell, 1997).

There are some exceptions to this general pattern, however. For example, Pashos (2000) asked adults from Germany and from urban and rural Greek communities to evaluate how much care each set of grandparents devoted to them as children. The patterns were consistent with the predictions based on paternity certainty and inclusive fitness theory for the adults from Germany and urban Greece but reversed in the rural Greece sample. In rural Greece, paternal grandparents provided the most care for grandchildren. Familial customs require children to live close to their paternal grandparents, often in the same house, and paternal grandparents are also socially obligated to help care for their grandchildren, especially their grandsons. These familial customs ensure a greater amount of physical closeness between paternal grandparents and their son's wife, leading to more paternity certainty. Paternal grandparents can monitor and control the movements of the daughters-in-law, lowering the risk of cuckoldry. Because paternity certainty is low, it behooves the paternal grandparents to invest care and resources in their grandchildren. As such, it is important to consider cultural and familial customs when evaluating a family investment, as they can influence what is predicted by paternity certainty and inclusive fitness theory, shifting patterns to the unexpected.

Similar to modern humans, our ancestors likely were marginally monogamous/marginally polygamous. This is reflected by (1) the degree of sexual dimorphism (differences in size between men and women), which is more similar to that observed in monogamous species than in polygamous species (Dewsbury, 1988); (2) the number and diversity of white blood cells reflective of susceptibility to sexually transmitted diseases (more similar to the polygamous, harem-based, gorilla [*Gorilla gorilla*] and the monogamous gibbon [*Hylobates lar*] than the promiscuous chimpanzee [*Pan troglodytes*], Nunn, Gittleman, & Antonovics, 2000); and (3) testes size, reflected of sperm competition (proportionally smaller than the promiscuous chimpanzee but larger than the polygamous gorilla, Smith, 1984). Females probably reached puberty and had their first births in their late teens or early 20s. Infants likely stayed in close physical contact with their mothers, nursed on demand, and were not weaned until 2 or 3 years of age. Birth intervals were likely 4 or 5 years, on average.

Once weaned, children spent an increasing amount of time in the company of other children. If modern hunter/gatherer groups are an indication of how our ancestors lived, they spent much of their day playing in mixed aged and sex groups with little adult supervision. Much of children's play involved imitating adult tasks, such as cooking, gathering, or hunting. They had few chores and little formal instruction from adults; there was no prescribed bedtime and no formal education. Children were loved and valued, and little was demanded of them. According to Lancy (2015):

> *There is a world in which children almost always feel 'wanted'... Infants are suckled on demand by their mothers and by other women in her absence. They are indulged and cosseted by their fathers, grandparents, and siblings. Children wean themselves over a long period and are given nutritious foods... They are subject to little or no restraint or coercion. Infants and toddlers are carried on long journeys and comforted when distressed. If they die in infancy, they may be mourned... They are rarely or never physically punished or even scolded... They are not expected to make a significant contribution to the household economy and are free to play until the mid to late teens... Their experience of adolescence is relatively stress free... This paradise exists among a globally dispersed group of isolated societies – all of which depend heavily on foraging for their subsistence (pp. 66, 67).*

Such rearing conditions can be viewed, by modern standards, as permissive parenting, which, in Western societies, are typically associated with poor psychological outcomes (e.g., Block, Block, & Keyes, 1988; Lamborn, Mounts, Steinberg, & Dornbusch, 1991). Given the substantial demands that a nomadic lifestyle entails, how could ancestral parents have had such a hands-off style of child-rearing? Lancy (2015) suggests that, unlike children in technologically developed societies, hunter/gatherer children had few options of what they could do when they grow up. They would become hunters and gatherers like their parents and pick up the skills necessary to survive, mainly through observation and play, or they would die.

The lifestyles of hunter/gatherers are quite different from those of modern people from what has been called WEIRD societies (Western, Educated, Industrialized, Rich, Democratic) (Henrich, Heine, & Norenzayan, 2010; Nielsen, Haun, Kärtner, & Legare, 2017). Most psychological research has been done with people from WEIRD cultures, with authors often assuming that the patterns found in their studies reflect species universals. This research disregards the fact that the conditions in which our species evolved are substantially different from the way people live today. Moreover, worldwide more people actually live in smaller farming and fishing communities or poor urban centers than in WEIRD cultures (Keller, 2017). The family and child-rearing practices in many of these cultures are substantially different not only from those found in WEIRD societies but also from those practiced by hunter/gatherers. The wide range of physical environments, cultural practices, and family structures found among groups of humans is only possible because children are able to adapt to broad variations in living conditions. In the next section, we discuss the evolved abilities necessary for children's flexibility and examine some of the diverse child-rearing practices that children are able to adjust to.

Developmental plasticity and variability in child-rearing practices in evolutionary perspective

Human family structure is more variable than that of most other mammals. Humans are a marginally monogamous/marginally polygamous species, with family structure and child-rearing practices varying across culture and with ecological conditions (Dewsbury, 1988; Whiting, 1963). Although most marriages in most cultures are monogamous (at least for some period of time, if not for a lifetime), polygyny (one man and several wives) is observed in many cultures, whereas polyandry (one woman and many husbands) is less common but nonetheless practiced. Families are sometimes nuclear, as is typical in the United States and much of Europe, or extended, including several generations in the same household. This wide range of family structures for humans is made possible by our species' large brains that afford a high degree of *plasticity* (ability to change) in response to environmental conditions. The diversity of ecological and social conditions in which humans live requires behaviorally flexible systems that are able to adjust to local conditions rather than highly constrained mating and parenting systems. Plasticity, which affords learning, is an evolved characteristic of *Homo sapiens*.

To some extent, each family within each culture is unique, with individual children from a single family being treated differently by their parents. Children typically have enough plasticity to adjust to the vagaries of the particular child-rearing practices of their family to reach maturity, although some might attain adulthood better prepared for success in their community than others. Despite the fact that differences in child-rearing practices may vary for different children within a given family, more substantial differences are found between children growing up in more-advantaged versus less-advantaged families within a society, and there is even more variation among the thousands of different world cultures.

In this section, we first briefly describe the importance of developmental plasticity that permits children to adapt to a wide range of parenting practices. We then describe some of the child-rearing practices among cultures that differ substantially from those typical of WEIRD cultures. We conclude the section by looking at ways of classifying, in a general sense, different forms of child-rearing practices.

The importance of developmental plasticity

Because cultures vary widely in how children are viewed and treated (see Keller, 2007, 2017; Lancy, 2015), children must be able to adapt to a wide range of rearing conditions. A species that lives in as diverse environments as humans do could not require a narrow range of parenting styles to produce competent adults lest it faces extinction. This point was made by developmental psychologist Sandra Scarr (1992, 1993), who proposed that there is an *average expectable environment*, with infants and children prepared by natural selection to respond to a relatively wide range of environmental conditions in order to acquire developmentally appropriate knowledge. This degree of plasticity does not mean that human children are born as blank slates, with an infinitely pliable nervous system that can adjust to all but the most atypical environments. Children and their parents are mammals, after all, and natural selection has produced mechanisms designed to foster the formation of social relationships with significant others in order to survive. Within these constraints, however, humans have evolved neural, cognitive, and behavioral plasticity, and this plasticity is greatest early in life (see Bjorklund & Rosenberg, 2005; Causey, Gardiner, & Bjorklund, 2008). One must consider both humans' evolved constraints and evolved plasticity in order to acquire a proper understanding of human parenting and family practices.

How variable are human family and child-rearing practices?

Western scientists and educations have extensively documented long-term effects of early parent-child interaction on children's intellectual, linguistic, and emotional development (e.g., Sameroff, Seifer, Baldwin, & Baldwin, 1993; Thompson, 2006). Despite the often seemingly substantial differences in the experiences of children from advantaged versus disadvantaged homes in WEIRD cultures, these differences are dwarfed when compared to some of the typical child-rearing practices found in traditional cultures. Below are few examples from David Lancy's (2015) book, *The Anthropology of Childhood*, of what, from a Western perspective, would be considered "extreme" parenting practices:

> *Yoruba mothers feed children barely visible scraps compared to portions they give themselves. Good food might spoil the child's moral character (p. 108).*
>
> *Pashtu mothers rarely make eye contact with their infants when nursing unless there's a problem... This seeming indifference may be reinforced by custom whereby a mother is chastised by peers if she is overly fond of her child (p. 121).*
>
> *The Zulu of South Africa use a more direct approach [to accelerate walking]; they place the child on an ant's nest to motivate it to stand and walk (p. 134).*
>
> *As they begin to become more and more children rather than babies, and begin to be a bit irritating and willful because they are "thinking for themselves," [Tahitians] begin to find children less amusing. Instead of being the center of the household stage, the child... becomes annoying (p. 135).*
>
> *[From medieval times in Europe] parents sent their five- to seven-year-old children (who were, in their view, no longer actually children) to the homes of master-craftsmen or merchants as apprentices, where the first decade of service might well be 'scut-work' rather than the acquisition of usable skills (p. 151).*
>
> *Like the Xhosa, [Gapun] mothers actively pit their three-year-olds (girls as well as boys) against each other, holding them in proximity and shouting orders to strike out at the opponent. Children are also encouraged and praised for hitting dogs and chickens (p. 192).*
>
> *Corporal punishment is, thus, often seen as a legitimate tool in shaping the child's behavior. The Rwala Bedu (Syria) utilize an arsenal of physical punishments ranging from spanking with a stick (small children) to slashing with a saber or dagger (older children) (p. 196).*
>
> *One of the consequences of treating young children as senseless, incomplete, and amoral is that adults feel no qualms about their exposure to sexual activity (p. 199).*
>
> *Qualities we [westerners] value, such as precocity, verbal fluency, independent and creative thought, personal expression, and ability to engage in repartee, would all be seen by [the lower Tapajós] villagers as defects to be curtailed as quickly as possible (p. 200).*

Some of these practices, and many others described in Lancy's book, may seem not only as examples of poor parenting, but downright cruel (*placing children on ants' nests to motivate walking* and *encouraging the abuse of animals!*). Yet, there are few if any species-universal good or bad parenting practices. Rather, different parenting practices seem to be well-suited (if not perfectly so) to the ecological niche in which children and their parents live. Children in all of these societies grow up to be functional members of their group, capable of making a living, forming meaningful relationships with others, and producing and raising children. In fact, according to Keller (2017), the particular ecological, social, and historical milieu of a culture "necessitate different psychologies to enable and facilitate competence in particular environments. Different psychologies are constituted along different developmental pathways" (p. 834). In other words, different child-rearing practices evolved (over historical time) to prepare children for life in a culture similar to the one that their parents and grandparents experienced. As we mentioned earlier, because the demands and practices of human cultures vary so greatly, a narrow range of parenting practices could not be required to raise competent adults across the wide range of human societies.

It is not practical to examine the child-rearing practices of every human culture; however, some generalities can be made based on shared common features among cultures. For example, Heidi Keller (Keller, 2007, 2017, 2018; Keller & Kärtner, 2013) contrasts child-rearing practices between Western middle-class families (i.e., WEIRD cultures) and traditional farmers. Most research on the effects of child-rearing practices on the development of children has been done in Western cultures, and there is a tendency to generalize the findings of such studies to all of humanity. However, Keller notes that Western middle-class families represent only a small minority of the world's population (about 5%), making generalizations inappropriate. In comparison, traditional farmers constitute between 30% and 40% of the world's population. Western middle-class families tend to be highly educated, marry late, have few children, and live in nuclear two-generation families. In contrast, traditional farmers tend to have little or no formal education, marry early, have many children, and live in extended multigenerational households. Given these differences, it is not surprising that to raise a successful child in a Western middle-class society requires different parental practices than those needed to raise a successful child in a traditional farming society.

Keller identified some generalities that can be made about the different child-rearing practices between Western middle-class and traditional farming cultures. For example, young infants in traditional cultures are rarely out of contact with their mothers or other caregivers, even when sleeping, whereas infants in Western cultures, although spending much face-to-face time with their mothers, also spend much time alone, which serves to foster greater independence. In many traditional cultures, parents place more of an emphasis on motor development than Western cultures, attempting to hasten walking (recall the Zulu of South Africa placing infants on ant hills). The earlier children can locomote on their own, the earlier they can do simple chores that are appropriate for youngsters in farming communities. In contrast, parents from Western middle-class cultures tend to place greater emphasis on cognitive development relative to traditional cultures, attempting to facilitate the "independent mental agency of the baby" (Keller, 2017, p. 836). Middle-class parents frequently talk to their preverbal infants about their feelings and what they know. Although the goal in both cultures is to raise an independent child, motor independence is more important for traditional farmers, whereas cognitive independence is more important for Western families.

These different perspectives on what is important early in life are associated with different patterns of development. For example, although Western children typically recognize themselves in a mirror by 18 or 19 months (mirror recognition being an indication of a developing sense of self), children from traditional Nso farming communities only do so several months later (Kärtner, Keller, Chaudhary, & Yovsi, 2012). Developmental patterns of attachment also differ between traditional and Western cultures, with children in traditional cultures frequently forming attachments to multiple caregivers and some not showing the typical fear of strangers, which is usually found in Western babies between 6 and 12 months of age (Lancy, 2015; Otto, 2014).

Taking a slightly different perspective, Lancy (2015) classifies different cultural child-rearing practices on a continuum. At one extreme are cultures in which children are highly valued. They are viewed as innocent, requiring substantial nurturing and protection, often at great expense to their parents. Lancy refers to such cultures as *neontocracies*, in which "children have authority – lording it over their valet parents" (Lancy, 2015, p. 72). Lancy argues that the Western middle-class cultures described by Keller are neontocracies but so are traditional hunter/gatherer societies, which likely reflected the environments of our ancient ancestors. Hunter/gatherer societies today, and likely those of our forebears, cherish their children and make few demands on them. Children are carried as infants until they can walk, and once weaned they spend most of their time playing with other children with little adult supervision; they have no formal education (adults rarely spend time directly teaching children), and they acquire the skills necessary for survival mainly through play and imitation. In fact, a number of theorists have proposed that children's propensities for imitation are evolved mechanisms for acquiring the skills and social rituals of one's culture (e.g., Legare & Nielsen, 2015; Nielsen, 2012, 2018).

At the other extreme are cultures Lancy calls *gerontocracies*, in which adults are the most valued members of society. Infants and children are recognized as a drain on resources, with children seen as owing their parents for parents' investment in them over the years. Infants and children may be valued only for their economic contribution to the family, last in line when it comes to receiving resources such as food, and viewed as property (chattel). Children may be expected to work and even be indentured to others to pay for parents' debts. Gerontocracies are characteristic of some of the traditional farming societies that Keller describes.

Compared to neontocracies and certainly to current Western standards, gerontocracies can be viewed as harsh and perhaps stultifying. Yet, gerontocracies may be well-suited to the type of adult life that children in these societies can expect, which can be seen with respect to the use of corporal punishment. In the research conducted in Western cultures, study after study has found that, on average, spanking does not lead to positive behavioral outcomes and is rather associated with negative psychological outcomes (e.g., Gershoff et al., 2018; Ma, Grogan-Kaylor, & Lee, 2018; Turns & Sibley, 2018). In fact, corporal punishment for children has been banned in at least 42 countries, including, Argentina, Austria, Bolivia, Brazil, Croatia, Cyprus, Denmark, Finland, Germany, Honduras, Israel, Italy, Latvia, Norway, Republic of the Congo, and Sweden (UN Tribune, 2014).

Some Western psychologists have referred to spanking as a form of terrorism. Although seemingly extreme, there is a truth reflected in this perspective; terrorism relies on evoking fear in which a terrorizing force acts aggressively in the knowledge that its victim will be unable to effectively retaliate. According to this perspective, a parent who spanks a child is intending to cease an undesired behavior by way of employing fear tactics and the anticipation that the child will passively succumb to the punishment. It is no wonder, then, that when children become old enough to fight back, this form of punishment ceases. Western researchers have concluded that corporal punishment can lead to explicit child-abusive behaviors and that spanking serves as a model for violent behavior, teaching children that violence is an appropriate way to handle frustration or anger (Curran, 2002). For example, Turns and Sibley (2018) found that spanking at 1 year of age positively predicted childhood bullying at age three for boys, though interestingly, a negative prediction was observed for girls.

These negative effects of corporal punishment are not found in all cultures, however. As noted, spanking is most commonly practiced in gerontocracies, which are marked by an unequal power dynamic in association with high levels of political hierarchy, social stratification, or by a colonial power, that often incorporate corporal punishment due to the embedded values of compliance and obedience (Levinson, 1989; Ripoll-Núñe & Rohner, 2006). In cultures where spanking is not frequently practiced as a normative means of punishment (i.e., neontocracies), the long-term effects of its usage are usually characterized by insecurity, increases in antisocial and criminal behavior, slowing of cognitive development (Straus, Douglas, & Medeiros, 2014), and overall maladjustment (Gershoff et al., 2018). In contrast, in cultures where spanking is the norm, developmental outcomes of the child have not been shown to be detrimental and even have been shown as beneficial according to a cultural normativeness perspective (Lancy, 2015). Differences in the consequences of corporal punishment are also found within a culture. For example, in a review of the effects of spanking on psychological outcomes, Lansford (2010) reported the expected patterns between use of corporal punishment and behavior problems for samples of White American children (higher levels of corporal punishment was associated with higher levels of current and future behavior problems); however, these relations were smaller or nonexistent for samples of Black and Hispanic American children.

There are evidently many differences in the lifestyles of children growing up in WEIRD cultures versus those growing up in more traditional cultures. Different cultures require different skills for their adults, and, regardless of culture, the social, emotional, intellectual, and practical skills necessary to survive and thrive in a culture take time to learn. The ability to acquire essentially any of the psychologies typical of the world's thousands of cultures necessitates a high level of neural, cognitive, and behavioral plasticity on the part of the child. As Keller (2017) states, "Humans are born with an enormous and universal array of developmental possibilities. Cultures select and reinforce the behavioral tendencies that are promising adaptive value in the particular environment in which the baby is born" (p. 837).

Explaining patterns of individual differences in child-rearing practices between and within cultures: Life history theory

Natural selection has provided children with the cognitive, neural, and behavioral plasticity to develop into functional members of any of the world's various cultures. *Life history theory* is an evolutionary framework for explaining how developmental plasticity in interaction with an animal's ecology produces adaptive individual differences (between species, between cultures, and between individuals within a culture) (e.g., Ellis, Figueredo, Brumbach, & Schlomer, 2009; Hill & Kaplan, 1999; Stearns, 1992).

Life involves trade-offs, and this becomes especially apparent once one becomes a parent. All sexually reproducing animals must make decisions (although likely not conscious ones) about how to distribute their limited resources. A central tenet of natural selection is that resources are limited—if not, all individuals would be able to get what they need to survive, eliminating the role of the environment in influencing evolution. Time can also be considered a limited resource, in that time spent foraging for food, for example, cannot be used for finding a mate or raising offspring. Because each of these activities contributes to an organism's fitness, devoting effort to one activity will usually involve both costs and benefits, and thus trade-offs between different fitness components (Williams, 1966). At perhaps the most basic level, there is a trade-off between reproduction and bodily growth, in that both, especially for females, require substantial energetic investment: having babies reduces somatic development. Somatic development includes not only growth and fitness of the physical body, but also developmental activity, which includes learning, exercise, play, and other activities that contribute to building *embodied capital*, such as strength, skills, and knowledge (Kaplan & Gangestad, 2005). Individuals make a series of decisions about fundamental trade-offs between mating and parenting effort, quality and quantity of offspring, and current and future reproduction, which constitute the individual's *life history strategy* (Ellis et al., 2009; Hill, 1993; Kaplan & Gangestad, 2005).

Depending on the species, its phylogenetic history, and the ecological context in which an animal lives, different trade-offs will produce different probable outcomes. That is, there is no ideal life history strategy, but rather the strategy that an individual develops is one suited to current (and anticipated) environmental conditions. At the broadest level of analysis, life history strategies exist on a continuum from *fast* to *slow* (see Fig. 2.1).

A *fast life history strategy* is typical of short-lived animals or animals living in harsh and unpredictable environments. Under these conditions, it is adaptive to take an opportunistic versus a futuristic orientation, to take risks, to mature and reproduce quickly, to invest little in long-term relationships, and to have many offspring. In comparison, a *slow life history strategy* is typical of long-lived animals or animals living in predictable and resource-rich environments with reliable relationships. Under these conditions, it is adaptive

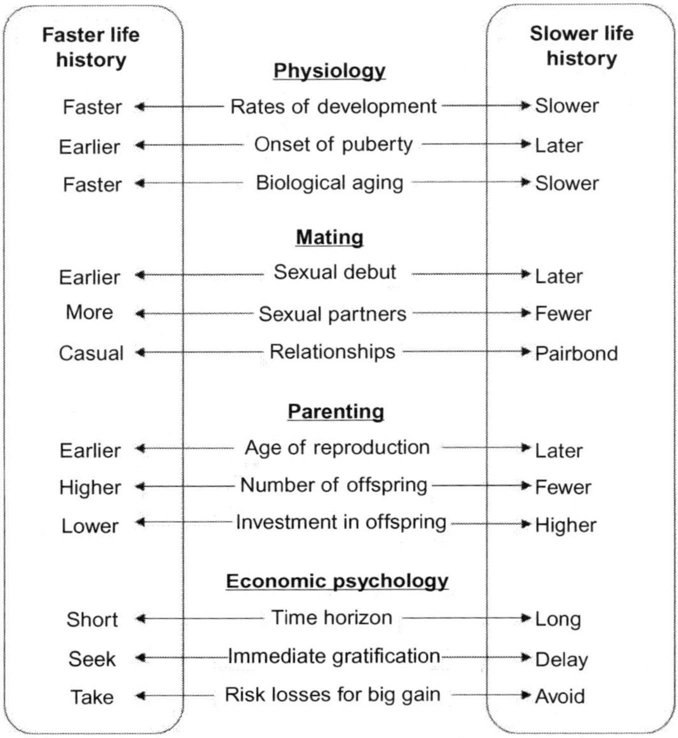

Fig. 2.1 The fast-slow continuum of life history variation. *(From Ellis, B. J., Del Giudice, M., Dishion, T. J., Figueredo, A. J., Gray, P., Griskevicius, V., et al. (2012). The evolutionary basis of risky adolescent behavior: Implications for science, policy, and practice.* Developmental Psychology, 48, *598–623. https://doi.org/10.1037/a0026220.)*

to take a more futuristic perspective, to develop more slowly, invest more in developing one's body and behavioral skills (embodied capital), be risk-prone, and invest more in relationships and in a smaller number of offspring. As a species, *Homo sapiens* are clearly at the extreme slow end of the life history continuum. However, there is substantial variability both between and within cultures in the factors that influence the life history strategy a person adopts, particularly in terms of harshness (few vs many resources, high vs low morbidity/mortality) and predictability, and thus variability in the life history strategies that people follow.

Consistent with the idea that plasticity is greatest early in life, children are sensitive to early environmental conditions and adjust important aspects of their developmental trajectory in anticipation of future environments. In an early study in developmental psychology that adopted a life history approach, Belsky, Steinberg, and Draper (1991) proposed that "a principal evolutionary function of early experience – the first 5–7 years of life – is to induce in the child an understanding of the availability and predictability of resources (broadly defined) in the environment, of the trustworthiness of others, and of the enduringness of close interpersonal relationships – all of which will affect how the developing person apportions reproductive effort" (p. 650). Essentially, Belsky and his colleagues proposed that children evolved the necessary plasticity to adapt to important aspects of their development to match current and anticipated environments.

Life history strategies in gerontocracies versus neontocracies

Compared to other species, humans follow a slow life history strategy. As noted earlier, humans have a prolonged period of immaturity and a long lifespan relative to other mammals, extending a trend observed in primates in general. Also compared to other mammals, humans are slow to attain puberty; women usually do not have their first births until the late teens or early 20s, they have relatively few offspring (mostly single births with 4 or 5 years between births), have enduring relationships—in part because of the need to cooperate with others in rearing a child—and invest substantially in each offspring. Yet, within this species-general pattern there can be great variability, with most societies described as gerontocracies (traditional farmers) adopting fast life history strategies and societies described as neontocracies (Western middle class) adopting slow life history strategies.

According to Lancy (2015), the earliest humans (our hunter/gatherer ancestors) were the first neontocracies, despite relatively high mortality (life expectancy at birth likely ranged from about 21 to 40 years, Gurven & Kaplan, 2007) and sometimes living in harsh conditions (e.g., possible conflict between neighboring groups, Diamond, 2013; Pinker, 2011). The nature of hunter/gatherer life allowed adults to treat children well and to follow a slow life history strategy. As we noted previously, hunter/gatherer children were highly valued with little expected of them; this was afforded by the obligatory features of foragers' lifestyles. According to Lancy (2015):

> *There is a cluster of factors that undergird this pattern of infant and childcare. First, unlike agrarian and industrialized societies, among most foragers, children's lack of strength and stamina renders them incapable of contributing very much to family subsistence or income... Second, foraging is, by definition, the active pursuit of food resources, and mothers will, of necessity, carry their newborn constantly unless they can pass it off to another caretaker. Third... foragers adopt a 'survivorship'*

reproductive strategy. Around-the-clock nursing and a postpartum sex taboo combine to insure long intervals between births, leading to lower fertility. Low fertility is offset by the attention bestowed on the few offspring, enhancing their chances of survival... Fourth, there are absolutely zero alternative life courses available within a foraging community. Children will either learn what they need to – including important social skills – to feed themselves and find a mate, or starve. So there is no need to rein them in, guide them, or teach them (p. 69).

With the advent of agriculture and the disappearance of a foraging way of life, many aspects of group life changed, including the lives of children. Mortality rates for the early farmers apparently decreased only slightly relative to hunter/gatherers, but women typically reduced their interbirth intervals relative to those of hunter-gatherers, resulting in larger families (Gage & DeWitte, 2009). Being sedentary and having to tend to small plots of land and care for domesticated animals brought with it many simple chores, some of which children could do. Children were expected to "pay back" parents for parents' investments. Children were valued for their usefulness, including child labor. Children were seen as chattel, providing benefits to parents and could be sold into slavery.

The extreme examples of contemporary gerontocracies described by Lancy (2015) are likely similar in many ways to the lifestyles of the early agriculturalists, and they are prototypical cases of environments that foster fast life history strategies. As noted previously, compared to children growing up in neontocracies, children growing up in gerontocracies have faster rates of physical development and earlier sexual debut and ages of marriage and first births; they have more children and are more likely to use corporal punishment as a form of childhood discipline. Harsh life conditions and fast life history strategies not only typify traditional farmers but also urban dwellers and children who lived in most of the world until the 20th century when affluence in Western societies resulted in the development of the middle class and contemporary neontocracies, which tend to follow slower life history strategies. The opportunistic nature of fast life history strategies fits well with the environmental conditions in which children in gerontocracies grow up. Resources are often scare and unpredictable, and there is every reason to believe children's lives as adults will be the same; postponing having children or anticipating conditions too far into the future is maladaptive for those living in gerontocracies. The lives of children and adults in these cultures may not be happy ones by modern standards, but they are well suited to the economic and social conditions in which they live.

We need to keep in mind that Lancy's (2015) neontocracy-gerontocracy distinction is a continuum. Not all traditional farming cultures are best described as gerontocracies, and even if they are, they are not necessarily at the extreme end of that continuum. Likewise, within modern Western societies, there are individual differences in parenting practices and thus in the life history strategies that children follow. These differences are apt to be most apparent in countries with substantial income inequality, with the socioeconomic status being a major multifaceted factor influencing one's life history strategy.

Life history strategies for explaining individual differences in developed societies

Over the past decade or so, much research has been done looking at differences in child-rearing practices and adult outcomes of children in developed, mostly WEIRD, cultures through the lens of life history theory. Belsky et al. (1991), discussed briefly earlier, proposed that children growing up in homes with low resources, high unpredictability, harsh child-rearing practices, exposure to violence, and who display insecure attachment to their parents, will adopt fast life history strategies, engaging in high levels of risky and aggressive behavior, forming unstable pair bonds, and providing less parental investment to their children, with girls achieving menarche and engaging in sex earlier than girls growing up in less harsh and more predictable environments. This pattern has generally been confirmed in a number of both retrospective and prospective studies (e.g., Chang & King Lu, 2018; Ellis, 2004; Ellis et al., 2009; Hentges & Wang, 2018; Nettle, 2010; Nettle & Cockerill, 2010; Placek & Quinlan, 2012; Szepsenwol, Simpson, Raby, & Griskevicius, 2018). Other research has similarly shown that children classified as insecurely attached at 12 months of age tend to display more negative emotions in their adult romantic relationships in early adulthood (e.g., Simpson, Collins, & Salvatore, 2011; Simpson, Collins, Tran, & Haydon, 2007).

With respect to perhaps, the most-studied factor predicted by life history theory—pubertal timing in girls as a function of their early family environments—several lines of research support that stress early in life is associated with an earlier age of pubertal onset. For example, greater familial warmth, positivity, closeness, and parental approval predict later pubertal maturation in girls (e.g., Ellis, McFadyen-Ketchum, Dodge, Pettit, & Bates, 1999; Graber, Brooks-Gunn, & Warren, 1995); higher levels of parent-child conflict and coercion predict earlier pubertal maturation in girls (e.g., Belsky et al., 2007; Costello, Sung, Worthman, & Angold, 2007); and growing up in families with positive spousal relationships and little conflict is associated with girls' later pubertal maturation (e.g., Ellis & Graber, 2000; Saxbe & Repetti, 2009).

In other research, Simpson, Griskevicius, Kuo, Sung, and Collins (2012) reported that children living in highly unpredictable environments (e.g., parental job changes, changes in residences, different nonbiologically related adult males living in the household) during their first 5 years of life had their first sexual encounter earlier, had more sex partners, and engaged in higher levels of risk-taking and delinquent behavior at age 23 than children growing up in more predictable homes. In a study examining the effects of unpredictable environments on cognitive development, adults who had grown up in unpredictable environments performed worse on cognitive inhibition tasks but better on tasks requiring cognitive flexibility (Mittal, Griskevicius, Simpson, Sung, & Young, 2015). The authors speculated that, although inhibition is important for attaining long-term goals, it can be maladaptive when ecological conditions favor opportunism. In comparison, the ability to shift effectively between tasks is critical for adapting to unpredictable environments.

From the perspective of Western culture, a fast life history strategy is associated with socially maladaptive outcomes—high levels of aggression, criminality, unsafe sex—often with poor mental health outcomes. However, from an evolutionary perspective, an opportunistic, fast life history strategy may be adaptive for a child growing up in a harsh and uncertain environment (Ellis, Bianchi, Griskevicius, & Frankenhuis, 2017; Ellis et al., 2012, 2009). Succeeding in an unpredictable environment is made possible by plasticity, allowing for adaptability, and likely requires prioritizing short-term gains over what may be achieved in an uncertain future.

Conclusion

Humans, perhaps more than any other species, depend on the prolonged actions of their parents and other significant individuals to survive and develop into functional members of their society. The complexity and diversity of humans' social lives necessitated the evolution of a highly plastic nervous system in its youngest members, permitting them to adapt to the particular features of their surroundings. Although the lifestyles of different groups of our ancient hunter/gatherer forebears were seemingly similar to one another—what Lancy (2015) describes as the original neontocracies—with the advent of the agricultural revolution and the dispersion of humans to all parts of the globe, the relative homogeneity of lifestyles and child-rearing practices (they were never all the same, of course) have disappeared. New lifestyles and diverse environments produced different child-rearing practices, adapted to the physical and social demands of a particular ecology. Children and their parents still retained mechanisms evolved in the environment of evolutionary adaptedness to promote attachment and investment, though the execution of these mechanisms varied from culture to culture, depending on the demands of the environment.

From this perspective, there are no universal "good" or "poor" parenting practices, in that different practices have developed over historical time to produce successful adults in particular cultures. Within a culture, some forms of parenting might yield better adjustment or more successful adults than others (e.g., the use of corporal punishment in agrarian societies). Similarly, some forms of parenting might result in maladaptation if a person's adult environment is substantially different from the one of his or her childhood. One must also consider what does being a "successful" adult mean. From a strictly biological perspective, success might mean reaching adulthood and bearing children and grandchildren, perhaps many of them; from the perspective of Western culture, it might mean having meaningful relationships and maintaining psychological well-being.

As we have mentioned previously, some theorists (e.g., Gray, 2013; Konner, 2010) have argued that the childhoods of hunter/gatherers should be the standard by which we judge the child-rearing practices of different cultures today and that deviations from the child-rearing practices of these original humans can result in poor mental health outcomes. In fact, although we cannot know the mental health status of our foraging ancestors, some scientists argue that despite higher mortality rates, modern hunter/gatherers—and thus likely our ancient ancestors—are happier and have better mental health (e.g., less anxiety and depression) than modern Western people (e.g., Suzman, 2017).

In many ways, the childhoods of children from WEIRD societies are likely more similar to those of the early hunter/gatherers than those of children from gerontocracies, which typified much of humans' last 10,000 years. But survival, rather than happiness and psychological well-being, was likely the major concern for most people throughout most of human (pre)history, and it is only recently that our species has had the luxury to worry about its members' psychological adjustment or happiness. A number of contemporary researchers and theorists have argued that increasing departures from the childhoods of hunter/gatherers for contemporary children are responsible, in part, for elevated levels of anxiety, depression, loneliness, and suicide among teenagers and young adults from WEIRD cultures, as well as a general unpreparedness for adult life (e.g., Gray, 2013; Konner, 2010; Lukianoff & Haidt, 2018; Twenge, 2017). For example, compared to children growing up in America during the middle part of the 20th century, today's American middle-class children engage in less free play, are overprotected by helicopter parents, and spend much time online playing video games and engaging with social media, often at the expense of face-to-face socializing with other people. Both correlational and experimental studies have found links between time spent on digital media and rates of depression, anxiety, suicide, loneliness, and unhappiness (e.g., Primack et al., 2017; Twenge, Martin, & Campbell, 2018; see Twenge, 2017 for an extensive review). At the same time, formal education itself is an evolutionarily novel experience, which often overlooks children's evolved ways of learning (Bjorklund, 2007; Gray, 2013). This may especially be the case for parents and educational programs that emphasize formal instruction during the preschool years (e.g., Bjorklund & Beers, 2016).

An evolutionary perspective provides a unique look at the differences in child-rearing practices seen in our ancestral past compared to today and across today's broad range of cultures. Such a perspective helps explain both the diversity among cultures as well as the species-general commonalities, or universals, that all humans share. Furthermore, an evolutionary approach provides not only (pre)historical context for understanding how different child-rearing approaches developed in different world cultures but can also offer insights into modern practices and perhaps suggest ways in which *Homo sapiens* can live happier, more fulfilled lives.

References

Alexander, R. D. (1989). Evolution of the human psyche. In P. Mellers & C. Stringer (Eds.), *The human revolution: Behavioural and biological perspectives on the origins of modern humans* (pp. 455–513). Princeton, NJ: Princeton University Press. https://doi.org/10.1002/ajhb.1310040119.

Anderson, K., Kaplan, H., & Lancaster, J. (1999). Paternal care by genetic fathers and stepfathers I: Reports from Albuquerque men. *Evolution and Human Behavior, 20*, 405–431. https://doi.org/10.1016/S1090-5138(99)00023-9.

Bardi, L., Regolin, L., & Simion, F. (2011). Biological motion preference in humans at birth: Role of dynamic and configural properties. *Developmental Science, 14*, 353–359. https://doi.org/10.1111/j.1467-7687.2010.00985.x.

Bardi, L., Regolin, L., & Simion, F. (2014). The first time ever I saw your feet: Inversion effect in newborns' sensitivity to biological motion. *Developmental Psychology*, *50*, 986–993. https://doi.org/10.1037/a0034678.

Belsky, J., Steinberg, L., & Draper, P. (1991). Childhood experience, interpersonal development, and reproductive strategy: An evolutionary theory of socialization. *Child Development*, *62*, 647–670. https://doi.org/10.2307/1131166.

Belsky, J., Steinberg, L., Houts, R. M., Friedman, S. L., DeHart, G., Cauffman, E., et al. (2007). Family rearing antecedents of pubertal timing. *Child Development*, *78*, 1302–1321. https://doi.org/10.1111/j.1467-8624.2007.01067.x.

Bjorklund, D. F. (1987). A note on neonatal imitation. *Developmental Review*, 7, 86–92. https://doi.org/10.1016/0273-2297(87)90006-2.

Bjorklund, D. F. (2007). *Why youth is not wasted on the young: Immaturity in human development.* Malden, MA: Blackwell.

Bjorklund, D. F. (2018). A metatheory for cognitive development (or "Piaget is dead" revisited). *Child Development*, *89*, 2288–2302. https://doi.org/10.1111/cdev.13019.

Bjorklund, D. F., & Beers, C. (2016). The adaptive value of cognitive immaturity: Applications of evolutionary developmental psychology to early education. In D. C. Geary & D. B. Berch (Eds.), *Evolutionary perspectives on education and child development* (pp. 3–32). New York, NY: Springer. https://doi.org/10.1007/978-3-319-29986-0_1.

Bjorklund, D. F., Hernández Blasi, C., & Periss, V. (2010). Lorenz revisited: The adaptive nature of children's supernatural thinking. *Human Nature*, *21*, 371–392. https://doi.org/10.1007/s12110-010-9099-8.

Bjorklund, D. F., & Myers, A. J. (2019). The evolution of parenting and evolutionary approaches to childrearing. In M. H. Bornstein (Ed.), *The biology of parenting: Vol. 1. Handbook of parenting*. (3rd ed., pp. 3–29). Mahwah, NJ: Erlbaum. https://doi.org/10.4324/9780429401459-1.

Bjorklund, D. F., & Pellegrini, A. D. (2002). *The origins of human nature: Evolutionary developmental psychology*. Washington, DC: American Psychological Association.https://doi.org/10.1037/10425-000.

Bjorklund, D. F., & Rosenberg, J. S. (2005). The role of developmental plasticity in the evolution of human cognition. In B. J. Ellis & D. F. Bjorklund (Eds.), *Origins of the social mind: Evolutionary psychology and child development* (pp. 45–75). New York, NY: Guilford.

Block, J., Block, J. H., & Keyes, S. (1988). Longitudinally foretelling drug usage in adolescence: Early childhood personality and environmental precursors. *Child Development*, *59*, 336–355. https://doi.org/10.2307/1130314.

Bogin, B. (2001). *The growth of humanity*. New York, NY: Wiley.

Borgi, M., Cogliati-Dezza, I., Brelsford, V., Meints, K., & Cirulli, F. (2014). Baby schema in human and animal faces induces cuteness perception and gaze allocation in children. *Frontiers in Psychology*, *5*, 411. https://doi.org/10.3389/fpsyg.2014.00411.

Bowlby, J. (1969). *Attachment and loss: Vol. 1: Attachment.* London: Hogarth.https://doi.org/10.1111/j.1939-0025.1982.tb01456.x.

Bugental, D. B., Beaulieu, D. A., & Silbert-Geiger, A. (2010). Increases in parental investment and child health as a result of an early intervention. *Journal of Experimental Child Psychology*, *106*, 30–40. https://doi.org/10.1016/j.jecp.2009.10.004.

Bugos, P., & McCarthy, L. (1984). Ayoreo infanticide: A case study. In G. Hausfaterand & S. Hrdy (Eds.), *Infanticide: Comparative and evolutionary perspectives*. New York, NY: Aldine.https://doi.org/10.1177/0022022190211001.

Buss, D. M. (2011). *The dangerous passion: Why jealousy is as necessary as love and sex* (2nd ed.). New York, NY: The Free Press.https://doi.org/10.1037/0003-066X.53.5.533.

Buss, D. M. (2016). *The evolution of desire: Strategies of human mating* (4th ed.). New York, NY: Basic Books.

Buss, D. M., Abbott, M., Angletimer, A., Asherian, A., Biaggio, A., et al. (1990). International preferences in selecting mates: A study of 37 cultures. *Journal of Cross-Cultural Psychology, 21*, 5–47. https://doi.org/10.1177/0022022190211001.

Buss, D. M., Haselton, M. G., Shackelford, T. K., Bleske, A. L., & Wakefield, J. C. (1998). Adaptations, exaptations, and spandrels. *American Psychologist, 53*, 533–548. https://doi.org/10.1037/0003-066X.53.5.533.

Campbell, A. (1999). Staying alive: Evolution, culture, and women's intrasexual aggression. *Behavioral and Brain Sciences, 22*, 203–252. https://doi.org/10.1017/S0140525X99001818.

Campbell, A. (2013). *A mind of her own: The evolutionary psychology of women*. New York, NY: Oxford University Press. https://doi.org/10.1093/acprof:oso/9780199609543.001.0001.

Camperio Ciani, A. S. C., & Fontanesi, L. (2012). Mothers who kill their offspring: Testing evolutionary hypothesis in a 110-case Italian sample. *Child Abuse & Neglect, 36*, 519–527. https://doi.org/10.1016/j.chiabu.2012.05.001.

Causey, K., Gardiner, A., & Bjorklund, D. F. (2008). Evolutionary developmental psychology and the role of plasticity in ontogeny and phylogeny. *Psychological Inquiry, 19*, 27–30. https://doi.org/10.1080/10478400701774089.

Chang, L., & King Lu, H. (2018). Resource and extrinsic risk in defining fast life histories of rural Chinese left-behind children. *Evolution and Human Behavior, 39*, 59–66. https://doi.org/10.1016/j.evolhumbehav.2017.10.003.

Clutton-Brock, T. H. (1991). *The evolution of parental care*. Princeton, NJ: Princeton University Press.

Costello, E. J., Sung, M., Worthman, C., & Angold, A. (2007). Pubertal maturation and the development of alcohol use and abuse. *Drug and Alcohol Dependence, 88*(Suppl. 1), S50–S59. https://doi.org/10.1016/j.drugalcdep.2006.12.009.

Curran, K. (2002). A cross-cultural comparison between Finland and the United States on the attitudes and practices of corporal punishment and abuse. *Dissertation Abstracts International Section A: Humanities and Social Sciences, 62*(11-A), 3688.

Daly, M., & Wilson, M. (1981). Abuse and neglect of children in evolutionary perspective. In R. D. Alexander & D. W. Tinkle (Eds.), *Natural selection and social behavior* (pp. 405–416). New York, NY: Chiron.

Daly, M., & Wilson, M. (1983). *Sex, evolution and behavior* (2nd ed.). Belmont, CA: Wadsworth.

Daly, M., & Wilson, M. (1988). *Homicide*. New York, NY: Aldine.

Danielsbacka, M., & Tanskanen, A. O. (2012). Adolescent grandchildren's perceptions of grandparents' involvement in UK: An interpretation from life course and evolutionary theory perspective. *European Journal of Ageing, 9*, 329–341. https://doi.org/10.1007/s10433-012-0240-x.

Danielsbacka, M., Tanskanen, A. O., Jokela, M., & Rotkirch, A. (2011). Grandparental child care in Europe: Evidence for preferential investment in more certain kin. *Evolutionary Psychology, 9*, 3–24. https://doi.org/10.1177/147470491100900102.

Del Giudice, M. (2009). Sex, attachment, and the development of reproductive strategies. *Behavioral and Brain Sciences, 32*, 1–21. https://doi.org/10.1017/S0140525X09000016.

DeVries, M. W. (1984). Temperament and infant mortality among the Masai of East Africa. *The American Journal of Psychiatry, 141*, 1189–1194. https://doi.org/10.1176/ajp.141.10.1189.

Dewsbury, D. A. (1988). The comparative psychology of monogamy. In D. W. Leger (Ed.), *Comparative perspectives in modern psychology: Nebraska Symposium on Motivation, 1987*. Lincoln, NE: University of Nebraska Press. https://doi.org/10.1037/0735-7036.102.2.124.

Diamond, J. (2013). *The world until yesterday: What can we learn from traditional societies?* New York, NY: Viking.

Dubas, J., Heijkoop, M., & Aken, M. (2009). A preliminary investigation of parent–progeny olfactory recognition and parental investment. *Human Nature, 20*, 80–92. https://doi.org/10.1007/s12110-009-9058-4.

Dunbar, R. I. M. (2003). The social brain: Mind, language, and society in evolutionary perspective. *Annual Review of Anthropology, 32*, 163–181. https://doi.org/10.1146/annurev.anthro.32.061002.093158.

Easterbrook, M. A., Kisilevsky, B. S., Hains, S. M. J., & Muir, D. W. (1999). Faceness or complexity: Evidence from newborn visual tracking of facelike stimuli. *Infant Behavior and Development, 22*, 17–35. https://doi.org/10.1016/S0163-6383(99)80003-4.

Ellis, B. J. (2004). Timing of pubertal maturation in girls: An integrated life history approach. *Psychological Bulletin, 130*, 920–958. https://doi.org/10.1037/0033-2909.130.6.920.

Ellis, B. J., Bianchi, J., Griskevicius, V., & Frankenhuis, W. E. (2017). Beyond risk and protective factors: An adaptation-based approach to resilience. *Perspectives on Psychological Science, 12*, 561–587. https://doi.org/10.1177/1745691617693054.

Ellis, B. J., Del Giudice, M., Dishion, T. J., Figueredo, A. J., Gray, P., Griskevicius, V., et al. (2012). The evolutionary basis of risky adolescent behavior: Implications for science, policy, and practice. *Developmental Psychology, 48*, 598–623. https://doi.org/10.1037/a0026220.

Ellis, B. J., Figueredo, A. J., Brumbach, B. H., & Schlomer, G. L. (2009). Fundamental dimensions of environmental risk: The impact of harsh versus unpredictable environments on the evolution and development of life history strategies. *Human Nature, 20*, 204–268. https://doi.org/10.1007/s12110-009-9063-7.

Ellis, B. J., & Graber, J. (2000). Psychosocial antecedents of variation in girls' pubertal timing: Maternal depression, stepfather presence, and marital and family stress. *Child Development, 71*, 485–501. https://doi.org/10.1111/1467-8624.00159.

Ellis, B. J., McFadyen-Ketchum, S., Dodge, K. A., Pettit, G. S., & Bates, J. E. (1999). Quality of early family relationships and individual differences in the timing of pubertal maturation in girls: A longitudinal test of an evolutionary model. *Journal of Personality and Social Psychology*, 77, 387–401. https://doi.org/10.1037/0022-3514.77.2.387.

Euler, H. A., & Michalski, R. (2007). Grandparental and extended kin relationships. In C. A. Salmon & T. K. Shackelford (Eds.), *Family relationships: An evolutionary perspective.* (pp. 230–255). New York, NY: Oxford University Press. https://doi.org/10.1093/acprof:oso/9780195320510.003.0011.

Euler, H. A., & Weitzel, B. (1996). Discriminative grandparental solicitude as reproductive strategy. *Human Nature*, 7, 39–59.

Fisher, H. (2004). *Why we love: The nature and chemistry of romantic love.* New York, NY: Holt.

Flinn, M. V. (1988). Step and genetic parent/offspring relationships in a Caribbean village. *Ethology and Sociobiology, 9*, 335–369. https://doi.org/10.1016/0162-3095(88)90026-X.

Franklin, P., & Volk, A. A. (2018). A review of infants' and children's facial cues' influence on adults' perceptions and behaviors. *Evolutionary Behavioral Sciences, 12*, 296–321. https://doi.org/10.1037/ebs0000108.

Fullard, W., & Reiling, A. M. (1976). An investigation of Lorenz's "babyness". *Child Development, 47*, 1191–1193. https://doi.org/10.2307/1128462.

Gage, T. B., & DeWitte, S. (2009). What do we know about the agricultural demographic transition? *Current Anthropology, 50*, 649–655. https://doi.org/10.1086/605017.

Gangestad, S. W., & Simpson, J. A. (2000). The evolution of human mating: Trade-offs and strategic pluralism. *Behavioral and Brain Sciences, 23*, 573–587. https://doi.org/10.1017/S0140525X0000337X.

Gaulin, S. J., McBurney, D. H., & Brakeman-Wartell, S. L. (1997). Matrilateral biases in the investment of aunts and uncles: A consequence and measure of paternity uncertainty. *Human Nature, 8*, 139–151. https://doi.org/10.1007/s12110-997-1008-4.

Geary, D. C. (2000). Evolution and proximate expression of human paternal investment. *Psychological Bulletin, 126*, 55–77. https://doi.org/10.1037/0033-2909.126.1.55.

Gershoff, E. T., Goodman, G. S., Miller-Perrin, C. L., Holden, G. W., Jackson, Y., & Kazdin, A. E. (2018). The strength of the causal evidence against physical punishment of children and its implications for parents, psychologists, and policymakers. *American Psychologist*, 7, 626–638. https://doi.org/10.1037/amp0000327.

Glocker, M. L., Langleben, D. D., Ruparel, K., Loughead, J. W., Gur, R. C., & Sachser, N. (2009). Baby schema in infant faces induces cuteness perception and motivation for caretaking in adults. *Ethology, 115*, 257–263. https://doi.org/10.1111/j.1439-0310.2008.01603.x.

Goetz, J. L., Keltner, D., & Simon-Thomas, E. (2010). Compassion: An evolutionary analysis and empirical review. *Psychological Bulletin, 136*, 351–374. https://doi.org/10.1037/a0018807.

Graber, J. A., Brooks-Gunn, J., & Warren, M. P. (1995). The antecedents of menarcheal age: Heredity, family environment, and stressful life events. *Child Development, 66*, 346–359. https://doi.org/10.2307/1131582.

Gray, P. (2013). *Free to learn: Why unleashing the instinct to play will make our children happier, more self-reliant, and better students for life*. New York, NY: Basic Books.

Gross, T. F. (1997). Children's perception of faces of varied immaturity. *Journal of Experimental Child Psychology, 66*, 42–63. https://doi.org/10.1006/jecp.1997.2373.

Gurven, M., & Kaplan, H. (2007). Longevity among hunter-gatherers: A cross-cultural examination. *Population and Development Review, 33*, 321–365. https://doi.org/10.1111/j.1728-4457.2007.00171.x.

Hamilton, W. D. (1964). The genetical theory of social behavior. *Journal of Theoretical Biology, 7*, 1–52. https://doi.org/10.1016/0022-5193(64)90038-4.

Hart, S. L. (2018). Jealousy *and* attachment: Adaptations to threat posed by the birth of a sibling. *Evolutionary Behavioral Sciences, 12*, 263–275. https://doi.org/10.1037/ebs0000102.

Hart, S. L., & Carrington, H. (2002). Jealousy in 6-month-old infants. *Infancy, 3*, 395–402. https://doi.org/10.1207/S15327078IN0303_6.

Hart, S. L., Carrington, H. A., Tronick, E. Z., & Carroll, S. R. (2004). When infants lose exclusive maternal attention: Is it jealousy? *Infancy, 6*, 57–78. https://doi.org/10.1207/s15327078in0601_3.

Henrich, J., Heine, S. J., & Norenzayan, A. (2010). The weirdest people in the world. *Behavioral and Brain Sciences, 33*, 61–135. https://doi.org/10.1017/S0140525X0999152X.

Hentges, R. F., & Wang, M.-T. (2018). Gender differences in the developmental cascade from harsh parenting to educational attainment: An evolutionary perspective. *Child Development, 89*, 397–413. https://doi.org/10.1111/cdev.12719.

Hernández Blasi, C., & Bjorklund, D. F. (2018). Adolescents' sensitivity to children's supernatural thinking: A preparation for adulthood? *Psicothema, 30*, 201–206. 2018-24698-009.

Hernández Blasi, C., Bjorklund, D. F., Agut, S., Lozano, F., & Martínez, M. A. (2018). Vocal cues as signaling behavior in early childhood. In *30th annual meeting of the Human Behavior and Evolution Society, Amsterdam, The Netherlands*.

Hernández Blasi, C., Bjorklund, D. F., & Ruiz-Soler, M. (2017). Children's supernatural thinking as a signaling behavior in early childhood. *British Journal of Psychology, 108*, 467–485. https://doi.org/10.1111/bjop.12211.

Hernández-Blasi, C., Bjorklund, D. F., & Ruiz-Soler, M. (2015). Cognitive cues are more compelling than facial cues in determining adults' reactions towards young children. *Evolutionary Psychology, 13*, 511–530. https://doi.org/10.1177/147470491501300212.

Hill, K. (1993). Life history theory and evolutionary anthropology. *Evolutionary Anthropology, 2*, 78–88. https://doi.org/10.1002/evan.1360020303.

Hill, K., & Kaplan, H. (1999). Life history traits in humans: Theory and empirical studies. *Annual Review of Anthropology, 28*, 397–430. https://doi.org/10.1146/annurev.anthro.28.1.397.

Hrdy, S. B. (1999). *Mother nature: A history of mothers, infants, and natural selection*. New York, NY: Pantheon Books.

Hrdy, S. B. (2009). *Mothers and others: The evolutionary origins of mutual understanding*. Cambridge, MA: Belnap Press.

Hrdy, S. B. (2016). Of marmosets, men, and the transformative power of babies. In W. R. Trevanthan & K. R. Rosenberg (Eds.), *Costly and cute: Helpless infants and human evolution* (pp. 177–203). Albuquerque, NM: University of New Mexico Press.

Huber, B. R., & Breedlove, W. L. (2007). Evolutionary theory, kinship, and childbirth in cross-cultural perspective. *Cross-Cultural Research, 41*, 196–219. https://doi.org/10.1177/1069397106298261.

Jasienska, G., Lipson, S. F., Ellison, P. T., Thune, I., & Ziomkiewicz, A. (2006). Symmetrical women have higher potential fertility. *Evolution and Human Behavior, 27*, 390–400. https://doi.org/10.1016/j.evolhumbehav.2006.01.001.

Kaplan, H., & Gangestad, S. W. (2005). Life history theory and evolutionary psychology. In D. M. Buss (Ed.), *The handbook of evolutionary psychology* (1st ed., pp. 68–95). Hoboken, NJ: Wiley. https://doi.org/10.1002/9780470939376.ch2.

Kaplan, H., Hill, K., Lancaster, J., & Hurtado, A. M. (2000). A theory of human life history evolution: Diet, intelligence, and longevity. *Evolutionary Anthropology, 9*, 156–185. https://doi.org/10.1002/1520-6505(2000)9:4<156::AID-EVAN5>3.0.CO;2-7.

Kärtner, J., Keller, H., Chaudhary, N., & Yovsi, R. (2012). The development of mirror self-recognition in different socio-cultural contexts. *Monographs of the Society for Research in Child Development*, 77(4) [serial no. 305].

Keller, H. (2007). *Cultures of infancy*. Mahwah, NJ: Erlbaum.

Keller, H. (2017). Culture and development: A systematic relationship. *Perspectives on Psychological Science, 12*, 833–840. https://doi.org/10.1177/1745691617704097.

Keller, H. (2018). Parenting and socioemotional development in infancy and early childhood. *Developmental Review, 50*, 31–41. https://doi.org/10.1016/j.dr.2018.03.001.

Keller, H., & Kärtner, J. (2013). Development—The culture-specific solution of universal developmental tasks. In M. L. Gelfand, C.-Y. Chiu, & Y.Y. Hong (Eds.), *Vol. 3. Advances in culture and psychology* (pp. 63–116). Oxford, England: Oxford University Press. https://doi.org/10.1093/acprof:oso/9780199930449.003.0002.

Konner, M. (2010). *The evolution of childhood: Relationships, emotions, mind*. Cambridge, MA: Belknap Press.

Kringelbach, M. L., Stark, E. A., Alexander, C., Bornstein, M. H., & Stein, A. (2016). On cuteness: Unlocking the parental brain and beyond. *Trends in Cognitive Science, 20*, 545–558. https://doi.org/10.1016/j.tics.2016.05.003.

Laham, S. M., Gonsalkorale, K., & von Hippel, W. (2005). Darwinian grandparenting: Preferential investment in more certain kin. *Personality and Social Psychology Bulletin, 31*, 63–72. https://doi.org/10.1177/0146167204271318.

Lamborn, S. D., Mounts, N., Steinberg, L., & Dornbusch, S. M. (1991). Patterns of competence and adjustment from authoritative, authoritarian, indulgent, and neglectful families. *Child Development, 62*, 1049–1065. https://doi.org/10.2307/1131151.

Lancy, D. (2015). *The anthropology of childhood* (2nd ed.). Cambridge, England: Cambridge University Press.

Langlois, J., Ritter, J., Casey, R., & Sawin, D. (1995). Infant attractiveness predicts maternal behavior and attitudes. *Developmental Psychology, 31*, 464–472. https://doi.org/10.1037/0012-1649.31.3.464.

Lansford, J. E. (2010). The special problem of cultural differences in effects of corporal punishment. *Law and Contemporary Problems, 73*, 89–106.

Lassek, W. D., & Gaulin, S. J. C. (2018). Do the low WHRs and BMIs judged most attractive indicate higher fertility? *Evolutionary Psychology*, 1–16. https://doi.org/10.1177/1474704918800063.

Lee, B. J., & George, R. M. (1999). Poverty, early childbearing and child maltreatment: A multinomial analysis. *Children and Youth Services Review, 21*, 755–780. https://doi.org/10.1016/S0190-7409(99)00053-5.

Legare, C. H., & Nielsen, M. (2015). Imitation and innovation: The dual engines of cultural learning. *Trends in Cognitive Sciences, 19*, 688–699. https://doi.org/10.1016/j.tics.2015.08.005.

Leibenluft, E., Gobbini, M. I., Harrison, T., & Haxby, J.V. (2004). Mothers' neural activation in response to pictures of their children and other children. *Biological Psychiatry*, *56*, 225–232. https://doi.org/10.1016/j.biopsych.2004.05.017.

Levinson, D. (1989). *Family violence in cross-cultural perspective*. Newbury Park, CA: Sage.https://doi.org/10.1007/978-1-4757-5360-8_18.

Lorenz, K. (1943). Die angeborenen Formen möglicher Erfahrung [The innate forms of possible experience]. *Zeitschrift fur Tierpsychologie*, *5*, 235–409. https://doi.org/10.1111/j.1439-0310.1943.tb00655.x.

Lucion, M. K., Oliveira, V., Bizarro, L., Bischoff, A. R., Silveira, P. P., & Kauer-Sant'Ana, M. (2017). Attentional bias toward infant faces—Review of the adaptive and clinical relevance. *International Journal of Psychophysiology*, *114*, 1–8. https://doi.org/10.1016/j.ijpsycho.2017.01.008.

Lukianoff, G., & Haidt, J. (2018). *The coddling of the American mind: How good intentions and bad ideas are setting up a generation for failure*. New York, NY: Penguin Press.

Luo, L., Li, L., & Lee, K. (2011). Are children's faces really more appealing than those of adults? Testing the baby schema hypothesis beyond infancy. *Journal of Experimental Child Psychology*, *110*, 115–124. https://doi.org/10.1016/j.jecp.2011.04.002.

Ma, J., Grogan-Kaylor, A., & Lee, S. J. (2018). Associations of neighborhood disorganization and maternal spanking with children's aggression: A fixed-effects regression analysis. *Child Abuse & Neglect*, *76*, 106–116. https://doi.org/10.1016/j.chiabu.2017.10.013.

Machluf, K., & Bjorklund, D. F. (2016). *Physical immaturity in infants triggers greater empathy in adults*. Paper presented at meeting of the Association of Psychological Science, Chicago, IL.

Mann, J. (1992). Nurture or negligence: Maternal psychology and behavioral preference among preterm twins. In J. Barkow, L. Cosmides, & J. Tooby (Eds.), *The adapted mind: Evolutionary psychology and the generation of culture* (pp. 367–390). New York, NY: Oxford University Press.

Marlowe, F. (1999). Showoffs or providers? The parenting effort of Hazda men. *Evolution and Human Behavior*, *20*, 391–404. https://doi.org/10.1016/S1090-5138(99)00021-5.

Meltzoff, A. N., & Moore, M. K. (1977). Imitation of facial and manual gestures by human neonates. *Science*, *198*, 75–78. https://doi.org/10.1126/science.198.4312.75.

Mittal, C., Griskevicius, V., Simpson, J. A., Sung, S., & Young, E. S. (2015). Cognitive adaptations to stressful environments: When childhood adversity enhances adult executive function. *Journal of Personality and Social Psychology*, *109*, 604–621. https://doi.org/10.1037/pspi0000028.

Mize, K. D., Pineda, M., Blau, A. K., Marsh, K., & Jones, N. A. (2014). Infant physiological and behavioral responses to a jealousy provoking condition. *Infancy*, 338–348. https://doi.org/10.1111/infa.12046.

Mondloch, C. J., Lewis, T. M., Budreau, D. R., Maurer, D., Dannemiller, J. L., Stephens, B. R., et al. (1999). Face perception during early infancy. *Psychological Science*, *10*, 419–422. https://doi.org/10.1111/1467-9280.00179.

Myers, A. J., & Bjorklund, D. F. (2018). An evolutionary perspective of rivalry in the family. In N. A. Jones & S. Hart (Eds.), *The psychology of rivalry* (pp. 1–33). Hauppauge, NY: Nova Science Publishers.

Nettle, D. (2010). Dying young and living fast: Variation in life history across English neighborhoods. *Behavioral Ecology*, *21*, 387–395. https://doi.org/10.1093/beheco/arp202.

Nettle, D., & Cockerill, M. (2010). Development of social variation in reproductive schedules: A study of an English urban area. *PLoS ONE*, *5*, e12690. https://doi.org/10.1371/journal.pone.0012690.

Nielsen, M. (2012). Imitation, pretend play, and childhood: Essential elements in the evolution of human culture? *Journal of Comparative Psychology*, *126*, 170–181. https://doi.org/10.1037/a0025168.

Nielsen, M. (2018). The social glue of cumulative culture and ritual behavior. *Child Development Perspectives, 12*, 264–268. https://doi.org/10.1111/cdep.12297.

Nielsen, M., Haun, D., Kärtner, J., & Legare, C. H. (2017). The persistent sampling bias in developmental psychology: A call to action. *Journal of Experimental Child Psychology, 162*, 31–38. https://doi.org/10.1016/j.jecp.2017.04.017.

Nunn, C. L., Gittleman, J. L., & Antonovics, J. (2000). Promiscuity and the primate immune system. *Science, 290*, 1168–1170. https://doi.org/10.1126/science.290.5494.1168.

Oostenbroek, J., Suddendorf, T., Nielsen, M., Redshaw, J., Kennedy-Costantini, S., Davis, J., et al. (2016). Comprehensive longitudinal study challenges the existence of neonatal imitation in humans. *Current Biology, 26*, 1334–1338. https://doi.org/10.1016/j.cub.2016.03.047.

Otto, H. (2014). Don't show your emotions! Emotion regulation and attachment in the Cameroonian Nso. In H. Otto & H. Keller (Eds.), *Different faces of attachment* (pp. 215–229). Cambridge, England: Cambridge University Press. https://doi.org/10.1017/CBO9781139226684.012.

Pashos, A. (2000). Does paternal uncertainty explain discriminative grandparental solicitude? A cross-cultural study in Greece and Germany. *Evolution and Human Behavior, 21*, 97–109. https://doi.org/10.1016/S1090-5138(99)00030-6.

Periss, V., & Bjorklund, D. F. (2011). Trials and tribulations of childhood: An evolutionary perspective. In C. Salmon & T. Shackelford (Eds.), *Oxford handbook of evolutionary family psychology* (pp. 149–168). Oxford, England: Oxford University Press. https://doi.org/10.1093/oxfordhb/9780195396690.013.0010.

Periss, V., Hernández Blasi, C., & Bjorklund, D. F. (2012). Cognitive "babyness": Developmental differences in the power of young children's supernatural thinking to influence positive and negative affect. *Developmental Psychology, 48*, 1203–1214. https://doi.org/10.1037/a0026979.

Pinker, S. (2011). *The better angels of our nature: Why violence has declined*. New York, NY: Viking.

Placek, C. D., & Quinlan, R. J. (2012). Adolescent fertility and risky environments: A population-level perspective across the lifespan. *Proceedings of the Royal Society B, 279*, 4003–4008. https://doi.org/10.1098/rspb.2012.1022.

Primack, B. A., Shensa, A., Sidani, J. E., Whaite, E. O., Lin, L.y., Rosen, D., et al. (2017). Social media use and perceived social isolation among young adults in the U.S. *American Journal of Preventive Medicine, 5*, 1–8. https://doi.org/10.1016/j.amepre.2017.01.010.

Ripoll-Núñe, K. J., & Rohner, R. P. (2006). Corporal punishment in cross-cultural perspective: Directions for a research agenda. *Cross-Cultural Research, 40*, 220–249. https://doi.org/10.1177/1069397105284395.

Sameroff, A. J., Seifer, R., Baldwin, A., & Baldwin, C. (1993). Stability of intelligence from preschool to adolescence: The influence of social risk factors. *Child Development, 64*, 80–97. https://doi.org/10.2307/1131438.

Saxbe, D. E., & Repetti, R. L. (2009). Brief report: Fathers' and mothers' marital relationship predicts daughters' pubertal development two years later. *Journal of Adolescence, 32*, 415–423. https://doi.org/10.1016/j.adolescence.2008.06.009.

Scarr, S. (1992). Developmental theories for the 1990s: Development and individual differences. *Child Development, 63*, 1–19. https://doi.org/10.2307/1130897.

Scarr, S. (1993). Biological and cultural diversity: The legacy of Darwin for development. *Child Development, 64*, 1333–1353. https://doi.org/10.2307/1131538.

Senese, V. P., De Falco, S., Bornstein, M. H., Caria, A., Buffolino, S., & Venutti, P. (2013). Human infant faces provoke implicit positive affective responses in parents and non-parents alike. *PLoS ONE, 8*, e80379. https://doi.org/10.1371/journal.pone.0080379.

Simpson, J. A., Collins, W. A., & Salvatore, J. E. (2011). The impact of early interpersonal experience on adult romantic relationship functioning: Recent findings from the Minnesota Longitudinal Study of Risk and Adaptation. *Current Directions in Psychological Science, 20*, 355–359. https://doi.org/10.1177/0963721411418468.

Simpson, J. A., Collins, W. A., Tran, S., & Haydon, K. C. (2007). Attachment and the experience and expression of emotions in adult romantic relationships: A developmental perspective. *Journal of Personality and Social Psychology, 92*, 355–367. https://doi.org/10.1037/0022-3514.92.2.355.

Simpson, J. A., Griskevicius, V., Kuo, S., Sung, S., & Collins, W. A. (2012). Evolution, stress, and sensitive periods: The influence of unpredictability in early versus late childhood on sex and risky behavior. *Developmental Psychology, 48*, 674–686. https://doi.org/10.1037/a0027293.

Singh, D. (1993). Adaptive significance of female physical attractiveness: Role of waist-to-hip ratio. *Journal of Personality and Social Psychology, 65*, 293–307. https://doi.org/10.1037/0022-3514.65.2.293.

Singh, D. (2002). Female mate value at a glance: Relationship of waist-to-hip ratio to health, fecundity and attractiveness. *Neuro Endocrinology Letters, 23*, 81–91.

Smith, R. L. (1984). Human sperm competition. In R. L. Smith (Ed.), *Sperm competition and the evolution of mating systems* (pp. 601–659). New York, NY: Academic Press. https://doi.org/10.1016/B978-0-12-652570-0.50026-9.

Soltis, J. (2004). The signal functions of early infant crying. *Behavioral and Brain Sciences, 27*, 443–458. https://doi.org/10.1017/S0140525X0442010X.

Sprengelmeyer, R., Perrett, D. I., Fagan, E. C., Cornwell, R. E., Lobmaier, J. S., Sprengelmeyer, A., et al. (2009). The cutest little baby face: A hormonal link to sensitivity to cuteness in infant faces. *Psychological Science, 20*, 149–154. https://doi.org/10.1111/j.1467-9280.2009.02272.x.

Stearns, S. (1992). *The evolution of life histories*. Oxford, England: Oxford University Press.

Straus, M. A., Douglas, E. M., & Medeiros, R. A. (2014). *The primordial violence: Spanking children, psychological development, violence, and crime*RoutledgeNew York, NYhttps://doi.org/10.1007/s10896-013-9560-7.

Sugiyama, L. S. (2005). Physical attractiveness in adaptationist perspective. In D. M. Buss (Ed.), *Handbook of evolutionary psychology*. (1st ed., pp. 292–343). Hoboken, NJ: John Wiley. https://doi.org/10.1002/9780470939376.ch10.

Suzman, J. (2017). *Affluence without abundance: The disappearing world of the Bushmen*. New York, NY: Bloomsbury.

Szepsenwol, O., Simpson, J. A., Raby, K. L., & Griskevicius, V. (2018). The effect of unpredictable early childhood environments on parenting in adulthood. *Journal of Personality and Social Psychology, 109*, 1045–1067. https://doi.org/10.1037/pspi0000032.

Thompson, R. A. (2006). The development of the person: Social understanding, relationships, conscience, self. In W. Damon, & R. M. Lerner (Eds.) & N. Eisenberg (Vol. Ed.), *Handbook of child psychology. Vol. 3, Social, emotional, and personality development* (6th ed.) (pp. 24–98). New York: Wiley.

Trivers, R. L. (1972). Parental investment and sexual selection. In B. Campbell (Ed.), *Sexual selection and the descent of man* (pp. 136–179). New York, NY: Aldine de Gruyter. https://doi.org/10.4324/9781315129266-7.

Trivers, R. L. (1974). Parent-offspring conflict. *American Zoologist, 14*, 249–264. https://doi.org/10.1093/icb/14.1.249.

Turns, B. A., & Sibley, D. S. (2018). Does maternal spanking lead to bullying behaviors at school? A longitudinal study. *Journal of Child and Family Studies, 27*, 2824–2832. https://doi.org/10.1007/s10826-018-1129-x.

Twenge, J. M. (2017). *iGen: Why today's super-connected kids are growing up less rebellious, more tolerant, less happy – and completely unprepared for adulthood*. New York, NY: Atria.

Twenge, J. M., Martin, G. N., & Campbell, W. K. (2018). Decreases in psychological well-being among American adolescents after 2012 and links to screen time during the rise of smartphone technology. *Emotion, 16*, 765–780. https://doi.org/10.1037/emo0000403.

UN Tribune. (2014). *The 42 countries that have banned corporal punishment*. http://untribune.com/42-countries-banned-corporal-punishment/. downloaded 12 November 2018.

van Schaik, C. P., & Burkart, J. M. (2010). Mind the gap: Cooperative breeding and the evolution of our unique features. In P. Kappeler & J. Silk (Eds.), *Mind the gap: Tracing the origins of human universals*. (pp. 477–496). Berlin, Germany: Springer Verlag. https://doi.org/10.1007/978-3-642-02725-3_22.

Volk, A. A., & Atkinson, J. A. (2013). Infant and child death in the human environment of evolutionary adaptation. *Evolution and Human Behavior, 34*, 182–192. https://doi.org/10.1016/j.evolhumbehav.2012.11.007.

Waller, K. L., Volk, A., & Quinsey, V. L. (2004). The effect of infant fetal alcohol syndrome facial features on adoption preferences. *Human Nature, 15*, 101–117. https://doi.org/10.1007/s12110-004-1006-8.

Weeden, J., & Sabini, J. (2005). Physical attractiveness and health in western societies: A review. *Psychological Bulletin, 131*, 635–653. https://doi.org/10.1037/0033-2909.131.5.635.

Weisfeld, G. E., Czilli, T., Phillips, K. A., Gal, J. A., & Lichtman, C. M. (2003). Possible olfaction-based mechanisms in human kin recognition and inbreeding avoidance. *Journal of Experimental Child Psychology, 85*, 279–295. https://doi.org/10.1016/S0022-0965(03)00061-4.

Whiting, B. B. (Ed.), (1963). *Six cultures: Studies of child rearing*. New York, NY: Wiley.

Williams, G. C. (1966). *Adaptation and natural selection*. Princeton, NJ: Cambridge University Press.

CHAPTER 3

Convergence and difference: Marriage and family life from a cross-cultural perspective

Belinda Hewitt and Brendan Churchill
School of Social and Political Sciences, The University of Melbourne, Parkville, VIC, Australia

Families are recognized across all societies and cultures, even though what defines "family" is often debated and contested. Arguably, the institution of marriage has been, and in many societies continues to be, the bedrock of family life (Coontz, 2004; Twamley, 2013). Globally, over the last few hundred years, the institution of marriage and family life have undergone significant changes (see Chapter 17). Some commentators have observed that these changes suggest a decline in the relative influence of the family (Goode, 2003) and the deinstitutionalization of marriage (Cherlin, 2004). From a global perspective, this conclusion may not hold true. The prevalence, nature, rhythm, and timing of changes in family life vary depending on the society and culture, and contemporary family life and marriage are marked by substantial cross-cultural variation that does not necessarily suggest the universal demise of the family in all countries or regions.

Family organization is a vital aspect of cultural identity (McDonald, 2000a). Family life, and marriage in particular, is governed by legal, social, cultural, and religious structures that constrain some activities and encourage others (Hennon & Wilson, 2008). One of the main roles of marriage in family life has been the regulation of who may or may not have sex with whom (Therborn, 2014). Negotiating these structures is a shared marker of what it is to be a member of a given society, and thus individual choice and social and cultural values that shape those choices show up in trends in marriage at an aggregate country level. While individuals do possess agency, no one possesses total agency where decisions are not constrained by legal, ethical, cultural, or economic restraint (Williams, 2010). Individual agency within a given society also varies depending on factors, such as gender, sexual orientation, and age, and we follow these themes throughout the chapter.

Cross-Cultural Family Research and Practice
https://doi.org/10.1016/B978-0-12-815493-9.00003-X

In this chapter, we undertake both a sociological and demographic investigation of cross-national differences in marriage and family trends. We highlight and discuss some of the key ways that contemporary family life, with a particular focus on marriage, varies across societies and cultures. In the first section of the chapter, we overview some of the explanations for the worldwide changes in marriage and differences between countries. We then examine cross-national data on marriage using a range of marriage indicators. First, we examine and discuss differences in crude marriage rates between countries. We then turn our attention to country differences in the average age at marriage, including an investigation of child marriage and a discussion of arranged marriage. One of the large and more contentious shifts in marriage since the turn of the 21st century has been the legalization of same-sex marriages, and we present and discuss some of the most recently available cross-national data on same-sex marriage. In the final section of the chapter, we discuss variation in the extent to which marriage is a breakable bond by examining divorce rates across countries. Throughout the chapter, we have a commitment to include as many countries as possible in our analysis. There are many studies which have examined cross-national variation in affluent or Western countries, and we wanted to incorporate less wealthy and non-Western countries into the discussion.

Global explanations for cross-national differences in marriage

There are several theoretical frameworks that have emerged to explain and understand differences in marriage and family life between countries. Some perspectives focus on similarities and emphasize worldwide trends in the convergence of marriage and family life, such as the World Society Theory, Developmental Idealism, and Second Demographic Transition Theory. In contrast, other frameworks, such as the Family Systems Theory, attempt to describe and explain differences between countries. Other perspectives have focused on the mechanisms such as the rise of individualism, secularism, and change in the gender order. In this section of the chapter, we provide a brief overview of some of the more dominant perspectives as a context for the data and discussion that follow.

World society theory

The world society theory (Meyer, 2010; Meyer, Boli, Thomas, & Ramirez, 1997) posits that since the end of World War II a world value system and culture has emerged. This value system and culture has shaped and influenced the policies and decisions of all nation- states across the globe. The diffusion and influence of this world culture and value system has occurred through global agencies, such as the United Nations (UN) and its subsidiary funding organizations (i.e., the United Nations International Children's Emergency Fund (UNICEF), United Nations High Commissioner for Refugees (UNHCR), United Nations Development Program (UNDP)), the World Bank, the International Monetary Fund (IMF), and countless other not-for-profit global organizations (Meyer et al., 1997). These agencies are globally active; some have enormous resources of money for aid and infrastructure, and they also provide expertise in technology, ideas, and services. Access to these valuable resources are used to reward and reinforce a nation-state's adherence to a common set of global norms (Meyer, 2010). While their membership includes countries from all over the globe, global agencies are dominated by wealthy Western countries that make the main financial and cultural contributions (Thornton, 2005).

At the core of world culture is the concept of modernization, and the premise that modernization is a good and achievable goal for all nation-states. These global agencies promulgate well-known, highly abstracted, and stylized theories of functional requirements of a modern society. These functional requirements are identified and defined by a professional class that mostly comes from wealthy Western countries (Meyer et al., 1997). Within this framework, the legitimate goals of properly constructed modern nation-states focus on two central platforms: (1) socioeconomic development of the nation-state, and (2) the human rights and entitlement to comprehensive self-development of the individuals within that nation-state. Importantly, in relation to the latter is the global rise of social movements and discourses legitimating the human rights of marginalized and excluded groups, for example, rise in the world discourse advocating for the rights of those in the LGBTQI community, which has produced national policy changes in often surprising ways (Meyer et al., 1997, p. 160). However, the world culture is also an external culture that is without borders and, more often than not, it cannot be simply or easily imported wholesale into a country whose culture and value systems are not closely aligned with it. Meyer et al. (1997) also argue that world cultural models are highly idealized and often internally inconsistent, making them impossible to achieve. Therefore, there is an eclectic adaption of these conflicting principles across the globe. Even countries in the wealthy West struggle to meet all the requirements and demands of the idealized modern state (Meyer, 2010).

Despite this, these values are culturally transcendent and causally important in that they shape expectations about the legitimate goals of nation-states across the world.

While the world society theory does not specifically address the impact of world culture and values on family life, there is increasing evidence that these organizations and the principles of economic development and human rights have had profound implications for family policy and practice in many nation-states, often with mixed results. One such example is the implementation of a series of successful programs to increase the education and retention of young girls in school in rural Bangladesh (Amin & Bajracharya, 2011; Kamal, Hassan, Alam, & Ying, 2015). Increasing levels of education delayed the age at marriage for the girls involved. In Bangladesh, it is customary that the bride's family pays a dowry to the groom's family, and the older a girl is when she marries, the greater the dowry demanded by the groom and his family (Amin & Bajracharya, 2011). Consequently, successful education programs placed a major financial burden on the brides' families because they were not accompanied by cultural change in the value and role of women in families in society.

Developmental idealism in family research

Thornton (2005, 2013) proposes that developmental idealism is the cultural schema that underpins much of the research and understanding of global trends and change in family life. Developmental idealism is closely aligned with modernization and is a narrative based on the belief that the shift from the traditional family to the modern family will accompany the further development of a culture and society (Allendorf, 2013). This developmental narrative dominates the belief systems of our global institutions (i.e., those identified by the world society theory) and influences the interpretation and understanding of country differences in family practices. The developmental narrative upholds wealthy western nations, mostly in North Western Europe, as the modern *ideal type* for society and the practice of family. The ideal modern society starts with the premise that all members of a society have political liberty and that nonfamilial institutions provide safety nets for individuals. It is underpinned by an individualistic rather than community orientation and there is a focus in policy and practice on the nuclear (not extended) family. The family life course is characterized by later marriage, high rates of nonmarriage (marriage is not universal), autonomous love marriages, which are considered superior to arranged marriages, and spousal equality (Thornton, 2013).

Developmental idealism is also underpinned by the belief that both modern society and modern families are good and attainable (Allendorf, 2013). Family practices outside of North West Europe, such as those in countries that are less wealthy and developed, are defined as traditional. Many of these traditional and customary family practices are not consistent with ideal modern family life or are considered harmful and therefore in need of reform. Developmental idealism implies that any differences observed between countries at any given time are due to differences in the location of a country on this trajectory from traditional to modern. Importantly, Thornton (2005) notes that the trajectory is not purely linear, as some countries may show good progress at certain times and then regress.

Despite the pervasiveness of developmental idealism in cross-national family research, it has a number of shortcomings. Not all countries are on the same trajectory, as they do not come from the same starting point or adhere to the same value systems that make the ideal family type upheld by the developmental idealism achievable. Nor do the assumptions about what are considered to be problematic family practices, such as arranged marriages, necessarily have the negative impacts that they are assumed to have (Thornton, 2005). Therefore, the expectations underpinning developmental idealism are problematic in predicting the future of countries that are viewed to be less developed.

Taken together, both world society and developmental idealism imply that marriage and family life are converging across the globe toward one modern, ideal type of marriage and family life. This argument has generally been referred to as the *convergence* theory (Pimentel, 2000). On the one hand, there is good evidence that there is some convergence in marriage and family life, with well-documented global declines in marriage, increases in the age at marriage, and increases in divorce (Goode, 2003). Nevertheless, there is not complete agreement with this view and there is evidence that other family systems remain evident and strong across the globe.

Family systems theory

Therborn (2004, 2006, 2014) has developed a theory of family systems. In contrast to the arguments of the world society theory and developmental idealism, Therborn emphasizes differences between family types. His family systems framework, developed on observations of family life across the globe between 1900 and 2000 (Therborn, 2004), identifies seven different systems of marriage and family life that vary between regions and countries. Most notably for this chapter, each system differs in the rhythm and timing of marriage and family formation within the life course, access to divorce, and in the gender power order within the family (Therborn, 2004).

The *Christian-European* family system is the dominant North Western European family type identified by the world society theory and developmental idealism. This family system is characterized by relatively low rates of marriage; marriage is not universal and occurs at later ages. There are also relatively high rates of divorce. The focus is on the nuclear family and household. Gender equity and equality are emphasized. This family type has extended to some nations settled by the North Western Europeans, such as the United States (although there is some US exceptionalism), Canada, Australia, and New Zealand (Lesthaeghe, 1991). It should also be noted that this family type is not consistent across Europe (Kalmijn, 2007). There are relatively large differences in family and marriage practices between East and West Europe. Some suggest that these differences stem from historical differences along the Hajnal line from Leningrad to Trieste (Hajnal, 1965), where Eastern Europe family formation is characterized by younger age at marriage, universal marriage, patriarchal dominance, and greater importance is placed on the extended family (Svab, Rener, & Kuhar, 2012; Therborn, 2006). Similarly, differences in family life between Southern European families and those of North Western Europe have been noted, also in terms of the timing of marriage and the importance of the extended family (Kalmijn, 2007; Lesthaeghe, 1991; McDonald, 2000a). Therefore, the Christian-European family system in essence refers to the family life typical of North Western Europe.

The *Islamic West/Asian/North African* family system is characterized by strongly entrenched gender inequality, where women and girls are subordinate to husbands and men. Marriage is universal and occurs at a young age, particularly for women, and child marriage is practiced. In *sub-Saharan African* family systems, men hold the power and the exercise of their power and authority is sometimes harsh. Mass polygyny is practiced and it is relatively common for men to have several wives in rural areas and/or partners in urban areas (Therborn, 2006). This system is characterized by young age at marriage, particularly for women, and high birth rates. Nevertheless, women's economic contributions to the household are highly valued; typically, families receive a bride price for the marriage of their daughters. In recent decades, family life has become unstable and uncertain, impacted by an HIV/AIDS epidemic and war and civil unrest in the region (Therborn, 2014).

The *South Asian* family system found in countries such as India, Pakistan, Bangladesh, and Afghanistan is predominantly characterized by strong and intractable patriarchy within the family and social institutions. Wifely obedience is a strong norm reinforced through Hindu and Muslim religious practices and traditions. Women marry young and have their children young, child marriage is common, and there are often large age differences between spouses when they marry. There is limited polygyny. Marriage is universal and is predominantly arranged by parents within castes, although arranged marriage practices have changed and adapted over time. Sex is tightly controlled and formalized, family households are large, and divorce is rare and largely inaccessible for women. There has been a growth in the number of highly educated women and girls and this is having an impact on family life.

Confucian East Asian family systems of Japan, China, Korea, and Taiwan are characterized by marriage as a contract between families. There are stronger male rights within families (Ji, 2015). For example, divorce is possible, but men have more economic power and a greater say in whether the marriage ends. There is also a patrilineal system for inheritance and living arrangements, filial piety is expected, and it is common for multiple generations of family to live together. This is a system in transition that has been disrupted by communism and colonization (Therborn, 2004).

The *Southeast Asian* family system found in countries such as Thailand, Malaysia, Vietnam, and Singapore is characterized by male superordination and a favoring of freedom in mate selection (Therborn, 2006). There has been a substantial increase in the education of girls over recent decades; as such, marriage is often not an attractive choice for educated women; nor is marriage to an educated woman attractive for a man. Therefore, this family system is in transition from young universal marriage to relatively low marriage rates at older ages, with relatively low fertility.

Creole families are mostly located in Latin America and the Caribbean and are a product of white male colonization, black slavery, and disenfranchised indigenous populations (Therborn, 2014). These families are characterized by matrilocality, low rates of marriage, high rates of fertility and union instability, and mother-centered families. However, conservative influences are also present, particularly in Latin America where there are legal limits on divorce and strong conservative religious beliefs (Therborn, 2006).

More recently, Therborn (2014) has argued that family systems in the world are not necessarily converging, as predicted by the other theories, and in fact finds evidence that for some country comparisons there is an increasing difference in family practice that suggests divergence. For example, comparing countries in North West and Eastern Europe, Therborn (2004) shows that the difference in the average age at first marriage for couples in France and Germany compared to couples in Russia and the Ukraine had increased from 5 or 6 years in 1900 to 8 years in the early 2000s. Other European studies have also found evidence that suggests divergence between marriage and family life in South and Eastern Europe compared to North Western Europe rather than convergence (Fokkema & Liefbrouer, 2008; Kalmijn, 2007).

Since the development and proposal of Therborn's (2004) theory based on his observations of marriage and family life during the 20th century, family practices have continued to change and shift early in the 21st century. There are some countries that fit within the family systems Therborn identified and some that do not. In later sections of the chapter, we will highlight some of these differences and departures. Nevertheless, many of these generalized family systems remain strong, and sit in tension with the more global forces that have been identified by the world society theory and developmental idealism. Meyer et al. (1997) allude to this tension when they emphasize that the world culture is an idealized type, in which many nation-states and countries do not have the capacity or the cultural background to emulate the idealized model. Therefore, even if countries with different family systems sign up to UN treaties or programs and have policies in place, in practice on the ground many of these undertakings and policies are difficult to implement, and unable to be enforced, due to the strength of traditional family and marriage practices (Kim, Longhofer, Heger Boyle, & Brehm Nyseth, 2013; Koski, Clark, & Nandi, 2017; Maswikwa, Richter, Kaufman, & Nandi, 2015).

The second demographic transition

The second demographic transition (SDT) theory (see Chapter 4) is a catchall often used to describe changes in family life and marriage since the end of WWII, particularly in the wealthy industrialized West (Goldscheider, Bernhardt, & Lappegård, 2015). The first demographic transition was characterized by declines in mortality and fertility, to replacement fertility levels, in Western and industrialized countries during the 18th and early 19th centuries (Davis, 1945). One particular characteristic of the second demographic transition is below replacement fertility, which was not predicted by the first demographic transition (Lesthaeghe, 1991). With respect to marriage, the second demographic transition is characterized by low rates of marriage, increased age with marriage, and relatively high divorce rates (Lesthaeghe, 1991). These demographic family factors result in a greater diversity of living arrangements for couples and families and suggest a general decline in the importance of marriage.

Lesthaeghe (1991) argues that these trends in family life are underpinned by ideational changes in value systems at a global level, which have shifted toward individualization and secularization, particularly since the end of World War II. As countries recovered from the war, modernization and industrialization increased their wealth, and the wealth of their citizens. With the increased wealth, countries and their citizens had their basic subsistence and survival needs met; this freed them up to focus on higher order needs and expressive social roles. These higher order needs, such as self-actualization, individual autonomy, and symmetrical gender relationships, are highly individualized, and therefore individual needs become more important than family and community. Of note is that SDT emphasizes that economic development created the space for higher order needs to be considered. This is strongly consistent with the expectations of properly formed nation-states in the world society theory, where legitimate goals of nation-states focus on economic development and meeting higher order needs such as human rights and equality of opportunity (Meyer, 2010). This, in part, may explain why poorer countries struggle with goals of human rights and self-determination while individuals in their nations struggle to meet basic needs. The second demographic transition theory has predominantly focused on explaining family trends in the wealthy West and highly industrialized countries, such as Japan, but the argument suggests that these same trends are evident in non-Western countries. Thus, the wealthy West is further along in terms of the second demographic transition, which will eventually extend to the rest of the world (Lesthaeghe, 2010).

Gender systems and the gender revolution?

One of the core functions of marriage is to regulate and define relationships between men and women. Therefore, a strong theme running through all explanations for cross-national differences in family life is gender relations and gender equity within the family, and in particular the position of women and girls within the family. Here we briefly outline two perspectives that emphasize the importance of gender relations in shaping contemporary marriage and family life.

McDonald (2000a, 2000b) highlights how familial gender systems within a society and their expectations prescribe a division of labor and responsibilities between women and men and grant different rights and obligations to them. Essential to MacDonald's argument is the notion that gender stratification and gender roles in different social institutions within a given society, particularly societies undergoing economic reform and transition, can be inconsistent with each other. For example, many countries have invested in the education of girls and improving women's access to employment; this is often promulgated as important for the economic development of nation-states. Thus, women's status and position within the broader society is relatively high compared to men's and they have greater access to economic independence and autonomy. However, within the family, women and girls continue to be viewed and treated as second-class citizens and are expected to take a subordinate role in the family and household. In countries undergoing this transition, women experience *institutional disjuncture*, where their value and status in the family is inconsistent with their value and status in other institutions (Ji, 2015). While women continue to have a subordinate status in the home, marriage rates will remain low. These insights are particularly useful for explaining marriage trends in regions and countries, such as Southern Europe and East and Southeast Asia, whose economies are developing and strengthening, women's socioeconomic status is increasing but patriarchal male-dominated family gender systems remain deeply entrenched (Svab et al., 2012; Tsutsui, 2013). Thus, family changes such as low rates of marriage and older age at marriage are being driven by the changing status of women in these countries.

Goldscheider et al. (2015) attribute the major changes in family behavior to structural changes in the labor market postindustrialization and a gender revolution in the relationships between men and women and household production. According to their theory, the gender revolution has two stages. The first stage began shortly after World War II, specifically in industrialized societies, where women's labor market participation increased. This accelerated during the 1960s and 1970s in industrialized western nations. Goldscheider et al. (2015) argue that increases in women's labor force participation meant that marriage and children became less attractive to educated women, and within marriage husbands and wives had to renegotiate their relationships. During the 1980s and 1990s and onward, it became apparent that the so-called gender revolution had stalled (England, 2010); men have not participated equally in the demands of family life to the extent that women in industrialized countries economically contribute to the household. Thus, working women often take on the double or second shift which involves being the primary carer in the home and family while also working to contribute income to the family (Hochschild & Machung, 1989). During this first stage of the gender revolution and the stalled revolution, women's increased participation in employment was linked to family processes such as delayed marriage and nonmarriage, divorce, and low fertility (Goldscheider et al., 2015, p. 211). Importantly, the framework proposed by Goldscheider et al. (2015) argues that the processes heralding the second stage of the gender revolution are beginning to occur. These processes entail the increased involvement of men in the domestic sphere, particularly in relation to caring for children, but as noted by Goldscheider et al. (2015) this second stage is only emerging, and changes in men's domestic activities are relatively small and mostly observed among younger and higher educated men.

In summary, most explanations for cross-cultural differences between countries in marriage and family life are often observing the same phenomenon but emphasize different elements of the process or different underlying causes. All theories reference major changes that have occurred since WWII that have resulted in some of the family and marriage trends we have today. All either directly or indirectly imply that the marriage and family trends of the wealthy West, and more specifically those that emerged in North West Europe, are more like the idealized family type that family life across the globe is benchmarked to. Gender and gender inequality within the family define and differentiate countries and equality is a major priority for the large global organizations referred to in the world society theory. There are elements of all theories that have relevance for what we observe in the following sections of the chapter where we examine the most recently available cross-national data on the prevalence of marriage, age at marriage, same-sex marriage, and divorce rates.

The prevalence of marriage

Some commentators have argued that there is a worldwide trend in the decline of marriage and the importance of marriage (Goode, 2003). Particularly in the wealthy West, marriage has become less central to family life and therefore these societies tend to be characterized by low marriage rates (Cherlin, 2009a, 2009b; Thornton, 2013). The expectation is, based on many of the theories' outlined in the previous section, that most other countries in the world will follow suit and eventually have equally low rates of marriage. Nevertheless, in some industrialized Western countries, such as Italy and the United States (Cherlin, 2009a, 2009b; Rosina & Fraboni, 2004) and many other cultures and societies, such as Russia, Bangladesh, and China (Kamal et al., 2015), marriage is a valued and core social and economic institution. Conversely, low marriage rates are already observed in many non-Western nations, such as Latin America (Liu, Esteve, & Trevin, 2017). To investigate these trends further, we examine the latest available UN data on Crude marriage rates (marriages per 1000 mean head of population). The data are presented in Fig. 3.1.[a]

The mean crude marriage rate averaged over all countries with available data was 5.4 marriages per 1000 mean head of population. From a close examination of Fig. 3.1, it quickly becomes apparent that it is problematic to make generalizations about marriage based on region or continent. For example, some Latin American countries, such as Argentina, Chile, and Venezuela have among the lowest crude marriage rates (all below 3.0 marriages per 1000 mean head of population), whereas Cuba has a crude marriage rate of 5.5, about double of these countries. In East Asia, which traditionally has a Confucianist patriarchal family system (Ji, 2015), Japan has a relatively low crude rate of marriage (4.9), whereas China (7.2) and Singapore (7.1) have relatively high crude marriage rates. Therborn (2014) suggests that the large differences in contemporary family life between Japan and China, reflected here in the crude marriage rates, may be due to the rise of communism in China and the influence of the US occupation in Japan following WWII.

Most wealthy Western countries of North West Europe, such as the Scandinavian countries, the United Kingdom, France, Germany, and the Netherlands, and European settler countries, such as Canada, Australia, and New Zealand, are concentrated around the middle to lower end of the distribution. Among wealthy Western nations, the United States is an outlier when it comes to marriage, with a crude marriage rate of 6.9 per 1000 head of population. It has been well documented that in the United States, couples tend to marry younger, divorce more often, and remarry more often than they do in most other countries in the world (Cherlin, 2009a, 2009b).

[a]We include all available data in the most recent UN Demographic Handbook (United Nations, 2017a, 2017b). There are still some notable limitations with the information in Fig. 3.1. First, the UN data relies on country level collection of data on registered marriages, and not all countries collect or collate data on marriage rates. Some notable exclusions from Fig. 3.1 are India, Pakistan, and South Africa. In addition, crude marriage rates may not adequately capture the prevalence of long-term committed relationships in some countries. For many countries, particularly those with large regional and rural populations, such as those in sub-Saharan Africa and South Asia, many marriages are not officially registered. Similarly, in countries with high levels of unmarried cohabitation, such as North Western Europe, official marriage rates do not adequately capture family formation and long-term committed relationships.

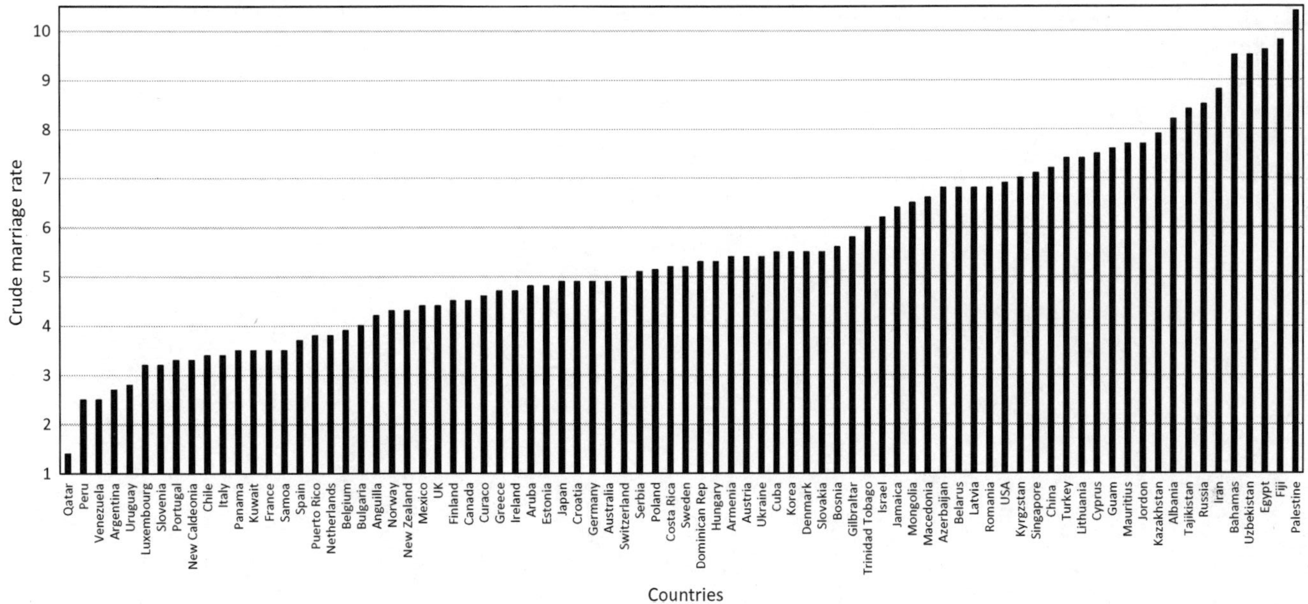

Fig. 3.1 Crude marriage rates, 2013–17, selected countries. *(Data from United Nations. (2017a). Table 23: Marriages and crude marriage rates, by urban/rural residence: 2013–2017. In* 2017 demographic yearbook*. New York: Department of Economic and Social Affairs, United Nations; Data for Canada taken from United Nations. (2011a). Table 23: Marriages and crude marriage rates, by urban/rural residence: 2000–2009. In* 2009–2010 demographic yearbook*. New York: Department of Economic and Social Affairs, United Nations.)*

As discussed in the previous section, family life in Europe is not uniform and there are several noteworthy differences. The distinction between East and West Europe, where Eastern European family life is characterized by near universal marriage (Thornton, 2005), is reflected in Fig. 3.1. For example, several former Soviet Union countries, including Uzbekistan, Russia, and Kazakhstan, all have relatively high crude marriage rates of over 8 per 1000 head of population. However, the large differences in marriage rates between countries in East and West Europe are inconsistent, with the other Eastern European and former Soviet countries having much lower crude marriage rates. For example, Bulgaria and Slovenia have crude marriage rates around 4 per 1000 head of population that are closer to countries in North West Europe than Eastern Europe.

In strong contrast to other Arab countries, which tend to have high or above average crude marriage rates (Palestine and Iran, for example, have some of the highest crude marriage rates), Qatar has the lowest crude rate of marriage of all countries listed. Qatar is an interesting case because it has experienced enormous economic growth over recent decades due to the growth in their oil and natural gas industries. This has attracted workers from poor neighboring countries. Consequently, 88% of the population are migrants (mostly male migrant workers from India, Nepal, Bangladesh, and Philippines) and only 25% of the total population are women (Alharahsheh & Almeer, 2018). In addition, the increased wealth of the country has facilitated the mass education of boys and girls. However, marriage remains strongly patriarchal and is influenced by traditional customs such as consanguineous marriage. There are restrictions on the marriages of Qatari national women, who are not easily able to marry non-Qataris, whereas Qatari men can have more than one wife (Alharahsheh & Almeer, 2018; Harkness & Khaled, 2014; State of Qatar, 2009). These particular economic, educational, demographic, cultural, and legal contexts combine to result in an extremely low crude marriage rate for Qatari nationals in Qatar.

In sum, our examination of crude marriage rates suggests a large variation between countries. There are a number of notable country exclusions from Fig. 3.1. We do not have crude marriage rates for large countries in South Asia, such as India and Pakistan, as well as Southeast Asian and African countries. This is because many of these countries do not have the systems in place to collect systematic country level data on marriages. The differences between the countries represented in Fig. 3.1 can be accounted for in terms of cultural traditions, demographic makeup, economic circumstances, and the influence of contemporary social change. The tensions between all these forces play an important part in shaping the prevalence of marriage in a society and contribute to the large differences between countries, even those with a similar or shared history. While crude rates are useful to indicate the overall rate of marriage in a country, they do not provide information about the rhythm and timing of marriage in the life course which also varies across countries.

The timing of marriage in the life course

The age at which a person gets married, or at which they are expected to marry, is one of the most fundamental events in the family life course. In many countries and cultures, marriage is considered a major life event in the transition to adulthood and as such determines access to important social and economic resources (Kamal et al., 2015). For other countries and cultures, marriage is not universal or essential to a good life; it is not the major signifier of transition to adulthood, but rather a symbolic capstone of adulthood to be achieved after economic stability and other milestones (Cherlin, 2004; Fussell, Guthier, & Evans, 2007). For women, in particular, these differences have major implications for their fertility, mortality, and health. In this section, we examine recent cross-national trends in age at marriage, child marriage, and arranged marriages.

The majority of countries in the world have a legal minimum age at marriage; only a handful of countries, including Gambia and Arabia, do not specify a minimum legal age for marriage, with or without parental consent (United Nations, 2013). For most countries, the legal age at marriage is the same for men and women; however, several countries have a gendered age difference in the minimum legal age for marriage. In all instances, it is women (or girls) who can marry at younger ages. About one quarter ($n=51$) of the 209 countries listed in the United Nations statistical database have a minimum legal age of marriage under 18 (United Nations Statistical Division, 2013). Generally, countries that have a minimum legal age under 18 are located in South, Central, or sub-Saharan Africa, South East Asia, the Arab World, or were former member states of the Soviet Union. In some countries, such as Venezuela and Lebanon, it is legal for girls to marry at the age of 12 without parental consent. The majority of wealthy Western countries have a minimum legal age of marriage of 18, although most of them also allow child marriage at the age of 16 with parental consent (United Nations, 2013). At the other end of the spectrum, around 10% of UN countries have a minimum legal age of marriage of 21 (without parental consent).

Interestingly, many of these countries are located in North, Central, South, and sub-Saharan Africa, regions that also have a concentration of countries with a legal minimum age at marriage below 18. These contrasts underscore the importance of not making generalizations about the legal restrictions on marriage based on regions, as there is a large variation between countries within the same region.

In Fig. 3.2, we present World Bank data on the average age of marriage for men and women in 160 countries (The World Bank, 2019a, 2019b). The majority of men and women in each country tend to marry later than the legal minimum age of marriage. Notably in two countries, Niger and the Central African Republic, the average age of first marriage for girls is less than 18 years; in contrast, the average age of first marriage for men in these countries is their early- to mid-20s. In several countries, women on average first marry under or around the age of 20, including Chad, South Sudan, Mali, Uganda, Bangladesh, India, and Afghanistan, which suggests that a large proportion of the female population in these countries are marrying under the age of 20, and possibly as children under the age of 18. Child marriage is addressed in more detail in the next section.

Overall, there is a large variation in the average age of first marriage for countries represented in Fig. 3.2, ranging from 17.2 for girls in Niger (see next section on child marriage) to 35.7 for men in Slovenia. The average age of first marriage across all countries was 24.8 years of age for women and 28.6 years for men. Men marry on average 3.8 years older than women. In general, Fig. 3.2 suggests that the age gap between men and women declines as the average age at marriage increases. In some countries, the gender age difference for first marriage is large, where men are on average over 7 or 8 years older than women when they first marry in countries such as Niger, Senegal, and Gambia. These large gender age gaps in marriage both reflect and reinforce gender inequality in a society (Alharahsheh & Almeer, 2018; Hotchkiss, Godha, Gage, & Cappa, 2016; Maswikwa et al., 2015). In contrast, in Bermuda, the average age difference between men and women when they marry is less than 1 year according to the data provided in Fig. 3.2.

There are large differences in the average age at marriage between wealthy Western nations and less wealthy Western and non-Western countries. Younger age at marriage is more typical in countries located in Africa, Southeast Asia (including India, Pakistan, and Bangladesh), the Arab World, and Latin America. The marriage patterns of South Asian countries are fairly consistent with the family system defined by Therborn (2004). The average

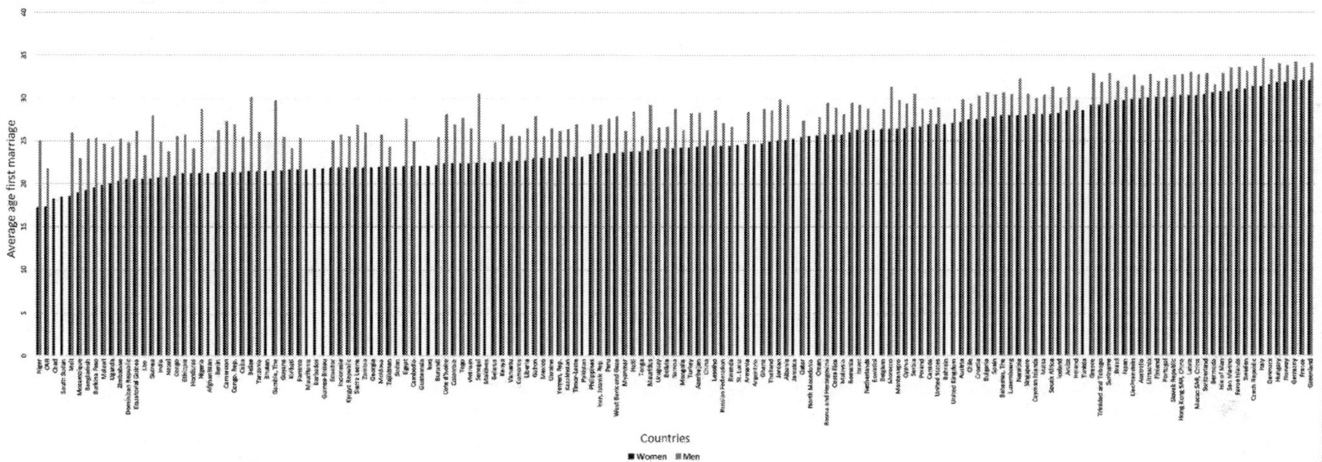

Fig. 3.2 Mean age at first marriage for women and men, selected countries United Nations 2009–14. *(Data from Gender statistics database (The World Bank. (2019a).* Data bank, gender statistics, men age at first marriage. *Retrieved from https//databank.worldbank.org; The World Bank. (2019b).* Data bank, gender statistics, women age at first marriage. *Retrieved from: https//databank.worldbank.org.))*

ages at marriage for women in Bangladesh, India, and Afghanistan are all around 20 years of age, suggesting that a large number of women marry at younger ages than that. Pakistan is a notable exception with an average age at marriage of 23.1. These countries also tend to have larger than average gender age gaps in marriage, particularly in Bangladesh where this gap is 6 years. Similarly, countries in sub-Saharan Africa also have low average ages at first marriage for women and relatively large gender age gaps, which is generally consistent with the male-dominated polygynous family systems identified by Therborn in the family systems theory. In these societies, women tend to have restricted access to social and economic resources outside of marriage. Therefore, marriage to older, more financially established men provides them with economic security and protection (Amin & Bajracharya, 2011; Maswikwa et al., 2015). Conversely, it may take men several years to accumulate the necessary capital to marry, also contributing to large gender age gaps in marriage.

In contrast, the differences between the Confucian-influenced East Asian countries of China and Japan are quite marked. Chinese men (26.2 years) and women (24.4 years) on average marry at relatively young ages by world standards, whereas Japan is more comparable to wealthy Western nations with an average of 29.7 for women and 31.2 for men, likely reflecting their diverging paths since the end of WWII, with the influence of communism in China and the United States on Japan, which were noted earlier.

In countries in Europe or European settler countries, such as Australia, New Zealand, the United States, and Canada, the age of first marriage for women tends to be older, ranging from 29 to 33 years. Not surprisingly, the Nordic countries and France, with historically high rates of cohabitation and correspondingly low rates of marriage (Kiernan, 2000), have among the highest average ages at first marriage, averaging over 31 for women and over 32 for men. In these countries, marriage, if it happens at all, often follows a long period of cohabitation and the birth of one or more children (Holland, 2017; Lappegard & Noak, 2015). Perhaps more surprisingly, Italy, which is a country known for its strong family system (Lesthaeghe, 2010; Rosina & Fraboni, 2004), is also among those countries with the highest average age of first marriage. Similarly, in contrast to most other postsocialist countries, Slovenia and the Czech Republic have much older ages at first marriage. Both countries are relatively wealthy, stable nations who experienced relatively smooth transitions from socialism to democracy (Mařiková, 2004;

Svab et al., 2012; Ule, 2004). Decisions around getting married and having a family hinge on completing education, achieving economic stability, finding stable housing out of the parental home, and potential compromises and losses of individual freedoms (Ule, 2004). In all three countries, there have been major gains in women's education and employment, but traditional family systems remain very strong and patriarchal (Mařiková, 2004; Rosina & Fraboni, 2004; Svab et al., 2012; Ule, 2004). While Italy, Czechia, and Slovenia differ markedly in terms of their history, in all countries contemporary young women deciding about marriage and family are likely to experience high levels of *institutional disjuncture*, where women's value and status in the family is low compared to their value and status in the labor market (McDonald, 2000a).

The practice of child marriage

Child marriage is globally defined as the marriage of boys and girls under the age of 18 (UNICEF, 2019), although we note that different countries and cultures define adulthood and marriageability in a multitude of ways (Kamal et al., 2015). While there is evidence that the rates of child marriage are declining worldwide (Kamal et al., 2015; Koski et al., 2017; Nguyen & Wodon, 2015), there are many countries where child marriage remains common. This is one contentious aspect of family life in poorer Western and non-Western countries that faces increasing scrutiny from global institutions dominated by wealthy Western nations, such as the World Health Organization and the United Nations. The 1989 UN Convention on the Rights of the Child forms the backdrop for this scrutiny around child marriage, as child marriage is seen as a fundamental violation of their human rights to self-determination. The following is a statement on the UNICEF website (UNICEF, 2019):

> *Marriage before the age of 18 is a fundamental violation of human rights. Many factors interact to place a child at risk of marriage, including poverty, the perception that marriage will provide 'protection', family honor, social norms, customary or religious laws that condone the practice, an inadequate legislative framework and the state of a country's civil registration system. While the practice is more common among girls than boys, it is a violation of rights regardless of sex.*

Child marriage is one aspect of traditional family life that is viewed as particularly problematic in the West, but it has remained highly resistant to the diffusion of the Western family. Fig. 3.3 shows the most recently

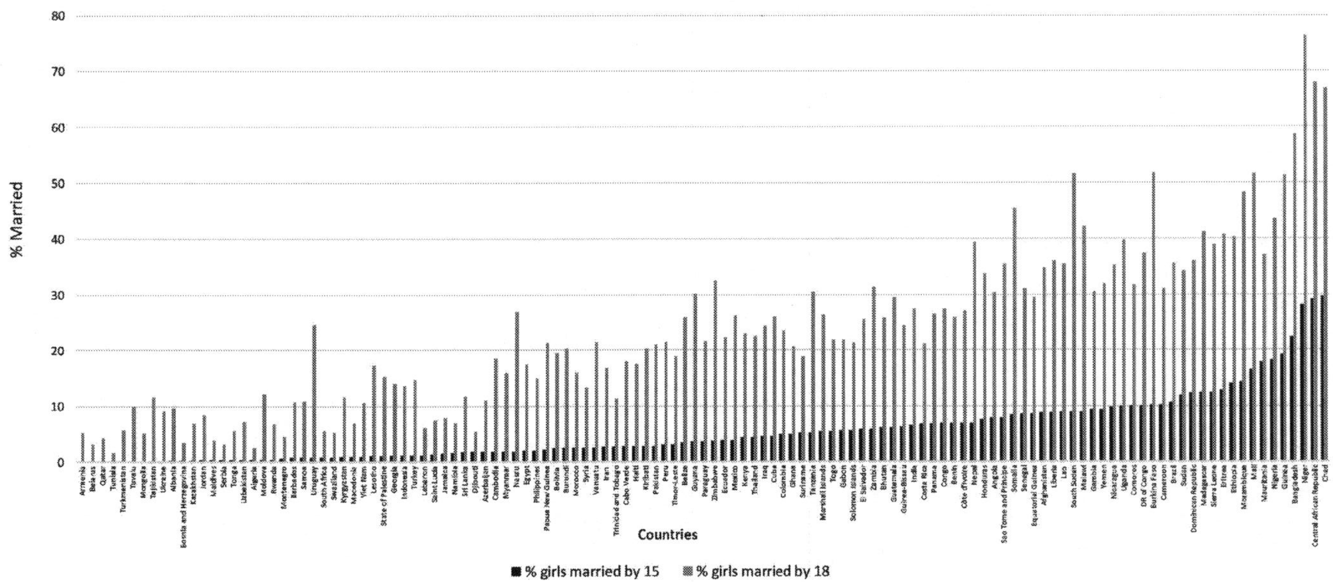

Fig. 3.3 Cross-national data on Child marriage, the proportion of girls aged 20–24 who were married by the age of 15 and 18. *(Data from UNICEF. (2019).* Child marriage. *Retrieved from https://data.unicef.org/topic/child-protection/child-marriage/. The term "child marriage" is used to refer to both formal marriages and informal unions in which a girl or boy lives with a partner as if married before the age of 18. An informal union is one in which a couple live together for some time, intending to have a lasting relationship, but do not have a formal civil or religious ceremony. The main sources of such data are national censuses and national household surveys, predominantly the Multiple Indicator Cluster Surveys (MICS) and Demographic and Health Surveys (DHS).)*

available UNICEF[b] data on Child Marriages among girls; the data are collected on girls because child marriage for boys is much less common. The data show the proportion of women aged 20–25 in each country who were legally married or in a marriage-like union either under the age of 15 or under the age of 18 (UNICEF, 2019). There are 120 countries where the marriage of girls under the age of 18 is relatively common, and in the majority of those countries the marriage of girls under the age of 15 also occurs. Most notably, in Chad, the Central African Republic, and Niger, just under one-third of women are married before their 15th birthday. In Fig. 3.3, there are several countries where more than 50% of girls are married under the age of 18, including Guinea, South Sudan, Mali, Bangladesh as well as Chad, Niger, and the Central African Republic. There are over 40 countries where 30% or more of girls are married before their 18th birthday. Many of these countries are relatively small West African countries, but in some with large populations, such as India, Pakistan, Ethiopia, and Nigeria, there are tens of millions of girls who are married each year (Kamal et al., 2015; Nguyen & Wodon, 2015; Nour, 2009).

Importantly, child marriages, particularly of girls under the age of 15, often happen outside of the legal frameworks in a nation. For example, Niger has a minimum age of 15 for marriage and Central African Republic has a minimum legal age of 18 for marriage (United Nations Statistical Division, 2019). This indicates that introducing legal minimum ages at marriage does not necessarily stop the practice of child marriage, because many countries that have a minimum legal age also allow children to marry with parental consent, according to tribal customary law or other extenuating circumstances, such as pregnancy (Nguyen & Wodon, 2015; Svanemyr, Scolaro, Blondeel, Chandra-Mouli, & Temmerman, 2013; Syrett, 2016). It has been argued that some nations engage in symbolic reform, where they introduce a legal minimum age at marriage to meet the requirements for aid and assistance from global agencies, but these policies are completely inconsistent with family practices on the ground and officials have limited motivation and capability to regulate or enforce the policies (Kim et al., 2013).

[b]The data are not comprehensive as many countries do not collect data on child marriage and many child marriages are not officially registered. The data mainly come from the Multiple Indicator Cluster Survey (MICS) run through UNICEF (http://mics.unicef.org/) and the Demographic Health Surveys (DHS) run through US AID (https://www.usaid.gov/), which involve directly surveying women and girls in each country. There may be reporting error or bias in that girls who marry are less likely to have completed education and may not know their age or their husband's age.

Nevertheless, recent research suggests that strong legislation may be an important part of the solution in reducing the practice of child marriage (Kim et al., 2013). For example, in their study of women in sub-Saharan Africa, Maswikwa et al. (2015) found that countries with consistent laws against child marriage had significantly lower rates of child marriage for girls than countries that did not have consistent laws.

The negative consequences of child marriage for the girls and women involved have been extensively documented; they are twofold. The first set of concerns are around the health and well-being of child brides, particularly those married under the age of 15, who have significantly poorer health and well-being outcomes than women married over the age of 18 (Koski et al., 2017; Nour, 2009). Research suggests that women who marry under the age of 18 have lower levels of control over their fertility, shorter birth spacings, and higher rates of major birth complications (Godha, Hotchkiss, & Gage, 2013; Maswikwa et al., 2015). Maternal mortality is the main cause of death for girls aged 15–19 in poor countries in Asia and Africa (Syrett, 2016). It is not only physical health that is negatively affected, but women who marry as girls have also been found to have poorer mental health outcomes and higher levels of social isolation (Erulkar, 2013). The other range of concerns focuses on social and economic outcomes. Child brides are less likely to complete their education, less likely to find employment, and more likely to live in poverty than those who marry over the age of 18 (Parsons et al., 2015).

Countries where child marriage is practiced are located all around the globe, including Africa (all regions, but as discussed earlier Central and sub-Saharan Africa have particularly high rates), Southern and Eastern Europe, Southern and Eastern Asia, Oceania, the Caribbean, Latin America, and the Arab World. Within countries, there is often quite a large variation between urban and rural areas, where the rates of child marriage in rural areas are much higher than those in urban areas (Maswikwa et al., 2015; Nguyen & Wodon, 2015). The forces influencing child marriage in these countries and regions are complex and varied, but there are some common factors. There are higher rates of child marriage in poor countries with low GDP, and within those countries the practice is more common among ethnic minorities, the poor, and lower educated groups (Sibanda, 2011). In some family systems, such as those in Africa, South Asia, and Eastern Europe, marrying a daughter as a child often brings money and/or status to the parents' household (in the form of a bride price or dowry) (Amin & Bajracharya, 2011). Conversely, it saves the household money because parents no longer have to pay for the girl's education and the costs of supporting the girl shift to her husband and husband's family (Hotchkiss et al., 2016). In addition, many of these countries and groups have strong religious and cultural traditions that reinforce the subordinate status of girls and women in the society, where the marriageability of women is based on their virginity, appearance, and reproductive capacity (Kamal et al., 2015; Machel, Pires, & Carlsson, 2013). Most child marriages are arranged (discussed in the next section) or occur with parental consent, and many parents believe that the marriage is in the best interests of the girl, protecting them from the loss of their honor and reputation by preventing premarital sex (Kamal et al., 2015). It also allegedly provides girls with ongoing financial security by marriage into a good family or to an older financially established husband (Tsutsui, 2013).

Child marriage is a major point of tension between the value systems of the West and the traditional customs, values, and economic systems of poorer non-Western nations. Nevertheless, child marriage is also legal and occurs in many Western countries. The majority of Western countries permit the marriage of children at the age of 16 with the consent of their parents, or in some circumstances with the court's consent (Syrett, 2016). Recent research in the United States, using the American Community Survey Data from 2010 to 2014, found that out of every 1000 children surveyed, 6.2 had been married. This varied depending on the state and ethnic background of the children, with those living in Southern rural areas and immigrant children more likely to have been married (Koski & Heymann, 2018). Other research has highlighted the ongoing practice of child marriage in some Central and Eastern Europe countries. For example, Hotchkiss et al. (2016) investigated child marriage amongst the Roma, an ethnic minority group in Serbia. Using data from the Serbian MICS,[c] they found that approximately 50% of Roma women had been married prior to 18 and 13% had been married before their 15th birthday. The factors associated with child marriage among the Roma were similar to those found in other parts of the world. Girls with lower levels of education, lower levels of wealth in the parental household, and who lived in rural areas were more likely to marry before their 18th birthday.

Arranged marriages and the role of love in marriage: Change and continuity

As highlighted in the world family theory (Therborn, 2004), there are two dominant approaches to mate selection in marriage, one where marriage is primarily *arranged* and approved by family members, and the second

[c]Multiple Indicator Cluster Survey (MICS) run through UNICEF (http://mics.unicef.org/).

is *autonomous* marriage where the couple choose each other and marry for love. Autonomous mate selection is a relatively recent trend that has largely emerged over the last 200 years, even in the West (Coontz, 2004). It is imbued with expectations of intimacy, passionate love, and sexual desire (Hatfield & Rapson, 1996). Arranged marriages are not necessarily without love, but there is no cultural expectation of passionate love. Although it is expected that companionate love will develop (Hatfield & Rapson, 1996), family connections and economic considerations are the primary considerations for arranged marriages. Lee and Stone (1980), in their study of 122 countries, found that arranged marriages were more common in countries with extended family systems. Conversely, they found that autonomous mate selection based on love was more likely in industrialized societies with nuclear family systems (Lee & Stone, 1980). They argue that this is because in family systems that emphasize the importance of extended family the whole family is impacted by a poor marriage choice, not just the couple involved.

The level of autonomy in mate selection for marriage is another point of tension between the global values of world society theory, such as human rights, individual autonomy and freedom. Through the lens of developmental idealism, autonomous marriages are more modern and aligned with the world cultural values, whereas arranged marriages are traditional and not aligned with world cultural values (Thornton, 2005). Consistent with this, scholars have argued that the processes of globalization, transnational migration, and modernization have contributed to changing understandings of mate selection for marriage and the role of love and intimacy within it (Twamley, 2013). In support of this observation, there is good evidence to suggest that there has been a decrease in arranged marriages in many countries, such as Nepal, China, Japan, and Korea, for recent generations (Ghimire, Axinn, Yabiku, & Thornton, 2006; Pimentel, 2000; Tsutsui, 2013). Nevertheless, many cultures and societies currently practice arranged marriage in one form or another, particularly in Africa, South and South East Asia, and parts of Eastern Europe. While there are no reliable data on the exact prevalence of arranged marriages, spouses who have marriages arranged by their parents tend to marry at younger ages (Ghimire et al., 2006, p. 1205), and arranged marriages are practiced in many of the countries with young age at marriage in Fig. 3.2. Most child marriages are also arranged marriages (Maswikwa et al., 2015).

The two approaches to mate selection have often been conceptualized as a simple dichotomy, with autonomous mate selection representing the "modern" and arranged "tradition." However, this dichotomy does not accurately capture how mate selection for marriage is practiced. Lee and Stone (1980) observed that arranged marriage practices have always varied by country, religion, and culture, where young people, parents, and elders exercise differential amounts of control over the mate selection process. In addition, there is a large and increasing body of evidence spanning decades that suggests arranged marriage practices are becoming less strict (Allendorf, 2013; Twamley, 2013). Several studies in Southeast Asia find that there has been an increase in "elopements," which end in marriage with parental approval (Allendorf, 2013). In addition, a large and growing body of research suggests that arranged marriage practices in India (Allendorf & Pandian, 2016; Deosthale & Hennon, 2008; Prakasa & Rao, 1979), China (Pimentel, 2000), and other countries in Southeast Asia (Tsutsui, 2013) are more inclusive of the young persons' choice, particularly among the elites.

Same-sex marriage: The emergence of a new marriage form

Although marriage is a centuries-old institution, found in all countries and cultures across the globe, it is a recent phenomenon for same-sex couples to marry. Indeed, the concept and practice of same-sex marriage is a relatively new one. There was a growing global movement toward the end of the 20th century to enshrine the rights of lesbian, gay, bisexual, transgender, intersex, and queer/questioning (LGBTIQ) peoples in some countries (Meyer et al., 1997). Prior to the turn of the 21st century, there were only six countries, in which same-sex couples could register a legal domestic partnership: Denmark, France, Iceland, the Netherlands, Norway, and Sweden (Chamie & Mirkin, 2011). The emergence of same-sex marriage rights (and rights for same-sex attracted people more generally) at the time of the new millennium is a significant break with the past (Winter, Forest, & Sénac, 2018). The 20th century was characterized by a number of events which were not favorable to the rights of LGBTIQ peoples, including efforts to ban same-sex marriage. As Chamie and Mirkin (2011, p. 529) observe, the emergence of same-sex marriage is surprising if not unexpected given this historical context: "[it is] extraordinary given that even during most of the twentieth century, homosexuals were closeted, and the concept of same sex marriage was inconceivable, perceived by nearly all as an oxymoron" (Chamie & Mirkin, 2011, p. 529).

In the first two decades of the third millennium, 29 countries have legislated marriage rights for same-sex couples (see Fig. 3.4). The first country

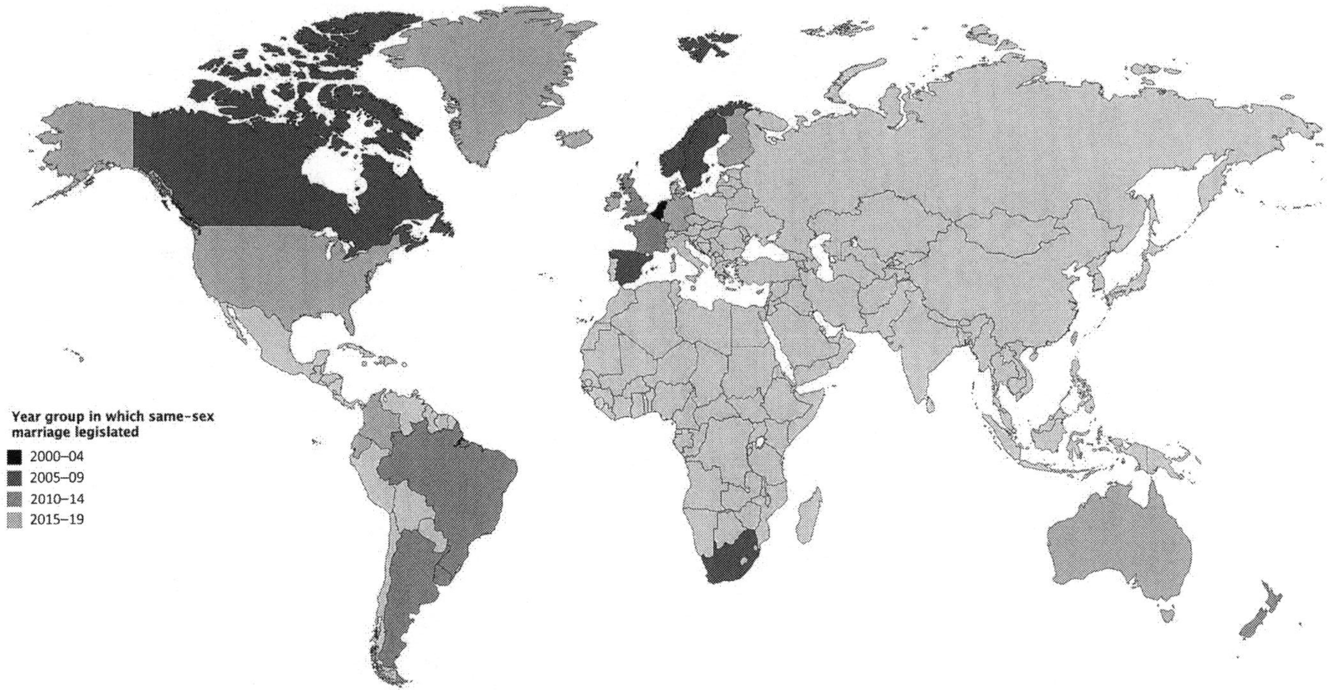

Fig. 3.4 Countries with same-sex marriage rights and year same-sex marriage was legislated. NB. *Large lightest gray shading* are countries that have not passed same-sex marriage legislation.

to do so was the Netherlands in 2000 and the most recent was Ecuador whose High Court ruled in favor of granting marriage rights to same-sex couples in June 2019. Notably, there is considerable cultural and economic diversity across the countries in which same-sex marriage is legal. The majority of countries are located in North West Europe: 17 countries as of 2019 allow same-sex marriage. Same-sex marriage is largely permissible in the countries that make up North America, although it is legal only in some Mexican jurisdictions. It is also becoming increasingly accepted in South America: at present six countries allow same-sex marriage: Argentina, Brazil, Colombia, Ecuador, French Guiana, and Uruguay. South Africa is the only African nation to legalize same-sex marriage, and similarly, Taiwan carries the torch in Asia. Australia and New Zealand as European settler colonies have followed the North West European lead, and are the only countries in the Oceanic region to allow same-sex couples to marry.

While most countries are Western or Western-aligned liberal democracies, there are some considerable differences. As Winter et al. (2018, p. 2) note, the types of countries that allow same-sex marriage include "recent or longstanding democracies, republics and monarchies, unitary and federal states, and reflect different positions with respect to religion and cultural foundations of the nations." For example, same-sex marriage is permissible in a secular republic like France, unitary monarchies like Spain or Sweden, and highly religiously conservative countries such as Brazil and Argentina. What is clear is that there is no discernible pattern identifiable among those countries that enjoy same-sex marriage rights. This is true of the reverse: "[c]ountries opposed to the legalisation of same-sex marriage, including those who have taken measures in recent years to legally reinforce the heterosexual character of marriage, present similar diversity" (Winter et al., 2018, p. 2).

What leads to the legislation of same-sex marriage in countries is the result of both domestic and global forces. For instance, some research has suggested that the decline in negative attitudes toward homosexuality has been the driving force in the acceptance of homosexual couples and same-sex marriage rights. Countries in which there are liberal attitudes toward homosexuality are more likely to recognize same-sex marriage and rights of same-sex couples than those countries in which attitudes are not liberal (Hooghe & Meeusen, 2013). However, attitudes are not the single driving factor in creating same-sex marriage. Political action and actors are needed, as Hooghe and Meeusen (2013, p. 268) hypothesize: "…a tolerant population creates the opportunity to introduce same-sex marriage, [but] political elite actors still have to be convinced to use this opportunity."

From a world society theory perspective, same-sex marriage is legalized in a nation-state through the actions of global actors (Meyer et al., 1997). Paternotte (2015) argues that some countries become regional leaders in the context of change and debate surrounding same-sex marriage. For example, Paternotte (2015) suggests that the inevitable albeit late adoption of same-sex marriage in France and Luxembourg was about both domestic politics and "cross-border emulation." Neighboring countries, in this instance, France, Belgium, and the Netherlands, can become "…caught in a 'rhetorical trap,' whereby their status as 'modern' and 'liberal' or 'European' depends on their extension of extensive rights to the LGBT community" (Siegel & Wang, 2018, p. 385). This idea of regional diffusion has other examples, for example, in Scandinavia: Norway legalized same-sex marriage in 2008, followed by Sweden in 2009 and Denmark in 2012. England and Wales legislated same-sex marriage in 2013 and Scotland followed in 2014. In South America, Uruguay and Brazil both legislated same-sex marriage in 2013 after Argentina in 2010. Same-sex debates in early adopter countries or other high-profile countries, which garner significant international public attention, subsequently set the agenda for other countries. In the case of Spain, legalization of same-sex marriage in 2005 meant that same-sex marriage came to be seen as a social and political issue not just limited to more progressive countries like the Netherlands or Scandinavian countries. Further, the historic nature of same-sex legalization in Spain influenced debates in France, Italy, and Portugal (Paternotte, 2015). Paternotte and Kollman (2013, p. 3) argue that the adoption of same-sex marriage in some countries has been "accelerated by processes of international learning via transnational networks, elite lesson-drawing, as well as—albeit to a much lesser extent—more formal processes of policy harmonization." Thus, Spain was not only influential in the legalization of same-sex marriage in neighboring countries of Europe, but also Latin America. For example, in Argentina, the Spanish model was used to push the same-sex debate and inevitable legalization in 2010 (Paternotte, 2015) and in a similar way, both Germany and Italy were encouraged to pass same-sex legislation after the Irish referendum (Ayoub & Garretson, 2017).

Country differences in same-sex marriage

The percentage of same-sex marriages as a proportion of all marriages contracted each year across 13 countries since their legalization is presented in Table 3.1. Using the average, we can see that same-sex marriage as a proportion of all marriages is largest in Ireland (3.8% of all marriages averaged

Table 3.1 Same-sex marriage as a percent of total marriages (%).

Year	Country												
	NL	BL	ES	SA	NO	DK	FR	NZ	BZ	SC	EG+W	IR	FL
2001	2.9												
2002	2.1												
2003	1.9												
2004	1.6	2.5											
2005	1.6	2.4	1.1										
2006	1.7	2.5	2.2										
2007	1.9	2.5	1.6										
2008	1.9	2.4	1.8	0.4									
2009	1.8	2.5	1.9	0.4	1.1								
2010	1.8	2.6	2.1	0.5	1.1								
2011	1.9	2.6	2.4	0.5	1.1								
2012	1.8	2.5	2.3	0.6	1.1	0.9							
2013	1.9	2.6	2.0	0.6	1.0	1.3	3.1	1.1					
2014	1.9	2.7	2.0	0.7	1.1	1.2	4.4	2.4	0.4		1.9		
2015	2.2	2.4	2.2	0.8	1.3	1.3	3.3	2.2	0.5	5.6	2.7	4.7	

Continued

Table 3.1 Same-sex marriage as a percent of total marriages (%)—cont'd

Year	Country												
	NL	BL	ES	SA	NO	DK	FR	NZ	BZ	SC	EG+W	IR	FL
2016	2.2	2.6	2.5	0.9	1.2	1.2	3.1	2.3	0.5	3.4	2.8	3.4	
2017	2.1	2.6	2.6	1.0	1.5	1.3	3.1	2.2	0.5	3.5		3.2	2.1
2018	2.3	2.9			1.5	1.4	2.6	2.3					
Average	2.0	2.5	2.1	0.7	1.2	1.3	3.3	2.1	0.5	3.5	2.5	3.8	2.1

Notes: Authors calculations based on available data. *BL*, Belgium; *BZ*, Brazil; *DK*, Denmark; *EG+W*, England and Wales; *ES*, Spain; *FI*, Finland; *FR*, France; *IR*, Ireland; *NL*, The Netherlands; *NO*, Norway; *NZ*, New Zealand; *SA*, South Africa; and *SC*, Scotland.

Data from Statistics Netherlands (CBS). (2019). *Lesbian couples likelier to break up than male couples*. Retrieved from https://www.cbs.nl/en-gb/news/2016/13/lesbian-couples-likelier-to-break-up-than-male-couples (Accessed 1 July 2019); Statistics Belgium. (2019). *Belgian people are getting married increasingly later and divorce less and less*. Retrieved from https://statbel.fgov.be/en/news/belgian-people-are-getting-married-increasingly-later-and-divorce-less-and-less (Accessed 1 July 2019); INE. (2019). *Demography and population*. Retrieved from https://www.ine.es/dyngs/INEbase/en/categoria.htm?c=Estadistica_P&cid=1254735572981; Statistics South Africa. (2017). *Marriage and divorces*. Retrieved from http://www.statssa.gov.za/publications/P0307/P03072017.pdf; Statistics Norway. (2019). *Marriages and divorce*. Retrieved from https://www.ssb.no/en/ekteskap/; Statistics Denmark. (2019). Marriages and divorces. Retrieved from https://www.dst.dk/en/Statistik/emner/befolkning-og-valg/vielser-og-skilsmisser; NSEE. (2019). *Mariages et pacs*. Retrieved from https://www.insee.fr/fr/statistiques/2381498; Stats NZ. (2019). *Marriages, civil unions, and divorces: Year ended December 2018*. Retrieved from https://www.stats.govt.nz/information-releases/marriages-civil-unions-and-divorces-year-ended-december-2018; IBGE. (2019). *Sistema de Estatísticas Vitais*. Retrieved from https://www.ibge.gov.br/estatisticas/sociais/populacao/9110-estatisticas-do-registro-civil.html?edicao=26178&t=resultados; National Records of Scotland. (2018). *Marriages and civil partnership—Time series data*. Retrieved from https://www.nrscotland.gov.uk/statistics-and-data/statistics/statistics-by-theme/vital-events/marriages-and-civil-partnerships/marriages-and-civil-partnership-time-series-data; Office for National Statistics (ONS). (2018). *Marriages in England and Wales*. Retrieved from https://www.ons.gov.uk/peoplepopulationandcommunity/birthsdeathsandmarriages/marriagecohabitationandcivilpartnerships/bulletins/marriagesinenglandandwalesprovisional/2016; Central Statistics Office. (2019). *Marriage and civil partnerships*. Retrieved from https://www.cso.ie/en/statistics/birthsdeathsandmarriages/marriagescivilpartnerships/; Statistics Finland. (2019). *Number of marriages contracted decreased considerably*. Retrieved from https://www.stat.fi/til/ssaaty/2018/ssaaty_2018_2019-06-18_tie_001_en.html.

over time), Scotland (3.5%), and France (3.3%). There are some interesting between-country differences over time. For example, in Spain, same-sex marriage as a proportion of total marriages has gone from 1.1% in 2005 to 2.6% in 2016. Similarly, same-sex marriage as a proportion of total marriages doubled in South Africa over a similar period; however, it remains only 1% of all marriages contracted in 2017. Same-sex marriages in England and Wales as a proportion of all marriages has also increased sharply over a smaller time period, going from 1.9% in 2014 to 2.8% in 2015. In other countries like Belgium and Denmark, there have been increases in same-sex marriage as a proportion of all marriages, but they are less marked than those in Spain or England and Wales. In the Netherlands and New Zealand, there have been small fluctuations over time, particularly in the Netherlands where same-sex marriages have gone from as high as 2.9% of all marriages in 2001, the first year of legal same-sex marriage, to as low as 1.6% in 2004 and back up to 2.3% in 2018. France, Scotland, and Wales all had relatively high proportions of same-sex marriage immediately after it was legalized, for example, 5.6% in Scotland and 4.7% in Ireland in 2015; however, in the years following, rates have declined a little. In the case of France and Ireland, these patterns of decline may reflect the significant public campaigns and public debate to legislate same-sex marriage in those countries. In the case of France, where same- sex marriage has gone from as high as 4.4% in 2014 to 2.6% in 2018, there was considerable social and political unrest following legalization in 2014 (Sénac, 2018).

Gender and age differences in same-sex marriage

The proportions of marriages that are between two men are presented in Table 3.2. Across the eight countries where data were available, marriages contracted between two men were highest over time in the Netherlands, Belgium, Spain, and Ireland. Just under three-quarters of all same-sex marriages in the first year of same-sex marriage being available in Spain were between men. This proportion has decreased considerably in the intervening years since with half of all same-sex marriages between men in 2017. Belgium experienced a similar decline where around 58% of all same-sex marriages following legalization in 2004 were between two men, steadily declining to around 47.6% in 2018. In the Netherlands, trends have also fluctuated over time. Initially, there were more marriages contracted between two women (55.5%); this changed from 2003 with a greater proportion of marriages between men, peaking at 57.8% in 2009. The trend reversed again in 2016, with more marriages between two women. In Denmark

Table 3.2 Two men marrying as a percent of total marriages (%).

Year	Country								
	NL	BL	ES	NO	DK	BZ	SC	EG+W	IR
2001	44.5								
2002	49.1								
2003	51.0								
2004	52.1	58.2							
2005	50.4	56.5	72.4						
2006	52.2	53.0	69.7						
2007	51.6	51.7	67.1						
2008	53.4	52.6	64.8						
2009	57.8	55.9	64.8	37.1					
2010	51.3	49.1	61.8	36.7					
2011	55.6	51.8	59.1	35.9					
2012	57.7	51.6	56.8	37.9	38.8				
2013	57.3	51.2	53.7	35.7	35.5				
2014	57.7	48.9	51.3	39.4	42.6	49.7	47.1	43.9	

2015	53.6	49.8	51.5	37.7	43.9		41.6	44.0	
2016	46.1	48.9	50.4	43.5	42.3		39.2	44.3	57.4
2017	45.1	50.1	50.3	35.7	43.6		41.4		55.9
2018	45.4	47.6		42.0	39.0				56.0
Average	51.8	52.2	59.5	38.2	40.8	49.7	42.3	44.1	56.4

Notes: Authors calculations based on available data. *BL*, Belgium; *BZ*, Brazil; *DK*, Denmark; *EG+W*, England and Wales; *ES*, Spain; *IR*, Ireland.; *NL*, The Netherlands; *NO*, Norway; and *SC*, Scotland.

Data from Statistics Netherlands (CBS). (2019). *Lesbian couples likelier to break up than male couples*. Retrieved from https://www.cbs.nl/en-gb/news/2016/13/lesbian-couples-likelier-to-break-up-than-male-couples (Accessed 1 July 2019); Statistics Belgium. (2019). *Belgian people are getting married increasingly later and divorce less and less*. Retrieved from https://statbel.fgov.be/en/news/belgian-people-are-getting-married-increasingly-later-and-divorce-less-and-less (Accessed 1 July 2019); INE. (2019). *Demography and population*. Retrieved from https://www.ine.es/dyngs/INEbase/en/categoria.htm?c=Estadistica_P&cid=1254735572981; Statistics Norway. (2019). *Marriages and divorce*. Retrieved from https://www.ssb.no/en/ekteskap/; Statistics Denmark. (2019). Marriages and divorces. Retrieved from https://www.dst.dk/en/Statistik/emner/befolkning-og-valg/vielser-og-skilsmisser; IBGE. (2019). *Sistema de Estatísticas Vitais*. Retrieved from https://www.ibge.gov.br/estatisticas/sociais/populacao/9110-estatisticas-do-registro-civil.html?edicao=26178&t=resultados; National Records of Scotland. (2018). *Marriages and civil partnership—Time series data*. Retrieved from https://www.nrscotland.gov.uk/statistics-and-data/statistics/statistics-by-theme/vital-events/marriages-and-civil-partnerships/marriages-and-civil-partnership-time-series-data; Office for National Statistics (ONS). (2018). *Marriages in England and Wales*. Retrieved from https://www.ons.gov.uk/peoplepopulationandcommunity/birthsdeathsandmarriages/marriagecohabitationandcivilpartnerships/bulletins/marriagesinenglandandwalesprovisional/2016; Central Statistics Office. (2019). *Marriage and civil partnerships*. Retrieved from https://www.cso.ie/en/statistics/birthsdeathsandmarriages/marriagescivilpartnerships/.

and Norway, there are higher rates of same-sex marriage between two women. Similarly, in the United Kingdom, same-sex marriage is contracted at higher rates between two women in Scotland, England, and Wales.

There is little information on age of same-sex couples at marriage; however, what evidence does exist suggests that same-sex couples marry at older ages than heterosexual couples. This is likely due to the fact that in many countries same-sex marriage rights are relatively recent and may reflect the fact that same-sex couples who get married were likely to have been in a long-term partnership prior to marriage. For example, in England and Wales and Ireland, the average age of men in a same-sex marriage between 2014 and 2018 was 40 years of age (Central Statistics Office, 2019; Office of National Statistics, 2019). The average age of women who were contracted into a same-sex marriage in England was 37.4 years in 2016, up from 36.9 in 2014. In Ireland, the average age of women who entered into a same-sex marriage went from 41 years of age in 2016 to 38.7 years of age in 2018 (Central Statistics Office, 2019).

In Table 3.3, we compare age-specific marriage rates for men and women in same-sex couples for Spain and the Netherlands in 2015. Both countries were among the first in the world to have same-sex marriage rights (the Netherlands in 2000, Spain in 2005) and these age-specific rates provide more depth to the age profile of same-sex married couples. Overall, same-sex attracted Spaniards are more likely to marry at younger ages than same-sex attracted Dutch people. There were some gender differences between and within countries. Dutch and Spanish women in same-sex couples were

Table 3.3 Age-specific marriage rates for men and women in same-sex couples, The Netherlands and Spain (2015) (%).

	Age group				
Country	**15–24**	**25–34**	**35–44**	**45–54**	**55+**
Netherlands					
Men	4.0	24.8	26.9	25	19.2
Women	5.9	37.6	25.1	19.2	12.2
Spain					
Men	4.1	32.3	32.3	20.2	11.3
Women	4.5	42.8	35.2	12.7	4.7

Notes: Authors calculations based on available data.
Data from Statistics Netherlands (CBS). (2019). *Lesbian couples likelier to break up than male couples*. Retrieved from https://www.cbs.nl/en-gb/news/2016/13/lesbian-couples-likelier-to-break-up-than-male-couples (Accessed 1 July 2019); IBGE. (2019). *Sistema de Estatísticas Vitais*. Retrieved from https://www.ibge.gov.br/estatisticas/sociais/populacao/9110-estatisticas-do-registro-civil.html?edicao=26178&t=resultados.

more likely to marry at younger ages than Dutch and Spain men. While there is little research on the topic, this gender gap in average at marriage for same-sex couples may reflect the general tendency for men to marry at older ages than women. Gender differences were also evident within countries. For instance, the age distribution of marriage among same-sex attracted men in the Netherlands was more evenly distributed, whereas marriage is concentrated in the younger years for women in the Netherlands. There were notable differences between countries. For example, over 42% of Spanish women in a same-sex couple were likely to marry between the ages of 25 and 34 compared to 37% of Dutch women.

Divorce: Cross-national differences in the enduring nature of marriage

One of the major global changes to marriage over recent years has been the rise in the incidence of divorce (Wang & Schofer, 2018). Traditionally, marriage was for life; however, the growth in divorce across the globe suggests that marriage is no longer an unbreakable bond. This trend, more than any other, has provoked concerns about the future of marriage and the family (van Acker, 2008; Willekens & Scheiwe, 2003). While the increase in the rate of divorce in wealthy Western countries commenced several decades ago in the 1960s and 1970s, there is evidence that the rest of the world has since followed this trend (Wang & Schofer, 2018). Nevertheless, the extent to which marriage is an enduring or breakable bond varies considerably by country. In Fig. 3.5, we present the net divorce rates in 72 countries. The net divorce rate,[d] which is the proportion of registered divorces of the married population, is a better measure of divorce rates for comparative purposes than the crude divorce rate, which is more commonly used, because it accounts for differences in the age and marriage rates in each country (Wang & Schofer, 2018). We note that there are many countries that do not

[d]The net divorce rate used in this chapter is the number of divorces per 1000 married women. The net rate is preferable to the crude divorce rate, which is the proportion of registered divorces per 1000 total population (not married population), because the denominator includes only those people at risk of divorce. Using the crude divorce rate may overestimate or underestimate the divorce rate depending on the distribution of marriage in the country. For example, countries where marriage is less common may have an underinflated divorce rate because most of the population is not married, but countries with high rates of marriage are going to have an overinflated divorce rate.

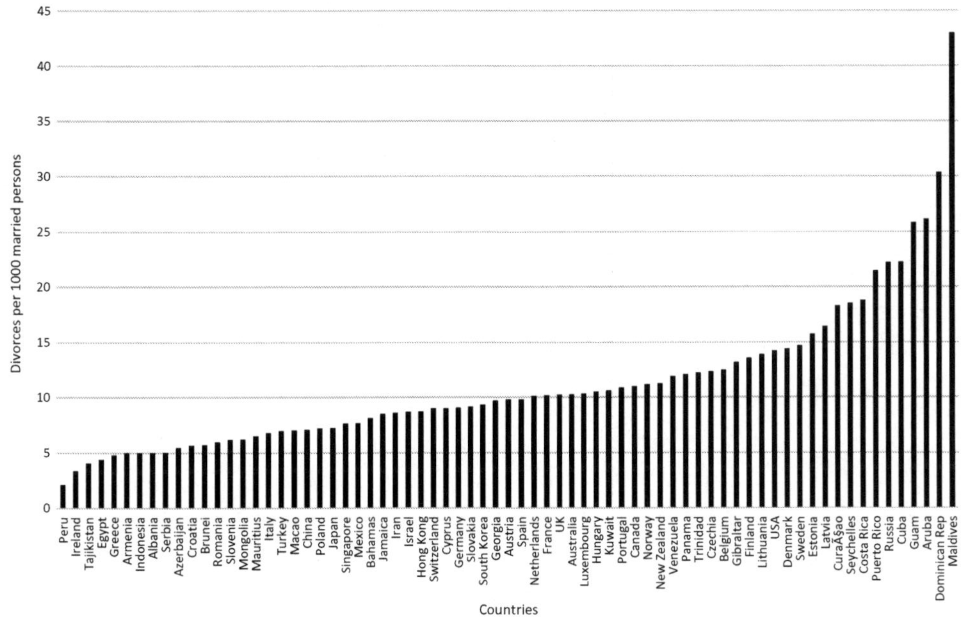

Fig. 3.5 National divorce rates, selected countries, 2010–17. (Some countries did not have data available within this date range including Peru and Canada (2007), and Egypt (2006).) *(Data from United Nations. (2011b). Table 25: Divorces and crude divorce rates, by urban/rural residence: 2005–2009. In 2009–2010 demographic yearbook. New York: Department of Economic and Social Affairs, United Nations; United Nations. (2013b). Table 25: Divorces and crude divorce rates, by urban/rural residence: 2009–2013. In 2013 demographic yearbook. New York: Department of Economic and Social Affairs, United Nations; United Nations. (2017b). Table 25: Divorces and crude divorce rates, by urban/rural residence: 2013–2017. In Demographic yearbook 2017. New York: Department of Economic and Social Affairs, United Nations; United Nations Statistical Division. (2019). UN demographic statistics database. Population by marital status, age, sex and urban/rural residence. http://data.un.org/Data.aspx?q=population+by+marital+status&d=POP&f=tableCode%3a23.)*

collect official data on registered divorces, particularly countries in Central, South, and sub-Saharan Africa, South Asia, and Southeast Asia.

As shown in Fig. 3.5, the divorce rate ranges from a low of 2.1 divorces per 1000 married population in Peru to a high of just under 43 for the Maldives. The overall average for the 72 countries was 10.9 divorces per 1000 married population. Most notably, the Maldives has the highest divorce rate and is an extreme outlier. The Maldives is a small Islamic country comprising 1160 small islands located in South Asia. Despite being an Islamic country in South Asia, divorce has traditionally been relatively fluid and easy (El-Horr & Prabha Pande, 2016). While gender inequality exists, women have relatively equal access to education and have similar levels of employment as men, but women's role in the family and household is restricted. There is little stigma attached to divorce (Walsh, 2013).

It is difficult to generalize about divorce rates based on region. Countries located in Latin America are at the extreme ends of the distribution. Some Latin American countries have very low rates of divorce, such as Mexico and Peru (which has the lowest rate of divorce of all countries). In contrast, Cuba, the Dominican Republic, Puerto Rico, and Costa Rica all have relatively high rates of divorce. Similarly, the divorce rates for former Soviet Union countries vary considerably. Russia, Estonia, Lithuania, and Latvia have relatively high divorce rates, but other former Soviet states such as Tajikistan and Azerbaijan have very low rates of divorce.

Divorce rates also vary among wealthy Western nations. Countries in Northern Europe, such as Finland, Denmark, and Sweden, and the United States tend to have higher rates of divorce compared to other European countries (Germany, France, the United Kingdom) and other former European settler countries in North America (Canada) and Oceania (Australia and New Zealand). The high divorce rates in Finland, Denmark, and the United States are not underpinned by similar marriage trends in these countries. Finland, Denmark, and Sweden have had relatively high rates of divorce for several decades, but they also have low rates of marriage and high rates of long-term unmarried cohabitation (Kalmijn, 2007). Whereas the United States has relatively low rates of cohabitation and relatively high rates of marriage, but they also have high rates of divorce (Cherlin, 2009a, 2009b), which is often referred to as US exceptionalism (Lesthaeghe, 2010). Interestingly, despite quite different patterns in the timing of marriage in China and Japan, with China having a higher rate of marriage at younger ages, the countries have similar rates of divorce of around 7 divorces per 1000 married population, which is below the country average in Fig. 3.5.

Same-sex divorce

There is relatively little information on divorce in same-sex marriages. Previous literature has found that marriage and partnership dissolution is greater among lesbian couples (Anderson, Noack, Seierstad, & Weedon-Fekjer, 2006). This largely mirrors the results shown in Table 3.2, which reports the percentage of two men divorcing as a percent of all same-sex divorces. Of the five countries where data were available, there were only two in which the proportion of men in same-sex couples has higher rates of divorce: Belgium and Spain. Men in same-sex relationships in Norway and Denmark are less likely to divorce; however, the proportion of divorces among this group has been steadily increasing over time: in Norway, the proportion has gone from 21% in 2011 to 31% in 2018, and in Denmark it has doubled from 17% in 2013 to 34% in 2018.

Conclusions: Convergence and difference in marriage and family

In this chapter, we examined recent cross-national differences in marriage practices. As marriage and family life change, some commentators on worldwide trends in family life have observed a convergence toward a dominant idealized family type represented by the North Western European family system described throughout this chapter (Lesthaeghe, 2010; Meyer et al., 1997; Thornton, 2005). There is some evidence for this convergence with global declines in the rate of marriage (Goode, 2003), increases in the average age at marriage, and an increase in the incidence of divorce (Wang & Schofer, 2018). Some traditional family practices in non-Western countries such as child marriage (Nguyen & Wodon, 2015) and arranged marriages (Ghimire et al., 2006; Pimentel, 2000; Twamley, 2013) have declined and undergone major transformations in recent years. At face value, it appears that there is a worldwide convergence in family life underpinned by a world cultural view of the family.

Despite this, there is overwhelming evidence in our analysis that marriage practices, measured in this study by crude marriage rates, marriage timing (age at marriage), and net divorce rates throughout the world are very diverse. More consistent with the arguments of Therborn (2014) and others (Fokkema & Liefbrouer, 2008; Kalmijn, 2007), we find that countries in our examination have their own unique marriage patterns that do not necessarily suggest a convergence of family life. For example, the United States and China have similar crude marriage rates which are high by world standards, but that is where the similarities end; marriage occurs on average a couple of years earlier for Chinese couples and net divorce rates in China are half of those in the United States. This is not surprising given that the United States and China have vastly different historical backgrounds, economies, political systems, and populations. Japan, by comparison, shares a traditional Confucian family system with China; however, it has much lower crude marriage rates and couples marry much later, but it has similar rates of divorce to China.

The high rates of marriage in the United States stand it apart from other wealthy western nations. For example, in North West Europe, countries such as Sweden, Norway, and Finland, tend to have substantially lower rates of marriage, and when couples marry, they are 3–4 years older on average than US couples, although divorce rates in these countries tended to be similar to those in the United States. In contrast, the Southern European countries of Spain, Greece, and Italy have very low crude marriage rates combined with older ages at first marriage, but also have quite low net divorce rates. Countries in Central and Eastern Europe also have some inconsistent patterns. Russia stands out with very high crude marriage rates, relatively young age at first marriage, and very high net divorce rates. Slovenia is also an interesting transition country with relatively low rates of marriage, older ages at marriage, but then relatively low rates of divorce and, according to recent research, relatively long marriage duration before divorce (Svab et al., 2012; Ule, 2004). These marriage patterns are much more like neighboring countries in Southern Europe than those in the Central or Western Europe. These countries in particular have high levels of *institutional disjuncture* among young educated women.

Countries such as Australia and the United Kingdom had almost identical crude marriage rates and net divorce rates; however, Australian couples were on average 3 years older than couples from the United Kingdom when they married. In contrast, Egypt, which is one of the few African/Arab nations that provide information on marriage and divorce, had very high crude marriage rates; women get married quite young (on average aged 22) and they were on average 5.5 years younger than their husband. Egypt also has very low net divorce rates. Iran, who also provided information on marriage and divorce, had a relatively high crude marriage rate, but compared to the women in Egypt, Iranian women were a couple of years older when they married, and the gender age gap was smaller, around 3 years difference between men and women. The net divorce rate in Iran is low by international standards, but Iran's net divorce rate is double that of Egypt.

These few examples and comparisons illustrate the diversity of family life across the globe and highlight the differences not only between regions with different family systems, but also between countries within regions that share similar family systems. While there may be some evidence for worldwide trends suggesting a decline in marriage (Goode, 2003), the stark differences between countries are also apparent. We were limited in examining these general patterns for many countries in Africa, Latin America, and Asia as most do not collect information on marriages and divorces. However, there were data available from many of those countries on age at marriage and rates of child marriage. There were pronounced differences between wealthy Western countries and poor or non-Western countries in terms of the average age at which couples marry. There were virtually no Western countries with an average age of marriage for women under 25 years. Most countries with young ages at marriage, in the late teens and early 20s, were located in Central and sub-Saharan Africa, South East Asia, the Arab World, and Latin America. Many of these regions and the countries within them also had high rates of child marriage, particularly among girls (Koski et al., 2017; Maswikwa et al., 2015; Nguyen & Wodon, 2015; Nour, 2009). There is evidence that worldwide rates of child marriage are declining, and most countries have a legal minimum age of marriage above 18 (Kim et al., 2013; Maswikwa et al., 2015). Nevertheless, in many countries and regions, poverty, customary practices, and the persistence of gender inequality mean that child marriage is one of the few limited options for girls. It is also noteworthy that most countries in the world, including those in the West, have legal provisions for children to marry at 16 and there is evidence to suggest that while relatively rare, it is more common than may be expected (Koski & Heymann, 2018; Syrett, 2016).

Overall, we conclude that many of these differences in the rhythm, timing, and practice of marriage and family life are likely to remain. This is because the functions, cultural, social, and economic, of marriage and family within different regions and countries differ. To illustrate, in high-income Western countries, marriage, if it takes place at all, is primarily an arrangement that meets the emotional and economic needs of the partners and children within a nuclear family context (Lee & Stone, 1980; Thornton, 2005). In low- to mid-income countries, a marriage often serves a broader social and economic function where couples are embedded in a network of extended family and community (Lee & Stone, 1980; Tsutsui, 2013). Thus, in the latter case, the Western model of family, which is often promulgated as the ideal standard by large international organizations, is unlikely to be fully embraced unless there are major changes in the functions of marriage and family in society.

One relatively new and emerging marriage trend that strongly differentiates countries is same-sex marriage. Prior to 2000, there were no countries in the world where same-sex couples could marry, but over the last two decades 29 countries have legalized same-sex marriage. The overwhelming majority of countries that have legalized same-sex marriage are located in North Western Europe, or are former European settler countries such as Canada, the United States, Australia, and New Zealand. South Africa is the only country in Africa to have legalized same-sex marriage and Taiwan is the only country in Asia. Several Latin American countries have also legalized same-sex marriage. On the one hand, this is not surprising. Therborn (2004) suggests that the Latin American family system is diverse, with low rates of marriage, which suggests it is not universal or as institutionalized as it is in other family systems. This may have created the space for same-sex marriage to be legalized. On the other hand, it is somewhat surprising because many Latin American countries where same-sex marriage has been legalized have conservative religious beliefs. Data on the prevalence of same-sex marriage were difficult to obtain. It was only available in a select number of European countries, and overall the proportion of same-sex couples marrying in each country was relatively small. We find some evidence that same-sex marriages have a different rhythm and timing during the life course compared to heterosexual marriage, where marriage occurs at older ages. It is not yet clear whether these are long-term trends or whether this is suggestive of long-term same-sex couples marrying. Either way, the numbers are so small that they are unlikely to have a major impact on worldwide marriage trends. The main unique contribution is the major cultural and social shifts that have occurred in the rights of minority groups with respect to marriage (Meyer et al., 1997)

We conclude with the general observation that marriage and family formation continue to evolve and change; however, these patterns are not necessarily suggesting a convergence to a uniform family system. Rather, different and new family systems are emerging and further diversification within and between countries is occurring (Therborn, 2014). Even if there is a shared or similar cultural family system, different political, social, and economic systems may influence marriage patterns in different ways. This fragmentation of family systems is underpinned by the tensions between tradition, economic development, and modernity, which all contribute to shaping marriage patterns within each country.

References

Alharahsheh, S. T., & Almeer, F. K. (2018). Cross-national marriage in Qatar. *Journal of Women of the Middle East and the Islamic World, 16*, 170–204. https://doi.org/10.1163/15692086-12341336.

Allendorf, K. (2013). Schemas of marital change: From arranged marriages to eloping for love. *Journal of Marriage and Family, 75*, 453–469.

Allendorf, K., & Pandian, R. K. (2016). The decline of arranged marriage? Marital change and continuity in India. *Population and Development Review, 42*(3), 435–464. https://doi.org/10.1111/j.1728-4457.2016.00149.x.

Amin, S., & Bajracharya, A. (2011). *Costs of marriage—Marriage transactions in the developing world* New York. Retrieved from www.popcouncil.org/publications/serialsbriefs/TABriefs.asp.

Anderson, G., Noack, T., Seierstad, A., & Weedon-Fekjer, H. (2006). The demographics of same-sex marriage in Norway and Sweden. *Demography, 43*, 79–98. https://doi.org/10.1353/dem.2006.0001.

Ayoub, P. M., & Garretson, J. (2017). Getting the message out: Media context and global changes in attitudes toward homosexuality. *Comparative Political Studies, 50*(8), 1055–1085.

Central Statistics Office. (2019). *Marriages and civil partnerships*. Central Statistics Office. Retrieved from https://www.cso.ie/px/pxeirestat/Database/eirestat/Marriages%20and%20Civil%20Partnerships/Marriages%20and%20Civil%20Partnerships_statbank.asp?SP=Marriages%20and%20Civil%20Partnerships&Planguage=0.

Chamie, J., & Mirkin, B. (2011). Same-sex marriage: A new social phenomenon. *Population and Development Review, 37*(3), 529–551. https://doi.org/10.1111/j.1728-4457.2011.00433.x.

Cherlin, A. J. (2004). The deinstitutionalization of American marriage. *Journal of Marriage and Family, 66*, 848–861. https://doi.org/10.1111/j.0022-2445.2004.00058.x.

Cherlin, A. (2009a). *The marriage-go-round: The state of marriage and the family in America today*. New York: Alfred A. Knopf.

Cherlin, A. J. (2009b). The origins of the ambivalent acceptance of divorce. *Journal of Marriage and Family, 71*, 226–229.

Coontz, S. (2004). The world historical transformation of marriage. *Journal of Marriage and Family, 66*(4), 974–979.

Davis, K. (1945). The world demographic transition. *The Annals of the American Academy of Political and Social Science, 237*, 1–11.

Deosthale, D. C., & Hennon, C. B. (2008). Family and tradition in modern India. In C. B. Hennon & S. M. Wilson (Eds.), *Families in a global context* (pp. 295–323). New York: Routledge/Taylor & Francis Group.

El-Horr, J., & Prabha Pande, P. (2016). *Understanding gender in Maldives: Toward inclusive development*. Washington, DC: World Bank.

Erulkar, A. (2013). Adolescence lost: The realities of child marriage. *Journal of Adolescent Health, 52*, 513–514. https://doi.org/10.1016/j.jadohealth.2013.03.004.

Fokkema, T., & Liefbrouer, A. (2008). Trends in living arrangements in Europe: Convergence or divergence? *Demographic Research, 19*, 1351–1418. https://doi.org/10.4054/DemRes.2008.19.36.

Fussell, E., Guthier, A. H., & Evans, A. (2007). Heterogeneity in the transition to adulthood: The cases of Australia, Canada and the United States. *European Journal of Population/Revue europeenne de demographie, 23*, 389–414. https://doi.org/10.1007/s10680-007-9136-4.

Ghimire, D. J., Axinn, W. G., Yabiku, S. T., & Thornton, A. (2006). Social change, premarital nonfamily experience, and spouse choice in an arranged marriage society. *American Journal of Sociology, 111*(4), 1181–1218.

Godha, D., Hotchkiss, D. R., & Gage, A. J. (2013). Association between child marriage and reproductive health outcomes and service utilization: A multi-country study from South Asia. *Journal of Adolescent Health, 52*, 552–558. https://doi.org/10.1016/j.jadohealth.2013.01.021.

Goldscheider, F., Bernhardt, E., & Lappegård, T. (2015). The gender revolution: A framework for understanding changing family and demographic behavior. *Population and Development Review*, *41*(2), 207–239. https://doi.org/10.1111/j.1728-4457.2015.00045.x.

Goode, W. J. (2003). Family changes over the long term: A sociological commentary. *Journal of Family History*, *28*(1), 15–30. https://doi.org/10.1177/0363199002238552.

Hajnal, J. (1965). European marriage patterns in perspective. In D.V. Glass & D. E. C. Eversely (Eds.), *Population in history* (pp. 101–143). London: Arnold.

Harkness, G., & Khaled, R. (2014). Modern traditionalism: Consanguineous marriage in Qatar. *Journal of Marriage and Family*, *76*, 587–603. https://doi.org/10.1111/jomf.12106.

Hatfield, E., & Rapson, R. L. (1996). *Love and sex: Cross-cultural perspectives*. Needham Heights, MA: Allyn & Bacon.

Hennon, C. B. & Wilson, S. M. (Eds.), (2008). *Families in a global context*. London: Routledge.

Hochschild, A., & Machung, A. (1989). *The second shift: Working parents and the revolution at home*. New York: Penguin Books.

Holland, J. (2017). The timing of marriage vis-à-vis coresidence and childbearing in Europe and the United States. *Demographic Research*, *35*, 609–626. https://doi.org/10.4054/DemRes.2017.36.20.

Hooghe, M., & Meeusen, C. (2013). Is same-sex marriage legislation related to attitudes toward homosexuality?: Trends in tolerance of homosexuality in European countries between 2002 and 2010. *Sexuality Research and Social Policy*, *10*(4), 258–268. https://doi.org/10.1007/s13178-013-0125-6.

Hotchkiss, D. R., Godha, D., Gage, A. J., & Cappa, C. (2016). Risk factors associated with the practice of child marriage among Roma girls in Serbia. *BMC International Health and Human Rights*, *16*(6), https://doi.org/10.1186/s12914-016-0081-3.

Ji, Y. (2015). Between tradition and modernity: "Leftover" women in Shanghai. *Journal of Marriage and Family*, *77*, 1057–1073. https://doi.org/10.1111/jomf.12220.

Kalmijn, M. (2007). Explaining cross-national differences in marriage, cohabitation, and divorce in Europe, 1990–2000. *Population Studies*, *61*(3), 243–263. https://doi.org/10.1080/00324720701571806.

Kamal, S. M. M., Hassan, C. H., Alam, G. M., & Ying, Y. (2015). Child marriage in Bangladesh: Trends and determinants. *Journal of Biosocial Science*, *47*, 120–139. https://doi.org/10.1017/S0021932013000746.

Kiernan, K. E. (2000). European perspectives on union formation. In L. J. Waite, C. Bachrach, M. Hindin, E. Thompson, & A. Thornton (Eds.), *The ties that bind: Perspectives on marriage and cohabitation* (pp. 40–58). New York: Aldine De Gruyter.

Kim, M., Longhofer, W., Heger Boyle, E., & Brehm Nyseth, H. (2013). When do laws matter? National minimum-age-of-marriage laws, child rights, and adolescent fertility, 1989–2007. *Law & Society Review*, *47*(3), 589–619. https://doi.org/10.1111/lasr.12033.

Koski, A., Clark, S., & Nandi, A. (2017). Has child marriage declined in sub-Saharan Africa? An analysis of trends in 31 countries. *Population and Development Review*, *43*(1), 7–29.

Koski, A., & Heymann, J. (2018). Child marriage in the United States: How common is the practice, and which children are at greatest risk? *Perspectives on Sexual and Reproductive Health*, *50*(2), 59–65. https://doi.org/10.1363/psrh.12055.

Lappegard, T., & Noak, T. (2015). The link between parenthood and partnership in contemporary Norway—Findings from focus group research. *Demographic Research*, *32*, 287–310. https://doi.org/10.4054/DemRes.2015.32.9.

Lee, G. R., & Stone, L. H. (1980). Mate-selection systems and criteria: Variation according to family structure. *Journal of Marriage and Family*, *42*(2), 319–326. https://doi.org/10.2307/351229.

Lesthaeghe, R. (1991). *The second demographic transition in western countries: An interpretation* International Demographic Program, working paper 1991–2. Brussels, Belgium. Retrieved from https://www.vub.be/demography/wp-content/uploads/2016/02/WP-IPD-1991-2.pdf.

Lesthaeghe, R. (2010). The unfolding story of the second demographic transition. *Population and Development Review, 36*(2), 211–251. https://doi.org/10.1111/j.1728-4457.2010.00328.x.

Liu, C., Esteve, A., & Trevin, R. (2017). Female-headed households and living conditions in Latin America. *World Development, 90*, 311–328. https://doi.org/10.1016/j.worlddev.2016.10.008.

Machel, G., Pires, E., & Carlsson, G. (2013). The world we want: An end to child marriage. *The Lancet, 382*(9897), 1005–1006. https://doi.org/10.1016/S0140-6736(13)61944-3.

Maříková, H. (2004). The Czech family at present and in the recent past. In M. Robila (Ed.), *Families in Eastern Europe* (pp. 29–47). Emerald Group Publishing Limited.

Maswikwa, B., Richter, L., Kaufman, J., & Nandi, A. (2015). Minimum marriage age laws and the prevalence of child marriage and adolescent birth: Evidence from Sub-Saharan Africa. *International Perspectives on Sexual and Reproductive Health, 41*(2), 58–68. https://doi.org/10.1363/4105815.

McDonald, P. (2000a). Gender equity in theories of fertility transition. *Population and Development Review, 26*(3), 427–439. https://doi.org/10.1111/j.1728-4457.2000.00427.x.

McDonald, P. (2000b). Gender equity, social institutions and the future of fertility. *Journal of Population Research, 17*(1), 1–16. https://doi.org/10.1007/BF03029445.

Meyer, J. W. (2010). World society, institutional theories, and the actor. *Annual Review of Sociology, 36*, 1–20. https://doi.org/10.1146/annurev.soc.012809.102506.

Meyer, J. W., Boli, J., Thomas, G. M., & Ramirez, F. O. (1997). World society and the nation-state. *American Journal of Sociology, 103*(1), 144–181.

Nguyen, M. C., & Wodon, Q. (2015). Global and regional trends in child marriage. *The Review of Faith & International Affairs, 13*(3), 6–11. https://doi.org/10.1080/15570274.2015.1075756.

Nour, N. M. (2009). Child marriage: A silent health and human rights issue. *Reviews in Obstetrics & Gynecology, 2*(1), 51–56.

Office of National Statistics. (2019). *Marriages in England and Wales: 2016. Number of marriages that took place in England and Wales analysed by age, sex, previous marital status and civil or religious ceremony*. Office of National Statistics. Retrieved from https://www.ons.gov.uk/peoplepopulationandcommunity/birthsdeathsandmarriages/marriagecohabitationandcivilpartnerships/bulletins/marriagesinenglandandwalesprovisional/2016.

Parsons, J., Edmeades, J., Kes, A., Petroni, S., Sexton, M., & Wodon, Q. (2015). Economic impacts of child marriage: A review of the literature. *The Review of Faith & International Affairs, 13*(3), 12–22. https://doi.org/10.1080/15570274.2015.1075757.

Paternotte, D. (2015). Global times, global debates? Same-sex marriage worldwide. *Social Politics, 22*(4), 653–674. https://doi.org/10.1093/sp/jxv038.

Paternotte, D., & Kollman, K. (2013). Regulating intimate relationships in the European polity: Same-sex unions and policy convergence. *Social Politics, 20*(4), 510–533. https://doi.org/10.1093/sp/jxs024.

Pimentel, E. E. (2000). Just how do I love thee?: Marital relations in urban China. *Journal of Marriage and Family, 62*, 32–47. https://doi.org/10.1111/j.1741-3737.2000.00032.x.

Prakasa, V. V., & Rao, V. N. (1979). Arranged marriages: An assessment of the attitudes of the college students in India. In G. Kurian (Ed.), *Cross-cultural perspectives of mate selection and marriage*. Connecticut: Greenwood Press.

Rosina, A., & Fraboni, R. (2004). Is marriage losing its centrality in Italy? *Demographic Research, 11*(6), 149–172. https://doi.org/10.4054/DemRes.2004.11.6.

Sénac, R. (2018). Same-sex marriage in France and Spain comparing resistance in centralized secular republics. In B. Winter, M. Forest, & R. Sénac (Eds.), *Global perspectives on same sex marriage: A neo-institutional approach* (pp. 105–125). London: Palgrave.

Sibanda, M. (2011). *Married too soon: Child marriage in Zimbabwe*. Research and Advocacy Unit. Retrieved from: http://archive.kubatana.net/html/archive/chiyou/111030rau.asp.

Siegel, S., & Wang, Y. A. (2018). Broken rainbows: The partisan politics of marriage equality in Europe. *European Politics and Society*, *4*, 377–395. https://doi.org/10.1080/23745118.2018.1429195.

State of Qatar. (2009). *Vital statistics annual bulletin, marriages and divorces.* . State of Qatar, Qatar (Table 4).

Svab, A., Rener, T., & Kuhar, M. (2012). Behind and beyond hajnal's line: Families and family life in Slovenia. *Journal of Comparative Family Studies*, *43*(3), 419–437. https://doi.org/10.3138/jcfs.43.3.419.

Svanemyr, J., Scolaro, E., Blondeel, K., Chandra-Mouli, V., & Temmerman, M. (2013). *The contribution of laws to change the practice of child marriage in Africa*. Inter-Parliamentarian Union. Retrieved from https://www.girlsnotbrides.org/wp-content/uploads/2018/04/IPU-WHO-Child-marriage_study-October-2013.pdf.

Syrett, N. L. (2016). *American child bride: A history of minors and marriage in the United States*. Chapel Hill: University of North Carolina Press.

The World Bank. (2019a). *Data bank, gender statistics, men age at first marriage* Retrieved from https//databank.worldbank.org.

The World Bank. (2019b). *Data bank, gender statistics, women age at first marriage* Retrieved from https//databank.worldbank.org.

Therborn, G. (2004). *Between sex and power: Family in the world, l900–2000*. London: Routledge.

Therborn, G. (2006). Families in today's world-and tomorrow's. *International Journal of Health Services*, *36*(3), 593–603. https://doi.org/10.2190/64UN-12LK-43HP-JXY3.

Therborn, G. (2014). Family systems of the world: Are they converging? In J. Treas, J. Scott, & M. Richards (Eds.), *The Wiley Blackwell companion to the sociology of families*. (1st ed., pp. 3–19). Chichester: John Wiley & Sons, Ltd.

Thornton, A. (2005). *Reading history sideways: The fallacy and enduring impact of the developmental paradigm on family life*. Chicago, IL: The University of Chicago Press.

Thornton, A. (2013). Historical and cross-cultural perspectives on marriage. In E. Peters & C. M. Kamp Dush (Eds.), *Marriage and family: Perspectives and complexities* (pp. 3–32). New York: Columbia University Press.

Tsutsui, J. (2013). The transitional phase of mate selection in East Asian countries. *International Sociology*, *28*(3), 257–276. https://doi.org/10.1177/0268580913484775.

Twamley, K. (2013). The globalisation of love? Examining narratives of intimacy and marriage among middle-class Gujarati Indians in the UK and India. *Families, Relationships and Societies*, *2*(2), 267–283. https://doi.org/10.1332/204674313X664923.

Ule, M. (2004). Changes in family life courses in Slovenia. In M. Robila (Ed.), *Families in Eastern Europe* (pp. 87–101): Emerald Group Publishing Limited.

UNICEF. (2019). *Child marriage* Retrieved from: https://data.unicef.org/topic/child-protection/child-marriage/.

United Nations. (2013). *2013 demographic yearbook* (64th ed.). New York: Department of Economic and Social Affairs, United Nations.

United Nations. (2017a). Table 23: Marriages and crude marriage rates, by urban/rural residence: 2013–2017. In *2017 demographic yearbook*. New York: Department of Economic and Social Affairs, United Nations.

United Nations. (2017b). Table 25: Divorces and crude divorce rates, by urban/rural residence: 2013–2017. In *Demographic yearbook 2017*. New York: Department of Economic and Social Affairs, United Nations.

United Nations Statistical Division. (2013). *Table 3a. Legal age for marriage*. Retrieved from https://unstats.un.org/unsd/demographic/products/indwm/June%202013/3a.xls.

United Nations Statistical Division. (2019). *UN demographic statistics database. Population by marital status, age, sex and urban/rural residence*. http://data.un.org/Data.aspx?q=population+by+marital+status&d=POP&f=tableCode%3a23.

van Acker, E. (2008). *Governments and marriage education policy perspectives from the UK, Australia and the US*. Basingstoke: Palgrave Macmillan.
Walsh, J. (2013). Maldives. In R. E. Emery (Ed.), *Cultural sociology of divorce: An encyclopedia*. Thousand Oaks: SAGE Publications.
Wang, C.-T.L., & Schofer, E. (2018). Coming out of the penumbras: World culture and cross-national variation in divorce rates. *Social Forces*, *97*(2), 675–704. https://doi.org/10.1093/sf/soy070.
Willekens, H., & Scheiwe, K. (2003). Introduction: The deep roots, stirring present and uncertain future of family law. *Journal of Family History*, *28*(1), 5–14. https://doi.org/10.1177/0363199002238551.
Williams, L. (2010). *Global marriage: Cross-border marriage migration in global context*. Hampshire: Palgrave McMillan.
Winter, B., Forest, M., & Sénac, R. (2018). Introduction. In B. Winter, M. Forest, & R. Sénac (Eds.), *Global perspectives on same sex marriage: A neo-institutional approach* (pp. 1–18). London: Palgrave.

CHAPTER 4

The second demographic transition: Cohabitation

Ron J. Lesthaeghe
Sociology Research Center, Vrije Universiteit of Brussels, Brussels, Belgium

The notion of a second demographic transition (SDT) was proposed in 1986 to account for: (i) emerging changes in forms of partnership formation, and (ii) the postponement of marriage and parenthood, jointly leading to structural long-term subreplacement fertility (Lesthaeghe & van de Kaa, 1986). At the time of writing that paper, now more than 30 years ago, these changes were only manifesting themselves in the Western world; with Scandinavian countries in the vanguard, then followed by France, Belgium, and the Netherlands, the province of Quebec in Canada, by the English-speaking world, and by the German-speaking countries (Germany, Austria, and Switzerland). It was then labeled as a "western idiosyncrasy" and was expected to stay that way (Coleman, 2004).[a,b]

During the following three decades there has been a fast development of both aspects of the SDT, with the diffusion of cohabitation to other major regions of the world, and further postponement of fertility. These two developments now proved to be highly context-specific. As cohabitation runs counter to often century-old cultural props of marriage, the limiting condition is of a moral and religious nature, whereas there are no such strong cultural opposing factors to fertility postponement. Especially in the context of increased female education (e.g., Neels, Murphy, Ni Brolchain, & Beaujouan, 2017; Ní Bhrolcháin & Beaujouan, 2012) and employment in advanced

[a] The SDT theory met with its first skeptics during the late 1980s and throughout the 1990s, the most important objection being that these changes constituted neither a transition nor a lasting structural modification. Instead, the changes would be merely temporary lifestyle changes, furthermore being solely confined to selected western populations. The "*not us, we're different*" reaction was particularly loud in the late 1980s with respect to cohabitation in Southern Europe and shortly thereafter in the 1990s in the former Communist countries as well.

[b] There has been an increase in the incidence of cohabitation in Central and Eastern Europe prior to 1989 (Hoem, Kostova, Jasilioniene, & Muresan, 2009), but the trend accelerated thereafter. Also, the economic crisis of the 1990s in these countries can hardly be blamed for the continuation of both high levels of cohabitation and low levels of fertility for recent cohorts which initiated partnerships after 2005.

Cross-Cultural Family Research and Practice
https://doi.org/10.1016/B978-0-12-815493-9.00004-1

sectors of the economy, fertility postponement has been more strongly conditioned by structural and economic factors. In the parlance of the "Ready, Willing, and Able" model (Coale, 1973; Lesthaeghe & Vanderhoeft, 2001), the limiting condition for cohabitation is the "Willingness" factor, or cultural acceptability, whereas this is not the case for fertility postponement.

In what follows, we shall be focusing mainly on the cohabitation aspect of the SDT. As a consequence, attention will be paid to two conditioning factors: (i) the historical background with respect to patterns of kinship organization, strength of the institution of marriage, and the concomitant position of women; and (ii) the current country position with respect to the "ethics revolution." The latter means the de-stigmatization of hitherto negatively sanctioned behaviors such as divorce, homosexuality, euthanasia, or suicide. Cohabitation acceptance increases broadly in step with this package of formerly stigmatized ethical issues (Lesthaeghe & Esteve, 2016; Surkyn & Lesthaeghe, 2004).

The disconnection of cohabitation and postponement of fertility

The Western European connection between the rise of cohabitation and the fertility postponement transition is not duplicated elsewhere. Procreation among European cohabitants did not become widespread until the 1970s onward. At that point, marriage postponement, which was associated with longer spells of cohabitation, stopped being a determinant of parenthood postponement.

In contrast, during the previous century, the Southern European countries exhibited the classic fertility postponement well before there were any signs of rising cohabitation. Instead, a longer period of co-living with the family of origin had become the rule and a cause of fertility postponement. Secondly, also in East Asia, the fertility postponement took on dramatic proportions from the 1970s and 1980s onward, whereas cohabitation followed only more timidly later on.[c] Thirdly, in Latin America,

[c] Japan, for instance, witnessed a drop in the Population Total Fertility Rate (PTFR) from 2.13 children per woman in 1970 to 1.57 in 1989, and 1.41 in 2012 (Fukuda, 2016), and a rise in the mean age at marriage of women of 5.0 years from 24.2 to 29.2 years, over the same period. Hong Kong and Singapore also had subreplacement fertility at the onset of the 1980s (PTFRs of 1.72 and 1.69, respectively), whereas Taïwan and South Korea recorded that about 5 years later. In the most recent period 2010–15, the PTFRs of these countries are all around 1.20 and even below that level in Taïwan (PTFR: 1.11) (United Nations, 2017). Over the last three decades the mean age at childbearing, as an indicator of the aging of overall fertility, rose as well. The smallest increase was recorded in Singapore (+ 2.8 years) and the largest in Taïwan (+ 4.5 years) (United Nations, 2017).

the opposite sequence prevailed. These countries not only have a historical tradition of cohabitation, and experienced a cohabitation boom starting during the 1980s and 1990s (Esteve, Lesthaeghe, & Lopez-Gay, 2012). Strikingly, during these decades there were no major accompanying shifts in the hitherto young age patterns of first marriages and of fertility (Esteve & Florez, 2014; Esteve, Lopez-Ruiz, & Spijker, 2013; Fussell & Palloni, 2004). However, at present, there may be first signs of some fertility postponement among the better educated Latin American women after 2000 (Berquo et al., 2014; Esteve, Garcia-Roman, Lesthaeghe, & Lopez-Gay, 2013; Rosero-Bixby, Castro-Martin, & Martin-Garcia, 2009).

From the above descriptions, it is obvious that there was a disconnection between the cohabitation and the fertility postponement parts of the SDT at the global level. This is the consequence of very different starting points and of century-long differences in kinship organization and socioeconomic development in the various regions.

The cohabitation boom: Second demographic transition (SDT) or pattern of disadvantage (POD)?

The initial view of the SDT as starting among the more educated and more secular parts of western societies was challenged by cross-sectional gradients of cohabitation by education. In several countries, and particularly in Latin America, cohabitation is characterized by a negative education gradient, meaning it is more common among people with lower education. On the basis of that association, it was suggested that the subsequent rises in the incidence of cohabitation had nothing to do with the SDT, but instead with a pattern of disadvantage (POD), and especially with the economic stagnation or crises that frustrated marriage among the lower social strata. In this view, cohabitation among the poor was, and continues to be, a substitute for marriage. The discussion took the form of opposition, that is, POD versus SDT as explanations for the rising rates of cohabitation.

The reality is more complex. Firstly, a clear distinction has to be made between preexisting historical or ethnicity-related education gradients and later (i.e., post-1970) rises following the POD pattern. Secondly, education gradients change over time and patterns of catching up of cohabitation are of major importance as well. Finally, there is not only the aspect of prevalence, but also of the duration of cohabitation, and this has methodological consequences as far as indicators are concerned.

Single cross-sectional associations between educational attainment and rates of cohabitation can be very misleading, and a time dimension is necessary to come up with the correct interpretation of both initiation and evolution. Hence, repeated cross sections or multiple cohort data are required to come up with a correct diagnosis. In the Latin-American setting, measurements over time and over successive cohorts are easily obtained since the censuses routinely included a category of consensual union along with the classic *current* marital status questions. Sample sizes are very large and the results are robust. However, probes about *ever* having experienced a period of premarital cohabitation are absent. The weakness then resides in the fact that the point prevalence of current cohabitation can be quite low despite the majority of the population having initiated a union with a period of cohabitation. This holds particularly when durations of the initial cohabitation are short. With long duration of cohabitation, as is common in Latin America, and cohabitation being an alternative to marriage, the absence of the "ever" question becomes less problematic, and the current-status data will yield a more trustworthy picture of overall rates of cohabitation.

The data for the European situation are different. At best only the most recent censuses—if any—record unmarried cohabitation. However, surveys routinely include questions about both the current and past statuses (currently cohabiting and ever-cohabited). The drawback is that the sample sizes in the relevant two or three age groups are only a fraction of what is available in the Latin American censuses. On the other hand, on the basis of the ever-cohabiting question, the gradients of older age groups can also be used to reconstruct the experience of successive cohorts. The drawback of the ever-cohabiting question is the lack of information about the duration: no distinction can be made between those who only cohabited for a few months prior to marriage and those who cohabited for a much longer period, often as a substitute for marriage. In what follows, we shall inspect the gradient of cohabitation according to education category in the European, American, and selected Pacific Asian settings (i.e., Japan, the People's Republic of China, and the Philippines).

The European patterns of the cohabitation boom

Although there were extant older patterns of premarital cohabitation in Europe (e.g., in Northern Sweden, Iceland, and Southern Portugal), the overwhelming part of the rise in this form of partnership formation constituted a completely new development. For example, in Belgium and The Netherlands, the acceptability of cohabitation first became evident in the 1950s among younger and better-educated cohorts. They had been secularized during their formative years, defended the political "postmaterialist" agenda (Felling, Peters, & Schreuder, 1983; Inglehart, 1990), stressed individual autonomy, challenged all forms of institutional authority, subscribed to expressive and egalitarian values, and articulated the quest for self-fulfillment. This resulted in keeping an "open future" for a more extended period of time, and in prolonging the child-free part of young adult life in particular. Furthermore, absolutist ethics were no longer widely accepted (i.e., stigmatization of divorce, abortion, homosexuality, euthanasia, or suicide) and gave way to circumstantial leniency in moral matters.[d] Such a de-stigmatization and ethical revolution spread very rapidly to all layers of the population as secularism had a historical foothold in all social strata to start with, and became even more pronounced during the 1960s in tandem with the overall rise in secondary and university education. The initiators of increasing cohabitation may have been concentrated among the better educated at the onset, but the acceptability of cohabitation, and even procreation among cohabitants, spread very rapidly across the strata of these societies, with only a religious core sticking to the sanctity of marriage.[e]

The education gradients of proportions of *ever-cohabiting* adults and their evolution over time for successive cohorts are available for several European countries that participated in the rounds of the European Gender and Generations Survey (GGS). The results for three education categories[f] are reported in Neels and Perelli-Harris (2012), and these results capture the experience starting with the birth cohorts of 1930–39 and ending with the cohorts of 1970–79. In this way, we can follow the evolution of the experience of cohabitation since the early 1960s until the first years of the new century.

Four patterns were found in the education gradients of women *ever-cohabiting.* First, the Netherlands, France, Germany, and Austria exhibit a predominantly *positive* education gradient throughout the entire process of rising cohabitation from low levels to percentages of *over* 75% in *all*

[d] For an empirical description of the association between living arrangements and a wide variety of ethical, political, expressive, and societal orientations using the European Values Studies, see Lesthaeghe and Moors (2002) and particularly Surkyn and Lesthaeghe (2004).

[e] For France, the rapidity of this diffusion from a young urban elite to the children of farmers is well documented in Leridon and Villeneuve-Gokalp (1988). Religious opposition was considerably stronger in the United States (cf. Thornton, Axinn, & Xie, 2007).

[f] The education categories are: low = International Standard Classification of Education (ISCED) codes 1 + 2, medium = ISCED codes 3 + 4, high = ISCED codes 5 + 6.

educational categories among the birth cohorts of 1970–79. These are clear cases of a rapid diffusion with the better educated continuously in the lead. Second, similar large progressions of percentages ever-cohabiting are found in Bulgaria and Estonia, but with a moderate *negative* gradient by education (i.e., more cohabitation among the less well educated). However, in these two countries, the better educated in the youngest cohort have completely caught up with the other two classes. Third, essentially *flat education gradients* are found for Belgium, Norway, Hungary, and Poland, but the first two countries have witnessed the "cohabitation boom" to its fullest extent, whereas the latter two had more moderate increases among the five cohorts considered. Especially Poland lagged further behind. Fourth, negative education gradients were found in Russia and Romania, but the overall rises were still moderate among all the cohorts considered. Also, these increases are by no means limited to the lowest education category but are equally witnessed among the better educated.

As already indicated, percentages of the population *ever-cohabiting* fail to differentiate between short and long duration cohabitations, that is, between mere premarital cohabitation as a prelude to marriage, and cohabitation replacing marriage. To remedy this to some extent, the following figures give the percentages of women of age 25–29 who experienced a cohabitation spell of *at least 1 year* relative to all women who have been in a union (marriage + cohabitation) in that age group for at least 1 year. To increase sample sizes, the results are grouped by decades. This indicator, calculated by Jorik Vergauwen on the basis of the GGSs data, gives a better idea of established longer-term cohabitation. The results are presented in Figs. 4.1 and 4.2.

The results depicted in Fig. 4.1 clearly indicate how longer-term cohabitation starts its explosive growth in the western countries from the 1970s onward to reach 60% or more shortly after the turn of the century. The former Communist countries start a decade later and the trend is less steep (see also Puur, Rahnu, Maslauskaite, Stankuniene, & Zakharov, 2012). Also noteworthy is the relatively high incidence of cohabitation in Russia at the onset, that is, prior to the 1990s.

Fig. 4.2 depicts the evolution of education gradients in these countries. Most of the previous conclusions are again substantiated for longer-term cohabitation. The rise occurs systematically in *all* education categories. In Belgium, Austria, France, Germany, especially the Netherlands, there is a *positive slope* with the better educated leading the way. In the United Kingdom, the older gradients are positive but the most recent data seem to indicate some stronger growth among the less educated (see also Ni

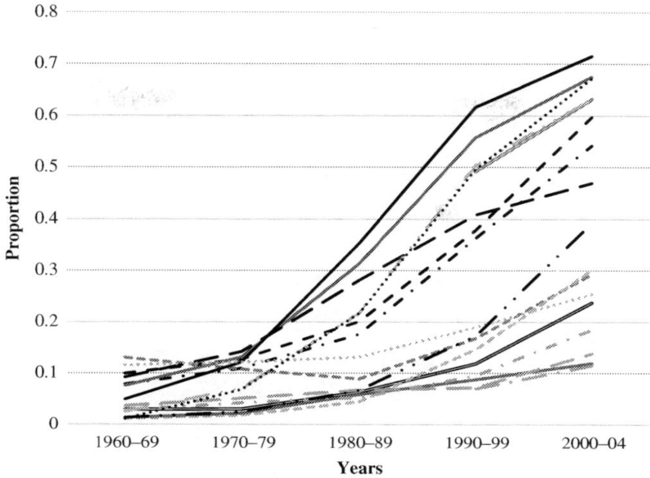

Fig. 4.1 Expansion of cohabitation in selected European countries—proportion of women aged 25–29 who cohabited for at least 1 year in that age group relative to women 25–29 who were in a union (married + cohabiting). The order of the countries at the most recent measurement is from top to bottom: Norway, France, United Kingdom, Austria, Netherlands, Estonia, Belgium, and Germany (leading group), then Spain, Hungary, Georgia, Russia, Bulgaria, Lithuania, Romania, Italy, and Poland (later group). *(Courtesy of Jorik Vergauwen, Gender and Generations Survey harmonized data.)*

Brolchain & Beaujouan, 2013). In Norway, as in the United Kingdom, the gradients are essentially flat. The *negative slope* is typical for Central and Eastern Europe, but mostly at the onset. Thereafter, however, the higher education group is catching up, and this produces a *V-shaped* gradient. In Lithuania and Bulgaria, the left arm of the V is still dominant, but in Estonia, Hungary, Romania, and Poland the right arm has taken over, whereas in Russia both sides of the V are of equal magnitude. Finally, Spain and Italy follow the western European pattern with positive gradients, but the Italian upward trend is one of the slowest moving ones in the western world and exhibit a slight catching up at the lower end of the educational spectrum.

The conclusion for the education gradients of ever-cohabiting and more permanent cohabitation is that there is no significantly large selection associated with the pattern of disadvantage (POD), but that *all* educational groups and social strata participated in the cohabitation boom since the 1970s. *In many countries, education-related differentials are dwarfed by the sheer overall rise of cohabitation.* Mainly the tempo of the overall expansion of cohabitation is different between the western and eastern parts, with Eastern and Central Europe taking off a decade later and moving more slowly. It

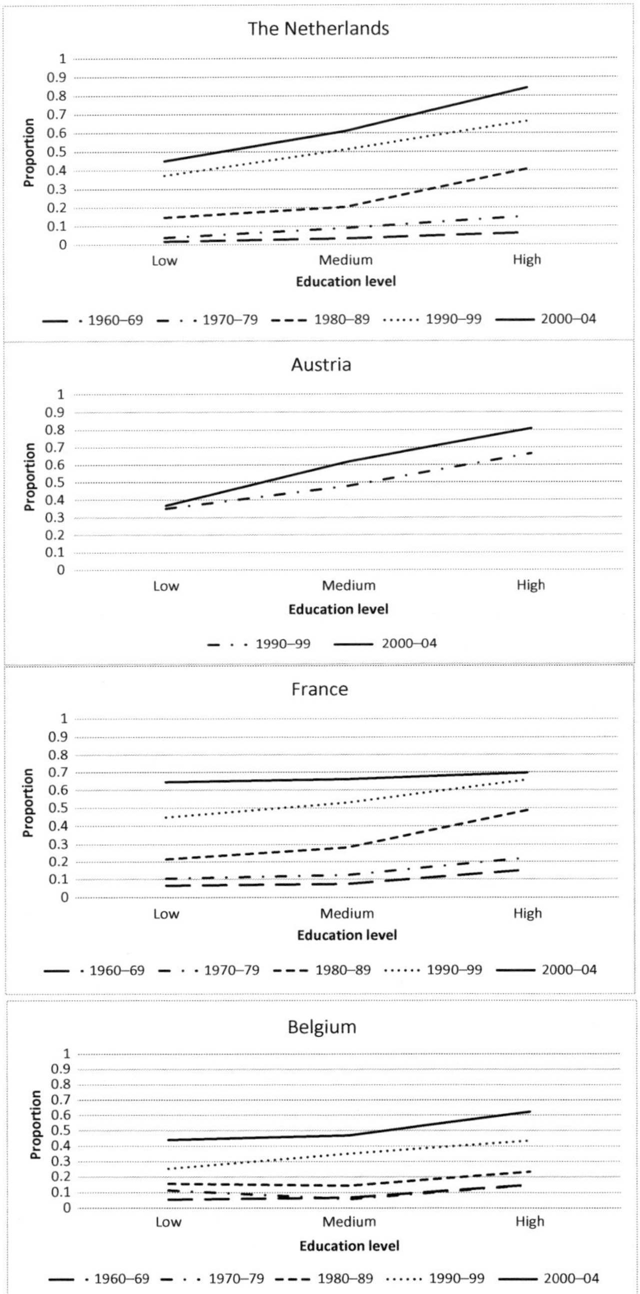

Fig. 4.2 Expansion of cohabitation by the level of education in selected European countries—proportion of women aged 25–29 who cohabited for at least 1 year at these ages relative to women 25–29 who were in a union (married + cohabiting) for at least 1 year. *(Courtesy of Jorik Vergauwen, Gender and Generation Survey harmonized data, and national surveys.)*

(Continued)

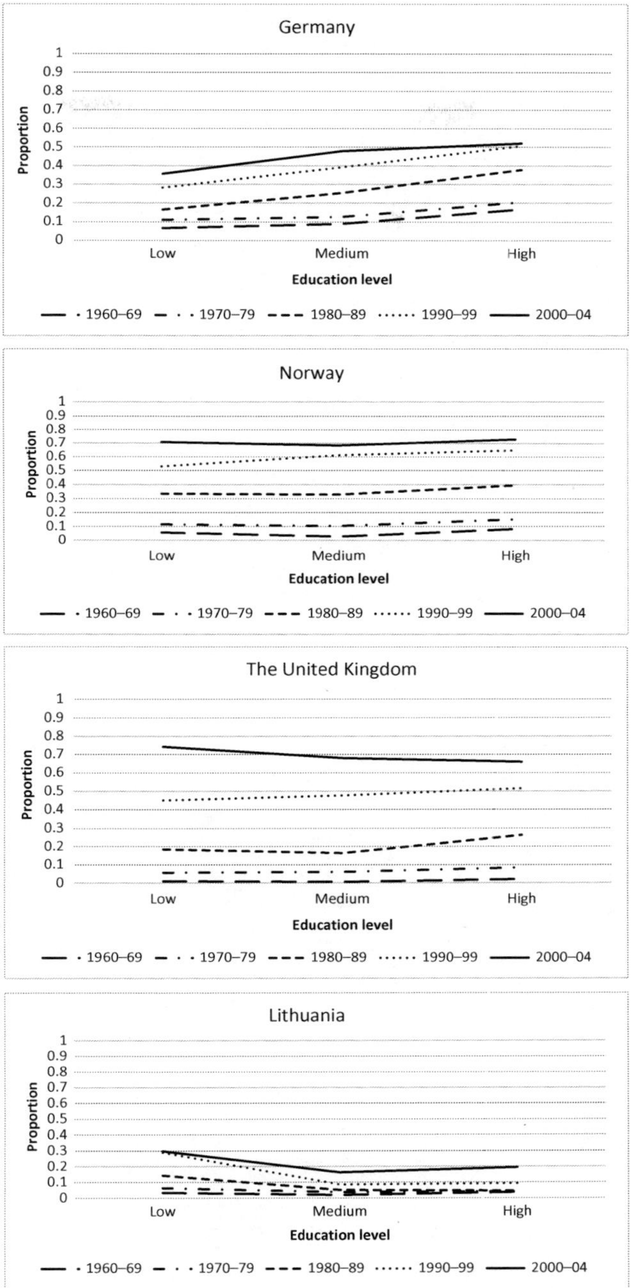

Fig. 4.2, Cont'd

(Continued)

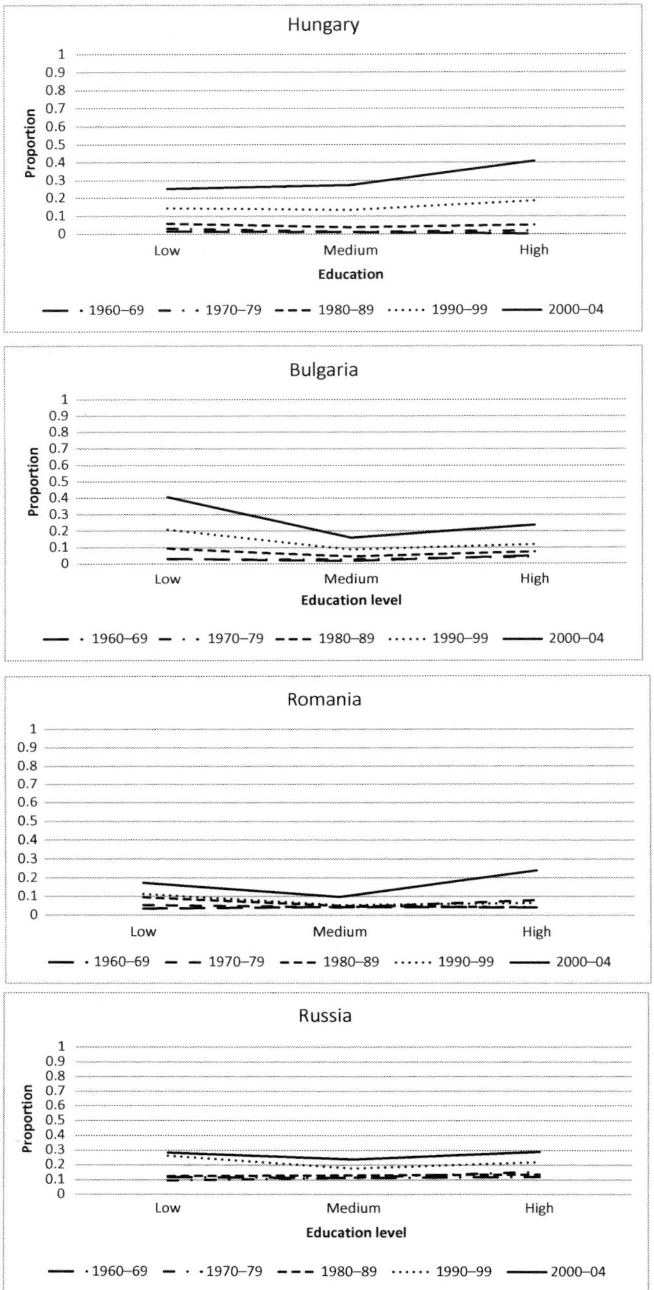
Hungary
Proportion
1
0.9
0.8
0.7
0.6
0.5
0.4
0.3
0.2
0.1
0
Low
Medium
High
Education
1960–69
1970–79
1980–89
1990–99
2000–04
Bulgaria
Proportion
Low
Medium
High
Education level
1960–69
1970–79
1980–89
1990–99
2000–04
Romania
Low
Medium
High
1960–69
1970–79
1980–89
1990–99
2000–04
Russia
Proportion
Low
Medium
High
Education level
1960–69
1970–79
1980–89
1990–99
2000–04

Fig. 4.2, Cont'd

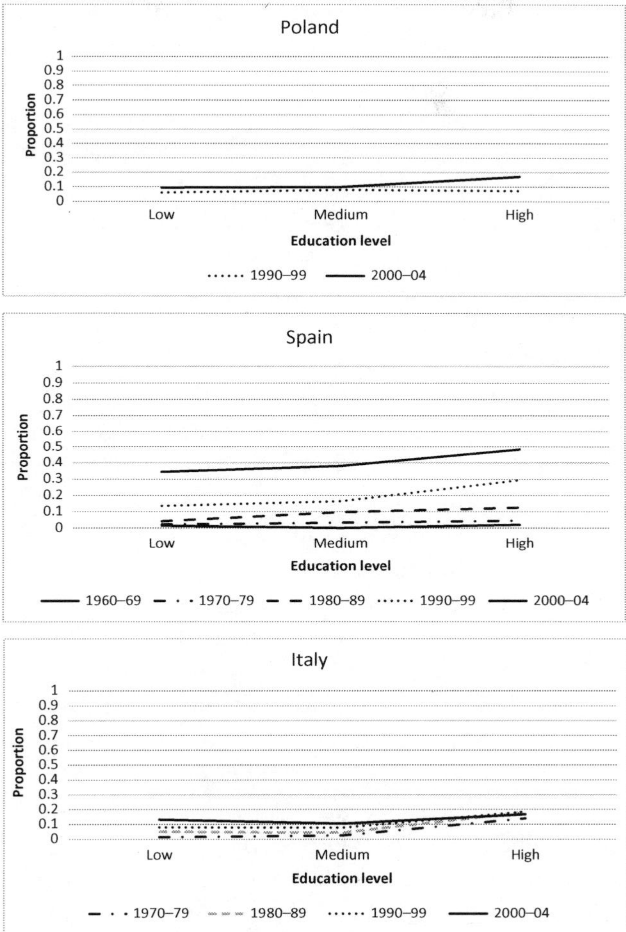

Fig. 4.2, Cont'd

should be noted, however, that the observations stopped in 2005 and that the subsequent evolution is not yet documented.

The second main observation is that initially negative gradients are altered over time into V-shaped ones in the former Communist countries as a result of catching up of the better-educated women. This once more illustrates how misleading conclusions can be when based on a single observation exhibiting negative education gradients, that is, at the onset of the process. The development of the V-shape, by contrast, suggests that both mechanisms are at work, a POD-type at the left side[g] and an SDT-type at the right side of the education axis. In this instance, the diagnosis is not one versus the other, but POD plus SDT.

Weak or no educational differences with respect to the rise of cohabitation do not imply that the *consequences* of the cohabitation boom would be neutral for further developments of family life and children. The story can be quite different as will be discussed in a later section on "diverging destinies."

Cohabitation in North America

Data on proportions of currently cohabiting women of ages 20–25 years among those women in a union (married + cohabiting) are available for the United States and for Canada. The data for the US stems from the censuses of 1990 and 2000 are basically best estimates,[h] but thereafter we can rely on direct questions in the American Community Surveys (ACS) for the years 2007–11 (pooled data). The Canadian data are census data for the years 1986, 1996, and 2006 (Laplante & Fostik, 2016).

On the basis of percentages, *ever-cohabited* the United States has experienced a transition from a slightly positive gradient to a slightly negative one, but the education differences are again not substantial (Neels & Perelli-Harris, 2012). The data for *currently* rather than ever-cohabiting women for the United States exhibit a more systematic negative gradient from ages 25 onward, mainly because the better educated have faster transitions to marriages and not because of a lower incidence at earlier ages (Lesthaeghe, Lopez-Colas, & Neidert, 2016). On the whole, the incidence of cohabitation is lower in the United States than in any of the other countries on the entire American continent and lower than in Western European countries.

[g] A remarkable finding is that respondents in interviews probing for the meaning of cohabitation in the Eastern European countries, with the exception of Russia, do not at all give a financial reason for preferring cohabitation. Among six possible answers, the item that would reflect the strength of the POD is "too poor to marry." The percentages of such answers amount only to 10.8% of respondents in Bulgaria, 2.7 in Hungary, 8.0 in Lithuania, and 8.7 in Romania. This is barely any different from 12.1 in France, 9.8 in Germany, or 6.8 in Norway. Only Russia stands out with 22.4%. Percentages answering "prelude to marriage" are also very similar for the Eastern and Western countries listed here, but those for the item "trial marriage" are lower in the former countries. See Hiekel, Liefbroer, and Poortman (2012), Table 3. Hence, the strict interpretation of the current negative POD education gradient as caused by current poverty or financial stress may not be valid, as such a gradient could just as well reflect a long-standing locked-in working class phenomenon.

[h] For a discussion of the data quality, see Lesthaeghe et al. (2016, pp. 101–104).

Table 4.1 Percentage currently cohabiting women 25–29 among all women 25–29 in a union (cohabiting and married), in the United States 1990–2011.

Education level	Census 1990	Census 2000	ACS 2007–11
Did not complete high school	13.6	16.2	24.3
Completed high school	9.9	16.4	24.4
University degree	9.7	15.3	20.5
Total	10.3	16.0	22.9

Source: Lesthaeghe, R., Lopez-Colas, J., & Neidert, L. (2016). The social geography of unmarried cohabitation in the USA, 2007–2011. In A. Esteve & R. Lesthaeghe (Eds.), *Cohabitation and marriage in the Americas—Geo-historical legacies and new trends* (pp. 101–132). Switzerland: Springer Open. Data: Censuses and American Communities Survey (ACS) samples available through the Integrated Public Use Microdata Series—USA database. https://usa.ipums.org/usa/.

At the 1990 census, the education gradient of current cohabitation exhibits a POD pattern among women aged 25–29 years. This pattern has been maintained over time[i] but is weakening in relative terms as the overall percentage of the whole population of 25- to 29-year-old women cohabiting more than trebled from 1990 to 2007–11 (see Table 4.1). The middle education group (complete high school, some university) has the largest gains in rates of cohabitation across this period and has come even with the lowest education group. The proportion among those with at least a university degree also doubled. As with European countries, the overall rise in rates of cohabitation in the United States is considerably more important than the educational differences. Another interesting feature for the United States is that the two higher education groups in the younger age group 20–24 years in the ACS data have a slightly *higher* incidence of cohabitation than those with less than complete high school, indicating that the better educated initiate their partnerships via cohabitation at least as frequently as the least educated (Lesthaeghe et al., 2016).

As age advances (i.e., as we examine cohorts of women of increasingly older ages than 25–29 years) the best-educated have a larger drop off of rates of cohabitation than the less educated, reflecting that the best-educated are more prone to convert their cohabiting union into a marriage. This is a "sorting effect" (Fürstenberg, 2013), probably reflecting the better financial

[i] It should be noted that the lower education category contains a disproportionally large group of blacks and native Americans who, of all ethnic groups, including Hispanics, have the highest proportions of cohabitation among women aged 25–29 years currently in a union, i.e., 31% and 33% against a national average of 23 in the period 2007–11 (Lesthaeghe et al., 2016, p. 108). Hence, the education gradient contains an ethnic component in the United States as well.

position of university degree holders relative to those with less education, in tandem with a cultural "embourgeoisement" that sets them apart as an educated upper middle class.

For the Canadian provinces, measurements are available for 1986, 1996, and 2006 (Laplante & Fostik, 2016) for percentages of currently cohabiting women of 25–29 years among all women of that age currently in a union (married plus cohabiting). The education gradients are essentially flat in the provinces, with the highest incidence of cohabitation in Quebec (above 60% in 2006), the North (mainly Inuït), and the Atlantic provinces. The provinces with the lowest incidence of cohabitation (Ontario, Manitoba, Saskatchewan, Alberta, and British Columbia) exhibit a slightly negative educational gradient in 2006 but there was no such negative gradient in 1986, and in Quebec a higher proportion of the better educated cohabitated than the proportion of less educated in 1986. Once more, in all provinces, the overall rise during the 20 years of observation is more substantial than any of the educational differences.

The Latin American patterns of the cohabitation boom

As is well known, many Latin American populations had significant proportions of consensual cohabiting unions well before the cohabitation boom that started in Europe and North America after 1980. The custom of cohabitation mainly concentrated in the poorest strata. There were also important differences in rates of cohabitation between ethnic groups, with notably high rates among Indigenous populations with hunter-gatherer or nomadic backgrounds, as well as the black populations. Also, racially mixed populations tend to have extant traditions of consensual unions. These high rates of cohabitation contrast with the low rates among better educated white populations, such as the late 19th and 20th century European immigrants in the Southern Cone of South America (Uruguay, Argentina, Chile) who strengthened the institution of marriage. Furthermore, European marriage became a mark of upward social mobility. The net result is that *all* Latin American countries, whether rich or poor, exhibited a negative education gradient relative to rates of cohabitation during the 1970s. In short, at the onset, there was a clear POD pattern, often with a strong ethnic component (Esteve, Lesthaeghe, Lopez-Gay, & Garcia-Roman, 2016; Esteve, Lesthaeghe, Quilodran, Lopez-Gay, & Lopez-Colas, 2016; Esteve, Lopez-Gay, et al., 2016) and this has remained a locked-in feature.

The question is whether this POD pattern has been strengthened during the boom years, or whether the better-educated strata have been catching up. After four decades of observation, the outcome is emerging quite clearly: in *all* countries, both the middle education groups and those with university education have experienced rapid rises in the proportions currently cohabiting at ages 25–29, that is, when most unions have passed the phase of dating. The outcome is not POD (where only the disadvantaged cohabit at high rates) versus SDT (where all education levels increase rates of cohabitation across time), but POD *and* SDT.

Of course, there are differences between the various countries as far as the orders of magnitude of POD and SDT are concerned. A few examples will illustrate this. Panama is at one extreme with the percentage of all women aged 25–29 years in cohabitation relative to all women 25–29 in a union (marriage + cohabitation) already very high among the least educated in the 1970s, so that any large magnitude rises could only be among those with higher levels of education (see Fig. 4.3A). A different pattern is evident in Brazil, where the largest increases in cohabitation between 1970 and 2010 concentrated among the less educated, with only a modest increase among the better educated. In the Brazilian case, the POD pattern accentuated over time (see Fig. 4.3B). A similar POD pattern also holds for Venezuela. In contrast, in Ecuador, Nicaragua, Dominican Republic, and especially in Colombia (Fig. 4.3C), the largest increases in rates of cohabitation were in the middle education groups, that is, among women who had completed primary or who completed secondary education. In another group made up of Costa Rica, Argentina, Mexico, and Chile, the initial rises in cohabitation began in the 1980s and 1990s, and tended to be concentrated among the least educated (i.e., a negative educational gradient), but there was a significant catching up of rates of cohabitation among the better educated after the turn of the 21st century. The case of Uruguay (see Fig. 4.3F) is particularly striking, where there was a huge increase in cohabitation between 1996 and 2010 in all educational strata, resulting in an almost flat education gradient by the later date.

To sum up, with the exception of Brazil and Venezuela, the POD gradients of the other Latin American cases have altered over time as the middle and better educated, including those with university degrees, have been catching up after 2000 with the high rates of cohabitation evident earlier among less-educated groups. However, the education gradients are still negative around 2010, as this catching up phase among the upper strata is presumably still in progress. Only in Uruguay has the new gradient become

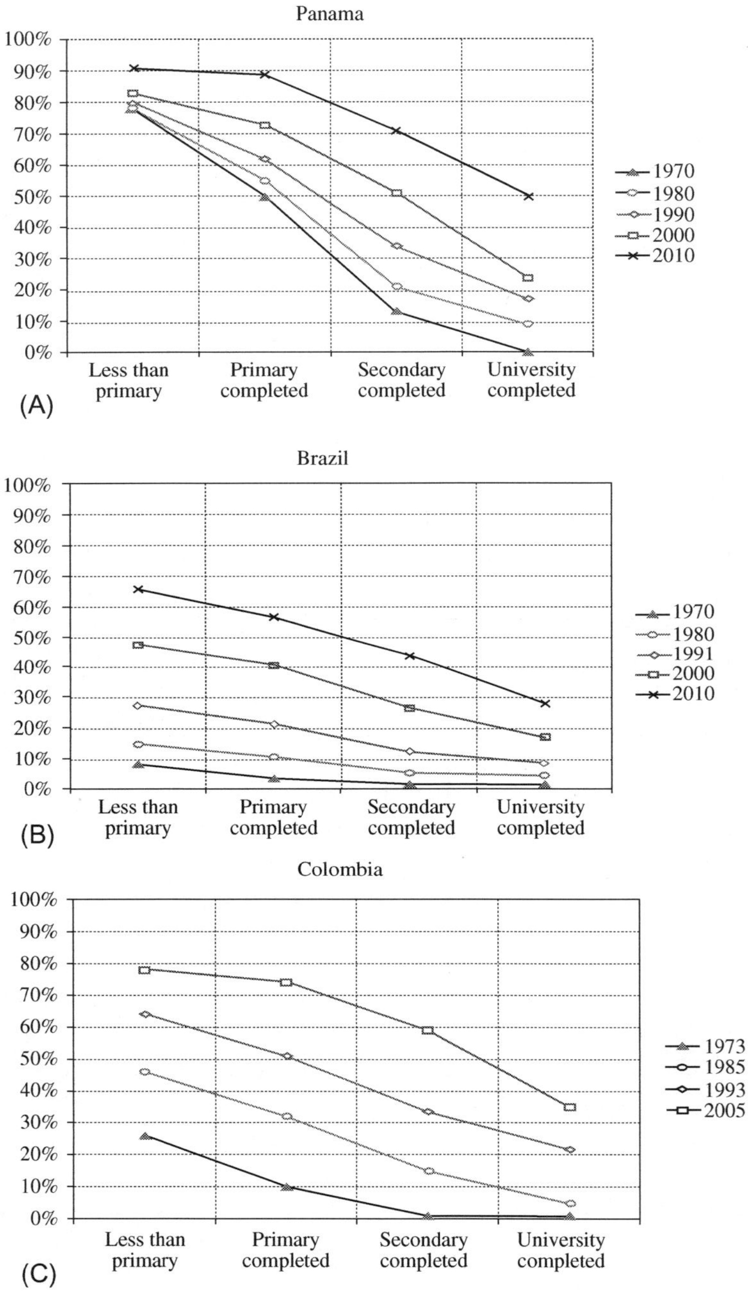

Fig. 4.3 The rise of cohabitation in selected Latin American countries. Percentage currently cohabiting among all women 25–29 in a union (cohabiting or married), 1970–2010. (A) Panama, (B) Brazil, (C) Columbia, (D) Argentina, (E) Mexico, and (F) Uruguay. *(Source: Esteve, A., Lesthaeghe, R., Lopez-Gay, A., & Garcia-Roman, J. (2016). The rise of cohabitation in Latin America and the Caribbean, 1970–2011. In A. Esteve & R. Lesthaeghe (Eds.),*

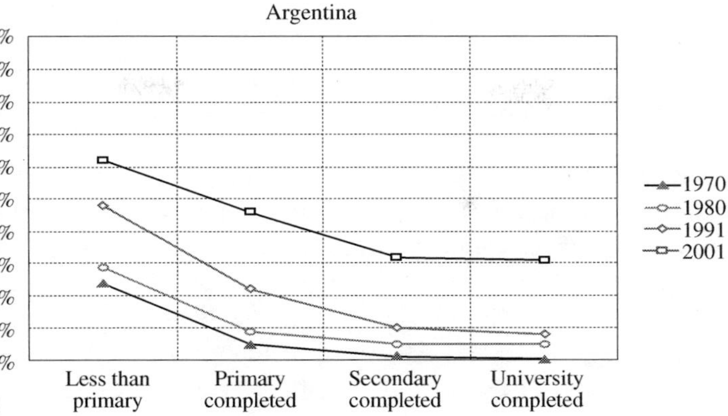

(D)

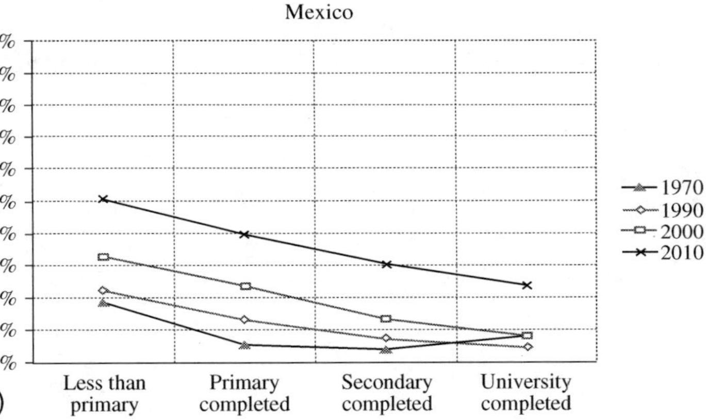

(E)

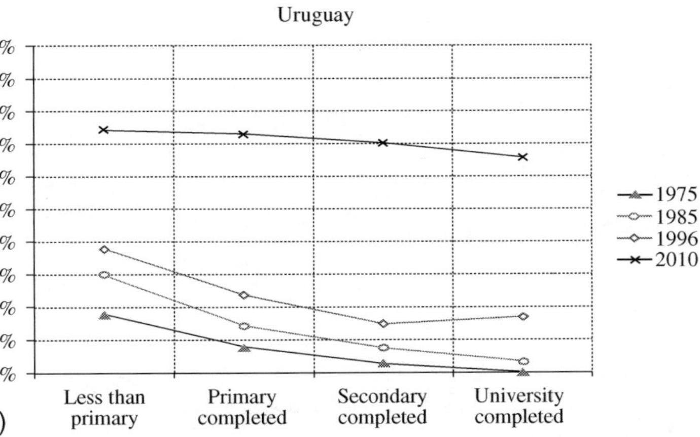

(F)

Fig. 4.3, Cont'd Cohabitation and marriage in the Americas (pp. 25–58). Switzerland: Springer Open; Esteve, A., Lesthaeghe, R., Quilodran, J., Lopez-Gay, A., & Lopez-Colas, J. (2016). The expansion of cohabitation in Mexico, 1930–2010: The revenge of history? In A. Esteve & R. Lesthaeghe (Eds.), Cohabitation and marriage in the Americas (pp. 133–156). Switzerland: Springer Open; Esteve, A., Lopez-Gay, A., Lopez-Colas, J., Permanyer, I., Kennedy, S., Laplante, B., … & Cusido, T. A. (2016). A geography of cohabitation in the Americas, 1970–2010. In A. Esteve & R. Lesthaeghe (Eds), Cohabitation and marriage in the Americas (pp. 1–24). Switzerland: Springer Open. Data: Census archives.)

education neutral by 2010 as a consequence of the rapid rise in cohabitation after 1996.[j]

The conclusion at this point is that starting from a historically founded POD and/or ethnic negative education gradient of cohabitation, a new component has been added in Latin America, resulting in major rises in rates of cohabitation after 1980, and particularly after 2000, among the better-educated strata. This increase in rates of cohabitation among the better educated is entirely consistent with the SDT prediction of a generalized diffusion of the "nonconformist" pattern of family formation. Finally, in several Latin American countries, there is consistency between rising rates of cohabitation and ideational changes from the 1990s to the 2000s reflecting greater social liberalism across all education levels, but the ideational changes were especially marked among those with higher education.[k] More specifically, as reported in the waves of the World Values Studies surveys (WVS), a distinct de-stigmatization occurred with respect to divorce, homosexuality, and euthanasia during that period (Lesthaeghe & Esteve, 2016). Although not measured in the WVS surveys, it is most likely that also the de-stigmatization of cohabitation among the better educated is part of this ideational evolution.

The Pacific Asian experience

The fact that premarital cohabitation spread to several Pacific Asian societies from the 1970s onwards might come as a surprise, as hitherto cohabitation had been viewed very negatively as it did not fit with the cohesively structured patriarchal traditions.[l] However, the proportions of the population married below age 30 had been falling since the 1970s and sometimes to very low levels (e.g., Jones, 2005). Given this trend, it was not improbable that, besides increases in prolonged living with parents, there also could be some of those unmarried individuals living alone, sharing with other singles,

[j] In this respect, Uruguay is the "Quebec of Latin America." Not only has Quebec the highest incidence of cohabitation in Canada, but also a flat education gradient. See Laplante and Fostik (2016), Figures 3.2a and 3.2b.

[k] This pertains to Brazil, Colombia, Mexico, Uruguay, and Chile for which several World Values surveys are available.

[l] In a 2000 International Union for the Scientific Study of Populations workshop held in Tokyo, it was still asserted that survey questions about cohabitation in Japan were ruled out since these would be most impolite.

and cohabitation could be a feasible alternative to marriage, particularly in situations with urban-bound migration and/or high housing costs.[m]

The first shock came from Japan in 2004 when the Mainichi Shimbun newspaper organized a survey asking the question: "have you ever cohabitated?" of single and married persons. The outcome was that over 20% of women born in the 1970s admitted to a period of cohabitation, which is double the figure for women who were 10 years older (Raymo, 2013). The Japanese Generations and Gender Survey (2004) confirmed the findings of increasing rates of cohabitation, and the 2009 Japanese Survey on Family and Economic conditions probed even further by inserting a question about the first step toward relationship commitment, being one of three possibilities of a pregnancy, a period of cohabitation, or direct transition to marriage without prior cohabitation.

A fuller layout of the patterns for the start of a couple union in Japan is given in Table 4.2 for the cohorts born prior to 1980 on the basis of the 2004 Japanese Generations and Gender Survey. The first three columns give the breakdown over the categories "started with cohabitation," "only ever-married without cohabitation," and "never in a union of either type." The fourth column is the classic indicator of the percentage ever-cohabiting for women ever in a union. The last column gives the percentages reporting a premarital pregnancy (or more rarely a premarital birth) among all

Table 4.2 Partnership formation and parenthood, Japanese women aged 25–69 years in 2004.

	% ever cohabitated	% married, not cohabitated	% neither cohabitated or married	% ever cohabitated of those ever in union	% premarital pregnancy of those ever having a child
25–29	20	34	47	37	43
30–34	17	62	21	22	28
35–39	16	77	8	17	19
40–44	12	8	5	13	16
45–49	8	87	5	8	14
50–54	7	89	4	7	10
55 +	4	93	3	4	9

Courtesy of Noriko Tsuya, data: Japan Generations and Gender Survey (2004).

[m] For a general update of trends in Far Eastern marriage and family patterns see Raymo, Park, Xie, and Yeung (2015).

ever-married and parous women. The results indicate that already before the turn of the Century a major change had occurred in the patterns of union formation. In the youngest cohort many had still not made it to a union, but of those who had more than a third had done so via cohabitation and 43% moved into marriage as a pregnant bride (or mother). Among the older cohorts, we effectively move back in history and note that there has been a steady rise in cohabitation among those in a union, and an even more marked rise in bridal pregnancies and marriages occurring when the bride is pregnant since the 1970s. The proportion of women ever-cohabiting among those who have formed a union with a partner was 37% of women aged 25–29, which is a similarly high proportion as that observed in Spain and several Eastern European countries in the 1990s.

There are some significant limitations on the available data on the education gradient of cohabitation in Japan. There is neither any data available on the widely used statistics of the percentages of women ever or currently cohabiting, nor is there longitudinal data for more than one period. Instead, we only have a single time point report of the proportion of all women cohabiting at the beginning of the century, and this proportion is relative to all women aged 25–69 years of all marital statuses. The broadness of the age range and the inclusion of those still single in the denominator is a problem when interpreting the results. Furthermore, these results reflect only the starting phase of cohabitation. There is evidence on the education gradient of the chance of a premarital pregnancy or birth for ever-married women in the same broad age range (Table 4.3). These data demonstrate a classic negative education gradient for both premarital cohabitation and pregnancy. That is, those women with lower education are more likely to cohabit or have a premarital pregnancy than those with higher education.

Table 4.3 Relative risk of ever-cohabiting and experiencing a premarital pregnancy or birth by education level for women aged 25–69 years in Japan in 2004.

Education	Relative risk ever cohabiting	Relative risk premarital pregnancy
Less than high school	1.39	1.52
High school[a]	1.00	1.00
Profession	1.18	1.11
Some university	0.44	0.71
4 + years university	0.39	0.58

[a]Reference category for estimation of relative risk is high school graduates.
Courtesy of Noriko Tsuya, data: Japan Generations and Gender Survey (2004).

Particularly among women with university-level education, the chance of either cohabitation or premarital pregnancy is only about half that for the female population overall. However, the inclusion of many single educated women in the denominator causes a downward bias in the result for this group. To sum up, there is probably a negative education gradient for cohabitation and premarital pregnancy in Japan, but the available figures on the downward slope may exaggerate the effect due to sampling biases.

Additional evidence on patterns of union formation is provided by the 2009 Japanese Survey on Family and Economic conditions, some of which is summarized in Table 4.4. Among women aged 20–49 years who had experienced at least one step in the family formation process (i.e., had a conception, initiated cohabitation, or entered a marriage), about 70% of women with a university degree (2 or 4 years) started with a marriage. In contrast, only 26% of women who did not finish high school had a marriage as a first step. The figure was about 50% for those women with high school or vocational school graduate education. Cohabitation, on the other hand, was the first step for 40% of women who had not completed high school education, and about 25% for high school or vocational school graduates. Hence, in the lower half of the educational spectrum, only a minority of women experienced a direct entry into marriage without prior pregnancy and/or cohabitation (Raymo, 2013). Furthermore, among highly educated women, cohabitation prior to marriage is associated with later childbearing within marriage. In contrast, among less-educated women, cohabitation appears to hasten marriage and parenthood via premarital conceptions and marriages when the woman is pregnant (referred to in some country as "shotgun marriages," and known in Japan as *dekichatta kekkon*) (Raymo, Iwasawa, & Bumpass, 2008). There needs to be a more recent analysis of social trends to find out what happened during the last decade to 2019, and to detect

Table 4.4 Percentage of Japanese women aged 20–49 years in a couple union who experienced union formation by conception, cohabitation or marriage stratified by education level.

Education level	Conception (%)	Cohabitation (%)	Marriage (%)
Junior high school	33.3	40.5	26.2
High school	18.6	24.9	56.5
Vocational training	18.7	30.5	50.8
College	11.9	16.5	71.6
University	6.8	23.9	69.3

Courtesy of James Raymo, data: Japanese Survey on Family and Economic Conditions (2009).

whether or not the new cohorts of college-educated Japanese women have been increasing rates of cohabitation relative to previous. This process of increasing rates of cohabitation among the better educated has happened in so many other countries, it might well be happening in Japan.

In the Philippines, the 2013 Demographic and Health Survey (DHS) reported record high percentages of cohabiting women in all age groups. This was a surprising finding in a largely Catholic country where both abortion and divorce are illegal.[n] The United Nations 2015 marriage data provided the results for earlier periods (United Nations Population Division, 2015). As shown in Fig. 4.4, the 2010 census results for percentages currently cohabiting among women in a union (married or cohabiting) show a modest increase compared to the two earlier surveys of 2003 and 2007, but the DHS (2013) outcomes are much higher than those in the census of 3 years

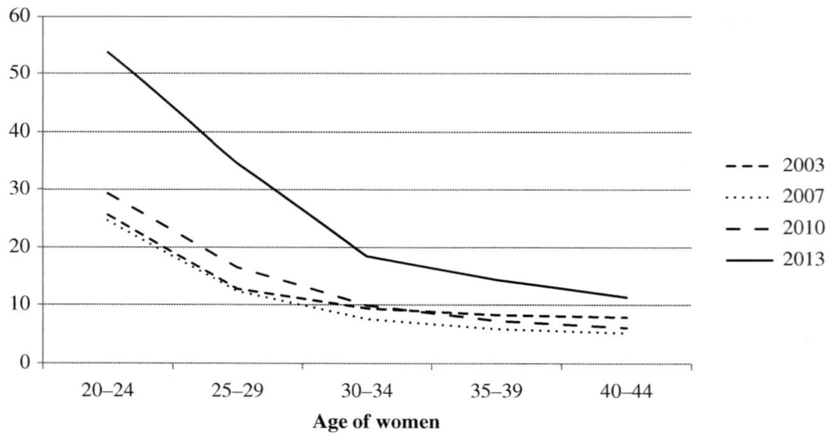

Fig. 4.4 Percentage of women currently cohabiting among women currently in a couple union (married or cohabitating) by age. *(Sources: Philippines survey data (2003, 2007), 2010 census as reported in United Nations, Department of Economic and Social Affairs, Population Division (2015). World marriage data 2015. New York: United Nations. http://www.un.org/en/development/desa/population/theme/marriage-unions/WMD2015.shtml.)*

[n] The Catholic Church is obviously worried about the rise of cohabitation. The Manilla Archdiocese and its Radio Veritas, for instance, commissioned a nation-wide "Truth" survey in 2012 among church attendants and found that 57.7% of the 2500 respondents was cohabiting instead of being married. Theoretically, cohabitants are banned from receiving communion. Also, church and official organizations are instrumental in helping poorer couples to arrange church marriages.

earlier. Such a large gap between observations only 3 years apart seems implausible. It is not clear whether the 2010 census underestimated the incidence of cohabitation, or whether the DHS (2013) was biased in the opposite direction. At any rate, whatever the source, it is clear that premarital cohabitation is quite common in the Philippines.

The education gradient recorded in the 2013 DHS exhibits the typical negative slope (Kuong, Perelli-Harris, & Padmadas, 2016) observed in many other countries, as described earlier in the current chapter. However, Kuong et al. use *all* women in the denominator. Given the much higher percentages of women who have never been in a couple union among the wealthier and more educated groups relative to less wealthy, less educated groups; the reported negative slope of the gradient is exaggerated. In fact, when only women already in a couple union (cohabiting or married) are included in the denominator, the share of cohabitation no longer follows a negative gradient by education and wealth in 2013 but turns into an inverted U-shaped pattern. This is illustrated in Table 4.5. Furthermore, given the strong Catholic stigmatization of cohabitation, there might be many women who falsely report being single rather than acknowledge their

Table 4.5 Percentage of women currently cohabiting by education and wealth quintiles in the Philippines among: (a) all women aged 25–29 years; and (b) among women in a union (married or cohabiting) 25–29 years.

	(a) Percentage of all women		(b) Percentage of women in a union	
	2003	**2013**	**2003**	**2013**
By education				
No education	10.6	23.0	12.5	26.0
Primary school	10.2	27.3	12.1	32.2
Secondary school	6.5	32.1	8.4	40.3
Postsecondary	2.9	13.5	5.1	26.3
Total	6.3	23.8	8.7	34.6
By wealth quintile				
Poorest	12.5	26.0	13.7	28.0
Poor	15.5	31.8	17.2	38.0
Middle	8.6	25.4	10.5	35.1
Rich	7.8	24.6	11.2	41.3
Richest	5.4	14.0	9.7	30.1
Total	9.7	23.8	12.7	34.6

Courtesy of Brienna Perelli-Harris on the basis of the Demographic and Health Survey data 2003 and 2013.

cohabitation, and this might be more common among the better educated, wealthier parts of the population. Finally, it should be noted that there may also be some historical effect of cohabitation being acceptable among some ethnic minorities as well,[o] which could explain a part of the incidence of cohabitation among the lower education groups. The Philippine's rate of cohabitation will be an interesting one to follow in the coming decade, as the best-educated women could catch up in their rate of cohabitation with other socioeconomic groups, despite strong Catholic opposition to premarital couple relations.

The Chinese experience of cohabitation is entirely different from most other countries. The "Reform & Opening Up" policy of modernization initiated tremendous transformations since 1978, and it was accompanied with strongly enhanced urbanization and industrialization. Although it was intended that modernization would occur for the entire country, the policies of rapid development were mainly implemented in coastal areas and provincial capitals. As a result, these areas of rapid economic development were exposed to Western culture (Hu & Scott, 2016; Tian, 2016, 2017) In addition, there were high rates of internal immigration from rural areas to the developing regions, which weakened traditional forms of social control in those from rural areas. The concomitant changes in Chinese family structures are characterized by a reduction in household size, a decline in patrilocal multigenerational co-residence, high divorce rates, increases in single motherhood (Wang & Zhou, 2010), and a rise of the so-called DINK couples (dual income, no kids) (Ma, Shi, Li, Wang, & Tang, 2011). Furthermore, urban China also experienced a sexual revolution (Feng & Quanhee, 1996; Parish, Laumann, & Mojola, 2007; Parish et al., 2007), and cultural change also manifested itself in a weakening of both the traditional filial duties and male-dominated gender relations (Liu, Karl, & Ko, 2013; Zhang, 2008).[o] For example, Liu et al. (2013) reported that the notions of economic independence and individual autonomy entered the discourse of gender dynamics. Finally, there also seems to be a shift from a collectivist outlook to much greater individual autonomy (Yan, 2009, 2010).

In the context of the above described social changes, it should not come as a surprise that rates of premarital cohabitation increased. Relative to traditional marriage, cohabitation offers freedom of partner choice, sexual

[o] However, Hu and Scott (2016) report that the better-educated still stress filial piety more than others, but are in the lead with respect to gender equality and the erosion of patrilineal values.

Table 4.6 Percentages ever-cohabiting among all women, by age and completed level of education, People's Republic of China, 2014.

Age	Illiterate[a]	Primary	Middle	High school	University
20–24	21.6	36.2	(22.7)	(15.6)	(5.9)
25–29	14.5	25.0	27.7	35.9	31.2
30–34	9.2	18.9	27.5	31.2	27.0
35–39	15.5	13.2	16.6	26.0	27.0
40–44	8.5	11.6	12.5	18.0	6.8

Note: figures within parentheses denote that many women 20–24 in these higher education categories are not yet in any union. As they progress to enter unions the rates of cohabitation are very likely to increase considerably.

[a]Includes some women who are partially literate.

Courtesy of Jia Yu and Yu Xie, data: China Family Panel Study (CFPS) 2014 round.

gratification, economies of scale in consumption and housing expenditures, and potentially lower exit costs if the relationship does not work out. Moreover, as there is no evidence of substantial rates of preexisting forms of cohabitation, there is presumably no significant locked-in effect of any historical legacy, as was the case of Latin America. The net result is the absence of the negative education gradient in the cohabitation pattern as shown in Table 4.6 with data from the 2014 round of the Chinese Family Panel Study (CFPS).[p]

It should be noted that the percentages ever-cohabiting reported in Table 4.6, and provided by Xie, are related to the *total* number of all women in the age groups, and not just the more commonly used number of women in a union (cohabiting or married). This means that percentages in the youngest two age groups are systematically lower than those reported previously since the larger numbers of single women in these age groups are inflating the denominators. Despite this drawback, it is still obvious that the percentages with a premarital cohabitation experience have been rising systematically in each successive cohort. The highest rate of cohabitation is among women with the most education: high school education or those with a university education (see also Xie, 2015). The lowest cohabitation rate is witnessed in women without any education, who are illiterate or only partially literate. Presumably, the least educated are disproportionally present in rural areas, which are more traditional and have not been exposed

[p] The Chinese Family Panel Study was conducted in 25 provinces, leaving out only the five most remote and sparsely populated ones. The baseline survey was organized in 2010 and covered 16,000 households. For further details see Xie and Lu (2015).

to outside influences to the same extent as the coastal cities or larger provincial urban areas. It should also be noted that there is a major shift in the educational composition of the five age groups presented in Table 4.6. For instance, in the oldest age group 40–44 years, almost 60% of women belonged to the illiterate/partially literate category, and only 6% had a university education. Among young women aged 20–24, the illiterate/partially literate category was only 23% of those women, whereas the university-educated were a full quarter of all those women. Hence in China, the combination of a positive education gradient of cohabitation and the upward shift in educational attainment across cohorts reinforces the expansion of cohabitation as a mode of family formation. Additional evidence for China also reveals that the prevalence of cohabitation is positively related to urbanity,[q] to fathers having a college education, and, as expected, particularly to being a rural to urban migrant (Xie, 2015).

From the analyses of trends just presented it is clear that the paths followed by rising cohabitation in Japan and China are very different, with the latter conforming more to the original SDT script of the 1980s. However, China and Japan still share "an orderly transition" to parenthood (Raymo et al., 2015) as premarital births are rare in both countries. Also, the mean duration of cohabitation was limited, it was about 1 year in China and about 18 months in Japan during the first decade of this century. This suggests that both countries were in a phase in which cohabitation was increasing in frequency and acceptability, and there might well be increasing duration of cohabitation in both countries, as had happened across many other countries. Finally, whether China is currently joining the Japanese pattern of more premarital pregnancies and subsequent abortions or shotgun marriages cannot be ruled out, but still needs to be documented.

Further globalization? Expectations for the next decade and for the other regions in the world

In formulating expectations about the future of cohabitation I consider two sets of factors. First, I consider factors related to ethnohistorical preexisting forms of marriage and partnership. Second, I examine factors associated with religions and with ethical conservatism in matters of individual decisions concerning matters of life and death.

[q] Particularly residents of the provinces of Shanghai and Guangdong have a higher incidence of cohabitation (Xie, 2015).

Legacies of the past

If one is to explain the current contrasts in cohabitation between countries on a global scale, one needs to pay attention to the preexisting ethnohistorical legacies with respect to kinship organization and partnership. At the most general level, four types of societies can be distinguished: (i) the patriarchal societies of Asia and North Africa; (ii) the conjugal societies of the "old and new Europes"; (iii) the ethnically mixed societies of Latin America and the Caribbean; and (iv) the lineage-based societies of sub-Saharan Africa with exogamous marriage.

The patriarchal societies are essentially made up of populations with Confucian, Buddhist, Hindu, and Muslim religious traditions. All have what Goody (1976) called "the Eurasian" patrilineal and male-dominated family and household organization.[r] These societies have or had arranged endogamous marriages protecting the lineage or caste prerogatives, and upon marriage, women were cutoff from their own kinship group to end up in a less favorable situation. Free partner choice was avoided, and of course, premarital cohabitation was an aberration.[s] Several features of these patriarchal structures have weakened in Japan (Ochiai, 1996), China, Taiwan, and Hong Kong, but are still strong in many other instances such as India, and the conservative countries of the Muslim world. But, as the Japanese case shows, partner selection still occurs in a context of social class endogamy, and marriages lead to persistent domestication of women in nuclear family households[t] (Matsuo, 2003). Overall the rates of cohabitation is expected to be very low for these societies. However, as the examples of the People's Republic of China, Japan, and the Philippines indicate, these countries are not immune to a nascent manifestation of premarital cohabitation in situations with weakening traditional forms of social control. In addition, the rapid growth of the number of singles in the age group 20–29 in many of these societies is an element that could trigger the emergence of higher rates of cohabitation.

[r] There are, however, examples of ethnic minorities who have maintained other patterns of kinship organization or who had these in the past.

[s] For an update on South and Southeast Asia, see Yeung, Desai, and Jones (2018).

[t] A low incidence of cohabitation is found in selected ethnic groups in the Central Volcanic region of Mexico or the former Aztec-dominated areas (Mazahua, Otomi, Nahuatl, Purpecha), the Maya of Yucatan and Guatemala, and the entire strip of central highland plains of the Andes. See Esteve et al. (2016, pp. 140–143) on Mexico and Esteve et al. (2016, pp. 19–21) on the Andes.

The second group of societies is mainly a European offshoot of the first one, but Christianity has weakened the patriarchal control over endogamous and arranged marriage (Goody, 1976). In fact, except for the nobility and other wealthy families, marriage evolved toward a pact between two individuals ("With this ring I thee wed") who could escape parental control. Marriage became a more strictly individual affair. This individualization was further advanced by Protestantism, and it comes as no surprise that Protestant regions started the cohabitation boom in Europe. Furthermore, differential secularization during the 19th and 20th centuries produced greater heterogeneity among the Catholic regions as well. Hence the more rapid rise of cohabitation in more secularized Spain and Portugal than in Italy or Poland. Furthermore, the conjugal system has been "exported" to the former overseas colonies with a majority of European settlers, that is, to the "new Europes" of Canada, the United States, Australia, New Zealand, Argentina, and Uruguay. Also there, further heterogeneity was added subsequently as a result of ethnic mixing or other more idiosyncratic causes. To sum up, the individualization and the strongly conjugal tradition are considerably more conducive to a shift from marriage to cohabitation. In these instances cohabitation initially is merely a temporary form of courtship, but subsequently it becomes a more permanent arrangement in which procreation can take place.

The third group of societies consist of the mixed-race Latin American and Caribbean populations, which have syncretic traditions of various origins: native American patterns of partnership formation, visiting unions among imported African slaves, and elite conjugal European marriage. White immigrants brought the European Christian marriage and nuclear family organization with them, and, more often than not, also the corresponding legal code about marriage and inheritance. The imposition of Catholicism strengthened this European model. By contrast, indigenous populations historically displayed a wide variety of partnership arrangements with varying strength of the institution of marriage, from weak among hunters/gatherers and nomads, to stronger marriage arrangements among the successors of the agriculture-based and stratified pre-Columbian civilizations.[t] Also in contrast with the Asian patriarchal societies and the concomitant dowry transfers, women rarely alienated property upon marriage. Instead, husbands rendered services to their wives' families (i.e., bridewealth) among most indigenous American populations. Added to that the importation of African slaves for whom marriage was not possible, and who had to resort to so-called visiting unions or cohabitation instead. Moreover, any of the above systems can be combined with co-residence with the wives' parents or other family members in most Latin American countries (Esteve, Garcia-Roman, & Lesthaeghe, 2012). Consequently, a plethora of customs and traditions came into existence in which women were not disadvantaged to the same degree as in Asian or the Muslim world. Hence, the coexistence of multiple forms of partnership not only resulted in a high incidence of cohabitation prior to the 1970s but also facilitated the subsequent cohabitation boom and its spread to the better educated. The SDT was, as it were, grafted on preexisting patterns of POD and ethnicity-related cohabitation. However, despite pronounced catching up, elites of European descent still have not reached the levels of cohabitation of the Mestizo, Pardo, or Black populations.

The fourth group is made up of sub-Saharan African populations. Their kinship structures are the opposite of those of the patriarchal Asian group: the dominant rule is that of exogamous marriage. Partners can never be chosen among members of one's own lineage. Furthermore, the transfers of goods and/or services go in the opposite direction of those in Asia: bridewealth is the rule, not a dowry (Goody, 1976).[u] In other words, women bring in wealth to their lineage, and it is the lineage of the husband which has to "pay" for the transferred procreative rights in patrilineal African societies. In matrilineal lineage societies, even these procreative rights are not transferred on marriage to the man's family, as children remain members of their maternal lineage. Husbands have the status of the visitor, whereas control over children remains in the hands of maternal uncles. Furthermore, sub-Saharan African systems had the highest incidence of polygamy, which means strong control of older men over women. In such gerontocratic settings, women enter into a sexual union at a young age, but the average size of the age gap between partners is large, and often in excess of 10 years. The younger men are in fact the "main victims" of the system as they have to wait for a partner or have to form illicit relationships. If various props of the system erode (rising education, urbanization, migration), room is created for

[u] Premarital cohabitation is not one of the items used by the WFS, but we take it that in a gradation of more to least acceptable, divorce would be the easiest item, probably followed by cohabitation, and then by the other four. At any rate, if there is a de-stigmatization of abortion, homosexuality, euthanasia, and suicide, it is likely that there has been one with respect to cohabitation as well. At the individual level, a strong correlation between current or ever-cohabiting and a liberal attitude toward these four ethics items could be established in the European societies on the basis of the European Values Surveys (EVS). See, for instance, Lesthaeghe and Surkyn (2002).

other forms of partnership with cohabitation being one of the possibilities. It should be noted though that the Islamized societies of the Sahel or East Coast, still have very low levels of education and have maintained the traditions better than the others. However, under the influence of Islam, some have shifted to arranged marriage and cross-cousin endogamous marriage. In these instances, these shifts operate against the drift toward more generalized cohabitation.

The differences in kinship structures and marriage traditions just outlined form a first and necessary backdrop for the understanding of subsequent global evolutions. Multiple forces have shaped these further on, but a related and much more recent evolution can also be singled out as a crucial player: the ethics revolution since the 1960s.

Cohabitation and the ethics revolution

The second factor facilitating the emergence and diffusion of cohabitation is the advancement of an ethics revolution. More specifically, I use indicators pertaining to the acceptability of divorce, abortion, homosexuality, euthanasia, and suicide[v] as indices of ethical conservatism versus liberalism. A high acceptability of these behaviors not only indicates a liberal position, it also implies a circumstantial evaluation rather than an absolutist one, as it reflects that under some circumstances these behaviors are acceptable rather than never justified. This equally implies that there is respect for *individual decisions* against strong normative institutional stigmatization. The degree of resistance to any circumstantial evaluation of these ethical features can be gleaned from the World Values Survey data (WVS) of the first decade of this century. Fig. 4.5 brings together the percentages of respondents aged 18 + years who consider the just mentioned five acts as *never justified* (NJ) for a variety of countries, except sub-Sahara ones. The length of the bars corresponds to the sum of these five percentages.

Obviously, the "ethics revolution" is in an advanced stage in Western Europe, with the exception of Italy, which incidentally, also corresponds with the slower secularization and the concomitant slow expansion of cohabitation in this country. The Anglo-Saxon world tends to be more conservative, but not to the same degree as Central and Eastern Europe. In

[v] Among Islamized populations, a distinction should be made between those who adopted Arab customs such as endogamy and cousin marriage, and the others which were often conquered tribes but maintained the essential characteristics of sub-Saharan kinship organization (e.g., exogamy, refusal of purdah, "market women" active in trade, sometimes even survival of female secret societies).

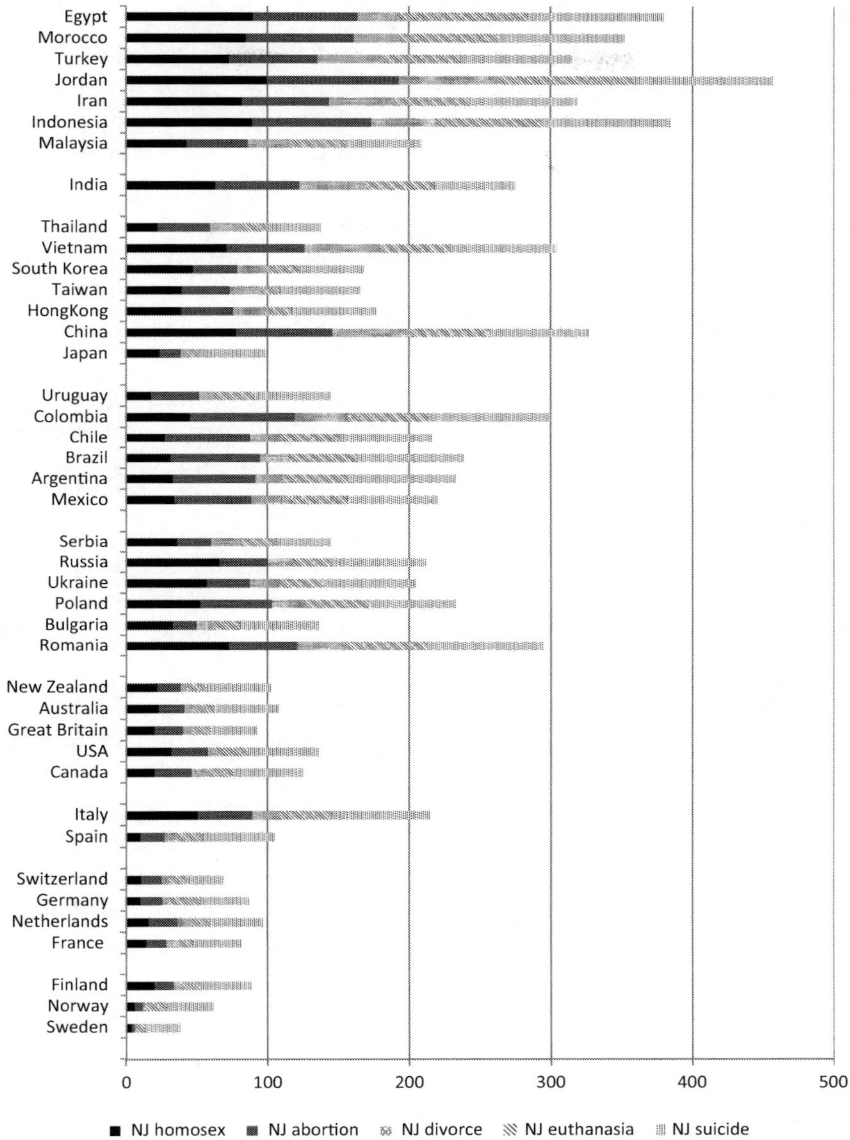

Fig. 4.5 Ethical conservatism—Cumulated percentages of the total population for whom homosexuality, abortion, divorce, euthanasia, suicide are NJ = never justified. (World Values Studies, 2005–07.) NB maximum possible score = 500, where higher scores reflect more ethical conservatism.

the latter group, Serbia and Bulgaria have witnessed a greater degree of de-stigmatization than the other countries reported here. The difference between Bulgaria and Romania mirrors that between Spain and Italy, and in both pairs the first mentioned countries have a noticeably higher incidence of cohabitation coupled to a significantly more liberal ethics profile.

The Latin American countries are currently still more conservative than their European counterparts, and this is equally connected to a lower degree of secularization. This also brings out once more the importance of the locked-in facilitating factor, that is, the ethnohistorical legacies of cohabitation in most of these countries. But, as mentioned before, the trend in ethical matters since the early 1990s goes in the direction of de-stigmatization. The diffusion of more tolerant attitudes is top-down, that is, spreading from the higher to the lower educated social strata. Also, Uruguay stands out as a leader in the "ethics revolution" just as it witnessed the strongest cohabitation boom. Given this top-down diffusion of liberal ethics, it is likely that catching up with respect to cohabitation among the better educated social strata may continue and result in a greater flattening of the education gradient. This trend may also be accompanied by a postponement of fertility among the more educated women, in which case a further step along the lines of the SDT would be taken.

Among the Confucian and Buddhist countries listed in Fig. 4.5, another contrast appears between China and Japan: the ethics revolution in the latter has advanced to the Anglo-Saxon levels, whereas there is no such a comparable erosion of traditional ethics in China. Also, Vietnam has maintained a more conservative position. South Korea, Hong Kong, and Taiwan are at par, and Thailand has advanced slightly further on the de-stigmatization path. These are all candidates for an incipient takeoff or further expansion of cohabitation as a prelude to marriage, but probably not yet for accepting unmarried motherhood.

As described previously, India and all the Muslim countries have strong patriarchal forms of family organization with weak female positions. In addition, large segments of their female populations are persistently characterized by low education. When paired to a rigorist ethical stance, the expectations for the acceptability of premarital cohabitation are slim, despite the fact that some countries have completed their fertility transition of delayed and reduced fertility (e.g., Turkey, Iran, and Malaysia). Hence, despite the marked heterogeneity of the Muslim world in terms of human development and advancement of fertility control, these countries still form a major monolithic block of opposition to the "nonconformist" family aspects of the SDT.

Sub-Saharan Africa. Sub-Saharan populations, including several Islamized ones,[w] have forms of social and lineage organization that are the complete opposite of those in Asia.[x] Christianity and urbanization have altered traditional arrangements. The former has tried to turn marriage into a single event, that is, a church marriage, rather than a prolonged process of negotiations, and into a closed monogamous relationship as well. Rapid urbanization and substantial labor migration flow, by contrast, have had the opposite effect with large numbers of men, both single and married, vying for sexual gratification. In such instances, classic *co*-residential polygyny is being replaced by illicit concubinage. More specifically in South Africa, cohabitation is increasing (Moore & Govender, 2013) and seasonal labor migrants can find a "little house" or a second family with children (Odimegwa et al., 2017). High separation rates lead, furthermore, to large numbers of single-mother families and to more out-fostering of children.

Within this context, the sub-Saharan historical legacy will be favorable to long-term cohabitation displacing the "Christian marriage" to some extent, but on the whole, partnerships will retain their multiple forms as is already the case today. The issue for sub-Sahara Africa is to advance its fertility transition, especially since many large West-African populations have barely started it. In other words, the "first demographic transition" (FDT) is still prominently on the agenda and likely to remain there for the next decades.

Cohabitation and diverging destinies

The current chapter has been devoted to the genesis and rise of cohabitation. But, without going into detail, it requires a major *postscriptum* with respect to its consequences. The rapid rise of cohabitation over and beyond possible historical locked-in effects is by definition the result of individuals placing their own choice above that of century-old cultural norms and practices favoring direct marriage. Individual freedom of choice gained the upper hand. But the outcomes of these individual choices are rarely neutral for the actors themselves, their children or for societies as a whole. Ever since McLanahan's (2004) article on "Diverging destinies" the impact of the alternative life course paths, also including rising separation and divorce, on the life chances of women and children has given ground for major concerns. At this point, the sheer demographics of the SDT enter

[w] See Goody (1976) for a detailed elaboration based on the distinction between societies with or without "devolution of property" through the marriage of women.

[x] For more details see Odimegwa et al. (2017).

into interaction with other major societal characteristics such as the degree of social class inequality, the strength of the welfare state and of income redistribution, older and more recent developments in ethnic composition, or with diverging family ideologies. In other words, not only the determinants of rising cohabitation are strongly context-specific, but so are the consequences as well.

At this point, the impact of the rise of cohabitation may not be that worrisome in Japan or China, mainly because there is little fertility out of wedlock (cf. Raymo & Iwasawa, 2017). The main culprit of adverse effects on women and children in these two nations may currently be the increasing incidence of divorce. At the other end of the spectrum are most Latin American countries and the United States. All are characterized by sharp social class and/or ethnic group contrasts and by weak or nonexistent income redistribution measures. This is a combination that puts the lower social strata at particularly high risk and produces the most adverse consequences of the rising trend in cohabitation, procreation among cohabitants, and especially separation. Eastern and Central European countries equally suffer from weak welfare state provisions. Differential patterns of family formation by education and income initiated after 1989 may lead to enhanced precariousness of women and children in poorer segments of the population. The United Kingdom also tends to go in that direction, given its social class contrasts and less developed welfare state provisions than those of the Western and Northern European countries. The better off in terms of the magnitude of the SDT-related negative consequences are, not surprisingly, the Scandinavian countries, but even there the issue of lone mothers entering and being trapped in poverty is on the agenda.

Conclusions

On a global scale, there are three patterns of connections between the nonconformist pattern of cohabitation being an alternative to marriage, which often involves procreation, and the postponement aspects of cohabitation in which cohabitation is used as a prelude to marriage and procreation. In much of Europe and overseas English speaking countries, these two changes were occurring simultaneously. In Latin America, the cohabitation boom largely preceded any fertility postponement. In the East Asian societies, the sequence is completely the reverse, with fertility postponement preceding the cohabitation boom. These various patterns are very closely related to historical forms of family and kinship organization and the concomitant differences in moral acceptability of cohabitation.

In the process of the expansion of cohabitation, especially among Latin American populations, there is a carry forward of the negative education gradient as a historical/ethnic feature and further growth of cohabitation among the less educated women. However, in all cases, there is a remarkable and more recent catching up of rates of cohabitation among the better educated, including those with a university degree. This evolution of increasing cohabitation among the better educated is presumably still in progress, and in this process, cohabitation may lose its social class stigmatization. Eastern European populations may have had a slight negative education gradient at the onset of the rise of cohabitation, but, as the percentages cohabiting are rising from very low to moderate degrees, a V-shaped gradient developed, signaling the existence of both a POD and SDT pattern of cohabitation. Western and Northern European populations generally started with a positive education gradient and follow the classic SDT pattern of development with either the maintenance of this positive gradient or with a pattern of a uniform rapid rise of cohabitation among all education strata simultaneously. In Canada and the United States, various forms coexist. Quebec and the Atlantic provinces of Canada follow the Western European pattern, whereas the rest of Canada follows a slower evolution with the maintenance of a slight negative education gradient. We also have detailed geographical information on cohabitation by education in the United States (see maps in Lesthaeghe et al., 2016), showing that college-educated women aged 25–29 fall in the top two quintiles of the national distribution in the whole of New England, New York state, the three Pacific states, and Colorado. Moreover, we know that all social classes start with very similar percentages cohabiting and that after age 25 the better-educated convert cohabiting partnerships into marriage to a greater extent than the less educated.

Japan and China have completely opposite patterns of development despite the fact that they have had strong cultures favoring marriage. In Japan, cohabitation starts as a lower class feature, but in China cohabitation is taking off among the better-educated women in the larger Eastern urban areas with the greatest exposure to Western media. At this point, evolution is still in the early stages in both countries, and the pace of progression is unknown. In Japan, this will hinge on the better educated catching up, and in China on the westward diffusion to provincial towns and countryside.

The next candidates for an incipient "nonconformist" transition are probably to be found in countries with a Buddhist tradition such as Vietnam or Thailand. The former is of course also the more secular one, but according to the WVS, it has the more conservative ethics profile of the two. The Hindu and Muslim world is most likely to remain on the other side of a Great Divide in rates of cohabitation, following strict traditional rules leading to direct marriage, often endogamous and arranged or brokered. A fertility postponement effect is, however, a possibility in economically more advanced countries such as Iran or Turkey or as a consequence of the postponement of marriage (e.g., Maghreb). In sub-Saharan Africa, by contrast, cohabitation may occur at the expense of monogamous Christian marriage, but it is likely to be just another variant of the already existing plethora of partnership arrangements. Fertility, however, is still unchecked or only partially under control, and further high population growth will be substantial as a result of a slow first demographic transition.

Finally, the consequences of more generalized cohabitation are just as diverse and context-specific as its causes. Cohabitation interacts with other demographic events, such as partner separation, social and family cohesion, and with major features of social class stratification, the absence or presence of welfare state provisions and amenities, or migration and changing ethnic composition. Not only individuals but also entire countries will face the diverging destinies in this respect of the impact of cohabitation rates on populations.

References

Berquo, E. S., Cunha-Waldvogel, B. C., Garcia, S., de Cavalho-Ferreira, C. E., Di Giacomo do Lago, T., & Batista, L. E. Reprodução apos os 30 anos no estade de São Paulo [Reproduction in women over 30 years of age in the state of Sao Paulo]. *Novos Estudias CEBRAP, 100*, (2014). 9–25. https://doi.org/10.1590/S0101-33002014000300002.

Coale, A. J. (1973). *The demographic transition reconsidered.* In: *Proceedings of the international population conference.*Vol.1. Liège, Belgium: International Union for the Scientific Study of Population. (pp. 53–72).

Coleman, D. (2004). *Why we don't have to believe without doubting in the 'second demographic transition': Some agnostic comments.* In Vol. 2. *Vienna yearbook of population research.*Vienna: Austrian Academy of Sciences. (pp. 11–24).

Esteve, A., & Florez, E. (2014). Edad a la primera union y al primer hijo en America Latina: estabilidad en cohortes mas educadas. *Notas de Poblacion, 99*, 39–65.

Esteve, A., Garcia-Roman, J., & Lesthaeghe, R. (2012). The family context of cohabitation and single motherhood in Latin America. *Population and Development Review, 38*(4), 699–720. https://doi.org/10.1111/j.1728-4457.2012.00533.x.

Esteve, A., Garcia-Roman, J., Lesthaeghe, R., & Lopez-Gay, A. (2013). *The second demographic transition features in Latin America: The 2010 update.* Unpublished manuscript, table 4Barcelona: Universitat Autònoma de Barcelona.

Esteve, A., Lesthaeghe, R., & Lopez-Gay, A. (2012). The Latin American cohabitation boom 1970–2007. *Population and Development Review, 38*(1), 55–81. https://doi.org/10.1111/j.1728-4457.2012.00472.x.

Esteve, A., Lesthaeghe, R., Lopez-Gay, A., & Garcia-Roman, J. (2016). The rise of cohabitation in Latin America and the Caribbean, 1970-2011. In A. Esteve & R. Lesthaeghe (Eds.), *Cohabitation and marriage in the Americas* (pp. 25–58). Switzerland: Springer Open.

Esteve, A., Lesthaeghe, R., Quilodran, J., Lopez-Gay, A., & Lopez-Colas, J. (2016). The expansion of cohabitation in Mexico, 1930-2010: The revenge of history? In A. Esteve & R. Lesthaeghe (Eds.), *Cohabitation and marriage in the Americas* (pp. 133–156). Switzerland: Springer Open.

Esteve, A., Lopez-Gay, A., Lopez-Colas, J., Permanyer, I., Kennedy, S., Laplante, B., … Cusido, T. A. (2016). A geography of cohabitation in the Americas, 1970-2010. In A. Esteve & R. Lesthaeghe (Eds.), *Cohabitation and marriage in the Americas* (pp. 1–24). Switzerland: Springer Open.

Esteve, A., Lopez-Ruiz, L. A., & Spijker, J. (2013). Disentangling how educational expansion did not increase women's age at union formation in Latin America from 1970 to 2000. *Demographic Research, 28*, 63–76. https://doi.org/10.4054/DemRes.2013.28.3.

Felling, A., Peters, J., & Schreuder, O. (1983). *Burgerlijk en onburgerlijk Nederland. [Civil and non-civil Netherlands (Bourgeios and non-Bourgeois Dutch): A national survey of value orientations on the thereshold of the 1980s]*. Deventer: Van Loghum-Slaterus.

Feng, W., & Quanhee, Y. (1996). Age at marriage and the first birth interval—The emerging change in sexual behaviour among young couples in China. *Population and Development Review, 22*(2), 299–320. https://doi.org/10.2307/2137436.

Fukuda, N. (2016). *Marriage and fertility behaviour in Japan—Economic status and value-orientation*. Singapore: Springer.

Fürstenberg, F. (2013). *Fifty years of family change—From consensus to complexity*. In: *Paper presented to the conference on family complexity, poverty and public policy*.Madison: University of Wisconsin, Institute for Research on Poverty.

Fussell, E., & Palloni, A. (2004). Persistent marriage regimes in changing times. *Journal of Marriage and the Family, 66*(5), 1201–1213.

Goody, J. (1976). *Production and reproduction—A comparative study of the domestic domain*. Cambridge: Cambridge University Press.

Hiekel, N., Liefbroer, A., & Poortman, A. (2012). *Understanding diversity in the meaning of cohabitation across Europe*. In: *Paper presented to the 2012 European population conference, Stockholm, Sweden*.

Hoem, J., Kostova, D., Jasilioniene, A., & Muresan, C. (2009). Traces of the second demographic transition in four selected countries in Central and Eastern Europe: Union formation as a demographic manifestation. *European Journal of Population Research, 25*(3), 239–255. https://doi.org/10.1007/s10680-009-9177-y.

Hu, Y., & Scott, J. (2016). Family and gender values in China—Generational, geographic and gender differences. *Journal of Family Issues, 37*(9), 1267–1293. https://doi.org/10.1177/0192513X14528710.

Inglehart, R. (1990). *Culture shift in advanced industrial society*. Princeton, NJ: Princeton University Press.

Jones, G. W. (2005). The flight from marriage in South-East and East-Asia. *Journal of Comparative Family Studies, 36*(1), 93–119. https://doi.org/10.3138/jcfs.36.1.93.

Kuong, B., Perelli-Harris, B., & Padmadas, S. (2016). *The unexpected rise of cohabitation in the Philippines—Evidence for a negative educational gradient*. In: *Paper presented at the 2016 annual meetings of the population association of America, Washington, DC, USA*.

Laplante, B., & Fostik, A. L. (2016). Cohabitation and marriage in Canada—The geography, law and politics of competing views on gender equality. In A. Esteve & R. Lesthaeghe (Eds.), *Cohabitation and marriage in the Americas—Geo-historical legacies and new trends* (pp. 59–100). Switzerland: Springer Open.

Leridon, H., & Villeneuve-Gokalp, C. Les nouveaux couples—Nombre, caractéristiques et attitudes [The new couples: Number, characteristics and attitudes]. *Population, 43*(2), (1988). 331–373.

Lesthaeghe, R., & Esteve, A. (2016). Cohabitation—The Pan-American view. In A. Esteve & R. Lesthaeghe (Eds.), *Cohabitation and marriage in the Americas—Geo-historical legacies and new trends* (pp. 269–291). Switzerland: Springer Open.

Lesthaeghe, R., Lopez-Colas, J., & Neidert, L. (2016). The social geography of unmarried cohabitation in the USA, 2007–2011. In A. Esteve & R. Lesthaeghe (Eds.), *Cohabitation and marriage in the Americas—Geo-historical legacies and new trends* (pp. 101–132). Switzerland: Springer Open.

Lesthaeghe, R., & Moors, G. (2002). Life course transitions and value orientations—Selection and adaptation. In R. Lesthaeghe (Ed.), *Meaning and choice: Value orientations and life course decisions* (pp. 1–44). The Hague: Netherlands Interdiciplinary Demographic Institute, NIDI-CBGS Publications.

Lesthaeghe, R., & Surkyn, J. (2002). *New forms of household formation in Central and Eastern Europe: Are they related to newly emerging value orientations?* In *Economic survey of Europe 02-1*. United Nations Economic Commission for Europe. (pp. 97–216).

Lesthaeghe, R., & van de Kaa, V. J. (1986). Twee demografische transities? [Two demographic transitions?]. In R. Lesthaeghe & D. van de Kaa (Eds.), *Groei en Krimp* (pp. 9–24). Deventer, The Netherlands: Van Loghum-Slaterus.

Lesthaeghe, R., & Vanderhoeft, C. (2001). Ready, willing and able—A conceptualisation of transitions to new behavioural forms. In J. Casterline (Ed.), *Diffusion process and fertility transition—Selected perspectives* (pp. 240–264). Washington, DC: National Research Council, National Academy Press.

Liu, L. H., Karl, R. E., & Ko, D. (2013). *The birth of Chinese feminism: Essential texts in transnational theory*. New York, NY: Columbia University Press.

Ma, C.-H., Shi, J.-Q., Li, Y.-H., Wang, Z.-Y., & Tang, C. (2011). The new trends of changing family forms in urban China. *Sociology Studies*, *2*, 182–216.

Matsuo, H. (2003). *The transition to motherhood in Japan*. Amsterdam: Rozenberg Publishers.

McLanahan, S. (2004). Diverging destinies: How children are faring under the second demographic transition. *Demography*, *41*(4), 607–627. https://doi.org/10.1353/dem.2004.00.

Moore, E., & Govender, J. (2013). Marriage and cohabitation in South Africa—An enriching explanation? *Journal of Comparative Family Studies*, *44*(5), 623–640. https://doi.org/10.3138/jcfs.44.5.623.

Neels, K., Murphy, M., Ni Brolchain, M., & Beaujouan, E. (2017). Rising educational participation and the trend to later childbearing. *Population and Development Review*, *43*(1), 1–27. https://doi.org/10.1111/padr.12112.

Neels, K., & Perelli-Harris, B. (2012). *Heterogeneity in demographic change—The educational gradient of "Second Demographic Transition"-behaviour in Europe and the US*. In: *Paper presented to the European population conference, Stockholm, Sweden*.

Ní Bhrolcháin, M., & Beaujouan, E. (2012). Fertility postponement is largely due to rising educational enrolment. *Population Studies*, *66*(3), 311–327. https://doi.org/10.1080/00324728.2012.697569.

Ni Brolchain, M., & Beaujouan, E. (2013). Education and cohabitation in Britain: A return to traditional patterns? *Population and Development Review*, *39*(3), 441–458. https://doi.org/10.1111/j.1728-4457.2013.00611.x.

Ochiai, E. (1996). *The Japanese family system in transition—A sociological analysis of family change in postwar Japan*. Tokyo: LTCB International Library Foundation.

Odimegwa, C., Wet, N. D., Adedini, S., Nzimande, M., Appuni, S., Hochfeld, T., & Dube, T. (2017). Family demography in sub-saharan Africa—A systematic review of family research. *African Populations Studies*, *31*(1), 3528–3572. https://doi.org/10.11564/31-1-1023.

Parish, W., Laumann, E., & Mojola, S. (2007). Sexual behaviour in China—Trends and comparisons. *Population and Development Review*, *33*(4), 729–756. https://doi.org/10.1111/j.1728-4457.2007.00195.x.

Parish, W., Luo, Y., Stolzenberg, R., Laumann, E., Farrer, G., & Pan, S. (2007). A population-based study of Chinese urban adults. *Archives of Sexual Behavior*, *36*(1), 5–20. https://doi.org/10.1007/s10508-006-9082-y.

Puur, A., Rahnu, L., Maslauskaite, A., Stankuniene, V., & Zakharov, S. (2012). Transformation of partnership formation in Eastern Europe—The legacy of the past demographic divide. *Journal of Comparative Family Studies*, *43*(3), 389–417. https://doi.org/10.3138/jcfs.43.3.389.

Raymo, J. (2013). *Cohabitation in Japan*. Minneapolis, USA: National Council of Family Relations. www.ncfr.org/ncfr-report/focus/family-focus-family-demography/cohabitation-japan.

Raymo, J., & Iwasawa, M. (2017). *Diverging destinies—The Japanese case*. Singapore: Springer.

Raymo, J., Iwasawa, M., & Bumpass, L. (2008). *Cohabitation and family formation in Japan*. Research report number 714Osaka, Japan: Institute for Social and Economic Research, Osaka University.

Raymo, J., Park, H., Xie, Y., & Yeung, W.-J.J. (2015). Marriage and family in East Asia—Continuity and change. *Annual Review of Sociology*, *41*, 471–492. https://doi.org/10.1146/aanurev.soc-073014-112428.

Rosero-Bixby, L., Castro-Martin, T., & Martin-Garcia, T. (2009). Is Latin America starting to retreat from early and universal childbearing? *Demographic Research*, *20*(9), 169–194. https://doi.org/10.4054/DemRes.2009.20.9.

Surkyn, J., & Lesthaeghe, R. (2004). Values orientations and the Second Demographic Transition (SDT) in Northern, Western and Southern Europe: An update. *Demographic Research*, *S3-3*, 45–86. https://doi.org/10.4054/DemRes.2004.S3.3.

Thornton, A., Axinn, W., & Xie, Y. J. (2007). *Marriage and cohabitation*. Chicago: University of Chicago Press.

Tian, F. F. (2016). Transition to adulthood in China 1982–2005—A structural view. *Demographic Research*, *34*, 451–466. https://doi.org/10.4054/DemRes.2016.34.16.

Tian, F. F. (2017). Global interactions and family formation pathways in Reform-era urban China. *Chinese Sociological Review*, *49*, 183–211. https://doi.org/10.1080/21620555.2016.1230011.

United Nations. (2017). *World population prospects—The 2017 revision*. Department of Economic and Social Affairs, Population Division. DVD edition, tables FERT/04 and /08.

United Nations, Department of Economic and Social Affairs, Population Division. (2015). *World marriage data 2015*. New York: United Nations. http://www.un.org/en/development/desa/population/theme/marriage-unions/WMD2015.shtml.

Wang, Q., & Zhou, Q. (2010). China's divorce and remarriage rates: Trends and regional disparities. *Journal of Divorce & Remarriage*, *51*, 257–267. https://doi.org/10.1080/10502551003597949.

Xie, Y. J. (2015). Cohabitation in China—Trends and determinants. *Population and Development Review*, *41*(4), 607–628. https://doi.org/10.1111/j.1728-4457.2015.00087.x.

Xie, Y. J., & Lu, P. (2015). The sampling design of the China Family Panel Study. *Chinese Journal of Sociology*, *1*(4), 471–484. https://doi.org/10.1177/2057150X15614535.

Yan, Y. (2009). *The individualisating Chinese society*. Oxford, UK: Berg Publishers.

Yan, Y. (2010). The Chinese path to individualisation. *British Journal of Sociology*, *61*(3), 489–512. https://doi.org/10.1111/j.1468-4446.2010.01323.x.

Yeung, W.-J.J., Desai, S., & Jones, G. (2018). Families in Southeast and South Asia. *Annual Review of Sociology*, *44*, 469–495. https://doi.org/10.1146/annurev-soc-073117-041124.

Zhang, Y. (2008). *Trends of divorce in today's China* [dangdai zhongguo lihun qushi]. China Academy of Social Sciences. Retrieved from http://www.sociology.cass.cn/shxw/shgz/shgz42/P020080218335825008116.pdf.

CHAPTER 5

Family law across cultures (comparative family law)

Arlette Gautier
Department of Sociology, University of Brest, LABERS, Brest, France

Studies of comparative family law in the 1950s sought to identify the "special genius of each system," rather than focusing on the details of their specific provisos (Rabel, 1958). For example, René David's famous *Treaty of Comparative Law* (1950), now translated into 11 languages, did not document the particularities of national law, but rather grouped systems of law according to methods of reasoning and philosophical, economic, and political frameworks. David identified four different types of provisos: a Romano-Germanic tradition, characterized by the unifying effect of codification; Common Law, which revolves around the binding force of precedent; a socialist tradition, where law is deeply influenced by political party; and traditional and religious laws, formed by Muslim, Hindu, African, and East Asian Laws. The latest edition of this classic provides an update of this typology, by taking into account the disappearance of the socialist model in favor of civil Russian law, which has undergone surprisingly few changes, except the elimination of the rhetoric of socialism (David, Goré, & Jauffret-Spinosi, 2016).

The David (1950) classification of legal systems is now questionable because there has been increasing recognition of human rights and the international mixing of legal models is developing, driven by the growth of individual mobility and international law based on human rights, which override the specific characteristics of these different legal systems. According to Anno (2015), the only difference that remains is between human rights–based legal systems and law based on divine will. Are there no longer any regional differences in family law?

A new approach to understanding comparative law focuses on the coproduction of law and gender (Cardi & Devreux, 2014). This approach is particularly appropriate in family law since it has long assigned specific rights and duties according to sex. Various international conventions, influenced

Cross-Cultural Family Research and Practice
https://doi.org/10.1016/B978-0-12-815493-9.00005-3

by the development of women's movements, now aim to eliminate this gender hierarchy. However, the move to promote gender equality has had varying degrees of success internationally. In this chapter, I review this line of research in family law, focusing on gender relations in family law (Banda, 2014; Wallbank, Choudhry, & Herring, 2012), rather than parenting law which is reviewed elsewhere in this book (Chapter 18).

The transformations that have occurred in family law are studied with different methods. Ethnologists and sociologists have studied the law-making process (Bereni, Debauche, Latour, & Revillard, 2009; Shalleck, 2011) and its social use (Gautier, 2017) through observations and interviews. A quantitative approach to human rights, in particular family rights, has also been developed since 2000. It is facilitated by the fact that legal provisos are very precise and therefore easily quantifiable (Gautier, 2005; Hallward-Driemeier, Hasan, & Rusu, 2013; Htun & Weldon, 2012, 2018; Hudson, Bowen, & Nielsen, 2011). Some authors have questioned the value of quantitative methods (Shalleck, 2011), but they are used in this chapter as they offer a unique possibility to compare all countries and not selected ones, with all the biases this selection may provoke.

Some quantitative analyses provide global and specific indicators, including family rights (Hallward-Driemeier et al., 2013; Organization for Economic Cooperation and Development. Social Institutions Gender Index (OECD-SIGI), 2019), to measure progress. It is useful for evaluating policies but lacks precision in exactly what legal rights exist to be useful from a scientific point of view. It seems more useful to document a typology of family rights. This was done on the basis of the provisos concerning the "rights and duties" prescribed to spouses, which make it possible to recognize four models: obedience of the wife, husband head of the family, inequalities, and equality (Gautier, 2005). Nevertheless, 20 years later, the obedience of the wife model has largely disappeared and the egalitarian model has become more complex. Thus, the model of nondiscrimination, promoted by feminist organizations at the end of the 19th century, showed its limits: it is necessary but insufficient. The Beijing Declaration and Platform for Action, adopted at United Nations Fourth World Conference on Women held in Beijing, China, included a broader perspective on women's rights including measures to combat violence against women and empowering women as a means of dissolving gender hierarchies.

According to Htun and Weldon (2018), family law and violence against women are two areas which raise the question of the different status of men and women. The first depends on state-religion institutional legacies and religious authority, and consequently doctrinal issues, but not the second. It should therefore be easier to change laws concerning violence against women than family laws. Yet analysis of the legislative process in the Yucatan Peninsula of Mexico has shown that, while it is difficult for legislators unilaterally to oppose the idea of combating violence against women, Yucatan elected representatives have resisted the effective implementation of Mexican federal law on domestic violence (Gautier, 2018). In Yucatan, the penalties for domestic violence are light and only related to second offenses. These laws give the illusion of aligning with international treaties and national Mexican law, as well as appeasing local feminist associations. At the same time, they are compatible with a paternalistic, even patriarchal, vision of gender relations (Lacombe, 2018). It is therefore essential to know whether specific provisos are enacted to make laws effective. Moreover, whatever is stated in the law, their efficacy is subject, in part, to the financial resources made available to enforce the laws.

The focus on gender equality within marriage might seem insufficient in the face of the continued prevalence of underage marriages and polygamy in many countries. Underage marriage (under age 18 years) is legal, and much more common for women than men, in many countries. Polygamy is almost always accessible only to men. These gender differences undermine the very existence of gender equality of rights in the couple even before marriage, so they must be taken into account in the evaluation of equality in marriage laws. In addition, the development, since 2001, of same-sex marriage in many countries has changed the conditions of marriage, undermining its traditional heteronormativity, and raising different issues about gender roles and gender equality.

This chapter describes the varying provisos of family rights across diverse countries, including marriage laws linked to the age, number, and gender of the spouses, as well as those of conjugality (civil and economic rights as well as laws against domestic violence). Based on these analyses, the chapter describes finally a more inclusive typology of matrimonial rights, specified at the regional level.

Quantifying family law

Authors generally use civil codes, legal sites, legal articles, and monographs, as well as monitoring reports from various human rights bodies, including the United Nations (1979), to study legal provisos, whether for rights and duties in the couple (Gautier, 2005; Htun & Weldon, 2012, 2018) or for the restriction of minimum age at marriage (Arthur et al., 2018). Information gathered with these methods tends to be uneven in reliability because sources often are not properly identified, and because different provisos might be covered with different levels of specificity in the legal structure of different countries. For example, the "European-style" codes, which regulate many aspects of married life and are very precise, exist in only half of the 142 countries studied by Gautier (2005). In the other countries, marriage laws range from partial to almost no regulations. Comprehensive laws covering marriage are most frequent in Europe and are least frequent in sub-Saharan Africa.

The Women, Business, and the Law database of the World Bank provides much more reliable data than other sources on the laws and regulations that constrain women as more than 10,000 professionals[a] (lawyers, judges, civil society representatives, and public officials) collected it in each of the 189 countries covered. The 2009–18 editions include a wide range of information, from political parity to professional equality, including the legal and economic effects of marriage and policies against gender-based violence. In addition, it is possible to verify not only the existence of legal provisos but also their effectiveness, such as what penalties are incurred, if any, for violations of the law. The data set and edition of *Women, Business and the Law* 2019[b] is more restricted than the earlier editions, as the index is structured around the economic decisions women make as they go through different stages of their working lives, and it does not differentiate between unmarried and married women, so it was not used, except for two provisos concerning divorce and remarriage.[c]

The *Women, Business, and the Law* database does not cover polygamy and same-sex marriages, so in the research for the current chapter I used data from various sources. For polygyny, articles from Wikipedia, which mainly uses data from the Organization for Economic Cooperation and Development. Social Institutions Gender Index (OECD-SIGI) (n.d.), family legal codes, and newspaper articles were compared with those of monographs, OECD SIGI country profiles, and the legal site of the United Nations High Commissioner for Refugees (2019), and sometimes with

[a]References to the local experts: https://wbl.worldbank.org/en/local-experts.

[b]http://pubdocs.worldbank.org/en/596151551985875100/WBL-Data-Notes-03-06.pdf.

[c]Can a woman obtain a divorce in the same way as a man? Do women have the same rights to remarry as men?

civil codes.[d] Data on same-sex marriages come from International Lesbian, Gay, Bisexual, Trans, and Intersex Association (ILGA) website (Mendos, 2019) and an article in Wikipedia, which refers to countries' official texts (downloadable) and more rarely to press articles from the countries themselves. Both Mendos (2019) and Wikipedia appear extremely reliable and match perfectly.

Some countries allow the coexistence of different rights according to ethnic and religious affiliations (e.g., Malaysia has different marriage laws for Muslims and non-Muslims), localities (e.g., United States), or provinces/states (e.g., Canada and Mexico). Several solutions are possible in the face of this legal pluralism. Gautier (2005) defined a country by the most unequal law, a solution chosen because the State accepts this type of inequality. In contrast, Hudson et al. (2011) and *Women, Business, and the Law* database choose the most widely used law, as that is the proviso that effects most people. In this chapter, I followed the procedure used by *Women, Business, and the Law* database.

Several databases, notably those of OECD-SIGI (2019) and Womanstats (2019), have chosen to code family law data not on the basis of written legal restrictions, but according to the prevalence of things like underage marriage or polygamy. They reason that there is little importance of having no legal age at marriage or accepting polygamy if the prevalence of these arrangements is very low. While this argument has some validity, this chapter describes the extant law, as that is what the government has determined is potentially acceptable.

In this chapter, I relate family laws to regions of the world to look for trends. The typology of regions used was that of the United Nations and particularly of the World Bank, as the data on the dispositions of family law come from it. It differentiates not only East Asia and Pacific, Europe and Central Asia (former USSR and popular democracies), Latin America and the Caribbean, Middle East and North Africa, South Asia and sub-Saharan Africa but also high-income OECD countries. This last category is not geographical like the other ones, but economic as it includes countries of the European Union (including some having been socialist), former British colonies (Australia, Canada, New Zealand, United States), and also Japan and the Republic of Korea. The strange thing in this typology is that it works

[d]Only data on Gabon seems inadequate as Wikipedia states that polyandry is possible while the Civil Code of the Gabonese Republic (1995) mentions only men and women in its articles.

very well statistically, all the correlations being very significant as opposed to income, a category often used but not significant in this case.

In this chapter, marriage laws were quantified across two broad domains: marriage rights (Who has the right to marry?) and conjugal rights (What are the rights and duties of the spouses? What provisos concern economic relations and the fight against domestic violence?) before comparing them. Are these different rights subject to different principles or do they agree? To do this, the provisos will be grouped together to cross-reference these different registers.

Right to marry laws

The law in many countries specifies the minimum age at marriage, the possibility or not of a polygynous marriage, and the sex of the partner. After presenting data on the conditions for marriage in these three areas, an attempt has been made to build models of legally acceptable marriage internationally.

Age at marriage

Age is a social construction, as are childhood and adolescence. In many societies, young people work to help their families as soon as they can in a family-based economy and they marry quite early. Economic transformations and the development of formal education changed this, but at different paces. As late as 1938, only 78% in the civil codes of the then 63 independent countries specified a minimum age for marriage, with an average minimum of 16.5 years for men and 14.8 years for women (Gautier, 2005). Still today, child marriage occurs in every region of the world and globally, one out of every five girls are married before the age of 18, while for boys the percentage is only 5% (United Nations, Child Fund, 2019). Marriage before the age of 18 carries implications for one's whole life, yet the person getting married is not legally recognized as an adult capable of performing many legal acts of lesser scope. In this sense, it is argued in many international texts that child marriage is contrary to the rights of the child, and even is a violence, notably by the United Nations (1989). According to the World Health Organization, when minors, particularly girls, marry and have children, their health can be adversely affected and their education is impeded. As a result the economic autonomy of the individual is restricted (Committee on the Elimination of Discrimination Against Women, 1994, art. 36). In addition, the legal age for marriage for girls is often less than that for boys, which constitutes discrimination.

International standards have gradually defended the need for a legal age of marriage, but without specifying an age. Thus, United Nations (1962) insists on consent to marriage, the enactment of a minimum age, and the obligatory registration.[e] This was changed in 1994 as the Committee on the Elimination of Discrimination against Women recommended setting the minimum marriageable age at 18 for both sexes (Committee on the Elimination of Discrimination Against Women, 1994, art. 36). A common declaration of The Convention on the Elimination of All Forms of Discrimination against Women (CEDAW) (United Nations, 2016) and the Convention on the Rights of the Child specified in 2016 is that marriage with judicial authorization for boys and girls under 18 years of age, but at least 16 years old, could be acceptable. Under the United Nations leadership (2016), countries around the world set 2030 as a target year for ending all child marriage, and removing provision for judicial or parental authorization of exemptions to this requirement.

At the beginning of the 21st century, 78% of countries had enacted a legal age for marriage as 18 years for boys and 60% of countries for girls (Gautier, 2005). Since then, many other countries have increased the legal age of marriage to be consistent with the international recommendation (Arthur et al., 2018). However, the extent to which exceptions to child marriage by parental and judicial assent are permitted has undermined the effectiveness of these legal changes (Arthur et al., 2018).

The *Women, Business, and the Law* database also provides data on the effectiveness of laws prohibiting child marriage, by collating measures such as penalties and invalidation of marriage before the legal age. Table 5.1 presents minimum age for marriage and provisions for exceptions across 189 countries. At first view, international pressure for a legal age of marriage of 18 years, identical for both sexes, seems to have worked; since in 2018, 168 countries (88%) have enacted a legal age of marriage of at least 18 years for girls and as did 180 countries for boys (95%). However, the figure of 23 countries with no legal age of marriage of 18 years for girls rises to 58 if one takes into account the absence of prohibition or invalidation of child marriages performed before 18 years of age. In all, 83 countries have no penalties for child marriage, and 157 countries still allow possible exceptions

[e]It is regrettable that the *Women, business, and the law* database does not provide data on legal obligations regarding consent to marriage and its registration, which is essential for the "rights and duties of marriage" to be respected. Nor are there any information on obligation and prevalence of marriage registration.

Table 5.1 Number and percent of countries with minimum age to marry by law, and by exemption of parental and judicial consent, by gender and age, $N = 189$ countries.

	No legal age, or < 10 years	10–15 years	16–17 years	18 years	19 years and more	Not applicable
Prohibited by law						
Boys	3 (2%)	2 (1%)	4 (2%)	141 (75%)	39 (20%)	
Girls	19[a] (10%)	0 (0%)	4 (2%)	132 (70%)	34 (18%)	
Allowed with parental consent						
Boys	12 (6%)	8 (4%)	46 (24%)	23 (12%)	2 (1%)	97 (51%)
Girls	13 (7%)	8 (4%)	60 (32%)	16 (8%)	0 (0%)	79 (42%)
Allowed with judge authorization						
Boys	30 (16%)	3 (2%)	55 (29%)	8 (4%)	1 (1%)	84 (44%)
Girls	30 (16%)	24 (13%)	51 (27%)	4 (2%)	0 (0%)	80 (42%)

[a]14 countries are in Middle East North Africa region, 2 in South Asia (Malaysia and Maldives), 3 in sub-Saharan Africa (Comoros, Mauritania, and Sudan).

Based on World Bank Group. (2018). *Women, business and the law 2018.* Washington, DC: World Bank. https://wbl.worldbank.org/.

Note: Percentages have been rounded to the nearest whole number.

to the minimum age of marriage. Parents, therefore, have little incentive to follow the law, and it can even be said that ineffectiveness is enshrined in it.

In summary, the legal age at marriage varies by gender in many countries, and even more so for child marriage with parental consent (one-third of all countries accept that possibility for boys and half of all countries for girls) than for judicial consent (respectively, 51% and 55% of countries), showing that marriage of girls is more a family controlled matter than marriage of boys. What is more, the accepted age for marriage often is younger for girls than for boys. So, if the countries that accept legal age of marriage as below 18 years are "only" 5% for boys and 9% for girls, it amounts to 49% for boys and 58% for girls with parental consent and to 56% for boys and 58% for girls with judicial consent. So, half of the countries allow child marriage under some circumstances.

We can therefore see that there are a number of inconsistencies between the various provisos, a sign of a highly contradictory legislative process, or even a purely rhetorical political will. For example, Iraq has a legal age of marriage as 18 years for both sexes, but parents can allow girls to marry from the age of 15 and there is no prohibition, invalidation, or penalty associated with such early marriages. On the other hand, in Afghanistan, provisos for punishing parents for allowing child marriage exist, but there is no legal age of marriage for girls, so they cannot be applied. It is a purely formal law without any effectiveness. Therefore, these provisos concerning the legal age of marriage will not be added together to produce a numerical indicator, but they will be added in a logical and disjunctive order.

Table 5.2 presents the percentage of countries with no legal restrictions on age of marriage by girls by region. In Table 5.2, the existence of a prohibition on marriage before the age of 18 is not classified as existing if a penalty for allowing marriage (e.g., invalidation of the marriage) under the legal age for marriage is not specified, or if the minimum allowable age for marriage is less than 18 years. The category "CEDAW" is based on a country adhering to the recommendations of the Convention on the Elimination of all forms of Discrimination against Women (CEDAW) and the Convention on the Rights of the Child (CRC), both contain recommendations in relation to the elimination of harmful practices affecting girls under 18 years of age. More specifically, these shared guidelines require a minimum age of 18 years for marriage, with possible judicially approved exceptions from age 16 for both sexes. Adhering to CEDAW principle 1 requires that a country enact a proviso for invalidation of a marriage if the age at marriage is less than 18 ***or*** a penalty for allowing such a marriage; while CEDAW principle

Table 5.2 Percent of countries by region having no laws prohibiting female child marriage, allowing exceptions to any such law, and adhering to CEDAW[a] principles 1 and 2 on child marriage.

Region	East Asia and Pacific $N = 25$	Europe and Central Asia $N = 24$	Rich countries of OECD[b] $N = 32$	Latin America and Caribbean $N = 32$	Middle East and North Africa (MENA) $N = 20$	South Asia $N = 8$	Sub-Saharan Africa $N = 47$	Total $N = 189$
No law (%)	8	12	10	0	70	25	11	14
Exceptions (%)	56	24	41	66	20	25	36	41
CEDAW 1 (%)	12	40	44	12	10	25	32	26
CEDAW2 (%)	24	24	16	22	0	25	21	19

[a]Women (CEDAW) principle 1 prohibits any marriage under age 18, at the exception of marriage from age 16 by judicial authorization and having provision for marriage invalidation ***or*** penalty if the law is broken; CEDAW principle 2 prohibits any marriage under age 18, or allowing marriage under 18 only by judicial authorization from age 16, and having provision for marriage invalidation ***and*** penalty if the law is broken.

[b]OECD = Organization for Economic Cooperation and Development. This category includes rich European countries.

Based on World Bank Group. (2018). *Women, business and the law 2018*. Washington, DC: World Bank. https://wbl.worldbank.org/.

2 requires that a country enact a proviso for invalidation of a marriage if the age at marriage is less than 18 ***and*** a penalty for allowing such a marriage.

As summarized in Table 5.2, 26 out of 189 countries for which data is available have no law concerning marriageable age, and only 36 have enacted all the legal conditions for the law prohibiting marriage before the age of 18 years to be effectively enforced. Nevertheless, these two groups are in the minority. The two largest groups allow both parental and judicial exceptions to permit underage marriage. If we add to this the countries that accept other exceptions, two-thirds of the countries allow some form of underage marriage.

Table 5.2 presents the differences in underage marriage laws within and between regions. The Middle East and North Africa region is characterized by a high degree of consistency. Only one-third of the countries, including three Maghreb countries, have legislated an age of marriage of 18 years for girls, and few countries penalize or invalidate marriages that are contrary to this requirement. Only Egypt accepts marriage at the age of 18 without exception and penalizes offenders. Others, such as Djibouti, have no legal age for marriage except for boys. Marriage remains largely determined by families and not by the state. The classical Islamic law only requires the occurrence of puberty to allow marriage but permits a great fluidity (Büchler & Schlatter, 2013). In contrast, the former communist countries of Europe and Central Asia are characterized by a high proportion of countries having legislation prohibiting marriage before the age of 18. However, many countries in that region allow exceptions, and relatively few penalize offenders and invalidate underage marriages. In this region, countries' laws ostensibly follow legal traditions of protecting children, particularly girls, but in practice these laws have little effectiveness.

Latin American countries, including Caribbean countries, and rich OECD countries (including rich European countries) have almost all legislated against marriage before the age of 18, but many countries do not prohibit the practice entirely, and accept a high level of exceptions.[f] Almost 70% of the Latin American and Caribbean countries have legal penalties for illegal underage marriages, while among the rich OECD countries less than half have such penalties. Common law countries and especially federations leave matrimonial matters largely in the hands of local authorities, but civil-law countries are also in this position.

[f]It is also the case of the United States, where only two states do not accept exceptions for marriage before 18 years ("Marriage," 2019).

East Asia, the Pacific, and sub-Saharan Africa not only have countries that are similar in having low percentages of absence of legalization restricting marriage to those under 18 years of age, but also have substantial proportions of countries that allow exceptions, and a very low proportion of countries that impose penalties for such marriages. Finally, most South Asian countries prohibit child marriage, but half do not follow the international definition of child marriage in terms of either specifying a minimum age for marriage of 18 years, or allowing exceptions.

Polygamy, polygyny, and polyandry

Polygamy is a very ancient practice which includes polygyny, in which co-wives share a husband, and polyandry in which co-husbands share a wife. Anthropologists have shown that polygyny exists in more than 83% of 849 cultures worldwide; in contrast polyandry is found in fewer cultures (Starkweather & Hames, 2012; Zeitzen, 2008). Where polyandry occurs, it always co-occurs with polygyny, such that high-status men in such cultures take multiple wives (Hudson et al., 2011). The proportion of women in polygamous marriage is declining everywhere, but it remains prevalent in some sub-Saharan countries, involving 48% of women in Guinea and 42% in Mali (United Nations, 2015).[g] Economist Esther Boserup (1970) and anthropologist Jack Goody (1973) have explained that this high incidence of polygyny in sub-Saharan Africa is rooted in sexual division of labor in hoe-farming and the large economic contribution of women. It is still a normative marital system in the majority of sub-Saharan societies, even in towns, where its economic utility is minimal (Marcoux & Antoine, 2014). In other places, polygyny may be rarely practiced, except by the richer men or professionals, who are also the ones who have political power.[h] In such contexts, young and poor men are often denied the right to be married.

Polyandry is never accepted by the law, even in the regions where it is de facto the norm (e.g., rural areas of Nepal or Tibet). Polygynous marriage contravenes a woman's right to equality with men as it defines marriage in a radically different way for women and men. For example, infidelity in the

[g]Demographic and health surveys indicate that six other sub-Saharan African countries have a prevalence of women in polygynous marriage more than 30%, while no non-African countries exceeds 5%, except Haiti (15%). However, Burundi, Eritrea, Namibia, and Rwanda do not exceed 10% (United Nations, 2015).

[h]Thus, the Moroccan Minister of State for Human Rights and senior leader of the Justice and Development Party, himself a polygamist, was questioned by the UN High Commissioner for Human Rights on the issue of polygamy in Moroccan law (Polygamie, 2017).

context of polygyny can only be perpetrated by women. Spousal inheritance for women has to be divided among multiple wives, whereas men inherit all of their spouse's inheritance. Thus, article 16 of CEDAW, which advocates equality in marriage, is generally understood as prohibiting polygyny. The Committee on the Elimination of Discrimination against Women stated more precisely in 1994 in its general recommendation about equality in marriage and family relations that "Polygamous marriage contravenes a woman's right to equality with men, and can have such serious emotional and financial consequences for her and her dependents that such marriages ought to be discouraged and prohibited" (Committee on the Elimination of Discrimination Against Women, 1994, art. 14). However, the extinction of polygamy is not included in the objectives and goals adopted by the international community, as it is a contentious subject influenced by religious doctrines. Specifically, most monotheist religions have explicitly rejected polygamy as unacceptable for some time, but polygyny is condoned within Islamic doctrine, and not condemned by Buddhism. Islamic movements very often refuse any change in polygamy rights as a matter of religious and cultural identity. The debate about the legalization of polygamy continues in several countries. For example, in Guinea, where a new family law had been under discussion since 2000, the President vetoed a provision making polygamy the norm. The code was finally ratified with the possibility of opting for polygamy at the time of marriage, without being able to change options (Pierret, 2019). "Accordingly, with regard to women in existing polygamous marriages, States parties should take the necessary measures to ensure the protection of the economic rights of women" (Committee on the Elimination of Discrimination Against Women, 2013, article 28).

The conditions of polygamy are quite diverse. In some countries, codifications have established the possibility of including a clause for monogamy in the marriage contract, or requiring consent from the first wife or authorization from a judge, both for polygamy and repudiation, thereby diminishing the traditionally preeminent position attributed to men. For example, since 2003, according to the Moroccan Personal Status Law, the *Mudawana*, polygamist marriage must be authorized by a judge and authorization is subordinated to the existence of an objective and exceptional justification, as well as the petitioner having sufficient financial resources (Article 41) (Naciri, 2014).

Criminalization of polygyny also varies. For example, in Nepal, it incurs 1–3 years of prison and fines. However, the second marriage is not annulled and once the sentence served, the second wife has the same rights as the first. Even the government pension provided to the wife of a retired government employee after his death is split between the spouses (Katmandu, 2015). Unfortunately, there is no data concerning the conditions posed by codes for polygyny and for its criminalization, so we will specify only if it is legalized for all the population or only for a part of it (Muslims, certain regions, or marriage following custom). Finally, the question of polygyny in states where monogamy is the norm (at least for marriage) arises from the arrival of immigrants from countries where polygyny is accepted and who are legally married to various wives. It provoked heated debates as it is illegal in most countries of Christian tradition (Dupret, 2012; Malik, 2014). Some countries, like Australia, the Netherlands, New Zealand, and Sweden, legally accept immigrants who are in such a union from a country where it is legal and recognize all the spouses (up to four in Sweden), but without giving residence or social security rights to spouses other than the first one. France in 1993 and Ireland in 2017 ruled that if someone had two legal marriages abroad, only the first is legal, "though that did not necessarily mean [the second] marriage can never have legal consequences" (Carolan, 2017). I did not classify countries that give some recognition to polygamous marriages established elsewhere, but do not accept such marriages be concluded on their soil (Table 5.3).

Polygyny is illegal in the majority (129) of countries in the world, mostly in Europe, America, and rich OECD countries. Only 47 countries (23%) allow polygyny for all their population. It is most often legal is the Middle East and North Africa (MENA), where it is allowed in 75% of the

Table 5.3 Percentage of countries in four regions and the world where polygyny is accepted by law for all the population, certain groups, or illegal.

Region	East Asia and Pacific *N* = 28	Middle East and Northern Africa *N* = 20	South Asia *N* = 8	Sub-Saharan Africa *N* = 47	Total world *N* = 189
Legal for all (%)	8	75	50	47	23
Legal for certain groups (%)	16	5	25	22	9
Illegal (%)	76	20	25	31	68

Based on Wikipedia. (2019a). *Legality of polygamy*. https://en.wikipedia.org/wiki/Legality_of_polygamy; United Nations High Commissioner for Refugees. (2019). *Report of the United Nations High Commissioner for Refugees*. https://www.refworld.org/docid/3b00f1d454.html.

countries for all citizens, while one country permits it only for Muslims. However, legality of polygyny is linked in most countries to Islamic beliefs being recognized in law, and this forms the basis for acceptance of polygyny. Countries of sub-Saharan Africa and South Asia accept polygyny but in a somewhat different way to MENA countries. Only about half of the countries of sub-Saharan Africa and South Asia accept polygyny as legal for all citizens, while almost a quarter accept polygyny only for parts of the population: Muslims or persons married within customary law. In East Asia, polygyny is not the legal norm for the population in most countries, and is only accepted for certain groups, notably Muslims, in a small proportion of countries. Nine countries that do not permit polygynous civil marriages do recognize polygynous marriages under customary law. Two countries recognize polygyny in certain regions (Eritrea and Nigeria) and five for Muslims only, notably India.

Same-sex marriage

Homosexuality exists in almost all human societies. "Cultures that openly accepted homosexuality, of which there were many, generally had non-marital categories of partnership through which such bonds could be expressed and socially regulated… (They) slowly succumbed to the homogenization imposed by colonialism" (Encyclopedia Britannica, 2019). The trend reversed only at the end of the 20th century.

The first legislation of unregistered cohabitation of same-sex couples was introduced in the Netherlands in June 1979, registered partnerships were first enacted in Denmark in June 1989, and the Netherlands was the first country to grant equal marriage rights to same-sex couples in 2001 (Merin, 2002). This recognition is partly the manifestation of the transformation of marriage. In the past, marriage was most often a socioeconomic alliance formed to serve the interests of tribes or families, whereas in many developed countries marriage is now seen as a voluntary partnership between adults associated with love and sexuality (Coontz, 2005). The recognition of same-sex marriage is also linked to the development of movements for the legal recognition of homosexuality, which had been important in the period between World Wars I and II, and resurfaced with new vigor from 1969 onwards. Initially focused on the abolition of discrimination, the gay rights movement then called for the right to marriage, parenthood, and adoption in the name of equal rights and nondiscrimination (Winter, Forrest, & Sénac, 2018). Gay rights are now supported by many human rights and civil rights organizations.

Same-sex marriages are recognized to varying degrees by two international courts of justice. On January 8, 2018, the Inter-American Court of Human Rights, which applies to countries which have signed the American Convention on Human Rights, ruled that this Convention mandates and requires the legalization of same-sex marriage. On June 5, 2018, the European Court of Justice ruled that European Union member states may choose whether or not to allow same-sex marriage, but they cannot obstruct the freedom of residence of an EU citizen and their spouse. Furthermore, the Court ruled that the term "spouse" is gender-neutral, and that it does not necessarily imply a person of the opposite sex. It said that governments "must recognize and guarantee all the rights that are derived from a family bond between people of the same sex." It also said that it was discriminatory for a separate legal provision to be established (such as civil unions) instead of same-sex marriage (Wikipedia, 2019b).

However, the issue of same-sex marriages remains controversial, and is not recognized as a right by many countries in the international community. The main United Nations documents on the topic of "human rights," the International Covenant on Civil and Political Rights, in article 23, provides only that "The right of men and women of marriageable age to marry and to found a family shall be recognized" (Human Rights Committee, 2000). There is no mention of rights to same-sex marriage.

Same-sex civil unions are accepted in 15 countries (Table 5.4). In some countries, civil unions or registered partnership provide all the same rights and responsibilities to same-sex couples as civil marriage and differ in name only; countries with such laws include Greece, Slovenia, and Switzerland. In other countries, civil unions provide some, but not all, of these rights.[i] These are mainly European countries (including five Eastern European countries) and Chile. Other states recognize same-sex unions, which have been legalized elsewhere. This is the case for European countries bound by the decision of the European Court of Justice, and also outside Europe in Cambodia, Hong Kong, Costa Rica, and Israel, as well as in dependent territories such as the Cayman Islands, Curacao, and St. Maarten.

As of July 13, 2019, 28 countries allow same-sex couples to marry (including two in only part of their territory). Same-sex marriages are most often recognized in Western and Southern Europe, where family life has changed the most, with 16 countries recognizing same-sex marriage. Seven American countries accept marriage for all (including the United States

[i]Croatia and Hungary accepted civil unions but prohibited same-sex marriages.

Table 5.4 Percent of countries by region with laws relevant to same-sex couple relationships.

Region	East Asia and Pacific $N = 25$	Europe and Central Asia $N = 25$	Rich countries [a]OECD $N = 32$	Latin America and Caribbean $N = 32$	Middle East and Northern Africa $N = 20$	South Asia $N = 8$	Sub-Saharan Africa $N = 47$	Total $N = 189$
Prohibit same-sex marriage (%)	8	48	9	19	0	0	6	14
Allow civil unions (%)	0	0	3	3	0	0	0	1
Allow same-sex marriage (%)	4	0	53	16	5	0	2	13

[a]OECD = Organization for Economic Cooperation and Development.

Based on Mendos, L. R. (2019). *State-sponsored homophobia 2019*. Geneva: International Lesbian, Gay, Bisexual, Trans and Intersex Association. https://ilga.org/downloads/ILGA_State_Sponsored_Homophobia_2019.pdf.

since 2015, 18 Mexican states and the Federal District). South Africa and Taiwan are the only states in Africa and Asia to allow same-sex marriage.

In total, 43 countries recognize some form of conjugal life for same-sex couples (23% of the 189 countries), which constitutes a significant and recent transformation of the conditions of marriage. The situation is changing rapidly. Thus, for example, the United States, which had adopted a federal law prohibiting the recognition of same-sex marriages, had to accept them in 2015 because their Supreme Court ruled that this law violated not only individual rights but also the right of local authorities to define their matrimonial laws (Carter, 2013; Van Eeden-Moorefield & Alvarez, 2015).

At the time of writing, 146 countries (77%) do not recognize same-sex marriage. Among them, 71 countries have prohibited homosexuality, many with harsh penalties. Also, 12 countries have even imposed the death penalty on consensual same-sex sexual acts (Mendos, 2019). Even among the countries which accept homosexuality, "As of 15 July 2019, thirty-two countries and four jurisdictions worldwide have passed constitutional amendments that explicitly prohibit the legal recognition of same-sex marriage and sometimes other forms of legal unions as well. Jamaica was the first country to prohibit same-sex marriage in July 1962. Paraguay, Honduras and Bolivia are the only where not only marriage is prohibited but also de facto unions" (Wikipedia, 2019a). Among these, seven countries, which accept polygamy, prohibited same-sex marriage. Croatia and Hungary, which prohibited same-sex marriages for their nationals, nonetheless accept them for nonnationals, because they are obliged to do so by the European court.

About half of European and Central Asian countries have laws prohibiting same-sex marriage. There also is a strong division of law in Latin American and Caribbean countries, where the proportion of countries with pro- and anti-same–sex marriage laws are almost equal (respectively, 16% and 19% of countries) and a majority of the countries' laws support heterosexuality as the only legally recognized basis for marriage. In the rich OECD countries, a small majority legalized same-sex marriages while heterosexual marriage is still the only legal norm in a third of the countries.

The lack of legal rights provided to same-sex couples tends to occur more often in regions and countries that accept polygamy. As presented in Table 5.3, three quarters of the countries of MENA, South Asia, and sub-Saharan Africa accept polygamy, but few if any of these countries recognize same-sex unions. Conversely, 56% of rich OECD countries and 16% of Latin American and Caribbean countries provide legal recognition to same-sex marriages, but no countries in these regions recognize polygamous marriages. Very few (n = 40 out 47 countries) sub-Saharan African countries have explicit laws about same-sex marriages, the exceptions being South Africa in favor and 6 other countries against. While same-sex relationships are mainly legal in East Asia and Pacific region, the question of same-sex marriage is largely ignored, as only 12% of the countries have laws relating to it (either prohibiting or allowing it). A quarter of these countries accept polygamy. Two-thirds of East Asian countries only recognize monogamous heterosexual marriage, not recognizing either polygamous or same-sex relationships. However, the legalization of same-sex marriages is a recent change and is moving very quickly.

In conclusion, with respect to the right to marry, two of these (polygyny and recognition of same-sex marriages) almost never occur together in one country. There is no clear relationship of recognition of polygyny or same-sex marriage with legal recognition of an effective minimum age of marriage of 18 years.

The law and relations within couples

The saying: "A woman is never free" exists in many languages (Gautier, 2003), and this applies particularly to wives. European family laws in the 19th century deprived wives of all the most basic civil rights she could have as a single woman, such as the right to leave the house, to travel, to choose where to live, to have identity documents, and to be able to sue (Gautier, 2005). Feminist organizations have been mobilizing since the 19th century to obtain these civil rights for women, irrespective of marital status. They also called for greater economic justice by the recognition of the work done in the family. It is interesting that the term "economy" etymologically refers to "the administration of the house." Yet the discipline of the same name, economics has usually ignored the economic production carried out in families, and focused attention to market-based production alone (Marçal, 2019). Family laws of the 19th-century support this occultation of domestic work that feminists tried to take into account, with varying success. It continues to affect wives' economic wealth where it is still accepted (Deere, Oduro, & Swaminathan, 2013).

Traditionally, family law entrenched men's rights within marriage over women. For example, the French civil code enacted by Napoleon in 1804, referred to as the Napoleonic code, like in many other forms of family law, allowed for and justified the husband's "right to correct" his wife. In recent years, such laws increasingly are being replaced by laws against violence by an intimate partner, although these laws are not always effective.

The civil rights of wives

A typology of family rights has been constructed from the laws concerning the rights and duties prescribed for husbands and wives and contrasts three prescriptions: obedience of the wife to the husband, the husband as head of the family, and sexual equality (Gautier, 2005). The wife's duty of obedience to her husband is often prescribed by religion, as in Muslim laws, as enacted in the majority Muslim countries when they achieved independence from European colonization. The wife's duty of obedience was also prescribed by civil law in England in 1753 and in France in 1804. Under these laws the wife had no right to administer her own property, including her salary, she had to get her husband's authorization to work outside the home, she was required to follow his instructions in what was regarded as his home, and she had no authority over her children. The French code was exported to Europe, and after independence it became widespread in Latin America's former Spanish and Portuguese colonies, and some former French colonies in Africa.

Subsequently, improvements emerged, but often contradictory provisos maintained a residual inequality model. Thus, in 1881, England abolished the duty of obedience and recognized the civil capacity of wives, but they still had no right to the property acquired during the marriage. The legal provision for the man to be the head of the family originated in 1900 in Germany. It admits the wife's civil capacity and recognizes her as a wise collaborator. Nevertheless, in case of disagreement the husband has the last word.

The adoption of certain articles of the Napoleonic Code in African countries removed the customary protections that existed in matrilineal ethnic groups. Indeed, the Ivorian code recognizes the husband as head of the household and administrator of common property and also of the wife's own property. The husband can garnish the wife's salary, whereas traditionally she kept her earnings. The Ivorian code sought to create a conjugal family by inventing a patronymic name that did not exist. The child bears the father's name, whereas the lineage chose among the names of the lineage, and the wife that of her husband, whereas traditionally she was not even supposed to pronounce it (Dumetz, 1975).

The egalitarian model of family law eliminated all previous sex-based discrimination, but it took two different forms. In the USSR, in 1917, gender equality was affirmed but there was no provision to require its implementation, and the legislative text was only a few pages long. This egalitarian communist law was extended to North Korea in 1946, China in 1950, North Vietnam in 1959, South Yemen in 1974, and Cuba in 1975 (Molyneux, 1985). In contrast, the Scandinavian countries established an egalitarian model of family law in 1918 as a rule of public policy and ensured the division of property in the event of separation. This model was adopted and further developed in 1979 by Article 16 of the Convention on the Elimination of Discrimination against Women (CEDAW), which is now the most ratified convention in the world. Article 16 specifies what a nondiscriminatory right to marriage would be, which is further elaborated in General Comment No. 8 of the International Covenant on Civil and Political Rights (Human Rights Committee, 2000).

We will use responses to 15 questions generated by World Bank researchers concerning the civil rights of wives to measure its application. In some countries, the civil rights of wives differ greatly from those of husbands and unmarried women, with marriage depriving married women of rights single women sometimes enjoy. As noted previously, many such laws reflect the influence of the French Civil Code of 1804, which codified married women's rights as often less than that of single women, and as less than those of a married man. Each variable is dichotomously coded to indicate whether or not the legal provisions in question disadvantages wives, or promotes equal rights to husbands. Table 5.5 presents the number and percentage of countries upholding equal rights for married women as married men.

As presented in Table 5.5, 10% of countries have laws that require wife's obedience to their husband. Such laws no longer exist in America and Europe, but the laws do exist in 11 countries in the Middle East, 5 sub-Saharan African countries (11% of these countries were mostly ex-French colonies with civil law), and Malaysia. In effect, this makes wives the slaves of husbands, as said the great French poet Victor Hugo. Sociologist Colette Guillaumin proposed in 1992 the concept of "sexage," which is the appropriation of women by men, to describe more adequately this situation which is different from slavery.

The vast majority of countries now extend similar civil rights to married women and men. But a significant minority of countries restricts women's rights in dramatic ways. The most sensitive issues are therefore the wife's freedom of movement and her ability to transmit her nationality, two issues that are essential for individual freedom and for the recognition of full citizenship. The question of divorce is also a very serious one, as it limits the ability of

Table 5.5 The number of countries and percentage (in parentheses) giving wives equal civil rights relative to husbands, $N = 189$ countries.

Right in law: Can a married woman do this in the same way as a married man?	No	Yes
Apply for a passport?	37 (20%)	152 (80%)
Obtain a national identity card?[a]	11 (7%)	146 (93%)
Travel outside the country?	6 (3%)	183 (97%)
Travel outside her home?	17 (9%)	172 (91%)
Get a job or pursue a trade or profession?	18 (9.5%)	171 (90.5%)
Sign a legal contract?	1 (1%)	188 (99%)
Register a business?	4 (2%)	185 (98%)
Open a bank account?	3 (2%)	186 (98%)
Choose where to live?	31 (16%)	158 (84%)
Confer citizenship to her children?	24 (13%)	165 (87%)
Confer citizenship to a nonnational spouse?	51 (27%)	138 (73%)
Give testimony carrying the same evidentiary weight?	16 (9%)	173 (92%)
Are wives required by law to obey their husbands?	170 (90%)	19 (10%)
Obtain a judgment of divorce?[b]	45 (24%)	142 (76%)
Do women have the same rights to remarry as men?[b]	66 (35%)	121 (65%)

[a]32 countries have no identity cards, hence numbers and percentages out of 157 countries.
[b]Source: World Bank Group (2018). Data on 187 countries.
Based on World Bank Group. (2018). *Women, business and the law 2018*. Washington, DC: World Bank. https://wbl.worldbank.org/.

women to live as they want, with whom they want. The difference in the possibility to sue for divorce is a serious discrimination.

The regions of the world differ significantly in the civil rights of spouses. Europe and Central Asia (except for the acquisition of a passport) and the rich OECD countries give the same civil rights to married women as to men. In all, 20% of East Asian and Pacific countries limit the ability of a married woman to get a passport in the same manner as a married man, and 8% of countries do not permit a married woman to travel outside her home or the country or to live where she chooses. Three countries do not permit a married woman to confer citizenship to her children in the same way as a married man. In total, 25% of Latin American and Caribbean countries limit the acquisition of a passport by a married woman and pose limits on three other rights.[j] One or two countries of the four South Asian countries

[j]The possibility to register a business, choose where to live and the transmission of citizenship to her children.

limit the freedom of movement of married women. Between 19% and 34% of the sub-Sahara African countries limit their possibility to get a passport, travel, register a business, and choose where to live or be head of a family.

Middle East and North African countries limit civil rights of wives the most. Around 20%–25% of Middle East and North African countries limit the ability of a married woman to obtain a national identity card, or to travel outside the country in the same way as a married man. In all, 45% of these countries place restrictions on the acquisition of a passport by married woman and her registration of a trade; and 65% limit her getting outside the home, choosing where to live, being head of a family, or transferring her citizenship to her children. So, a majority of these countries seriously limit the freedom of movement of a married woman.

To summarize these data, I counted the different responses, adding one for each "yes," except for the obedience question where a "no" added a point. This yields a scale with a maximum possible score of 15, which reflects equality of rights of wives and husbands across all areas assessed. As presented in Table 5.6, 27 countries (14%) grant 2–10 civil rights to wives, 26 countries grant 10–12 civil rights (14%), 57 countries (30%) grant 13–14 civil rights, and 77 (41%) grant wives the maximum of 15 civil rights. Some Middle East countries had scores below 10 (notably Bahrein and Iran which got 2 points, and Lebanon got 13[k]), as did nine sub-Saharan African countries (predominantly former French colonies), as well as Afghanistan, Pakistan, and Malaysia. The Maghreb countries of Northern Africa got civil rights scores of 11 or 12.[l] All but six rich OECD countries and five European and Central Asian countries had high scores.

The economic consequences of marriage

Women's economic rights are severely limited when the wife cannot decide to work or manage her property. They can also be reduced by the rules of family inheritance. Girls' inheritance rights are not identical to boys in 39 countries (21%). This was the case in sharia regimes, whereas the Napoleonic Code established an egalitarian inheritance between all children. The English

[k]However, Lebanon is a pluralist country where each community has its own personal status even if a code civil does exist, so the general situation may be not so equalitarian (Edde, 2018).

[l]However, these rights may be limited in the current context of civil war. For example, in 2017, "the military authorities in eastern Libya (from Derna to Ben Jawwad) announced a ban on women under 60 years of age travelling without a 'male guard' outside Libyan territory" (Delatouche, 2017).

Table 5.6 Number of countries (and percent) with different numbers of provisos relating to civil and economic rights and policies to combat domestic violence in 189 countries.

No. of provisos	Civil rights		Economic relations		Policies against intimate partner violence	
	Number	%	Number	%	Number	%
1			1	1	40	21
2	2	1	25	13	3	2
3			10	5	2	1
4	2	1	8	4	2	1
5	4	2	24	13	1	1
6	6	3	36	19	9	5
7	2	1	62	33	15	8
8	5	3	23	12	14	7
9	6	3			37	20
10	2	1				
11	7	4			66	35
12	17	9				
13	26	14				
14	31	16				
15	77	41				
Total	189	100	189	100	189	100

Source: [dataset] World Bank Group. 2018. *Women, Business and the Law 2018*. Washington, DC: World Bank. https://wbl.worldbank.org/.

legal system was egalitarian in the sense that it allowed for the exclusion of anyone, irrespective of sex, even if English parents most often left wealth to the oldest son. The Committee on the Elimination of Discrimination Against Women (2013) explained in its general recommendation number 29 the extent of the economic consequences for wives of marriage, family relations, and their dissolution and the necessity to eliminate all these discriminations.

Marital property systems provide what properties are in common and what are individual, how and by whom they are managed, and how they will be divided and inherited at the end of the marriage. These provisions are applied by the operation of law, which generally defines a default marriage system to those who do not make a prenuptial agreement in civil-law countries. The laws also depend on the *lex domicilii* (the legal jurisdiction and laws deemed by a court to be the most relevant to hear the case when there are different laws in different jurisdictions). The *lex domicilii* is usually determined as being the legal jurisdiction of the spouses at the time of, or immediately following, the wedding.

Marital property systems have long and often been unequal for husbands and wives (Miles & Scherpe, 2014). The Napoleonic Civil Code, British law since 1753, and Islamic law all assume that the wife dedicates herself to the family, including in the context of farming or commerce, and is supported in return by the husband. However, the way they organize this dependency is different. The Napoleonic Code establishes the community of property between spouses, which favors the unity of family patrimony, and is managed by the husband without the need to seek the advice of his wife. The husband even has the enjoyment and administration of the wife's property, including her wages and her own property when she has not made a property separation agreement. Even in this case, because of her civil incapacity, the husband represents the wife. On the other hand, English and Islamic law follows the legal regime of separation of property; wives manage their own property without being able to oversee common property, and they do not derive any recognition from their unpaid domestic work, except that which they may have provided for in their marriage contracts or which the husband decides. In British law, the husband has complete freedom of will and can disinherit his wife. In 1888, Costa Rica invented the system of conjugal administration, which allows each spouse to administer his or her property as he or she sees fit, and provides the division of common property into two halves upon dissolution of the marriage, thus recognizing the domestic labor of wives and mothers[m] (Gautier, 2005).

In 2018, countries were divided into three groups in terms of the default marital property regimes. There are about the same number of countries which follow limited or partial Community Property laws (77 countries, 41%) as countries that follow the separation of property laws (75 countries, 40%). The separation of ownership of property to each spouse is the only matrimonial regime of most Common law and Islamic jurisdictions. Other default marital property regimes are quite variable. The Napoleonic full community principle is now the default marital property regime of only six countries (3%). A total of 14 countries in America, Western Europe, and Israel (7.4% of countries) mix community and separation property regimes

[m]The Latin American version of this model reduces common property to acquisitions, while the Scandinavian version accepts the universal community, i.e., it includes property acquired before marriage or given in a personal capacity.

as marital property is separately owned during the marriage, but after marriage (divorce and death of a spouse), the net assets are lumped together as property in joint tenancy and divided. In all, 12 countries have a different default marital property regime (3%) and 6 do not have any.

Mostly matrimonial property regimes differ from region to region. Europe and Central Asia overwhelmingly use legal regimes in which property acquired after marriage is deemed to be shared community property of the spouses, with the exception of Cyprus which has chosen separation of property. The situation is exactly the opposite in the Middle East and North Africa, where the separation of property regime is the rule. The other regions of the world have a mix of partial community and separation of property regimes.

The division of property under matrimonial legal regimes is profoundly modified in some countries by taking into account the sharing of household responsibilities and the recognition of nonfinancial contributions. The Committee on the Elimination of Discrimination against Women defined that in 2013 as "household and family care, lost economic opportunity and tangible or intangible contributions to either spouse's career development and other economic activity and to the development of his or her human capital" (article 47). The last two measures are being taken in all European, Latin American (except three Caribbean islands), and rich OECD countries. The sharing of household responsibilities is accepted as a way to contribute to the household's wealth by more than 80% of countries in East Asia and sub-Saharan Africa, but only by one-third of South Asian countries and by one quarter of countries in the Middle East and North Africa. Two-thirds of countries in East Asia and sub-Saharan Africa recognize nonfinancial contributions in marriage. In total, only 39% of countries in Africa and the Middle East accept the relevance of nonfinancial contributions and 47 countries refuse to accept the relevance of these contributions.

Moreover, the joint decisions on the family home are guaranteed to the nonprimary owner in only 43 countries (23%), but 31% of countries where property acquired after marriage is held in common, whereas this is the case in only 8% of countries under separation of property. So, in the last set of countries, the family home belongs to the primary owner. As to financial decisions, married couples have to make financial decisions together in 86% of countries under the community of property regimes, while in contrast the original owner decides alone in 98% of countries that follow the separation of property (Table 5.7).

Table 5.7 Number and percent (in parentheses) of laws on economic relationships of families in and out of marriage in 2018, $N = 189$ countries.

Law	No	Yes
Do married couples jointly share legal responsibility for financially maintaining the family's expenses?	29 (15%)	160 (85%)
Are there special provisions for major transactions concerning the marital home?	145 (77%)	43 (23%)
Do married men and married women have equal ownership rights to property? (7 nonresponses)[a]	13 (7%)	169 (93%)
Do female and male surviving spouses have equal rights to inherit assets? (6 nonresponses)[a]	36 (20%)	147 (80%)
Do married couples jointly share legal responsibility for financially maintaining the family's expenses?	29 (15%)	160 (85%)
Does the law provide for the valuation of nonmonetary contributions?	55 (29%)	134 (71%)
Do sons and daughters have equal rights to inherit assets from their parents? (4 nonresponses)[a]	39 (21%)	146 (77%)

[a]For these questions some countries did not have available data, and percentages are reported of countries that did respond.
Based on World Bank Group. (2018). *Women, business and the law 2018*. Washington, DC: World Bank. https://wbl.worldbank.org/.

In total, 36 countries grant 1 or 2 economic rights to wives (19%), 68 admit 3–5 rights (36%), and 85 accept 6 or 7 rights (45%)[n] (see Table 5.6). The legal recognition of this right is more heterogeneous across countries than is the case for wives' civil rights. The lowest values are found in countries belonging to the African and Asian regions, but it is mainly 75% of the Middle Eastern countries that are in this category. Thus, among the countries that grant 3–5 economic rights to wives, we find half of the East African and Asian countries and a quarter of the OECD countries. Among the countries which grant 6–7 economic rights, we found 24% of the East Asian countries, 32% of the African countries, 72% of the OECD countries, and 92% of the East European and central Asian countries.

Laws against intimate partner violence

Another area of law that has undergone a profound transformation is that of violence against women by their male partners. The French Civil Code of 1804 stated that: "The conjugal duty authorizes the husband to use violence within the limits set by nature, morals and the law, as long as they are not acts contrary to the legitimate purpose of the marriage"

[n]Each response in favor of women's rights is counted as 1.

(Arnaud-Duc, 1992, p. 105). Even though this violence[o] had been criticized for a long time, it was the autonomous feminist movements of the 1970s that brought about a change, more than the action of left-wing parties, women legislators, or even national wealth (Htun & Weldon, 2018). In 1993, the United Nations General Assembly ratified the Declaration on the Elimination of Violence against Women, which is the first international instrument to explicitly combat it, emphasizing that "violence against women reflects historically unequal power relations between men and women, which have led to domination and discrimination by the former and hindered the promotion of the latter, and that it is one of the main social mechanisms to which women are subordinated to men." The struggle against physical or sexual violence by an intimate partner became the 17th Sustainable Development Goal adopted by the General Assembly of the United Nations in September 2015.

Local, national, and transnational mobilizations (Akhami, Erturk, & Mayer, 2018) have not been in vain. The number of countries with laws against violence by an intimate partner and other family violence has increased significantly. Moreover, the scope of the laws has been broadened, so that behaviors deemed as violent have become more inclusive (e.g., to include sexual and emotional abuse), and safety measures (e.g., order to prohibit contact, or removal of a spouse from the family home) have been introduced. Table 5.8 presents the number of countries which have enacted each of 14 provisos against intimate partner violence.

Intimate partner violence, whether physical and/or emotional, is now illegal in three quarters of countries. The possibility of issuing protection orders makes these laws potentially effective. Sexual violence is less often illegal, with 119 countries criminalizing and only 78 countries having specific criminal penalties against marital rape. Among the 111 countries without explicit laws against marital rape, 34 accept a complaint of rape against the husband. In total, 77 countries explicitly maintain the right of husbands to have sex with their wives whenever the man desires it, and wives have no recognized right to decline to have sex with their husband, essentially conveying ownership of the wife's body to husbands. Countries that make marital rape a special category are Latin America and the Caribbean (69% of these countries), rich OECD countries (50%), East Asia, Europe, and Central Asia (44%), but such laws are very scarce in the Middle East and North Africa (5%).

[o]Although universal, violence against women by an intimate partner during the year varies from 0.2% in Denmark to 40% in Equatorial Guinea and Zambia (United Nations, 2015).

Table 5.8 Number of countries and percentages with each of 14 provisos against domestic violence, $N = 189$ economies.

Law has provision	Yes	%
Is there domestic violence legislation?	144	76
If there is no domestic violence legislation, are there penalties for crimes against a spouse or family member?	9	5
Are there clear **criminal penalties** for domestic violence?	117	62
Does domestic violence legislation cover **emotional** violence?	142	75
Does domestic violence legislation cover **physical** violence?	142	75
Does domestic violence legislation cover **sexual** violence?	119	63
Does legislation explicitly criminalize **marital rape**?	78	41
If there is no law against marital rape, can a woman file a criminal complaint against her husband for rape?[a]	34	30
Does domestic violence legislation cover **economic** violence?	95	50
Do **protection orders** for domestic violence exist?	141	75
Do protection orders provide for **removal** of the perpetrator from the home?	130	69
Do protection orders cover **prohibition of contact** and maintaining distance from the survivor?	131	69
Does domestic violence legislation protect **former spouses**?	117	62
Does domestic violence legislation protect **unmarried** intimate partners?	69	37

[a]111 countries have no law against marital rape and the % in this row is of those 111 countries.
Based on World Bank Group. (2018). *Women, business and the law 2018*. Washington, DC: World Bank. https://wbl.worldbank.org/.

In total, 117 countries have laws protecting women against violence perpetrated by a previous intimate partner from whom the woman is separated, which is important as former partners often are perpetrators of intimate partner violence. However, laws protecting unmarried women cohabiting with a male partner exist in only 69 countries. This leaves a significant number of women without any legal protection, which is a major limit of the existing laws as rates of cohabitation have increased dramatically in many countries (Chapter 4), and rates of interpartner violence are often higher among cohabiting than married couples (Chapter 8).

In conclusion, there is a clear improvement in laws protecting women against violence by an intimate partner, but there is still a long way to go to provide global consistency of adequate protections. A quarter of countries have little or no legislation against domestic violence (5 or less of the 14 legal provisions against violence), 40% of countries have average policies (between 6 and 9), while only one-third of countries have strong legal frameworks protecting against violence (between 11 and 14 legal provisions).

Legal regimes about conjugal relationships

In order to better compare national achievement of the various types of legal rights cross-nationally, Table 5.9 presents the mean number of laws by countries of the different regions of the world guaranteeing married women civil rights, economic rights, and freedom from intimate partner violence. There is great variability in the number of these laws internationally. Some countries have very few laws protecting these rights. Four countries have four or less legal provisos relating to civil rights, while this is the case for 14% of countries for economic relations and 23% for policies against domestic violence. On the other hand, 68% of countries achieved the two highest values for civil rights, 50% for economic rights, and 35% for policies against domestic violence. However, there is no linear evolution from one right to another. Pakistan, for example, has implemented 7 civil rights provisos, 11 provisos against domestic violence, and only 1 economic rights measure. For the same total number of provisos, the distribution between the types of rights may be different from one country to another, some giving wives civil rights but not economic rights (Lebanon), or civil and economic rights but no protection against violence (Chile).

The number of legal provisos guaranteeing civil rights, economic rights, and freedom from intimate partner violence show medium to large correlation, $r = 0.653$ between civil rights and economic relations between spouses, $r = 0.480$ between civil rights and rights protecting against violence, and $r = 0.413$ between rights against violence and economic relations between spouses. In other words, countries that legally recognize one set of rights also tend to recognize other rights, but the statistical association is stronger for civil and economic rights than for provisos against intimate partner violence.

In all, 27 countries (14% of the 189 countries) do not recognize any rights for wives, while 55 (29%) implemented, based on existing data, nondiscriminatory marital policies. The most frequent situation (109 countries, or 58%) is some mix of discrimination against women in some areas, and legal provision for equality in other areas. More specifically, 14 countries (7.5%) have granted mainly civil rights, 29 countries (15%) promote economic equality for spouses, 48 other countries (25%) have measures against violence, and 18 countries (9.5%) allow measures for legal equality and measures to combat violence.

Table 5.9 Mean number of legal provisos providing wives with civil and economic rights, and freedom from interpartner violence, by region, $N = 189$ countries.

	East Asia	Europe and Central Asia	Rich countries of OECD	Latin America and Caribbean	Middle East and North Africa	South Asia	Sub-Saharan Africa	Total
Civil rights (15 measures assessed)	13.2	14.4	14.7	14.0	7.8	11.7	12.2	12.8
Economic rights (8 measures assessed)	4.2	6.2	5.9	5.5	1.5	2.2	4.4	4.6
Freedom from violence (11 measures assessed)	8.4	8.6	8.6	9.9	2.8	9.5	5.2	7.4
Total (of 34 measures assessed)	25.8	29.2	29.2	29.4	12.1	23.4	21.8	24.8

Based on World Bank Group. (2018). *Women, business and the law 2018*. Washington, DC: World Bank. https://wbl.worldbank.org/.

Different regions show marked differences in the legal status of married women's rights. Countries affording women the least overall legal protection of their rights are mainly in the Middle East and North Africa, although North African countries have more protection against domestic violence. Countries with coherent and strong legal systems to combat domestic violence exist across most regions, but there are notably low mean numbers of provisos in the MENA and sub-Saharan African regions. Countries that have taken the most legal measures for economic equality between spouses are mainly the former communist countries of Europe and Central Asia, OECD countries, and Latin America. Jamaica, Cambodia, and Laos have notably high legal protections for wives' economic equality. Finally, the countries with most laws protecting civil rights equality are mainly in Europe and Central Asia, OECD, countries and Latin America, with none of the regions being first in all types of rights.

Of course, these rankings are only for the rights included in the current study. Reproductive rights, notably the right of a woman to terminate a pregnancy, are often out of line with other legal protections of women's rights. For example, the United States ranks higher on legal protections of most women's rights, but has patchy legal protection of women's right to abortion.

Matrimonial rights

Laws on polygyny (the number allowed) and gender of spouses define two regimes of rights that seem mutually incompatible in all countries, except in South Africa. How do these rights to marriage combine with the rights of conjugality? Are there relationship between polygamy and same-sex marriages and the provisos admitting civil and economic rights to wives, laws against domestic violence, or total legal equality in marriage? Such relations would justify a classification of matrimonial rights combining those different elements. Would this typology be specific to regions as the preceding legal classifications (i.e., African or East Asian)?

Seven regimes of matrimonial laws in 2018

Table 5.10 presents the mean number of legal provisions supporting wives' rights by the acceptance of polygyny and same-sex marriage in those countries. As is evident, the legal acceptance of polygyny negatively correlates with the laws supporting women's civil, economic, and protection from violence rights. Moreover, those countries that allow polygyny for the whole

Table 5.10 Mean number of civil, economic, and domestic violence rights legal provision according by the legal recognition of polygynous marriage and same-sex union for $N = 189$ countries.

Laws on who can marry	Civil rights	Economic rights	Freedom from violence rights	Total rights	Number of countries
Polygyny	9.4	2.5	4.0	17.4	43
Partial polygyny	12.5	4.3	6.0	24.0	17
Monogamy	13.9	5.1	8.4	28.0	103
Same-sex union	14.7	6.0	8.7	30.0	26
Total	12.9	4.6	7.4	–	189

N.B. There is only country (South Africa) that gives any recognition to polygyny ***and*** to same-sex unions, in all other countries it is either one or the other recognition. In this table, monogamy refers to laws that recognize only monogamous heterosexual marriage; same-sex union refers to countries that recognize same-sex unions but not polygyny.

Based on World Bank Group. (2018). *Women, business and the law 2018*. Washington, DC: World Bank. https://wbl.worldbank.org/.

population tend to have fewer legal protections for women's rights than those countries that accept polygyny only for some of the population (e.g., laws that apply only to Muslims or customary marriages). The absence of laws specifying a minimum legal age of marriage are also negatively correlated with legal recognition of women's rights more broadly, but to a lesser extent than the negative correlation with recognition of polygamy. Countries that accept same-sex marriages typically have enacted more legal protections supporting wives' rights, and in particular half of those countries have legal recognition of a large number of economic rights for wives. These countries generally follow an approach in which marital property is viewed as belonging to individual partners. However, as the differences between countries with and without marriage for all are not huge, we will set them aside so as not to overcomplicate our presentation.

Based on the current review of legal provisos to support married women's rights, the 62 countries accepting polygyny fall into the categories of "no law favorable to wives," "civil rights," or "laws against violence against women." Only those countries where polygyny is partially accepted achieve a higher recognition of economic rights relative to those that accept polygyny universally. Countries where polygyny is recognized therefore combine the discrimination directly related to it with the low recognition of women's rights in marriage.

So, I suggest a typology of seven classes of matrimonial laws, as defined by the number of spouses allowed and "the rights and duties of spouses," including the combat against domestic violence. Two are "pure": totally "traditional polygamist patriarchal" or admitting all legal rights mentioned ("quasilegally equalitarian"). The other ones mix polygyny or monogamy with civil rights, economic rights, and combat against violence.

The "polygamist patriarchal" class consists of 27 countries (14% of the 189 countries, but 42% of legally polygamous countries), which recognize polygyny for the whole population and do not recognize any rights for wives. The term "intractable gender," the title of the book by the Moroccan philosopher Nadia Tazi (2018), seems suited to evoke this group. However, it is not specific to the countries of the Middle East, descendants of the desert lords, since it also applies to Malaysia, Afghanistan, and 11 countries in sub-Saharan Africa. The most populous of these countries is Iran. Among the states allowing polygyny, 21 countries grant wives civil rights (as in Indonesia), and are called "polygamous civil." A total of 13 other countries called "polygamous protective" states (Bangladesh, India, Pakistan, Philippines, or Nigeria) have taken significant legal measures for violence against women, which may be more protective of the family than transformative in the absence of civil rights for wives.

In contrast, no country rejecting polygamy has kept all its antiwomen discriminatory laws. They have more often laws that support wives' rights, even though, as we have just seen, they are very diverse. In all, 35 "monogamous civil" states admit a significant number of civil rights (among them Japan, United States, United Kingdom), they are more often of liberal traditions. In total, 19 "monogamous protective" countries have effective laws against violence. A total of 41 countries have a significant number of economic rights (e.g., China, Ethiopia, Germany). Many of these "monogamous economic" countries came from a communist tradition but many accept also same-sex marriage and seem to follow a more individualistic conception of marriage. A total of 32 "quasilegally equalitarian" countries (17%) have implemented nondiscriminatory conjugal laws. Of these, only eight got the highest score on the three types of laws. Except for Timor Leste, which is located in South East Asia, all countries are in the regions of Europe and Central Asia, OECD, or Latin America and the Caribbean (notably Brazil and Mexico).

Regional cultures of family law

Throughout this chapter, I have identified the regions of the world where different legal provisos exist about married women's rights. The differences between regions are always significant both for the variables taken one by one (e.g., wife's duty of obedience to the husband, acceptance of polygynous marriage, etc.) and for the typologies of legal system. This does not mean that all countries in the same region follow exactly the same legal provisos, as there are large differences between regions in the mean number of legal provisos protecting women's rights.

The Middle East and North Africa are the regions with the most homogenous legal provisos, maintaining its patriarchal model in all areas: age at marriage, polygamy, absence of civil rights, economic rights, and liberty from violence. It may help than gender inequalities seem endorsed by divine will but some provisos are not made obligatory by traditional Islamic law (Büchler & Schlatter, 2013). It must be added that this region is the one with greatest social inequalities (Piketty, 2019), so gender inequality, although specific, is part of this unequal share of power and wealth. However, North Africa distinguishes itself from the Middle East as three North African countries grant civil rights to wives, and Tunisia has abolished polygyny. Only two other countries distinguishes themselves depart from it: Djibouti granting civil rights to spouses and Lebanon enacting a comprehensive policy against violence, mainly, though, in a domestic and not gender framework (Edde, 2018).

In all, 75% of the eight very populous countries of South Asia (including India and Pakistan) have accepted polygamy but do have legal rights for protection against violence for wives. None of the South Asian countries provide any economic or total rights. So South Asian states seem to reflect a patriarchal but protective approach to women.

In contrast to the homogeneity of the Middle East and North Africa, and South Asian regions, the East Asia region is very diverse in its legal provisos protecting women's rights. A quarter of the countries accept polygynous marriage, and are split between countries that afford almost no rights to women at all, those that have legal provisos protecting civil rights, and those which have enacted laws against intimate partner violence. In the nonpolygynous states, the matrimonial system, the most accepted legal recognition of civil rights (with 36%), is followed by less frequent enactment of laws supporting wives' economic rights or laws against violence.

Sub-Saharan Africa countries also are heterogeneous with respect to women's rights. Polygyny is legal in half the countries of the region, and 21% of the countries have almost none or no laws favorable to wives' rights, while a few more countries extend civil rights to wives (28%) than countries that have laws protecting women against violence (15%). In the nonpolygynous African states, laws against violence (17%) are more frequent than economic rights (13%). There is a clear divide between countries in the west of this region, which tend to be more patriarchal, and countries in the eastern region, which tend to be more egalitarian.

The last three regions (Europe and Central Asia, the OECD, and Latin America and the Caribbean) all reject polygyny, but their laws are quite heterogeneous on other aspects of women's rights. The European and Central Asian region are split between two positions: 44% of countries have legal recognition of the economic rights of wives, and a similar percentage extends to recognition of all rights. This makes this region the highest ranking region for equality in family law, and four countries in this region got the maximum number of legal provisos that were assessed. The rich OECD countries got the same mean percentage of legal provision supporting women's economic rights as the European and Central Asian countries, but the OECD countries are more heterogeneous with respect to the recognition of other women's rights. Even Nordic countries, which often are celebrated for their gender equality, are not doing so well in some aspects of family law.

These countries were among the first to legislate for civil and economic rights for wives, but have been tardy in legislating to protect women against intimate partner violence. About one-third of the Latin American countries have legislated to support all the rights considered in the current study.

Conclusion

This chapter analyzed the cross-national patterns of legal recognition of married women's rights, based on the World Bank database of 2018. The analysis shows that the typology of a divide between divine law (often patriarchal) and humanist (egalitarian) law does not reflect the complexity of laws concerning women's rights. Even in the 27 countries where laws still give all power to the husband, only half claim to follow laws inspired by God while the other half, mostly in West African states, use civil law, whose origin is human and not divine.

"A special genius of each system" is no longer tenable as profound transformations are under way: limits have been placed on the age at marriage, same-sex marriage has been allowed in some countries, wives have acquired civil rights and, to a lesser extent, economic rights, as well as rights against domestic violence. Different developmental paths of laws have been taken, which give seven different regimes of family law, depending on whether they are polygamous or monogamous and the civil, economic, and antiviolence rights granted to wives. Countries that have the same total number of provisos favorable to women may be in different groups, which show the interest of this classification for a better understanding of family laws (in the broad acceptation of it). The impact of polygamy as a direct discrimination and as a factor in discriminative regimes has also been assessed. In contrast, same-sex marriage pertains to the monogamous economic rights group and does not form a group per se.

Overall, family law changes have moved toward more gender equality, and these changes are influenced by international migration and international conventions. However, only laws on the minimum age for marriage and the fight against domestic violence are among the sustainable development objectives that are supposed to guide the action of all governments. The other themes dealt with here are not part of those sustainable development objectives, even though all countries have ratified CEDAW and therefore its article 16 on equality in marriage. As an exemplar of this, low international interest in these areas of family law, data and research on polygamy laws are lacking. In addition, the World Bank has significantly reduced the size of its database on family law in 2019, further eroding the available information on family law across countries.

There is also a lot of resistance to framing laws that protect women's rights effectively. Some of the principles of protecting women's rights agreed to by international organizations include caveats and exemptions, or countries that enact laws lacking any penalties for infraction, so many countries have laws that are essentially empty frameworks. For example, in many countries the specified minimum legal age of marriage has provision for exceptions by way of parental or judicial permission, and there often are no penalties when laws about minimum marriageable age are broken. Another example is the laws against domestic violence, which in many countries limits the scope of behaviors classified as violence, reinforce patriarchal norms in the name of protecting the family, or again fail to provide for meaningful penalties for breaking the laws that do exist.

References

Akhami, A., Erturk, Y., & Mayer, A. E. *Feminist advocacy, family law and violence against women: International perspective*. (2018). London: Taylor & Francis Ltd.

Anno, C. Droit comparé: Vers une nouvelle classification des systèmes juridiques [Comparative law: Towards a new classification of legal systems]. *Le petit juriste*, (2015). 25 June www.lepetitjuriste.fr/droit-compare-vers-une-nouvelle-classification-des-systemes-juridiques/.

Arnaud-Duc, N. Les contradictions du droit [The contradictions of Law]. In Duby, G. & Perrot, M. (Eds.), [Women's history]Tome 4. *Histoire des femmes*. (1992). (pp. 87–116). Paris: Plon.

Arthur, M., Earle, A., Raub, A., Vincent, I., Atabay, E., Latz, I., … Heyman, J. (2018). Child marriage laws around the world: Minimum marriage age, legal exceptions, and gender disparities. *Journal of Women, Politics & Policy*, *39*(1), 51–74. https://doi.org/10.1080/1554477X.2017.1375786.

Banda, F. (2014). Gender discrimination and the right to family life. In J. Eekelaar (Ed.), *Routledge handbook of family law and policy* (pp. 297–313). London: Taylor and Francis.

Bereni, L., Debauche, A., Latour, E., & Revillard, A. (Eds.), (2009). Le droit à l'épreuve du genre: Les lois du genre (I). [The right to the test of gender: Gender laws]*Nouvelles Questions Féministes*, *28*(2), 4–79.

Boserup, E. (1970). *Woman's role in economic development*. London: Cromwell Press.

Büchler, A., & Schlatter, C. (2013). Marriage age in Islamic and contemporary Muslim family laws: A comparative survey. *Electronic Journal of Islamic and Middle Eastern Law*, *1*(2), 37–74. https://doi.org/10.5167/uzh-78204.

Cardi, C., & Devreux, A. Le genre et le droit: Une coproduction: Introduction [Gender and the law: A co-production: Introduction]. *Cahiers du Genre*, *57*(2), (2014). 5–18. https://doi.org/10.3917/cdge.057.0005.

Carolan, M. (2017). First marriage of Lebanese man with two wives recognised under Irish law. *The Irish Times*. 15 June https://www.irishtimes.com/news/crime-and-law/courts/supreme-court/first-marriage-of-lebanese-man-with-two-wives-recognised-under-irish-law-1.3120790.

Carter, W. B. (2013). The "Federal Law of Marriage": Deference, deviation, and DOMA. *American University Journal of Gender Social Policy and Law*, *21*(4), 705–795.

Code civil de la République gabonaise [Civil code of the Gabonese Republic]. (1995). *Libreville: Journal Officiel de la République gabonaise*. https://tbinternet.ohchr.org/Treaties/CEDAW/Shared%20Documents/1_Global/A_49_38(SUPP)_4733_E.pdf.

Committee on the Elimination of Discrimination Against Women. (1994). *General recommendation No. 21—Thirteenth session, 1994 equality in marriage and family relations*. https://tbinternet.ohchr.org/Treaties/CEDAW/Shared%20Documents/1_Global/A_49_38(SUPP)_4733_E.pdf.

Committee on the Elimination of Discrimination Against Women. (2013). *General recommendation on article 16 of the Convention on the Elimination of All Forms of Discrimination against Women (Economic consequences of marriage, family relations and their dissolution)*. New York. https://tbinternet.ohchr.org/_layouts/15/treatybodyexternal/Download.aspx?symbolno=CEDAW/C/GC/29&Lang=en.

Coontz, S. (2005). *Marriage, a history: From obedience to intimacy or how love conquered marriage*. NY:Viking.

David, R. *Traité élémentaire de droit civil comparé: introduction à l'étude des droits étrangers et à la méthode comparative* [Basic treatise on comparative civil law: Introduction to the study of foreign law and the comparative method]. (1950). Paris: Dalloz.

David, R., Goré, M., & Jauffret-Spinosi, C. *Les grands systèmes de droit contemporains* [The major contemporary legal systems]. (2016). Paris: Dalloz.

Deere, C. D., Oduro, A. D., & Swaminathan, H. (2013). Property rights and the gender distribution of wealth in Ecuador, Ghana and India. *The Journal of Economic Inequality*, *11*(2), 249–265. https://doi.org/10.1007/s10888-013-9241-z.

Delatouche, M. Les droits des femmes en pleine régression dans la Libye post-Kadhafi [Women's rights in regression in post-Gaddafi Libya]. *Orient-le Jour*, (2017). 03/03/2017 https://www.lorientlejour.com/article/1038348/les-droits-des-femmes-en-pleine-regression-dans-la-libye-post-kadhafi.html.

Dumetz, M. *Le droit du mariage en Côte d'Ivoire*. [Marriage law in Ivory Coast]. (1975). Paris: Librairie générale de droit et de jurisprudence.

Dupret, B. (Ed.), (2012). *La charia aujourd'hui. Usages de la référence au droit islamique*. [Sharia law today. Uses of the reference to Islamic law]Paris: La Découverte.

Edde, R. La lutte contre les violences conjugales au Liban: Une première avancée pour le droit des femmes [The fight against domestic violence in Lebanon: A first step forward for women's rights]. *Autrepart*, *85*, (2018). 109–132.

Encyclopedia Britannica. (2019). *Same-sex marriage*. https://www.britannica.com/topic/same-sex-marriage.

Gautier, A. Femmes et colonialisme [Women and colonialism]. In Ferro, M. (Ed.), *Le livre noir du colonialisme. XVIe-XXIe siècle: de l'extermination à la repentance*. [The black book of colonialism. 16th-21st century: From extermination to repentance](2003). (pp. 569–607). Paris: éditions Robert Laffont.

Gautier, A. (2005). Legal regulation of marital relations: Historical and comparative approach. *International Journal of Law, Policy and the Family*, *19*, 47–72. https://doi.org/10.1093/lawfam/ebi003.

Gautier, A. Las relaciones conyugales según las leyes y las prácticas en Yucatán [Marital relations according to law and practice in Yucatan]. In Guadarrama, M. E. & Mass, L. B. (Eds.), *Cuidado, violencia y desilguadad. Las mujeres entre el hacer y el deber ser*. [Care, violence and inequalities. Women between doing and duty to be](2017). (pp. 252–283). Veracruz et Oaxaca: Centro de Estudios de Género, Universidad Veracruzana; Instituto de Ciencias Educativas, Universidad Benito Juárez de Oaxaca.

Gautier, A. Cultura y contra cultura de la violencia de género en Yucatán [Culture and counter culture of gender violence in Yucatan]. In Ramos, T. M. (Ed.), *Ruralidades, cultura laboral y feminismos en el sureste de México*. [Ruralities, work culture and feminism in southeastern Mexico](2018). (pp. 251–270). San Cristobal de las Casas: ediciones del CESMECA, universidad de Ciencias y Artes de Chiapas.

Goody, J. (1973). Polygyny, economy and the role of women. In J. Goody (Ed.), *The character of kinship* (pp. 180–190). London: Cambridge University Press.
Guillaumin, C. (1992). *Sexe, race et pratique du pouvoir. L'idée de nature* [Gender, race and power practice. The idea of nature]. Paris: Côté-femmes.
Hallward-Driemeier, M., Hasan, T., & Rusu, A. B. (2013). *Women's legal rights over 50 years: Progress, stagnation or regression?* Policy research working paper Number WPS 6616Washington, DC: World Bank Group. http://documents.worldbank.org/curated/en/312581468338398063/Womens-legal-rights-over-50-years-progress-stagnation-or-regression.
Htun, M. S., & Weldon, L. (2012). *Sex equality in family law: Historical legacies, feminist activism, and religious power in 70 countries*. Washington, DC: World Bank. World Development Report 2012, Background Paper, 2011 https://openknowledge.worldbank.org/bitstream/handle/10986/9204/WDR2012-0006.pdf?sequence=1.
Htun, M. S., & Weldon, L. (2018). *The logics of gender justice: State action on women's rights around the world*. Cambridge: Cambridge University Press.
Hudson, V. M., Bowen, D. L., & Nielsen, P. L. (2011). What is the relationship between inequity in family law and violence against women? Approaching the issue of legal enclaves. *Politics and Gender, 7*, 453–492. https://doi.org/10.1017/S1743923X11000328.
Human Rights Committee. (2000). *General comment 28, Equality of rights between men and women (article 3)*. [Sixty-eighth session, 2000), United Nations Document CCPR/C/21/Rev.1/Add.10 (2000]Geneva: Office of the United Nations High Commissioner for Human Rights. http://hrlibrary.umn.edu/gencomm/hrcom28.htm.
Katmandu. (August 22, 2015). Katmandu sees rise in polygamy cases. *Himalaya Times*.
Lacombe, D. Légiférer Sur les « violences de genre » tout en préservant l'ordre patriarcal. L'exemple du Nicaragua (1990-2017) [Legislate on gender violence while preserving the patriarchical order. The example of Nicaragua]. *Droit et société, 99*(2), 287–303. https://www.cairn.info/revue-droit-et-societe-2018-2-page-287.htm.
Malik, M. (2014). Family law in diverse societies. In J. Eekelaar (Ed.), *Routledge handbook of family law and policy* (pp. 424–442). London: Taylor and Francis.
Marçal, K. *Le dîner d'Adam Smith. Comment le libéralisme a zappé les femmes et pourquoi c'est un gros problème* [Adam Smith's dinner. How liberalism has overlooked women and why it is a big problem]. (2019). Paris: Les arènes.
Marcoux, R., & Antoine, P. *Le mariage en Afrique: Pluralité des formes et des modèles matrimoniaux* [Marriage in Africa: Plurality of matrimonial forms and patterns]. (2014). Laval: Presses de l'Université du Québec.
Mendos, L. R. (2019). *State-sponsored homophobia 2019*. Geneva: International Lesbian, Gay, Bisexual, Trans and Intersex Association. https://ilga.org/downloads/ILGA_State_Sponsored_Homophobia_2019.pdf.
Merin, Y. (2002). *Equality for same-sex couples*. Chicago: University of Chicago Press.
Miles, J., & Scherpe, J. (2014). The legal consequences of dissolution: Property and financial support between spouses. In J. Eekelaar (Ed.), *Routledge handbook of family law and policy* (pp. 138–152). London: Taylor and Francis.
Molyneux, M. (1985). Family reform in socialist states: The hidden agenda. *Feminist Review, 21*(1), 47–64. https://doi.org/10.1057/fr.1985.28.
Naciri, R. Le mouvement des femmes au Maroc [Women's movement in Morocco]. *Nouvelles Questions Féministes, 33*(2), (2014). 43–64. https://doi.org/10.3917/nqf.332.0043.
Organization for Economic Cooperation and Development. Social Institutions Gender Index (OECD-SIGI). (2019). *Global report: Transforming challenges into opportunities, Social Institutions and Gender Index*. Paris: OECD. https://doi.org/10.1787/bc56d212-en.
Organization for Economic Cooperation and Development. Social Institutions Gender Index (OECD-SIGI). *Database*. https://www.genderindex.org/.
Pierret, C. (2019). *La Guinée adopte un nouveau Code civil faisant de la monogamie la règle* [Guinea adopts a new Civil Code making monogamy the rule]http://www.rfi.fr/afrique/20190510-guinee-le-nouveau-code-civil-introduit-article-monogamie.

Piketty, T. (2019). *Capital et ideologie* [Capital and ideology]. Paris: Le Seuil.

Polygamie. (2017). Polygamie – Avec ses 2 épouses, le ministre d'Etat Mustapha Ramid embarrasse le Maroc, *article 19.ma*, 12 avril 2017. http://article19.ma/accueil/archives/70384 (Accessed 15 October 2019).

Rabel, E. (1958). *The conflict of laws: A comparative study*. Ann Arbor, MI: University of Michigan Press. https://repository.law.umich.edu/michigan_legal_studies/11/.

Shalleck, A. (2011). Introduction to comparative family law: What is the global family? Family law in decolonization, modernization and globalization. *American University Journal of Gender Social Policy and Law*, *19*(2), 449–458.

Starkweather, K. E., & Hames, R. (2012). A survey of non-classical polyandry. *Human Nature*, *23*(2), 149–172. https://doi.org/10.1007/s12110-012-9144-x.

Tazi, N. (2018). Le Genre intraitable. Politiques de la virilité dans le monde musulman. Arles, Actes Sud, coll. Questions de société.

United Nations. (1962). *Convention on consent to marriage, minimum age for marriage and registration of marriages opened for signature and ratification by General Assembly resolution 1763 A (XVII) of 7 November 1962. Entry into force: 9 December 1964, in accordance with article 6*. New York: Office of the United Nations High Commissioner for Human Rights. https://www.ohchr.org/Documents/ProfessionalInterest/convention.pdf.

United Nations. (1979). *Convention on the elimination of all forms of discrimination against women New York, 18 December 1979*. New York: Office of the United Nations High Commissioner for Human Rights. https://www.ohchr.org/EN/ProfessionalInterest/Pages/CEDAW.aspx.

United Nations. (1989). *Convention on the Rights of the Child Adopted and opened for signature, ratification and accession by General Assembly resolution 44/25 of 20 November 1989 entry into force 2 September 1990, in accordance with article 49*. New York: Office of the United Nations High Commissioner for Human Rights. https://www.ohchr.org/en/professionalinterest/pages/crc.aspx.

United Nations. (2015). *The world's women 2015: Trends and statistics*. New York: United Nations, Department of Economic and Social Affairs, Statistics Division. https://unstats.un.org/unsd/gender/downloads/worldswomen2015_report.pdf.

United Nations. (2016). *The Sustainable Development Goals Report 2016*. New York: Department of Economic and Social Affair. https://unstats.un.org/sdgs/report/2016/The%20Sustainable%20Development%20Goals%20Report%202016.pdf.

United Nations, Child Fund. (2019). *2018 annual report: UNFPA-UNICEF global programme to accelerate action to end child marriage*. https://www.unicef.org/media/60276/file.

United Nations High Commissioner for Refugees. (2019). *Report of the United Nations High Commissioner for Refugees*. https://www.refworld.org/docid/3b00f1d454.html.

Van Eeden-Moorefield, B., & Alvarez, A. (2015). The social, political, and judicial contexts of queer marriage in the United States. *Journal of GLBT Family Studies*, *11*(1), 57–81. https://doi.org/10.1080/1550428X.2014.883560.

Wallbank, J., Choudhry, S., & Herring, J. (2012). *Rights, gender and family law*. London: Routledge-Cavendish.

Wikipedia. (2019a). *Legality of polygamy*. https://en.wikipedia.org/wiki/Legality_of_polygamy.

Wikipedia. (2019b). *Same sex marriage*. https://en.wikipedia.org/wiki/Same-sex_marriage.

Winter, B., Forrest, M., & Sénac, R. (2018). *Global perspectives on same-sex marriage. A neo-institutional approach*. New York: Palgrave Macmillan.

Womanstats. (2019). http://www.womanstats.org/.

World Bank Group. (2018). *Women, business and the law 2018*. Washington, DC: World Bank. https://wbl.worldbank.org/.

Zeitzen, M. K. (2008). *Polygamy: A cross-cultural analysis*. New York: Oxford University Press.

CHAPTER 6

The intra-spousal balance of power within the family: Cross-cultural evidence☆

Pierre-André Chiappori[a] **and José Alberto Molina**[b,c]
[a]Department of Economics, Columbia University, New York, NY, United States
[b]Department of Economics, University of Zaragoza, Zaragoza, Spain
[c]IZA Institute of Labor Economics, Bonn, Germany

This chapter adopts an economic perspective to analyze cross-cultural evidence of the intra-family spousal balance of power. The family unit is, according to the Nobel Laureate economist Gary Becker, the decision-making unit linked through kinship that constitutes society's greatest treasure in terms of its capacity to generate individual and social welfare. The economic studies of the family were originally predominantly descriptive, but in recent decades we utilized rigorous economic models that develop detailed analyses of both internal and external influences on family power and distribution of resources.

We wish to emphasize that the detailed model-based study of the socio-economic behavior of families in society is a way of looking at the social consequences of economic constraints on individual behavior. In this chapter, we concentrate on the basic idea that there are substantial benefits to individuals from living in couples. Individuals gain from sharing goods within the household and can take advantage of specialization of roles, resulting in a comparative advantage in home and/or market production. For example, individuals in a household can pool income risk, share in each other's gains and insure one another against shocks (e.g., poor health or job loss). Most importantly, individuals in a household can jointly invest in their children.

☆This chapter was written while Jose Alberto Molina was a visiting professor at the Department of Economics of Boston College in the United States. He expresses his thanks for the hospitality and facilities provided. The chapter writing has been partially supported by the Government of Aragon (Project S32-17R) and FEDER 2014–20.

Cross-Cultural Family Research and Practice
https://doi.org/10.1016/B978-0-12-815493-9.00006-5

Family empirical analyses traditionally have been based on the so-called unitary model, which assumes that family members maximize a single utility function for the family unit (Becker, 1981; Samuelson, 1956). That is, the modeling of the distribution of family resources is based on the assumption that the family acts as a single unit with specific and well-defined preferences. Various justifications have been given for that assumption. For Samuelson, individual preferences are merged at marriage into a common family index. However, as pointed out by Becker, the common index is simply postulated as existing in a somewhat ad hoc fashion; nothing is said about how individuals generate that common index. In particular, Samuelson explicitly posits that the family utility index is independent of individual incomes, meaning that how much one spouse earns relative to the other has no effect on the expenditure of the shared family income. This assumption is central to the traditional models of consumer theory that apply at the household level but is hard to reconcile with any representation of family decision-making process that entails bargaining between members. Becker himself, in his well-known "Rotten Kid" theorem, provides an alternative justification of the unitary model based on game theory. However, his approach relies on two key assumptions regarding the distribution of power within the family. First, there is a "benevolent patriarch" who is the ultimate decision-maker, which seems increasingly doubtful as a model in many cultures in the 21st century. Second, utility is transferable in such a way that one individual can transfer utility or value to another one in a costless way, which requires that individuals have a common currency that is valued equally by all (see Bergstrom, 1989).

The unitary model has numerous implications for how families behave economically. In particular, it requires that individual members pool their income in achieving family outcomes in such a way that the identity of the income recipient is of no consequence for the family's expenditure. In this traditional unitary model, neither the identity of the income recipient nor the income source (e.g., paid employment, government welfare payment) has any effect on the family's expenditure allocation; all that matters is total family income. This restriction has raised some questions about the validity of the unitary model. Empirically, many modeling studies of actual family expenditure have found the unitary model does not fit patterns of observed family expenditure (see, e.g., Attanasio & Lechene, 2002; Duflo, 2003; Hoddinott & Haddad, 1995; Kanbur & Haddad, 1994; Lundberg & Pollak, 1994; Lundberg, Pollak, & Wales, 1997; Maitra & Ray, 2002, 2003; Phipps & Burton, 1998; Quisumbing & Maluccio, 2003; Schultz, 1990; Thomas, 1990). On the theory side, it essentially ignores the existence of potentially complex decision processes within the household and disregards intra-family inequality.

Consequently, the unitary model has been increasingly replaced in recent decades by attempts to model individual utility to incorporate the different preferences of family members, with the relative power of the individual being crucial to predicting how family income is spent in these nonunitary models (Pollak, 1994). Recent economic theory about family power typically includes attempts to model individual utility (i.e., to incorporate the divergent and conflicting preferences of different family members), and include either explicit or axiomatic representation of intra-household decision processes. More general models of family decision-making share a common characteristic: they take the individual as the basic element and treat family decisions as the outcomes of interactions among the members.

Two strands of economic models of family economic decision-making have emerged in the literature. The first strand, the so-called collective approach[a] (Browning & Chiappori, 1998; Chiappori, 1988, 1992; Donni & Chiappori, 2011; Donni & Molina, 2018), is based on cooperative game theory. It either models the decision as the solution to an explicitly described bargaining game (Manser & Brown, 1980; McElroy & Horney, 1981), or simply postulate it as the outcome of negotiations that achieve efficiency, in the sense that no alternative choice would have been preferred by *all* family members. Note that, since bargaining games typically postulate economic efficiency in the distribution of resources, the first specification is a particular case of the second. A second direction refers to noncooperative game theory, in which it is assumed individuals act in self-interest and there is no possibility to manage alliances, with this theory being specifically applied to the topic of private contributions to public goods (Doepke & Tertilt, 2018; Konrad & Lommerud, 1995). Lastly, some works try to establish a bridge between the two approaches, for instance by taking the noncooperative solution as the status quo point for a cooperative bargaining model (Basu, 2006; Kanbur & Haddad, 1994; Lundberg & Pollak, 1994).

There is a significant literature, both historical and recent, in sociology and anthropology arguing that spousal power is important in the analysis of family economic behavior (see, e.g., Blumberg & Coleman, 1989; Desai & Jain, 1994; Kousta, 2019; Pollak, 1994; Riley, 1997; Slatton & Brailey, 2019). Here, we focus on the economic empirical analyses that have estimated the determinants of spousal power and have quantified the magnitude of its impact on the various expenditure patterns of the family.

This chapter reviews evidence of the intra-spousal balance of power in the nonunitary models, with the collective models *à la* Chiappori being

[a]See Browning, Chiappori, and Weiss (2014) for a general overview.

the most used. An immediate advantage of the collective approach is that it provides a direct and tractable translation of the notion of power. Theory shows that any efficient decision must maximize a weighted sum of individual utilities; if one accepts this representation, then there exists a direct relationship between the corresponding weights (often referred to as Pareto weights) and the respective powers of household members. Hence, unlike the traditional unitary model, any change in a variable that affects the balance of power in the family can impact economic decisions of the family (consumption, the extent of participation of individual family members in the paid labor force, etc.), even when the family's total income and all prices remain unchanged. In these collective models, the relative power of family members plays a central role in shaping family decisions. Spouses derive power from multiple sources, many of which reflect the options an individual has outside the family. We will see during this chapter that many studies have considered the income-generating capacity of husbands and wives as measures of their relative power. Other studies have relied on nonlabor income, or the value of assets, or other physical or social characteristics of individuals, as alternative determinants of spousal power. Finally, a complementary approach investigates how the intra-household allocation of power is related to global factors, such as legislation governing divorce or property rights, or the sex ratio of men to women which can be seen as reflecting the available supply of potential spouses on the relevant "market for marriage."

Given the objective of analyzing cross-cultural evidence of the intra-family balance of power, we have identified both socioeconomically highly developed and less developed areas around the world, using the United Nations Human Development Index (HDI) (2018). According to assessment on the HDI, the United States, Canada, Australia, New Zealand, most of Western Europe, South Korea, Chile, Mexico, and Colombia are included in the Very High Human Development category; South Africa and Indonesia are at the top of the Medium Human Development category, while countries from South-Asia, such as Nepal and Pakistan, and from sub-Saharan Africa are either at the bottom of the Medium Human Development group, or in the Low Human Development category.

The rest of the paper is organized as follows. The next section summarizes the most relevant international evidence in developed cultural areas. The section after that is dedicated to evidence of the intra-spousal balance of power in developing regions. The studies that form the evidence base to be presented in these two sections are summarized in Table 6.1. Finally, the paper includes a summary of the most important conclusions.

Table 6.1 Chronological evidence on the intra-marriage balance of power.

Year	Authors	Country	Data	Keywords
2002	Basu and Ray	Nepal	NLSS/1995	Balance of Power, Child Labor, Woman's Power
2002	Rubalcava, Teruel, and Thomas	Mexico	PROGRESA	Woman's Power, Consumption Time, Welfare
2002	Thomas, Contreras, and Frankenberg	Indonesia	IFLS/1993–97	Relative Power, Intra-Family, Child Health
2003	Andreoni, Brown, and Rischall	United States	Gallup Organization/1992–94	Charitable Giving, Bargaining, Relative Power
2003	Duflo	South Africa	SALDRU/1993	Public Transfers, Programs, Spousal Recipient
2003	Maitra and Ray	Australia	AHES/1998–99	Spousal Power, Expenditure, Intra-Family
2003	Quisumbing and Maluccio	Bangladesh, South Africa, Ethiopia, Indonesia,	Various/1996–98	Marriage, Intra-family, Bargaining Power
2004	Lundberg and Ward-Batts	United States	HRS/1992	Bargaining, Spouses' Control, Net Worth
2005	Maitra and Ray	Australia	AHES/1998–99	Simultaneous Expenditure, Bargaining Power
2006	Lancaster, Maitra, and Ray	India	NSS/1993–94	Endogeneity, Bargaining Power, Expenditure
2007	Lee	South Korea	KHPS/1993–98	Consumption, Spouses Power, Dynamic
2011	de Palma, Picard, and Ziegelmeyer	Germany	Sessions/2005	Balance of Power, Experiments, Intra-Family
2013	Browning, Chiappori, and Lewel	Canada	FAMEX/1974–92	Consumption, Economies of Scale, Bargaining Power
2013	Fujii and Ishikawa	Japan	JPSC/1994–2002	Childbirth, Intra-household, Expenditure Share
2013	Martínez	Chile	CASEN/1990–92–94–96–2000	Intra-family, Bargaining Power
2015	Couprie and Ferrant	United Kingdom	UKTUS/2000	Welfare, Time Use, Bargaining Power
2015	Yilmazer and Lich	United States	HRS/1992–2006	Portfolio Choice, Risky Assets, Bargaining Power
2017	Porter	China	CLHLS/1998–2009	Spousal Bargaining Care, Elderly Parents
2018	Campaña, Giménez-Nadal, and Molina	Mexico Colombia	Time Use Surveys: Mexico (2009) Colombia (2012)	Efficiency, Labor Supply, Time Use, Bargaining Power

The intra-family balance of power: Cross-cultural evidence in developed areas

Cultural regions are geographical areas in which individuals share common cultural characteristics (Gupta & Ferguson, 1997), and we have identified seven developed areas which show a very high or a top medium HDI; namely, the United States and Canada, Australia, Europe, East-Asia, Latin America, South Africa, and Indonesia, where economic analyses of spousal power in nonunitary frameworks have been developed.

The United States and Canada

The United States and Canada are often considered as belonging to the same cultural region. The majority language in both countries is English, and both countries were once British colonies. The majority in both countries follows the Christian religion, both were once home to Native Americans, and had very large spaces in the West which were gradually settled by pioneers. Citizens in both countries follow similar occupations and enjoy similar ways and standards of living.

In the case of the United States, numerous authors have found evidence against the hypothesis that a married couple behaves as if governed by a single utility function, and instead found couple behavior is more consistent with a household influenced by two individual utility functions (Andreoni, Brown, & Rischall, 2003; Lundberg & Ward-Batts, 2004; Yilmazer & Lich, 2015). Expenditures on consumer goods by the family depend on the bargaining between spouses about their different preferences (utilities) for expenditure in different categories of goods. Some goods show a clear gender difference in their utility (e.g., in the United States on average women tend to value and spend more on clothing, men tend to value and spend more on leisure activities). However, the bulk of family consumption goods take the form of collective or shared goods (that economists describe as public goods) that can be seen to benefit both spouses. One such public good is the family's charitable giving. Andreoni et al. (2003) have identified systematic gender differences with respect to charitable giving. Relative to single men, single women tend to give more money overall to charities, give to a greater variety of charitable entities, and give less to each charity. Andreoni and colleagues have shown that, when a particular spouse has sole control, the family decisions seem to reflect his or her own tastes. But, when decisions are made jointly, husbands seem to be getting more of what they want than wives. The authors also found that the probability that the wife controls decisions decreases when her husband earns more than she does, or when he is more highly educated than her, with these two measures (income/earnings and education) being generally associated with high bargaining power. Although the interspousal bargaining on charity mostly favors men when it comes to total giving, when women do become the decision-makers, they wield their power to influence the disbursement of the family's charity. Thus, the authors provide direct evidence to support the growing feeling among fundraisers that men and women behave very differently with respect to charitable giving.

Husbands and wives can also have different individual preferences in savings and wealth accumulation, which must be resolved through the household decision process. Lundberg and Ward-Batts (2004) use a collective framework, which includes indicators of the relative control over resources of husbands and wives, to show that the wife's long-run relative power over household decisions is positively associated with household net worth (i.e., wives have more say in wealthy than less wealthy households). Interestingly, households in which the husband has substantially more education than his wife appear to have significantly lower net worth than households in which both spouses have high levels of education.

There is limited research on how intra-family financial decisions are made. Prior studies suggest that women invest their asset portfolios more conservatively than men, but it is not gender alone that drives the investment decisions of men and women, but rather a combination of gender and marital status. Yilmazer and Lich (2015) present an analysis of the allocation of assets in family financial wealth, using a nonunitary model where the spouses differ in their risk preferences. Using data from the Human Retirement Survey (HRS) they show that families invest more in risky assets when the spouse who has greater bargaining power has higher risk tolerance. As there seems to be no evidence that bargaining power is related to risk tolerance, this study provides evidence that the risk preference of the decision-maker in the family plays an important role in household portfolio decisions. The decision-maker is, most of the time, the more educated partner. However, the literature does not indicate whether the spouse who makes the decision consults, or not, with the other spouse. It can be argued that couples simply think it is wise to rely on the person with the higher education, and that this is not just reflecting power or bargaining power of the spouses. However, when the more educated partner gets to make a decision that is in line with his or her preferences, it is still a distributional factor that results in more weight being placed on his or her preferences.

This would matter for welfare analysis and predictions regarding the investment behavior of two-person households.

In the case of Canada, evidence using consumption data on couples and singles supports the collective model of couple economic decision-making, and also demonstrate the economic efficiencies of being in a couple relative to being single. In a recent study by Browning, Chiappori, and Lewbel (2013) of Canadian data they show that couples spend about one-third less in total expenditure per head than singles, with the savings coming from efficiencies brought about by the couples' shared consumption of goods and services. For example, heating, lighting, food, rent, and transport all tend to be cheaper per head when sharing with a partner relative to the costs for a single person. Single adults need to spend between one half to three quarters more per person to attain the same standard of living for themselves as for the members of a two-person household.

Australia

Australia is similar in many ways to the United States and Canada, although it is unique in occupying an entire continent, and with having an Indigenous population who have occupied the land for more than 60,000 years. British settlers established the first colony in 1788, and even today the arid interior of Australia is sparsely populated. The South and East coasts have large cities, with these areas having a prosperous way of life.

As has been demonstrated in research in other regions, female income-share has significant effects on family expenditure. Maitra and Ray (2005a, 2005b) provide evidence from Australia of the effect of spousal power on family expenditure patterns, exploiting a useful feature of the collective model, which allows both weights of the spousal utilities to be identified from the data, rather than assumed a priori. In this context, weights of spousal power were estimated to depend not only on the share of total household earnings each spouse contributed, which has been conventionally used, but also on additional variables such as the educational experience of the adults, family income, and family size. The overall message from these results is that the bargaining power of each spouse does affect household expenditure patterns substantially and that the income-pooling hypothesis underlying the unitary model does not fit the data. The paper provides evidence that the bargaining power of a spouse does not show a simple, consistent monotonic relationship to decisions about all expenditure items, but rather the association of spousal power with decisions can vary from one specific good to another. For example, the male spouse might have a relatively high power on decisions over expenditure on cars and holidays, and the female spouse has high power over expenditure on food.

Europe

The Greek culture has had a lasting impact on the West. Much of Europe was later united as part of the Roman Empire, but following its decline and fall, Europe became divided into a series of warring Kingdoms, with their own languages and cultures. Nevertheless, Europe remained united by versions of the Christian religion. The late 15th century saw Europe become a global leader during the Renaissance, and the region remained the most powerful and technologically advanced for the next five centuries. European nations colonized vast regions of the world and spread their technology and many aspects of their way of life. Europeans then fought two major global wars, but today most of the constituent nations are united in the European Union. European evidence favors the collective model, with relevant countries analyzed in terms of the intra-family balance of power (see, e.g., Couprie & Ferrant, 2015, for the United Kingdom; de Palma, Picard, & Ziegelmeyer, 2011, for Germany).

The process by which individual partner's use of time (e.g., paid work, unpaid work like household task and child care, leisure time) are transformed into family welfare (subjective well-being) implies negotiation between the spouses. Couprie and Ferrant (2015) define a family utility function as a weighted sum of family members' individual subutility functions, with the family behavior taken to be Pareto-efficient. Pareto-efficient means that the pattern of expenditure is optimized to address the utilities of the spouses, with the utilities weighted by spousal power. Any change from Pareto-efficiency in expenditure could only improve the situation of one spouse by making the situation of the other spouse worse in a way that detracted from the optimization of utilities weighted by spousal power. The authors model economies of scale in time use, based on a UK subsample that is restricted to singles and couples without children. The results bring important insights on how time use and intra-family bargaining could affect the analysis of the individual well-being of each individual. Two singles living apart require more spare time a day to achieve the same well-being as when living in a couple. The well-being derived from individual time use activities tends to be increased by living as a couple, on average. The authors also demonstrate that the effects on individual well-being due to interactions in time use are important, even when considering pure leisure.

A single woman (man) requires 110% (104%) of the time a partnered woman (man) in a couple time-resources to attain the same level of self-reported well-being as those in a couple relationship. Individuals can benefit unequally from economies of scale, such that living in a couple allows time to be saved. By living in a couple a female (male), on average, requires 22 (20) minutes per day of housework and 72 (60) minutes of leisure less to attain the same level of well-being as a single person, with these savings depending on the characteristics of the couple.

A growing literature using experimental research in economics shows that family financial investments are significantly affected by how decision-making power is allocated between men and women. de Palma et al. (2011) report German results on individual and couple choices, in an experiment involving risk. Individuals and couples make binary choices between a large payoff that might occur from a lottery or a smaller guaranteed payoff. In the first part of the experiment, spouses are separated and independently choose the payoff they prefer. Individual choices express individual risk preferences. In the second part of the experiment, spouses rejoin their partner, and they make joint choices. In most cases, the couple implemented the man's choices, which express collective risk preferences. The authors investigate the evolution of the balance of power within the family via the individual risk aversions estimated from choices made in the first part of the experiment, to explain the couple's risk aversions estimated from choices made in the second part. The man is initially more successful than the woman in influencing couple choices, but that woman progressively gains power over the course of protracted decision-making processes, and the authors speculate that part of this power increase relates to the ultimate control over the implementation of joint choices. This evidence on the dynamics of the power balance does not identify whether the man or the woman is particularly powerful in specific decisions and, additionally, it suggests that actual decision processes within the family are adaptive and depend on the context.

East-Asia

While culturally different from the West, East-Asia is far from homogeneous; different groups in East-Asia speak different languages, use different systems of writing, and share distinctive religious beliefs (Confucianism, Taoism, Buddhism, Islam, among others). Since opening their economies to foreign investment, this area has undergone an economic revolution, placing the region among the world's fastest-growing economies.

Prior empirical evidence using cross-sectional data has suggested that the interspousal balance of power within couples is relatively stable over time. Lee (2007) uses longitudinal data on assignable private consumption in South Korea, to examine the marital balance of power between spouses in a dynamic collective model. The data allows the author to account for unobserved heterogeneity at the family level and time-constant unobserved bargaining power. A comprehensive framework of marriage and intra-family resource allocation suggests that the balance of power is partly determined by spousal matching and that it is robust over time as long as the marriage is sustained. Results also suggest that relative earning capacities at the time of marriage may be a good proxy for the spouses' balance of power. However, temporal changes in relative earning capacities within marriage (e.g., the woman being promoted to a higher pay rate than her husband across time) do not appear to induce any significant resource transfer between spouses. This finding indicates that relationship commitment plays a significant role in the dynamics of marital relationships, and allows, in particular, a significant degree of mutual insurance within the family. It also suggests that, at least in the case of South Korea, as individuals sort themselves into marriages, the balance of power is largely shaped at that moment, and continues to affect intra-family allocations in the long run.

Given that childbirth is an important event for many couples, it is important to analyze how the couple manages the new expenses from childbirth, and how both spouses modify their own expenditures. Fujii and Ishikawa (2013) investigated the impact of a new child on intra-household allocation for married Japanese couples, estimating a collective model of the household by using nine rounds of the Japanese Panel Survey of Consumers (JPSC) between 1994 and 2002. The wife's share of private expenditure by the family is lower in households with more children. Specifically, each additional child is associated with a decrease of at least 2% points in a woman's share of the private goods purchased for her by the family.

Women in China play a substantial role in caring for aging parents. Porter (2017) uses the Chinese Longitudinal Healthy Longevity Survey (CLHLS) from 1998 to 2009 to estimate a collective model where the bargaining issue is the time devoted to one's own parents. Evidence shows increases in women's bargaining power within couple relationships due to demographic changes, which has altered social support time use of couples. The demographic changes are the ratio of men to women in the population, which has got higher, and the economic power of men within relationships has declined, presumably because their capacity to find an alternative spouse has declined. The increase in female spousal power is associated with couples providing greater support time for the wife's parents and less support time for the husband's parents.

Developed Latin America

We review evidence on economic decision-making in three developed Latin American countries, Mexico, Chile, and Colombia, with all three countries being included in the High Human Development category of the UN Human Development Index (2018). The Latin American cultural area received the name because it was once colonized by Spain and Portugal, whose languages are based on Latin. Thus, most Latin Americans speak Spanish or Portuguese, follow the Roman Catholic religion, and share a common history and many traditions.

Rubalcava, Teruel, and Thomas (2002) measured the effect in Mexico of a large, exogenous increase in resources distributed to women by government policy on family expenditure. This exogenous increase avoided the complexities associated with modeling increases in resources to women that might result from the woman changing her level of involvement in paid employment. PROGRESA (*Programa de Educación, Salud y Alimentación [Education, Health, and Food Program]*) is a large national program in Mexico aimed at developing the human capital of poor households, which provides money and in-kind benefits to reduce poverty and increase family investments in health, education, and nutrition. PROGRESA payments are made to women. Holding total household resources constant, an increase in income from PROGRESA can be interpreted as an increase in the share of total household resources that are attributed to the woman who receives the PROGRESA income. The authors have interpreted this exogenous shift in the attribution of income within the family as an increase in the bargaining power of the woman, relative to other family members. Estimation of the effect of this income on spending indicates that, as the share of household resources from PROGRESA increases, the share of the budget spent on children's clothing, education, and a higher quality diet increase. The share of the budget spent on adult clothing, transport, and staples declines. The authors conclude that there has been a shift in the balance of power within PROGRESA families that has resulted in greater investment in the human capital of the next generation. The results suggest that the impact of the income transfers to families has been greater than it would otherwise have been if the income had been given to a male in the family.

The PROGRESA intervention has also been studied by Attanasio and Lechene (2014). They find that the unitary model of family expenditure did not fit the data; the PROGRESA program affected the share of expenditure going to different family members even after controlling for the effects of the additional total resources given to the eligible households. Importantly, the authors also find that the relative strength of the family network of household members shapes household choices, supporting the view that several factors, including the importance of social networks, may affect the intra-family balance of power. Last but by no means least, the presence of two distribution factors (variables that affect expenditure shares while not affecting preferences or budget constraints) allows direct testing the efficiency hypothesis of the collective model, which predicts that the two distribution factors should affect expenditure shares in a proportional fashion (Browning, Bourguignon, & Chiappori, 2009). These authors find that this efficiency hypothesis underlying the collective framework fits the data.

In a similar political change that influenced family dynamics to that of Mexico, in 1999 Chile increased government-funded child support for children of nonmarried couples. As in Mexico, the Chilean Government directed payments to mothers. Martínez (2013) took advantage of this political change as a source of exogenous variations in bargaining power within families to estimate a nonunitary bargaining model using a series of cross-sectional data. Interestingly, Martnez found as child support increased, there was a decrease of 1.8% points in father's rate of employment, as well as a decrease in the fathers' mean hours worked. These results highlight the importance of considering changing dynamic family interactions when analyzing the impact of public policies, as there might be unintended consequences from changes in incentives and power within the family.

After the introduction of the Chilean law increasing child support payments, there was an increase in school attendance of children between 14 and 15 years (Martínez, 2013). This can be interpreted as a movement toward the preferences of women in cohabiting relationships as their bargaining power increased assuming that mothers tend to prioritize education for children more than fathers. The 1999 law increased women's options if the relationship ended, and the evidence is consistent with women using this increase in power to further their children's education. It is important to point out that, although the legal change increased school attendance, it did not affect either mothers' participation in the labor market or their hours of paid work to any significant extent. This lack of effect on women's paid work could be caused by labor market conditions or by cultural roles that preclude female labor force participation. If children are more likely to be in school, and mothers do not increase their work participation, it is likely mothers would have more free time, which would increase their well-being, providing that they value leisure. This research shows that legal change can increase children's well-being through indirect channels, specifically by improving women's economic bargaining power within their families.

Campaña, Giménez-Nadal, and Molina (2018) use time diary data from Mexico (2009) and Colombia (2012) to evaluate whether intra-household bargaining power, among other factors, is an important variable influencing the paid work participation of spouses. In doing so, they use the gender ratio (the number of men per 100 women) as a distribution factor, i.e., as an external factor that may directly affect the balance of power within households. The underlying idea, dating back to Becker (at least), is that individuals of the gender in short supply find themselves in a better bargaining position within a relationship. Hence, as the ratio of men to women increases, women's economic bargaining power within the family is likely to increase. Labor supply decisions appear to be efficient (in the Pareto sense: no alternative choice would have been preferred by both spouses). Higher female salaries are related to more labor market hours of female workers. The more men earn, the less hours women typically work, and conversely, the more women earn, the less hours men typically work—a standard income effect whereby a wealthier household tends to consume more leisure. Interestingly, gender-ratios are related to intra-household transfers: when there are (many) more men than women in the area, women consume more and men consume less as a proportion of the family resources, everything else being equal than in areas where the sex ratio is more balanced. Finally, the presence of children is negatively related to the paid workforce participation by female workers in Mexico, a standard result reflecting the time constraints on workforce participation by women generated by child care. Interestingly, in Colombia the presence of other household members over age 18 is positively related to the female workforce participation, likely reflecting that these additional adults need less direct care themselves and/or can provide care for younger children.

South Africa

Another strong support of a collective model of economic family decision-making is provided by Duflo (2003), who analyzed a reform of the South African social pension program for the elderly. The reform extended benefits to a large, previously not covered aging sector of the black population.

Due to eligibility criteria, the coverage is not universal; in some households, only one of the grandparents receives the benefit. Duflo showed that the recipient's gender is of considerable importance for the impact of the transfers on children's health. Specifically, a payment to the grandfather has no significant effect on grandchildren's health, whereas the same amount paid to the grandmother results in a huge improvement in the health status of girls in the family.

Indonesia

Southeast Asia has been strongly influenced by the large influx of peoples from China and India, as well as the Middle East and remnants from the European Colonial powers. The islands of Southeast Asia export important spices—pepper, cinnamon, nutmeg, and so on—used in cooking all over the world. In earlier times, these spices were highly prized in Europe and the Middle East, as they provided a means of preserving food that was more flavorful than salt alone. Although much of the region continue to follow traditional ways of life, the growth of cities, the spread of technology, and government modernization programs are making inroads into traditional village life. It is well-known that control of economic resources is an important source of power in negotiating family decisions.

In this context, Thomas, Contreras, and Frankenberg (2002) examined the effects of maternal and paternal assets in marriage in Indonesia on whether children experience coughs, fevers, diarrhea, or other symptoms in the month prior to the interview using the Indonesian Family Life Survey (IFLS). The authors argue that conditional on income, the assets that husbands and wives bring to the marriage are indices of the power each partner wields within the marriage. Although both the morbidity and assets at marriage data likely have some reporting errors, the authors address these errors and find that mothers who are more powerful (i.e., bring more assets to the marriage) allocate resources toward goods and services that they value differently from their husbands. In particular, mothers spend more on their son's welfare (e.g., clothing and nutrition) than fathers, and this is reflected in their sons having fewer episodes of illness than their daughters. The authors also examined the assumption that the allocations are Pareto efficient, which predicts that the effect of the ratio of paternal assets to maternal assets will be constant across morbidities, and found the data fitted that model of Pareto efficiency. The authors conclude that although spouses may not agree about resource allocations, they cannot reallocate resources in such a way that one family member is made better-off without making at least one other family member worse-off.

The intra-family balance of power: Cross-cultural evidence in nondeveloped areas

The analysis of the intra-family balance of power is particularly important in developing regions, given the underlying cultural norms that persist that tend to disadvantage women. We have identified two cultural nondeveloped areas that exhibit an HDI at the bottom of the medium range, or a low HDI (South-Asia and sub-Saharan Africa) where relevant economic analyses of spouses' power have been developed from a nonunitary perspective.

South-Asia

The Himalayas are a natural barrier dividing South-Asia, which separated the populations and allowed people on each side of the mountains to develop their own distinct languages, customs, and cultures. In ancient times, invaders introduced the Hindu religion and caste system. In the 18th century, much of India fell under British rule. When India became independent in 1947 it was partitioned into Hindu India and Moslem Pakistan, and then former East Pakistan became Bangladesh. The majority of the populations speak English, as well as Hindi and other local languages.

Koolwal and Ray (2002), using data from Nepal, conclude that the woman's share of family earnings derived from paid employment, excluding the unearned income, understates her true power in influencing family outcomes. Other interesting results indicate that rural women enjoy greater power within the family than their urban counterparts and that education plays an effective role in enhancing the power of women inside the family. These estimates of female power provide considerable support for the hypothesis that welfare weights should be estimated simultaneously with household outcomes. The statistical significance of the impact of female power on an item's budget share provides a convenient test of the income-pooling hypothesis underlying the unitary model. The results provide little evidence in support of income pooling, especially for items that have the characteristics of a private good. A private good is something that primarily is consumed by particular family members (e.g., in Nepal tobacco and alcohol are consumed, almost exclusively, by the adult male). However, there is some support for income pooling existing for expenditure items that are collectively consumed within the family (e.g., flour and rice, eggs and milk, and meat are jointly consumed within the household). The results reveal some nonmonotonic relationships between female power and budget share, which vary a good deal from item to item.

Sen (1990) argued that there were a million adult women missing from the Indian population relative to what would be expected from birth rates. Calvi (2020) analyzed the possible source of this well-known "missing women" phenomenon. The starting point of the investigation is the fact that a majority of missing Indian women die in adulthood. Calvi's explanation is that women of postreproductive age experience a decline in their health, and a consequent surge in mortality risk that is directly related to changes affecting their intra-household bargaining power. She exploited amendments to the Indian inheritance law as a natural experiment to test the proposition that declining intra-household power, and the resulting deprivation of individual access to resources, underpinned excess mortality in older women. The Hindu Succession Act amendments equalized women's inheritance rights to men's in several Indian states between 1976 and 2005. By granting women the right to inherit their natal family property, these reforms likely strengthened the women's economic bargaining position in their marital families. Using data from the 2005 to 2006 National Family Health Survey and the 2004 NSS Morbidity and Health Care Survey, Calvi showed that women living in states with these reforms improved their nutritional outcomes (measured by body mass index and anemia), reduced their likelihood to suffer from a variety of ailments, and lowered their risk of death. Thus, economic equality of opportunity enhanced the life expectancy of older women.

An important consequence of power relationships within the family is the measure of intra-family inequality. Brown, Calvi, and Penglase (2018) argue that standard antipoverty policies, which typically target poor households, may be in part misdirected because many poor individuals reside in nonpoor households. Using Bangladeshi data, they show that undernourished individuals are spread across the household per-capita expenditure distribution. In particular, women, children, and the elderly face significant probabilities of living in poverty even in households with per-capita expenditure above the poverty threshold. These inequities often reflect the lack of economic power of certain family members in determining the allocation of family resources.

Basu and Ray (2002) use the collective framework to add evidence that, in Nepal, there is a relationship between the balance of power between the husband and the wife in the family and child labor. Child labor is less likely when power is evenly balanced between the spouses. While prior studies showed that higher parental education makes child labor less likely, Basu and Ray's study shows that there are additional subtleties involved in the association of education with child labor. The incidence of child labor is kept at its lowest when there is a balanced level of high education among parents. Since most developing countries are characterized by a disproportionately low level of education and literacy among women, this study suggests that, among other reasons, female education ought to be emphasized in developing countries to benefit child outcomes.

Maitra and Ray (2005a, 2005b) use data from Pakistan to revisit the notion of resource-pooling from different sources, accruing to different members, in order to achieve a range of family outcomes. The study extends an earlier work (Maitra & Ray, 2003) by introducing gender-based disaggregation of resource inflows. The most important results of this paper include the following. First, each income stream is distinguished by the gender of the income recipient, in line with the spirit of collective family models that has characterized much of the recent literature on family behavior. Second, the authors distinguish between the different sources of nonlabor income, namely, unearned income and private remittances that have been lumped together in previous studies. Third, following from the two issues raised above, the authors test the hypothesis of income pooling, not only between men and women, as several studies have recently done, but also, pooling of the two components of nonlabor earnings, mentioned above, separately for men and for women. In particular, they find that private transfers received by the household significantly increase the budget shares of necessities like food and clothing, while public pensions do not have much of an impact on household expenditure patterns.

Does the power of an individual making family decisions have an impact on that family's expenditure pattern? To answer this question, Lancaster, Maitra, and Ray (2006) use the collective approach to examine whether relative spousal power, measured by her/his income share, has an effect on family expenditure patterns. The authors propose and estimate a framework of this type in Indian families. Empirical results using family-level unit record datasets provide considerable support for the idea that expenditure reflects most strongly the preferences of the high power spouse. However, the bargaining power of the spouses is made more even by increases in the family's socioeconomic status.

Consistent with recent empirical evidence, the Lancaster et al. (2006) study rejects an important implication of the unitary household model, namely that the identity of the income recipient is irrelevant in the determination of the family's expenditure outcomes. One caveat is that identity of income recipient has marginal or no effect on some of the smaller items of expenditure, especially those (e.g., alcohol and tobacco) that are privately consumed (typically by adult males) within the household. However, an even spread of power in decisions about major household expenditure is most likely to be found in a family where educational level is spread evenly between the spouses. Thus, promoting both male and female levels of education within the family, as a way to improve both partners' income earnings opportunities, should help to prevent an overwhelming dominance by one partner in the decision-making process.

Sub-Saharan Africa

The area of Africa south of the Sahara Desert, with its different climate, topography, and non-Arab populations, forms a separate and distinct cultural region from the rest of Africa. Most of the region has a warm climate, and many peoples in this area were once afflicted by the slave trade, and/or subject to European colonial rule. Populations remain divided into a large number of separate ethnic groups, or tribes, each with its own language and culture, although they often share similar religious beliefs. Africans have been influenced by both Islam and Christianity, as well as by local animist traditions.

The first line of research addresses intra-family inequality. The share of household resources devoted to children is typically hard to identify because consumption is usually observed at the household and not the individual level. The collective model, however, allows identification of how total household resources are divided up among household members by observing how each family member's expenditures on a single private good like clothing varies with income and family size. Dunbar, Lewbel, and Pendakur (2013) show, in data from Malawi, resources devoted to wives and children are generally much less than what a standard, equivalence for each family member would predict, and that the extent of inequality within families varies by family size and structure. Standard poverty indices based on family income implicitly assume a fair allocation of resources within the household and tend to seriously understate the incidence of child poverty.

Finally, while most studies support the Pareto efficiency hypothesis, Udry (1996) reaches a different conclusion analyzing agricultural production in Burkina Faso. Using plot-level agricultural data from households, he concludes that the allocation of resources within these African households is Pareto inefficient. Specifically, he finds that among plots planted with the same crop in the same year within a given household, those controlled by women produce lower yields than the men's plots, suggesting an inefficient allocation of inputs (notably fertilizers). These conclusions have however been qualified by Akresh (2008), who, using an alternative nationally representative dataset, finds that only households in regions geographically proximate to those studied by Udry exhibit Pareto inefficient intra-household allocations, while the rest of the country reveals no evidence of Pareto inefficiencies. That is, fertilizer use was not dependent on the gender of the person doing the planting of the plot. Moreover, households in regions experiencing very low rainfall are on average less likely to exhibit Pareto inefficient intra-household allocations, and low rainfall shocks are correlated with increases in labor resources allocated to the wife's plots. This suggests that, under pressure to produce adequate food when rainfall is low, households avoid losses by becoming Pareto efficient. More recently, Rangel and Thomas (2019) provide a detailed investigation of rural families in Burkina Faso and argue that direct estimates of productivity may be contaminated by unobserved heterogeneity that is correlated with the identity of the plot owner/manager. This point was also made by Goldstein and Udry (2008), who document that plots managed by females are less likely to be followed and are, therefore, less productive. However, using land ownership as an indicator of intra-household power, decisions regarding consumption functionally are Pareto efficient (Rangel & Thomas, 2019).

Discussion and conclusions

This chapter provides an overview of the cross-cultural evidence for the intra-marriage balance of power on the basis of the nonunitary approach. In addition to the literature in sociology and anthropology arguing that spousal power is important in the analysis of family behavior, the economic evidence has recently increased in the context of the collective models. This framework assumes that a family maximizes a weighted average of the wife's and husband's utilities, where the weights capture each agent's power in the decision-making process.

Throughout the paper, we have presented recent international evidence on the intra-family balance of power for a variety of cultural areas. We have identified cultural regions as geographical areas in which individuals share common cultural characteristics, focusing on seven developed cultural areas: United States and Canada, Australia, Europe, East-Asia, Latin America, South Africa and Indonesia, and two broad nondeveloped areas: South-Asia, and sub-Saharan Africa, where important economic analyses of spousal power have been carried out from a collective perspective.

A number of conclusions emerge from this broad overview and the evidence base is summarized in Table 6.1.

1. Empirical investigations clearly indicate that, in both developed and non-developed cultural areas, the "unitary" representation of the household, which assumes that the household behaves as a single decision-maker, does not provide an accurate description of actual behavior. In particular, income pooling, a central prediction of the unitary framework, has been repeatedly rejected. Consequently, collective models, in which intra-family bargaining and the respective powers of spouses play important roles, are by now widely accepted
2. The Pareto-optimality hypothesis of collective models has been generally supported by empirical studies. While some violations have been pointed out (e.g., Udry, 1996), they seem to constitute exceptions; efficiency not only appears to be a natural benchmark but also a largely acceptable description of actual behavior.
3. Regarding the (few) violations of Pareto-efficiency, not much is known on the ultimate cause of these observed inefficiencies. A natural but somewhat ad-hoc justification could invoke social norms. For example, men seek to perpetuate their power in allocating resources that suit them and reinforce their power. Alternatively, inefficiencies may stem from informational problems (e.g., asymmetric information, imperfect monitoring) within the household; yet another cause could be limitations to the spouses' ability to commit to future behavior. Clearly, more work is needed on this important issue.[b]
4. Within the efficiency paradigm, the collective model allows the researcher to analyze effects on each individual family member of issues such as household formation and dissolution, or the impact of policies targeting specific household members (e.g., women or children). The widely used standard unitary approach was essentially disregarding different outcomes for different family members. Of particular importance is the notion of *intra-household inequality*. The traditional analysis, based on equivalence scales, was essentially assuming an equal (or "fair") allocation of resources within the family—a view that is both unrealistic and dangerously misleading. On the contrary, the collective approach—and specifically the notion of *indifference scale*—recognizes the endogenous nature of resource allocation within the family and provides empirical tools aimed at quantifying the resulting, intra-family inequality.

[b]On the latter point, see for instance Chiappori and Mazzocco (2017).

5. In both developed and developing countries, the respective bargaining powers of the adult spouses affect the family's economic decisions. Bargaining power relates to a number of family economic decisions (expenditure, time use at home, paid work, charity donations, wealth accumulation, childbirth and elder care, among many others). Moreover, some aspects of family-based economic decision-making vary widely across cultures. For instance, increasing women's power through targeted transfers ultimately benefits daughters in South Africa (Duflo, 2003); in Ethiopia, however, mothers with more assets invest preferentially in sons, while in Bangladesh, fathers' and mothers' assets do not have differential effects on daughters relative to sons (Quisumbing & Maluccio, 2003).
6. The determinants of the spouses' respective bargaining powers strongly depend on the cultural and socioeconomic context. In Western economies, human capital, wealth and incomes, as well as global factors such as sex ratios or laws governing divorce and separation, have been empirically linked to an individual spouse's decision power. In developing countries, relevant factors also include religion, ethnicity, cast, family background, and many other determinants. In fact, the same general mechanism—individual bargaining powers playing a crucial role in determining household behavior—operates in very specific ways depending on the cultural context.
7. The cultural variations do not mean there has not been some universal influences on economic decision-making within families. In all cultural contexts studied, education appears to play an important role in economic power and decision-making. In particular, a fair allocation is more likely to occur in a family where education and human capital is spread evenly between the spouses.

On this last point, recent evolutions provide strong motives for optimism. Basically, in all developed economies, and in an increasing number of developing ones, women are now more educated than men. We confidently predict that this evolution will deeply affect, for the better, economic, and social welfare of all family members as resource allocations become more equally influenced by mothers and fathers.

References

Akresh, R. (2008). *(In)Efficiency in intrahousehold allocations*. IZA Discussion paper 1858.

Andreoni, J., Brown, E., & Rischall, I. (2003). Charitable giving by married couples. Who decides and why does it matter? *Journal of Human Resources*, *38*, 111–133. https://doi.org/10.3368/jhr.XXXVIII.1.111.

Attanasio, O., & Lechene,V. (2002). Tests of income pooling in household decisions. *Review of Economic Dynamics, 5*, 720–748. https://doi.org/10.1006/redy.2002.0191.

Attanasio, O., & Lechene, V. (2014). Efficient responses to targeted cash transfers. *Journal of Political Economy, 122*(1), 178–222.

Basu, K. (2006). Gender and say: A model of household behaviour with endogenously determined balance of power. *Economic Journal, 116*(511), 558–580. https://doi.org/10.1111/j.1468-0297.2006.01092.x.

Basu, K., & Ray, R. (2002). *The collective model of the household and an unexpected implication for child labor.* Policy Research Working Paper 2813. The World Bank eLibrary.

Becker, G. S. (1981). *A treatise on the family.* English Edition. Cambridge: Cambridge University Press.

Bergstrom, T. C. (1989). A fresh look at the rotten kid theorem and other household mysteries. *Journal of Political Economy, 97*, 1138–1159.

Blumberg, R., & Coleman, M. (1989). A theoretical look at the gender balance of power in the American couple. *Journal of Family Issues, 10*, 225–250. https://doi.org/10.1177/019251389010002005.

Brown, C., Calvi, R., & Penglase, J. (2018). *Sharing the pie: Undernutrition, intra-household allocation, and poverty.* SSRN online paper. https://doi.org/10.2139/ssrn.3199129.

Browning, M., & Chiappori, P. A. (1998). Efficient intra-household allocations: A general characterization and empirical tests. *Econometrica, 66*, 1241–1278. https://doi.org/10.2307/2999616.

Browning, M., Bourguignon, F., & Chiappori, P.-A. (2009). Efficient intra-household allocations and distribution factors: Implications and identification. *Review of Economic Studies, 76*, 503–526.

Browning, M., Chiappori, P. A., & Lewbel, A. (2013). Estimating consumption economies of scale, adult equivalence scales, and household bargaining power. *The Review of Economic Studies, 80*, 1267–1303. https://doi.org/10.1093/restud/rdt019.

Browning, M., Chiappori, P. A., & Weiss, Y. (2014). *Economics of the family.* Cambridge: Cambridge University Press.

Calvi, R. (2020). Why are older women missing in India? The age profile of bargaining power and poverty. *Journal of Political Economy* [in press].

Campaña, J. C., Giménez-Nadal, J. I., & Molina, J. A. (2018). *Efficient labor supply for Latin families: Is the intra-household bargaining power relevant?* IZA Institute of Labor Economics Discussion Paper No. 11695. https://ssrn.com/abstract=3234218.

Chiappori, P. A. (1988). Rational household labor supply. *Econometrica, 56*, 63–89. https://doi.org/10.2307/1911842.

Chiappori, P. A. (1992). Collective labour supply and welfare. *Journal of Political Economy, 100*, 437–467. https://doi.org/10.1086/261825.

Chiappori, P. A., & Mazzocco, M. (2017). Static and intertemporal household decisions. *Journal of Economic Literature, 55*(3), 1–61. https://doi.org/10.1257/jel.20150715.

Couprie, H., & Ferrant, G. (2015). Welfare comparisons, economies of scales and equivalence scale in time use. *Annales d'Économie et de Statistique, 117–118*, 185–210. https://doi.org/10.15609/annaeconstat2009.117-118.185.

de Palma, A., Picard, N., & Ziegelmeyer, A. (2011). Individual and couple decision behavior under risk: Evidence on the dynamics of power balance. *Theory and Decision, 70*, 45–64. https://doi.org/10.1007/s11238-009-9179-6.

Desai, S., & Jain, D. (1994). Maternal employment and family dynamics: The social context of women's work in rural South India. *Population and Development Review, 20*, 115–136. https://doi.org/10.2307/2137632.

Doepke, M., & Tertilt, M. (2018). Women's empowerment, the gender gap in desired fertility, and fertility outcomes in developing countries. (Vol. 108, pp. 358–362). *AEA Papers and Proceedings, American Economic Association.*

Donni, O., & Chiappori, P.A. (2011). Non-unitary models of household behavior: A survey of the literature. In J. A. Molina (Ed.), *Household economic behaviors* (pp. 1–40). Berlin: Springer Verlag.

Donni, O., & Molina, J. A. (2018). *Household collective models: Three decades of theoretical contributions and empirical evidence* IZA Discussion Paper No. 11915. https://papers.ssrn.com/sol3/papers.cfm?abstract_id=3286175.

Duflo, E. (2003). Grandmothers and granddaughters: Old age pension and intra-household allocation in South Africa. *World Bank Economic Review*, *17*(1), 1–25. https://doi.org/10.1093/wber/lhg013.

Dunbar, G. R., Lewbel, A., & Pendakur, K. (2013). Children's resources in collective households: Identification, estimation, and an application to child poverty in Malawi. *American Economic Review*, *103*, 438–471. https://doi.org/10.1257/aer.103.1.438.

Fujii, T., & Ishikawa, R. (2013). How does childbirth alter intra-household resource allocation? Evidence from Japan. *Oxford Bulletin of Economics and Statistics*, *75*(3), 362–387. https://doi.org/10.1111/j.1468-0084.2012.00699.

Goldstein, M., & Udry, C. (2008). The profits of power: Land rights and agricultural Investment in Ghana. *Journal of Political Economy*, *116*(6), 981–1022.

Gupta, A., & Ferguson, J. (1997). *Culture, power, place: Explorations in critical anthropology*. Durham, NC: Duke University Press.

Hoddinott, J., & Haddad, L. (1995). Does female income share influence household expenditures? Evidence from Côte d'Ivoire. *Oxford Bulletin of Economics and Statistics*, *57*, 77–96.

Kanbur, R., & Haddad, L. (1994). Are better off households more unequal or less unequal: A bargaining theoretic approach to "Kuznets" effects at the micro level. *Oxford Economic Papers*, *46*(3), 335–458.

Konrad, K. A., & Lommerud, K. E. (1995). Family policy with non-cooperative families. *The Scandinavian Journal of Economics*, *97*, 581–601. https://doi.org/10.2307/3440544.

Koolwal, G., & Ray, R. (2002). *Estimating the endogenously determined intrahousehold balance of power and its impact on expenditure pattern.* Policy Research Working Paper 2814. eLibrary The World Bank. https://elibrary.worldbank.org/doi/abs/10.1596/1813-9450-2814.

Kousta, S. (2019). Why gender gaps persist. *Nature Human Behaviour*, *3*, 418. https://doi.org/10.1177/0001839219832310.

Lancaster, G., Maitra, P., & Ray, R. (2006). Endogenous intra-household balance of power and its impact on expenditure patterns: Evidence from India. *Economica*, *73*, 435–460. https://doi.org/10.1111/j.1468-0335.2006.00502.x.

Lee, J. (2007). Marriage, the sharing rule, and pocket money: The case of South Korea. *Economic Development and Cultural Change*, *55*, 557–581. https://doi.org/10.1086/511194.

Lundberg, S., & Pollak, R. (1994). Noncooperative bargaining models of marriage. *American Economic Review Papers & Proceedings*, *84*, 132–137. https://www.jstor.org/stable/2117816.

Lundberg, S., Pollak, R. A., & Wales, T. J. (1997). Do husbands and wives pool their resources? Evidence from the United Kingdom child benefit. *Journal of Human Resources*, *32*, 463–480.

Lundberg, S., & Ward-Batts, J. (2004). *Saving for retirement: Household bargaining and household net worth.* Claremont McKenna College Robert Day School of Economics and Finance Research Paper No. 2004-03. https://papers.ssrn.com/sol3/papers.cfm?abstract_id=532822.

Maitra, P., & Ray, R. (2002). *Household resources, expenditure patterns and resource pooling: Evidence from South Africa.* Unpublished Manuscript. University of Adelaide.

Maitra, P., & Ray, R. (2003). The effect of transfers on household expenditure patterns and poverty in South Africa. *Journal of Development Economics*, *71*(1), 23–49. https://doi.org/10.1016/S0304-3878(02)00132-3.

Maitra, P., & Ray, R. (2005a). *An integrated test of the unitary household model: Evidence from Pakistan.* ABERU Discussion Paper No. 7. Melbourne: Monash University. http://citeseerx.ist.psu.edu/viewdoc/download?doi=10.1.1.411.1298&rep=rep1&type=pdf.

Maitra, P., & Ray, R. (2005b). The impact of intra-household balance of power on expenditure pattern: The Australian evidence. *Australian Economic Papers, 44*, 15–29. https://doi.org/10.1111/j.1467-8454.2005.00246.x.

Manser, M., & Brown, M. (1980). Marriage and household decision making: A bargaining analysis. *International Economic Review, 21*(3), 31–44. https://doi.org/10.2307/2526238.

Martínez, C. (2013). Intrahousehold allocation and bargaining power: Evidence from Chile. *Economic Development and Cultural Change, 61*, 577–605. https://doi.org/10.1086/669260.

McElroy, M., & Horney, M. (1981). Nash bargained household decisions. *International Economic Review, 22*(2), 333–350. https://doi.org/10.2307/2526280.

Phipps, S. A., & Burton, P. (1998). What's mine is yours? The influence of male and female incomes on patterns of household expenditure. *Economica, 65*, 599–613. https://doi.org/10.1111/1468-0335.00148.

Pollak, R. A. (1994). For better or worse: The roles of power in models of distribution within marriage. *American Economic Review: Papers and Proceedings, 84*(2), 148–152. https://www.jstor.org/stable/2117819.

Porter, M. (2017). Spousal bargaining over care for elderly parents in China: Imbalances in sex ratios influence the allocation of support. *The Journal of Development Studies, 53*(4), 514–529. https://doi.org/10.1080/00220388.2016.1156093.

Quisumbing, A. R., & Maluccio, J. A. (2003). Resources at marriage and intrahousehold allocation: Evidence from Bangladesh, Ethiopia, Indonesia and South Africa. *Oxford Bulletin of Economics and Statistics, 65*, 283–327. https://doi.org/10.1111/1468-0084.t01-1-00052.

Rangel, M., & Thomas, D. (2019). *Decision-making in complex households.* Working paper, Duke University.

Riley, N. (1997). Gender, power and population change. *Population Bulletin, 52*, 1–46.

Rubalcava, L., Teruel, G., & Thomas, D. (2002). *Welfare design, women's empowerment and income pooling.* Documento de Trabajo Numero 244. CIDE (Centro de Investigatión y Docencia Económicas). http://aleph.academica.mx/jspui/bitstream/56789/3833/1/DOCT2065350.PDF.

Samuelson, P. (1956). Social indifference curves. *Quarterly Journal of Economics, 70*, 1–22. https://doi.org/10.2307/1884510.

Schultz, T. P. (1990). Testing the neoclassical model of family labor supply and fertility. *Journal of Human Resources, 25*, 599–631. https://doi.org/10.2307/145669.

Sen, A. (1990). More than 100 million women are missing. *New York Review of Books*, 61–66. December 20.

Slatton, B. C. & Brailey, C. D. (Eds.), (2019). *Women and inequality in the 21st century.* New York: Routledge.

Thomas, D. (1990). Intra-household resource allocation: An inferential approach. *Journal of Human Resources, 25*, 635–664. https://doi.org/10.2307/145670.

Thomas, D., Contreras, D., & Frankenberg, E. (2002). *Distribution of power within the household and child health.* Munich Personal RePEc Archive Paper No. 80075. https://mpra.ub.uni-muenchen.de/80075/.

Udry, C. (1996). Gender, agricultural production, and the theory of the household. *The Journal of Political Economy, 104*, 1010–1046.

United Nations Human Development Index. (2018). *Global human development indicators.* United Nations Development Programme. http://hdr.undp.org/en/content/human-development-index-hdi.

Yilmazer, T., & Lich, S. (2015). Portfolio choice and risk attitudes: A household bargaining approach. *Review of Economics of the Household, 13*, 219–241. https://doi.org/10.1007/s11150-013-9207-8.

CHAPTER 7

Couple communication from a cross-cultural perspective

Shuangyue Zhang[a] and Susan Kline[b]
[a]Department of Communication Studies, Sam Houston State University, Huntsville, TX, United States
[b]School of Communication, The Ohio State University, Columbus, OH, United States

Across cultures, one of the strongest factors that contribute to life satisfaction is a loving marriage (Diener, Suh, Lucas, & Smith, 1999). According to Halford's (2011) ecological model of influences on couple relationships, four groups of variables predict relationship outcomes. They include sociocultural variables such as family and friendship support, life events such as the birth of a child, individual characteristics such as personality variables, and couple interaction such as couple communication. Of the four groups of variables, couple communication has been the most powerful predictor of couple relational outcomes (p. 29).

Although couple communication is strongly associated with functional relationship outcomes, it is also "the common pathway to relationship dysfunction across theories, therapists, and clients" (Heyman, 2001, p. 6). Both scholars and couples rate communication as a top problem area and both researchers and therapists stress the importance of the role of communication (see Heyman, 2001). Longitudinal data has shown that negative communication behaviors of husbands and wives predict lower marital satisfaction for both husbands and wives (Karney & Bradbury, 1995). Negative behaviors from husbands also consistently predict lower marriage stability. Couple communication has also been associated with marital satisfaction in non-Western cultures, such as in Filipino (Vazhappily & Reyes, 2016), Argentinian (Falconier & Epstein, 2011), Iranian (Sadeghi, Mazaheri, Motabi, & Zahedi, 2012), Chinese (Li et al., 2018), Mexican (Bravo & Martinez, 2016), Indian (Yelsma & Athappilly, 1988), Pakistan (Rehman & Holtzworth-Munroe, 2007), and Turkish (Celenk & van de Vijver, 2013) cultures.

Cross-Cultural Family Research and Practice
https://doi.org/10.1016/B978-0-12-815493-9.00007-7

Recently, several authors have outlined the advantages and importance of studying couple communication from a cross-cultural perspective (e.g., Hiew, Halford, van de Vijver, & Liu, 2016; Rehman & Holtzworth-Munroe, 2007). First, people from different cultures hold very different views on the roles of romantic love and extended family networks in marriages. Romantic love and mutual attraction are considered to be necessary for marriages in Western countries such as the United States (Dion & Dion, 1993; Sastry, 1999). Across the world, however, the majority of marriages are still by arrangement (Goodwin & Pillay, 2006). The majority of marriages in India and other South Asian cultures are based on material criteria such as age, education, and earning potential (Rehman & Holtzworth-Munroe, 2006; Sastry, 1999). According to Yelsma and Athappilly (1988), love comes after marriage for arranged Indian couples, and romantic love comes before marriage for Western couples. When a person marries, he/she not only marries another person but also that person's family. Sadeghi et al. (2012) found that distressed Iranian couples are more critical and insulting toward each other's family than nondistressed couples. They suggest the meaning and the importance of family-of-origin is different among different cultures and societies. Similarly, social networks of family members and close friends predicted Chinese young adults' marital intentions and relationship commitment (Zhang & Kline, 2009).

A second reason for studying couple communication from a cross-cultural perspective is that intercultural relationships are increasing and common throughout the world (Hiew, Halford, van de Vijver, & Liu, 2015). In Australia, Hawaii, Singapore, and Taiwan, one in three marriages are interracial (see Hiew et al., 2015 for a review). Even in nonmetropolitan areas such as the Southern and Midwestern parts of the United States, intercultural relationships are on the increase due to recent growing number of immigrants (Donato, Tolbert, Nucci, & Kawano, 2007). Compared with same-culture couples, intercultural couples face unique stressors and are less satisfied in marriage due to cultural differences (Fu, Tora, & Kendall, 2001). Intercultural couples have to cope with differences in childrearing practices, time orientation, gender role expectations, affection expressions, religion, importance of food, and residence decisions (Bustamante, Nelson, Henrilksen, & Monakes, 2011; Tien, Softas-Nall, & Barritt, 2017). In addition, due to language barriers, communicating between partners and extended family members can lead to frustration and marital problems (Tien et al., 2017).

It is also important to study which aspects of communication in intimate relationships are universal and which aspects are culturally specific because such work can clarify "the boundary conditions that govern the breadth or our models of intimate functioning" (Williamson, Ju, Bradbury, Karney, & Fang, 2012, p. 308). Examining and comparing marital interactions in different countries makes it possible to investigate the role of culture in understanding couple communication (Bodenmann, Kaiser, Hahlweg, & Fehm-Wolfdorf, 1998). Yet the interventions that are designed for improving relationships based on samples from one culture may not work well in another culture. For example, couples in different cultures may have different views about the top priorities in their marriages. Su et al. (2015) found that raising children was the most frequently reported problem in marriages for both Taiwanese husbands and wives. Interestingly, it was the least reported marital problem for US couples. The number one problem for both US husbands and wives was communication. Su et al. believe such cultural differences in problem ratings reflect different values about marriage. In individual cultures, couples emphasize dyadic relationships for the purpose of fulfilling personal autonomy and freedom of choice (Schwartz, 1990). In contrast, in collectivistic Chinese culture, continuing the family heritage is the primary goal of marriage. Sacrificing for the family is considered the most cherished virtue for Chinese parents (Su et al., 2015).

Finally and importantly, there are theoretical reasons for studying couple communication, or personal relationships in general, from a cross-cultural perspective (see Goodwin & Pillay, 2006 for a review). First, cultures may differ in the *relative magnitude* of different relationship phenomena. For example, a few cross-cultural studies on couple communication found that both Chinese husbands and wives reported lower marital satisfaction than their Western counterparts (e.g., Williamson et al., 2012). Culture may also differ in the importance and salience of couple communication on marital satisfaction. As indicated earlier, Taiwanese couples see rearing children as the top priority in marriage, but American couples see communication as the number one issue in marriage (Su et al., 2015). Finally, even when there is universal consistency in certain relational phenomena, the way in which these influence actual behaviors may differ across cultures. For example, negative communication behaviors have been associated with marital distress in both American and Pakistani couples, but the association is stronger for American couples than for Pakistani couples (Rehman & Holtzworth-Munroe, 2007).

The purpose of this chapter is to synthesize empirical research on couple communication from a cross-cultural perspective. We organize the chapter into five sections. We first review how couple communication is conceptualized and measured in cross-cultural research. We then review theoretical frameworks for studying culture, particularly the individualism-collectivism framework, and cultural relationship beliefs. Then we review and integrate the findings of couple communication from a cross-cultural perspective. Next, we discuss practical implications for couple therapy and couple relationship education (CRE). Further, we describe areas of research deficits and opportunities for future directions.

Conceptualizations and measurements of couple communication

Couple communication refers to communication between romantic partners in dating, cohabitating, or married relationships. As couple communication has been studied mainly in married relationships, it is often used interchangeably with marital communication and spousal communication. The study of marital communication is an interdisciplinary endeavor, but the two disciplines of communication and psychology are at the forefront of research (Fincham, 2003a, 2003b). Within the communication discipline, research on marital communication has been informed by family communication researchers and interpersonal communication scholars employing family systems theory and interactionist theory, among others. Within psychology, the study of marital communication has been mainly informed by social exchange theory and social learning theory. Many recent investigations have been based on observations of couple interactions.

Kelly, Fincham, and Beach (2003) argue for the necessity of distinguishing communication behavior from communication skills in couple relationships. For Kelly et al., the former refers to any verbal or nonverbal behavior that occurs during the course of interaction, whereas the latter implies "the ability to realize communicative goals while behaving in a socially appropriate manner" (p. 724). Communication behavior may be analyzed descriptively without a focus on the enactment of skills, but the study of communication skills involves the normative analysis of communicative behavior. Theorists such as Burleson and Denton (1997) have argued that problem-solving discussions in most couple communication research may be a better operationalization for communication behaviors than for communication skills. Kelly et al. (2003) cautioned that "the extent to which behavioral observation accurately taps communication skills" (p. 726) remains a question. Distressed couples may not be motivated to use communication skills they do have, and indeed they might use them to hurt their partner.

Kelly et al.'s (2003) perspective fit the stance taken by many interpersonal communication scholars who study couple communication and close relationships. Two approaches have typically been employed (Burleson, Metts, & Kirch, 2003). One has been a systems-functional approach in which communication is seen as individual intentions enacted as strategies, motivations, skills, and behaviors, all in service of accomplishing situated tasks and general functions. Within this view, scholars discriminate between several primary functions that communication serves for relational partners, such as engaging in social influence, creating identities, maintaining relationships, and engaging in discussion and decision-making. The other approach has been to employ a sociocultural approach to study the social structure and culture within which relational partners construct their joint meanings. Speech codes, narratives and idioms, interaction routines and rituals, and relational norms and rules are all topics for study with this second approach to learning how couples construct their understandings and ways of speaking with each another.

Interpersonal communication scholars like Caughlin and Scott (2010) have combined aspects of constructivist and interactionist theories to explicate skilled communication behavior for relational partners. Four bases for skill can be identified. First, the situation may call for a particular message design logic for handling the situation. Design logics reflect ways about communicating in any situation. O'Keefe (1988) has documented three message design logics in US interpersonal communication: expressive (say what you think or feel); conventional (speak appropriately considering the social context); and rhetorical (create new situations and identities). Interpersonal skill reflects a communicator's ability to use one or more of these design logics. Second, because the functional task can be difficult communicators may also differ in their strategic knowledge about what best to say or do in the situation. Third, communicators can also differ in the number of subsidiary aims they may have in the situation, such as crafting a particular self-presentation or relationship state, or maintaining face through politeness. Communicators' ability to manage these multiple aims in interaction can differ (Brown & Levinson, 1987; Weinstein, 1969). Finally, communicators differ in the way they facilitate the process of communication, with some focusing on establishing common ground through shared ways of speaking, others engaged in openness and attentive communication, and still others engaged in cooperative practices like empathizing and helping.

In sum, individual differences in couple communication skill may involve knowing effective practices for general tasks like influencing others; knowing ways of designing specific messages such as knowing appropriate ways of speaking; knowing ways of pursuing multiple goals and knowing ways of ensuring understanding (such as identifying with the other, asking questions, or empathizing).

These approaches to studying couple communication have produced a variety of methods. The focus here is on methods used in published research on couple communication. Of these, most studies in marital communication have been based on self-report methods (Heyman, 2001). Common questionnaires measuring marital communication include the Communication Patterns Questionnaire (CPQ) (Christensen & Sullaway, 1984), the Marital Communication Inventory (Bienvenu, 1970), and the Primary Communication Inventory (Navran, 1967). Among these questionnaires, the CPQ has been most often used for cross-cultural comparisons of couple communication (Bodenmann et al., 1998; Christensen, Eldridge, Catta-Preta, Lim, & Santagata, 2006). This questionnaire examines marital communication patterns in three stages of marital conflict: when the problem arises, when couples discuss the conflict, and after the conflict discussion. It measures five aspects of marital communication: mutual constructive communication, mutual discussion avoidance, total demand/withdrawal, husband demand/wife withdrawal, and wife demand/husband withdrawal. Bodenmann et al. (1998) found evidence of criterion validity of the CPQ using a German and a Swiss sample. Recently, Crenshaw, Christensen, Baucom, Epstein, and Baucom (2017) used exploratory and confirmatory factor analyses, reexamined the factor structure, and revised the CPQ for improved reliability. The reliability and validity of the scale have been assessed in samples from United States, Germany, Sweden, Taiwan, Brazil, and Italy (Bodenmann et al., 1998; Christensen et al., 2006).

Aside from scales that assess the general construct of couple communication, scales that measure how couples handle conflict have also been used in the study of couple communication. They include the Marital Coping Inventory (Bowman, 1990), the Conflict and Problem-Solving Scales (Kerig, 1996), the Problem-Solving subscale of the Marital Satisfaction Inventory Revised (Snyder, 1997) the Conflict Tactics Scale (Straus, Hamby, Boney-McCoy, & Sugarman, 1996), and the Perceived Criticism Measure (Chambless & Blake, 2009).

Finally, scales have been used to measure social support and relational maintenance within couples. These include validated instruments for measuring relational maintenance which operationalize maintenance as engaging in particular positive behaviors, openness, assurances about the relationship, sharing tasks, and utilizing social networks (Canary & Stafford, 1992; Stafford, 2011). Scales used to measure social support among couples have included the Spousal Support Scale (Xu & Burleson, 2001), Comforting Message Evaluation Scale (Burleson & Mortenson, 2003), and Support Goals Inventory (Burleson & Mortenson, 2003).

Although couples' reports of marital problems and communication are useful in understanding their own conceptualizations of distress, it does not present the whole picture for the clinicians to design the best treatment plans. Heyman (2001) believes that outsiders' observation of couple communication behaviors can add "valuable nonredundant information" (p. 6) to that gathered from self-reports, which is important for researchers who study marital functions and clinicians who design treatment plans. Rehman and Holtzworth-Munroe (2006) contend that using observational data to compare couple communication across cultures has advantages over using self-report methods. That is, people from different cultures may employ different understanding and interpretations of terms used in questionnaire studies, in addition to attributional biases.

Thus, for behaviorally oriented researchers the most common method to measure communication behavior and skills is direct observation (Kelly et al., 2003). This method typically consists of videotaping couples discussing issues in their relationship for a short period of time (mostly around 10 minutes) about a problem (Halford, Hahlweg, & Dunne, 1990; Rehman & Holtzworth-Munroe, 2007) or a neutral topic (Williamson et al., 2012), or a positive topic (Hiew et al., 2015), or both a problem and a supportive topic (Cao et al., 2015). Trained coders then use a certain coding system to assess verbal and nonverbal communication behaviors as functional or dysfunctional. Different coding systems used in cross-cultural research of couple communication include the Marital Interaction Coding System (MICS) (e.g., Hooley & Hahlweg, 1989), the Kategoriensystem Partnerschaftliche Interaction (KPI) (e.g., Halford et al., 1990; Hiew et al., 2016), the Iowa Family Interaction Rating Scales (IFIRS) (e.g., Williamson et al., 2012), the Iranian Couples Interaction Coding System (ICICS) (Sadeghi et al., 2012), the Social Support Interaction Rating System (SSIRS) (Rehman & Holtzworth-Munroe, 2007), and the Couple Interaction Rating System (CIRS) (Rehman & Holtzworth-Munroe, 2006).

Although numerous coding systems have been used, and different content categories have resulted from different coding systems, researchers have seemed to agree in their reduction of categories to summary categories. In most studies, researchers construct two final summary categories of positive communication behaviors (i.e., functional) and negative communication behaviors (i.e., dysfunctional), and these summary categories are used in data analyses (e.g., Williamson et al., 2012). Examples of positive communication behaviors include compliments, encouraging statements, and agreement, while examples of negative communication behaviors are insults, interruptions, denying, excuse, and disagreement. In Halford et al. (1990), a third category was named "neutral" communication behaviors, which includes problem descriptions and metacommunication like clarification requests.

Kelly et al. (2003) caution readers about the findings obtained through observational studies of couple communication. Couple communication tends to vary with contextual factors, and research settings may alter participants' typical communication behavior. Furthermore, coding systems "inevitably restrict the researcher's field of vision with respect to the complexities of couple interaction" (p. 725). Sillars and Vangelisti (2006) contend that since communication conveys multiple messages simultaneously on different levels of analysis, coding communication behavior into "singular, or even primary, meanings, functions, or outcomes" (p. 345) may be problematic. In addition, coding systems are different in terms of coding unit and unit complexity (Floyd, 1989), making it harder to compare findings across different studies.

Although there are limitations for observational studies on couple communication, there is evidence to support the validity of structured research laboratory observations. Couple communication in laboratory settings seems to be similar to marital interactions at home (Kelly & Halford, 1993). Observed behaviors in laboratory settings reliably distinguish happy and unhappy couples. Moreover, there is evidence to show that observation is as valid as or even more valid than self-report measures. For example, Sanford (2010) compared the validity of self-report, partner-report, and observer ratings of conflict communication using a sample of 83 married couples in the United States. He found all three different methods demonstrated good validity, with partner-report ratings and observer ratings showing high correlation and higher validity than self-report ratings. Similarly, using a sample of 81 couples in Germany, Hahlweg, Kaiser, Christensen, Fehm-Wolfsdorf, and Groth (2000) found high correlations between the Communication Patterns Questionnaire and the KPI observation system. Another study by

Markman, Rhoades, Stanley, Ragan, and Whitton (2010) found that premarital self-report negative communication predicted divorce 5 years after marriage slightly better than did premarital observational data.

Aside from laboratory observations and self-report questionnaires, a few studies on cross-cultural couple communication have used interviews and diary methods. For example, Ingersoll-Dayton, Campbell, and Mattson (1998) interviewed 11 Japanese (age ranging from 54 to 77 years) and 13 American (age ranging from 62 to 89 years) older couples. The participants were asked to tell a joint story about their marriage. Interviews were tape-recorded, transcribed, and coded into five forms of communication. Three of the forms (prompting, questioning, and echoing) were regarded as supportive and harmonious exchanges. The other two forms (teasing and contradicting) were seen as acrimonious and barbed exchanges.

Incorporating culture into a theoretical framework

Cross-cultural researchers of personal relationships often acknowledge difficulties in defining and categorizing culture (Goodwin & Pillay, 2006). Various scholars have differentiated cultures based upon their scores on key values. The most influential typologies of this endeavor have been Hofstede's (1980) four cultural dimensions: power distance, masculinity-femininity, uncertainty avoidance, and individualism-collectivism. Of these four dimensions, the individualism-collectivism has been the most widely used framework to study close relationships across cultures. Collectivism refers to "the subordination of individual goals to the goals of the collective, and a sense of harmony, interdependence and concern of others" (Hui & Triandis, 1986, pp. 244–245). In contrast, individualistic cultures value the subordination of group goals to individual goals by encouraging the expression of unique ideas, feelings, and preferences, and the development of independence (Markus & Kitayama, 1991).

Individualistic and collectivistic cultural orientations can directly influence one's behavior through social norms/rules or indirectly through personality factors (Gudykunst & Matsumoto, 1996). One such influence that mirrors the individualism-collectivism dimension is the distinction of high-context and low-context communication. In high-context communication, people rely more on nonverbal cues and context for decoding meanings, whereas in low-context communication people express their thoughts and feelings explicitly (Ting-Toomey, 1999). In the cross-cultural study of couple communication, especially when Western cultures such as the United States and Australia are compared with Eastern cultures such as China, Taiwan, Korea, and Japan, the framework of individualism-collectivism and high-context versus low-context has been most often used (e.g., Hiew et al., 2016; Williamson et al., 2012).

Aside from the dimensions of individualism and collectivism, and high-context and low-context communication, researchers have identified cultural beliefs about marriage and close relationships and cultural concepts to examine cross-cultural similarities and differences in close relationships. For example, Rehman and Holtzworth-Munroe (2006) argue that patriarchal values are still prevalent in Pakistani culture. Compared with men, women are given low status and fewer opportunities in "educational, economic, political and legal arenas" (p. 757). From their analyses, they proposed that women from patriarchal cultures are less likely to engage in a direct, aggressive manner and more likely to use unassertive methods to influence their husbands than women from egalitarian cultures. In another study, Zhang and Kline (2009) discussed two Chinese relational concepts of *xiao* (filial piety) and *guanxi* (network) and found that these concepts still influence mate selection behaviors in China. Similarly, Hiew et al. (2015, 2016) show the importance of *enqing* (feeling of gratitude and admiration) in understanding Chinese marriage relationships.

While these broad value orientations play a role in explaining cultural differences in couple communication, a core theoretical problem is to determine how these broad orientations exactly affect the expression and interpretation of couple communication behaviors. Drawing from his research Burleson (2003) contends that cultural differences may shape communication in three ways. First, cultural differences can affect the value couples place on particular kinds of communication skills (like emotional support), which have implications for how relationships might function. For instance, using the Communication Functions Questionnaire Xu and Burleson (2001) found that while Chinese and American married individuals both valued emotional support from their spouse over other types of support, Chinese individuals valued informational and network support more from their spouses than did Americans. Cultural differences can also affect the way people evaluate their partners' messages. For instance, Burleson (2003) found that while both Chinese and Americans evaluate highly person-centered messages as more effective than low person-centered messages, Chinese see low person-centered messages as more sensitive than Americans, while Americans see highly person-centered messages as more sensitive than Chinese. Finally, Burleson (2003) has found that cultural differences may shape what goals are pursued in support situations. Using his Support Goals Inventory, Burleson and his colleagues have found that Americans and Chinese differ in the types of support goals they pursue in support situations, with Americans viewing emotion management to be more important than problem management for providing support while Chinese consider avoidance and problem management to be more important for handling support situations than Americans. In sum, while each area needs substantial replication, there is initial support for cultural differences in support-related values, message preferences, and goal orientations.

Besides value orientations, another concept relevant to the study of couple communication has been relationship standards. Campbell, Simpson, Kashy, and Fletcher (2001) found that perception of discrepancies between the partner and ideal standards predicted relational satisfaction. Similarly, Chi, Epstein, Fang, and Lam (2013) found that similarity of relationship standards such as on sharing values, or having equal outcomes between Chinese couples affected their marital satisfaction indirectly through couple communication patterns (more mutual constructive communication and less avoidance and demand-withdrawal). A few recent studies have indicated that Chinese and Western couples hold different relationship standards (Epstein, Chen, & Beyder-Kamjou, 2005; Halford et al., 2018; Hiew et al., 2015), with Chinese couples holding standards on family responsibilities (e.g., respect for elders), sharing values, giving in, instrumental investments, trying to change the other, and seeking to make the relationship perfect more than US couples, who held standards on expressing intimacy more than Chinese couples. Such differences may affect couple communication behaviors and pose challenges for couple therapy. For example, partners who have different levels of Family Responsibility Standards are likely to have conflict over upholding family's face and to what extent they should support their parents financially. Halford et al. (2018) suggest that therapists for intercultural couples should pay particular attention to the similarities and differences in partners' relationship standards.

Review of couple communication from a cross-cultural perspective

Marital communication has been studied extensively in Western cultures (see review in Heyman, 2001). For this chapter, we have limited our review to studies using non-Western samples and their cross-cultural comparisons. We first review findings of behavioral research from both observational studies and survey research, specifically on cross-cultural differences and similarities on communication behavior, relational satisfaction, and links between communication behavior and satisfaction. Then we review findings related to migration, sexual communication, and sources of relational problems. We finally review findings from interview studies.

Couple communication behavior: Cultural similarities and differences

Culture influences how people display and interpret marital communication behaviors (Williamson et al., 2012). Studies have found both similarities and differences across cultures in observed couple communication behaviors. For instance, Halford et al. (1990) compared the problem-solving behaviors of happy and unhappy Australian and West Germany couples. They found that German happy and unhappy couples expressed and reciprocated significantly less neutral communication (problem descriptions and metacommunication) and more critiques and refusals than did Australian couples. Across the two cultures, happy couples were more likely to use positive verbal or nonverbal behaviors such as acceptance-agreement and neutral communication and less likely to use critiques and refusals. By contrast, unhappy couples were more likely to engage in verbal coercive escalation than happy couples, but the German couples engaged in more verbal coercive escalation than did the Australian couples. Halford et al. suggest that the cultural differences in verbal behaviors between Germans and Australians might be due to different understandings of the categories used by the coders in the two countries.

Two other studies found differences between Chinese and Western couples in their communication behaviors. Williamson et al. (2012) analyzed two videotaped discussions of a social support task, and after controlling for relational satisfaction, found that American and Chinese couples did not differ in their levels of positive behaviors (shared enjoyment, positive mood, warmth/support, humor, affection, endearment, listener responsiveness), contrary to their predictions. Yet Chinese couples were generally more negative than American couples, using more coercion, contempt, dominance, disruption, denial, hostility, interrogation, and verbal attacks. Specifically, though, American and Chinese husbands did not differ in their levels of negativity, but Chinese wives were significantly more negative than American wives. It is possible, as explained by Williamson et al., that seeking support from spouses might be seen as self-centered and inappropriate in collectivistic cultures, but as functional and appropriate behavior for individualistic Americans.

Similar to Williamson et al.'s (2012) study, Hiew et al. (2016) showed that gender affected rates of positive and negative behaviors. They analyzed couple problem discussions and positive reminiscence discussions for their level of positive (validating and discussion) and negative (invalidate and conflict) behaviors. Compared with Australian couples and Australian-Chinese couples, Chinese couples displayed the lowest level of validating behaviors. Couples with an Australian female partner showed more positive verbal and fewer negative verbal behaviors than couples with a Chinese female partner. By contrast, couples with a Chinese female partner showed higher rates of conflict and invalidation, and lower levels of discussion (p. 198). Overall, no cultural differences were observed in nonverbal communication. Hiew et al. concluded that most cultural differences were a function of the female partner's culture (p. 200).

Other studies have shown that couple communication behavior differences may be a result of differences in patriarchal cultural values. Comparing Pakistani, immigrant Pakistani in the United States, and US American couples, Rehman and Holtzworth-Munroe (2006) analyzed marital problem discussions for the frequency of spouses' aggressive demands, discussion, and withdrawal behaviors, and found that for all three groups, the wife-demand-husband-withdraw pattern was negatively associated with marital satisfaction. American couples were more likely to engage in the wife-demand-husband-withdraw pattern and less likely to engage in husband-demand-wife-withdraw than Pakistani and immigrant Pakistani couples. American wives were more likely to show aggressive demands than Pakistani and immigrant Pakistani wives. Within cultures, American wives were more likely to engage in aggressive demands than their husbands. The opposite was true for Pakistani couples. Pakistani husbands were more likely to show aggressive demand behaviors than their wives. In a similar study, Rehman and Holtzworth-Munroe (2007) rated marital problem discussions for the positive global affect and behaviors as well as negative global affect and behaviors each spouse showed in the interactions. They found that American wives were more negative than Pakistani wives, but American husbands were less negative than Pakistani husbands.

Further, couple communication behaviors change over time and vary according to contexts. For example, Markman et al. (2010) demonstrated a time factor in couples' communication behaviors, in that levels of negative communication decreased among all couples in their first 5 years of marriage, but the decrease was significantly larger for nondistressed couples than for distressed couples. This may suggest that happy couples handle their negative communication better than unhappy couples. Other studies have found that the tasks used in the studies affect what communication behaviors couples use. Christensen and Heavey (1990) found couples tended to demand when requesting a change they wanted and withdraw when discussing a change their partner wanted. Similarly, Pasch and Bradbury (1998) found that couples showed more positive behaviors and fewer negative behaviors in support interactions than in conflict resolution interactions. Hiew et al. (2016) extended these findings among Western, Chinese, and Chinese-Western couples, finding higher negative behaviors and lower validation and positive affect in problem discussion tasks than in positive reminiscence tasks.

While much research has focused on the couple's discussion and resolution of problems, communication researchers contend that two other functions of communication are important for marriage. One is relationship maintenance, or the behaviors that sustain desired definitions of the relationship. Canary and Stafford's (1992) program of research has identified the behaviors couples used to maintain their relationships; positivity, openness, assurances, sharing tasks, and social networks are behaviors linked to relationship qualities such as relational satisfaction, perceived equity, and commitment. Yum and Canary (2009) used cultural modernization theory (Inglehart & Welzel, 2005) to argue that particular countries would value equity and maintenance strategies more due to their endorsement of self-expression values as opposed to traditional/secular values. Their comparison of equity and relational maintenance strategy used in six countries found that relationships between equity and relationship maintenance strategies occur as predicted for the United States, Spain, and Japanese samples, but not for China, South Korea, and the Czech Republic. Using the same theoretical framework, in a second study Yum, Canary, and Baptist (2015) examined the same hypotheses with romantic partners in the United States, Singapore, and Malaysia. Consistent with modernization theory hypotheses the United States and Malaysians reported using more relational maintenance strategies than did Singapore partners (self-expression vs survival), but only the US sample reported associating relational maintenance strategies with perceived equity.

Relational satisfaction

Most studies on couple communication have examined relational or marital satisfaction to be an outcome that correlates with communication behaviors. Similar to cross-cultural variations in communication behaviors, cultures may also differ in terms of reported relational satisfaction. In the studies that reported statistical means for relationship satisfaction, only one study did not find differences between the cultural groups that were compared. In Rehman and Holtzworth-Munroe's (2006) study, Pakistani, immigrant Pakistani, and American couples reported similar levels of satisfaction.

Hiew et al.'s (2015, 2016) two studies used the same sample of 123 Chinese, Western, and Chinese-Western couples. They found that relationship satisfaction was higher in Western couples than in Chinese couples, with intercultural couples (Chinese-Western) in between. Interestingly, they also found that both male and female partners' satisfaction was higher in couples when they married a Westerner than when they married a Chinese. Similarly, in Williamson et al.'s (2012) study comparing Chinese and American couples, both American husbands and wives reported higher relational satisfaction than did Chinese couples. Hiew et al. (2015) suggested some possible reasons for the lower relational satisfaction in the Chinese sample. First, the Chinese typically reported a lower level of happiness with life than Westerners (Epstein et al., 2005). In addition, there could be a cultural bias in how relational satisfaction was measured, as the scale items they used seemed to focus more on romanticism (e.g., "For me, my partner is the perfect romantic partner") than they reflected Chinese relational concepts like *enqing* (feeling of gratitude and admiration). Hiew et al. (2015, 2016), though, did not explain why relational satisfaction was lower for Westerners who married Chinese.

Yelsma and Athappilly (1988) compared marital communication among US American couples and Indian couples in arranged and free-choice marriages. Contrary to their predictions, they found Indian couples in arranged marriages reported higher levels of relational satisfaction than for US free-choice married couples. Yelsma and Athappilly suggested that for arranged marriages, a sense of life-long commitment and cultural tradition seems to engender marital satisfaction, while for American couples the effectiveness of communication is a more important influence on marital satisfaction.

The cultural variations in reported marital satisfaction from the limited studies seem to be very interesting. More research is needed on how marital satisfaction is conceptualized and operationalized across cultures.

The communication behavior-satisfaction link: Cultural differences and similarities

Although different coding systems have been used in different observational studies, a consistent finding based on Western samples is that negativity and withdrawal are negatively associated with relational satisfaction

(Heyman, 2001). This association is verified in most cross-cultural comparisons and studies that have used non-Western samples.

Findings from observational studies

The behavior-satisfaction link has mostly received support from observational cross-cultural studies. Hooley and Hahlweg (1989) examined the association between marital satisfaction and marital communication among two samples: a group of Western German couples seeking treatment for marital problems and a group of depressed patients in Oxford, England, and their spouses. In both samples, they found that marital satisfaction was positively related to positive verbal and nonverbal behaviors, and negatively correlated with negative verbal and nonverbal behaviors. Hooley and Hahlweg suggested that certain patterns of dysfunctional communication may transcend cultures and diagnostic groups, and bad marriages may be very much alike within Western cultures.

More recent cross-cultural studies have also lent support to the communication behavior-satisfaction link, but the magnitude of such a link seems to be moderated by culture, gender, and contexts. For example, comparing Pakistani and American couples, Rehman and Holtzworth-Munroe (2006) found a significant relationship between the wife-demand-husband-withdraw pattern and lower marital satisfaction. The relationship was stronger for American couples than for Pakistani couples, in that both demand-withdraw and discussion behaviors were more strongly associated with marital satisfaction for American couples than for Pakistani couples. Similarly, in Williamson et al.'s (2012) study American husbands' positive behaviors, such as warmth and responsiveness, were positively related to their own marital satisfaction, but Chinese husbands' positive behaviors were not related to their marital satisfaction. Interestingly, Chinese husbands' negative behaviors, such as hostility and dominance, were negatively associated with their own marital satisfaction, but such an association was not found for American husbands. For both American and Chinese wives, their own positive or negative communicative behaviors were not related to their own marital satisfaction. Wives' positivity was significantly associated with husbands' marital satisfaction for American husbands, but not for Chinese husbands. In contrast, wives' negative behaviors were significantly related to husbands' marital satisfaction for Chinese husbands, but not for American husbands. For both American and Chinese samples, husbands' positive and negative behaviors were not related to wives' level of marital satisfaction.

The most recent cross-cultural observation study by Hiew et al. (2016) also found significant relationships between communication behaviors and relational satisfaction in problem discussions regardless of culture, in that higher level of negative communication behaviors (invalidate, conflict, and negative affect) were associated with lower relational satisfaction, and higher level of positive behaviors (validating, positive affect) were related to higher relational satisfaction. Different from Rehman and Holtzworth-Munroe (2006) and Williamson et al.'s (2012) studies, Hiew et al. (2016) did not find culture to be a moderator between communication behaviors and satisfaction, but they did find a gender effect in the association between communication behaviors and marital satisfaction. The shared variance between male and female's negative affect in problem discussions and positive affect in both problem discussion and positive reminiscence predicted marital satisfaction. Females' invalidation behaviors in both problem and positive reminiscence discussions predicted low male and female satisfaction, but males' invalidation behaviors were not found to be a predictor for male or female relational satisfaction. Hiew et al. (2016) concluded that cultural differences existed more likely in levels of communicative behaviors, especially those of female partners, and less likely in the associations between behaviors and relational satisfaction (p. 200).

The communication behavior-satisfaction link also received support from an observational study in Iran. Using the Iranian Couples Interaction Coding System, Sadeghi et al. (2012) had couples complete four interaction tasks and found that similar to Western couples, distressed Iranian couples exhibited more negative behaviors such as criticism, contempt, defensiveness, dominance, and condemning the relationship than did well-adjusted Iranian couples, who showed more interest, and engaged in more validation, the use of "we," affection, humor, and self-disclosure. Distressed Iranian couples were also observed using insulting and abusive language when referring to the partner's family members and relatives (Sadeghi et al., 2012). Sadeghi et al. explain that the Iranian family structure is different from the Western families, and in that, extended families in Iran are economically and socially bonded.

Findings from self-report studies

The communication behavior-satisfaction link also received support from survey studies of cross-cultural comparisons. Christensen et al. (2006) compared samples from four cultures: Taiwan, Brazil, Italy, and the United States in three areas of communication: mutual constructive communication, woman demand/man withdrawal, and man demand/woman withdrawal. Across cultures, they found that mutual constructive communication was positively related to relational satisfaction, whereas demand/withdrawal communication was negatively associated with relational satisfaction.

Bodenmann et al. (1998) also used the same questionnaire to compare German and Swiss couples. They found similar findings as Christensen et al. (2006). One cultural difference was noted in Bodenmann et al. (1998) in that the number of children was negatively associated with constructive communication and marital satisfaction, and positively related to negative communication such as avoidance and demand/withdrawal in Swiss couples, but not for German couples. Although the German couples had more children, the children of the Swiss sample tended to be younger. It could be that children had a negative impact on marital quality for the Swiss sample. Another study also found that culture moderates the association between communication and satisfaction. Yelsma and Athappilly (1988) found that all three modes of communication (verbal, nonverbal, and sexual) were more influential in affecting marital satisfaction for American couples than for Indian couples in arranged marriages.

The communication behavior-satisfaction link has also received support from self-report studies using non-Western samples. As indicated in the introduction, couple communication has also been found to be associated with marital satisfaction in non-Western cultures, such as in Filipino (Vazhappily & Reyes, 2016), Argentinian (Falconier & Epstein, 2011), Chinese (Li et al., 2018), Mexican (Bravo & Martinez, 2016), Indian (Yelsma & Athappilly, 1988), and Turkish (Celenk & van de Vijver, 2013) cultures.

However, one study for Israeli couples did not find support for the behavior-satisfaction link. Winkler and Doherty (1983) compared Israeli couples living in the United States and US American couples. They found that Israeli couples reported higher levels of verbal conflict but less physical violence than did American couples. For Israeli couples, high level of verbal conflict was not associated with marital satisfaction, as it was significantly correlated for American couples. Winkler and Doherty (1983) explained that the seemingly contradictory findings regarding verbal and physical violence in Israeli couples reflected the "cultural confinement of violent behavior to nonfamily context" (p. 226). They suggest that marital therapists for Israeli couples should be cautious when using American communication skills approaches.

Directionality: Communication to satisfaction or satisfaction to communication or a third variable affecting both?

Although numerous studies have supported the relationship between communication behaviors and relational satisfaction, Lavner, Karney, and Bradbury (2016) have raised the question as to the directionality of this relationship. That is, does couple communication predict marital satisfaction or does marital satisfaction predict communication? Or does a third variable like life stress predict both communication and satisfaction? Although they acknowledge that overall both effects (communication-to-satisfaction and satisfaction-to-communication) were similar in magnitude, Lavner et al. contend that their longitudinal data lends more support to the satisfaction-to-communication path than for the reverse path (communication-to-satisfaction). However, an early metaanalysis has found that behavioral couple therapy based on communication training does produce sustained benefits (Giblin, Sprenkle, & Sheehan, 1985).

Consistent with Giblin et al.'s (1985) metaanalysis, Li et al. (2018) examined daily communication, conflict resolution, and marital quality among newlywed Chinese couples in a 3-year longitudinal study. Their findings support the behavioral model of marital relationship in that marital satisfaction is a function of daily interaction, not vice versa. Li et al. explained that their different findings from Lavner et al. (2016) may be due to sampling and data collection methods. Li et al.'s samples had relatively high socioeconomic status, while Lavner et al.'s sample consisted of low-income couples. Li et al. used self-report surveys, and Lavner et al. used observational methods for communication behaviors. Lavner et al. (2016) acknowledge that neither the effect from communication to satisfaction nor the effect from satisfaction to communication, was particularly strong or consistent. They suggest that some other factors may predict communication and satisfaction or may mediate the association between communication and satisfaction. In fact, a couple of studies have lent support to this speculation. Sanford (2003) found that topic difficulty affected communication behavior and relationship satisfaction at the distal level. Couples who experienced conflict over a highly difficult topic reported low relational satisfaction and used negative forms of communication behavior. Another study by Williamson, Karney, and Bradbury (2013) found that financial strain and stressful events were the best predictors of negative communication behaviors in newlywed couples, whereas relational satisfaction was the best predictor of positive couple communication behaviors.

Couple communication stability

Many behavioral studies have examined communication behaviors from cross-sectional survey and observational studies, and focused on the frequency and/or intensity of communication behaviors. Cao et al. (2015) have advocated the importance of also examining couple communication stability, and suggest that couple communication stability includes two distinct but overlapping aspects: the stability of communication behaviors and patterns over the course of interaction, and the consistency of communication behaviors across different contexts (p. 852). They contend that the consistency of negative communication behaviors across contexts may be particularly detrimental to relationships.

Cao et al. (2015) examined the communication patterns of 144 Chinese newlywed couples across 4 different contexts (2 problem-solving interactions and 2 social support interactions with 1 initiated by wife and 1 initiated by husband). Cao et al. (2015) identified three types of couples: consistently supportive (high levels of positivity such as physical affection, warm support, and humor), consistent emotional quarreling (high levels of negativity such as dominance, hostility, and interrogation), and traditionally undemonstrative (relatively low levels of both hostility and positivity, but high levels of communication and listening responsiveness). In total, the consistently supportive group constituted around 17% of the sample, the consistently quarreling group 33%, and the traditional group 50% of the sample. They found that the consistently supportive group showed the highest marital satisfaction, the emotionally quarreling group the lowest satisfaction, and the traditional undemonstrative group in the middle. Both the supportive and traditional groups reported significantly higher marital satisfaction than the quarreling group. There were no significant differences between the supportive and the traditional group in terms of marital satisfaction. Interestingly, the identification of the consistently supportive and quarreling groups seems to be inconsistent with traditional norm of Chinese interpersonal interactions. However, Cao et al. (2015) suggested the vast social and economic reform during the last few decades in China has changed the dynamics of Chinese marriages, especially for Chinese urban newlywed young couples.

Cao et al.'s (2015) findings of communication patterns share some similarities to typologies identified by Fitzpatrick (1988) and Gottman (1994). By analyzing couple interaction patterns, Fitzpatrick identified three types of couples: traditional (less confrontational and more conciliatory), independent (assertive and confrontational), and separate (avoidant). Similarly,

Gottman classified three types of couples: validators (validating each other), volatiles (passionate, confrontational but with warmth and love), and avoiders (avoiding conflicts). More cross-cultural research seems to be needed to compare whether similar typologies exist across cultures and how this might inform couple relationship therapy.

Migration and acculturation

Migration and acculturation might also influence couple communication behavior and perceptions of marital satisfaction. For example, Rehman and Holtzworth-Munroe (2006) found that communication behaviors of both immigrant Pakistani husbands and wives were related to levels of acculturation in the United States. More acculturated wives were more assertive, and more acculturated husbands were less likely to engage in aggressive demands than their native counterparts. Rehman and Holtzworth-Munroe (2007) also found that the association between global positive affect and behaviors and marital satisfaction was similar for both immigrant Pakistani and native Pakistani couples, but the association between global negative affect and behaviors and marital satisfaction was stronger for immigrant than for native Pakistani couples. They suggest that future research should examine the impact of acculturation on relationship function and stability.

In addition to the influence of communication behaviors, migration and acculturation might change one's relationship beliefs. Halford et al. (2018) found that immigrant Chinese women showed alignment with western standards for Couple Bond, but a divergence of Family Responsibility standards from western standards, indicating that migration influences women's relationship standards more than it does for men's relationship standards. Future research could focus on how standards might affect communication behaviors.

Sexual communication

Sexual communication has been found to be associated with both sexual and relational satisfaction (see Sprecher, Christopher, & Cate, 2006, for a review). A more recent study that used US samples found that greater amount of sexual communication was associated with increased female orgasm frequency, and greater sexual and relational satisfaction for both partners (Jones, Robinson, & Seedall, 2017). Another study by Rehman et al. (2011) found that positive communication behaviors displayed during a sexual conflict discussion were associated with relational satisfaction for both partners, and negative communication behaviors significantly predicted female's relational satisfaction. Cross-cultural research on sexual communication still remains scarce. One study by Yelsma and Athappilly (1988) found that couples in arranged marriages in India were found to use as much sexual communication as US American couples. Both groups used more sexual communication than couples in free-choice marriages in India. Yelsma and Athappilly explained that in arranged marriages, wives might be socialized to be deferent to husbands to avoid conflicting issues on matters of sexual interest.

Different from Yelsma and Athappilly's (1988) findings, discussing sexual topics remains a taboo in Asian cultures (Kim & Ward, 2007). Xie and Galliher (2018) found that bilingual Chinese living in the United States felt more comfortable using English for sexual communication due to the lack of vocabulary in Chinese. Similarly, traditional norms restrict sexual communication in Vietnamese marriages (Huong, 2010). Huong found that couples used more nonverbal than verbal means to communicate about sex and that both the quantity and quality of sexual communication decreased with the length of marriage.

Sources of relational problems

Using an evolutionary perspective, Dillon et al. (2015) studied the sources of marital conflict in the United States, United Kingdom, China, Russia, and Turkey. Across cultures, all four issues (i.e., sex, finances, division of labor, and raising children) were significantly related to marital problems in almost all cases. All four issues predicted marital problems for British and Chinese wives, and for British, Chinese, and Turkish husbands. For American husbands and wives, childrearing was not a significant predictor for their marital problems. For Russian husbands and wives, finance was not a significant predictor. Across all five cultures, wives perceived more problems and disagreements than their husbands.

Another study by Su et al. (2015) found that US couples reported more marital problems than Taiwanese couples in five of six areas: communication, intimacy/sexual relations, financial matters, division of labor, and power. Raising children was the most frequently reported problem in marriages for both Taiwanese husbands and wives, but it was the least reported problem for US couples. Su et al. (2015) suggest that *Ren* (tolerance) in Chinese couple relationships is an important element for conflict resolution and maintaining relational harmony. *Ren* is the Confucian relational concept that focuses on self-control, resilience, and concerns for others (Li & Hsiao, 2008), which is different from the Western concept of conflict avoidance.

According to Confucian teachings, a family is the fundamental unit of society, country, and the world; being nonconfrontational in conflict resolution helps maintain relational harmony (Su et al., 2015). While Westerners emphasized dyadic relationships for personal fulfillment, the Eastern Asians see parent-child relationships and the achievement of children as major issues in maintaining a happy marriage.

Findings of interview studies

Compared with studies using observations and self-report surveys, the number of cross-cultural studies on couple communication using in-depth interviews still remains scarce. However, a few interview studies have compared the similarities and differences of communication behaviors between couples from different cultures, or between the spouses of intercultural couples. Using a phenomenological approach, Tien et al. (2017) interviewed eight intercultural couples to learn how culture affects couple interactions. Couples found it difficult to express their feelings or to understand specific terms related to cultural concepts. Couples were sometimes "lost in translation" because they did not understand the cultural concept behind particular linguistic terms. Moreover, it was not easy for one partner to communicate with the partner's extended family members due to language barriers. Couples did express their desire to learn their partner's language. In another phenomenological study, Bustamante et al. (2011) interviewed five intercultural couples about how they coped with culturally related stressors in their relationship. Primary stressors included cultural differences in child-rearing practices, orientations toward time, gender role expectations, and extended family; for instance, one partner reported experienced marginalization from in-laws and extended family members while the other partner conversed with their in-laws in their native language. Bustamante et al. (2011) found that couples developed their relationships by creating flexible gender roles, using humor, showing cultural deference to one's partner, recognizing similarities, developing blended values and expectations and expressing a genuine appreciation for other cultures.

Similar to Tien et al.'s (2017) study, Tili and Barker (2015) interviewed nine intercultural couples in the United States with one partner Caucasian and one partner Asian. They found three factors that influenced couple communication: the motivation to learn about the partner's culture, language fluency, and different conceptions of high-context and low-context communication. They concluded that intercultural competence was displayed when partners were aware of cultural similarities and differences, were open-minded, showed respect, expressed their emotions, and provided face support to one's partner.

Different from the above interview studies, Ingersoll-Dayton et al. (1998) used a narrative approach to examine communication behaviors among older couples. Through in-depth interviews with 11 Japanese (aged from 54 to 77) and 13 American couples (aged from 62 to 89), Ingersoll-Dayton et al. challenged some of the stereotypes related to communication patterns in Japan and the United States. Both Japanese and American couples asked each other questions, embellished each other's points, contradicted, and teased each other in the interviews. Contrary to the view of Japanese being indirect and Americans being direct in communication, Japanese couples were more direct with each other than were American couples, who used indirect contradictions and teasing more often to address toxic topics. Ingersoll-Dayton et al. explained that older American couples seem to have learned how to "optimize positive experience and minimize negative emotional experiences" (p. 75). On the other hand, Japanese husbands appeared to use direct communication to express control in the relationship, while Japanese wives were more open about their feelings. Consistent with traditional views, Ingersoll-Dayton et al. also found that American couples talked significantly more to each other than did the Japanese couples, indicating the value of talk in US culture and the value of "not talking too much" in Japan.

In sum, these interview studies corroborated past findings and also provided insights about other kinds of communicative practices that may be profitably examined in couple communication. However, it should be noted that these studies employed small samples, which necessitates replication with consistent instrumentation and procedures.

Practical implications for couple therapy and relationship education

Although cross-cultural comparisons of couple communication have been conducted among limited countries and cultures, the findings have significant implications for couple therapy and relationship education. Results from both observational and self-report studies have generally supported the communication behavior-satisfaction link across cultures, lending support to applications of Western behavioral couple therapy, and couple education programs to non-Western and intercultural couples.

One primary purpose of recent behavioral couple therapy is to boost relational satisfaction by enhancing communication skills (Kelly et al., 2003).

There are two common types of behavioral couple therapy programs: traditional behavioral couple therapy (BCT) and modified cognitive-behavioral couples therapy (CBCT). The traditional BCT has focused on behavioral exchange, communication skills training, negative affect management, and problem-solving training. The CBCT adds an additional focus on helping clients change their beliefs about marriage, the relationship, and the partner (Epstein, Baucom, & Daiuto, 1997). Different from behavioral couple therapy programs that mostly target couples with relational problems, CRE focuses on helping couples develop relationship knowledge, attitude, and skills to enhance relationship satisfaction and commitment and to prevent relationship erosion (Halford, 2011).

Although both BCT and CBCT programs can be applied to cross-cultural couple therapy, studies suggest such applications could be modified in several ways. For instance, Hiew et al. (2016) found significant associations between communication behaviors (invalidation/validation, conflict/discussion, negative/positive affect) and relationship satisfaction. Although the findings were based on a cross-sectional study that did not indicate causation, Hiew et al. suggest that increasing positive behaviors and decreasing negative communication behaviors may help Chinese couples and Chinese-Western couples in couple therapy. Williamson et al. (2012), however, found that gender moderated the association between communication behaviors and marital satisfaction, so educational interventions may not work the same way for Chinese couples. Reducing negative behaviors may benefit Chinese husbands more, but increasing positive behaviors may benefit American husbands more. In addition, Rehman and Holtzworth-Munroe (2006) warn that the same behavior may have different meanings in different cultural contexts. In the Western world, withdrawal may be seen as an attempt to avoid escalation of a conflict or to resist change, but the same behavior may indicate a lack of power for women in Pakistani culture. Thus, couple therapists should be both culturally sensitive and knowledgeable.

In terms of the cognitive beliefs of CBCT, therapists should be aware that relationship standards are different across cultures. Halford et al. (2018) suggest that relationship therapists should be aware of Family Responsibility Standards (such as relations with extended family members, and preserving face) held by Eastern Asians. Immigrants may selectively maintain relationship standards from their culture of origins or endorse standards of the majority culture. In addition, couples from different cultures may have different relational problems. In Chinese societies, for example, the most frequent problem reported in marriage and family therapy clinics is about the child's academic performance (Sze, Hou, Lan, & Fang, 2011). Therapists in Western societies should be aware that Asian couples may face different problems and challenges in their marriages. Thus, couple education and therapy should be culturally sensitive.

For CRE, Halford (2011) suggests that three things need to be considered when providing CRE: to whom CRE is appropriate, what education resources are provided, and when to provide CRE. Therapists should identify couples who may have high risks for future relationship problems, provide appropriate educational materials, and select the best time such as entry to marriage, major life events such as the birth of a child, or relocation to a different residence for CRE. There is empirical evidence to show the importance of CRE before marriage and during the first few years of marriage. Markman et al. (2010) have shown that premarital communication can affect marital adjustment and significantly predicts divorce in the first 5 years of marriage. Further, Cao et al. (2015) suggest that the associations between marital communication and marital outcomes may be particularly salient in the early years of marriage when couples establish interaction patterns that could affect long-term outcomes (Lavner & Bradbury, 2010).

Lastly, we acknowledge the various challenges that face cross-cultural couple therapy. Bhugra and Silva (2000) have listed seven common pitfalls for inexperienced therapists in cross-cultural couple therapy: color blindness (assuming the minority culture client is the same as the majority culture client), color consciousness (assuming that problems come from minority status), cultural transference (client's feelings result from therapist's race), cultural countertransference (therapist's feelings toward clients result from their race), cultural ambivalence (therapists wish to help but needs to control to maintain power), overidentification (minority therapist overidentifies everything in terms of racism and defines problems as racially based), and identification with an oppressor (minority therapist denies his/her status by virtue of power; p. 191). Besides, couple therapy may not be accepted by some cultural or ethnic groups. For example, little research has studied the willingness or prevalence of South Asian immigrant groups seeking couple therapy (Rehman & Holtzworth-Munroe, 2006). Couples may also not be able to articulate culturally based problems due to language barriers (Tien et al., 2017). In addition, therapists need to be aware that extended family members of couples seeking help may not approve the therapeutic interventions (Bhugra & Silva, 2000).

In general, counselors and therapists should develop their cultural competence, and be aware of their own worldviews and potential biases.

They should also help couples understand and appreciate joys and challenges of relationships, assist couples with exploring their core values and beliefs, and remind couples about their feelings when they first met (Bustamante et al., 2011; Tien et al., 2017).

Limitations and future directions

This chapter illuminates the need for further research on couple communication from a cross-cultural perspective. Below we describe the areas in which we see limitations and opportunities for future research.

First, one of the limitations of observational studies on couple communication is related to the coding process. Coders in different cultures may operationalize the coding schemes in different ways (Halford et al., 1990). In addition, the coding system developed by Western researchers based on Western samples may not work well in other cultures. Hiew et al. (2016) noticed that Chinese partners would often say things like "You have been busy lately, but..." before they offer criticism. The coding system failed to capture the disclaimer and politeness strategies that can mitigate criticism (Brown & Levinson, 1987), and affect perceived relationship satisfaction. Furthermore, certain marital dysfunctional features may be related to certain aspects of a particular culture's communication expectations. For example, in Sadeghi et al.'s (2012) study of Iranian couples, they noticed that many Western coding systems were neither culturally sensitive to nor representative of Iranian culture. After they found that attacking a spouse's family of origin was common for distressed Iranian couples, they modified their coding systems by creating a category of "family contempt/criticizing" because communicating about spouses' family members is very important for effective interpersonal relations. Finally, in most of the observational studies, coding was done by the authors involved in the project or trained coders such as graduate students (e.g., Hiew et al., 2016). Future research should also include member checking, or including insiders' (i.e., both spouses) perspectives to fully understand the thoughts and feelings underlying behaviors over the course of conversation (Noller & Feeney, 2004).

Second, marital satisfaction was the most common relational outcome variable that was examined in the literature in correlation with couple communication. However, Karney and Bradbury (1995) suggest that marital satisfaction does not strongly predict marital stability. Other marital outcomes such as relational commitment, divorce, and closeness need to be examined as well for cross-cultural comparisons. For example, quite a few studies found that Chinese couples reported lower levels of marital satisfaction and higher levels of negative communication behaviors than did Western couples (e.g., Hiew et al., 2016). It does not mean Chinese couples are less committed or more likely to divorce. In fact, one longitudinal study found that individual risk factors predicted divorce, whereas observed conflict management correlated with marital satisfaction (Rogge & Bradbury, 1999). Thus, future research should examine how couple communication is associated with other indices of relationship quality, such as relational commitment.

Third, related to the outcome variable, the measurement of relational satisfaction that was developed by Western researchers using Western samples may not be a valid scale to measure relational satisfaction among couples from Chinese (Hiew et al., 2015) or other cultures. For instance, Chinese couples may employ different relationship standards. Besides using verbal expressions Chinese couples commonly express love for each other through pursuing activities together (Russell & Yik, 1996).

Fourth, almost all cross-cultural comparisons on couple communication have used cross-sectional designs, but it is hard to draw causal inferences from such findings (Hiew et al., 2016). Studies have shown that premarital communication among American couples can predict divorce during the first 5 years of marriage (Markman et al., 2010). Li et al. (2018) have also found that Chinese couples' daily communication and conflict management predicted marital quality over a 3-year period. Future research should continue to use longitudinal designs to examine how the behavioral-satisfaction link varies across cultures. However, as Heyman and Hunt (2007) suggest, we caution interpretation of these findings from longitudinal research, as the results might be inflated by making postdiction rather than predictions. Postdiction refers to fitting prediction models post hoc once outcomes are known, such as using multiple indices of communication to predict known outcomes like relationship satisfaction. Cross-validation and replication in observational couple research are needed to ensure that prediction is reliable (Heyman & Slep, 2001).

Fifth, although it is not possible to find the perfect comparable samples for cross-cultural studies, future research has to avoid using samples that are apparently not comparable. For example, samples from small towns in the United States may not be a good comparison with samples from metropolitan Beijing. Further, quite a few studies on Chinese marital communication used samples with relatively higher levels of social economic status (SES) than the general population. People with higher levels of education and income in collectivistic cultures may be more likely to hold individualistic marital values that emphasize equality, intimacy, and happiness (Xu, Xie, Liu, Xia, & Liu, 2007).

Sixth, there are cultural differences in how people respond to questionnaires (Harzing, 2006). For example, Chinese tend to endorse the midpoint more than Westerners. Van de Vijver and Tanzer (1997) suggested that the midpoint response style (MRS) can be estimated by assessing the proportion of midpoint responses across a diverse range of areas. Future research in cross-cultural comparisons should always use the MRS to check if such response bias exists and control it in the analysis if it does occur.

Seventh, for survey research, most questionnaires were designed in Western cultures. Although these questionnaires might demonstrate good reliability, their validity needs to be examined. For example, a study showed that Taiwanese couples have different marital problems (Sze et al., 2011). Another study showed that for Iranian couples one of the most important relationship issues is about extended family members (Sadeghi et al., 2012). Hence people in different cultures have different relationship standards (Hiew et al., 2015) that may not be reflected in Western designed questionnaires. An ideal way for cross-cultural research to proceed would be to combine Western and non-Western relational concepts in the questionnaire design (Kline & Zhang, 2009; Zhang & Kline, 2009). Banville, Desrosiers, and Genet-Volet (2000) contend that it is crucial for cross-cultural studies to use comparable, yet culture-specific measurements.

Eighth, most couple communication studies have focused on married relationships. Similar studies should extend to other relationship contexts such as dating relationships, cohabitating relationships, gay/lesbian/transgender relationships, and remarried relationships. Studies have shown that partners can face unique challenges in cohabitating relationships (see Kline, Pleasant, Whitton, & Markman, 2006 for a review), in same-sex relationships (Diamond, 2006), and in remarried relationships (Afifi & Schrodt, 2003). In addition, most studies on cross-cultural comparisons of couple communication have used newlywed couples. Future research should also focus on older age couples, as older couples face different marital problems and exhibit different communication styles from younger couples (Zietlow & Sillars, 1988).

Ninth, mixed methods research designs are needed in cross-cultural research on couple communication. Most communication behaviors from observational studies were observed in a lab setting. It is possible that couples' interactions may be different in natural settings (Cao et al., 2015).

Fincham (2003a, 2003b), for instance, suggests that couple conflicts videotaped in a lab setting may be less severe than at home. Thus, it is important to use additional methods such as diary or in-depth interviews to obtain a full picture of couple communication across cultures.

Tenth, as indicated earlier, only a handful of countries have been examined in cross-cultural comparisons of couple communication. Of the studies that have been conducted most have focused on Western (the United States and Australia) countries and Chinese cultures in mainland China, Hong Kong, and overseas Chinese. Future research should examine this relational phenomenon across different continents. This requires collaborations between scholars across disciplines and continents. Further, most research on this topic has ignored within-culture variations. Future research should examine regional differences within specific cultures.

Conclusion

In this chapter, we have provided an overview of findings on couple communication from a cross-cultural perspective, highlighted the implications for couple therapy and relationship education, and described areas for potential future research in the area of couple communication across cultures. The majority of studies from both observational and self-report methods have generally supported the behavioral link between communication behaviors and relational satisfaction, providing support for couple therapy and relationship education across cultures. Specifically, positive communication behaviors such as positive affect, validation, and warmth are positively related to relational satisfaction. In contrast, negative communication behaviors such as hostility, invalidate, dominance are negatively associated with relational satisfaction. The magnitude of the association between communication and relational satisfaction is moderated by culture, gender, and contexts. Overall, the relationship between communication and satisfaction is stronger for Western couples than for non-Western couples. A couple of studies have also shown that the communication behaviors of the female partner affect couple relationship satisfaction more so than those of the male partner (e.g., Hiew et al., 2016).

Although we admit that certain communication behaviors and behavioral sequences are more likely to either increase relational satisfaction or lead to relationship problems, we acknowledge that "there is no simple formula for functional communication in couples" (Kelly et al., 2003, p. 743). In addition, cross-cultural research on couple communication still remains scarce, due to limited cultures being studied and compared around the world. It is our hope that this chapter will inspire more research in this exciting area of study. We anticipate that the development of cross-cultural research on couple communication will progress further through "accidental and purposeful collaborations" (Sillars & Vangelisti, 2006) among scholars who will come from across different disciplines and nations.

References

Afifi, T. D., & Schrodt, P. (2003). Uncertainty and the avoidance of the state of one's family/relationships in stepfamilies, post-divorce single parent families, and first marriage families. *Human Communication Research*, *29*, 516–533. https://doi.org/10.1111/j.1468-2958.2003.tb00854.x.

Banville, D., Desrosiers, P., & Genet-Volet, Y. (2000). Translating questionnaires and inventories using a cross-cultural translation technique. *Journal of Teaching in Physical Education*, *19*, 374–387. https://doi.org/10.1123/jtpe.19.3.374.

Bhugra, D., & Silva, P. D. (2000). Couple therapy across cultures. *Sexual and Relationship Therapy*, *15*, 183–192. https://doi.org/10.1080/14681990050010763.

Bienvenu, M. J. (1970). Measurement of marital communication. *The Family Coordinator*, *19*, 26–31. https://doi.org/10.2307/582142.

Bodenmann, G., Kaiser, A., Hahlweg, K., & Fehm-Wolfdorf, G. (1998). Communication patterns during marital conflict: A cross-cultural replication. *Personal Relationships*, *5*, 343–356. https://doi.org/10.2307/j.ctt1xp3vn1.9.

Bowman, M. L. (1990). Coping efforts and marital satisfaction: Measuring marital coping and its correlates. *Journal of Marriage and the Family*, *52*, 463–474. https://doi.org/10.2307/353040.

Bravo, C. S., & Martinez, A. W. (2016). Profiles using indicators of marital communication, communication styles, and marital satisfaction in Mexican couples. *Journal of Sex & Marital Therapy*, *43*, 361–376. https://doi.org/10.1080/0092623x.2016.1168332.

Brown, P., & Levinson, S. C. (1987). *Politeness: Some universals in language usage*. Cambridge: Cambridge University Press.

Burleson, B. R. (2003). The experience and effects of emotional support: What the study of cultural and gender differences can tell us about close relationships, emotion, and interpersonal communication. *Personal Relationships*, *10*, 1–23. https://doi.org/10.1111/1475-6811.00033.

Burleson, B. R., & Denton, W. H. (1997). The relationship between communication skill and marital satisfaction: Some moderating effects. *Journal of Marriage and Family*, *59*, 884–902. https://www.jstor.org/stable/353790.

Burleson, B. R., Metts, S., & Kirch, M. W. (2003). Communication in close relationships. In C. Hendrick & S. S. Hendrick (Eds.), *Close relationships: A sourcebook* (pp. 245–258). Thousand Oaks, CA: Sage Publications.

Burleson, B. R., & Mortenson, S. R. (2003). Explaining cultural differences in evaluations of emotional support behaviors exploring the mediating influences of value systems and interaction goals. *Communication Research*, *30*(2), 113–146. https://doi.org/10.1177/0093650202250873.

Bustamante, R. M., Nelson, J. A., Henrilksen, R. C., Jr., & Monakes, S. (2011). Intercultural couples: Coping with culture-related stressors. *The Family Journal: Counseling and Therapy for Couples and Families*, *19*, 154–164. https://doi.org/10.1177/1066480711399723.

Campbell, L., Simpson, J. A., Kashy, D. A., & Fletcher, G. J. O. (2001). Ideal standards, the self, and flexibility of ideals in close relationships. *Personality and Social Psychology Bulletin*, *27*, 447–462. https://doi.org/10.1037/e413792005-117.

Canary, D. J., & Stafford, L. (1992). Relational maintenance strategies and equity in marriage. *Communication Monographs*, *59*, 243–266. https://doi.org/10.1080/03637759209376268.

Cao, H., Fang, X., Fine, M., Ju, X., Lan, J., & Liu, X. (2015). Beyond the average marital communication: Latent profiles of the observed interaction among Chinese newlywed couples. *Journal of Family Psychology*, *29*, 850–862. https://doi.org/10.1037/fam0000126.

Caughlin, J. P., & Scott, A. M. (2010). Toward a communication theory of the demand/withdraw pattern of interaction in interpersonal relationships. In S. W. Smith & S. R. Wilson (Eds.), *New directions in interpersonal communication research* (pp. 180–200). Beverly Hills, CA: Sage.

Celenk, O., & van de Vijver, F. J. R. (2013). Perceived antecedents of marital satisfaction among Turkish, Turkish-Dutch, and Dutch couples. *International Journal of Psychology*, *48*, 1165–1175. https://doi.org/10.1080/00207594.2012.741242.

Chambless, D. L., & Blake, K. D. (2009). Construct validity of the perceived criticism measure. *Behavior Therapy*, *40*, 155–163. https://doi.org/10.1016/j.beth.2008.05.005.

Chi, P., Epstein, N. B., Fang, X., & Lam, D. O. B. (2013). Similarity of relationship standards, couple communication patterns, and marital satisfaction among Chinese couple. *Journal of Family Psychology*, *27*, 806–816. https://doi.org/10.1037/a0034113.

Christensen, A., Eldridge, K., Catta-Preta, A. B., Lim, V. R., & Santagata, R. (2006). Cross-cultural consistency of the demand/withdraw interaction pattern in couples. *Journal of Marriage and Family*, *68*, 1029–1044. https://doi.org/10.1111/j.1741-3737.2006.00311.x.

Christensen, A., & Heavey, C. L. (1990). Gender and social structure in the demand/withdraw pattern of marital conflict. *Journal of Personality and Social Psychology*, *59*, 73–81. https://doi.org/10.1037//0022-3514.59.1.73.

Christensen, A., & Sullaway, M. (1984). *Communication patterns questionnaire*. Unpublished manuscript. Los Angeles, CA: University of California.

Crenshaw, A. O., Christensen, A., Baucom, D. H., Epstein, N. B., & Baucom, B. R. W. (2017). Revised scoring and improved reliability for the Communication Patterns Questionnaire. *Psychological Assessment*, *29*(7), 913–925. https://doi.org/10.1037/pas0000385.

Diamond, L. M. (2006). The intimate same-sex relationships of sexual minorities. In A. L. Vangelisti & D. Perlman (Eds.), *The Cambridge handbook of personal relationships* (pp. 293–312). New York, NY: Cambridge University Press.

Diener, E., Suh, E. M., Lucas, R. E., & Smith, H. L. (1999). Subjective well-being: Three decades of progress. *Psychological Bulletin*, *125*(2), 276–302. https://doi.org/10.1037/0033-2909.125.2.276.

Dillon, L. M., Novak, N., Weisfeld, G. E., Weisfeld, C. C., Shattuck, K. S., Imagoglu, O. E., et al. (2015). Sources of marital conflict in five cultures. *Evolutionary Psychology*, *13*, 1–15. https://doi.org/10.1177/147470491501300101.

Dion, K. K., & Dion, K. L. (1993). Individualistic and collectivistic perspective on gender and the cultural context of love and intimacy. *Journal of Social Issues*, *49*, 53–59. https://doi.org/10.1111/j.1540-4560.1993.tb01168.x.

Donato, K. M., Tolbert, C. M., Nucci, A., & Kawano, Y. (2007). Recent immigrant settlement in the nonmetropolitan United States: Evidence from internal census data. *Rural Sociology*, *72*, 537–559. https://doi.org/10.1526/003601107782638666.

Epstein, N. H., Baucom, D. H., & Daiuto, A. (1997). Cognitive-behavioral couples therapy. In W. K. Halford & H. J. Markman (Eds.), *Clinical handbook of marriage and couples intervention* (pp. 415–449). West Sussex, England: Wiley.

Epstein, N. B., Chen, F., & Beyder-Kamjou, I. (2005). Relationship standards and marital satisfaction in Chinese and American couples. *Journal of Marital and Family Therapy*, *31*, 59–74. https://doi.org/10.1111/j.1752-0606.2005.tb01543.x.

Falconier, M. K., & Epstein, N. B. (2011). Female-demand/male-withdraw communication in Argentinian couples: A mediating factor between economic strain and relationship distress. *Personal Relationships*, *18*, 586–603. https://doi.org/10.1111/j.1475-6811.2010.01326.x.

Fincham, F. D. (2003a). Marital conflict: Correlates, structure, and context. *Current Directions in Psychological Science*, *12*, 23–27. https://doi.org/10.1111/1467-8721.01215.

Fincham, F. D. (2003b). Communication in marriage. In A. L. Vangelisti (Ed.), *Handbook of family communication* (pp. 83–103). Mahwah, NJ: Lawrence Erlbaum Associates, Publishers.

Fitzpatrick, M. (1988). A typological approach to marital interaction. In P. Noller & M. Fitzpatrick (Eds.), *Perspectives on marital interaction* (pp. 98–120). Philadelphia, PA: Multilingual.

Floyd, F. J. (1989). Segmenting interactions: Coding units for assessing marital and family behaviors. *Behavioral Assessment*, *11*(1), 13–29.

Fu, X., Tora, J., & Kendall, H. (2001). Marital happiness and interracial marriage: A study in a multi-ethnic community in Hawaii. *Journal of Comparative Family Studies*, *31*, 45–61.

Giblin, P., Sprenkle, D. H., & Sheehan, R. (1985). Enrichment outcome research: A meta-analysis of premarital, marital, and family interventions. *Journal of Marital and Family Therapy*, *11*, 257–271. https://doi.org/10.1111/j.1752-0606.1985.tb00619.x.

Goodwin, R., & Pillay, U. (2006). Relationships, culture, and social change. In A. L. Vangelisti & D. Perlman (Eds.), *The Cambridge handbook of personal relationships* (pp. 695–708). New York, NY: Cambridge University Press.

Gottman, J. M. (1994). *What predicts divorce? The relationship between marital processes and marital outcomes*. Hillsdale, NJ: Erlbaum.

Gudykunst, W. B., & Matsumoto, Y. (1996). Cross-cultural variability of communication in personal relationships. In W. B. Gudykunst, S. Ting-Toomey, & T. Nishida (Eds.), *Communication in personal relationships across cultures* (pp. 19–56). Thousand Oaks, CA: Sage.

Hahlweg, K., Kaiser, A., Christensen, A., Fehm-Wolfsdorf, G., & Groth, T. (2000). Self-report and observational assessment of couples' conflict: The concordance between the Communication Patterns Questionnaire and the KPI observation system. *Journal of Marriage and the Family*, *62*, 61–67. https://doi.org/10.1111/j.1741-3737.2000.00061.x.

Halford, W. K. (2011). *Marriage and relationship education: What works and how to provide it*. New York: Guilford Press.

Halford, W. K., Hahlweg, K., & Dunne, M. (1990). The cross-cultural consistency of marital communication associated with marital distress. *Journal of Marriage and the Family*, *52*, 487–500. https://doi.org/10.2307/353042.

Halford, W. K., Leung, P., Hung-Cheung, C., Chau-Wan, L., Hiew, D., & van de Vijver, F. (2018). Couple relationships standards and migration: Comparing Hong Kong Chinese with Australian Chinese. *Family Process*, *57*(4), 996–1011. https://doi.org/10.1111/famp.12337.

Harzing, A. W. (2006). Response style in cross-national survey research: A 26-country study. *International Journal of Cross-Cultural Management*, *6*, 243–266. https://doi.org/10.1177/1470595806066332.

Heyman, R. E. (2001). Observation of couple conflicts: Clinical assessment applications, stubborn truths, and shaky foundations. *Psychological Assessment*, *13*, 5–35. https://doi.org/10.1037//1040-3590.13.1.5.

Heyman, R. E., & Hunt, A. N. (2007). Replication in observational couples research: A commentary. *Journal of Marriage and Family*, *69*, 81–85. https://doi.org/10.1111/j.1741-3737.2006.00345.x.

Heyman, R. E., & Slep, A. M. S. (2001). The hazards of predicting divorce without cross-validation. *Journal of Marriage and the Family*, *63*, 473–479. https://doi.org/10.1111/j.1741-3737.2001.00473.x.

Hiew, D. N., Halford, W. K., van de Vijver, F. J., & Liu, S. (2015). Relationships standards and satisfaction in Chinese, Western, and intercultural Chinese-Western couples in Australia. *Journal of Cross-Cultural Psychology*, *46*, 684–701. https://doi.org/10.1177/0022022115579936.

Hiew, D. N., Halford, W. K., van de Vijver, F. J., & Liu, S. (2016). Communication and relationship satisfaction in Chinese, Western, and intercultural Chinese-Western couples. *Journal of Family Psychology*, *30*, 191–202. https://doi.org/10.1037/fam0000144.

Hofstede, G. (1980). *Culture's consequences: International difference in work-related values*. Newbury Park, CA: Sage.

Hooley, J. M., & Hahlweg, K. (1989). Marital satisfaction and marital communication in German and English couples. *Behavioral Assessment*, *11*, 119–133.

Hui, C. H., & Triandis, H. C. (1986). Individualism-collectivism: A study of cross-cultural researchers. *Journal of Cross-Cultural Psychology*, *17*, 225–248. https://doi.org/10.1177/0022002186017002006.

Huong, B. T. (2010). 'Let's talk about sex, baby': Sexual communication in marriage in contemporary Vietnam. *Culture, Health & Sexuality*, *12*, S19–S29. https://doi.org/10.1080/13691050903072025.

Ingersoll-Dayton, B., Campbell, R., & Mattson, J. (1998). Forms of communication: A cross-cultural comparison of older couples in the USA and Japan. *Journal of Cross-Cultural Gerontology*, *13*, 63–80.

Inglehart, R., & Welzel, C. (2005). *Modernization, cultural change and democracy: The human development sequence*. New York, NY: Cambridge University Press.

Jones, A. C., Robinson, D., & Seedall, R. B. (2017). The role of sexual communication in couples' sexual outcomes: A dyadic path analysis. *Journal of Marital and Family Therapy*, *44*, 606–623. https://doi.org/10.1111/jmft.12282.

Karney, B. R., & Bradbury, T. N. (1995). The longitudinal course of marital quality and stability: A review of theory, method, and research. *Psychological Bulletin*, *118*, 3–34. https://doi.org/10.1037/0033-2909.118.1.3.

Kelly, A. B., Fincham, F. D., & Beach, S. R. H. (2003). Communication skills in couples: A review and discussion of emerging perspectives. In J. O. Green & B. R. Burleson (Eds.), *Handbook of communication and social interaction skills* (pp. 723–751). Mahwah, NJ: Laurence Erlbaum Associates Publishers.

Kelly, A. B., & Halford, W. K. (1993). Adapting behavioural marital therapy to the needs of a secondary marriage: A case evaluation. *Behavioural Psychotherapy*, *21*(2), 115–125. https://doi.org/10.1017/s0141347300018085.

Kerig, P. K. (1996). Assessing the links between interparental conflict and child adjustment: The conflicts and problem-solving scales. *Journal of Family Psychology*, *10*, 454–473. https://doi.org/10.1037//0893-3200.10.4.454.

Kim, J. L., & Ward, L. M. (2007). Silence speaks volumes: Parental sexual communication among Asian American emerging adults. *Journal of Adolescent Research*, *22*, 3–331. https://doi.org/10.1177/0743558406294916.

Kline, G. H., Pleasant, N. D., Whitton, S. W., & Markman, H. J. (2006). Understanding couple conflict. In A. L. Vangelisti & D. Perlman (Eds.), *The Cambridge handbook of personal relationships* (pp. 445–462). New York: Cambridge University Press.

Kline, S., & Zhang, S. (2009). The role of relational communication characteristics and filial piety in mate preferences: Cross-cultural comparisons of Chinese and US college students. *Journal of Comparative Family Studies*, *40*(3), 325–353.

Lavner, J. A., & Bradbury, T. N. (2010). Patterns of change in marital satisfaction over the newlywed years. *Journal of Marriage and the Family*, *58*, 945–957. https://doi.org/10.1111/j.1741-3737.2010.00757.x.

Lavner, J. A., Karney, B. R., & Bradbury, T. N. (2016). Does couples' communication predict marital satisfaction or does marital satisfaction predict communication. *Journal of Marriage and Family*, *78*, 680–694. https://doi.org/10.1111/jomf.12301.

Li, X., Cao, H., Zou, N., Ju, X., Lan, J., Zhu, Q., et al. (2018). Daily communication, conflict resolution, and marital quality in Chinese marriage: A three wave, cross-lagged analysis. *Journal of Family Psychology*, *32*, 733–742. https://doi.org/10.1037/fam0000430.

Li, T., & Hsiao, Y. (2008). Huaren hunyin pinzhi de weixi: Chongtu yu renrang de zhongjie xiaoguo [Maintaining quality in marriage: The mediating effect of conflict and ren (tolerance) for Taiwanese marriage couples]. *Indigenous Psychological Research in Chinese Societies*, *29*, 77–116.

Markman, H. J., Rhoades, G. K., Stanley, S. M., Ragan, E. P., & Whitton, S. W. (2010). The premarital communication roots of marital distress and divorce: The first five years of marriage. *Journal of Family Psychology*, *24*, 289–298. https://doi.org/10.1037/a0019481.

Markus, H. R., & Kitayama, S. (1991). Culture and the self: Implications for cognition, emotion, and motivation. *Psychological Review*, *98*, 224–253. https://doi.org/10.1037//0033-295x.98.2.224.

Navran, L. (1967). Communication and adjustment in marriage. *Family Process*, *6*, 173–184. https://doi.org/10.1111/j.1545-5300.1967.00173.x.

Noller, P., & Feeney, J. A. (2004). Studying family communication: Multiple methods and multiple sources. In A. L. Vangelisti (Ed.), *Handbook of family communication* (pp. 31–50). Mahwah, NJ: Erlbaum.

O'Keefe, B. J. (1988). The logic of message design: Individual differences in reasoning about communication. *Communication Monographs*, *55*(1), 80–103. https://doi.org/10.1080/03637758809376159.

Pasch, L. A., & Bradbury, T. N. (1998). Social support, conflict, and development of marital dysfunction. *Journal of Consulting and Clinical Psychology*, *66*, 219–230. https://doi.org/10.1037/0022-006x.66.2.219.

Rehman, U. S., & Holtzworth-Munroe, A. (2006). A cross-cultural analysis of the demand-withdraw martial interaction: Observing couples from a developing country. *Journal of Consulting and Clinical Psychology*, *74*, 755–766. https://doi.org/10.1037/0022-006x.74.4.755.

Rehman, U. S., & Holtzworth-Munroe, A. (2007). A cross-cultural examination of the relation of marital communication behavior to marital satisfaction. *Journal of Family Psychology*, *21*, 759–763. https://doi.org/10.1037/0893-3200.21.4.759.

Rehman, U. S., Janssen, E., Newhouse, S., Heiman, J., Holtzworth-Munroe, A., Fallis, E., et al. (2011). Marital satisfaction and communication behaviors during sexual and non-sexual conflict discussions in newlywed couples: A pilot study. *Journal of Sex and Marital Therapy*, *37*, 94–103. https://doi.org/10.1080/0092623x.2011.547352.

Rogge, R. M., & Bradbury, T. N. (1999). Recent advances in the prediction of marital outcomes. In R. Berger & M. T. Hannah (Eds.), *Preventive approaches in couple therapy* (pp. 331–360). Philadelphia, PA: Brunner/Mazel.

Russell, J. A., & Yik, M. S. (1996). Emotion among the Chinese. In M. H. Bond (Ed.), *The handbook of Chinese psychology* (pp. 166–188). Oxford: Oxford University Press.

Sadeghi, M. S., Mazaheri, D. A., Motabi, D. F., & Zahedi, K. (2012). Marital interaction in Iranian couples: Examining the role of culture. *Journal of Comparative Family Studies*, *43*, 281–300. https://doi.org/10.3138/jcfs.43.2.281.

Sanford, K. (2003). Problem-solving conversations in marriage: Does it matter what topics couples discuss? *Personal Relationships*, *10*, 97–112. https://doi.org/10.1111/1475-6811.00038.

Sanford, K. (2010). Assessing conflict communication in couples: Comparing the validity of self-report, partner-report, and observer ratings. *Journal of Family Psychology*, *24*, 165–174. https://doi.org/10.1037/a0017953.

Sastry, J. (1999). Household structure, satisfaction, and distress in India and the United States: A comparative cultural examination. *Journal of Comparative Family Studies*, *30*, 135–152.

Schwartz, S. H. (1990). Individualism-collectivism: Critique and proposed refinements. *Journal of Cross-Cultural Psychology*, *21*, 139–157. https://doi.org/10.1177/0022022190212001.

Sillars, A. L., & Vangelisti, A. L. (2006). Communication: Basic properties and their relevance to relationship research. In A. L. Vangelisti & D. Perlman (Eds.), *The Cambridge handbook of personal relationships* (pp. 33–351). New York, NY: Cambridge University Press.

Snyder, D. K. (1997). *Martial satisfaction inventory, revised*. Los Angeles, CA: Western Psychological Services.

Sprecher, S., Christopher, F. S., & Cate, R. (2006). Sexuality in close relationships. In A. L. Vangelisti & D. Perlman (Eds.), *The Cambridge handbook of personal relationships* (pp. 463–482). New York, NY: Cambridge University Press.

Stafford, L. (2011). Measuring relationship maintenance behaviors: Critique and development of the revised relationship maintenance behavior scale. *Journal of Social and Personal Relationships*, *28*, 278–303. https://doi.org/10.1177/0265407510378125.

Straus, M. A., Hamby, S. L., Boney-McCoy, S., & Sugarman, D. B. (1996). The revised Conflict Tactics Scales (CTS2): Development and preliminary psychometric data. *Journal of Family Issues*, *17*, 283–316. https://doi.org/10.1177/019251396017003001.

Su, L. P., Miller, R. B., Canlas, J. M., Li, T.-S., Hsiao, Y.-L., & Willoughby, B. J. (2015). A cross-cultural study of perceived marital problems in Taiwan and the United States. *Contemporary Family Therapy*, *37*, 165–175. https://doi.org/10.1007/s10591-015-9337-7.

Sze, Y. T., Hou, J., Lan, J., & Fang, X. (2011). Profiling family therapy users of a therapy center in Beijing. *American Journal of Family Therapy*, *39*, 299–306. https://doi.org/10.1080/01926187.2010.542073.

Tien, N. C., Softas-Nall, L., & Barritt, J. (2017). Intercultural/multilingual couples: Implications for counseling. *The Family Journal: Counseling and Therapy for Couples and Families*, *25*, 156–163. https://doi.org/10.1037/e741382011-239.

Tili, T. R., & Barker, G. G. (2015). Communication in intercultural marriages: Managing cultural differences and conflicts. *Southern Communication Journal*, *80*, 189–210. https://doi.org/10.1080/1041794x.2015.1023826.

Ting-Toomey, S. (1999). *Communicating across cultures*. New York, NY: Guilford Press.

Van de Vijver, F. J. R., & Tanzer, N. K. (1997). Bias and equivalence in cross-cultural assessment: An overview. *European Review of Applied Psychology*, *47*, 263–280.

Vazhappily, J. J., & Reyes, M. E. (2016). Couples' communication as a predictor of marital satisfaction among selected Filipino couples. *Psychological Studies*, *61*, 301–306. https://doi.org/10.1007/s12646-016-0375-5.

Weinstein, E. A. (1969). The development of interpersonal competence. In E. A. Goslin (Ed.), *Handbook of socialization theory and research* (pp. 753–775). Chicago, IL: Rand-McNally.

Williamson, H. C., Ju, X., Bradbury, T. N., Karney, B. R., & Fang, X. (2012). Communication behavior and relationship satisfaction among American and Chinese newlywed couples. *Journal of Family Psychology*, *26*, 308–315. https://doi.org/10.1037/a0027752.

Williamson, H. C., Karney, B. R., & Bradbury, T. N. (2013). Financial strain and stressful events predict newlywed' negative communication independent of relationship satisfaction. *Journal of Family Psychology*, *27*, 65–75. https://doi.org/10.1037/a0031104.

Winkler, I., & Doherty, W. J. (1983). Communication style and marital satisfaction in Israeli and American couples. *Family Process*, *22*, 221–228. https://doi.org/10.1111/j.1545-5300.1983.00221.x.

Xie, T., & Galliher, R. V. (2018). Acculturation and language use in intimate and sexual relationships among Chinese bilinguals. *Asian American Journal of Psychology*, *9*, 227–236. https://doi.org/10.1037/aap0000116.

Xu, Y., & Burleson, B. R. (2001). Effects of sex, culture and social type on spousal social support: An assessment the "support gap" hypothesis in early marriage. *Human Communication Research*, *27*(4), 533–566. https://doi.org/10.1111/j.1468-2958.2001.tb00792.x.

Xu, A., Xie, X., Liu, W., Xia, Y., & Liu, D. (2007). Chinese family strengths and resiliency. *Marriage & Family Review*, *41*, 143–164. https://doi.org/10.1300/j002v41n01_08.

Yelsma, P., & Athappilly, K. (1988). Marital satisfaction and communication practices: Comparisons among Indian and American couples. *Journal of Comparative Studies*, *19*, 37–54.

Yum, Y. O., & Canary, D. J. (2009). Cultural differences in equity theory predictions of relational maintenance strategies. *Human Communication Research*, *35*, 384–406. https://doi.org/10.1111/j.1468-2958.2009.01356.x.

Yum, Y. O., Canary, D. J., & Baptist, J. A. (2015). The role of culture and fairness in maintaining relationships in Malaysia, Singapore, and the United States. *International Journal of Intercultural Relations*, *44*, 100–112. https://doi.org/10.1016/j.ijintrel.2014.12.003.

Zhang, S., & Kline, S. (2009). Can I make my own decision? A cross-cultural study of social network influence in mate selection. *Journal of Cross-Cultural Psychology*, *40*, 3–23. https://doi.org/10.1177/0022022108326192.

Zietlow, P. H., & Sillars, A. L. (1988). Life stage differences in communication during martial conflicts. *Journal of Social and Personal Relationships*, *5*, 223–245. https://doi.org/10.1177/026540758800500206.

CHAPTER 8

Intimate partner violence and child maltreatment: Definitions, prevalence, research, and theory through a cross-cultural lens

Aleja M. Parsons, Richard E. Heyman, Danielle M. Mitnick, and Amy M. Smith Slep
Family Translational Research Group, New York University, New York, NY, United States

What constitutes intimate partner abuse and neglect? What constitutes child abuse and neglect? Are these phenomena socially determined by the norms of cultures and subcultures, or is it possible to establish worldwide standards that delineate the line between acceptable and abusive/neglectful family behavior? This chapter will take on these issues.

On the one hand, widespread recognition of the existence and pervasive nature of family maltreatment[a] has gained momentum in the past few decades and is now prioritized as a global issue with significant health implications (e.g., World Health Organization, 2016). Researchers internationally—including those at the World Health Organization (WHO), the United Nations agency charged with monitoring and improving global public health—have established that family maltreatment is common, and guidelines for maltreatment were included in clinical frameworks (e.g., 10th International Classification of Diseases [ICD-10], World Health Organization, 1992).

On the other hand, those creating worldwide guidelines are well aware that variability in norms and traditions (including those tied to religious beliefs) make a single standard agreed on by all people in all lands impossible.

[a]In the child abuse and neglect field, "child maltreatment" is the umbrella term for "child abuse" and "child neglect," with "abuse" referring to harm due to acts by the parent and "neglect" referring to acts or omissions that failed to protect the child from harm. In a parallel fashion, "partner maltreatment" is used as an umbrella term for partner abuse and neglect.

Cross-Cultural Family Research and Practice
https://doi.org/10.1016/B978-0-12-815493-9.00008-9

From a research perspective, such heterogeneity led to wide-ranging operationalizations of maltreatment, making interpretation of prevalence difficult, or operationalizations (such as those that measure acts but not impacts) that do not match evolving guidelines for maltreatment (Saltzman, 2004).

Thus, this chapter will provide a brief consideration of maltreatment from a cross-cultural perspective, given the current status of the field. First, we discuss recent attempts at sharpening the global guidelines for both partner and child maltreatment (and the research underpinning these guidelines), which we consider a necessary prerequisite to understanding what aspects of maltreatment may be universal and which may be shaped by culturally bound factors. Despite the current limitations, we also review the research on cross-cultural prevalence. Finally, we highlight how cultural factors must be integrated into theoretical conceptualizations of the etiological factors and mechanisms contributing to the occurrence and maintenance of partner and child abuse and neglect to fully understand the impact of family maltreatment globally. It is our view that both standardizing definitions for maltreatment and investigating the culturally bound nuance in experiences of maltreatment are of vital importance.

Emerging global definitions of partner and child maltreatment

The evolution of global criteria used in clinical, research, and public health forums occurred in concert with broader international declarations, treaties, and legal obligations. For example, the United Nations (1989) enacted the Convention on the Rights of the Child, which was ratified by 196 countries and asserted that children have the right to be free from all forms of mental and physical violence, injury, abuse, and neglect from any caregiver. This helped establish international standards for appropriate treatment of children and set the stage for more uniform standards defining maltreatment. Although no similar single convention for partner maltreatment exists, the United Nations (UN) (2009, p. 1) declared that: "States have clear obligations under international law to enact, implement and monitor legislation addressing all forms of violence against women. Over the past two decades, many States have adopted or revised legislation on violence against women. However, … [m]any States still do not have in place legislative provisions that specifically address violence against women and, even where legislation exists, it is often limited in scope and coverage, or is not enforced." Although the UN language is framed in terms of violence against women, laws in many countries are gender-neutral (thus encompassing violence by any partner in an intimate relationship).

Operationalizations for health professionals are found in the World Health Organization's International Classification of Diseases (ICD). ICD provides international guidelines (i.e., criteria) for diseases and health conditions, making it the most critical global diagnostic standard for both clinical and research use. ICD-10 (World Health Organization, 1992) included listings for each of the varied forms of maltreatment (physical, psychological, sexual, neglect) but did not provide detailed guidelines, leaving the meaning of that form of maltreatment up to the discretion of the assessor.

During the ensuing decades, however, a foundational set of maltreatment criteria were developed and tested for reliability and validity. Criteria were: (a) adopted by some key entities (the US Department of Defense and the psychiatric "bible" [the Diagnostic and Statistical Manual of Mental Disorders, DSM-5; APA, 2013]); and (b) are being field-tested for the ICD-11 (see Heyman et al., 2018). Because of their incorporation into these major systems, possibly including the ICD-11, we focus on these systems in this chapter. We first provide an overview of the research that developed and tested the criteria that DSM and ICD adapted and adopted, then detail the criteria for each form of family maltreatment, and discuss the cross-cultural considerations and implications of attempting to make worldwide criteria for such phenomena.

Foundational criteria for family maltreatment

The criteria were originally developed and field-tested in a five-study program (Heyman & Slep, 2006) that included (1) a content validity study, (2) a mixed-methods study regarding clinical utility with treatment professionals for family maltreatment, (3) development of operationalized criteria, (4) evaluation of the interrater agreement of the criteria under typical usage in field settings, and (5) evaluation of the interrater agreement of the criteria using a computerized decision support tool. Agreement between field-users at five sites and master reviewers using prior definitions—similar to the vague ones used in the laws in the United States and in international guidelines—was 50%; in the final field trial of the development study, agreement was 92% (Heyman & Slep, 2006) and was maintained at 91% when the criteria were disseminated to 41 sites worldwide (Heyman & Slep, 2009).

Evidence of content and criterion validity of maltreatment has been documented across multiple studies (Heyman & Slep, 2019). These studies established the consistency (reliability) of decisions made using the criteria; correctness (validity) of the criteria was established through several means. First, content validity of the proposed maltreatment criteria can be inferred given: (a) the content validity studies conducted during the development process (on the preceding criteria; Heyman & Slep, 2006, Study 1), and the correction of deficits based on expert (research and clinician) and field-clinician recommendations; and (b) the reasonably high concordance between the proposed criteria and that of (1) the US Centers for Disease Control and Prevention (Breiding, Basile, Smith, Black, & Mahendra, 2015; Leeb, Paulozzi, Melanson, Simon, & Arias, 2008) which were approved by expert researchers, representatives from the United States and international governments and nongovernmental organizations, the US state health department representatives (professional associations, medical organizations, and advocacy coalitions) and (2) the WHO Multicountry Study on Women's Health and Domestic Violence against Women Study (Garcia-Moreno et al., 2006). Second, construct validity was established via relations in questionnaire studies (with over 100,000 participants) of abuse classifications (using the criteria) to a variety of hypothesized outcomes such as relationship satisfaction, depressive symptoms, and alcohol problems (see Wojda et al., 2017). Third, clinical validity was indicated by US Air Force users' ratings that the criteria made determinations fair to both alleged victims and offenders and that they were generally easy to use (Heyman, Collins, Slep, & Knickerbocker, 2010). Similar results came from a study of social workers and supervisors in the US state of Alaska (which adopted the child maltreatment criteria), who found them to be clear, useful, efficient, and a significant improvement over previous practices (Perkins, Slep, & Heyman, 2017).

What constitutes maltreatment: Creating a harm criterion

To best understand the conceptual framework used in the foundational criteria and put them in a cross-cultural context, it is critical to understand their incorporation of a "harm criterion." Since DSM-III-R (APA, 1987), DSM individual diagnoses required (a) a signature syndrome and (b) harm (e.g., "present distress … or disability … or significantly increased risk of suffering death, pain, disability, or an important loss of freedom;") so as "to reduce the risk of 'false positives' that might otherwise arise by diagnosing cases that are too mild to be properly considered disorders" (Cooper, 2015, p. 87). The maltreatment criteria harnessed the harm criterion concept by requiring that all criteria sets include (a) at least one signature act (or, in the case of neglect, an act or omission) and (b) a clinically significant impact or a more than the reasonable potential for a clinically significant impact. Given the intensely personal violation involved with sexual abuse and the brighter line drawn legally and in (most) cultures around acts of coercive or nonconsensual sexual contact, only a qualifying act was required. (That is, harm or potential for harm was presumed for sexual abuse.)

Thus, in this chapter, "abuse" requires both an aggressive act of a particular type *and* a qualifying impact; similarly, "neglect" requires an egregious act/omission *and* a qualifying impact. (We sometimes use the phrase "clinically significant abuse" to emphasize that we mean incidents that meet the criteria to distinguish such operationalizations from more nebulous usage of "abuse.")

Harm criterion in a cross-cultural context

As with any behavioral health criterion, creating a threshold between "socially sanctioned" and "problematic" is an inherently culture-bound decision. Some operationalizations of family maltreatment measure only act and operationalize all acts as abusive (which not only sets the threshold at a low level that does not have international consensus but also ignores contextual factors, such as protection of self). In contrast, the advantage of the harm criterion is that it requires acts to be damaging (or have a more than reasonable potential to be damaging) for them to be considered abusive (and not just ill-advised). In summary, although people from various cultures worldwide can disagree on whether causing injury to an intimate partner is "wrong," the harm criterion itself is a fairly objective way to draw a bright line between abusive and nonabusive acts.

Note that the notion of a harm criterion does not resolve all arguments regarding a bright-line distinction. For example, there is considerable evidence indicating that physical punishment of children is detrimental to their well-being (e.g., Gershoff, 2002; Gershoff & Grogan-Kaylor, 2016); 54 countries worldwide prohibit it (Global Initiative to End All Corporal Punishment of Children, 2018). As indicated below, the harm criterion does not consider local illegality or future, agglomerated harm, but instead operationalizes harm from an act (or acts) as (a) injury, (b) more than the reasonable potential for injury, or (c) significant fear.

In the subsequent sections, we discuss the foundational criteria that were found in the proposed ICD-11 guidelines. Table 8.1 shows the two criteria sets.

Table 8.1 Comparison of foundational criteria developed and tested by Heyman and Slep (2006) and implemented by the US Department of Defense (DoD) and the proposed criteria for ICD-11.

	DoD	ICD-11 (proposed)
Spouse or Intimate Partner Violence, Physical		
Act	Nonaccidental use of physical force • Physical force includes, but is not limited to, at least one of the following: hitting with open hand or slapping, including spanking, dropping, pushing or shoving; grabbing or yanking limbs or body, poking; hair-pulling; scratching; pinching; restraining; or squeezing shaking; throwing; biting; kicking; hitting with fist; hitting with a stick, strap, belt, electric cord or other object; scalding or burning; poisoning; stabbing; applying force to throat; strangling or cutting off air supply; holding underwater; brandishing or using a weapon.	At least one nonaccidental act of physical force (e.g., push/shove, scratch, slap, throw something that could hurt, punch, bite)
Impact	Significant impact on the intimate partner involving any of the following: 1. Any physical injury (including, but not limited to, pain that lasts at least 4 hours, bruises, cuts, sprains, broken bones, loss of consciousness, death) 2. Reasonable potential for more than inconsequential physical injury[a] given the inherent dangerousness of the act; the degree of force used; and the physical environment in which acts occurred; 3. More than inconsequential fear reaction.[b]	Act causes (or exacerbates) at least one of the following impacts: • Any physical injury; • Significant fear; • Reasonable potential for significant physical injury.
Exclusions	1. Acts committed to protect self from imminent physical harm a. Act occurred while partner was in the act of using physical force. "In the act" begins with the initiation of motoric behavior that typically would result in an act of physical force such as charging at the partner to hit him/her and ends when the use of force is no longer imminent. b. Sole function of act was to stop partner's use of physical force. c. Act used minimally sufficient force to stop partner's use of physical force.	Act for physical protection of self (e.g., to ward off partner's punches) or partner (e.g., prevent partner from attempting suicide)

	2. Act committed to protect self from imminent physical harm where (a) act followed partner's threat (verbal or nonverbal) to imminently inflict more than inconsequential physical injury and (b) at least one previous incident of partner inflicting more than inconsequential physical injury.[a] 3. Acts committed to protect partner, others, or pet	
Spouse or Intimate Partner Violence, Psychological		
Act	Nonaccidental verbal or symbolic act(s) (excluding physical and sexual abusive acts) such as those listed below. Acts not listed, but of similar severity, are also eligible. • Interrogating partner • Berating, disparaging, humiliating partner (or other similar behavior) • Isolating partner from family, friends, or social support resources • Interfering with the partner's cultural adaptation • Restricting partner's access to or use of economic resources • Obstructing partner from receiving medical, mental health, or dental services • Restricting the partner's ability to come and go freely • Trying to make partner think that s/he is mentally ill or make others think that the partner is mentally ill • Threatening to harm partner directly or indirectly (including, but not limited to, threatening to: inflict physical or sexual abuse; harm the partner's children, pets, or people that the partner cares about; damage or destroy the partner's property) • Harming the partner's children, pets, or property • Stalking partner • Obstructing partner's access to protective assistance	Verbal or symbolic acts with the potential to cause psychological harm to the victim, e.g.: • Berating, disparaging, degrading, humiliating partner; • Interrogating partner; • Restricting partner's ability to come and go freely; • Obstructing partner's access to assistance (example: police aid, legal help, protective resources, medical resources, mental health resources); • Threatening partner; • Harming, or threatening to harm, people/things that partner cares about; • Restricting partner's access to or use of economic resources; • Isolating partner from family, friends, or social support resources; • Stalking partner; • Trying to make people think that s/he is crazy (or make others think that partner is crazy)

Continued

Table 8.1 Comparison of foundational criteria developed and tested by Heyman and Slep (2006) and implemented by the US Department of Defense (DoD) and the proposed criteria for ICD-11—cont'd

	DoD	ICD-11 (proposed)
Impact	Significant impact on the intimate partner involving any of the following: 1. Psychological harm, including any of the following a. More than inconsequential fear reaction[b] b. Significant psychological distress (major depressive disorder, posttraumatic stress disorder, acute stress disorder, or a constellation of symptoms similar to other psychiatric disorders) related to, or exacerbated by, the act(s) c. Fear of emotionally abusive act(s) that significantly interferes with the ability to carry out any of these five major life activities: work; education; religion; necessary medical or mental health services; or contact with family/friends 2. Stress-related somatic symptoms (related to or exacerbated by the acts) that significantly interfere with normal functioning	Act causes (or exacerbates) at least one of the following impacts: • Significant fear; • Significant psychological distress; • Somatic symptoms that interfere with normal functioning; • Fear of the recurrence of emotionally abusive act(s) that cause victim to significantly limit any of these five major life activities: (a) work; (b) education; (c) religion; (d) medical or mental health services; (e) contact with family/friends.
Spouse or Intimate Partner Violence, Sexual		
Act	1. The use of physical force to compel an intimate partner to engage in a sex act against his or her will, whether or not the act is completed. • A sex act is defined as contact between the penis and the vulva or the penis and the anus involving penetration, however slight; contact between the mouth and the penis, vulva, or anus; or penetration of the anal or genital opening by a hand, finger, or other object. 2. The use of a physically or emotionally aggressive act or use of one's body, size, or strength, or an emotionally aggressive act to coerce an intimate partner to engage in a sexual act whether or not the sexual act is completed. 3. An attempted or completed sex act involving an intimate partner who is unable to provide consent. That is, the partner is unable to understand the nature or conditions of the act, to decline participation, or to communicate unwillingness to engage in the sexual act (including, but not limited to, illness; disability; being asleep or under the influence of alcohol or other drugs).	• The use of physical force to compel participation in a sex act against the partner's will or when partner is incapable of consent (whether or not the act is completed); • The use of physical or emotional aggression to coerce the partner to participate in a sex act; • Physical contact of a sexual nature (e.g., kissing, fondling) that is against the expressed wishes of the partner and that causes considerable distress to the partner.
Impact	None required	None required

Spouse or Intimate Partner Violence, Neglect		
Act	When a partner is incapable of self-care due to substantial physical, psychological, or cultural limitations, the withholding, or threatening to withhold, the partner's access to, any of the following: • Appropriate, medically indicated health care, including, but not limited to, appropriate medical, mental health, or dental care; • Appropriate nourishment, shelter, clothing, or hygiene; or • Caregiving for more than 24 hours without having arranged for an appropriate surrogate caregiver.	At least one egregious act(s) or omission(s) by an adult's caregiver that deprives an intimate partner who is incapable of self-care of needed or adequate food, shelter, hygiene, or necessary services.
Impact	Significant deprivation-related impact involving any of the following: • More than inconsequential physical injury[a] • Reasonable potential for more than inconsequential physical injury[a]	Significant physical injury; reasonable potential for significant injury.
Child Abuse, Physical		
Act	Nonaccidental use of physical force • Physical force includes, but is not limited to, hitting with the open hand or slapping, including spanking; dropping; pushing or shoving; grabbing or yanking limbs or body; poking; hair-pulling; scratching; pinching; restraining or squeezing; shaking; throwing; biting; kicking; hitting with fist; hitting with a stick, strap, belt, electrical cord, or other object; scalding or burning; poisoning; stabbing; applying force to throat; strangling or cutting off air supply; holding under water; brandishing or using a weapon	Confirmed or suspected verbal or symbolic
Impact	Significant impact on the child involving any of the following: 1. More than inconsequential physical injury[a] 2. Reasonable potential for more than inconsequential physical injury[a] given the inherent dangerousness of the act; the degree of force used; and the physical environment in which acts occurred; 3. More than inconsequential fear reaction.[b]	Act causes (or exacerbates) at least one of the following impacts: any physical injury; significant fear; reasonable potential for significant physical injury

Continued

Table 8.1 Comparison of foundational criteria developed and tested by Heyman and Slep (2006) and implemented by the US Department of Defense (DoD) and the proposed criteria for ICD-11—cont'd

	DoD	ICD-11 (proposed)
Exclusions	1. Acts committed to protect self from imminent physical harm a. Act occurred while child/adolescent was in the act of using physical force. "In the act" begins with the initiation of motoric behavior that typically would result in an act of physical force such as charging at the partner to hit him/her and ends when the use of force is no longer imminent. b. Sole function of act was to stop child/adolescent's use of physical force. c. Act used minimally sufficient force to stop child/adolescent's use of physical force 2. Acts committed to protect child, others, or pet from imminent harm, as evidenced by both of the following: a. Sole function of act was to protect child, another person, or pet from imminent physical harm. b. Act used minimally sufficient force to stop child's potentially harmful behavior. 3. Acts committed during developmentally appropriate physical play (including, but not limited to, horseplay, wrestling, tackle football).	Act was not for physical protection of self (e.g., to ward off child/adolescent's aggressive acts) or child (e.g., prevent child from being injured from environmental danger) or during developmentally appropriate physical play.
Child Abuse, Psychological		
Act	Nonaccidental verbal or symbolic act(s) (excluding physical and sexual abusive acts) such as those listed below. Acts not listed, but of similar severity, are also eligible. • Berating, disparaging, degrading, scapegoating, or humiliating child (or other similar behavior)	Nonaccidental verbal or symbolic acts with the potential to cause psychological harm to the child. Examples: • Berating, disparaging, degrading, humiliating child;

	• Threatening child (including, but not limited to, indicating/implying future physical harm, abandonment, sexual assault) • Harming/abandoning—or indicating that alleged abuser will harm/abandon—people/things that child cares about, such as pets, property, loved ones • Confining child (a means of punishment involving restriction of movement, as by tying a child's arms or legs together or binding a child to a chair, bed, or other object, or confining a child to an enclosed area [such as a closet]) • Coercing the child to inflict pain on him/herself (including, but not limited to, ordering child to kneel on split peas/rice for long periods or ordering child to ingest highly spiced food) • Disciplining child (through physical or nonphysical means) excessively (i.e., extremely high frequency or duration, though not meeting physical abuse criteria)	• Threatening child (including, but not limited to, indicating/implying future physical harm, abandonment, sexual assault); • Harming/abandoning—or indicating that the parent/caregiver will harm/abandon—people/things that child cares about, such as pets, property, loved ones (including exposing child to criteria-meeting or subthreshold partner violence); • Confining child (e.g., typing a child's arms or legs together; binding a child to a chair, bed, or other object; or confining a child to an small enclosed area [such as a closet]); • Scapegoating child, i.e., blaming child for things for which they cannot possibly be responsible; coercing the child to inflict pain on him/herself; • Disciplining child (through physical or nonphysical means) excessively (i.e., extremely high frequency or duration, though not meeting physical abuse criteria); • Purposefully indoctrinating child to consider a parent evil, dangerous, or not worthy of affection.

Continued

Table 8.1 Comparison of foundational criteria developed and tested by Heyman and Slep (2006) and implemented by the US Department of Defense (DoD) and the proposed criteria for ICD-11—cont'd

	DoD	ICD-11 (proposed)
Impact	Significant impact on the child involving any of the following: 1. Psychological harm, including any of the following a. More than inconsequential fear reaction[b] b. Significant psychological distress (major depressive disorder, posttraumatic stress disorder, acute stress disorder, or a constellation of symptoms similar to other psychiatric disorders) related to, or exacerbated by, the act(s) 2. Reasonable potential for psychological harm a. The act (or pattern of acts) creates reasonable potential for the development of a psychiatric disorder (at or near diagnostic thresholds) related to, or exacerbated by, the act(s). The child's level of functioning and the risk and resilience factors present should be taken into consideration. b. The act (or pattern of acts) carries a reasonable potential for significant disruption of the child's physical, psychological, cognitive, or social development. 3. Stress-related somatic symptoms (related to or exacerbated by the acts) that significantly interfere with normal functioning	• Psychological harm (e.g., significant fear of abusive parent or other psychological distress); • Reasonable potential for significant psychological harm (e.g., for developing significant psychological problems or for significant disruption of the child's physical, psychological, cognitive, or social development); • Stress-related somatic symptoms that interfere with normal functioning.
Exclusions	Excluded are generally accepted practices such as child car seats, safety harnesses, swaddling of infants, and discipline involving "grounding" a child or restricting a child to his/her room.	

Child Abuse, Sexual		
Act	1. Forcing, tricking, enticing, threatening or pressuring a child to participate in acts for anyone's sexual gratification without direct physical contact between child and the alleged abuser. • Acts include, but are not limited to, exposing child's or alleged abuser's genitals, anus, or (female) breasts; having child masturbate or watch masturbation; having child participate in sexual activity with a third person (including child prostitution); having child pose, undress or perform in a sexual fashion (including child pornography); exposing child to pornography or live sexual performance; "peeping" or other prurient watching (i.e., voyeurism) without the child's knowledge. 2. The caregiver's use of force, emotional manipulation, trickery, threatening, or taking advantage of the child's youth or naiveté to engage in penetration of the vagina, however slight, by the penis or by a hand or finger or any object with the intent to abuse, humiliate, harass, or degrade the child; or to arouse or gratify the sexual desire of the alleged abuser, the child, or any other person. 3. The caregiver's engaging in any of the following: a. Placing the alleged abuser's sexual organ in the mouth or anus of a child, however slight the penetration; or b. Taking into the alleged abuser's mouth or anus the sexual organ of a child, however slight the penetration.	Sexual acts involving a child that are intended to provide sexual gratification to an adult: (a) Physical contact of a sexual nature between child and adult. For example, vaginal or anal penetration (or attempted penetration), oral-genital or oral-anal contact, fondling (directly or through clothing); (b) Noncontact exploitation—an adult forcing, tricking, enticing, threatening, or pressuring a child to participate in acts for anyone's sexual gratification without direct physical contact between child and offender. For example, exposing child's genitals, anus, or breasts; having child masturbate or watch masturbation; having child participate in sexual activity with a third person (including child prostitution); having child pose, undress or perform in a sexual fashion (including child pornography).

Continued

Table 8.1 Comparison of foundational criteria developed and tested by Heyman and Slep (2006) and implemented by the US Department of Defense (DoD) and the proposed criteria for ICD-11—cont'd

	DoD	ICD-11 (proposed)
	4. Physical contact of a sexual nature not involving rape, intercourse, or sodomy between the child and the caregiver, including, but not limited to any of the following: a. The fondling or stroking of the genitals or buttocks, directly or through clothing, with the intent to abuse, humiliate, harass, or degrade the child, or to arouse or gratify the sexual desire of the alleged abuser, the child, or any other person. b. The fondling or stroking of a female's breast, directly or through clothing, with the intent to abuse, humiliate, harass, or degrade the child, or to arouse or gratify the sexual desire of the alleged abuser, the child, or any other person. c. The attempted penile penetration of the vagina, anus, or mouth. d. The attempted penetration of the vagina, with a hand or finger or any object with the intent to abuse, humiliate, harass, or degrade the child, or to arouse or gratify the sexual desire of the alleged abuser, the child, or any other person.	
Impact	None required	None required
Child Abuse, Neglect		
Act	Egregious act(s) or omission(s) on the part of the child's caregiver that deprives the child of needed age-appropriate care. 1. Lack of supervision: Egregious absence or inattention. Child's age and level of functioning should be considered in making determination about level of supervision required. Note: Leaving children 10 or older unattended in a vehicle for brief periods of time in a safe area *does not* meet this criterion.	Confirmed or suspected egregious act(s) or omission(s) that by deprives the child of needed, age-appropriate care (e.g., abandonment, lack of appropriate supervision; exposure to physical hazard; failure to provide necessary education, health care, nourishment, shelter, clothing).

	2. Exposure to physical hazards: inattention to child's safety by exposing child to physical dangers or home hazards (including, but not limited to, exposed electrical wiring; broken glass; nonsecured, loaded firearms in home; illegal drugs in home; dangerous or unhygienic pets; asking child to perform dangerous activities; driving while intoxicated with child in vehicle, hazardous chemicals, unhygienic living conditions dangerous to health; caregivers known to be abusive or neglectful; domestic violence close enough in proximity to the child to create risk of injury to the child). 3. Educational neglect: Knowingly allowing the child to have extended or frequent absences from school, neglecting to enroll the child in some type of home schooling or public or private education, or preventing the child from attending school for other than justifiable reasons (when education is compulsory by law). 4. Neglect of health care: Refusal or failure to provide appropriate health care including, but not limited to, failure to obtain appropriate professionally indicated medical, mental health, or dental services, procedures, or medications, although the parent was financially able to do so or was offered other means to do so. It includes withholding of medically indicated treatment for a child with life-threatening conditions. 5. Deprivation of necessities: The failure to provide age-appropriate nourishment, shelter, and clothing to the child. Includes nonorganic failure to thrive as determined by a competent medical authority. (FTT is a type of child neglect evidenced by an infant's or young child's failure to adequately grow and develop to or above the third percentile in height and weight when no organic basis for this deviation is found.) 6. Abandonment: The caregiver is absent and does not intend to return or is away from the home for more than 24 hours without having arranged for an appropriate surrogate caregiver. *No impact ("B") criterion is necessary for abandonment to be deemed neglect.*	

Continued

Table 8.1 Comparison of foundational criteria developed and tested by Heyman and Slep (2006) and implemented by the US Department of Defense (DoD) and the proposed criteria for ICD-11—cont'd

	DoD	ICD-11 (proposed)
Impact	Significant deprivation-related impact involving any of the following: 1. More than inconsequential physical injury[a] 2. Psychological harm, including any of the following a. More than inconsequential fear reaction[b] b. Significant psychological distress (major depressive disorder, posttraumatic stress disorder, acute stress disorder, or a constellation of symptoms similar to other psychiatric disorders) related to, or exacerbated by, the act(s) 3. Stress-related somatic symptoms (related to or exacerbated by the acts) that that significantly interfere with normal functioning. 4. Reasonable potential for more than inconsequential physical injury[a] 5. Reasonable potential for psychological harm a. The act (or pattern of acts) creates reasonable potential for the development of a psychiatric disorder (at or near diagnostic thresholds) related to, or exacerbated by, the act(s). The child's level of functioning and the risk and resilience factors present should be taken into consideration. b. The act (or pattern of acts) carries a reasonable potential for significant disruption of the child's physical, psychological, cognitive, or social development	• Physical injury or reasonable potential for injury; • Significant fear or psychological distress; • Stress-related somatic symptoms; • Reasonable potential for the development of a psychiatric disorder; • Reasonable potential for significant disruption of the child's physical, psychological, cognitive, or social development.

[a]More than inconsequential physical injury: An injury involving any of the following:
- Any injury to the face or head
- A more than superficial bruise. The bruise was a color other than very light red or had a total area exceeding that of the partner's hand or was tender to light touch
- More than superficial cut or scratch. The cut or scratch was bleeding and required pressure to stop the bleeding
- Bleeding internally or from mouth or ears
- A welt (a bump or ridge raised on the skin)
- Loss of consciousness

- A burn
- Heat exhaustion or heat stroke
- Loss of functioning (including, but not limited to, sprains, broken bones, detached retina, a loose or chipped tooth)
- Damage to internal organs
- Disfigurement (including, but not limited to scarring)
- Swelling lasting at least 24 hours
- Pain felt in the course of normal activities *and* at least 24 hours after the physical injury was suffered. Note: At times, the partner is (a) unable to report verbally about pain or (b) inaccessible for assessment. In such cases, the criterion is met if the nature of the injuries would typically result in pain as defined above.
- Death.

[b]More than inconsequential fear reaction:

(a) Fear (verbalized or displayed) of bodily injury to self or others *and*

(b) At least one of the following signs of fear or anxiety lasting at least 48 hours:

1. Persistent intrusive recollections of the incident
2. Marked negative reactions to cues *related to incident*, as evidenced by (a) avoidance of cues; (b) subjective or overt distress to cues; or (c) physiological hyperarousal to cues (Note: Perpetrator can be a cue)
3. Acting or feeling as if incident is recurring
4. Marked symptoms of anxiety (any of the following):
    - Difficulty falling or staying asleep
    - Irritability or outbursts of anger
    - Difficulty concentrating
    - Hypervigilance (i.e., acting overly sensitive to sounds and sights in the environment; scanning the environment expecting danger; feeling keyed up and on edge)
    - Exaggerated startle response.

Intimate partner maltreatment

Intimate Partner Violence (IPV)[b] exerts a considerable public health burden internationally, including demonstrated degradation of myriad markers of morbidity (e.g., physical, sexual/reproductive, and psychological/behavioral impacts) and mortality (Heise & Garcia-Moreno, 2002). Additionally, IPV exerts a financial burden internationally, with trillions of dollars in medical and mental health services costs, lost productivity, and lost wages over a lifetime (Peterson et al., 2018). Further, exposure to partner physical IPV also affects children in the household, three-quarters of whom witness or hear it (e.g., O'Brien, John, Margolin, & Erel, 1994); such children exhibit a host of deleterious outcomes (e.g., internalizing problems, externalizing problems, social problems, academic problems, negative affect/distress, and negative cognitions; Kitzmann, Gaylord, Holt, & Kenny, 2003).[c]

Partner physical abuse

Physical aggressive acts involve the nonaccidental use of physical force against a spouse or intimate partner. Physical force includes, but is not limited to, at least one of the following: hitting with open hand or slapping, including spanking, dropping, pushing or shoving; grabbing or yanking limbs or body, poking; hair-pulling; scratching; pinching; restraining; or squeezing or shaking; throwing; biting; kicking; hitting with fist; hitting with a stick, strap, belt, electric cord or other object; scalding or burning; poisoning; stabbing; applying force to throat; strangling or cutting off air supply; holding underwater; brandishing or using a weapon. Note that these criteria refer to "nonaccidental acts," as opposed to the oft-used "intentional acts"

[b]Intimate Partner Violence is now the most accepted usage for partner abuse, following the lead of the US Centers for Disease Control and Prevention (CDC; e.g., Saltzman, Fanslow, McMahon, & Shelley, 1999). Although CDC's definitions of IPV are very close to the maltreatment criteria discussed below that require an act and an impact, IPV is typically used in the field to mean any act of violence. In this chapter, we will use IPV generically and will use partner abuse specifically when we mean acts and impacts that meet criteria for clinical significance.

[c]Research on the impact of IPV on children has been conducted almost exclusively in high-income countries. A United Nations report on the impact of violence on children worldwide (Pinheiro, 2006) cited this research base as evidence for the deleterious impact of witnessing IPV, implying that the extant findings are likely generalizable to all children. While this expectation of generalizability is one that we share, obviously studies in various countries around the globe would be necessary to support it more definitively.

(e.g., CDC's definitions; Breiding et al., 2015). "Nonaccidental" was used throughout the tested criteria set because (a) it does not require knowledge of alleged perpetrators' private thoughts/motives, (b) is thus easier for partners and assessors to infer, and (c) field testing indicated superior results when anchoring users to a threshold just above "accidental."

Impacts include (a) any physical injury (e.g., pain that lasts at least 4 hours, bruises, cuts, sprains, broken bones, loss of consciousness, and death); (b) reasonable potential for significant physical injury given the inherent dangerousness of the act(s), the degree of force used, and the physical environment in which the act(s) occurred; or (c) significant fear.

Exclusions

In some situations, acts of physical aggression that result in a significant impact (or have the potential to) are still not considered to be physical abuse. These include protection of self from imminent physical harm ("self-defense") and protection of partner from imminent physical harm (e.g., while partner is attempting to commit suicide or self-harm). In the full criteria developed for the US Department of Defense (Heyman & Slep, 2006), protection of self requires all three of the following: (a) the physically aggressive act occurred while partner was in the act of using physical force; (b) the sole function of the act was to stop partner's use of physical force; and (c) the act used minimally sufficient force to stop partner's use of physical force. Another criterion excluded acts committed in preemptive protection of self when (a) the protective acts were in response to a threat to imminently inflict significant physical injury, and (b) there had been at least one prior incident of significant physical injury.

Partner emotional abuse

Psychologically/emotionally aggressive acts involve verbal or symbolic acts that have the potential to adversely affect the psychological well-being of a partner including, but not limited to, berating, disparaging, degrading, humiliating the partner; interrogating the partner; restricting the partner's ability to come and go freely (when unwarranted); obstructing the partner's access to assistance (e.g., law enforcement, legal, protective, or medical resources); threatening the partner (e.g., indicating/implying future physical harm or sexual assault); harming, or indicating that the alleged perpetrator will harm, people/things the partner cares about, such as children, self, other people, pets, or property; restricting the partner's access to or use of economic resources (when unwarranted); isolating the partner from family, friends, or social support resources; stalking the partner; and trying to make partner think that s/he is mentally ill (or make others think that the partner is mentally ill).

Finally, psychological aggressive acts include those that use coercive control. A specific type of coercive control that has received increased recent attention—and one with considerable global resonance—is reproductive coercion (e.g., Grace & Anderson, 2017). CDC (Breiding et al., 2015, p. 15) operationalizes it as "control of reproductive or sexual health (e.g., refusal to use birth control and coerced pregnancy terminations);" other examples in the literature include tampering with birth control devices and interfering with the partner's capacity to use birth control. We concur with CDC's inclusion of this form of violence in psychological aggression.

Impacts include psychological harm (i.e., significant fear; significant psychological distress related to or exacerbated by the act; fear of the recurrence of emotionally abusive act(s) that cause the partner to significantly limit any of five major life activities: (a) work; (b) education; (c) religion; (d) medical or mental health services; and (e) contact with family/friends) and somatic symptoms that interfere with normal functioning.

Partner sexual abuse

Sexually aggressive acts involve a sexual act with the intimate partner without the consent of the partner. A sexual act involves contact between the penis and the vulva or the penis and the anus involving penetration, however slight; masturbation by self or partner; contact between the mouth and the penis, vulva, or anus; or penetration of the anal or genital opening by a hand, finger, or an object. A lack of consent may involve using physical force to compel the partner to engage in a sexual act; using physically or emotionally aggressive acts to coerce the partner to engage in a sexual act; or a partner who is unable to provide consent (i.e., the partner is unable to understand the nature or conditions of the act, to decline participation, or to communicate unwillingness to engage in the sexual act because of illness, disability, being asleep, being under the influence of alcohol or other drugs, or other reasons). As noted earlier, because of the high potential impact of sexual abuse, no demonstrated impact is required to meet criteria.

Partner neglect

Partner neglectful acts/omissions involve withholding necessary care or assistance for a partner who is incapable of self-care physically, psychologically, or culturally, despite the caregiver's financial ability to do so or the caregiver having other means to do so. Impacts include (a) significant physical injury and (b) reasonable potential for significant physical injury.

Cross-cultural differences in applying the ICD-11 partner physical and psychological abuse criteria

Using English-speaking mental health professionals ($N=738$) from 65 nations drawn from the WHO's Global Clinical Practice Network, Heyman et al. (2018) tested if the more operationalized ICD-11 criteria performed better and had higher clinical utility than the unelaborated ICD-10 listings. The study used standardized clinical vignette conditions with experimentally manipulated case presentations of presence or absence of (a) individual mental health diagnoses and (b) relational problems or maltreatment. Of note to the focus of this chapter, clinician factors (e.g., gender, native language, and world region) largely did not impact classification performance, implying that clinicians who can complete studies in English largely do not differ on how they apply maltreatment guidelines. As hypothesized, ICD-11 guidelines, compared with ICD-10, resulted in significantly better classification accuracy although only in the presence of comorbid mental health problems.

Furthermore, in not yet published data from the same study, which included $N=1225$ mental health professionals who completed the study in English, Spanish, French, and Japanese, we also found few differences across clinicians and hypothesized superiority of the ICD-11 maltreatment guidelines.

Not only do the more operationalized ICD-11 guidelines generally function better than those in ICD-10 but health professionals across the globe also appear to apply them in relatively similar ways. Although not conclusive, this is initial evidence for the cross-cultural clinical utility of the criteria.

Child maltreatment

Child maltreatment (physical, emotional, and sexual abuse, and neglect) has been recognized as a significant global public health and social problem for decades. Consequences include death, increased risk of later premature mortality, changes in brain development and physiological responding, physical injury, internalizing and externalizing psychopathology, future victimization and perpetration of violence, and lessened future earning potential (Del Vecchio & Slep, 2012).

Child physical abuse

Physical abuse is the most concrete and easily agreed-upon type of child maltreatment. However, cultural norms vary regarding the use of corporal punishment. For example, multiple countries (e.g., Sweden, Croatia, and United Kingdom) have specifically outlawed all forms of child corporal punishment, whereas all 50 states of the United States legally permit disciplinary spanking (Ripoll-Núñez & Rohner, 2006). This variability and the fact that parenting young children inherently involves physically intervening at times can make even this type of abuse difficult to define. As with partner physical abuse, the act criterion requires a nonaccidental act of physical force (which could include corporal punishment or physically moving a child). Impacts include a more than inconsequential injury, a reasonable potential for such an injury, or significant fear and anxiety. There are a few specifically defined exclusions that are relevant. First, if the parent is acting to protect the child from imminent harm (e.g., grabbing a child running into the street to prevent them from being hit by an approaching vehicle, resulting in a dislocated elbow), the incident would not be considered abuse. Similarly, if while a child was using force against the parent, the parent used minimally sufficient force to stop the child's use of force, it would be excluded as nonabusive. Finally, if the force and injury occurred in the context of developmentally appropriate physical play, it would be excluded as nonabusive. Notably, whether the force was intended as a discipline is not considered as an exclusion. Thus, these criteria implement the harm criterion (i.e., children should not be injured or traumatized) for parental aggression, even if it might be considered corporal punishment. Thus, the criteria are neither as restrictive as the illegality of all corporal punishment in 54 countries nor as permissive as the laws of many US states and countries that require an intentional act that causes injury.

Child psychological abuse

Psychological abuse is similarly operationalized as including a qualifying verbal or symbolic act, including behaviors such as berating, humiliating, exposing the child to violence against someone they care about (e.g., IPV), destroying prized objects, threatening to harm or abandon. The impacts considered for psychological abuse are emotional rather than physical. They include psychological harm (significant fear or significant psychological distress), reasonable potential for harm (either likelihood to lead to a psychiatric disorder or a significant disruption of development), and somatic symptoms. The potential for disorders and developmental disruption, although arguably difficult to assess, were intended to allow for the determination that particular acts or patterns of acts of psychological abuse are likely to result in harm for a given child, but that sufficient time might not have passed to demonstrate that impact at the time of the assessment, either due to the acute and severe nature of the act or the developmental stage of the child. In these situations, the potential for future psychological distress or disrupted development would be more likely to be apparent later as the child progresses through different developmental stages.

Child sexual abuse

Sexual abuse is defined only in terms of the act. Although there is evidence in the literature that sexual abuse is not always dramatically impactful (e.g., Rind, Tromovitch, & Bauserman, 1998), most definitions of sexual abuse used in clinical, research, and legal settings do not include a need for short-term, demonstrable harm to the child. Rather, these acts are seen as such a violation of society's standards that they are considered endangering and abusive regardless of apparent impact. The proposed definitions included three types of acts: use of physical force, emotional manipulation, and a child's youth or naïveté to engage in (1) forced sexual contact, (2) other sexual contact, and (3) exploitation of the child for anyone's sexual gratification in a manner that involves no direct contact with the child (e.g., photographs).

Child neglect

The definition of child neglect was the longest and most complex of the criteria sets due to the variety of significant neglecting acts and omissions that can occur. Qualifying omissions are categorized by type and include (1) lack of supervision, (2) exposure to physical hazards, (3) educational neglect, (4) health care neglect, (5) deprivation of necessities, and (6) abandonment. It is important to note that the principle underlying these categories was that the act or omission must have been egregious and have deprived the child of needed, age-appropriate care. This principle allowed flexibility in applying these standards in different cultures where standards for age-appropriate education, supervision, and healthcare might differ tremendously. Neglect requires both an egregious act/omission and a significant impact; because of the diversity of the types of neglect, all categories of impact included in both physical and emotional abuse were included under neglect.

Prevalence across cultures

Partner abuse

The literature provides substantial evidence that IPV is a widespread global phenomenon with considerable individual, familial, and societal implications (e.g., Heise & Garcia-Moreno, 2002; World Health Organization, 2005). In addition to the complexities in interpreting any cross-cultural research, the vast variability in how IPV is conceptualized and measured across studies makes it difficult to infer meaningful conclusions by comparing results across studies. Methodological variability also makes it challenging to narrow a reliable estimate of prevalence. Furthermore, despite growing evidence that IPV occurs all over the world, data from low-income countries are particularly scarce. The absence of these data undermines attempts to identify and understand disparities and develop interventions. Given these limitations, the following content reviews findings from the World Health Organization's (2005) Multicountry Study on Women's Health and Violence Against Women and data from the US Agency for International Development Demographic and Health Surveys (DHS) (Hindin, Kishor, & Ansara, 2008). Both of these surveys have attempted to capture a global picture of rates of IPV by using standardized measurement and methodology across countries. Of note, given the limitations outlined above and the focus on violence perpetrated against women in the few multisite cross-cultural datasets available, the current chapter precludes a discussion of violence against men. This omission is a result of the limitations in available data rather than the implication that male partners are not victimized.

The WHO (2005) study collected data from over 24,000 women from 15 sites in 11 countries (Samoa, Peru, Brazil, Namibia, United Republic of Tanzania, Ethiopia, Serbia and Montenegro, Bangladesh, Thailand, Japan, and New Zealand) and examined forms and patterns of violence as well as their links to women's health to inform prevention efforts. Measurement tools underwent an intensive and iterative refinement process that included review by the core research team, content experts, and collaborators in participating countries. In half of the countries included in the study, data were collected from both the capital or other large city and a province or region with both urban and rural populations. The WHO study assessed physical, psychological (including controlling behaviors), and sexual IPV. Results indicate while IPV existed in all settings surveyed, there was considerable variability in prevalence both between countries and across sites within countries. The lifetime prevalence of physical IPV ranged from 13%

in Japan to 61% in provincial Peru. Estimates of physical violence were based on questions of participants experiencing explicit acts (e.g., slapped, kicked, choked). The most common form of physical IPV was being slapped followed by being struck with a fist, with 11%–21% of women endorsing being hit with a fist during their lifetimes at most study sites. When acts likely to cause injury were considered, lifetime clinically significant physical abuse rates varied from 4% in Japan to 49% in provincial Peru with most sites ranging between 13% and 26%. Specific acts were grouped by likelihood to cause injury into two groupings—Moderate physical violence included being slapped, pushed, or shoved. Severe physical violence included being hit with a fist, kicked, dragged, threatened with a weapon, or having a weapon used against her. In most countries between 13% and 26% of women reported experiencing severe abuse. Further, in almost all countries included in analyses, if a woman endorsed ever experiencing partner violence, a severe act was highly likely to have occurred. Reports of sexual violence ranged from 6% in Serbia, Montenegro city, and Japan, to 59% in provincial Ethiopia, with the majority of sites ranging from 10% to 50%. Sexual IPV was operationalized as being physically forced to engage in intercourse, having intercourse because of fear of consequences, or being forced to do something sexual the respondent found to be degrading or humiliating. Approximately half of the reported sexual IPV was via physical force rather than via fear. Generally, sexual IPV was much less prevalent than physical IPV except in Bangladesh, Ethiopia, and urban Thailand. Other than these settings, most women reported either experiencing only physical IPV or experiencing both physical and sexual IPV. When comparing urban and provincial settings, provincial settings, which included rural populations, reported higher rates of IPV. Of note, the majority of women who reported experiencing physical IPV reported that it happened more than once, suggesting that IPV for women involves repeated victimization across their lifetimes, rather than an isolated incident. Whether this involves patterns of victimization within a single relationship, repeated sporadic incidents across relationships, or signs of repeated bidirectional conflict (within or across relationships) cannot be determined from this study.

A particular challenge for the designers of the WHO study was operationalizing emotional IPV in a way that would be consistently applicable in various cultural contexts (and administrable in a relatively brief survey). Reports of lifetime emotional IPV ranged from 20% to 75%. Similar to patterns observed in physical and sexual IPV, most women who experienced an act of emotional IPV had experienced it repeatedly.

Hindin et al. (2008) included data from married or cohabiting women who participated in 10 recent DHS (Bangladesh, Bolivia, the Dominican Republic, Haiti, Kenya, Malawi, Moldova, Rwanda, Zambia, and Zimbabwe). Of note, the countries included in DHS do not overlap with those in the WHO Study, except Bangladesh. IPV was measured using a modified version of the Conflict Tactics Scale (CTS; Straus, 1979) comprising physical (e.g., pushed, slapped, or hit) and sexual IPV (e.g., forced to have intercourse or perform sexual acts against will). Women who endorsed any of the items included in the scale were categorized as experiencing IPV. Reports of physical IPV ranged from 12% in Haiti to 71% in Bangladesh. Similar to results of the WHO Study, the most common acts of physical IPV in almost all countries were being slapped, having arms twisted, or hair pulled with the highest ratings from women in Bangladesh, 64% of whom endorsed her partner as having twisted their arm. The more severe acts of physical violence (punching with a fist, kicking, dragging, choking, and threatening or use of a weapon) were less common. Being punched with a fist ranged from 16% in Bangladesh to 6% in Malawi. Being kicked, dragged, or beaten up fell below 10% for all participating countries except for Bangladesh (12%). The least common acts were being threatened by a weapon, which ranged from less than 1% in Malawi to 6% in Kenya, and attack with a weapon which fell below 2% in most countries aside from the Dominican Republic, where close to 4% of women endorsed these items. Rates of sexual IPV ranged from 3% in Moldova to 26% in Bangladesh. Mirroring patterns found in the WHO Study, physical IPV was far more prevalent than sexual IPV and a considerable number of women reported lifetime victimization of both physical and sexual IPV although there is considerable variability across countries.

While broader studies of types of IPV like those cited above typically result in comparatively lower ratings of extreme forms of violence, a body of research has also focused on arguably the most severe type of partner violence, homicides. The United Nations Office on Drugs and Crime released a Global Study On Homicide (2018) with a focus on gender-related killings of women and girls both within and outside of the family sphere. According to their data, of the total number of women killed in 2017, more than a third of them were killed by an intimate partner, a number that appears to be an increase from previous data waves. Their report ranks the rates of homicides by intimate partners as the highest in Africa, followed by the Americas, Oceania, Europe, and Asia. As a supplement to the range of acts of IPV reviewed above, these data demonstrate that IPV exists on a global scale even in the extreme and severe forms.

When considering differences in prevalence in IPV across cultures, it is vital to also examine trends in general attitudes toward or against IPV in various contexts. The degree of social acceptability contributes to perpetration rates, as well as bystander responses, rates of reporting incidences of violence, and attention and resources devoted to subsequent supports and services (Waltermaurer, 2012). In general, individuals in underdeveloped nations tend to hold attitudes that are more supportive of IPV than those living in developed nations (Rodriguez Martínez & Khalil, 2013). When considering specific behavioral antecedents, a literature review of 23 studies spanning 61 countries found women neglecting her children or going out without permission from her husband as the highest-ranking scenarios for which individuals felt IPV was justifiable (Waltermaurer, 2012). In contrast, the same review found poor care of household duties was the lowest ranking justification. Of note, women in the included studies were more likely to report IPV was justifiable than their male counterparts (Waltermaurer, 2012). Younger women were also found to be significantly more likely to justify IPV than older women (Waltermaurer, 2012). When considering patterns within and between countries, the authors found little to no consistency in pro or antijustification attitudes within each country. Rather, there appears to be quite a bit of variability in the endorsement of each scenario best explained by demographics such as age, education, relationship status, income, and family history of witnessing violence rather than notable between country patterns. When considering how these attitudes may shift as individuals move from one context to another, in a study of value acculturation comparing natives and immigrants in five Western countries, Rodriguez Martínez and Khalil (2017) demonstrated that attitudes of immigrants tended to be more positive toward IPV (in alignment with those of their home countries) than natives though these differences disappear across generations acculturation occurs. Studies investigating similar patterns in initial differences between immigrants and natives found similar patterns in Spain (Gracia, Herrero, & Lila, 2008) and Israel (Wallach, Weingram, & Avitan, 2010).

Child maltreatment

Child maltreatment is endemic, although prevalences vary based on type, location, and the process by which studies sample and measure maltreatment. The best estimates of cross-cultural prevalences are gleaned from reviews and metaanalyses. Systematic review of population-based surveys has included data from 112 studies in 96 developing and developed countries, estimating rates of abuse for children living in nearly half the countries in the world; this review indicates the prevalence of past-year physical acts against children to be highest in Asia (64%), Africa (50%), and North America (56%), with lower rates in Latin America (34%) and Europe (12%). Globally, over half of all children experienced past-year physical acts (Hillis, Mercy, Amobi, & Kress, 2016). This translates to an estimation that globally, at least half of, or 1 billion children experienced physical acts in the past year. Concerning fatal abuse, the WHO (2016) estimated 41,000 deaths attributed to homicide among children under 15 years old.

Rates of subtypes of child maltreatment included in metaanalyses show consistency across the globe. The rate of child physical abuse from 168 independent samples is estimated to be 22.6% (Stoltenborgh, Bakermans-Kranenburg, van IJzendoorn, & Alink, 2013) and the rate of child emotional abuse from 46 independent samples was 36.3% (Stoltenborgh, Bakermans-Kranenburg, Alink, & van IJzendoorn, 2012). In both cases, rates did not significantly vary for a geographical region, level of country's economic development, or child's gender; within North American samples, predominant ethnicity did not significantly influence rates. A deeper dive into cross-cultural comparison of physical and emotional abuse does reveal differences in how common various abusive acts are; for example, spanking on the bottom is typically the most common behavior reported across a variety of countries, but reports in Egypt indicate that shaking, pinching, or slapping on the face or head were more frequently used as punishment. Similarly, for emotional abuse, shouting at children is the most common behavior in most locations, but threatening children with abandonment is common in the Philippines and much less so in Chile (Runyan et al., 2002).

In contrast, child sexual abuse does seem to vary by gender and location. A metaanalysis of 331 samples of child sexual abuse globally indicated an overall rate of 12.7%, with a 7.6% rate for boys and an 18% rate for girls (Stoltenborgh, van IJzendoorn, Euser, & Bakermans-Kranenburg, 2011). The highest reported rates were found in Australia for girls and in Africa for boys, with the lowest rates in Asia for both boys and girls. Generally, sexual abuse against boys was more frequent in low-income countries than in high-income countries. Within the United States and Canada, sexual abuse against boys is more prevalent within African-American samples.

For child neglect, the metaanalysis of studies is more preliminary, due to the relatively smaller number of studies that include child neglect rates. Global estimates find a rate of 16.3% for physical neglect and 18.4% for emotional neglect (Stoltenborgh, Bakermans-Kranenburg, & van IJzendoorn, 2013).

Rates did not vary by gender, but other cross-cultural comparisons were not possible due to the small number of studies that included these figures, and in particular, the fact that most studies of child neglect came from North America or other high-income countries.

One important finding of child maltreatment prevalence is that rates based on informant data (i.e., what is known to community "sentinels") are considerably lower than those based on self-report. Although the above prevalences are based on metaanalyses of self-report data, a metaanalysis of global informant research indicated rates of 0.3%, 0.3%, and 0.4%, for physical, psychological, and sexual abuse, respectively (Stoltenborgh, Bakermans-Kranenburg, Alink, & van Ijzendoorn, 2015). Given the "behind closed doors" nature of the phenomenon, it is not surprising that the amount of child maltreatment known or suspected by the community provides vast underestimates of the actual prevalence of the phenomenon.

Cultural considerations

Etic and emic lenses to IPV

As we have argued (e.g., Heyman & Slep, 2019), research-tested criteria for partner and child maltreatment are an essential step toward agreement across research, policies, and laws focused on the measurement, study, treatment, and prevention of maltreatment. We also recognize that although this foundational work is a necessary component of understanding how maltreatment is experienced across settings, it is not sufficient. With the significant strides the maltreatment fields have taken in their ability to classify and measure abuse and neglect, they are still met with questions of how to increase agreement that these behaviors are indeed problematic as they often collide with cultural and religious norms and rights. For example, women's ability to move in public freely and without the permission of a male relative is considered normative in some countries but counter religious and social norms in others. Responding to these differences requires addressing cultural nuance in the underlying meaning, impact, and ramifications of these experiences. Ignoring these factors is a sure misstep in attempts to develop culturally responsive interventions with maintainable impact, as being able to deem maltreatment as culturally deviant provides leverage to motivate change. Thus, in addition to creating global criteria to standardize the way we identify maltreatment, we must also critically examine the cultural norms, standards, and beliefs that justify and help perpetuate these acts/omissions. By both providing bright-line thresholds for what is and is not maltreatment *and* probing the ideologies and societal structures that determine what is or is not culturally accepted, we broaden our range of impact and potential for actualizing lasting change.

With this goal in mind, feminist theories, in particular, have detailed how societal factors, such as variability in patriarchal informed laws, gender roles, and normalized views of aggression and violence are linked to the origins and continued perpetuation of violence against women. For example, the socially normed, and in some countries, legally sanctioned power differentials between men and women vary widely. These aspects of women's role in society, which then carry over to their role in the home, are known risk factors for IPV (Fernandez, 2006). Illustrating this point, a study exploring the determinants of IPV in Bangladesh found that autonomous women were significantly more likely to experience violence in highly conservative areas known for stricter gender roles (Koenig et al., 2003). The authors conclude that the risk of violence is likely linked to the societal normalcy of women's empowerment and autonomy. The differences across countries in acceptability of gender inequities are indeed stark. For example, Morrisson and Jutting (2005) found significant disparities between Latin American and East Asian countries and countries in sub-Saharan Africa, South Asia, and Middle East and North Africa with the latter grouping having higher rates of markers of systematic gender discrimination to include constraints on education, access to health care, and opportunities to work. While scholars certainly argue that patriarchal gender inequities exist globally, it is clear that differences exist in countries whose laws legally sanction these inequities and those that do not, a necessary consideration in examining cross-cultural differences in IPV. To exemplify the impact of social norms on individuals, researches have examined differences in attitudes toward IPV. Indeed, patriarchal and pro-aggression attitudes by both men and women are not only associated with increased rates of IPV but also blur lines on whether acts that surpass clinical thresholds of IPV are culturally defined as abusive, criminal, or outside normative expectations (Hayes & Boyd, 2017). Interestingly, research demonstrating that the gap in the endorsement of IPV attitudes between immigrants and natives disappears in the second generation of immigration, suggests these beliefs about the acceptability of violence against women may be shaped, in part through socialization (Rodriguez Martinez & Khaliil, 2017). This highlights an important opportunity for intervention as research has shown that shifts in global views of the acceptability of partner maltreatment ultimately shifted individual attitudes. For instance, Pierotti (2013) documents how, following prominent messages of the women's movement including messages decrying violence against women and promoting women's rights and gender equity, there were coinciding shifts in women's attitudes toward IPV such that women in 23 of 26 countries surveyed became more likely to reject IPV as a justifiable form of social control. These changes in ideological views began to inform institutional policies and laws, likely shifted interpersonal behaviors, and ultimately were internalized as changes in attitudes about deserving to be victimized.

As another example of the influence of societal factors on IPV, in their review of key macro factors that help, in part contextualize the varying prevalence rates outline in this chapter, Larkin and Morris (2009) highlighted the significance of poverty. While it is clear that IPV occurs across socioeconomic status, economic stress contributes to the disproportionately higher rates of IPV found among people living in poverty. In addition to the direct impact on limited material resources, poverty may also impact relationship dynamics such as feeling constrained to stay in a violent relationship because of economic necessity (Larkin & Morris, 2009). Further, poverty on a macro level differentially affects countries' fiscal ability to implement and execute prevention and intervention programming (Larkin & Morris, 2009). Thus, individual and societal wealth disparities are vital factors in contextualizing cross-cultural experiences of IPV. Taken together, while there is a need to standardize a universal measurement of violence as is proposed in this chapter, culturally bound, contextual factors are also important in the consideration of the rates of acts of violence, as well as individual and communal responses to such acts.

Challenges in cross-cultural research

One of the primary challenges in gaining a culturally nuanced understanding of how IPV operates globally is the restricted way in which culture is traditionally defined. Cross-cultural research typically compares prevalence, outcomes, or mechanisms across people with different identities (e.g., racial groups) or in different geographic regions. There are clear benefits of this approach, such as its ability to highlight group differences and, as in the case with family maltreatment, document constructs as significant global phenomena. Furthermore, understanding these group differences helps to expose disparities and inequities in access, as well as guide interventions to populations or regions that are most in need.

However, prevalence is but a first step in a comprehensive understanding of the impact culture has on why variabilities in maltreatment exist in the first place. If this type of rudimentary comparison is accepted as a sufficient picture of cultural variability, it promotes the erroneous assumption that culture is a singular concept that can be fully measured by the selected grouping variable. Moreover, we must recognize that cultural characteristics are not stagnate. Rather, we must consider that many norms, beliefs, and practices are embedded in a rich sociohistorical context and in response to the constantly evolving allocation of societal and global power and resources. Last, the emphasis on *between*-group differences in traditional cross-cultural research de-emphasizes the importance of heterogeneity within populations. A *within*-group focus, including considerations of subcultural and intersectional factors (i.e., interconnected systems of power), provides greater clarity on variability in how family violence is experienced and how cultural factors may help explain why these differences occur. Reducing the broad, multivariate nature of culture to a narrowed and restricted construct overlooks opportunities to accurately depict both (a) the universal factors of maltreatment that help scaffold the global picture of these constructs and (b) the nuanced differences in how and why maltreatment is perpetuated in particular cultural contexts.

As reviewed above, researchers are moving toward operationalizing coordinated criteria sets for family violence that, if further reliability and validity work is supportive, may improve global research and response to maltreatment. In addition, cultural research must also consider how and why maltreatment is defined differently within a particular context to effectively intervene and affect change. For example, in attempts to understand what cultural factors inform and perpetuate normative attitudes toward IPV in sub-Saharan African, researchers highlight that egalitarian decision-making and contribution to the home is associated with reduced beliefs about the acceptability of IPV (Mann & Takyi, 2009). This type of research sheds light on how culturally bound experiences shape the perpetuating of IPV and aids in providing culturally informed clinical and prevention efforts.

To move in this direction, it is of paramount importance to integrate discourse on social identities and how they are positioned within systems of oppression into theoretic frameworks of IPV (Mitchell & Vanya, 2009). A social identity is the component of one's self-concept derived from membership in a social group (e.g., race, gender, sexual orientation). Social identities are relational in that they are embedded in hierarchies of privilege and marginalization. The sociopolitical history and societal norms within which one is situated dictate both the boundaries of how identities are defined as well as the associated assignment of superior or inferior status. The systems that maintain these inequities operate at three levels: ideologically as a foundational dichotomized view of "us and them"; institutionally, in-laws, policies, and regulations that create or maintain disparities; interpersonally and intrapersonally as internalized messages about self-worth. Although there are some global social identities (e.g., sex), there are also culture-bound social identities (e.g., tribe affiliation), as well as constructs that are operationalized differently across regions (e.g., race). Furthermore, intersectional frameworks emphasize that individuals experience a layered form of marginalization at crossroads of their marginalized identities (Crenshaw, 1989). For example, when considered separately, one's race is impacted by racism and one's sex is afflicted by sexism. However, that same individual also experiences a distinct fusion of racism and sexism distinct from the experiences of people outside of that overlapping territory.

Given the layered, multifaceted nature of these identities, it is clear that a comprehensive understanding of culture quickly becomes quite complex. Nonetheless, the examination of how power, privilege, and oppression operate both within cultures and as a determinant of inequitable distributions of resources across nations is an essential step in understanding the individual and macro impact these systems have on maltreatment. Through this framework, we can better understand how these larger systemic factors contribute to variability in cross-cultural prevalences of IPV and child maltreatment.

Summary and future directions

The Institute of Medicine's (now part of the US National Academy of Sciences) Prevention Intervention Research Cycle described five steps for effectively intervening in a problem (Mrazek & Haggerty, 1994). The first step is to define the problem and measure its prevalence. Subsequent steps involve identifying risk and protective factors, and developing and testing interventions. Because the steps are recursive, advances in research feedback to earlier stages to inform improvements. This chapter essentially describes the beginning of one such recursive loop, describing research-based improvements in defining the problems of partner and child physical, psychological, and sexual abuse and neglect. This research—organized around "harm criteria" that create a bright line between acts that are and are not maltreatment—may lead to improved international communication about maltreatment, especially if they are eventually incorporated into ICD-11.

Vignette-based field trials for ICD-11 document their improvement over those in ICD-10 and the consistency in which clinicians around the world use them.

This work may lead to improvements in knowledge about prevalence and causes of maltreatment and through improvements in the research literature, eventual improvements in the intervention. Yet the current research base still clearly indicates that maltreatment is endemic worldwide, yet with considerable variability within and across populations. Ultimately, worldwide reductions in maltreatment will require integrating cultural considerations into all stages of the Prevention Intervention Research Cycle to help differentiate what aspects of maltreatment are universal and which others are culturally specific. As there is currently a relative dearth in such emic research, future studies must consider cultural difference from a much more nuanced perspective, examining the varying levels at which culturally bound constructs influence the experiences of maltreatment as well as the vast heterogeneity within cultures that may help shed light on why these differences occur and on key targets for intervention. Cultural differences are not unidimensional. Thus, research examining the intersection of social identities—informed by an understanding of how systems of power, privilege, and oppression operate within specific cultural contexts—will be needed to better understand and create improved interventions for these global phenomena.

References

American Psychiatric Association. (1987). *Diagnostic and statistical manual of mental disorders* [3rd ed., rev.]. Arlington, VA: American Psychiatric Publishing.

American Psychiatric Association. (2013). *Diagnostic and statistical manual of mental disorders* (5th ed.). Washington, DC: American Psychiatric Publishing.

Breiding, M. J., Basile, K. C., Smith, S. G., Black, M. C., & Mahendra, R. R. (2015). *Intimate partner violence surveillance: Uniform definitions and recommended data elements, version 2.0.* Atlanta: National Center for Injury Prevention and Control, Centers for Disease Control and Prevention.

Cooper, R. V. (2015). Must disorders cause harm? The changing stance of the DSM. In S. Demazeux & P. Singy (Eds.), *The DSM-5 in perspective* (pp. 83–96). New York, NY: Springer. https://doi.org/10.1007/978-94-017-9765-8_5.

Crenshaw, K. (1989). Demarginalizing the intersection of race and sex: A black feminist critique of antidiscrimination doctrine, feminist theory, and antiracist politics. *University of Chicago Legal Forum, 1*, 139–167.

Del Vecchio, T., & Slep, A. M. S. (2012). The impact of child maltreatment: Psychological, physical, and economic consequences. In H. M. Foran, S. R. H. Beach, A. M. S. Slep, R. E. Heyman, & M. Z. Wamboldt (Eds.), *Family violence and family problems: Reliable assessment and the ICD-11* (pp. 149–158). New York, NY: Springer Publishing.

Fernández, M. (2006). *Cultural beliefs and domestic violence.* Blackwell Publishing.

Garcia-Moreno, C., et al. (2006). Prevalence of intimate partner violence: Findings from the WHO multi-country study on women's health and domestic violence. *Lancet, 368*(9543), 1260–1269.

Gershoff, E. T. (2002). Corporal punishment by parents and associated child behaviors and experiences: A meta-analytic and theoretical review. *Psychological Bulletin, 128*, 539–579. https://doi.org/10.1037/0033-2909.128.4.539.

Gershoff, E. T., & Grogan-Kaylor, A. (2016). Spanking and child outcomes: Old controversies and new meta-analyses. *Journal of Family Psychology, 30*, 453–469. https://doi.org/10.1037/fam0000191.

Global Initiative to End All Corporal Punishment of Children. (2018). *Global progress toward prohibiting all corporal punishment.* Retrieved from http://endcorporalpunishment.org/wp-content/uploads/legality-tables/Global-progress-table-commitment.pdf.

Grace, K. T., & Anderson, J. C. (2017). Reproductive coercion. *Trauma, Violence & Abuse, 121*, 1–20. https://doi.org/10.1177/1524838016663935.

Gracia, E., Herrero, J., & Lila, M. (2008). Explaining the immigration-partner violence link: Attitudes towards partner violence among Latin-American immigrants in Spain. *The Open Family Studies Journal, 1*(1), 31–38. https://doi.org/10.2174/1874922400801010031.

Hayes, B. E., & Boyd, K. A. (2017). Influence of individual-and national-level factors on attitudes toward intimate partner violence. *Sociological Perspectives, 60*, 685–701. https://doi.org/10.1177/0731121416662028.

Heise, L., & Garcia-Moreno, C. (2002). Violence by intimate partners. In E. G. Krug, L. L. Dahlberg, J. A. Mercy, A. B. Zwi, & R. Lozano (Eds.), *World report on violence and health* (pp. 87–121). Geneva: World Health Organization.

Heyman, R. E., Collins, P. S., Slep, A. M. S., & Knickerbocker, L. (2010). Evidence-based substantiation criteria: Improving the reliability of field decisions of child maltreatment and partner abuse. *Protecting Children, 25*, 35–46.

Heyman, R. E., Kogan, C. S., Foran, H. M., Burns, S. C., Slep, A. M. S., … Reed, G. M. (2018). A case-controlled field study evaluating ICD-11 proposals for relational problems and intimate partner violence. *International Journal of Clinical and Health Psychology, 18*, 113–123. https://doi.org/10.1016/j.ijchp.2018.03.001.

Heyman, R. E., & Slep, A. M. S. (2006). Creating and field-testing diagnostic criteria for partner and child maltreatment. *Journal of Family Psychology, 20*, 397–408. https://doi.org/10.1037/0893-3200.20.3.3.

Heyman, R. E., & Slep, A. M. S. (2009). Reliability of family maltreatment diagnostic criteria: 41 site dissemination field trial. *Journal of Family Psychology, 23*, 905–910. https://doi.org/10.1037/a0017011.

Heyman, R. E., & Slep, A. M. S. (2019). Relational disorders and beyond. In B. H. Fiese, M. Celano, K. Deater-Deckard, E. N. Jouriles, & M. A. Whisman (Eds.), *Family therapy and training*: Vol. 3. APA handbook of contemporary family psychology (pp. 19–34). Washington, DC: American Psychological Association. https://doi.org/10.1037/0000101-002.

Hillis, S., Mercy, J., Amobi, A., & Kress, H. (2016). Global prevalence of past-year violence against children: A systematic review and minimum estimates. *Pediatrics, 137*, e20154079.

Hindin, M. J., Kishor, S., & Ansara, D. L. (2008). *Intimate partner violence among couples in 10 DHS countries: Predictors and health outcomes.* DHS Analytical Studies No. 18 Calverton, MD: Macro International Inc..

Kitzmann, K., Gaylord, N. K., Holt, A. R., & Kenny, E. D. (2003). Child witnesses to domestic violence: A meta-analytic review. *Journal of Consulting and Clinical Psychology, 71*, 339–352. https://doi.org/10.1037/0022-006X.71.2.339.

Koenig, M. A., Ahmed, S., Hossain, M. B., et al. (2003). Women's status and domestic violence in rural Bangladesh: Individual- and community-level effects. *Demography, 40*, 269–288. https://doi.org/10.1353/dem.2003.0014.

Larkin, G. L., & Morris, S. C. (2009). *International perspectives on intimate partner violence.* In *Intimate partner violence: A health-based perspective.* (pp. 61–78).

Leeb, R. T., Paulozzi, L. J., Melanson, C., Simon, T. R., & Arias, I. (2008). *Child maltreatment surveillance: Uniform definitions for public health and recommended data elements.* Atlanta: Center for Disease Control and Prevention, National Center for Injury Prevention and Control. Retrieved from http://www.cdc.gov/violenceprevention/pdf/CM_Surveillance-a.pdf.

Mann, J. R., & Takyi, B. K. (2009). Autonomy, dependence or culture: Examining the impact of resources and socio-cultural processes on attitudes towards intimate partner violence in Ghana, Africa. *Journal of Family Violence, 24*(5), 323–335. https://doi.org/10.1007/s10896-009-9232-9.

Mitchell, C., & Vanya, M. (2009). Explanatory frameworks of intimate partner violence. In C. Mitchell & D. Anglin (Eds.), *Intimate partner violence: A health-based perspective* (pp. 39–52). New York, NY: Oxford University Press.

Morrisson, C., & Jütting, J. P. (2005). Women's discrimination in developing countries: A new data set for better policies. *World Development, 33*(7), 1065–1081.

Mrazek, P. J. & Haggerty, R. J. (Eds.), (1994). *Reducing risks for mental disorders: Frontiers for preventive intervention research.* Washington, DC: National Academy Press.

O'Brien, M., John, R. S., Margolin, G., & Erel, O. (1994). Reliability and diagnostic efficacy of parents' reports regarding children's exposure to marital aggression. *Violence and Victims, 9*, 45–62.

Perkins, D. F., Slep, A. M. S., & Heyman, R. E. (2017, August). *Report prepared for Alaska's Office for Children's Services: Summary report of implementation evaluation of the maltreatment assessment protocol.* State College, PA: Penn State University.

Peterson, C., Kearns, M. C., McIntosh, W. L., Estefan, L. F., Nicolaidis, C., … Florence, C. (2018). Lifetime economic burden of intimate partner violence among U.S. adults. *American Journal of Preventive Medicine, 55*, 433–444. https://doi.org/10.1016/j.amepre.2018.04.049.

Pierotti, R. S. (2013). Increasing rejection of intimate partner violence evidence of global cultural diffusion. *American Sociological Review, 78*, 240–265. https://doi.org/10.1177/0003122413480363.

Pinheiro, P. S. (2006). *Report of the independent expert for the United Nations study on violence against children.* Retrieved from https://www.unicef.org/violencestudy/reports/SG_violencestudy_en.pdf.

Rind, B., Tromovitch, P., & Bauserman, R. (1998). A meta-analytic examination of assumed properties of child sexual abuse using college samples. *Psychological Bulletin, 124*, 22–53. https://doi.org/10.1037/0033-2909.124.1.22.

Ripoll-Núñez, K. J., & Rohner, R. P. (2006). Corporal punishment in cross-cultural perspective: Directions for a research agenda. *Cross-Cultural Research, 40*(3), 220–249. https://doi.org/10.1177/1069397105284395.

Rodriguez Martínez, P., & Khalil, H. (2013). Battery and development: Exploring the link between intimate partner violence and modernization. *Cross-Cultural Research, 47*(3), 231–267. https://doi.org/10.1177/1069397112463329.

Rodriguez Martinez, P., & Khalil, H. (2017). Changing values: Attitudes about intimate partner violence in immigrants and natives in five western countries. *Deviant Behavior, 38*(3), 241–253.

Runyan, D., Wattam, C., Ikeda, R., Hassan, F., Ramiro, L., Krug, E. G., … Lozano, R. (2002). *World report on violence and health. Child abuse and neglect by parents and other caregivers.* Geneva: World Health Organization.

Saltzman, L. E. (2004). Definitional and methodological issues related to transnational research on intimate partner violence. *Violence Against Women, 10*(7), 812–830. https://doi.org/10.1177/1077801204265553.

Saltzman, L. E., Fanslow, J. L., McMahon, P. M., & Shelley, G. A. (1999). *Intimate partner violence surveillance: Uniform definitions and recommended data elements, version 1.0.* Atlanta: National Center for Injury Prevention and Control, Centers for Disease Control and Prevention.

Stoltenborgh, M., Bakermans-Kranenburg, M. J., Alink, L. R., & van IJzendoorn, M. H. (2012). The universality of childhood emotional abuse: A meta-analysis of worldwide prevalence. *Journal of Aggression, Maltreatment & Trauma, 21*, 870–890. https://doi.org/10.1080/10926771.2012.708014.

Stoltenborgh, M., Bakermans-Kranenburg, M. J., Alink, L. R., & van Ijzendoorn, M. H. (2015). The prevalence of child maltreatment across the globe: Review of a series of meta-analyses. *Child Abuse Review, 24*, 37–50. https://doi.org/10.1002/car.2353.

Stoltenborgh, M., Bakermans-Kranenburg, M. J., & van IJzendoorn, M. H. (2013). The neglect of child neglect: A meta-analytic review of the prevalence of neglect. *Social Psychiatry and Psychiatric Epidemiology, 48*, 345–355. https://doi.org/10.1007/s00127-012-0549-y.

Stoltenborgh, M., Bakermans-Kranenburg, M. J., van IJzendoorn, M. H., & Alink, L. R. (2013). Cultural–geographical differences in the occurrence of child physical abuse? A meta-analysis of global prevalence. *International Journal of Psychology, 48*, 81–94. https://doi.org/10.1080/00207594.2012.697165.

Stoltenborgh, M., van IJzendoorn, M. H., Euser, E. M., & Bakermans-Kranenburg, M. J. (2011). A global perspective on child sexual abuse: Meta-analysis of prevalence around the world. *Child Maltreatment, 16*, 79–101. https://doi.org/10.1177/1077559511403920.

Straus, M. A. (1979). Measuring intrafamily conflict and violence: The conflict tactics (CT) scale. *Journal of Marriage and the Family, 41*, 75–78.

United Nations. (1989). *Convention on the rights of the child.* Retrieved from http://www2.ohchr.org/english/law/crc.htm.

United Nations. (2009). *Handbook for legislation on violence against women.* New York, NY: United Nations.

Wallach, H. S., Weingram, Z., & Avitan, O. (2010). Attitudes toward domestic violence: A cultural perspective. *Journal of Interpersonal Violence, 25*(7), 1284–1297. https://doi.org/10.1177/0886260509340540.

Waltermaurer, E. (2012). Public justification of intimate partner violence: A review of the literature. *Trauma, Violence, & Abuse, 13*(3), 167–175. https://doi.org/10.1177/1524838012447699.

Wojda, A. K., Heyman, R. E., Smith Slep, A. M., Foran, H. M., Snarr, J. D., & Oliver, M. (2017). Family violence, suicidality, and substance abuse in active duty military families: An ecological perspective. *Military Behavioral Health, 5*, 300–312.

World Health Organization. (1992). *The ICD-10 classification of mental and behavioural disorders: Clinical descriptions and diagnostic guidelines.* Geneva: World Health Organization. https://doi.org/10.1080/21635781.2017.1343698.

World Health Organization. (2005). *WHO multi-country study on women's health and domestic violence against women: Initial results on prevalence, health outcomes and women's responses.* Geneva: World Health Organization.

World Health Organization. (2016). *Global plan of action to strengthen the role of the health system within a national multisectoral response to address interpersonal violence, in particular against women and girls, and against children.* Geneva: World Health Organization.

CHAPTER 9

Lesbian, gay, bisexual, and transgender families across the globe

Christopher A. Pepping[a], Jennifer Power[b], Adam Bourne[b], and Anthony Lyons[b]
[a]School of Psychology & Public Health, La Trobe University, Melbourne, VIC, Australia
[b]Australian Research Centre in Sex Health and Society, La Trobe University, Melbourne, VIC, Australia

There has been increasing awareness and acceptance of same-sex couples and LGBT families in high-income countries over the past 30 years. For example, the Netherlands held the first legally recognized same-sex marriage in 2001; since this time, same-sex marriages are legally recognized and performed in over 25 countries, with the most recent country legalizing same-sex marriage being Taiwan. Here we review research pertaining to LGBT families across the globe. We begin with an overview of definitions and terminology, and then highlight the similarities and differences between LGBT couples and heterosexual couples. We then report on research pertaining to the numbers of LGBT relationships and families across the globe, and the legal and social context of diverse relationships and families across countries. Next, we review the diversity of family forms and structures, and provide an overview of the methods used by LGBT families to become parents. We then review evidence pertaining to the psychological and social outcomes of children raised in LGBT families. Finally, we conclude by outlining some challenges and research limitations pertaining to LGBT couples and families, and offer recommendations for future research.

Definitions, terminology, and prevalence

Sexual orientation relates to an enduring pattern of sexual, romantic, and emotional attraction. Individuals who are attracted primarily to people of a different sex or gender are typically referred to as *heterosexual*. Those who are primarily attracted to people of the same sex or gender may refer to

Cross-Cultural Family Research and Practice
https://doi.org/10.1016/B978-0-12-815493-9.00009-0

themselves as *homosexual* or *gay*, and many gay women prefer the term *lesbian*. People who are attracted to both women and men may refer to themselves as *bisexual*; however, some may identify with other sexual orientations if their attractions span a number of diverse gender identities such as female, male, intersex, and transgender. Such individuals may use a number of terms to describe their sexual orientation, though common terms include *pansexual* and *queer*. Individuals who may be uncertain about their sexual orientation may refer to themselves as *questioning*. Population-based studies in Western contexts find that about 3.5% of adults identify as gay, lesbian, or bisexual (Gates, 2011), though less is known about the percentage of people who use other terms to describe their sexual orientation, or who are questioning.

It is important to note that being gay or lesbian is often not just about who someone is attracted to or has sex with; in contemporary life, sexuality can also be a social category. People perceive a gay, lesbian, or bisexual identity as a marker of who someone is and the communities with which they are aligned. Further, there are many people who are attracted to, and who have sex with, people of the same sex, but who identify as heterosexual. This is important, for example, in areas such as health promotion and HIV prevention. Specifically, many men who have sex with men, but do not identify as gay or bisexual, are unlikely to engage with health promotion messages targeting gay men.

Regarding gender identity, it is important to differentiate between *sex* and *gender*. *Sex* refers to the physical characteristics made up of genetics and hormones that lead to differences in primary (genitalia) and secondary (voice, body shape, body hair) characteristics. Sex is typically assigned at birth on the basis of an individual's genitalia. In contrast, the term *gender* refers to a person's identity in relation to societal and cultural assumptions about gender roles; these have traditionally been thought of as binary (i.e., boy/girl, man/woman), and inherently dictated by one's sex assigned at birth. The term *cisgender* refers to those whose gender identity aligns with their birth-assigned sex, whereas the term *transgender* refers to those with a gender identity that differs from the sex they were assigned at birth.

The term *transgender* has predominantly referred to individuals who identify with the gender identity opposite to their birth-assigned sex. However, many individuals do not identify with binary constructions of gender (i.e., male and female), and their identities fall somewhere in between, or outside of, binary concepts of male and female. Specifically, some individuals identify with a gender that may be a blend of male and female, or an identity that falls along a continuum but that is not represented by the binary identities of male and female. Finally, some may not identify with any specific gender. There are numerous terms people may use to describe a gender identity that is not binary, though commonly used terms include *nonbinary* and *gender-queer*. Although the term *transgender* includes those with *any* gender identity other than that assigned at birth, many who hold a non-binary gender identity do not feel that the term *transgender* is inclusive of their experiences (Hendricks & Testa, 2012; Kuper, Nussbaum, & Mustanski, 2012). Therefore, we will use the term *transgender and gender diverse* (TGD) throughout this chapter when focusing specifically on this group. We use the term LGBT when describing gender and sexual minorities as a group given that this is the most commonly used term internationally. Estimating the prevalence of TGD adults has been difficult given the variation in definitions (Collin, Reisner, Tangpricha, & Goodman, 2016). However, estimates of the prevalence of TGD adults often range from 0.03% to 0.05% of the population (Conron, Scott, Stowell, & Landers, 2012; Gates, 2011; Reisner et al., 2014). A systematic review of 27 studies that included both self-reports of gender identity from community surveys, as well as numbers of treatment seeking TGD individuals, estimated the prevalence to be approximately 355 in 100,000 (Collin et al., 2016).

LGBT couple relationships

Most LGBT individuals desire to be in a satisfying romantic relationship (D'Augelli, Rendina, Sinclair, & Grossman, 2007; Frost, 2011), and the relationships of same-sex couples are more similar to than they are different from those of heterosexual couples (Kurdek, 2004; Peplau & Fingerhut, 2007). Specifically, research tends to find no difference in reported relationship satisfaction between same-sex and married heterosexual couples (Peplau & Fingerhut, 2007). In addition, the predictors of relationship outcomes, such as communication patterns, conflict resolution skills, dyadic coping, social support, and the psychological adjustment of each partner, are the same as in heterosexual relationships (e.g., Gottman et al., 2003; Khaddouma, Norona, & Whitton, 2015; Kurdek, 2004). Relationship standards—referring to the beliefs people hold about what makes for a fulfilling relationship—are quite similar among heterosexuals and same-sex attracted individuals (Baker & Halford, 2019). For instance, as with heterosexuals, same-sex attracted individuals endorse the importance of the couple bond (e.g., expressing love, caring, and intimacy), family responsibility (e.g., maintaining positive relationships with extended family), relationship effort (working to improve the relationship), and religiosity (Baker & Halford, 2019). Nonetheless, there were some distinctive relationship standards that may be particularly relevant to same-sex couples, such as being "out" as a couple, and dyadic coping with discrimination (Baker & Halford, 2019). Despite the similarities, there are some important differences between same-sex and heterosexual relationships.

Some evidence suggests that same-sex couples are at greater risk of relationship breakup compared to heterosexuals (e.g., Andersson, Noack, Seierstad, & Weedon-Fekjar, 2006; Kurdek, 2004). This pattern may be somewhat accounted for by differences in marriage and marriage-like commitments between same-sex and heterosexual couples (Rosenfeld, 2014). Specifically, once marriage and marriage-like commitments were controlled for, Rosenfeld (2014) found that these disparities were no longer statistically significant over a 4-year period. However, studies conducted in Europe, where same-sex marriage has been legalized for a longer period of time compared to other countries such as the United States and Australia, found that same-sex couples were at higher risk of breakup over 8–10 year follow-up periods (Andersson et al., 2006; Kalmijn, Loeve, & Manting, 2007). Further, this finding remained when comparing only cohabiting same-sex couples with cohabiting heterosexual couples, or same-sex and heterosexual couples who are either married or legally registered (Andersson et al., 2006; Kalmijn et al., 2007).

One possible explanation for these higher rates of breakup may be that same-sex couples are less likely to have children, and the presence of children born within a marriage is associated with lower risk of relationship breakup (Heaton, 1990; Waite & Lillard, 1991). Higher rates of breakup may also, for some couples, be caused by external influences on the relationship, such as minority stress and stigmatization (Otis, Rostosky, Riggle, & Hamrin, 2006) and, for some, a lack of support for the relationship from family and friends (Khaddouma et al., 2015; Martell, Safren, & Prince, 2004). An Australian study of adults in a same-sex relationship found that some couples were hesitant to seek couple therapy due to fears of discrimination, and concerns that therapists might not be equipped to work with same-sex couples (Pepping, Lyons, Halford, Cronin, & Pachankis, 2017). Such barriers to accessing couple services are likely to prevent relationship help-seeking, which may also partially explain the higher rates of breakup among same-sex couples.

According to the minority stress theory (Meyer, 2003; Meyer, Schwartz, & Frost, 2008), LGB individuals are at higher risk of psychological and relationship difficulties because of living in a heteronormative environment. These experiences undermine the quality of life (Mays & Cochran, 2001) and predict reduced relationship satisfaction (Otis et al., 2006), which likely contributes to the disparities in relationship stability between same-sex couples and heterosexual couples. Some LGBT individuals internalize negative societal attitudes toward gender and sexual minorities, a process known as internalized stigma. Internalized stigma is associated with low couple satisfaction, and this appears to be mediated by the tendency to hide and conceal one's relationship (Pepping et al., 2017).

There are also important relationship differences that do not pertain to minority stress. Nonmonogamous relationships are more common, and more accepted, in gay male relationships compared to heterosexual (and lesbian) couples (Bailey, Gaulin, Agyei, & Gladue, 1994; Lyons & Hosking, 2014). Importantly, those who are in an open, nonmonogamous relationship fare just as well as their monogamous counterparts in terms of relationship satisfaction (Bricker & Horne, 2007). However, studies from the United States and Australia suggest that a particularly important determinant of relationship functioning is the extent to which the couple effectively negotiate and clarify rules and expectations about sexual exclusivity (Hoff, Beougher, Chakravarty, Darbes, & Neilands, 2010; Hosking, 2013a, 2013b; Parsons, Starks, DuBois, Grov, & Golub, 2013). In other words, gay male couples who are in an open, nonmonogamous relationship are just as satisfied in their relationship as those who are in monogamous arrangements. What appears to be most critical in terms of relationship functioning is the extent to which both partners can agree on rules, expectations, and boundaries pertaining to extra-dyadic sex.

Remarkably little research has investigated the romantic relationships of TGD individuals. However, there are some important and unique issues facing relationships with one or more TGD partners. For instance, many TGD individuals will affirm their gender identity while in a relationship (heterosexual or same-sex). Coming out to partners can be associated with a range of challenges, including fear of judgment, concern that the relationship will end, and sometimes concerns for safety. Further, for many TGD individuals, affirming their identity can involve taking medical and nonmedical steps to change their physical appearance to be more consistent with their gender identity. Taking such steps can often be a relief for TGD individuals, but can be met with complex reactions from their romantic partners (Bethea & McCollum, 2013; Platt & Bolland, 2018). For instance, partners may grieve the loss of their partner's body, and in many instances, one partner affirming their gender identity can change the sexual orientation labels of the relationship (Platt & Bolland, 2018). Thus, there are unique challenges facing relationships with one or more TGD individuals. Although there is limited research on this topic, one US study found that about half of those who were in a relationship before they had begun to transition and affirm their gender were no longer in the relationship (Meier, Sharp, Michonski, Babcock, & Fitzgerald, 2013). There is great need for research investigating the predictors of relationship outcomes for TGD couples, including factors that might be protective against some of the common challenges facing TGD couples.

LGBT families across the globe

It is difficult to determine the number of LGBT people who are parents. However, in high- and middle-income countries, it is increasingly common for questions about sexuality to be included in population-based data surveys (or national census data collection). Although there are still limitations with these data, this provides a more accurate picture of the number of LGBT people who are parenting children—showing that the number is in the hundreds of thousands. For instance, in the United States, 887,456 same-sex couple households were identified in the 2016 American Community Survey (U.S. Census Bureau, 2016). Of these, approximately 16.5% had children living in their household—23.3% of female couples and 9.5% of male couples. However, other data sources indicate that up to 48% of LGBT women and 20% of LGBT men are a parent to a child or children aged under 18 years (Gates, 2013). In the United Kingdom, the 2017 census data indicated that there were 190,000 same-sex couple households in the United Kingdom of which 18,000 (approximately 9.5%) had dependent children. In both the United States and the United Kingdom, same-sex couples who were legally married (or in a civil partnership in the United Kingdom) were more likely to have dependent children than those who were unmarried/cohabiting (Office for National Statistics, 2017).

Data from the 2016 Australian Census indicate that there were 46,800 cohabiting same-sex couples in Australia. Of these, 15% had children living with them in their household, 25% of female couples and 5% of male couples. In total, there were 10,500 children aged under 25 years living in same-sex couple households (Australian Bureau of Statistics (ABS), 2018). However, Australian census data are acknowledged by the Australian Bureau of Statistics to be a conservative estimate of the number of LGBT parents in Australia as it only includes same-sex couples (sexual identity is not recorded). As such, these data do not include single parents who identify as LGBT, nonresident parents (parents who do not live with their children), or people who were reluctant to declare their relationship within the census. There is also no record of transgender parents. A 2012 survey of over 3000 LGBT-identified Australians indicated that 22% of respondents were parents or stepparents, while 40% intended to have children (or more children) in the future (Leonard et al., 2012).

Legal context and legal recognition across the globe

The above estimates are, of course, from countries in which homosexuality is legal, LGBT people are protected by antidiscrimination laws, same-sex marriage is legal, and—although many LGBT people still face significant discrimination—there is wide social acceptance of LGBT people. However, like all dimensions of LGBT rights and experiences, the global picture of LGBT families is complex. It ranges from full parenting and adoption rights with explicitly LGBT-supportive adoption and surrogacy services, through to prosecution, imprisonment, and even the death penalty in some parts of the world for LGBT people. Even within those countries that are supportive or inclusive, the social and legal situation can vary significantly by state or province, and the circumstances vary considerably by gender identity and expression. For instance, a metaanalysis of 138 studies found high rates of victimization among gay, lesbian, and bisexual individuals, with between 55% and 80% having experienced verbal harassment, and 28% physical assault (Katz-Wise & Hyde, 2012).

Sex between men remains illegal in 73 countries around the world, and is subject to imprisonment or capital punishment (Equaldex, 2018). These countries are predominantly in sub-Saharan Africa and Southern and South-Eastern Asia and typically (but not exclusively) in countries previously colonized by the United Kingdom, which adopted penal codes relating to homosexuality. Others criminalize homosexuality under the auspices of Islamic Sharia Law. The situation for lesbian or bisexual women is often harder to decipher given a lack of specific reference to women in many homosexuality related laws or statutes, although many such laws have been used to prosecute or threaten women regardless. However, the Human Dignity Trust (2016) estimates that sex between women is criminalized in 34 jurisdictions, either by the expressed mention of female-female sexual conduct or by using neutral language when outlawing same-sex experience that can apply to both men and women. In many countries and cultures, women face considerable pressure to have children which can lead to being—or feeling—forced into marriages to conceive (Human Dignity Trust, 2016).

In such contexts, there is no legal recognition of, and little possibility for, LGBT people to formally or openly adopt children, and there are significant social and safety barriers to more informal co-parenting relationships. That is not to say that LGBT families do not exist in these countries, but rather that there is an abject lack of research detailing the number and nature of these. Even in those countries that do *not* criminalize same-sex activity, there is a mixed pattern of legal rights relating to adoption and surrogacy which impacts LGBT people. In total, there are 44 countries of the world where adoption by same-sex couples is explicitly legal (i.e., specifically legislated for, or was never illegal), and 27 countries where it is expressly illegal (Equaldex, 2018). The countries in the former category largely include Western Europe and large parts of South America, while those in the latter category (where it is illegal) are more often found in Africa, Eastern and Southern Asia, and certain parts of Eastern Europe.

For a further 153 countries and jurisdictions, the picture is considerably more complex, ambiguous, or uncertain. This complexity includes a requirement in some countries (such as Hungary) that same-sex couples must be married in order to be eligible to adopt, or that same-sex attracted people can adopt, but only as a single person and not as part of a same-sex couple (e.g., Russia, Chile, Romania). In a small number of countries (e.g., Finland and, until recently, Germany), adoption is permitted only for stepchildren. Specifically, if one partner in a same-sex couple already has a child, then the other partner has the legal right to become their adoptive parent; this is also known as *successive adoption*.

In certain countries, especially those with devolved legal jurisdictions, legal recognition and permission for same-sex adoption has been further complicated by differing legal perspectives and compliance. Within the United States, for example, prior to 2017 there was considerable variation across states regarding the ability of same-sex couples or individuals to adopt, with certain states explicitly legislating against it. A series of Supreme Court challenges (notably those in Mississippi and Arkansas) have facilitated an environment in which same-sex adoption cannot be made illegal by individual states (i.e., it struck down such laws). In reality, however, there still exist exemptions for religious adoption agencies that can decline to work with same-sex couples and there are still numerous reports of state-based adoption agencies that are reluctant to work with this population.

Only seven states (plus the District of Columbia [DC]) have laws that expressly prohibit discrimination in adoption based on sexual orientation, and three states (plus DC) have laws that prohibit discrimination based on gender identity (Movement Advancement Project, 2018).

The legal situation regarding other forms of LGBT parenting (see later section) are often harder to determine. The legal context of surrogacy varies considerably across jurisdictions and a full review is beyond the scope of this chapter. Within Australia (with a similar legal approach in this regard to the United Kingdom and many parts of Western Europe), commercial surrogacy is illegal in all states. While in vitro fertilization (IVF) is legal for all female same-sex couples, this is not legal for surrogates of male couples in Western Australia or the Northern Territory. Similarly, while altruistic surrogacy is legal in most states, it is illegal for single people and same-sex couples in Western Australia (although a bill to remove this restriction is pending).

Family forms and structures

There are various pathways to parenthood for LGBT people. Since the 1980s, there has been a gradual increase in availability and accessibility of assisted reproductive technologies and donor sperm, enhancing the options for lesbians to pursue pregnancy (Crouch, McNair, Waters, & Power, 2013). More recently, commercial surrogacy and egg donor services have emerged as an option for some gay men to become parents. These technologies, along with increasing social and political recognition of LGBT people, mean it is increasingly common for LGBT people to pursue parenthood in the context of a same-sex relationship or as a single parent. However, use of technology is only part of the story. Many LGBT people form families through other means including adoption, fostering, home-based insemination, or through a prior heterosexual relationship. Note that in this section, we describe reproductive technologies that are commonly used by lesbian and bisexual women (e.g., sperm donation), or gay and bisexual men (e.g., egg donor services). For this reason, we refer primarily to gay, lesbian, and bisexual individuals when describing these technologies. However, it is important to note that these technologies are not necessarily specific to gender. For instance, TGD men can use donor sperm to carry a pregnancy.

For many years, a heterosexual relationship was the primary method by which LGBT people became parents—often this was a relationship or marriage that people entered into before coming out as gay or lesbian

(Crouch et al., 2013), or prior to gender affirmation (von Doussa, Power, & Riggs, 2017). Recent years have seen a significant increase in the number of same-sex couples entering into parenthood together, but a heterosexual union is still a common pathway to parenthood for LGBT people, and especially gay men who face significant legal or financial barriers to becoming a parent in many parts of the world (Carneiro, Tasker, Salinas-Quiroz, Leal, & Costa, 2017). It is also worth remembering that there are people within LGBT communities, particularly bisexual and transgender people, who are in long-term opposite-sex relationships in which they conceive and parent children (Hines, 2006; Power et al., 2012).

There is limited research on the experiences of LGBT people who come to parenting as a stepparent to the children of their same-sex partner, though it is likely that this is a common experience. A 2008 Australian survey of LGBT parents found that, of the 445 respondents, 36% were parenting at least one child from their partner's former heterosexual relationship, while 10% were parenting a child from their partner's former same-sex relationship (Power et al., 2010). There are many challenges associated with step, or blended, families that are common to both heterosexual and same-sex parented families (Lynch, 2000). These may include tensions in adjusting to new family structures and cohabiting arrangements, as well as difficulties for the stepparent in adjusting to their new identity and role as a parent. In addition, there can be challenges relating to conflict between biological parents and one or more stepparents, including conflict about parenting practices, major life decisions that may affect children and the larger family, and difficulties agreeing on family rituals and celebrations. These challenges are likely to apply to both heterosexual and same-sex parented families. However, same-sex parented families may face unique challenges associated with social stigma and, for some families, decisions about 'coming out' that affect both children and parents (Robitaille & Saint-Jacques, 2009). For instance, younger children, who are likely to have been less exposed to societal heteronormativity, may accept a same-sex stepparent in much the same way as they would a stepparent of the opposite sex of their biological parent. Older children, on the other hand, may hold varying levels of bias as a result of socialization, and may also be at greater risk of peer ridicule and ostracism (Gold, 2017; Robitaille & Saint-Jacques, 2009). Finally, there may be varying levels of support for the same-sex relationship from extended family, including, at times, from heterosexual ex-partners (Gold, 2017; Robitaille & Saint-Jacques, 2009).

Adoption is another common pathway to parenthood for same-sex couples in countries or regions where adoption is legal for same-sex couples, and where adoption is a common practice. In some countries, such as Australia, adoption by same-sex couples is legal in all states and territories, but within-country adoption is not a common practice in Australia; most adoption is international. In these cases, the legality of same-sex couple adoption is determined by the country in which the child was born. In other countries, adoption by same-sex couples is much more common. For example, the 2011 US census indicated that over 16,000 same-sex couples were raising an estimated 22,000 adopted children, as same-sex couple adoption is legal throughout the United States (Gates, 2013). In fact, in 2011, same-sex couples were four times more likely than opposite-sex couples to be raising adopted children. However, adoption can be difficult for LGBT people even in jurisdictions where it is legal. Studies indicate that it can be complex to navigate a system designed to accommodate heterosexual parents. Further, discrimination against same-sex couples seeking to adopt children is not uncommon, with some "gatekeepers" in the adoption system potentially holding discriminatory attitudes (Brown, Smalling, Groza, & Ryan, 2009; Carneiro et al., 2017; Hicks, 2005).

Fostering of children is also common among LGBT communities, and has a longer history than adoption given that many countries allowed same-sex couples to become foster parents many years, or decades, before they were able to adopt. In the United Kingdom, LGBT people have been foster parents since the 1980s. While there are no official statistics on the number of same-sex couples who are foster parents, there are multiple established support organizations for LGBT foster carers. In the United States, fostering among LGBT couples is less common than adoption. However, similarly to adoption, same-sex couples are more likely to be raising foster children than opposite-sex couples. In 2011, it was estimated that approximately 2600 same-sex couples were raising 3400 foster children in the United States (Gates, 2013).

There is evidence that lesbians have been pursuing pregnancy through supportive medical clinics or home-based artificial insemination (self-insemination of donor sperm, the proverbial "turkey baster" method) since the 1960s or possibly even earlier. However, from the 1970s onward, advice about home-insemination for lesbians began to be distributed via lesbian and women's health networks. For example, the 1975 Conference on Women and Health held at Harvard Medical School included a workshop titled "Lesbians: Artificial Insemination at Home" (Agigian, 2004). Still today, many lesbians will become mothers through home-insemination with the sperm of a known donor—often a gay male friend. A 2014 Australian survey of 500 children with at least one LGBT parent found that 32% of those with female parents had been conceived via home-based insemination (27% of total sample) (Crouch, Waters, McNair, Power, & Davis, 2014). Research shows that many lesbians choose home-based insemination as a means to become pregnant because they want to use a known sperm donor and/or because home-insemination is less expensive and less invasive than fertility services (Hayman, Wilkes, Halcomb, & Jackson, 2015).

While home-insemination is generally associated with lesbian pregnancy, it is actually a pathway to parenthood for many gay men as well through their role as a sperm donor/father. There are variations on the role that gay men may play in their lives of children for whom they are a known sperm donor (Dempsey, 2015). Some donor/fathers will have no involvement in their children's everyday life. In others, the donor/father plays an "uncle-like" role in the family, spending time with the children occasionally or regularly. Some families engage in active co-parenting arrangements. Typically, this will involve a lesbian couple or single women co-parenting with a gay man or gay couple. All parents will be actively involved in the care of the children and take equal responsibility for decision-making about their needs. For some families, the children will be cared for across each of the parents' households (Power et al., 2010).

Since the 1980s, laws have gradually changed to allow female same-sex couples and single women access to sperm banks and fertility clinics. This has significantly enhanced the opportunities for lesbians to become parents. There are no statistics on the number of lesbian or bisexual women who become pregnant through assisted reproductive technology, though several studies suggest it is a relatively common practice within lesbian communities. A 2014 Australian survey of 500 children with at least one LGBT parent found that 48% of those with female parents were conceived using assisted reproductive technology. Of these, 75% had conceived using a clinic-acquired donor (a sperm bank) (Crouch et al., 2014). There are different types of fertility services that may be required by women using donor sperm. Intrauterine insemination (IUI) is a clinical procedure whereby sperm is placed into the uterus. IVF involves retrieval of eggs from a woman so they can be fertilized with sperm outside the body and transferred back into the uterus. Women utilizing donor sperm will generally undergo IUI or IVF depending on a range of factors, including the presence of conditions affecting her fertility. Research shows that many lesbians utilize sperm banks to become pregnant as they prefer an unknown sperm donor for legal or social reasons. In addition, some lesbians feel more comfortable using a fertility clinic than home-based insemination for health or safety reasons (Hayman et al., 2015).

Surrogacy services are an option for gay men to become parents. Surrogacy involves a woman carrying a pregnancy to term on behalf of another person or couple. There are two forms of surrogacy. Gestational surrogacy involves two women, an egg donor, and a surrogate. The egg from the donor is fertilized with sperm (usually the sperm of the intended gay father) and the embryo is carried by the surrogate. With traditional surrogacy, the surrogate also supplies the egg (Norton, Hudson, & Culley, 2013). Commercial surrogacy is legal in some jurisdictions, including several states in the United States. However, the high cost of commercial surrogacy means it is prohibitive for many gay men (Tuazon-McCheyne, 2010). There is an emerging body of research on the motivations of gay men to pursue surrogacy over other pathways to parenthood, including adoption. Some gay men report that biogenetic paternity is important to them because it confers a sense of validity to their family due to the high social value placed on biological kinship (Dempsey, 2013). Men may also have an emotional connection to their biogenetic heritage, which they seek to extend to their children. Gay men may also see surrogacy as more secure than other forms of parenting, including adoption or fostering. Surrogacy ensures the fathers will be able to raise the child from birth and that they are clearly the legal parents of the child (Lev, 2006).

The above issues regarding becoming a parent may apply to many TGD people (Stotzer, Herman, & Hasenbush, 2014). There are many TGD people who become parents prior to affirming their gender and continue to parent their children in their affirmed gender (von Doussa et al., 2017). Each TGD parent and their family will have a different experience of this. For some, it may deepen family connections, whereas for others it may lead to alienation or rejection. For some TGD people, being a parent may affect their process of gender affirmation, with some people delaying this for many years for fear of potential negative effects on their children or partner. For TGD people, the experience of gender affirmation and living as a TGD person does not happen in isolation to family life and family connections (von Doussa, Power, & Riggs, 2015). Parenthood, or the desire to become a parent, can be an important (but often overlooked) part of a TGD person's life narrative (Hines, 2006). Some TGD people will become parents after they affirm their gender, using the reproductive capacity of their body.

For transgender men, this may involve becoming pregnant as a man. This can be profoundly challenging for transgender men in terms of their emotional connection to their body and their masculine identity. However, the desire to become a parent means transgender men may see this as a "functional sacrifice" that is worth making (Charter, Ussher, Perz, & Robinson, 2018). Reproductive healthcare systems and fertility services are often not familiar with, or well equipped to deal with, transgender men. Thus, on top of the personal challenge of becoming pregnant, transgender men may encounter stigma or discrimination within fertility services as well as within maternal healthcare and childbirth facilities (Charter et al., 2018; Obedin-Maliver & Makadon, 2016).

As with all family life, the family formation of LGBT families is often complex, and many will be formed through a combination of methods. One study of 445 same-sex attracted parents based in Australia or New Zealand found that there were 6 typical family forms: (1) a two-parent same-sex couple based family; (2) families in which a lesbian couple were the primary parents but a known sperm donor/father was involved in the child's life; (3) families in which an LGBT parent was co-parenting with a former heterosexual partner (which may involve a current same-sex partner as well); (4) separated same-sex parented families in which an LGBT parent was co-parenting with their former same-sex partner; (5) sole parent families; and (6) multiparent families, which typically involved a lesbian couple co-parenting with a gay man or gay couple to raise children across their two households (Dempsey, 2015; Power et al., 2010).

For LGBT people, becoming a parent requires access to emotional, social, and financial resources. Surrogacy and fertility services are expensive, often costing tens if not hundreds of thousands of dollars (US). Adoption and fostering are geared toward middle-class households. While home-based insemination is an inexpensive way for lesbians to pursue pregnancy, it still requires a level of resourcefulness and access to social networks to identify a sperm donor and facilitate the insemination. This type of social capital is generally aligned with higher levels of income and education (Taylor, 2009). For these reasons, parenthood among LGBT people tends to be more common among people who are white, middle class, and highly educated. Therefore, while the increased rates of LGBT people becoming parents are evidence of significant social and political change toward greater acceptance of LGBT families, it is important to acknowledge that there is marginalization within this community, and that it is not simple or feasible for many LGBT people to become parents. In addition, most research on LGBT parenting has been framed within Western models of kinship, centered on similarities to, and difference from, the biogenetically linked nuclear family (Berkowitz, 2009; Cahill, Battle, & Meyer, 2003). Non-Western models of kinship—including those with different patterns of intergenerational connection and care for children within extended family systems—may be inclusive or exclusive of LGBT parents in ways that are not considered in this body of research. There is a need for more research on the family experiences of LGBT parents in diverse community settings that explores the intersections of class, race, sexuality, and kinship (Battle & Ashley, 2008; Berkowitz, 2009).

Finally, it is important to consider the rights of children born via donor conception to have knowledge about their conception, as well as access to information about the identity of the donor. Laws vary considerably across nations, and across states within nations, in regard to the number of families donor's sperm may be provided to, the number of births allowed per donor, who may access such methods (e.g., married couples only; heterosexual couples only; single people), and whether the donor may remain anonymous. Within Australia, for instance, all states and territories have laws that support the rights of children to have access to information about the identity of their donor once they reach the age of 18. However, research conducted in Australia, the United Kingdom, Spain, Italy, the Netherlands, and the United States suggests that most heterosexual parents with children conceived via donor conception do not disclose this fact to their children (Golombok et al., 2011; Kovacs, Wise, & Finch, 2012; Readings, Blake, Casey, Jadva, & Golombok, 2011; van Berkel, van der Veen, Kimmel, & te Velde, 1999). Nonetheless, professional guidelines emphasize the importance of children being informed about their donor conception, and whether they are being raised in LGBT families or heterosexual families (Victorian Assisted Reproductive Treatment Authority, 2014).

Psychosocial outcomes of children raised in LGBT families

As the visibility, legal status, and societal acceptance of LGBT families has increased, some have expressed concerns about the potential impact of diverse family structures on children (American Psychological Association, 2004). One area of concern that has been expressed is that children raised by same-sex parents will experience disturbances in their own identity development, and may be more likely to identify as LGBT themselves compared to children raised by different-sex parents. This concern rests on the assumption that being LGBT is somehow problematic, and research has largely moved away from addressing such questions. Some have expressed concern that children raised by same-sex parents will be at greater risk of poor mental health, psychological maladjustment, and display greater behavioral problems. Further, some have gone as far as to suggest that children raised by lesbian women or gay men are more likely to be sexually abused by their parent, or acquaintances close to their parent. Finally, there have been concerns raised about the potential for children raised by same-sex parents to experience discrimination, bullying, and teasing from peers and others.

There is a growing body of empirical research designed to address these concerns. However, it is important to note that most studies have investigated such questions in US populations, and there is more research pertaining to outcomes of children raised by lesbian parents than gay male parents (Patterson, 2006). Regarding the concern that children raised by same-sex couples may experience difficulties in their own identity development, research suggests that identity development, including sexual and gender identity, develops in similar ways for children of heterosexual parents and children raised by lesbian mothers (Patterson, 2006). Similarly, multiple studies conducted in the 1980s and 1990s examined whether children of same-sex parents were more likely to be gay or lesbian themselves; results of these studies did not suggest elevated rates of homosexuality among those raised by same-sex parents (see Patterson, 2006, for a review).

Regarding child social development, parental sexual orientation is unrelated to the social competence of their children, measured using parent and teacher reports of the Child Behavior Checklist (Chan, Raboy, & Patterson, 1998; Fulcher, Sutfin, Chan, Scheib, & Patterson, 2005; Patterson, 2006). Similarly, adolescent relationship involvement, sexual behavior, and school performance are not associated with parental sexual orientation (Wainright, Russell, & Patterson, 2004). Rather, the quality of the parent-child relationship is a stronger predictor of child and adolescent adjustment than is parental sexual orientation; parent-child relationships characterized by warmth and security consistently predict positive adjustment (Chan et al., 1998; Fulcher et al., 2005; Patterson, 2006; Wainright et al., 2004). The possibility that children of same-sex couples may be at greater risk of bullying and discrimination has also been investigated. Although children of same-sex parents sometimes do report having experienced homophobic or antigay abuse from their peers (Farr, Crain, Oakley, Cashen, & Garber, 2016; Gartrell, Rodas, Deck, Peyser, & Banks, 2005), comparative research found that children of same-sex parents were at no greater risk of peer victimization compared to children of opposite-sex parents (Wainright & Patterson, 2006). In fact, in one US study, the children of same-sex parents reported a *stronger* connection to their peers at school compared to those children raised by opposite-sex couples (Wainright & Patterson, 2006).

Perhaps the most widely researched topic has been the psychological adjustment and mental health outcomes of children raised by same-sex parents, relative to those raised by heterosexual parents. Reviews of the empirical literature on this topic have consistently concluded that children of lesbian mothers or gay fathers do not differ from those raised by heterosexual parents in emotional adjustment, psychological adjustment, behavioral functioning, and cognitive functioning (e.g., Anderssen, Amlie, & Ytterøy, 2002; Biblarz & Stacey, 2010; Patterson, 2006). Further, when significant differences between groups have been observed, these differences tend to show better outcomes for children raised by same-sex parents. A recent study conducted in Italy investigated the psychological adjustment of children (aged 3–11 years) raised by 70 gay fathers who became parents via surrogacy, 125 lesbian mothers who became parents via donor insemination, and 195 heterosexual parents who became parents via spontaneous conception (Baiocco, Carone, Ioverno, & Lingiardi, 2018). Children raised by gay fathers and lesbian mothers were less likely to have psychological problems compared to the children raised by heterosexual parents, though once family process factors were accounted for, the type of family was not predictive of child health outcomes. In brief, parental sexual orientation clearly cannot be seen as a risk factor for poor psychological adjustment in children.

There is a consistent finding in the literature that children raised by same-sex parents fare just as well as children raised by heterosexual parents. In one notable and widely publicized exception, Regnerus (2012) claimed to find evidence that children of parents who had been in a same-sex relationship were at higher risk of maladaptive outcomes. The author assessed a random sample of 2988 American adults (aged 18–39 years) who reported on their family of origin experiences and well-being. Regnerus (2012) reported that individuals who had one parent who had a same-sex relationship displayed worse outcomes on a range of psychological, social, and economic measures. However, the methodology and handling of the statistical analyses has been widely criticized (Cheng & Powell, 2015; Perrin, Cohen, & Caren, 2013), and it is important to note that neither parental sexual orientation, nor the family structure, was assessed. Rather, participants were asked "From when you were born until age 18 (or until you left home to be on your own), did either of your parents ever have a romantic relationship with someone of the same sex?" Those who responded "yes" to this question were classified as having a lesbian mother or gay father, and were then compared to those from intact biological families and respondents from other family environments (e.g., adopted, stepfamily). Whether this same-sex relationship was a long-term relationship, the result of an affair, or a casual sexual relationship was not assessed. Indeed, only 52% of those responding that their mother had a lesbian relationship had ever lived in the household with their mother and her lesbian partner.

The Regnerus (2012) study therefore does not shed light on the effects of same-sex parenting, because the study did not assess whether participants were raised by same-sex parents. Further, regarding the statistical concerns of the study, Cheng and Powell (2015) reanalyzed the data correcting for data-analytic errors and found that the originally reported disadvantage of having a parent who had been in a same-sex relationship largely disappeared. In brief, a large body of evidence suggests that children raised by same-sex parents fare just as well as those raised by heterosexual parents. Importantly, what appears to be most predictive of psychological adjustment is the quality of the parent-child relationship, rather than the sexual orientation of parents.

It is important to consider the determinants of well-being among LGBT individuals who have children. While the well-being of adults is inherently important, there are also clear benefits for children if their parents are coping well. Research has shown that the legal and political context relating to LGBT rights can affect the well-being of parents. For instance, a 2009 study comparing lesbian mothers in Canada and the United States, conducted prior to changes to federal law allowing same-sex marriage across the United States, found that lesbian mothers in the United States experienced greater worries about discrimination and were more likely to report depressive symptoms than Canadian lesbian mothers (Shapiro, Peterson, & Stewart, 2009). Heterosexual mothers in each country reported no differences. Another US-based study on well-being among gay and lesbian adoptive parents during their transition to parenthood found that higher levels of perceived support from families, colleagues, and friends were related to lower anxiety symptoms among adoptive parents (Goldberg & Smith, 2011). Adoptive parents with high levels of internalized homophobia, who also lived in states with legal climates that were less favorable to gay adoption, experienced a sharp increase in depressive or anxiety symptoms during their transition to parenthood.

Most LGBT individuals are raised by heterosexual parents, and thus it is important to consider the experience of LGBT individuals and their families when they come out to their family. First, LGBT individuals are often faced with a decision of whether to come out to their family and, if they do, to whom they wish to disclose. Some LGBT individuals remain "in the closet" and attempt to conceal their identity from family members, which is often done in order to avoid potentially negative consequences and reactions from family (LaSala, 2000; Malterud & Bjorkman, 2016; Valentine, Skelton, & Butler, 2003). For those who decide to come out to their families, it is often reported to be a highly stressful experience, with substantial fear about the potential for negative reactions and consequences (LaSala, 2000). For instance, reactions from parents can be positive, accepting, and affirming, but can also include negative reactions such as self-blame, grief, anger, rejection, and even violence (D'Augelli, Hershberger, & Pilkington, 1998; LaSala, 2000; Merighi & Grimes, 2000). Research suggests that negative reactions to disclosure of a child's sexual orientation is predicted by parental political conservatism and religiosity, as well as rigid family functioning and enmeshment (Baiocco et al., 2015, 2016), though there is substantial variation in familial acceptance, and the level of acceptance often changes over time.

Practice implications

LGBT individuals are at higher risk for poor mental health compared to their heterosexual and cisgender counterparts (King et al., 2008). However, many LGBT individuals, couples, and families who require assistance do not access appropriate interventions in high-income countries (Mayer et al., 2008; Steele et al., 2017; Williams & Chapman, 2011), and this pattern is likely to be even more marked in countries with less acceptance. As with the general population, LGBT individuals not only face a range of practical barriers to accessing services, such as a shortage of appropriate professionals, and the high costs of services (Bischoff, Hollist, Smith, & Flack, 2004; Cronin, Pepping, Halford, & Lyons, 2019; Shipherd, Green, & Abramovitz, 2010), but they also face some minority stress-related barriers that are likely to undermine help-seeking (Hatzenbuehler, Phelan, & Link, 2013; Willging, Salvador, & Kano, 2006). For instance, fear of disclosing one's sexual orientation to health professionals, and fears of discrimination from healthcare providers, are commonly reported barriers among gender and sexual minorities (e.g., Cronin et al., 2019; Lisy, Peters, Schofield, & Jefford, 2018; Pepping et al., 2017). In relation to couple-based services, a recent Australian study of 363 adults in a same-sex relationship found that there was very low uptake of couple services such as relationship counseling (4.41%) and couple workshops (0.55%); further, among those who had accessed these services, satisfaction with the service was low (Pepping et al., 2017). A range of minority stress-related barriers to accessing services were also identified, including fear of discrimination, and a lack of practitioners who are trained to work with same-sex couples.

How do interventions need to be tailored and modified to be appropriate for LGBT couples and families? First, existing couple and family interventions may need to incorporate material designed to assist people to cope with discrimination effectively in order to buffer the effects of stigmatization on individual and relational well-being (Buzzella, Whitton, & Tompson, 2012; Martell et al., 2004; Pepping et al., 2017; Whitton & Buzzella, 2012). For instance, same-sex couple interventions might focus on enhancing dyadic coping to effectively cope with discrimination and stigma as a couple (Buzzella et al., 2012; Martell et al., 2004; Pepping et al., 2017; Whitton & Buzzella, 2012). Such a focus may be especially useful for couples living in countries with less acceptance of their relationship, though to our knowledge, no research has addressed this question in such contexts.

For some couples and families, it may be important to assess for, and address, the impact of internalized stigma, which refers to the internalization of negative societal attitudes toward LGBT people. Internalized stigma predicts poor individual mental health (Newcomb & Mustanski, 2010) and lower couple relationship satisfaction (Frost & Meyer, 2009; Pepping, Cronin, Halford, & Lyons, 2018). Assisting individuals who are high in internalized stigma to recognize the impact and external origins of internalized stigma may be a potentially useful addition to couple and family interventions (Green & Mitchell, 2015; Greenan & Tunnell, 2003; Pepping & Halford, 2014; Whitton & Buzzella, 2012).

As mentioned earlier, there are characteristics of same-sex relationships that may need to be accommodated in couple interventions. For instance, with regard to nonmonogamous relationships, it is likely to be helpful to include a focus on effective negotiation surrounding sexual exclusivity, safe sex, and rules and expectations for sex outside of the relationship (Hosking, 2013a, 2013b; Parsons et al., 2013; Pepping et al., 2017). For LGBT couples who may not yet be out to their families, friends, or even partner, incorporating a focus on whether, when, and how to come out may be useful (Pepping, Cronin, et al., 2018). Similarly, if there are discrepancies between partners in terms of comfort with disclosure, or level of outness, therapy might include a specific focus on identifying and managing such differences effectively (Pepping, Cronin, et al., 2018). Thus, it has been recommended that interventions for LGBT populations include some content about coming out, offer guidance on coping with harassment and discrimination if appropriate, and highlight specific and unique issues that might be particularly relevant to same-sex couples (Lyons, Rozbroj, Pitts, Mitchell, & Christensen, 2015).

It is also important to note that existing couple and family interventions have tended to be designed for, and evaluated with, cisgender, heterosexual individuals, and thus these existing programs are likely to contain some level of heteronormative and cis-normative bias. It is therefore important that program materials are inclusive of LGBT families and relationships (Pepping & Halford, 2014; Whitton & Buzzella, 2012). For instance, Lyons et al. (2015) recommend that when interventions include vignettes and example characters, it is best to avoid the exclusive use of heterosexual and cisgender examples and vignettes, and the inclusion of vignettes depicting nontraditional family structures and gender roles will likely enhance the applicability and appropriateness of the intervention to a broader audience.

There is only one published trial of couple therapy with same-sex couples (Fals-Stewart, O'Farrell, & Lam, 2009), and this trial focused on couples where one partner had an alcohol use disorder. Fals-Stewart et al. (2009) found that behavioral couple therapy was more effective than individual treatment for alcohol misuse, and the intervention also led to increased couple satisfaction. Regarding relationship education, there have been only three published trials with same-sex couples (Buzzella et al., 2012; Newcomb et al., 2017; Whitton et al., 2017). Buzzella et al. (2012) found that relationship education led to increased relationship satisfaction and positive relationship processes, such as communication, following the program among same-sex male couples, though the sample was small ($n=7$ couples received relationship education immediately). Newcomb et al. (2017) found that a couple-based relationship education and HIV prevention program led to increased relationship investment and decreased HIV risk behavior among 57 young male couples. Finally, Whitton et al. (2017) tested relationship education in 37 female same-sex couples. Relationship education was associated with improved couple communication and relationship satisfaction immediately after the intervention, and the effects of improved communication, but not satisfaction, were maintained 3 months later.

Finally, how might service providers work to support families who may be having difficulties accepting their LGBT child or family member? It may be helpful to provide education about LGBT-related issues, including dispelling myths and stereotypes about gender and sexual minorities, as doing this will likely address potential concerns that the parents (or other family members) may have (Merighi & Grimes, 2000). It may also be important to work with the LGBT individual to plan how they wish to come out, and when they wish to do so. For instance, there may be circumstances in which delaying the disclosure of gender identity or sexual orientation may be preferable, if for instance they hold realistic fears that their safety or financial security may be jeopardized upon disclosure (Merighi & Grimes, 2000). Therapists might also work with the LGBT individual to enhance their communication skills to effectively have these conversations with family members, prepare for potential reactions from family, and identify methods to cope with any negative reactions or consequences (Merighi & Grimes, 2000).

It is important to note that service providers are often not equipped to address the specific needs of gender and sexual minorities, which may not be surprising given that they often receive very little training in LGBT-affirmative therapy and intervention (e.g., Obedin-Maliver et al., 2011; Pepping, Lyons, & Morris, 2018). The term *affirmative therapy* refers to the practice of adopting LGBT-affirmative attitudes, knowledge, and skills when delivering services (Davies, 1996; Pachankis & Goldfried, 2013). Pepping, Lyons, et al. (2018) examined whether training mental health professionals in LGBT-affirmative therapy would increase their knowledge and skills relating to therapeutic work with gender and sexual minorities, and decrease any negative attitudes toward this population. The protocol was a full-day workshop that included content and skills training in therapeutic work with gender sexual minorities, including issues associated with LGBT couples and families. Therapists' knowledge and skills increased following the training, and they also displayed reductions in negative attitudes toward gender and sexual minorities.

Research challenges and limitations, and recommendations for future research

Although research involving LGBT couples and families has grown over recent years, much work remains to be done, as a significant challenge is a lack of data in many countries on the numbers of LGBT couples and families. National census surveys in the vast majority of countries do not ask citizens about their sexual orientation and tend to ask about gender identity in a binary format (male or female). As mentioned earlier, the Australian census is now able to provide data on cohabiting same-sex couples but is not able to report on single LGBT people who may be in noncohabiting relationships or be raising children as a single parent. Many other countries around the world lack any data on sexual identity or TGD identities. Thus, representative or accurate data are unavailable in many cases, not only of the numbers of LGBT families and couples, but also the different ways in which LGBT families are structured. Advocating for the inclusion of questions on sexual and gender identity, with the inclusion of diverse gender identities, would enable a more complete assessment of family diversity in a given country than is presently the case.

Many would argue that forming relationships and raising children are universal rights, with exceptions in extreme cases such as child abuse. As we have noted thus far, the pathways for LGBT people to establish cohabiting relationships that are regarded as equal to those of heterosexual cisgender couples, and to establish families, have grown in recent decades in some countries. However, severe restrictions and prohibitions still exist in many parts of the world. Research is needed to identify the specific ways in which different legal frameworks and sociocultural environments impact LGBT relationships and families both positively and negatively.

Much of the research conducted thus far comprises single cross-sectional studies. Longer-term longitudinal studies that track changes over time would be especially useful; only a few have been conducted (e.g., Gartrell, Bos, & Koh, 2019; Power et al., 2015). Studies could, for example, track legislative changes and their impact on LGBT relationships and families. Research conducted with large cohorts over many years or decades would be particularly valuable. Such studies could follow individuals and families over significant periods of the life course as legislative and social contexts change. Even in countries where LGBT individuals enjoy many of the rights of the general population, much remains unknown about how couples and families navigate changes across the life course such as new opportunities for conceiving children, undergoing gender affirmation, changes in family structures, divorce and separation, raising children through key life stages such as commencing and graduating from school, becoming a grandparent, and so on. Charting the many ways in which LGBT couples and families navigate key milestones in relation to the legal and sociocultural contexts in which they live would deliver much-needed knowledge of the predictors of a range of psychosocial outcomes to assist in identifying specific support needs and legislative and policy changes.

There is a particularly acute need for additional research involving TGD relationships and families. While research on same-sex parents has grown, there is a distinct shortage of similar research involving TGD populations. It is especially important to recognize that the experiences and challenges for TGD couples and families can vary considerably compared to those of same-sex couples and families, as we have noted in this chapter. Research ought to be designed with sufficient nuance that allows the unique experiences of the various subpopulations within the broader LGBT group to be given consideration, and to therefore ensure that research findings can be utilized in legislative, policy, health, and social support initiatives in ways that target areas of particular need.

Quantitative data, such as data from large-scale national surveys, provide a picture of broader patterns, such as family structures, barriers, and facilitators for conceiving children, and persistent issues related to stigma and discrimination. However, qualitative research is also required. For example, in-depth interviews and qualitative longitudinal studies are valuable in exploring specific lived experiences such as how LGBT couples and families use resources to overcome challenges in conceiving and raising children, what it means for them to be a family in the sociocultural context of the community in which they live, and how multiple experiences, challenges, and life-stage milestones intersect in ways that affect them and their families. Some qualitative studies have been conducted that explore facets of the above (e.g., von Doussa et al., 2017; Welsh, 2011), but are specific to particular jurisdictions. Further work across a range of sociocultural contexts would deliver a comprehensive understanding of the diversity of lived experiences.

One limitation faced by researchers is access to populations. This is especially challenging in countries where LGBT people may feel that their safety is at risk if they disclose their identity and may therefore be reluctant to participate in research. In many cases, researchers will put in place protocols to ensure that participants can take part anonymously, such as completing an online survey. Ensuring confidentiality and the de-identification of data are also ways to help overcome this issue. However, as noted in this chapter, some people are further marginalized. For example, they may not be connected to LGBT communities and/or they may be additionally marginalized due to socioeconomic status, ethnic background, or religion. Where necessary, particular efforts may be needed to reach these populations to ensure that research findings are genuinely representative of LGBT couples and families. In addition, there may be legal or political barriers in some countries in conducting research on these topics, which prohibits researchers from carrying out research involving LGBT populations. Aside from these limitations, there is much scope for further research in countries where it is safe to do so.

It seems likely that at least some of the findings from research conducted in Western countries are likely to also be applicable to other parts of the world, though we cannot assume that Western research findings will apply in exactly the same ways. For instance, the research suggesting that the quality of the parent-child relationship is a better predictor of child adjustment than the sexual orientation of parents is likely to be applicable across cultures. However, given that societal acceptance of such relationships differs so markedly across cultures and countries, the findings from Western countries that children of same-sex parents are not subject to higher rates of bullying and stigma may not consistently hold across countries. These experiences of bullying and stigma are then highly likely to undermine the psychological adjustment of children raised by same-sex parents in such contexts. Further, it seems likely that examining these research questions in family structures that do not adhere to Western models of kinship may add greater complexity, and offer unique perspectives on the well-being of LGBT families that Western research has not explored. In brief, it is likely that some of the Western research findings will be applicable across cultures, but this needs to be explicitly tested. There are additional cultural and contextual factors that may moderate findings from Western research, and thus we cannot assume that Western research findings will transfer to other cultures, contexts, and family structures in the same way.

Finally, while much of the research that has been carried out has identified the challenges experienced by LGBT couples and families, as we have noted in this chapter, there are now a range of research findings showing positive experiences depending on the country or jurisdiction, particularly in high-income countries, as well as findings demonstrating that the children of same-sex couples are just as likely to thrive as those of heterosexual cisgender couples. Continued research involving LGBT couples and families is highly encouraged. In achieving a thorough and accurate picture, not only are myths and misconceptions likely to be corrected, but pathways toward overcoming challenges are also likely to be identified.

References

Agigian, A. (2004). *Baby steps: How lesbian alternative insemination is changing the world.* Middletown, CT: Wesleyan University Press.

American Psychological Association. (2004). *Sexual orientation, parents, & children: Research summary.* Retrieved from https://www.apa.org/about/policy/parenting.aspx.

Anderssen, N., Amlie, C., & Ytterøy, E. A. (2002). Outcomes for children with lesbian or gay parents. A review of studies from 1978 to 2000. *Scandinavian Journal of Psychology, 43*(4), 335–351. https://doi.org/10.1111/1467-9450.00302.

Andersson, G., Noack, T., Seierstad, A., & Weedon-Fekjar, H. (2006). The demographics of same-sex marriages in Norway and Sweden. *Demography, 43*(1), 79–98. https://doi.org/10.1353/dem.2006.0001.

Australian Bureau of Statistics (ABS). (2018). *2071.0—Census of population and housing: Reflecting Australia—Stories from the census, 2016: Same-sex couples in Australia, 2016.* Canberra: Australian Bureau of Statistics.

Bailey, J. M., Gaulin, S., Agyei, Y., & Gladue, B. A. (1994). Effects of gender and sexual orientation on evoluntionarily relevant aspects of human mating psychology. *Journal of Personality and Social Psychology, 66*, 1081–1093. https://doi.org/10.1037/0022-3514.66.6.1081.

Baiocco, R., Carone, N., Ioverno, S., & Lingiardi, V. (2018). Same-sex and different-sex parent families in Italy: Is parents' sexual orientation associated with child health outcomes and parental dimensions? *Journal of Developmental & Behavioral Pediatrics, 39*(7), 555–563. https://doi.org/10.1097/DBP.0000000000000583.

Baiocco, R., Fontanesi, L., Santamaria, F., Ioverno, S., Baumgartner, E., & Laghi, F. (2016). Coming out during adolescence: Perceived parents' reactions and internalized sexual stigma. *Journal of Health Psychology, 21*(8), 1809–1813. https://doi.org/10.1177/1359105314564019.

Baiocco, R., Fontanesi, L., Santamaria, F., Ioverno, S., Marasco, B., Baumgartner, E., … Laghi, F. (2015). Negative parental responses to coming out and family functioning in a sample of lesbian and gay young adults. *Journal of Child and Family Studies, 24*(5), 1490–1500. https://doi.org/10.1007/s10826-014-9954-z.

Baker, N. A., & Halford, W. K. (2019). Assessment of couple relationships standards in same-sex attracted adults. *Family Process.* https://doi.org/10.1111/famp.12447.

Battle, J., & Ashley, C. (2008). Intersectionality, heteronormativity, and Black lesbian, gay, bisexual, and transgender (LGBT) families. *Black Women, Gender & Families, 2*(1), 1–24.

Berkowitz, D. (2009). Theorizing lesbian and gay parenting: Past, present, and future scholarship. *Journal of Family Theory & Review, 1*(3), 117–132. https://doi.org/10.1111/j.1756-2589.2009.00017.x.

Bethea, M. S., & McCollum, E. E. (2013). The disclosure experiences of male-to-female transgender individuals: A systems theory perspective. *Journal of Couple & Relationship Therapy, 12*(2), 89–112. https://doi.org/10.1080/15332691.2013.779094.

Biblarz, T. J., & Stacey, J. (2010). How does the gender of parents matter? *Journal of Marriage and Family, 72*(1), 3–22. https://doi.org/10.1111/j.1741-3737.2009.00678.x.

Bischoff, R. J., Hollist, C. S., Smith, C. W., & Flack, P. (2004). Addressing the mental health needs of the rural underserved: Findings from a multiple case study of a behavioral telehealth project. *Contemporary Family Therapy, 26*(2), 179–198. https://doi.org/10.1023/B:COFT.0000031242.83259.fa.

Bricker, M. E., & Horne, S. G. (2007). Gay men in long-term relationships. *Journal of Couple & Relationship Therapy, 6*(4), 27–47. https://doi.org/10.1300/J398v06n04_02.

Brown, S., Smalling, S., Groza, V., & Ryan, S. (2009). The experiences of gay men and lesbians in becoming and being adoptive parents. *Adoption Quarterly, 12*, 229–246. https://doi.org/10.1080/10926750903313294.

Buzzella, B. A., Whitton, S. W., & Tompson, M. C. (2012). A preliminary evaluation of a relationship education program for male same-sex couples. *Couple and Family Psychology: Research and Practice*, *1*(4), 306–322. https://doi.org/10.1037/a0030380.

Cahill, S., Battle, J., & Meyer, D. (2003). Partnering, parenting, and policy: Family issues affecting Black lesbian, gay, bisexual, and transgender (LGBT) people. *Race and Society*, *6*(2), 85–98. https://doi.org/10.1016/j.racsoc.2004.11.002.

Carneiro, F. A., Tasker, F., Salinas-Quiroz, F., Leal, I., & Costa, P. A. (2017). Are the fathers alright? A systematic and critical review of studies on gay and bisexual fatherhood. *Frontiers in Psychology*, *8*, 1636. https://doi.org/10.3389/fpsyg.2017.01636/full.

Chan, R. W., Raboy, B., & Patterson, C. J. (1998). Psychosocial adjustment among children conceived via donor insemination by lesbian and heterosexual mothers. *Child Development*, *69*, 443–457. https://doi.org/10.1111/j.1467-8624.1998.tb06201.x.

Charter, R., Ussher, J. M., Perz, J., & Robinson, K. (2018). The transgender parent: Experiences and constructions of pregnancy and parenthood for transgender men in Australia. *International Journal of Transgenderism*, *19*(1), 64–77. https://doi.org/10.1080/15532739.2017.1399496.

Cheng, S., & Powell, B. (2015). Measurement, methods, and divergent patterns: Reassessing the effects of same-sex parents. *Social Science Research*, *52*, 615–626. https://doi.org/10.1016/j.ssresearch.2015.04.005.

Collin, L., Reisner, S. L., Tangpricha, V., & Goodman, M. (2016). Prevalence of transgender depends on the "case" definition: A systematic review. *The Journal of Sexual Medicine*, *13*(4), 613–626. https://doi.org/10.1016/j.jsxm.2016.02.001.

Conron, K. J., Scott, G., Stowell, G. S., & Landers, S. J. (2012). Transgender health in Massachusetts: Results from a household probability sample of adults. *American Journal of Public Health*, *102*(1), 118–122. https://doi.org/10.2105/AJPH.2011.300315.

Cronin, T. J., Pepping, C. A., Halford, W. K., & Lyons, A. (2019). Mental health help-seeking and barriers to service access in sexual minority Australians. (submitted for publication).

Crouch, S., McNair, R., Waters, E. B., & Power, J. (2013). What makes a same-sex parented family. *Medical Journal of Australia*, *199*(2), 94–96. https://doi.org/10.5694/mja12.11234.

Crouch, S. R., Waters, E., McNair, R., Power, J., & Davis, E. (2014). Parent-reported measures of child health and wellbeing in same-sex parent families: A cross-sectional survey. *BMC Public Health*, *14*(1), 635. https://doi.org/10.1186/1471-2458-14-635.

D'Augelli, A. R., Rendina, J., Sinclair, K. O., & Grossman, A. H. (2007). Lesbian and gay youth's aspirations for marriage and raising children. *Journal of LGBT Issues in Counseling*, *1*, 77–98. https://doi.org/10.1300/J462v01n04_06.

D'Augelli, A. R., Hershberger, S. L., & Pilkington, N. W. (1998). Lesbian, gay, and bisexual youth and their families: Disclosure of sexual orientation and its consequences. *American Journal of Orthopsychiatry*, *68*(3), 361–371. https://doi.org/10.1037/h0080345.

Davies, D. (1996). *Pink therapy: A guide for counsellors and therapists working with lesbian, gay, and bisexual clients*. (Vol. 1). Philadelphia, PA: Open University Press.

Dempsey, D. (2013). Surrogacy, gay male couples and the significance of biogenetic paternity. *New Genetics and Society*, *32*(1), 37–53. https://doi.org/10.1080/14636778.2012.735859.

Dempsey, D. (2015). Familiarly queer? Same-sex relationships and family formation. In G. Heard & D. Arunachalam (Eds.), *Family formation in 21st century Australia* (pp. 225–240). Dordrecht: Springer.

Equaldex. (2018). *LGBT inclusion indices*. Retrieved from http://www.equaldex.com/.

Fals-Stewart, W., O'Farrell, T. J., & Lam, W. K. K. (2009). Behavioral couple therapy for gay and lesbian couples with alcohol use disorders. *Journal of Substance Abuse Treatment*, *37*(4), 379–387. https://doi.org/10.1016/j.jsat.2009.05.001.

Farr, R. H., Crain, E. E., Oakley, M. K., Cashen, K. K., & Garber, K. J. (2016). Microaggressions, feelings of difference, and resilience among adopted children with sexual minority parents. *Journal of Youth and Adolescence*, *45*(1), 85–104. https://doi.org/10.1007/s10964-015-0353-6.

Frost, D. M. (2011). Similarities and differences in the pursuit of intimacy among sexual minority and heterosexual individuals: A personal projects analysis. *Journal of Social Issues*, *67*, 282–301. https://doi.org/10.1111/j.1540-4560.2011.01698.x.

Frost, D. M., & Meyer, I. H. (2009). Internalized homophobia and relationship quality among lesbians, gay men, and bisexuals. *Journal of Counseling Psychology*, *56*(1), 97–109. https://doi.org/10.1037/a0012844.

Fulcher, M., Sutfin, E. L., Chan, R. W., Scheib, J. E., & Patterson, C. J. (2005). Lesbian mothers and their children: Findings from the Contemporary Families Study. In A. Omoto & H. Kurtzman (Eds.), *Recent research on sexual orientation, mental health, and substance abuse* (pp. 281–299). Washington, DC: American Psychological Association.

Gartrell, N., Bos, H., & Koh, A. (2019). Sexual attraction, sexual identity, and same-sex sexual experiences of adult offspring in the US National Longitudinal Lesbian Family Study. *Archives of Sexual Behavior*, *48*(5), 1495–1503. https://doi.org/10.1007/s10508-019-1434-5.

Gartrell, N., Rodas, C., Deck, A., Peyser, H., & Banks, A. (2005). The national lesbian family study: 4. Interviews with the 10 year old children. *American Journal of Orthopsychiatry*, *75*(4), 518–524. https://doi.org/10.1037/0002-9432.75.4.518.

Gates, G. (2011). *How many people are lesbian, gay, bisexual, and transgender?* Retrieved from-http://www.ciis.edu/Documents/PDFs/PublicPrograms/ExpandingtheCircle/How-many-people-are-LGBT-Final.pdf.

Gates, G. (2013). *LGBT parenting in the United States*. Los Angeles, CA: The Williams Institute.

Gold, J. M. (2017). Honoring the experiences of gay stepfamilies: An unnoticed population. *Journal of Divorce & Remarriage*, *58*(2), 126–133. https://doi.org/10.1080/10502556.2016.1268020.

Goldberg, A. E., & Smith, J. Z. (2011). Stigma, social context, and mental health: Lesbian and gay couples across the transition to adoptive parenthood. *Journal of Counseling Psychology*, *58*(1), 139–150. https://doi.org/10.1037/a0021684.

Golombok, S., Readings, J., Blake, L., Casey, P., Mellish, L., Marks, A., & Jadva, V. (2011). Children conceived by gamete donation: Psychological adjustment and mother-child relationships at age 7. *Journal of Family Psychology*, *25*(2), 230–239. https://doi.org/10.1037/a0022769.

Gottman, J. M., Levenson, R. W., Gross, J., Fredrickson, B. L., McCoy, K., Rosenthal, L., … Yoshimoto, D. (2003). Correlates of gay and lesbian couples' relationship satisfaction and relationship dissolution. *Journal of Homosexuality*, *45*(1), 23–43. https://doi.org/10.1300/J082v45n01_02.

Green, R.-J., & Mitchell, V. (2015). Gay, lesbian, and bisexual issues in couple therapy. In A. S. Gurman, J. L. Lebow, & D. K. Snyder (Eds.), *Clinical handbook of couple therapy* (5th ed, pp. 489–511). New York: Guilford.

Greenan, D., & Tunnell, G. (2003). *Couple therapy with gay men*. New York: Guilford.

Hatzenbuehler, M. L., Phelan, J. C., & Link, B. G. (2013). Stigma as a fundamental cause of population health inequalities. *American Journal of Public Health*, *103*(5), 813–821. https://doi.org/10.2105/AJPH.2012.301069.

Hayman, B., Wilkes, L., Halcomb, E., & Jackson, D. (2015). Lesbian women choosing motherhood: The journey to conception. *Journal of GLBT Family Studies*, *11*(4), 395–409. https://doi.org/10.1080/1550428X.2014.921801.

Heaton, T. B. (1990). Marital stability throughout the child-rearing years. *Demography*, *27*(1), 55–63. https://doi.org/10.2307/2061552.

Hendricks, M. L., & Testa, R. J. (2012). A conceptual framework for clinical work with transgender and gender nonconforming clients: An adaptation of the Minority Stress Model. *Professional Psychology: Research and Practice*, *43*(5), 460–467. https://doi.org/10.1037/a0029597.

Hicks, S. (2005). Lesbian and gay foster care and adoption: A brief UK history. *Adoption and Fostering*, *29*(3), 42–56. https://doi.org/10.1177/030857590502900306.

Hines, S. (2006). Intimate transitions: Transgender practices of partnering and parenting. *Sociology*, *40*(2), 353–371. https://doi.org/10.1177/0038038506062037.

Hoff, C. C., Beougher, S. C., Chakravarty, D., Darbes, L. A., & Neilands, T. B. (2010). Relationship characteristics and motivations behind agreements among gay male couples: Differences by agreement type and couple serostatus. *AIDS Care*, *22*, 827–835. https://doi.org/10.1080/09540120903443384.

Hosking, W. (2013a). Agreements about extra-dyadic sex in gay men's relationships: Exploring differences in relationship quality by agreement type and rule-breaking behaviour. *Journal of Homosexuality*, *60*, 711–733. https://doi.org/10.1080/00918369.2013.773819.

Hosking, W. (2013b). Satisfaction with open sexual agreements in Australian gay men's relationships: The role of perceived discrepancies in benefit. *Archives of Sexual Behavior*, *42*, 1309–1317. https://doi.org/10.1007/s10508-012-0005-9.

Human Dignity Trust. (2016). *Breaking the silence: Criminalisation of lesbians and bisexual women and its impacts*. London: Human Dignity Trust.

Kalmijn, M., Loeve, A., & Manting, D. (2007). Income dynamics in couples and the dissolution of marriage and cohabitation. *Demography*, *44*(1), 159–179. https://doi.org/10.1353/dem.2007.0005.

Katz-Wise, S. L., & Hyde, J. S. (2012). Victimization experiences of lesbian, gay, and bisexual individuals: A meta-analysis. *Journal of Sex Research*, *49*(2–3), 142–167. https://doi.org/10.1080/00224499.2011.637247.

Khaddouma, A., Norona, J. C., & Whitton, S. W. (2015). Individual, couple, and contextual factors associated with same-sex relationship instability. *Couple and Family Psychology: Research and Practice*, *4*, 106–125. https://doi.org/10.1037/cfp0000043.

King, M., Semlyen, J., Tai, S. S., Killaspy, H., Osborn, D., Popelyuk, D., & Nazareth, I. (2008). A systematic review of mental disorder, suicide, and deliberate self harm in lesbian, gay and bisexual people. *BMC Psychiatry*, *8*(1), 70. https://doi.org/10.1186/1471-244X-8-70.

Kovacs, G. T., Wise, S., & Finch, S. (2012). Functioning of families with primary school-age children conceived using anonymous donor sperm. *Human Reproduction*, *28*(2), 375–384. https://doi.org/10.1093/humrep/des403.

Kuper, L., Nussbaum, R., & Mustanski, B. (2012). Exploring the diversity of gender and sexual orientation identities in an online sample of transgender individuals. *Journal of Sex Research*, *49*(2–3), 244–254. https://doi.org/10.1080/00224499.2011.596954.

Kurdek, L. A. (2004). Are gay and lesbian cohabiting couples really that different from heterosexual couples? *Journal of Marriage and the Family*, *66*(4), 880–890. https://doi.org/10.1111/j.0022-2445.2004.00060.x.

LaSala, M. C. (2000). Lesbians, gay men, and their parents: Family therapy for the coming out crisis. *Family Process*, *39*(1), 67–81. https://doi.org/10.1111/j.1545-5300.2000.39108.x.

Leonard, W., Pitts, M., Mitchell, A., Lyons, A., Smith, A., Patel, S., & Couch, M. (2012). *Private lives 2. The second national survey on the health and wellbeing of Gay, Lesbian, Bisexual, Transgender (GLBT) Australians*. Melbourne, VIC: La Trobe University.

Lev, A. I. (2006). Gay dads: Choosing surrogacy. *Lesbian & Gay Psychology Review*, 7, 72–76.

Lisy, K., Peters, M. D., Schofield, P., & Jefford, M. (2018). Experiences and unmet needs of lesbian, gay, and bisexual people with cancer care: A systematic review and meta-synthesis. *Psycho-Oncology*, *27*(6), 1480–1489. https://doi.org/10.1002/pon.4674.

Lynch, J. M. (2000). Considerations of family structure and gender composition: The lesbian and gay stepfamily. *Journal of Homosexuality*, *40*(2), 81–95. https://doi.org/10.1300/J082v40n02_06.

Lyons, A., & Hosking, W. (2014). Prevalence and correlates of sexual partner concurrency among Australian gay men aged 18-39. *AIDS and Behavior*, *18*, 801–809. https://doi.org/10.1007/s10461-013-0613-y.

Lyons, A., Rozbroj, T., Pitts, M., Mitchell, A., & Christensen, H. (2015). *Improving E-therapy for mood disorders among lesbians and gay men: A practical toolkit for developing tailored web and mobile phone-based depression and anxiety interventions.* Retrieved from http://arrow.latrobe.edu.au:8080/vital/access/manager/Repository/latrobe:37676.

Malterud, K., & Bjorkman, M. (2016). The invisible work of closeting: A qualitative study about strategies used by lesbian and gay persons to conceal their sexual orientation. *Journal of Homosexuality, 63*(10), 1339–1354. https://doi.org/10.1080/00918369.2016.1157995.

Martell, C. R., Safren, S. A., & Prince, S. E. (2004). *Cognitive-behavioural therapies with Lesbian, Gay, and Bisexual clients.* New York: The Guilford Press.

Mayer, K. H., Bradford, J. B., Makadon, H. J., Stall, R., Goldhammer, H., & Landers, S. (2008). Sexual and gender minority health: What we know and what needs to be done. *American Journal of Public Health, 98*(6), 989–995. https://doi.org/10.2105/AJPH.2007.127811.

Mays, V. M., & Cochran, S. D. (2001). Mental health correlates of perceived discrimination among lesbian, gay, and bisexual adults in the United States. *American Journal of Public Health, 91*(11), 1869–1876. https://doi.org/10.2105/AJPH.91.11.1869.

Meier, S. C., Sharp, C., Michonski, J., Babcock, J. C., & Fitzgerald, K. (2013). Romantic relationships of female-to-male trans men: A descriptive study. *International Journal of Transgenderism, 14*(2), 75–85. https://doi.org/10.1080/15532739.2013.791651.

Merighi, J. R., & Grimes, M. D. (2000). Coming out to families in a multicultural context. *Families in Society, 81*(1), 32–41. https://doi.org/10.1606/1044-3894.1090.

Meyer, I. H. (2003). Prejudice, social stress and mental health in lesbian, gay and bisexual populations: Conceptual issues and research evidence. *Psychological Bulletin, 129*, 674–697. https://doi.org/10.1037/0033-2909.129.5.674.

Meyer, I. H., Schwartz, S., & Frost, D. M. (2008). Social patterning of stress and coping: Does disadvantaged social status confer more stress and fewer coping resources? *Social Science & Medicine, 67*(3), 368–379. https://doi.org/10.1016/j.socscimed.2008.03.012.

Movement Advancement Project. (2018). *Foster and adoption laws (US).* Retrieved from-http://www.lgbtmap.org/equality-maps/foster_and_adoption_laws.

Newcomb, M. E., Macapagal, K. R., Feinstein, B. A., Bettin, E., Swann, G., & Whitton, S. W. (2017). Integrating HIV prevention and relationship education for young same-sex male couples: A pilot trial of the 2GETHER intervention. *AIDS and Behavior, 21*, 2464–2478. https://doi.org/10.1007/s10461-017-1674-0.

Newcomb, M. E., & Mustanski, B. (2010). Internalized homophobia and internalizing mental health problems: A meta-analytic review. *Clinical Psychology Review, 30*(8), 1019–1029. https://doi.org/10.1016/j.cpr.2010.07.003.

Norton, W., Hudson, N., & Culley, L. (2013). Gay men seeking surrogacy to achieve parenthood. *Reproductive Biomedicine Online, 27*(3), 271–279. https://doi.org/10.1016/j.rbmo.2013.03.016.

Obedin-Maliver, J., Goldsmith, E. S., Stewart, L., White, W., Tran, E., Brenman, S., … Lunn, M. R. (2011). Lesbian, gay, bisexual, and transgender–related content in undergraduate medical education. *JAMA, 306*(9), 971–977. https://doi.org/10.1001/jama.2011.1255.

Obedin-Maliver, J., & Makadon, H. J. (2016). Transgender men and pregnancy. *Obstetric Medicine, 9*(1), 4–8. https://doi.org/10.1177/1753495X15612658.

Office for National Statistics. (2017). *Trends in living arrangements including families (with and without dependent children), people living alone and people in shared accommodation, broken down by size and type of household.* Retrieved from https://www.ons.gov.uk/peoplepopulationandcommunity/birthsdeathsandmarriages/families/bulletins/familiesandhouseholds/2017.

Otis, M., Rostosky, S., Riggle, E., Hamrin. (2006). Stress and relationship quality in same-sex couples. *Journal of Social and Personal Relationships, 23*(1), 81–99. https://doi.org/10.1177/0265407506060179.

Pachankis, J., & Goldfried, M. (2013). Clinical issues in working with lesbian, gay, and bisexual clients. *Psychology of Sexual Orientation and Gender Diversity*, *1*, 45–58. https://doi.org/10.1037/2329-0382.1.S.45.

Parsons, J. T., Starks, T. J., DuBois, S., Grov, C., & Golub, S. A. (2013). Alternatives to monogamy among gay male couples in a community survey: Implications for mental health and sexual risk. *Archives of Sexual Behavior*, *42*(2), 303–312. https://doi.org/10.1007/s10508-011-9885-3.

Patterson, C. J. (2006). Children of lesbian and gay parents. *Current Directions in Psychological Science*, *15*(5), 241–244. https://doi.org/10.1111/j.1467-8721.2006.00444.x.

Peplau, L. A., & Fingerhut, A. W. (2007). The close relationships of lesbians and gay men. *Annual Review of Psychology*, *58*, 405–424. https://doi.org/10.1146/annurev.psych.58.110405.085701.

Pepping, C. A., Cronin, T. J., Halford, W. K., & Lyons, A. (2018). Minority stress and same-sex relationship satisfaction: The role of concealment motivation. *Family Process*, *58*(2), 496–508. https://doi.org/10.1111/famp.12365.

Pepping, C. A., & Halford, W. K. (2014). Relationship education and therapy for same-sex couples. *Australian and New Zealand Journal of Family Therapy*, *35*, 431–444. https://doi.org/10.1002/anzf.1075.

Pepping, C. A., Lyons, A., Halford, W. K., Cronin, T. J., & Pachankis, J. E. (2017). Couple interventions for same-sex couples: A consumer survey. *Couple and Family Psychology: Research and Practice*, *6*(4), 258–273. https://doi.org/10.1037/cfp0000092.

Pepping, C. A., Lyons, A., & Morris, E. M. (2018). Affirmative LGBT psychotherapy: Outcomes of a therapist training protocol. *Psychotherapy*, *55*(1), 52–62. https://doi.org/10.1037/pst0000149.

Perrin, A. J., Cohen, P. N., & Caren, N. (2013). Are children of parents who had same-sex relationships disadvantaged? A scientific evaluation of the no-differences hypothesis. *Journal of Gay & Lesbian Mental Health*, *17*(3), 327–336. https://doi.org/10.1080/19359705.2013.772553.

Platt, L. F., & Bolland, K. S. (2018). Relationship partners of transgender individuals: A qualitative exploration. *Journal of Social and Personal Relationships*, *35*(9), 1251–1272. https://doi.org/10.1177/0265407517709360.

Power, J., Perlesz, A., Brown, R., Schofield, M., Pitts, M., McNair, R., & Bickerdike, A. (2010). Diversity, tradition and family: Australian same-sex attracted parents and their families. *Gay and Lesbian Issues and Psychology Review*, *6*(2), 66–81.

Power, J., Perlesz, A., Brown, R., Schofield, M. J., Pitts, M. K., McNair, R., & Bickerdike, A. (2012). Bisexual parents and family diversity: Findings from the work, love, play study. *Journal of Bisexuality*, *12*(4), 519–538. https://doi.org/10.1080/15299716.2012.729432.

Power, J., Schofield, M. J., Farchione, D., Perlesz, A., McNair, R., Brown, R., … Bickerdike, A. (2015). Psychological wellbeing among same-sex attracted and heterosexual parents: Role of connectedness to family and friendship networks. *Australian and New Zealand Journal of Family Therapy*, *36*(3), 380–394. https://doi.org/10.1002/anzf.1109.

Readings, J., Blake, L., Casey, P., Jadva, V., & Golombok, S. (2011). Secrecy, disclosure and everything in between: Decisions of parents of children conceived by donor insemination, egg donation and surrogacy. *Reproductive Biomedicine Online*, *22*(5), 485–495. https://doi.org/10.1016/j.rbmo.2011.01.014.

Regnerus, M. (2012). How different are the adult children of parents who have same-sex relationships? Findings from the New Family Structures study. *Social Science Research*, *41*(4), 752–770. https://doi.org/10.1016/j.ssresearch.2012.03.009.

Reisner, S. L., Conron, K. J., Tardiff, L. A., Jarvi, S., Gordon, A. R., & Austin, S. B. (2014). Monitoring the health of transgender and other gender minority populations: Validity of natal sex and gender identity survey items in a US national cohort of young adults. *BMC Public Health*, *14*(1), 1224. https://doi.org/10.1186/1471-2458-14-1224.

Robitaille, C., & Saint-Jacques, M. C. (2009). Social stigma and the situation of young people in lesbian and gay stepfamilies. *Journal of Homosexuality*, *56*(4), 421–442. https://doi.org/10.1080/00918360902821429.

Rosenfeld, M. J. (2014). Couple longevity in the era of same-sex marriage in the United States. *Journal of Marriage and the Family*, *76*, 905–918. https://doi.org/10.1111/jomf.12141.

Shapiro, D. N., Peterson, C., & Stewart, A. J. (2009). Legal and social contexts and mental health among lesbian and heterosexual mothers. *Journal of Family Psychology*, *23*(2), 255–262. https://doi.org/10.1037/a0014973.

Shipherd, J. C., Green, K. E., & Abramovitz, S. (2010). Transgender clients: Identifying and minimizing barriers to mental health treatment. *Journal of Gay & Lesbian Mental Health*, *14*(2), 94–108. https://doi.org/10.1080/19359701003622875.

Steele, L. S., Daley, A., Curling, D., Gibson, M. F., Green, D. C., Williams, C. C., & Ross, L. E. (2017). LGBT identity, untreated depression, and unmet need for mental health services by sexual minority women and trans-identified people. *Journal of Women's Health*, *26*(2), 116–127. https://doi.org/10.1089/jwh.2015.5677.

Stotzer, R. L., Herman, J. L., & Hasenbush, A. (2014). *Transgender parenting: A review of existing research*. Los Angeles, CA: The Williams Institute.

Taylor, Y. (2009). *Lesbian and gay parenting: Securing social and educational capital*. Basingstoke: Palgrave McMillan.

Tuazon-McCheyne, J. (2010). Two dads: Gay male parenting and its politicisation—A cooperative inquiry action research study. *Australian and New Zealand Journal of Family Therapy*, *31*(4), 311–323. https://doi.org/10.1375/anft.31.4.311.

U.S. Census Bureau. (2016). *Characteristics of same-sex couple households: 2005 to present*. Retrieved from https://www.census.gov/data/tables/time-series/demo/same-sex-couples/ssc-house-characteristics.html.

Valentine, G., Skelton, T., & Butler, R. (2003). Coming out and outcomes: Negotiating lesbian and gay identities with, and in, the family. *Environment and Planning D: Society and Space*, *21*(4), 479–499. https://doi.org/10.1068/d277t.

van Berkel, D., van der Veen, L., Kimmel, I., & te Velde, E. (1999). Difference in the attitudes of couples whose children were conceived through artificial insemination by donor in 1980 and in 1996. *Fertility and Sterility*, *71*, 226–231. https://doi.org/10.1016/S0015-0282(98)00448-8.

Victorian Assisted Reproductive Treatment Authority. (2014). *Why, when and how to tell children about donor conception*. Melbourne, VIC : VARTA. Retrieved from https://www.varta.org.au/resources/publications/why-when-and-how-tell-children-about-donor-conception.

von Doussa, H., Power, J., & Riggs, D. (2015). Imagining parenthood: The possibilities and experiences of parenthood among transgender people. *Culture, Health and Sexuality*, *17*(9), 1119–1131. https://doi.org/10.1080/13691058.2015.1042919.

von Doussa, H., Power, J., & Riggs, D. W. (2017). Family matters: Transgender and gender diverse peoples' experience with family when they transition. *Journal of Family Studies*, 1–14. https://doi.org/10.1080/13229400.2017.1375965.

Wainright, J. L., & Patterson, C. J. (2006). Delinquency, victimization, and substance use among adolescents with female same-sex parents. *Journal of Family Psychology*, *20*(3), 526–530. https://doi.org/10.1037/0893-3200.20.3.526.

Wainright, J. L., Russell, S. T., & Patterson, C. J. (2004). Psychosocial adjustment, school outcomes, and romantic relationships of adolescents with same-sex parents. *Child Development*, *75*(6), 1886–1898. https://doi.org/10.1111/j.1467-8624.2004.00823.x.

Waite, L. J., & Lillard, L. A. (1991). Children and marital disruption. *American Journal of Sociology*, *96*(4), 930–953. https://doi.org/10.1086/229613.

Welsh, M. G. (2011). Growing up in a same-sex parented family: The adolescent voice of experience. *Journal of GLBT Family Studies*, 7, 49–71. https://doi.org/10.1080/1550428X.2010.537241.

Whitton, S.W., & Buzzella, B. A. (2012). Using relationship education programs with same-sex couples: A preliminary evaluation of program utility and needed modifications. *Marriage & Family Review*, *48*, 667–688. https://doi.org/10.1080/01494929.2012.700908.

Whitton, S. W., Scott, S. B., Dyar, C., Weitbrecht, E. M., Hutsell, D. W., & Kuryluk, A. D. (2017). Piloting relationship education for female same-sex couples: Results of a small randomized waitlist-control trial. *Journal of Family Psychology*, *31*(7), 878–888. https://doi.org/10.1037/fam0000337.

Willging, C. E., Salvador, M., & Kano, M. (2006). Unequal treatment: Mental health care for sexual and gender minority groups in a rural state. *Psychiatric Services*, *57*, 867–870. https://doi.org/10.1176/ps.2006.57.6.867.

Williams, K. A., & Chapman, M.V. (2011). Comparing health and mental health needs, service use, and barriers to services among sexual minority youths and their peers. *Health & Social Work*, *36*(3), 197–206. https://doi.org/10.1093/hsw/36.3.197.

PART 3

Influences on family functioning

CHAPTER 10

The work-family interface

Erin Kramer Holmes[a], Clare R. Thomas[a], Richard J. Petts[b], and E. Jeffrey Hill[a]
[a]School of Family Life, Brigham Young University, Provo, UT, United States
[b]Department of Sociology, Ball State University, Muncie, IN, United States

In this chapter, we focus our attention on the work-family interface. We ground our discussion in an ecological systems framework (Bronfenbrenner, 1986; Grzywacz & Marks, 2000; Hill & Holmes, 2018; Voydanoff, 2002). Within this framework, we address three other related theories of the work-family interface: boundary theory, role theory, and gender theory. We then introduce the literature on the work-family interface and focus our attention on the experiences of work-to-family and family-to-work conflict, work-to-family and family-to-work spillover and crossover, workplace flexibility, parental leave policies, and day-care arrangements. We highlight these theories and experiences because they represent defining features of the work-family interface within the United States (Hill & Holmes, 2018; Holmes, Baumgartner, Marks, Palkovitz, & Nesteruk, 2010) and globally (Behson, Holmes, Hill, & Robbins, 2018; Shockley, French, & Yu, 2018). We close with a brief conclusion summarizing our review. Though we have tried to emphasize research in multiple countries and regions, more research has focused on developed countries such as those found in Australia, North America, and Western Europe than in Africa, Asia, and South America. We recognize that cultural characteristics imbedded in different regions and countries moderate dimensions of the work-family interface, and we invite our readers to consider how their own cultural practices and assumptions shape their research regarding the work-family interface as they ponder this chapter (see House, Hanges, Javidan, Dorfman, & Gupta, 2004; Ollier-Malaterre & Foucreault, 2018).

Ecological systems framework

Though researchers use several prominent theories to study the work-family interface, we focus on an ecological systems framework because of its interdisciplinary breadth. We also introduce boundary theory, role theory, and

Cross-Cultural Family Research and Practice
https://doi.org/10.1016/B978-0-12-815493-9.00010-7

gender theories that are often embedded in ecological systems exploring the work-family interface (see also Hill & Holmes, 2018).

According to Bronfenbrenner (1986) individuals develop within the context of several environmental systems: the microsystem, mesosystem, exosystem, macrosystem, and chronosystem. In Voydanoff's (2002) application of this theory, work and family are microsystems, which interact in unique ways based on the larger cultural context (or macrosystem) in which microsystems occur. Interactions between work and family are identified as the mesosystem. In this theoretical context, features of family life may influence the relationship between work and family characteristics, work-family conflict, and work and family outcomes in diverse ways based on the macrosystem in which each is embedded, such as the extent of industrialization or development in a given country (Aryee, Fields, & Luk, 1999; Joplin, Shaffer, Francesco, & Lau, 2003), or other cultural dimensions such as institutional collectivism, future orientation, power distance, etc. (House et al., 2004; Ollier-Malaterre & Foucreault, 2018). Attention to cultural dimensions also requires us to acknowledge that the *type* of work one does differs by context. For example, while workplace flexibility is a well-studied support, the type of work (e.g., childcare, farming, retail, food services, manual labor, etc.) may make flexibility options less likely, or may alter the nature of the options available to the worker.

This framework also proposes links between work, family, and child development. Adults' work environments and the family are each part of systems affecting children's development (Holmes, Holladay, Hill, & Yorgason, 2018). This perspective suggests that adults' work—an exosystem—affects child development through its influence on family processes, such as spillover between work and family and its effect on parent-child relationship quality (Bumpus, Crouter, & McHale, 2006; St. George & Fletcher, 2011), or parent-child interactions in other contexts such as school (Holmes et al., 2018). The exosystem also exists in diverse macrosystemic contexts. Thus, the cultural systems of childcare may include grandparents caring for grandchildren (Jappens & Van Bavel, 2012), or other issues in adult development such as caring for aging parents (Hill, Erickson, Fellows, Martinengo, & Allen, 2014).

Boundary theory

As stated above, processes that occur within the home, workplace, and other overlapping systems occur within a broader socioeconomic context called a macrosystem. This macrosystem encompasses any group whose members "share resources, hazards, lifestyles, opportunity structures, life course options, and patterns of social interchange" (Bronfenbrenner, 1993, p. 23).

Boundary theory enhances our understanding of the interplay between macrosystems and other ecological systems by acknowledging that boundaries between a person's family roles and roles in the workplace are often situated within this broader macrosystem context (Clark, 2000). These boundaries include location, time, and psychological presence (Ashforth, Kreiner, & Fugate, 2000).

In general, the more permeable the boundary between work and family, the greater the integration of roles. Conversely, the less permeable the boundary, the greater the segmentation of roles. The degree to which the boundary between family and workplaces is permeable is a key point of analysis with boundary theory. For example, when the physical boundary between work and family is very permeable (e.g., working in one's family room while the children play nearby in the same room), there is greater potential for work-family conflict.

Role theory

To further enhance the understanding of the work-family interface, scholars may combine ecological systems theory with role theory. Role theory posits two perspectives relevant to the work-family interface: role strain and role enhancement (Barnett & Gareis, 2006b). On the one hand, role strain suggests that individuals experience conflict and stress as they try to meet the demands of multiple roles (Goode, 1960). The larger the number of roles, the more intense the conflict that exists. From this role strain perspective, time and energy spent fulfilling the duties of one role reduce time and energy available for other roles; stressors associated with one role negatively impact performance in other roles (Greenhaus & Beutell, 1985).

On the other hand, role enhancement proposes that participation in one role brings about rewards and privileges that enhance performance in other roles (Greenhaus & Powell, 2006; Marks, 1977; Voydanoff, 2002). Being an employee and being a parent both require significant amounts of time and energy. It is important to understand the relationship between these competing roles; whether it is one of conflict or enhancement; and how it affects home life, including family relationships (Buehler, O'Brien, & Walls, 2011).

Gender theories

For many years, conversations about gender difference in the work-family interface were focused primarily on biological sex differences (Livingston, 2018). Key theorizing about gender in psychology, sociology, and family studies expanded these theoretical models (e.g., Clark, Rudolph, Zhdanova, Michel, & Baltes, 2017; Pedulla & Thébaud, 2015). Now many theories acknowledge that gender is socially constructed (West & Zimmerman, 1987), and that social constructions of gender include gender role ideologies (e.g., traditional vs egalitarian), traits (e.g., masculine vs feminine vs androgynous), social expectations, and social norms (Holmes et al., 2010; Livingston, 2018). As such, gender becomes another core feature of an ecosystemic approach to studying the work-family interface. Based on the permeability of boundaries and roles (as discussed above), gender expectations and norms at home and at work will likely intersect, contributing to a complex system of expectations for individuals in their roles as parents, spouses, employees, and coworkers.

Gender ideologies and attitudes are impacted by a number of processes at the national level including the average educational levels in the country (Judge & Livingston, 2008), the gender ideologies of individuals within the country (Sjöberg, 2010), and the extent to which egalitarian behaviors exist between men and women within a given country (Grove, 2005). For example, countries with a higher education level per individual (on average) are likely to have more culturally egalitarian attitudes. With this in mind, some scholarly attention has focused on assessing national-level views of gender in families and workplaces (Grove, 2005; Hofstede, 1983; House et al., 2004) and has also focused on household structure and income inequality between the sexes (World Economic Forum, 2013). At these national levels, gender views have differential impact on wage gaps (Fortin, 2005; Weichselbaumer & Winter-Ebmer, 2005).

We offer one caution to our readers: while gender theories have expanded our view away from biological sex differences alone, we also acknowledge that biological sex still informs features of the work-family interface. For example, biological aspects of pregnancy, childbirth, and breastfeeding intersect with cultural expectations and norms about who will care for children, when parents will return to work after having a child, etc. In our sections on parental leave and day care, we discuss ways these issues may differentially impact men and women.

In summary, understanding and conceptualizing gender requires acknowledging the complexity of social constructions of gender and the realities of biology as features of the work-family interface at individual, family, and national levels (Livingston, 2018).

Research on work-family interface

With these theoretical models in place, we begin our review of the literature. Our review explores work-to-family and family-to-work conflict, work-to-family and family-to-work spillover and crossover, workplace flexibility, parental leave policies, and day-care arrangements. We end with an exploration of gender as a core feature of the work-family interface. Throughout, we highlight cross-cultural findings when possible to acknowledge that the work-family interface is a global phenomenon; we also address cross-cultural differences acknowledging that culture is an important moderator of associations between work and family.

What is the work family interface?

The work-family interface is a broad concept referring to the myriad ways families and workplaces intersect. Scholars make the key assumption that work and family influence one another in mutual and bidirectional ways, flowing from work to family and from family to work (see, e.g., Frone, Yardley, & Markel, 1997). These processes can be negative or positive, either inhibiting or facilitating processes between and within the two spheres of work and family. Finally, the way these spheres impact each other has implications for parenting (e.g., Behson et al., 2018; Holmes et al., 2010; Petts & Knoester, 2018), child development (e.g., Bumpus et al., 2006; Holmes et al., 2018; St. George & Fletcher, 2011), romantic relationships (e.g., Amstad, Meier, Fasel, Elfering, & Semmer, 2011; McAllister, Thornock, Hammond, Holmes, & Hill, 2012), individual well-being (e.g., Holmes, Erickson, & Hill, 2012; Petts, 2018), and capacity to care for extended family members such as aging parents (e.g., Hill et al., 2014).

Work-to-family and family-to-work conflict

Conflict has been historically defined as "a form of inter-role conflict in which the role pressures from the work and family domains are mutually incompatible so that participation in one role [such as family] is made more difficult by participation in another role [such as work]" (Greenhaus & Beutell, 1985, p. 77). Conflict between work and family is viewed as bidirectional. Thus, any conversation about conflict in the work-family interface includes both work-to-family conflict (WFC) and family-to-work conflict (FWC). WFC and FWC are two of the most frequently studied features of the work-family interface (see also Hill & Holmes, 2018).

Reports from the National Study of the Changing Workforce suggest that work-family conflict *increased* among men and women in the United States over the past 30 years (Aumann, Galinsky, & Matos, 2011; Kelly et al., 2014). One report suggested that 70% of employees' experience conflict between their work and nonwork lives (Schieman, Milkie, & Glavin, 2009). The rise in WFC may be due to the ideal worker schema (Moen & Roehling, 2005).

The ideal worker schema assumes that employees must be highly committed full-time workers whose work demands are set by the employer's needs (Moen & Roehling, 2005). Individuals who meet this ideal worker schema are assumed to be the best employees; those who deserve wage increases and work promotions. As one may imagine, demographic shifts in the labor market and in the family may collide with the ideal worker schema to create increases in WFC and FWC. For example, in the US women's labor force participation increased—particularly among women with infants and small children (Toossi & Morisi, 2017)—at the same time that men have increased their desire to be warmer, more nurturing, and more actively involved parents than men in prior generations (Kuo, Volling, & Gonzalez, 2017). Shifting expectations for parents in their family roles have collided with both mothers' and fathers' expectations to conform to the ideal worker schema. This produces tension for both mothers and fathers between the potentially competing demands of being an ideal employee and being an actively involved parent (Blair-Loy, 2003; Holmes et al., 2010). Interestingly, parents are not the only individuals feeling these tensions. WFC is increasing among single adults in the United States (Schieman et al., 2009).

Four of the most commonly studied predictors of WFC are work hours (Buehler & O'Brien, 2011; Byron, 2005; Ng & Feldman, 2008), workload (Jin, Ford, & Chen, 2013; Schieman & Young, 2015), schedule flexibility (Poelmans & Chenoy, 2008; see below section for a fuller accounting of flexibility), and workplace support from supervisors and colleagues (Allen, 2001; Frone et al., 1997; Kelly et al., 2014; Kossek & Michel, 2011). Most research supports the claim that longer work hours (Byron, 2005; Ng & Feldman, 2008), and heavier workloads (Hill, Märtinson, & Ferris, 2004; Jin et al., 2013; Schieman & Young, 2015), increase WFC and FWC (Ng & Feldman, 2008). Most research also supports the claim that schedule flexibility, and workplace support, reduce WFC (Allen & Shockley, 2009; Byron, 2005; Gajendran & Harrison, 2007; Mesmer-Magnus & Viswesvaran, 2006).

Cultural and social norms influence the values, beliefs, and roles associated with WFC and FWC (Luk & Shaffer, 2005; Yang, Chen, Choi, & Zou, 2000). Some have wondered how the differing conception of time between western and Asian countries may yield differences in the experience of WFC and FWC, respectively. These comparisons yield mixed findings with some detecting no difference in WFC or FWC (Jin et al., 2013; O'Brien, Del Pino, Yoo, Cinnamon, & Han, 2014), others finding significantly higher FWC in Asian societies (Hassan, Dollard, & Winefield, 2010; Lu, Gilmour, Kao, & Huang, 2006), and still others discovering lower WFC in Asia

(Hassan et al., 2010; Masuda et al., 2012). Luk and Shaffer (2005) found evidence that time commitment and work role expectations increased WFC, and parental demands increased FWC in Hong Kong.

A recent metaanalysis tried to sort out these mixed findings, and the authors concluded that there were few significant mean differences in the experiences of WFC and FWC cross-culturally (Allen, French, Dumani, & Shockley, 2015). Their metaanalysis further clarified that significant differences resulted from gender inequality (e.g., higher FWC in countries with more inequality), and from being in a more collectivistic rather than individualistic culture. For example, employees from Confucian Asian countries (e.g., China, Hong Kong, and Taiwan) report more FWC compared to those in Anglo countries (e.g., Australia, United Kingdom, and United States), but findings regarding WFC are less conclusive (Shockley et al., 2018).

Work-to-family and family-to-work spillover and crossover

Like WFC and FWC, spillover and crossover between work and family are mutual, bidirectional, potentially negative, or potentially positive (Grzywacz & Marks, 2000). Spillover and crossover represent:

> *"processes whereby stress, strain, or other emotions (positive or negative) are carried over within and across individuals and domains. Spillover is the within-person transmission of experiences and/or emotions, from work to home or from home to work, or the same individual. Crossover is an interpersonal process that occurs when job stress, psychological strain, or other emotions experienced by one person affect the level of strain or outcomes experience by another person in the same social environment" (Brough & Westman, 2018, p. 630).*

Discussions of boundaries and roles become critical to assessing the correlates of work-to-family and family-to-work spillover and crossover. Boundary theories help scholars theorize about how the permeability, flexibility, and strength of boundaries between home and work can contribute to within person spillover and between persons crossover (Clark, 2000; Lourel, Ford, Gamassou, Guégen, & Hartmann, 2009). Below we distinguish between the research findings focused on negative and positive spillover, and negative and positive crossover. Role theories help scholars consider how performing one role well may enhance one's mood or one's sense of efficacy in one domain (either work or family), and through permeable boundaries may spillover into one's personal sense of efficacy in the roles one fills in another domain (Greenhaus & Powell, 2006). One's mood or efficacy may also crossover, impacting relationships with coworkers, supervisors, children, or romantic partners (Brough & Westman, 2018).

Negative spillover

WFC and negative work-family spillover are closely related with the main difference being that work-family conflict is more focused on roles, while spillover emphasizes an individual's emotions (Cho & Tay, 2016). In affective spillover models, an individual's negative moods and experiences in one domain are assumed to impact their performance in another domain (Edwards & Rothbard, 2000) and may influence one's perceptions of subsequent events, fueling the negative emotional response across domains (Judge & Ilies, 2004; Rusting & DeHart, 2000). For example, a bad experience at work may make a mother feel discouraged, and her discouraged affect may make it more challenging for her to feel positive when she comes home for dinner. The more negative emotions she experiences across domains, the more likely these multiple negative experiences are to reinforce her bad mood. Because moods may fluctuate within and across days, some scholars have explored the transmission of negative spillover across days using daily diaries (Heller & Watson, 2005; Sonnentag & Binnewies, 2013). These studies have confirmed that spillover can occur within the same day and across days, and have explored how trying to regulate emotions in one domain may make it more difficult to continue to regulate similar emotions in another (Muraven, Tice, & Baumeister, 1998).

Cultural ideas about the expression (Mesquita & Delvaux, 2012) and the interpretation of emotion (Chun, Moos, & Cronkite, 2006) also play a role in negative spillover, though considerably more work needs to be conducted to better understand cultural attitudes about emotional expression. For example, in comparisons between Asian Americans and European Americans, researchers found that it was more difficult and potentially more maladaptive for European Americans to suppress their emotions than for Asian Americans or East Asians to do so (Mauss & Butler, 2010; Soto, Perez, Kim, Lee, & Minnick, 2011).

Negative spillover becomes an important part of the work-family interface because it is correlated with depression and anxiety (Frone, Russell, & Cooper, 1992; Grzywacz & Bass, 2003), physical health (Grzywacz, 2000), FWC (Rotondo & Kincaid, 2008), general life stress (Lourel et al., 2009), job satisfaction (Ilies, Wilson, & Wagner, 2009), and family relationships (Matias et al., 2017). While spillover from work to family and family to work is assumed to be bidirectional, more work continues to be done to better understand how these processes mutually impact one another in families and workplaces across cultural contexts.

Positive spillover

Although a great deal of research in the work-family realm focuses on work-family conflict and negative spillover, some positive spillover from work-to-family and family-to-work also exists. According to Edwards and Rothbard (2000), positive work-to-family spillover has been viewed as "work satisfaction that enhances family functioning," while family-to-work spillover is considered something that enhances work functioning. When the effects of the spillover are beneficial for the other domain, it is considered positive spillover, as opposed to negative (Cho & Tay, 2016; Edwards & Rothbard, 2000). Whether the spillover is positive or negative, it is focused on the relationship between work satisfaction and family (Edwards & Rothbard, 2000), as opposed to role conflict. The concept of spillover acknowledges that family and work are intricately connected and can either be at odds or support one another. While negative spillover explains the effects of the domains being at odds, positive spillover shows the benefits of each domain supporting the other.

When the domains of family and work support each other, both domains benefit. For example, positive work-family spillover is associated with greater life satisfaction (Cho & Tay, 2016), improved work attitudes (Kirchmeyer, 1992), and commitment (Orthner & Pittman, 1986). Thus, both domains improve when the spillover is positive, resulting in decreased psychological distress and greater family satisfaction (Haar & Bardoel, 2008).

Positive spillover is associated with individual personality (Polatcia & Akdoğan, 2014) mood (Edwards & Rothbard, 2000), older age (Grzywacz, Almeida, & McDonald, 2002), and work and family environments. When flexibility and support (Grzywacz & Marks, 2000) as well as improved resources (Hakanen, Peeters, & Perhoniemi, 2011) are available in work and families, the spillover is positively associated in both directions and enhances well-being. Because almost no family can escape the obligation to work, and many individuals who work also have families, understanding the interplay between both domains is essential for each one to function properly and benefit each other.

Positive and negative crossover

As noted above, spillover and crossover are related but distinct constructs. Our search of the literature suggests that crossover is less studied than spillover, thus we combine our review of positive and negative crossover effects into one section.

Processes whereby intrapersonal crossover occurs can be complex. Crossover may occur as the result of hostile marital interactions (Westman, 2002; Westman & Etzion, 2005), a partner's mental health (Hammer, Cullen, Neal, Sinclair, & Shafiro, 2005), a partner's positive affect (Hammer et al., 2005), a partner's empathy (Liu & Cheung, 2015a), children's misbehavior (Frone et al., 1997), or a partner's WFC (Bakker, Demerouti, & Dollard, 2008). Crossover in these situations may result in either negative effects such as worsened depression and physical health for one's partner (Shimazu, Bakker, & Demerouti, 2009), or positive effects such as happiness (Rodriguez-Muñoz, Sanz-Vergel, Demerouti, & Bakker, 2014) and marital satisfaction (Liu & Cheung, 2015b; Liu, Ngo, & Cheung, 2016). The effects of crossover appear to be significant and meaningful even after accounting for within-persons effects (Bakker et al., 2008; Hammer, Allen, & Grigsby, 1997; Liu et al., 2016; Shimazu et al., 2009). More recent models are considering the importance of gender role ideology when sex differences are detected in crossover processes (Liu & Cheung, 2015a, 2015b; Liu et al., 2016; Westman, Vinokur, Hamilton, & Roziner, 2004) and are also considering how each partner impacts the other to produce crossover effects over time (Muller & Brough, 2017).

Finally, acknowledging spillover and crossover between families and workplaces across the globe is becoming more and more important as current demographic data shows that dual-earner couples with children represent the most common employment pattern in the OECD (2018), and in China (Ho, Chen, Cheung, Liu, & Worthington Jr., 2013). For example, between 50% and 66% of couples in Denmark, Finland, Israel, Poland, Portugal, and Slovenia both work full-time (OECD, 2018). Another 35% of couples in Austria, Australia, the United Kingdom, and the Netherlands consist of one partner who works full time and another who works part-time.

Influences on the work-family interface

Workplace flexibility

Workplace flexibility is "the ability of workers to make choices influencing when, where, and for how long they engage in work-related tasks" (Hill et al., 2008, p. 152). Though its specific implementation may differ by industry, size of employer, wage group, age of the employee, and gender of employee, variations of workplace flexibility are ubiquitous (Galinsky, Sakai, & Wigton, 2011). Numerous studies reveal that workplace flexibility is generally associated with outcomes beneficial to individuals, families, and employers (Hill & Holmes, 2018). In a metaanalysis comprised of 58 studies, Allen, Johnson, Kiburz, and Shockley (2013) found workplace flexibility was associated with lower work-to-family conflict, but unrelated to family-to-work conflict. In an international study including 75 countries (N = 24,436), Hill, Erickson, Holmes, and Ferris (2010) found that employees with workplace flexibility were able to work an average of 8 hours or more per week more before reporting work-family conflict than those without workplace flexibility. Though objectively beneficial, research also suggests that some employees choose not to use workplace flexibility, even when offered, out of fear of being stigmatized (Williams, Blair-Loy, & Berdahl, 2013).

Workplace flexibility includes three basic domains: *schedule flexibility* (when work is done), *location flexibility* (where work is done), and *reduced workload* (for how long work is done). We will now briefly summarize some of the current research related to workplace flexibility in each of these three specific categories.

Schedule flexibility

Schedule flexibility permits employees to adapt their work schedules. Flextime allows workers to choose when they work from among a range of acceptable work hours (Coenen & Kok, 2014). In his metaanalysis, Byron (2005) revealed that flextime predicted lower work-to-family conflict and lower family-to-work conflict. Grzywacz, Carlson, and Shulkin (2008) found that flextime was associated with less stress and burnout for individuals and less work-family conflict for employees. Flextime was also associated with less negative work-family spillover and stress symptoms, especially among female workers, single parents, and/or employees with higher family workloads (Jang, Zippay, & Park, 2012).

Other research reveals that offering schedule flexibility may relate to positive outcomes for all employees involved, even those who do not use the flexibility (Hayman, 2009). An employee's perceived flexibility predicted work-family fit even better than used flexibility (Brown & Pitt-Catsouphes, 2016; Clark et al., 2017; Jones et al., 2008). "Indeed, just the perception that the flexibility would be available when needed appears to be sufficient" (Jones et al., 2008, p. 781).

Another type of schedule flexibility is the compressed work-week, where the employee works 40 hours per week in fewer than 5 days. Common arrangements include four 10-hour days or three 12-hour days (with an extra day of work every other week). In a metaanalysis of 40 studies, utilization of compressed work-week was associated with improved work-life balance (Bambra, Whitehead, Sowden, Akers, & Petticrew, 2008). Grzywacz et al. (2008) found that using a compressed work-week was related to less individual stress and burnout.

Location flexibility

Location flexibility is defined as working outside the conventional workplace and communicating through telecommunications or computer-based technology (Nilles, 1994). One way to offer flexibility to employees is to telecommute, or work remotely from their traditional places of employment (generally home), on a part-time or full-time basis (Coenen & Kok, 2014; Hill, Hawkins, Ferris, & Weitzman, 2001). The empirical research generally supports the positive influence of telecommuting on individuals, families, and employers. In a metaanalysis of 46 empirical telecommuting studies, Gajendran and Harrison (2007) found that telecommuting was associated with higher perceived autonomy, lower work-family conflict, higher job performance, lower turnover intent, and less role stress. Golden (2006) discovered that telework was positively related to commitment and negatively related to turnover intentions, and that work exhaustion mediated these relationships. In a metaanalysis of 22 studies conducted by Martin and MacDonnell (2012), telework was shown to have a positive relationship to four organizational outcomes (productivity, employee retention, organizational commitment, and performance). Age was a significant moderator of the relationship between telecommuting and organizational commitment, with younger employees being more committed to organizations where telecommuting was an option. When examining gender differences, Clark et al. (2017) found that men specifically benefit from telecommuting, feeling a stronger commitment to their jobs and intention for longevity, and Kurowska (2018) discovered that men more easily balance work and family responsibilities while working from home.

There have been ambiguous results about whether or not telecommuting has a positive or negative influence on overall job satisfaction. However, recent studies indicate that telecommuting is positively associated with job satisfaction (Clark et al., 2017; Gajendran & Harrison, 2007; Golden & Veiga, 2005), but that this relationship may be curvilinear based on the number of hours worked virtually per week (Golden & Veiga, 2005). Such a finding may be due to the fact that employees who telecommute exclusively tend to have less positive relationships with coworkers (Gajendran & Harrison, 2007) and/or because it may limit their career advancement opportunities (McCloskey & Igbaria, 1998).

Reduced workload

Reduced workload involves working fewer hours and having fewer job responsibilities than what is typically expected for fulltime employment. There is increasing pressure for employers to allow employees to work part time in order to meet work and family demands (Barnett & Gareis, 2000). Historically, part-time jobs have been lower status jobs with less pay and fewer opportunities for advancement than full-time jobs (Barnett & Hall, 2001). However, more recently, many part-time workers have chosen (with the support of their employers) to remain in high-status, high-paid employment but work fewer hours per week (Hill et al., 2004). Because of the negative connotations of *part-time* employment, some prefer to label it *reduced-hours* employment (Barnett & Gareis, 2000).

Wadsworth and Facer (2016) discovered that employees working a 4/40 schedule (allowing employees to work 40 hours per week across 4 days) had lower levels of work-family conflict and greater productivity than those who do not work 4/40 schedules. Some research suggests that individual benefits of part-time work are associated with transition to parenthood. Part time work benefits both the mother and the father as they begin to raise children by allowing them more time with their children for enrichment activities or daily routines, improving the parent child relationship (Kim, 2018). Holmes et al. (2012) found that part-time work was the most common employment preference for mothers during the 3 years following childbirth. In a study of mothers with young children, Hill et al. (2004) compared those working full-time with those working regular part-time schedules. The regular part-time group reported significantly lower work-to-family interference, work-to-family strain, job travel, and unnecessary work, and greater work-family fit, job flexibility, and family success. A common perceived drawback is that working reduced-hours leads to reduced career opportunities—a so-called "Mommy track" (Hill et al., 2004). However, in a study of professionals and managers working in part-time positions, Lee and MacDermid (1998) found that 35% of the participants were actually promoted during the time of their study. Reduced work hours may benefit dual-earning families, allowing more time to care for responsibilities within the home (Schooreel & Verbruggen, 2016).

One final issue to consider regarding reduced workload is whether or not employers should use age-specific practices such as reduced work hours, additional leave, or early retirement benefits for those above age 50 (Bal & Jansen, 2016; Johnson, 2011). Theoretically, flexibility options in the workplace would be moderated by the age of the worker because work-family demands change based on different constraints and opportunities that accompany age. Practically, however, less formal practices are being implemented (Matz-Costa & Pitt-Catsouphes, 2010). There is some evidence that reduced work hours for aging employees may benefit employers by reducing turnover, enhancing recruitment, and improving overall job performance in their organizations (Bal & Jansen, 2016), and that these changes represent the changing needs and capacities of aging employees. Unfortunately, it appears that regardless of the benefits, flexibility is used to attract applicants, but not always implemented based on applicants' individual needs (Way et al., 2015).

Parental leave

Parental leave is a key policy that may help to facilitate work-family balance by allowing workers to take time off work after the birth of a child to be home and adjust to having a new child. In contrast to other work-family issues in which extensive research has been conducted in the United States, more is known about parental leave internationally than within the United States. This is due, at least in part, to the United States being one of only a handful of countries (the others are Marshall Islands, Micronesia, Palau, Papua New Guinea, Suriname, and Tonga), and the only OECD country, that does not have a national paid parental leave policy (World Bank Group, 2018). The only national leave policy in the United States is the Family and Medical Leave Act (FMLA), which provides eligible employees with up to 12 weeks of unpaid leave following childbirth (Blum, Koslowski, Macht, & Moss, 2018). There is variation between states though, with six US states (California, Rhode Island, Massachusetts, New Jersey, New Work, and Washington) and Washington DC having passed and/or implemented paid parental leave policies (National Partnership for Women and Families, 2018). Due to the lack of statutory paid leave policies, US workers are often reliant on employers to provide paid parental leave, but only 16% of workers have access to paid family leave at their workplace (Bureau of Labor Statistics, 2018b).

In contrast to the United States in which access to paid parental leave is uncommon, paid maternity leave is available in the vast majority of countries throughout the world, and 42.5% of countries (94% of OECD countries) have policies that provide fathers access to either paid paternity leave or paid parental leave that can be shared by mothers and fathers (International Labour Organization, 2014; Raub et al., 2018; World Bank Group, 2018). Globally, the median length of paid maternity leave provided is 14 weeks, matching the International Labour Organization (ILO) minimum standard for promoting gender equality, but well-paid parental leave benefits (i.e., wage replacement of at least 66%) extend as long as 24 months in some countries (Blum et al., 2018). The Nordic countries (Finland, Iceland, Norway, and Sweden), in particular, have served as the model for parental leave policies. Policies in the Nordic countries provide 35–59 weeks of paid parental leave at 70%–100% wage replacement, and this time is divided into a reserved part for mothers, a reserved part for fathers, and time that may be shared between parents (Blum et al., 2018). Despite the widespread availability of paid parental leave, less than half of countries meet the ILO standard of providing well-paid parental leave (International Labour Organization, 2014; World Bank Group, 2018).

According to the ILO, providing paid parental leave is a fundamental human right and necessary to maintain work-family balance, particularly for women (International Labour Organization, 2014). Parental leave is believed to be beneficial to families for a number of reasons. First and foremost, parents are more likely to juggle multiple roles today as women's labor force participation remains high. Consequently, the majority of parents (71% of mothers and 93% of fathers in the United States) work in paid employment (Bureau of Labor Statistics, 2018a). Access to parental leave can help to alleviate role conflict by allowing parents to transition parenthood without having to also manage their roles as workers (Bianchi & Milkie, 2010).

Second, parental leave may help to promote maternal and child health. By allowing mothers to take time off to physically and mentally recover from pregnancy and childbirth, and fathers to take time off to help out at home, parental leave may reduce mothers' stress, risk of depression, and other negative health outcomes (Chatterji & Markowitz, 2012; Petts, 2018; Staehelin, Bertea, & Stutz, 2007). Indeed, longer periods of maternity leave are associated with reduced stress and a lower risk of depression for mothers (Avendano, Berkman, Brugiavini, & Pasini, 2015; Chatterji & Markowitz, 2012; McGovern et al., 1997; Petts, 2018). There is also some evidence that paternity leave may be positively associated with mothers' health outcomes (Bratberg & Naz, 2014; Redshaw & Henderson, 2013).

Parental leave may also promote child health and development. Spending time with young children is fundamental to child development; by providing time for parents to bond with their new child, parental leave may promote parental sensitivity and responsiveness and consequently encourage healthy child development (Waldfogel, 2006). Indeed, longer periods of parental leave are associated with greater engagement by both mothers and fathers (Chatterji, Markowitz, & Brooks-Gunn, 2013; Clark, Shibley Hyde, Essex, & Klein, 1997; Huerta et al., 2014; Petts, 2018; Petts & Knoester, 2018). Greater access to parental leave is also associated with lower infant mortality, a reduced risk of low birth weight, higher immunization rates, and more frequent breastfeeding (Baker & Milligan, 2010; Hajizadeh, Heymann, Strumpf, Harper, & Nandi, 2015; Huang & Yang, 2015; Ruhm, 2000; Tanaka, 2005). There is also some evidence from Norway suggesting that greater access to parental leave has long-term consequences for children including higher academic performance and higher wages in adulthood (Carneiro, Løken, & Salvanes, 2015; Cools, Fiva, & Kirkebøen, 2015).

Parental leave may also help to prevent discrimination toward women and promote gender equality. Cultural norms emphasizing that women should be primarily responsible for children continue to persist, which leads women to be especially likely to take time off for childbirth (even if they do not have access to parental leave). Consequently, women experience a wage penalty due to motherhood and employers may also discriminate against women (Budig & England, 2001). Providing parental leave (with job protection) helps to protect women against this discrimination, and may also help to promote gender equality. Indeed, access to generous parental leave policies is associated with improved labor market outcomes for women such as increased labor market participation, higher-income, and a smaller motherhood wage penalty (Baum & Ruhm, 2016; Berger & Waldfogel, 2004; Budig, Misra, & Boeckmann, 2016; Rossin-Slater, Ruhm, & Waldfogel, 2013; Waldfogel, 1998). Furthermore, allowing both mothers and fathers to take time off for the birth of a child may help to facilitate gender equality by encouraging and enabling both parents to share in childcare tasks.

Indeed, fathers who take longer periods of paternity leave are more engaged in their child's life and are also more likely to share in childcare and housework tasks than fathers who take shorter leaves or no leave at all (Almqvist & Duvander, 2014; Bünning, 2011; Haas & Hwang, 2008; Huerta et al., 2014; Kotsadam & Finseraas, 2011; Nepomnyaschy & Waldfogel, 2007; Petts & Knoester, 2018; Pragg & Knoester, 2017). Employer support for working fathers, such as leave and flexibility, also supports better work-life balance for both men and their working spouses (Harrington, Van Deusen, Sabatini Fraone, Eddy, & Haas, 2014).

Finally, access to parental leave may also help to reduce socioeconomic inequality. Current parental leave policies both internationally and in the USUnited States often have eligibility requirements that may limit access to leave for workers who are socioeconomically disadvantaged (International Labour Organization, 2014; McKay, Mathieu, & Doucet, 2016; Petts, Knoester, & Li, 2018; World Bank Group, 2018). Low socioeconomic status workers are also often unable and/or unwilling to take unpaid leaves or leaves with low wage replacement due to income constraints, and this is particularly true for fathers (Harrington et al., 2014; Moss, Duvander, & Koslowski, 2019; O'Brien, 2009; Petts et al., 2018). Providing well-paid parental leave may particularly benefit more disadvantaged populations and help to reduce the disparities between high- and low-income workers. Indeed, more generous parental leave policies are associated with lower poverty rates and improved health particularly for disadvantaged populations (Lichtman-Sadot & Bell, 2017; Misra, Moller, Strader, & Wemlinger, 2012; Petts, 2018). Overall, evidence suggests that greater access to paid parental leave—and more generous parental leave policies—provides benefits to families that may enable them to more effectively balance their work and family lives.

Daycare arrangements

Global daycare practices

Laws and policies are strongly influenced by culture and values (Lewis, 2008; Lokteff & Piercy, 2012, pp. 104–105). Thus, childcare policies often reflect the values of the individual society. Our review of global daycare policies suggests that current daycare policies across the globe are lacking in accessibility and affordability (Szelewa & Polakowski, 2008). While global parental leave policies after childbirth have gained traction, less priority has been given to childcare after parental leave ends.

Countries across the world largely differ in their childcare access and policies (Davis & Powell, 2003; Fagnani, 2007; Szelewa & Polakowski, 2008). UNESCO and UNICEF have historically worked to increase child access to quality care and education in lower income countries and countries in socioeconomic transition (Rosemberg, 2004). However, countries like Uganda have not only struggled with policy implementation, but with tracking the actual needs of children in working families (Pence et al., 2004). Childcare and education is different in lower income countries because of the struggles inherent in rural, unskilled areas; they often rely less on government assistance and more on community programs (Leroy, Gadsden, & Guijarro, 2012), as well as families, and individuals within villages (Pence et al., 2004). In many low-income countries, families are less concerned with daycare policies and more concerned with whether or not their child will live to adulthood (Kinney et al., 2012).

Many higher-income countries have implemented policies supporting working parents and especially working mothers (Davis & Powell, 2003). For example, France and Germany have legally recognized the family as a key societal institution (Fagnani, 2007); and Finland has developed policies focused on the well-being of children and the importance of having women actively participate in the workforce (Lammi-Taskula, 2008). Japan is leading the world in their publicly funded childcare as well as specific policies for childcare and parental leave (Lambert, 2007), and the United Kingdom has revamped their policies to increase public expenditure specifically for childcare purposes (Lewis, 2008). This new global focus on childcare increases employment opportunities for parents, which many believe will result in better economic growth and improved quality care for children across the socioeconomic spectrum.

However, even with these policies, many individuals still do not have access to the formal childcare opportunities that they need. According to an Australian report, even when policies focus on increasing opportunities for single-parent families, many of these policies still fall short (Whiteford & Adema, 2007). Although Germany has recognized the family as a legal institution, it still maintains the view that children are best cared for in the home, which has resulted in a value system that limits childcare opportunities and reduces parents' capacity to participate in the workforce and care well for their children (Fagnani, 2007).

Some countries continue to struggle to implement institutionalized childcare, most notably the United States (Davis & Powell, 2003), Korea (Won & Pascall, 2004), and Ireland (Murphy-Lawless, 2000). Although some employers offer childcare options (Bordone, Arpino, & Aassve, 2017), the lack of institutionalized polices harms low-income, single-parent families who cannot afford to pay out of pocket for care, and limits work opportunities. Without assistance for long-term childcare, many are forced to choose between employment and parenthood. For example, Korea has one of the lowest rates of state expenditures for childcare in the OECD (2001). As women's workforce presence has increased, the average number of women bearing children has decreased (Won & Pascall, 2004). Mexico follows similar trends (Joplin et al., 2003).

Countries such as Australia, Sweden, and England have developed hybrid systems for childcare using privatized methods to meet demand for daycare (Brennan, Cass, Himmelweit, & Szebehely, 2012). Although this marketization can lead to improved quality in private childcare organizations (White & Friendly, 2012), the competitive nature of private childcare also increases the cost of quality care sometimes reducing quality or limiting access for low-income children (Mahon, 2002). Whether countries institute strict government led policies, a hybrid of competitive private childcare, or no institutional childcare at all, most daycare systems have been inadequate for providing quality childcare to working families across socioeconomic levels.

Daycare practices and the work-family interface

When childcare is available, fathers and mothers are more likely to utilize it (Lammi-Taskula, 2008). Without policies in place, however, it is often more cost-effective for one parent to remain at home and care for the child(ren); based on societal values, that parent is often the mother (Barrientos & Kabeer, 2004; Whiteford & Adema, 2007). The lack of childcare policies often results in disadvantages for women in the workforce (Martin, 2007), or negative child development outcomes (Hughes-Belding, Hegland, Stein, Sideris, & Bryant, 2012). When workforce opportunities become limited because of lack of childcare, families are forced to acknowledge the limits of what they can accomplish in the constraints of the current workforce priorities and institutional availabilities (Lewis, 2008).

When institutional childcare policies are lacking and privatized care is too expensive, extended family members often take part in informal daycare arrangements for children. Grandparents in particular have begun to play a pivotal role in childcare (Arpino, Pronzato, & Tavares, 2014; Bordone et al., 2017; Da, 2003), with up to 50% of grandparents across Europe (Glaser et al., 2010) and the United States (Hank & Buber, 2009) providing childcare for relatives. In countries like India, broader family units including aunts, uncles, nieces, nephews, and grandparents are all integral for helping raise children (Sharma & LeVine, 1998). However, as grandparents age, they can become less of a resource and more of a problem as adults struggle to manage both the care of their children and their parents. Many of these less structured childcare provisions are often based less on the *quality* of the childcare and more on the *necessity* of childcare (Hank & Buber, 2009). Although high-quality care can have lasting, positive effects on child cognitive development (Hughes-Belding et al., 2012), most countries either do not have the necessary childcare policies in place (Davis & Powell, 2003; Murphy-Lawless, 2000), or the policies are inadequate for the current demand (Fagnani, 2007; Szelewa & Polakowski, 2008; Whiteford & Adema, 2007).

Current researchers argue that quality childcare should be viewed as a human right (Davis & Powell, 2003). Childcare provides opportunities for steady economic growth (Barrientos & Kabeer, 2004; Whiteford & Adema, 2007), and healthy child development. In our estimation, attention to childcare policies and practices can provide equal opportunities in the workforce for multiple family forms and across diverse socioeconomic lines.

Conclusion

Ecological systems theory alerts scholars to the complexity of the work-family interface, emphasizing that work and family have the potential to mutually influence one another, and that this influence exists across multiple levels (Bronfenbrenner, 1986; Hill & Holmes, 2018; Voydanoff, 2002). Ecological systems theory further asserts that this influence exists in the larger context of cultural and social norms (Allen et al., 2015), permeable and impermeable boundaries (Clark, 2000), family and work roles (Greenhaus & Powell, 2006), and gender roles, traits, and expectations (Livingston, 2018). We believe that understanding these multiple features of an ecological systems perspective will offer deeper insight into these complex processes between workplaces and families (Hill & Holmes, 2018).

Our review demonstrates that processes between work and family have important implications for individual mental and physical health (e.g., Hajizadeh et al., 2015; O'Brien et al., 2014), family relationships (e.g., Ho et al., 2013; Liu et al., 2016; Matias et al., 2017), children's development (e.g., Holmes et al., 2018; Hughes-Belding et al., 2012), job satisfaction (e.g., Ilies et al., 2009; Masuda et al., 2012), organizational functioning (e.g., Clark et al., 2017), and economic growth (e.g., Whiteford & Adema, 2007). Much of this work is being tested cross-culturally, giving us a better understanding of how cultural attitudes and beliefs shape the work-family interface (e.g., Hofstede, 2001; Shockley et al., 2018). Finally, it is apparent from our review that single-parent families, low-income families, less educated families, families with young children, and families in countries or in workplaces with fewer resources (such as institutionalized leave and childcare practices) are at greater risk and deserve continued attention (e.g., Craig & Habgood, 2018). We hope this will contribute to the reduction of inequality around the globe.

References

Allen, T. D. (2001). Family-supportive work environments: The role of organizational perceptions. *Journal of Vocational Behavior, 58*(3), 414–435. https://doi.org/10.1006/jvbe.2000.1774.

Allen, T. D., Johnson, R. C., Kiburz, K. M., & Shockley, K. M. (2013). Work–family conflict and flexible work arrangements: Deconstructing flexibility. *Personnel Psychology, 66*(2), 345–376. https://doi.org/10.1111/peps.12012.

Allen, T. D., French, K. A., Dumani, S., & Shockley, K. M. (2015). Metaanalysis of work-family conflict mean differences: Does national context matter? *Journal of Vocational Behavior, 90*, 90–100. https://doi.org/10.1016/j.jvb.2015.07.006.

Allen, T. D., & Shockley, K. (2009). Flexible work arrangements: Help or hype? In D. R. Crane & E. J. Hill (Eds.), *Handbook of families and work: Interdisciplinary perspectives* (pp. 265–284). Lanham, MD: University Press of America.

Almqvist, A., & Duvander, A. (2014). Changes in gender equality? Swedish fathers' parental leave, division of childcare and housework. *Journal of Family Studies, 20*, 19–27. https://doi.org/10.5172/jfs.2014.20.1.19.

Amstad, F. T., Meier, L. L., Fasel, U., Elfering, A., & Semmer, N. K. (2011). A metaanalysis of work-family conflict and various outcomes with a special emphasis on cross-domain versus matching-domain relations. *Journal of Occupational Health Psychology, 16*(2), 151–169. https://doi.org/10.1037/a0022170.

Arpino, B., Pronzato, C., & Tavares, L. (2014). The effect of grandparental support on mothers' labour market participation: An instrumental variable approach. *European Journal of Population, 30*(4), 369–390. https://doi.org/10.1007/s10680-014-9319-8.

Aryee, S., Fields, D., & Luk, V. (1999). A cross-cultural test of a model of the work-family interface. *Journal of Management, 25*(4), 491–511. https://doi.org/10.1177/014920639902500402.

Ashforth, B. E., Kreiner, G. E., & Fugate, M. (2000). All in a day's work: Boundaries and micro role transitions. *The Academy of Management Review, 25*(3), 472–491. https://doi.org/10.5465/AMR.2000.3363315.

Aumann, K., Galinsky, E., & Matos, K. (2011). *The new male mystique*. New York: Families and Work Institute.

Avendano, M., Berkman, L. F., Brugiavini, A., & Pasini, G. (2015). The long-run effect of maternity leave benefits on mental health: Evidence from European countries. *Social Science & Medicine, 131*, 45–53. https://doi.org/10.1016/j.socscimed.2015.02.037.

Baker, M., & Milligan, L. (2010). Evidence from maternity leave expansions of the impact of maternal care on early child development. *The Journal of Human Resources, 45*, 1–32. https://doi.org/10.3368/jhr.45.1.1.

Bakker, A. B., Demerouti, E., & Dollard, M. F. (2008). How job demands affect partners' experience of exhaustion: Integrating work-family conflict and crossover theory. *Journal of Applied Psychology, 93*(4), 901–911. https://doi.org/10.1037/0021-9010.93.4.901.

Bal, P. M., & Jansen, P. G. W. (2016). Workplace flexibility across the lifespan. *Research in Personnel and Human Resources Management, 34*, 43–99. https://doi.org/10.1108/S0742-730120160000034009.

Bambra, C., Whitehead, M., Sowden, A., Akers, J., & Petticrew, M. (2008). "A hard day's night?" The effects of compressed working week interventions on the health and work-life balance of shift workers: A systematic review. *Journal of Epidemiology and Community Health, 62*(9), 764–777. https://doi.org/10.1136/jech.2007.067249.

Barnett, R., & Gareis, K. (2000). Reduced-hours employment. The relationship between difficulty of trade-offs and quality of life. *Work and Occupations, 27*, 168–187. https://doi.org/10.1177/0730888400027002003.

Barnett, R. C., & Gareis, K. C. (2006b). Role theory perspectives on work and family. In M. Pitt-Catsouphes, E. E. Kossek, & S. Sweet (Eds.), *The work and family handbook: Multi-disciplinary perspectives, methods, and approaches* (pp. 209–222). Mahwah, NJ: Lawrence Erlbaum Associates.

Barnett, R. C., & Hall, D. T. (2001). How to use reduced hours to win the war for talent. *Organizational Dynamics, 29*, 192–210. https://doi.org/10.1016/S0090-2616(01)00024-9.

Barrientos, S., & Kabeer, N. (2004). Enhancing female employment in global production: Policy implications. *Global Social Policy, 4*(2), 153–169. https://doi.org/10.1177/1468018104045107.

Baum, C. L., & Ruhm, C. J. (2016). The effects of paid family leave in California on labor market outcomes. *Journal of Policy Analysis and Management, 35*, 333–356. https://doi.org/10.1002/pam.21894.

Behson, S. J., Holmes, E. K., Hill, E. J., & Robbins, N. L. (2018). Fatherhood. In W. Shen, K. Shockley, & R. Johnson (Eds.), *The Cambridge handbook of the global work-family interface*. (pp. 614–628). Cambridge: Cambridge University Press. https://doi.org/10.1017/9781108235556.03.

Berger, L. M., & Waldfogel, J. (2004). Maternity leave and the employment of new mothers in the United States. *Journal of Population Economics*, *17*, 331–349. https://doi.org/10.1007/s00148-003-0159-9.

Bianchi, S. M., & Milkie, M. A. (2010). Work and family research in the first decade of the 21st century. *Journal of Marriage and Family*, *72*, 705–725. https://doi.org/10.1111/j.1741-3737.2010.00726.x.

Blair-Loy, M. (2003). *Competing devotions: Career and family among women executives*. Cambridge, MA: Harvard University Press.

Blum, S., Koslowski, A., Macht, A., & Moss, P. (2018). *International review of leave policies and research 2018*. Available at: http://www.leavenetwork.org/lp_and_r_reports/.

Bordone, V., Arpino, B., & Aassve, A. (2017). Patterns of grandparental child care across Europe: The role of the policy context and working mothers' need. *Ageing and Society*, *37*(4), 845–873. https://doi.org/10.1017/S0144686X1600009X.

Bratberg, E., & Naz, G. (2014). Does paternity leave affect mothers' sickness absence? *European Sociological Review*, *30*, 500–511. https://doi.org/10.1093/esr/jcu058.

Brennan, D., Cass, B., Himmelweit, S., & Szebehely, M. (2012). The marketisation of care: Rationales and consequences in Nordic and liberal care regimes. *Journal of European Social Policy*, *22*(4), 377–391. https://doi.org/10.1177/0958928712449772.

Bronfenbrenner, U. (1986). Ecology of the family as a context for human development: Research perspectives. *Developmental Psychology*, *22*, 723–742. https://doi.org/10.1037/0012-1649.22.6.723.

Bronfenbrenner, U. (1993). The ecology of cognitive development: Research models and fugitive findings. In R. H. Wozniak & K.W. Fischer (Eds.), *The Jean Piaget symposium series. Development in context: Acting and thinking in specific environments* (pp. 3–44). Lawrence Erlbaum Associates, Inc.

Brough, P., & Westman, M. (2018). Crossover, culture, and dual-earner couples. In W. Shen, K. Shockley, & R. Johnson (Eds.), *The Cambridge handbook of the global work-family interface* (pp. 629–645). Cambridge: Cambridge University Press. https://doi.org/10.1017/9781108235556.034.

Brown, M., & Pitt-Catsouphes, M. (2016). A mediational model of workplace flexibility, work–family conflict, and perceived stress among caregivers of older adults. *Community, Work & Family*, *19*(4), 379–395. https://doi.org/10.1080/13668803.2015.1034656.

Budig, M. J., & England, P. (2001). The wage penalty for motherhood. *American Sociological Review*, *66*, 204–225. https://doi.org/10.4054/DemRes.2017.37.56.

Budig, M. J., Misra, J., & Boeckmann, I. (2016). Work-family policy trade-offs for mothers? Unpacking the cross-national variation in motherhood earnings penalties. *Work and Occupations*, *43*, 119–177. https://doi.org/10.1177/0730888415615385.

Buehler, C., & O'Brien, M. (2011). Mothers' part-time employment: Associations with mother and family well-being. *Journal of Family Psychology*, *25*(6), 895–906. https://doi.org/10.1037/a0025993.

Buehler, C., O'Brien, M., & Walls, J. K. (2011). Mothers' part-time employment: Child, parent, and family outcomes. *Journal of Family Theory & Review*, *3*, 256–272. https://doi.org/10.1111/j.1756-2589.2011.00110.x.

Bumpus, M. F., Crouter, A. C., & McHale, S. M. (2006). Linkages between negative work-to-family spillover and mothers' and fathers' knowledge of their young adolescents' daily lives. *The Journal of Early Adolescence*, *26*(1), 36–59. https://doi.org/10.1177/0272431605282652.

Bünning, M. (2011). What happens after the 'daddy months'? Fathers' involvement in paid work, childcare, and work after taking parental leave in Germany. *European Sociological Review, 31*, 738–748. https://doi.org/10.1093/esr/jcv072.

Bureau of Labor Statistics. (2018a). *Employment characteristics of families—2017*. Available at: https://www.bls.gov/news.release/famee.nr0.htm.

Bureau of Labor Statistics. (2018b). *Access to paid personal leave* December 2017. Available at: https://www.bls.gov/ebs/paid:personal_leave_122017.htm.

Byron, K. (2005). A metaanalytic review of work-family conflict and its antecedents. *Journal of Vocational Behavior, 67*, 169–198. https://doi.org/10.1016/j.jvb.2004.08.009.

Carneiro, P., Løken, K.V., & Salvanes, K. G. (2015). A flying start? Maternity leave benefits and long-run outcomes of children. *Journal of Political Economy, 123*, 365–412. https://doi.org/10.1086/679627.

Chatterji, P., & Markowitz, S. (2012). Family leave after childbirth and the mental health of new mothers. *The Journal of Mental Health Policy and Economics, 15*, 61–76. https://doi.org/10.3386/w14156.

Chatterji, P., Markowitz, S., & Brooks-Gunn, J. (2013). Effects of early maternal employment on maternal health and well-being. *Journal of Population Economics, 26*, 285–301. https://doi.org/10.1007/s00148-012-0437-5.

Cho, E., & Tay, L. (2016). Domain satisfaction as a mediator of the relationship between work–family spillover and subjective well-being: A longitudinal study. *Journal of Business Psychology, 31*, 445–457. https://doi.org/10.1007/s10869-015-9423-8.

Chun, C., Moos, R. H., & Cronkite, R. C. (2006). Culture: A fundamental context for the stress and coping paradigm. In P. T. P. Wong & L. C. J. Wong (Eds.), *Handbook of multicultural perspectives on stress and coping* (pp. 29–53). Dallas, TX: Springer. https://doi.org/10.1007/0-387-26238-5 2.

Clark, M.A., Rudolph, C.W., Zhdanova, L., Michel, J. S., & Baltes, B. B. (2017). Organizational support factors and work–family outcomes: Exploring gender differences. *Journal of Family Issues, 38*(11), 1520–1545. https://doi.org/10.1177/0192513X15585809.

Clark, R., Shibley Hyde, J., Essex, M. J., & Klein, M. H. (1997). Length of maternity leave and quality of mother-infant interactions. *Child Development, 68*, 364–383. https://doi.org/10.2307/1131855.

Clark, S. C. (2000). Work/family border theory: A new theory of work/family balance. *Human Relations, 53*(6), 747–770. https://doi.org/10.1177/0018726700536001.

Coenen, M., & Kok, R. A. W. Workplace flexibility and new product development performance: The role of telework and flexible work schedules. *European Management Journal, 32*(4), (2014). 564–576. https://doi.org/10.1016/j.emj.2013.12.003.

Cools, S., Fiva, J. H., & Kirkebøen, L. J. (2015). Causal effects of paternity leave on children and parents. *The Scandinavian Journal of Economics, 117*, 801–828. https://doi.org/10.1111/sjoe.12113.

Craig, L., & Habgood, R. (2018). Cultural considerations in the division of labor. In W. Shen, K. Shockley, & R. Johnson (Eds.), *The Cambridge handbook of the global work-family interface* (pp. 646–660). Cambridge: Cambridge University Press. https://doi.org/10.1017/9781108235556.035.

Da, W. W. (2003). Transnational grandparenting: Child care arrangements among migrants from the People's Republic of China to Australia. *Journal of International Migration and Integration, 4*(1), 79–103. https://doi.org/10.1007/s12134-003-1020-4.

Davis, M. F., & Powell, R. (2003). The international convention on the rights of the child: A catalyst for innovative child care policies. *Human Rights Quarterly, 25*, 689–719. https://doi.org/10.1353/hrq.2003.0031.

Edwards, J. R., & Rothbard, N. P. (2000). Mechanisms linking work and family: Clarifying the relationship between work and family constructs. *The Academy of Management Review, 25*, 178–199. https://doi.org/10.2307/259269.

Fagnani, J. (2007). Family policies in France and Germany: Sisters or distant cousins? *Community, Work, and Family*, *10*, 39–56. https://doi.org/10.1080/13668800601110769.
Fortin, N. M. (2005). Gender role attitudes and the labour-market outcomes of women across OECD countries. *Oxford Review of Economic Policy*, *21*(3), 416–438. https://doi.org/10.1093/oxrep/gri024.
Frone, M. R., Russell, M., & Cooper, M. L. (1992). Prevalence of work-family conflict: Are work and family boundaries asymmetrically permeable? *Journal of Organizational Behavior*, *13*, 723–729. https://doi.org/10.1093/oxrep/gri024.
Frone, M. R., Yardley, J. K., & Markel, K. S. (1997). Developing and testing an integrative model of the work-family interface. *Journal of Vocational Behavior*, *50*, 145–167. https://doi.org/10.1006/jvbe.1996.1577.
Gajendran, R. S., & Harrison, D. A. (2007). The good, the bad, and the unknown about telecommuting: Metaanalysis of psychological mediators and individual consequences. *Journal of Applied Psychology*, *97*, 1524–1541. https://doi.org/10.1037/0021-9010.92.6.1524.
Galinsky, E., Sakai, K., & Wigton, T. (2011). Workplace flexibility: From research to action. *The Future of Children*, *21*(2), 141–161. https://doi.org/10.1353/foc.2011.0019.
Glaser, K., Ribe, E., Waginger, U., Price, D., Stuchbury, R., & Tinker, A. (2010). *Grandparenting in Europe*. London, UK: Grandparents Plus.
Golden, T. D. (2006). Avoiding depletion in virtual work: Telework and the intervening impact of work exhaustion on commitment and turnover intentions. *Journal of Vocational Behavior*, *69*, 176–187. https://doi.org/10.1016/j.jvb.2006.02.003.
Golden, T. D., & Veiga, J. F. (2005). The impact of extent of telecommuting on job satisfaction: Resolving inconsistent findings. *Journal of Management*, *31*(2), 301–318. https://doi.org/10.1177/0149206304271768.
Goode, W. J. (1960). A theory of role strain. *American Sociological Review*, **25**(4), 483–496. https://doi.org/10.2307/2092933.
Greenhaus, J., & Beutell, N. (1985). Sources of conflict between work and family roles. *Academy of Management Review*, *10*, 76–88. https://doi.org/10.2307/258214.
Greenhaus, J. H., & Powell, G. N. (2006). When work and family are allies: A theory of work-family enrichment. *The Academy of Management Review*, *31*(1), 72–92. https://doi.org/10.5465/AMR.2006.19379625.
Grove, C. N. (2005). Worldwide differences in business values and practices: Overview of GLOBE research findings. In *Global leadership solutions*. Grovewell LLC.
Grzywacz, J. G. (2000). Work-family spillover and health during midlife: Is managing conflict everything? *American Journal of Health Promotion*, *14*(4), 236–243. https://doi.org/10.4278/0890-1171-14.4.236.
Grzywacz, J. G., & Bass, B. L. (2003). Work, family, and mental health: Testing different models of work-family fit. *Journal of Marriage and Family*, *65*(1), 248–261. https://doi.org/10.1111/j.1741-3737.2003.00248.x.
Grzywacz, J. G., Almeida, D. M., & McDonald, D. A. (2002). Work–family spillover and daily reports of work and family stress in the adult labor force. *Family Relations*, *51*(1), 28–36. https://doi.org/10.1111/j.1741-3729.2002.00028.x.
Grzywacz, J. G., Carlson, D., & Shulkin, S. (2008). Schedule flexibility and stress: Linking formal flexible arrangements and perceived flexibility to employee health. *Community, Work & Family*, *11*(2), 199–214. https://doi.org/10.1080/13668800802024652.
Grzywacz, J. G., & Marks, N. F. (2000). Reconceptualizing the work-family interface: An ecological perspective on the correlates of positive and negative spillover between work and family. *Journal of Occupational Health Psychology*, *5*, 111–126. https://doi.org/10.1037/1076-8998.5.1.111.
Haar, J. M., & Bardoel, E. A. (2008). Positive spillover from the work–family interface: A study of Australian employees. *Asia Pacific Journal of Human Resources*, *46*(3), 275–287. https://doi.org/10.1177/1038411108095759.

Haas, L., & Hwang, C. P. (2008). The impact of taking parental leave on fathers' participation in childcare and relationships with children: Lessons from Sweden. *Community, Work & Family*, *11*, 85–104. https://doi.org/10.1080/13668800701785346.

Hajizadeh, M., Heymann, J., Strumpf, E., Harper, S., & Nandi, A. (2015). Paid maternity leave and childhood vaccination update: Longitudinal evidence from 20 low-and-middle-income countries. *Social Science & Medicine*, *140*, 104–117. https://doi.org/10.1016/j.socscimed.2015.07.008.

Hakanen, J. J., Peeters, M. C., & Perhoniemi, R. (2011). Enrichment processes and gain spirals at work and at home: A 3-year crosslagged panel study. *Journal of Occupational and Organizational Psychology*, *84*(1), 8–30. https://doi.org/10.1111/j.2044-8325.2010.02014.x.

Hammer, L. B., Allen, E., & Grigsby, T. D. (1997). Work-family conflict in dual-earner couples: Within-individual and crossover effects of work and family. *Journal of Vocational Behavior*, *50*(2), 185–203. https://doi.org/10.1006/jvbe.1996.1557.

Hammer, L. B., Cullen, J. C., Neal, M. B., Sinclair, R. R., & Shafiro, M. V. (2005). The longitudinal effects of work–family conflict and positive spillover on depressive symptoms among dual-earner couples. *Journal of Occupational Health Psychology*, *10*, 138–154. https://doi.org/10.1037/1076-8998.10.2.138.

Hank, K., & Buber, I. (2009). Grandparents caring for their grandchildren: Findings from the 2004 Survey of Health, Aging, and Retirement in Europe. *Journal of Family Issues*, *30*, 53–73. https://doi.org/10.1177/0192513x08322627.

Harrington, B., Van Deusen, F., Sabatini Fraone, J., Eddy, S., & Haas, L. (2014). *The new dad: Take your leave*. Boston, MA: Boston College Center for Work and Family.

Hassan, Z., Dollard, M. F., & Winefield, A. H. (2010). Work-family conflict in East vs. Western countries. *Cross Cultural Management: An International Journal*, *17*, 30–49. https://doi.org/10.1108/13527601011016899.

Hayman, J. R. (2009). Flexible work arrangements: Exploring the linkages between perceived usability of flexible work schedules and work/life balance. *Community, Work & Family*, *12*(3), 327–338. https://doi.org/10.1080/13668800902966331.

Heller, D., & Watson, D. (2005). The dynamic spillover of satisfaction between work and marriage: The role of time and mood. *Journal of Applied Psychology*, *90*, 1273–1279. https://doi.org/10.1037/0021-9010.90.6.1273.

Hill, E. J., Erickson, J. J., Fellows, K. J., Martinengo, G., & Allen, S. M. (2014). Work and family over the life course: Do older workers differ? *Journal of Family and Economic Issues*, *35*, 1–13. https://doi.org/10.1007/s10834-012-9346-8.

Hill, E. J., Erickson, J. J., Holmes, E. K., & Ferris, M. (2010). Workplace flexibility, work hours, and work-life conflict: An extra day or two. *Journal of Family Psychology*, *24*(3), 349–358. https://doi.org/10.1037/a0019282.

Hill, E. J., Hawkins, A. J., Ferris, M., & Weitzman, M. (2001). Finding an extra day a week: The positive effect of job flexibility on work and family life balance. *Family Relations*, *50*(1), 49–58. https://doi.org/10.1111/j.1741-3729.2001.00049.x.

Hill, E. J., & Holmes, E. K. (2018). Families and workplaces. In B. H. Fiese, M. Celano, K. Deater-Deckard, E. N. Jouriles, & M. A. Whisman (Eds.), *Applications and broad impact of family psychology*: Vol. 2. American Psychological Association handbook of contemporary family psychology (pp. 379–395). Washington, DC: American Psychological Association.

Hill, E. J., Jacob, J. I., Shannon, L. L., Brennan, R. T., Blanchard, V. L., & Martinengo, G. (2008). Exploring the relationship of workplace flexibility, gender, and life stage to family-to-work conflict, and stress and burnout. *Community, Work & Family*, *11*, 165–181. https://doi.org/10.1080/13668800802027564.

Hill, E. J., Märtinson, V., & Ferris, M. (2004). New-concept part-time employment as a work-family adaptive strategy for women professionals with small children. *Family Relations*, *53*(3), 282–292. https://doi.org/10.1111/j.0022-2445.2004.0004.x.

Ho, M.Y., Chen, X., Cheung, F. M., Liu, H., Worthington, E. L., Jr. (2013). A dyadic model of the work-family interface: A study of dual-earner couples in China. *Journal of Occupational Health Psychology*, *18*(1), 53–63. https://doi.org/10.1037/a0030885.

Hofstede, G. (1983). The cultural relativity of organizational practices and theories. *Journal of International Business Studies*, *14*(2), 75–89. https://doi.org/10.1057/palgrave.jibs.8490867.

Hofstede, G. (2001). *Culture's consequences: Comparing values, behaviors, institutions, and organizations across nations* (2nd ed.). Thousand Oaks, CA: Sage.

Holmes, E. K., Baumgartner, J., Marks, L. D., Palkovitz, R., & Nesteruk, O. (2010). Contemporary contradictions and challenges facing married fathers and mothers. In K. S. Pearlman (Ed.), *Marriage: Roles, stability, and conflict* (pp. 157–171). Hauppage, NY: Nova Science.

Holmes, E. K., Erickson, J. J., & Hill, E. J. (2012). Doing what she thinks is best: Maternal psychological wellbeing and attaining desired work situations. *Human Relations*, *65*(4), 501–522. https://doi.org/10.1177/0018726711431351.

Holmes, E. K., Holladay, H. M., Hill, E. J., & Yorgason, J. (2018). Are mothers' work-to-family conflict, school involvement, and work status related to academic achievement? *Journal of Child and Family Studies*, *27*(6), 1881–1898. https://doi.org/10.1007/s10826-018-1021-8.

House, R. J., Hanges, P. J., Javidan, M., Dorfman, P. W., & Gupta, V. (Eds.), (2004). *Culture, leadership, and organizations: The GLOBE study of 62 societies*. Thousand Oaks, CA: Sage.

Huang, R., & Yang, M. (2015). Paid maternity leave and breastfeeding practice before and after California's implementation of the nation's first paid family leave program. *Economics and Human Biology*, *16*, 45–59. https://doi.org/10.1016/j.ehb.2013.12.009.

Huerta, M., Adema, W., Baxter, J., Han, W., Lausten, M., Lee, R., & Waldfogel, J. (2014). Fathers' leave and fathers' involvement: Evidence from four OECD countries. *European Journal of Social Security*, *16*, 308–346. https://doi.org/10.1177/138826271401600403.

Hughes-Belding, K., Hegland, S., Stein, A., Sideris, J., & Bryant, D. (2012). Predictors of global quality in family child care homes: Structural and belief characteristics. *Early Education & Development*, *23*(5), 697–712. https://doi.org/10.1080/10409289.2011.574257.

Ilies, R., Wilson, K. S., & Wagner, D. T. (2009). The spillover of daily job satisfaction onto employees' family lives: The facilitating role of work-family integration. *Academy of Management Journal*, *52*(1), 87–102. https://doi.org/10.5465/AMJ.2009.36461938.

International Labour Organization (ILO). (2014). *Maternity and paternity at work: Law and practice across the world*. Geneva: International Labour Organization.

Jang, S. J., Zippay, A., & Park, R. (2012). Family roles as moderators of the relationship between schedule flexibility and stress. *Journal of Marriage and Family*, *74*(4), 897–912. https://doi.org/10.1111/j.1741-3737.2012.00984.x.

Jappens, M., & Van Bavel, J. (2012). Regional family norms and child care by grandparents in Europe. *Demographic Research*, *27*(4), 85–120. https://doi.org/10.4054/DemRes.2012.27.4.

Jin, J. F., Ford, M. T., & Chen, C. C. (2013). Asymmetric differences in work-family spillover in North America and China: Results from two heterogeneous samples. *Journal of Business Ethics*, *113*, 1–14. https://doi.org/10.1007/s10551-012-1289-3.

Johnson, R. W. (2011). Phased retirement and workplace flexibility for older adults: Opportunities and challenges. *The Annals of the American Academy of Political and Social Science*, *638*, 68–85. https://doi.org/10.1177/0002716211413542.

Jones, B. L., Scoville, D. P., Hill, E. J., Childs, G., Leishman, J. M., & Nally, K. S. (2008). Perceived versus used workplace flexibility in Singapore: Predicting work-family fit. *Journal of Family Psychology*, *22*(5), 774. https://doi.org/10.1037/a0013181.

Joplin, J. R. W., Shaffer, M. A., Francesco, A. M., & Lau, T. (2003). The macro-environment and work-family conflict: Development of a cross cultural comparative framework. *International Journal of Cross Cultural Management*, *3*(3), 305–328. https://doi.org/10.1177/1470595803003003004.

Judge, T. A., & Ilies, R. (2004). Affect and job satisfaction: A study of their relationship at work and at home. *Journal of Applied Psychology*, *89*, 661–673. https://doi.org/10.1037/0021-9010.89.4.661.

Judge, T. A., & Livingston, B. A. (2008). Is the gap more than gender? A longitudinal analysis of gender, gender role orientation, and earnings. *Journal of Applied Psychology*, *93*(5), 994–1012. https://doi.org/10.1037/0021-9010.93.5.994.

Kelly, E. L., Moen, P., Oakes, J. M., Fan, W., Okechukwu, C., Davis, K. D., … Casper, L. (2014). Changing work and work-family conflict: Evidence from the work, family, and health network. *American Sociological Review*, *79*(3), 485–516. https://doi.org/10.1177/0003122414531435.

Kim, J. (2018). Workplace flexibility and parent–child interactions among working parents in the U.S. *Social Indicators Research*, https://doi.org/10.1007/s11205-018-2032-y.

Kinney, M.V., Kerber, K. J., Black, R. E., Cohen, B., Nkrumah, F., Coovadia, H., … Lawn, J. E. (2012). Sub-Saharan Africa's mothers, newborns, and children: Where and why do they die? *PLoS Medicine*, 7(6). https://doi.org/10.1371/journal.pmed.1000294.

Kirchmeyer, C. Nonwork participation and work attitudes: A test of scarcity vs. expansion models of personal resources. *Human Relations*, *45*, (1992). 775–795. https://doi.org/10.1177/001872679204500802.

Kossek, E. E., & Michel, J. (2011). Flexible work schedules. In S. Zedeck (Ed.), *APA handbook of industrial and organizational psychology* (pp. 535–572). Washington, DC: American Psychological Association.

Kotsadam, A., & Finseraas, H. (2011). The state intervenes in the battle of the sexes: Causal effects of paternity leave. *Social Science Research*, *40*, 1611–1622. https://doi.org/10.1016/j.ssresearch.2011.06.011.

Kuo, P. X., Volling, B. L., & Gonzalez, R. (2017). Gender role beliefs, work–family conflict, and father involvement after the birth of a second child. *Psychology of Men & Masculinity*, *19*(2), 243–256. https://doi.org/10.1037/men0000101.

Kurowska, A. (2018). Gendered effects of home-based work on parents' capability to balance work with nonwork: Two countries with different models of division of labour compared. *Social Indicators Research*, https://doi.org/10.1007/s11205-018-2034-9.

Lambert, P. A. (2007). The political economy of postwar family policy in Japan: Economic imperatives and electoral incentives. *Journal of Japanese Studies*, *33*(1), 1–29.

Lammi-Taskula, J. (2008). Doing fatherhood: Understanding the gendered use of parental leave in Finland. *Fathering*, *6*(2), 133–148. https://doi.org/10.3149/fth.0602.133.

Lee, M. D., & MacDermid, S. M. (1998). *Improvising new careers: Accommodation, elaboration, transformation*. West Lafayette, IN: The Center for Families at Purdue University.

Leroy, J. L., Gadsden, P., & Guijarro, M. (2012). The impact of daycare programmes on child health, nutrition, and development in developing countries: A systemic review. *Journal of Development Effectiveness*, *4*(3), 472–496. https://doi.org/10.1080/19439342.2011.639457.

Lewis, J. (2008). Child care policies and the politics of choice. *The Political Quarterly*, *79*(4), 499–507. https://doi.org/10.1111/j.1467-923X.2008.00962.x.

Lichtman-Sadot, S., & Bell, N. P. (2017). Child health in elementary school following California's paid family leave program. *Journal of Policy Analysis and Management*, *36*, 790–827. https://doi.org/10.1002/pam.22012.

Liu, H., & Cheung, F. M. (2015a). The moderating role of empathy in the work-family crossover process between Chinese dual-earner couples. *Journal of Career Assessment*, *23*(3), 442–458. https://doi.org/10.1177/1069072714547612.

Liu, H., & Cheung, F. M. (2015b). Testing crossover effects in an actor-partner interdependence model among Chinese dual-earner couples. *International Journal of Psychology, 50*(2), 106–114. https://doi.org/10.1002/ijop.12070.

Liu, H., Ngo, H.Y., & Cheung, F. M. (2016). Work-family enrichment and marital satisfaction among Chinese coupes: A crossover-spillover perspective. *International Journal of Stress Management, 23*(2), 209–231. https://doi.org/10.1037/a0039753.

Livingston, B. A. (2018). Gender, gender norms, and national culture: Global work-family at multiple levels. In W. Shen, K. Shockley, & R. Johnson (Eds.), *The Cambridge handbook of the global work-family interface* (pp. 599–613). Cambridge: Cambridge University Press. https://doi.org/10.1017/9781108235556.032.

Lokteff, M., & Piercy, K.W. (2012). "Who cares for the children?" Lessons from a global perspective of child care policy. *Journal of Child and Family Studies, 21*(1), 120–130. https://doi.org/10.1007/s10826-011-9467-y.

Lourel, M., Ford, M.T., Gamassou, C. E., Guégen, N., & Hartmann, A. (2009). Negative and positive spillover between work and home. *Journal of Managerial Psychology, 24*, 438–449. https://doi.org/10.1108/02683940910959762.

Lu, L., Gilmour, R., Kao, S. F., & Huang, M. T. (2006). A cross-cultural study of work/family demands, work/family conflict, and wellbeing: The Taiwanese vs. British. *Career Development International, 11*, 9–27. https://doi.org/10.1108/13620430610642354.

Luk, D. M., & Shaffer, M. A. (2005). Work and family domain stressors and support: Within and cross-domain influences on work-family conflict. *Journal of Occupational and Organizational Psychology, 78*, 489–508. https://doi.org/10.1348/096317905X26741.

Mahon, R. (2002). Child care: Toward what kind of "social Europe"? *Social Politics, 9*(3), 344–379. https://doi.org/10.1093/sp/9.3.343.

Marks, S. R. (1977). Multiple roles and role strain: Some notes on human energy, time, and commitment. *American Sociological Review, 42*(6), 921–936. https://doi.org/10.2307/2094577.

Martin, B. H., & MacDonnell, R. (2012). Is telework effective for organizations? A metaanalysis of empirical research on perceptions of telework and organizational outcomes. *Management Research Review, 35*(7), 602–616. https://doi.org/10.1108/01409171211238820.

Martin, C. (2007). A baby-friendly state: Lessons from the French case. *Pharmaceuticals, Policy and Law*, 9, 203–210.

Masuda, A. D., Poelmans, S. A.Y., Allen, T. D., Spetor, P. E., Lapierre, L. M., Cooper, C. L., … Moreno-Velazquez, I. (2012). Flexible work arrangements availability and their relationship with work-to-family conflict, job satisfaction, and turnover intentions: A comparison of three country clusters: Work-to-family conflict across countries. *Applied Psychology. An International Review, 61*, 1–29. https://doi.org/10.1111/j.1464-0597.2011.00453.x.

Matias, M., Ferreira, T., Vieira, J., Cadima, J., Leal, T., & Matos, P. M. (2017). Work-family conflict, psychological availability, and child emotion regulation: Spillover and crossover in dual-earner families. *Personal Relationships, 24*(3), 623–639. https://doi.org/10.1111/pere.12198.

Matz-Costa, C., & Pitt-Catsouphes, M. (2010). Workplace flexibility as an organizational response to the aging of the workforce: A comparison of nonprofit and for-profit organizations. *Journal of Social Service Research, 36*, 68–80. https://doi.org/10.1080/01488370903333645.

Mauss, I. B., & Butler, E. A. (2010). Cultural context moderates the relationship between emotion control values and cardiovascular challenge versus threat responses. *Biological Psychology, 84*, 521–530. https://doi.org/10.1016/j.biopsycho.2009.09.010.

McAllister, S., Thornock, C., Hammond, J., Holmes, E. K., & Hill, E. J. (2012). The influence of couple emotional intimacy on job perceptions and work-family conflict. *Family and Consumer Sciences Research Journal, 40*(4), 330–347. https://doi.org/10.1111/j.1552-3934.2012.02115.x.

McCloskey, D.W., & Igbaria, M. (1998). A review of the empirical research on telecommuting and directions for future research. In *The virtual workplace* (pp. 338–358). https://doi.org/10.4018/978-1-87828-947-6.

McGovern, P., Dowd, B., Gjerdingen, D., Moscovice, I., Kochevar, L., & Lohman, W. (1997). Time off work and the postpartum health of employed women. *Medical Care, 35*, 507–521. https://doi.org/10.1097/00005650-199705000-00007.

McKay, L., Mathieu, S., & Doucet, A. (2016). Parental-leave rich and parental-leave poor: Inequality in Canadian labour market based leave policies. *Journal of Industrial Relations, 58*, 543–562. https://doi.org/10.1177/0022185616643558.

Mesmer-Magnus, J. R., & Viswesvaran, C. (2006). How family friendly work environments affect work-family conflict: A metaanalytic examination. *Journal of Labor Research, 4*, 555–574. https://doi.org/10.1007/s12122-006-1020-1.

Mesquita, B., & Delvaux, E. (2012). A cultural perspective on emotional labor. In A. Grandey, J. Diefendorff, & D. Rupp (Eds.), *Emotional labor in the 21st century: Diverse perspectives on emotion regulation at work* (pp. 251–272). New York: Psychology Press/Routledge.

Misra, J., Moller, S., Strader, E., & Wemlinger, E. (2012). Family policies, employment and poverty among partnered and single mothers. *Research in Social Stratification and Mobility, 30*, 113–128. https://doi.org/10.1016/j.rssm.2011.12.001.

Moen, P., & Roehling, P.V. (2005). *The career mystique: Cracks in the American dream*. Boulder, CO: Rowman & Littlefield.

Moss, P., Duvander, A., & Koslowski, A. (2019). *Parental leave and beyond: Recent international developments, current issues, and future directions*. Bristol: Policy Press.

Muller, W., & Brough, P. (2017). Work, stress, and love: Investigating crossover through couples' lived experiences. In J. Fitzgerald (Ed.), *Foundations for couples therapy: Research for the real world* (pp. 464–474). New York, NY: Routledge.

Muraven, M., Tice, D. M., & Baumeister, R. F. (1998). Self-control as a limited resource: Regulatory depletion patterns. *Journal of Personality and Social Psychology, 74*, 774–789. https://doi.org/10.1037/0022-3514.74.3.774.

Murphy-Lawless, J. (2000). Changing women's lives: Child care policy in Ireland. *Feminist Economics, 6*(1), 89–94. https://doi.org/10.1080/135457000337697.

National Partnership for Women and Families. (2018). *State paid family and medical leave laws, July 2018*. Washington, DC. Available at: http://www.nationalpartnership.org/research-library/work-family/paid-leave/state-paid-family-leave-laws.pdf.

Nepomnyaschy, L., & Waldfogel, J. (2007). Paternity leave and fathers' involvement with their young children: Evidence from the American ECLS-B. *Community, Work & Family, 10*, 427–453. https://doi.org/10.1080/13668800701575077.

Ng, T. W. H., & Feldman, D. C. (2008). Long work hours: A social identity perspective on metaanalysis data. *Journal of Organizational Behavior, 29*(7), 853–880. https://doi.org/10.1002/job.536.

Nilles, J. (1994). *Making telework happen: A guide for telemanagers and telecommuters*. New York: Van Nostrand Reinhold.

O'Brien, K. M., Del Pino, H.V., Yoo, S. K., Cinnamon, R. G., & Han, Y. J. (2014). Work, family, support, and depression: Employed mothers in Israel, Korea, and the United States. *Journal of Counseling Psychology, 61*, 461–472. https://doi.org/10.1037/a0036339.

O'Brien, M. (2009). Fathers, parental leave policies, and infant quality of life: International perspectives and policy impact. *The Annals of the American Academy of Political and Social Science, 624*, 190–213. https://doi.org/10.1177/0002716209334349.

OECD. (2001). *Starting strong*. Retrieved from https://www.oecd.org/newsroom/earlychildhoodeducationandcare.htm.

OECD. (2018). *OECD family database: The labour market position of families* Retrieved December. 2018 from http://www.oecd.org/els/family/database.htm.

Ollier-Malaterre, A., & Foucreault, A. (2018). GLOBE cultural dimensions: Implications for global work-family research. In W. Shen, K. Shockley, & R. Johnson (Eds.), *The Cambridge handbook of the global work-family interface*. (pp. 69–85). Cambridge: Cambridge University Press. https://doi.org/10.1017/9781108235556.003.

Orthner, D. K., & Pittman, J. F. (1986). Family contributions to work commitment. *Journal of Marriage and the Family*, *48*, 573–581. https://doi.org/10.2307/352043.

Pedulla, D. S., & Thébaud, S. (2015). Can we finish the revolution? Gender, work-family ideals, and institutional constraint. *American Sociological Review*, *80*(1), 116–139. https://doi.org/10.1177/0003122414564008.

Pence, A. R., Amponsah, M., Chalamanda, F., Habtom, A., Kameka, G., & Nankunda, H. (2004). ECD policy development and implementation in Africa. *International Journal of Educational Policy, Research, and Practice*, *5*(3), 13–29.

Petts, R. J. (2018). Time off after childbirth and mothers' risk of depression, parenting stress, and parenting practices. *Journal of Family Issues*, *39*, 1827–1854. https://doi.org/10.1177/0192513X17728984.

Petts, R. J., & Knoester, C. (2018). Paternity leave-taking and father engagement. *Journal of Marriage and Family*, *80*, 1144–1162. https://doi.org/10.1111/jomf.12494.

Petts, R. J., Knoester, C., & Li, Q. (2018). Paid paternity leave-taking in the United States. *Community, Work & Family*, https://doi.org/10.1080/13668803.2018.1471589.

Poelmans, S. A. Y., & Chenoy, R. (2008). Investigating workplace flexibility using a multi-organization database: A collaboration of academics and practitioners [Special issue]. *Community, Work & Family*, *11*, 1–242. https://doi.org/10.1080/13668800802172576.

Polatcia, S., & Akdoğan, A. (2014). Psychological capital and performance: The mediating role of work family spillover and psychological well-being. *Business and Economics Research Journal*, *5*(1), 1–15.

Pragg, B., & Knoester, C. (2017). Parental leave use among disadvantaged fathers. *Journal of Family Issues*, *38*, 1157–1185. https://doi.org/10.1177/0192513X15623585.

Raub, A., Nandi, A., DeGuzman Chorny, N., Wong, E., Chung, P., Batra, P., … Heymann, J. (2018). *Paid parental leave: A detailed look at approaches across OECD countries*. Los Angeles, CA: World Policy Analysis Center.

Redshaw, M., & Henderson, J. (2013). Fathers' engagement in pregnancy and childbirth: Evidence from a national survey. *BMC Pregnancy and Childbirth*, *13*, 70. https://doi.org/10.1186/1471-2393-13-70.

Rodriguez-Muñoz, A., Sanz-Vergel, A. I., Demerouti, E., & Bakker, A. B. (2014). Engaged at work and happy at home: A spillover-crossover model. *Journal of Happiness Studies*, *15*(2), 271–283. https://doi.org/10.1007/s10902-013-9421-3.

Rosemberg, F. (2004). Early child-care and educational policies for the developing world: The Brazilian case. *Prospects: Quarterly Review of Comparative Education*, *34*(4), 481–491. https://doi.org/10.1007/s11125-005-2739-9.

Rossin-Slater, M., Ruhm, C., & Waldfogel, J. (2013). The effects of California's paid family leave program on mothers' leave-taking and subsequent labor market outcomes. *Journal of Policy Analysis and Management*, *32*, 224–245. https://doi.org/10.1002/pam.21676.

Rotondo, D. M., & Kincaid, J. F. (2008). Conflict, facilitation, and individual coping styles across the work and family domains. *Journal of Managerial Psychology*, *23*(5), 484–506. https://doi.org/10.1108/02683940810884504.

Ruhm, C. J. (2000). Parental leave and child health. *Journal of Health Economics*, *19*, 931–960. https://doi.org/10.1016/S0167-6296(00)00047-3.

Rusting, C. L., & DeHart, T. (2000). Retrieving positive memories to regulate negative mood: Consequences for mood-congruent memory. *Journal of Personality and Social Psychology*, *78*, 737–752. https://doi.org/10.1016/S0167-6296(00)00047-3.

Schieman, S., Milkie, M., & Glavin, P. (2009). When work interferes with life: The social distribution of work-nonwork interference and the influence of work-related demands and resources. *American Sociological Review*, *74*, 966–987. https://doi.org/10.1177/000312240907400606.

Schieman, W., & Young, M. (2015). Who engages in work-family multitasking? A study of Canadian and American workers. *Social Indicators Research*, *120*, 741–767. https://doi.org/10.1007/s11205-014-0609-7.

Schooreel, T., & Verbruggen, M. (2016). Use of family-friendly work arrangements and work–family conflict: Crossover effects in dual-earner couples. *Journal of Occupational Health Psychology*, *21*(1), 119–132. https://doi.org/10.1037/a0039669.

Sharma, D., & LeVine, R. A. (1998). Child care in India: A comparative developmental view of infant social environments. *New Directions for Child and Adolescent Development*, *81*, 45–67. https://doi.org/10.1002/cd.23219988005.

Shimazu, A., Bakker, A. B., & Demerouti, E. (2009). How job demands affect an intimate partner: A test of the spillover-crossover model in Japan. *Journal of Occupational Health Psychology*, *51*(3), 239–248. https://doi.org/10.1539/joh.L8160.

Shockley, K. M., French, K. A., & Yu, P. P. (2018). A comprehensive review and synthesis of the cross-cultural work-family literature. In W. Shen, K. Shockley, & R. Johnson (Eds.), *The Cambridge handbook of the global work-family interface* (pp. 9–68). Cambridge: Cambridge University Press. https://doi.org/10.1017/9781108235556.002.

Sjöberg, O. (2010). Ambivalent attitudes, contradictory institutions ambivalence in gender-role attitudes in comparative perspective. *International Journal of Comparative Sociology*, *51*(1–2), 33–57. https://doi.org/10.1177/0020715209347064.

Sonnentag, S., & Binnewies, C. (2013). Daily affect spillover from work to home: Detachment from work and sleep as moderators. *Journal of Vocational Behavior*, *83*, 198–208. https://doi.org/10.1016/j.jvb.2013.03.008.

Soto, J. A., Perez, C. R., Kim, Y. H., Lee, E. A., & Minnick, M. R. (2011). Is expressive suppression always associated with poorer psychological functioning? A cross-cultural comparison between European Americans and Hong Kong Chinese. *Emotion*, *11*, 1450–1455. https://doi.org/10.1037/a0023340.

St. George, J. M., & Fletcher, R. J. (2011). Fathers online: Learning about fatherhood through the internet. *Journal of Perinatal Education*, *20*(3), 154–162. https://doi.org/10.1891/1058-1243.20.3.154.

Staehelin, K., Bertea, P. C., & Stutz, E. Z. (2007). Length of maternity leave and health of mother and child—A review. *International Journal of Public Health*, *52*, 202–209. https://doi.org/10.1007/s00038-007-5122-1.

Szelewa, D., & Polakowski, M. P. (2008). Who cares? Changing patterns of childcare in Central and Eastern Europe. *Journal of European Social Policy*, *18*, 115–131. https://doi.org/10.1177/0958928707087589.

Tanaka, S. (2005). Parental leave and child health across OECD countries. *The Economic Journal*, *115*, F7–F28. https://doi.org/10.1111/j.0013-0133.2005.00970.x.

Toossi, M., & Morisi, T. L. (2017). *U. S. Bureau of Labor Statistics, spotlight on statistics: Women in the workforce before, during, and after the great recession.* https://www.bls.gov/spotlight/2017/women-in-the-workforce-before-during-and-after-the-great-recession/pdf/women-in-the-workforce-before-during-and-after-the-great-recession.pdf.

Voydanoff, P. (2002). Linkages between the work-family interface, and work, family, and individual outcomes: An integrative model. *Journal of Family Issues*, *23*(1), 138–164. https://doi.org/10.1177/0192513X02023001007.

Wadsworth, L. L., & Facer, R. L. (2016). Work–family balance and alternative work schedules: Exploring the impact of 4-day workweeks on state employees. *Public Personnel Management*, *45*(4), 382–404. https://doi.org/10.1177/0091026016678856.

Waldfogel, J. (1998). The family gap for young women in the United States and Britain: Can maternity leave make a difference? *Journal of Labor Economics*, *16*, 505–545. https://doi.org/10.1086/209897.

Waldfogel, J. (2006). *What children need.* Cambridge: Harvard University.

Way, S. A., Tracey, J. B., Fay, C. H., Wright, P. M., Snell, S. A., Chang, S., & Gong, Y. (2015). Validation of a multidimensional HR flexibility measure. *Journal of Management*, *41*, 1098–1131. https://doi.org/10.1177/0149206312463940.

Weichselbaumer, D., & Winter-Ebmer, R. (2005). A metaanalysis of the international gender wage gap. *Journal of Economic Surveys*, *19*(3), 479–511. https://doi.org/10.1111/j.0950-0804.2005.00256.x.

West, C., & Zimmerman, D. H. (1987). Doing gender. *Gender & Society*, *1*(2), 125–151. https://doi.org/10.1177/0891243287001002002.

Westman, M. (2002). Crossover of stress and strain in the family and in the workplace. In P. L. Perrewé & D. C. Ganster (Eds.), Vol. 2. *Research in occupational stress and well-being* (pp. 143–181). Greenwich, CT: JAI Press/Elsevier.

Westman, M., & Etzion, D. (2005). The crossover of work-family conflict from one spouse to the other. *Journal of Applied Social Psychology*, *35*(9), 1936–1957. https://doi.org/10.1111/j.1559-1816.2005.tb02203.x.

Westman, M., Vinokur, A. D., Hamilton, V., & Roziner, I. (2004). Crossover of marital dissatisfaction during military downsizing among Russian army officers and their spouses. *Journal of Applied Psychology*, *89*(5), 769–779. https://doi.org/10.1037/0021-9010.89.5.769.

White, L. A., & Friendly, M. (2012). Public funding, private delivery: States, markets, and early childhood education and care in liberal welfare states—A comparison of Australia, the UK, Quebec, and New Zealand. *Journal of Comparative Policy Analysis: Research and Practice*, *14*(4), 292–310. https://doi.org/10.1080/13876988.2012.699789.

Whiteford, P., & Adema, W. (2007). What works best in reducing child poverty: A benefit or work strategy. *OECD Social Employment and Migration Working Papers No. 51*. http://www.oecd.org/els.

Williams, J. C., Blair-Loy, M., & Berdahl, J. L. (2013). Cultural schemas, social class, and the flexibility stigma. *Journal of Social Issues*, *69*(2), 209–234. https://doi.org/10.1111/josi.12012.

Won, S., & Pascall, G. (2004). A Confucian war over childcare? Practice and policy in childcare and their implications for understanding the Korean gender regime. *Social Policy and Administration*, *38*(3), 270–289. https://doi.org/10.1111/j.1467-9515.2004.00390.x.

World Bank Group. (2018). *Women, business and the law 2018*. Washington, DC: World Bank.

World Economic Forum. (2013). *The global gender gap report*. Geneva: World Economic Forum.

Yang, N., Chen, C. C., Choi, J., & Zou, Y. (2000). Sources of work–family conflict: A Sino–US comparison of the effects of work and family demands. *Academy of Management Journal*, *43*(1), 113–123. https://doi.org/10.5465/1556390.

CHAPTER 11

Families and physical health

Haleama Al Sabbah
College of Natural and Health Sciences, Zayed University, Dubai, United Arab Emirates

The World Health Organization (WHO) defined health as "state of complete physical, mental, and social well-being, and not merely the absence of disease or infirmity" (WHO, 1948). Health is a dynamic state resulting from a body's constant adjustment and adaptation in response to changes in the environment. This chapter will focus on the physical health of families. Physical health is a state of well-being when all internal and external body parts, organs, tissues, and cells can function properly as they are supposed to function. Also, it includes a state of physical well-being in which a person is physically fit to perform their daily activities without restrictions. Good physical health means that, for example, our ears can normally hear, our eyes have normal vision, our legs can walk, jump, run, and perform many other normal activities without problems. The various components of physical health that are discussed below include diet and nutrition, physical activity (PA), sleep patterns, medical self-care, smoking as well as alcohol and drugs. Health is also associated with several biopsychosocial factors (such as genetics, personal behaviors, and socioeconomic status) and therefore, it is essential to develop health promotion strategies targeting these factors in order to improve the well-being of each individual in the community (Ennis & Bunting, 2013).

Effects of illness on family members

Illness is a distressing experience for the individual as well as for the family members and can result in adverse effects on the mental and physical well-being of the family, creating a family burden (Persson & Benzein, 2014). Family burden is a term that encompasses all the challenges a family experiences as a result of having a sick member. It includes subjective elements like emotional difficulties as well as practical/objective challenges (Ennis & Bunting, 2013). An individual's quality of life (QoL) is closely knit with the QoL of those around them, be it their spouse, children or parents.

Cross-Cultural Family Research and Practice
https://doi.org/10.1016/B978-0-12-815493-9.00011-9

As a result, any kind of chronic sickness present within the family has the potential to affect the lives of the other members (Golics, Basra, Finlay, & Salek, 2013). The effect that a patients' illness has on families is extensive and profound. There are several ways in which the QoL of family members living with a chronically ill individual is affected. Golics, Basra, Finlay, et al. (2013) conducted semistructured interviews on $n = 133$ family members who had having a chronically ill close relative. They identified the following key themes related to the impact of living with an ill person on the family: emotional impact, relationships, sleep, work, financial load, and social life. Out of these themes, emotional burden was the most widely reported, 92% of the participants mentioned it. The emotional impact had a great influence on various areas of the participants' lives, such as on their sleep and overall health. The family members of patients suffering from illness, particularly cancer, have been found often to be more emotionally disturbed than the patients themselves (Golics, Basra, Salek, & Finlay, 2013).

Persson and Benzein (2014) reported similar findings to Golics, Basra, Finlay, et al. (2013) and Golics, Basra, Salek, et al. (2013). Persson and Benzein found that chronic illnesses have an effect on the family members as well as the individual diagnosed with the illness, effects including emotional, financial, family relationships, education and work, leisure time, and social activities. Similarly, Ennis and Bunting (2013) found that health problems in any family member were associated with a high incidence of mental health issues in other family members. Some of the emotions experienced by the immediate relatives of an ill family member include helplessness, suffering, vulnerability, strain, and difficulties in relationships as well as a struggle to preserve normality in family living (Persson & Benzein, 2014).

Family influences on health

There is a moderate concordance among family members' health, most notably there is a medium correlation between spouses' health status. For example, there are well-replicated associations between spouses' overall longevity as well as between spouses' risk for cardiovascular disease (Meyler, Stimpson, & Peek, 2007).

There are many different pathways that account for the concordance in family members' health outcomes. It is well established that a mutually satisfying couple relationship is associated with good health, (e.g., low rates of cardiovascular disease, diabetes, and other chronic health problems) (Robles, Slatcher, Trombello, & McGinn, 2014). Of course association does not establish causality, and leading researchers propose there is a bi-directional influence between couple relationship satisfaction and health (Robles, 2014). As noted in the previous section, health issues in one family member impact upon the whole family, and severe illness in a partner often erodes marital satisfaction in both spouses (Valle, Weeks, Taylor, & Eberstein, 2013). Conversely, marital distress impacts negatively on spouses' health (Robles, 2014).

The negative health effects of marital distress can be classified as direct or indirect effects of marital distress. Direct effects of marital distress refer to when marital distress has specific physiological effects. For example, couple conflict increases cortisol and vasopressin levels, and these increases predict later risk of cardiovascular disease (Martire, Schulz, Helgeson, Small, & Saghafi, 2010; Robles & Kiecolt-Glaser, 2003). There are also indirect effects of marital distress, meaning marital distress often is associated with unhealthy behaviors, such as heavy alcohol consumption, lack of exercise, and smoking (Robles & Kiecolt-Glaser, 2003), and changes in these behaviors in response to marital distress often erode future health of all family members.

Family members' health behaviors (e.g., smoking, alcohol consumption, eating, exercise, and sleep) show more similarity than would be expected by chance, particularly the association of spouses' health-related behaviors (Di Castelnuovo, Quacquaruccio, Donati, de Gaetano, & Iacoviello, 2009; Hasler and Troxel, 2010). This similarity is partly due to assortative mating in how people select their marriage partner. For example, men who are heavy drinkers, smokers, or who do not exercise much, tend to partner with women with similar habits, and vice versa (Ask, Rognmo, Torvik, Røysamb, & Tambs, 2012; Di Castelnuovo et al., 2009). In addition to the effects of assortative mating, spouses' seem to influence each other's health-related behaviors, as husbands and wives behaviors become more similar across time (Ask et al., 2012). This convergence of health-related behaviors seems understandable as spouses are likely to be spending leisure time together, which is likely to be reflected in a shared active or sedentary lifestyle, eating many of the same foods, sleeping in the same bed so sleep problems of one affect the other and so forth.

Spouses also influence attempts to lead healthier lifestyles. When one spouse has a positive health behavior habit (e.g., exercises regularly, eats a healthy diet), then the other person is likely over time to also take on those same behaviors (Jackson, Steptoe, & Wardle, 2015). Even more striking is that when one spouse adopts a new health enhancing behavior (e.g., stops smoking), the other spouse is often influenced to also change their behavior (Jackson et al., 2015).

Components of physical health and families

Diet and nutrition

Frequent meals consumed together as a family support the development of healthy relationships among children and young people. Family meals provide an opportunity for families to prepare healthy meals while continuously strengthening their emotional bonds. Shared meals provide a regular window of time wherein parents and children have the opportunity to talk, connect and therefore develop a feeling of family unity (Giray & Ferguson, 2018). Research also shows greater family connection as well as parental monitoring and communication with adolescents which may help in prevention of certain health risk behaviors and moreover promote emotional well-being (Utter et al., 2018). Also, children that consume frequent meals with their family have reported to have better nutrition as well as eating behaviors (Berge et al., 2015).

Family meals have also shown to have beneficial effects on parents. A study conducted in the United States found a significant association between regular family meals and reduced body mass index (BMI) among parents (Sobal & Hanson, 2011). Another study conducted by Utter et al. (2018) found that parents who consumed regular family meals consumed more vegetables and fruits, fathers consumed less junk food, and mothers displayed fewer dieting behaviors. Such parents are also reported be have better social and emotional well-being, which positively affects the overall well-being of their children (Utter et al., 2018).

Eating meals together as a family has been suggested as a factor which may be protective against obesity in adolescence as well as later on in adulthood. More frequent consumption of meals together with the family is associated with an increased consumption of calcium, whole grains, fruits and vegetables, reduction in excessive weight control behavior, and binge eating as well as improved psychological behavior in adolescents. These associations have been proven effective across diverse racial/ethnic backgrounds and genders (Berge et al., 2015).

Childhood obesity poses another challenge with its prevalence increasing at an alarming rate (Al-Domi et al., 2019). It is linked to several adverse health outcomes in adulthood such as cardiovascular diseases (CVD), diabetes and other chronic illnesses. Hence, prevention of childhood obesity is essential (Al-Domi et al., 2019). The rate of obesity is increasing at an alarming rate worldwide with about one third of the population being obese and 2 billion overweight. The Global Burden of Metabolic Risk Factors of Chronic Diseases Collaborating Group reviewed data on the prevalence of obesity between 1980 and 2008 among 199 countries and found the prevalence of obesity to have doubled worldwide (Seidell & Halberstadt, 2015).

Physical activity

Physical activity is recommended by healthcare professionals as an essential aspect of our daily lives and the resulting health benefits are well recognized. The WHO recommends at least 150 min of moderate-intensity physical activity per week or 75 min of vigorous-intensity physical activity per week or an equal combination of vigorous and moderate intensity physical activity for adults aged 18–64. The recommended physical activity for children ages 5–17 years is a minimum of 60 min of moderate to high intensity physical activity daily (WHO, 2015). Family-based interventions pose a great opportunity to increase and support physical activity among children. The support of parents has always shown to be positively associated with an increase in children's physical activity levels (Brown, Schiff, & Sluijs, 2015). There is also growing evidence linked with regular physical activity providing mental health benefits among youth. A cross-sectional survey conducted by Wagner (2015) on a sample of 8256 children and adolescents found that individuals who performed a minimum of 60 min activity per day were at lower risk of developing depressive symptoms; similarly, those with lower levels of screen use (i.e., video games, television, etc.) were also less prone to having feelings of depression.

Aside from health benefits, physical activity helps in cognitive development as well as development of certain skills such as discipline, persistence, and confidence (Knaus, Lechner, & Reimers, 2018). There is substantial evidence showing the association between chronic illness and physical inactivity (Kohl et al., 2012). Physical inactivity is considered as the key risk factor related to major noncommunicable diseases in adults, including breast and colon cancer (Lee et al., 2012). The long-term impacts of these diseases have an increasing burden on national economies around the globe (Kahan, 2015).

Rapid changes in lifestyle including eating habits and inadequate physical activity have resulted in an increase in overweight and obesity prevalence across the world. A study conducted in Jordan revealed that boys (24%) and girls (25%) aged less than 20 years were overweight or obese, with 8% being obese (Al-Domi et al., 2019). In order to reduce the growing global burden of noncommunicable diseases in developed countries, there is a strong movement toward increasing community participation in physical activity (Kohl et al., 2012). Daily physical activity has shown to have a reduction in the health care costs by increasing the usage of primary preventive health services and reduced usage of secondary and tertiary health care (Kang & Xiang, 2017). Promoting daily physical activity has become a public health priority.

Investing in regular physical activity during childhood can prove to be beneficial in prevention of chronic diseases (Black, Johnston, Propper, & Shields, 2019), reducing morbidity later on during adulthood as well as preventing premature mortality (Kang & Xiang, 2017). Instilling the mindset of regular physical activity during childhood has shown to predict positive physical activity behavior during adulthood (Black et al., 2019). Schools provide a key platform to encourage physical activity among the youth through compulsory physical education courses and sports (Morton, Atkin, Corder, Suhrcke, & Sluijs, 2015). The development of such mindset of daily physical activity and knowledge of its resulting health benefits is carried on into adulthood. A study conducted by Black et al. (2019), reported that despite the strategies to promote physical activity only 21% of boys and 16% of girls in England are in the process of reaching the WHO recommendations of 60 min of daily physical activity for health. Cooper et al. (2015) assessed the association between young adults' physical activity and demographic characteristic among Australia, Belgium, Brazil, Denmark, Estonia, Norway, Portugal, Switzerland, United Kingdom, and United States, and found that the United States have the least active young adults, followed by the United Kingdom, and Australia to be the most active. A report published by the WHO (2019) revealed that on a global scale, 23% adults and 81% adolescents do not meet the physical activity recommendations.

There are also certain groups in the society that face challenges in performing physical activity. Individuals with physical and/or sensory disabilities were found to be more physically inactive, due to certain activity limitations, as compared to individuals that did not have any physical and/ or sensory disability. Therefore, physical activity strategies must be designed with promoting inclusion of people with disabilities into physical exercise (Hollander & Proper, 2018).

Culture also moderates exercise levels. In Arab countries, other Islamic majority countries, India, and in certain parts of Africa, separation of genders often precludes female participation in exercise, and that gender also affects the rate of physical activity. Kahan (2015) found that in Arab countries, females showed higher rates of inactivity relative to males, whereas that gender gap in activity was less marked in non-Arab countries. Similarly, the WHO's (2019) Global Health Observatory reported women to be less physically active than males by 6% globally. The magnitude of the lower female activity among Arab countries was higher (11.6%) than in non-Arab countries (5.3%). Multiple factors contribute to the marked physical inactivity among Arab females including lack of social support, lack of funding for women's sport, and lack of suitable facilities for exercise (Benjamin & Donnelly, 2013).

Recently the United Nations' (UN) sustainable development goals (SDGs) have included the promotion of mental well-being, and the prevention of mental disorders and substance abuse, as well as promoting the availability of effective treatments for mental health problems (Rebar & Taylor, 2017). The UN has aimed to achieve a total of 17 SDG's by 2030, out of which the third goal is aimed at "Good health and well-being." It accounts for increasing economic and social inequalities, growing rates of urbanization and the continuing burden of infectious diseases (United Nations Development Program, 2017). Increasing physical activity is an effective intervention to enhance many aspects of mental health, which is supported by strong evidence suggesting its benefits on mental health for both clinical and nonclinical populations (Rebar et al., 2015). One of the most common mental health concerns, which can greatly impair an individual's QoL and daily functioning, is anxiety.

Anxiety involves symptoms such as nervousness, pessimism, and pervasive thoughts of worry which if severe and debilitating is associated with diagnosis of an anxiety disorder (Stubbs et al., 2017). Anxiety according to international data is associated with less physical activity and promotion of physical activity could be an important aspect of health care in order to improve the overall physical and mental well-being of an individual (Stubbs et al., 2017). The results of 400 randomized trials consisting of more than 14,000 subjects revealed that physical activity had a moderately reductive effect on depression. A strong body of evidence suggests that increasing physical activity in the general population can have a reductive effect on depression, anxiety as well as their prevention (Rebar et al., 2015). Furthermore, a study reported that changes in the physical activity of a person's spouse are directly related to changes in the individual's own physical activity. Therefore, promotion efforts for physical activity should consider potentially targeting couples (Cobb et al., 2015).

Sleep

Sleep is an essential aspect of one's daily lives and has an impact on our health and well-being. According to Mireku et al. (2019), the duration and quality of sleep is essential for children and adolescents to promote and maintain their continuous physical and mental development. A healthy night's sleep is important for the brain and the nervous system as it helps the individual cope with the various sensory, cognitive, and various components of complex environmental stimuli when they are awake. On the other hand, poor sleep quality is linked to various mental health problems including those defined by increased emotional reactivity. Cross-sectional studies have persistently revealed that poor emotional states such as hostility, high negative and low positive effect as well as psychological distress are linked to poor sleep quality whereas, constructs such as well-being, self-esteem, and optimism are related to good sleep quality. Moreover, longitudinal studies and experimental studies indicate that poor psychological and emotional states can negatively affect an individual's sleep quality (Simor, Krietsch, Köteles, & Mccrae, 2015).

A 9–11 h sleep duration is recommended for school-aged children (6- to 13-year olds) by the United States National Sleep Foundation (NSF) (Hirshkowitz et al., 2015). However, it was found that more than 90% of students in high school slept for less than the recommended minimum of 9 h (Basch, Basch, Ruggles, & Rajan, 2014). Insufficient sleep has been found to have harmful health consequences on adolescents as evidenced by several studies (Gariépy, Doré, Whitehead, & Elgar, 2018). Furthermore, Gariépy et al. (2018) state that there is enough evidence suggesting that chronotype (the timing of our daily activities controlled by the circadian rhythm) or the timing of sleep play an important role in adolescent health.

The increasing use of technology has serious harmful effects and one of them includes adverse sleep outcomes. Nighttime usage of any electronic device especially mobile phones and television is associated with negative sleep outcomes including poor sleep quality, insufficient sleep duration, abnormal catch-up sleep, late midpoint of sleep, and abnormal social jetlag among adolescents (Mireku et al., 2019). According to Kalmbach, Pillai, Roth, and Drake (2014) and João, Jesus, Carmo, and Pinto (2018), there is substantial evidence suggesting the relationship between sleep deprivation and depression, anxiety and stress.

Sleep is also affected by age; older people tend to sleep less as compared to younger individuals. For the elderly, the main aim is to improve their QoL which can be achieved by increasing the years of healthy aging through identifying modifiable factors that are linked to high-level well-being. This step is important in order to optimize healthy aging. A study conducted in Australia revealed that maladaptive lifestyle behaviors such as excessive consumption of alcohol, poor diet, smoking, lack of physical activity, and inconsistent sleep patterns were associated with mortality in middle aged and above adults (Chasens, Yang, Baniak, Choi, & Imes, 2018). A key human requirement for health and well-being is receiving a "good sleep" and is considered as a primary factor linked to healthy aging. One aspect of obtaining good sleep involves receiving sufficient sleep duration, which in older adults is about 7–9 h every night (Chasens et al., 2018). Sleep is an important aspect of healthy physical and cognitive development. Lack of sleep can make children at risk for various physical and mental illnesses including obesity, overweight, and subsequent chronic illnesses. With the increasing reliance on electronic devices by adolescents and children, the presence of the same in bedrooms poses a barrier for children to achieve sufficient sleep. Parental rules and efforts along with modifications to improve home and bedroom environment in order to reduce bedroom screen use should be established in order to ensure children achieve adequate sleep for their optimum health. Moreover, persistent parental support for children of all ages is essential to ensure the child's overall sleep hygiene. Public health efforts to advocate healthy sleep patterns in children should focus on the importance of a good night's sleep in children and the role of parents in achieving it (Pyper, Harrington, & Manson, 2017).

Alcohol and smoking

Adolescence is a period when individuals take up new habits which often are shaped by social factors. Based on a report published by U.S. Department of Health and Human Services (2015), the ages between 12 and 17 years is regarded as a critical risk period wherein adolescents are most likely to initiate the use of alcohol. There are multiple lines of evidence suggesting an association between initiation of substance during this period and high risk for the development of later substance abuse. For instance, data published by National Survey on Drug Use and Health in 2012 shows that adults who initiated the use of marijuana at 14 years or younger were 6 times more likely to later develop illicit substance dependence than those who first used marijuana at 18 years of age or above (U.S. Department of Health and Human Services, 2015).

Adolescents are strongly influenced by their peers, who can make them feel pressured to adopt a certain behavior in order to "fit in" with everyone else around them and can result in the adoption of substance use behaviors (Schuler, Tucker, Pedersen, & Damico, 2019). Substance use, including alcohol consumption and smoking, among adolescents has been found to be a major health concern worldwide and contributes significantly toward the global burden of disease (Sajjadi, Shushtari, Mahboubi, Rafiey, & Salimi, 2018). A study conducted in Iran among high school students revealed a prevalence of drug use among 26.5% of boys and 11.5% of girls (Sajjadi et al., 2018). Similar findings were revealed in another study of university students, which showed 67% of male students and 42% of female students had tried drugs (Chinwong, Mookmanee, Chongpornchai, & Chinwong, 2018).

Also, when children are frequently exposed to alcohol drinking adults, they can get the impression that heavy consumption of alcohol is a common human behavior and is socially normative, which shapes their acceptance of alcohol and other harmful behaviors such as smoking (Kuntsche & Kuntsche, 2019). Due to the rapid physical and mental changes in adolescents, they are one of the most sensitive groups that can be easily influenced into adopting harmful behaviors (such as alcohol use and smoking) and therefore require more attention from school, society, and especially family members. Several studies have reported that family history of substance abuse and smoking is linked with adolescent's substance use (Sajjadi et al., 2018). The positive or negative health behaviors that adults have are usually initiated during adolescence therefore, it is important to target these vulnerable youths while developing health promotion programs. However, studies have shown that parental monitoring of where adolescents are, and who they are with, serves as a preventive barrier against adolescent substance use. Kristjansson, James, Allegrante, Sigfusdottir, and Helgason (2010) reported that parental supervision plays a crucial role in reducing the likelihood of substance abuse in children along with increasing their likelihood of selecting peers who abstain from substance.

Assessment of family physical health

Community assessment of the prevalence of substance use is essential to inform rates of use, and to assist identify individuals who might be offered access to services. A study conducted by Sacco, Cagle, Moreland, and Camlin (2017) in the United states found that almost 14% of people suffered from alcohol use disorder in the past year, and 29% of the population had a lifelong history of alcohol use disorder. Reliable and concise screening tools for substance use exist and are extensively used in various areas of healthcare. An approach recently developed combines screening with brief intervention models and is called Screening, Brief Intervention, and Referral for treatment (SBIRT). This model is being used widely in healthcare including emergency departments, primary care, and many other setting (Sacco et al., 2017). Even though smoking rates are declining worldwide in several countries, its prevalence is on the rise in the Eastern Mediterranean Region and African Region countries (World Health Organization, 2016). This relentless increase in untreated or undertreated substance use disorders in certain parts of Asia and Africa depicts the intensity of the challenge the community is facing. Therefore, there is a critical need to find evidence-based, practical approaches to effectively screen individuals, who are at a high risk for unhealthy tobacco, drug and alcohol use, increase detection of substance use disorders at an earlier stage as well as provide the appropriate care required. A brief questionnaire, called the family CAGE, is currently being used for the diagnosis of substance use disorders including alcohol (Mannelli & Wu, 2016).

Several successful public health strategies help to reduce the consumption of cigarettes. The first being price increase on tobacco products, statistics have shown, that by increasing the price of tobacco by 10%, has resulted in a decrease in smoking by 2.5%–5% (World Health Organization, 2003). Moreover, this price increase was assessed to reduce around six hundred thousand to 1.8 million premature deaths in various regions such as central Asia and Eastern Europe. The government also plays a crucial role in increasing taxes on cigarettes. Some countries took the initiative to increase tax percentages to be around 65%–80% of the price of a pack of cigarettes. This has resulted in the effectiveness to control tobacco consumption. Countries such as the United Kingdom, Canada, and South Africa have increased the tax on cigarettes, contributing to reduction in cigarette consumption. Other efficacious strategies include mass media campaigns specifically targeting the youth and implementing legal restrictions on consumption of tobacco in public places and work settings. Results from a systematic review on implementation of smoking restrictions in public settings, proved to be effective in reducing it by almost 10% (World Health Organization, 2003). In regards to the multimedia campaigns, examples of such initiatives taken in Norway and United States have shown to improve individual's attitudes toward tobacco use, a reduced prevalence and consumption among youth (World Health Organization, 2003).

An assessment conducted by Dave, Nimbalkar, Vasa, and Phatak (2017) reported that despite having sufficient playground equipment, most schools lacked a supporting environment that promoted physical activity. Therefore, similar assessments need to be conducted among various regions in order to improve physical activity in schools. Also, apart from assessing physical activity in children, other strategies such as including exercise in the school as part of the curriculum, developing public health policies to promote walking, cycling, and recreational activities can cumulatively aid in meeting the UN's 2030 Sustainable Development Goal 3. In order to reverse the current global trends of reduced physical activity, a system-based national action aimed at enhancing the factors that facilitate physical activity (i.e., economic, environmental, cultural, and social factors) is required (World Health Organization, 2019). For example, the measure taken by Japan Sports Agency in 2015 to promote physical activity in companies and academic organizations comprised of organized sport participation and active transportation, proved to be a favorable step toward increasing the level of physical activity among the community (Tanaka et al., 2019).

Factors associated with health-related families QoL

Following unhealthy lifestyle habits (i.e., consumption of unhealthy foods, alcohol, drugs, irregular sleep, and minimal physical activity) puts individuals at a high risk of developing chronic diseases, and negatively impacting their QoL. In cases wherein the patient has a history of substance use disorder or at-risk substance use, pain and symptom management become more difficult. Alcohol use disorder is a condition wherein problems arise from uncontrolled alcohol use reaching a level of severity which leads to impairment of overall body functioning (Sacco et al., 2017). A study conducted among undergraduate students in Serbia by Gazibara, Pekmezovic, Popovic, Paunic, and Kisic-Tepavcevic (2018) reported the patterns of chronic diseases in the university students and confirmed that the presence of chronic diseases is a constant factor linked to worse health-related quality of life (HRQoL). HRQoL refers to the ability of an individual to function in their life and their perceived well-being in social, mental, and physical aspects of health (Karimi & Brazier, 2016). Along with the individuals themselves, their families too are tremendously affected. Be it living with a chronically ill individual, an alcoholic, a smoker, or a drug addict, all of it leads to the families living an extremely stressful life, which in turn affects their QoL as well. Substance use can disrupt the basic social and family ties. Therefore, it is important to develop restorative strategies that will help preserve the individual's social and family networks and help limit the harmful effects of drug and alcohol use (Mannelli & Wu, 2016).

Improving family physical health

Noting the distinct components of physical health, each of them individually needs to be targeted in order to improve overall family physical health. It is essential to create awareness in the community on the importance of physical activity, improvement of physical education courses in school, development of infrastructure, and the creation of national policies as well as the political commitment to improve physical activity in the youth, as it will have a positive impact on the health condition of future generations (Dave et al., 2017). Several studies have reported that in individuals suffering from alcohol abuse, QoL can be used as an indicator for treatment efficacy, as various therapeutic interventions have led to an improvement in the QoL of such individuals (Prakash & Singh, 2018). Regular follow up with substance use patients, after receiving group therapy, ensured that the patient achieved complete abstinence from substance use and therefore, improved their QoL (Prakash & Singh, 2018).

Involvement of families can be effective in health promotion. For example, there are numerous programs targeted on parents that are effective in enhancing child health, such as parents engaging in and supporting child breast feeding, immunization, washing, injury prevention, exercise, and good nutrition (see Chapter 18 for a review). Similarly, involvement of spouses in couple-based interventions can assist the whole family when one partner is suffering from major illness, such as cancer, heart disease, and diabetes (Baucom, Porter, Kirby, & Hudepohl, 2012; Sher et al., 2014).

The family plays a central role in developing and maintaining its members behaviors related to health promotion. The mother and/or wife considered as the major health care provider in most families. The family is the source of physical and psychological disorders. The provision of basic needs (shelter, food, health, and illness care) supports the promotion of health. Health promotion of the family should focus on the outcomes of unity, flexibility, communication, and quality of health. These outcomes can be achieved through clear identification of expectations and related consequences, effective communication and flexible allocation of roles, effective problem solving, adaptability, resilience, effective interactions with health system, and recognition of developmental needs of family members (Loveland-Cherry, 2011).

Health professionals should work with families to promote healthy outcomes. The family is responsible for providing basic resources that support the promotion of health of its members. Building the recourses in families and promoting changes in families that use these resources is important in the development and maintaining the healthy lifestyles of family members.

References

Al-Domi, H. A., Faqih, A., Jaradat, Z., Al-Dalaeen, A., Jaradat, S., & Amarneh, B. (2019). Physical activity, sedentary behaviors and dietary patterns as risk factors of obesity among Jordanian schoolchildren. *Diabetes & Metabolic Syndrome: Clinical Research & Reviews*, *13*(1), 189–194. https://doi.org/10.1016/j.dsx.2018.08.033.

Ask, H., Rognmo, K., Torvik, F. A., Røysamb, E., & Tambs, K. (2012). Non-random mating and convergence over time for alcohol consumption, smoking, and exercise: The Nord-Trøndelag health study. *Behavior Genetics*, *42*(3), 354–365. https://doi.org/10.1007/s10519-011-9509-7.

Basch, C. E., Basch, C. H., Ruggles, K. V., & Rajan, S. (2014). Prevalence of sleep duration on an average school night among 4 nationally representative successive samples of American high school students, 2007–2013. *Preventing Chronic Disease*, *11*, e216. https://doi.org/10.5888/pcd11.140383.

Baucom, D. H., Porter, L. S., Kirby, J. S., & Hudepohl, J. (2012). Couple-based interventions for medical problems. *Behavior Therapy*, *43*, 61–76. https://doi.org/10.1016/j.beth.2011.01.008.

Benjamin, K., & Donnelly, T. T. (2013). Barriers and facilitators influencing the physical activity of Arabic adults: A literature review. *Avicenna*, *2013*(1), 1–16. https://doi.org/10.5339/avi.2013.8.

Berge, J. M., Wall, M., Hsueh, T.-F., Fulkerson, J. A., Larson, N., & Neumark-Sztainer, D. (2015). The protective role of family meals for youth obesity: 10-year longitudinal association. *The Journal of Pediatrics*, *166*(2), 296–301. https://doi.org/10.1016/j.jpeds.2014.08.030.

Black, N., Johnston, D. W., Propper, C., & Shields, M. A. (2019). The effect of school sports facilities on physical activity, health and socioeconomic status in adulthood. *Social Science & Medicine*, *220*, 120–128. https://doi.org/10.1016/j.socscimed.2018.10.025.

Brown, H. E., Schiff, A., & Sluijs, E.M.F.V. (2015). Engaging families in physical activity research: A family-based focus group study. *BMC Public Health*, *15*(1), https://doi.org/10.1186/s12889-015-2497-4.

Chasens, E. R., Yang, K., Baniak, L. M., Choi, J., & Imes, C. C. (2018). Sleep and other correlates of high-level health in older adults. *Geriatric Nursing*, *39*(3), 344–349. https://doi.org/10.1016/j.gerinurse.2017.11.002.

Chinwong, D., Mookmanee, N., Chongpornchai, J., & Chinwong, S. (2018). A comparison of gender differences in smoking behaviors, intention to quit, and nicotine dependence among Thai university students. *Journal of Addiction*, *2018*, 1–8. https://doi.org/10.1155/2018/8081670.

Cobb, L. K., Godino, J. G., Selvin, E., Kucharska-Newton, A., Coresh, J., & Koton, S. (2015). Spousal influence on physical activity in middle-aged and older adults. *American Journal of Epidemiology*, *183*(5), 444–451. https://doi.org/10.1093/aje/kwv104.

Cooper, A. R., Goodman, A., Page, A. S., Sherar, L. B., Esliger, D. W., Sluijs, E. M. V., … Ekelund, U. (2015). Objectively measured physical activity and sedentary time in youth: The International Children's Accelerometry Database (ICAD). *International Journal of Behavioral Nutrition and Physical Activity*, *12*(1), 113. https://doi.org/10.1186/s12966-015-0274-5.

Dave, H., Nimbalkar, S. M., Vasa, R., & Phatak, A. G. (2017). Assessment of physical activity among adolescents: A cross-sectional study. *Journal of Clinical and Diagnostic Research*, *11*(1), 21–24. https://doi.org/10.7860/jcdr/2017/27080.10870.

Di Castelnuovo, A., Quacquaruccio, G., Donati, M. B., de Gaetano, G., & Iacoviello, L. (2009). Spousal concordance for major coronary risk factors: A systematic review and meta-analysis. *American Journal of Epidemiology*, *169*(1), 1–8. https://doi.org/10.1093/aje/kwn234.

Ennis, E., & Bunting, B. P. (2013). Family burden, family health and personal mental health. *BMC Public Health*, *13*(1), e255. https://doi.org/10.1186/1471-2458-13-255.

Gariépy, G., Doré, I., Whitehead, R. D., & Elgar, F. J. (2018). More than just sleeping in: A late timing of sleep is associated with health problems and unhealthy behaviors in adolescents. *Sleep Medicine*, *56*, 66–72. https://doi.org/10.1016/j.sleep.2018.10.029.

Gazibara, T., Pekmezovic, T., Popovic, A., Paunic, M., & Kisic-Tepavcevic, D. (2018). Chronic diseases among university students: Prevalence, patterns and impact on health-related quality of life. *Vojnosanitetski Pregled Military Medical and Pharmaceutical Journal of Serbia*, *75*(12), 1178–1184. https://doi.org/10.2298/vsp160920050g.

Giray, C., & Ferguson, G. M. (2018). Say yes to "Sunday Dinner" and no to "Nyam and Scram": Family mealtimes, nutrition, and emotional health among adolescents and mothers in Jamaica. *Appetite*, *128*, 129–137. https://doi.org/10.1016/j.appet.2018.05.132.

Golics, C. J., Basra, M. K. A., Finlay, A.Y., & Salek, S. (2013). The impact of disease on family members: A critical aspect of medical care. *Journal of the Royal Society of Medicine*, *106*(10), 399–407. https://doi.org/10.1177/0141076812472616.

Golics, C. J., Basra, M. K., Salek, M. S., & Finlay, A.Y. (2013). The impact of patients' chronic disease on family quality of life: An experience from 26 specialties. *International Journal of General Medicine*, *6*, 787–798. https://doi.org/10.2147/ijgm.s45156.

Hasler, B. P., & Troxel, W. M. (2010). Couples' night time sleep efficiency and concordance: Evidence for bidirectional associations with daytime relationship functioning. *Psychosomatic Medicine*, *72*(8), 794–801. https://doi.org/10.1097/PSY.0b013e3181ecd08a.

Hirshkowitz, M., Whiton, K., Albert, S. M., Alessi, C., Bruni, O., Doncarlos, L., … Hillard, P. J. (2015). National Sleep Foundation's sleep time duration recommendations: Methodology and results summary. *Sleep Health*, *1*(1), 40–43. https://doi.org/10.1016/j.sleh.2014.12.010.

Hollander, E. L., & Proper, K. I. (2018). Physical activity levels of adults with various physical disabilities. *Preventive Medicine Reports*, *10*, 370–376. https://doi.org/10.1016/j.pmedr.2018.04.017.

Jackson, S. E., Steptoe, A., & Wardle, J. (2015). The influence of partner's behavior on health behavior change: The English Longitudinal Study of Aging. *JAMA Internal Medicine*, *175*(3), 385–392. https://doi.org/10.1001/jamainternmed.2014.7554.

João, K. A., Jesus, S. N., Carmo, C., & Pinto, P. (2018). Sleep quality components and mental health: Study with a non-clinical population. *Psychiatry Research*, *269*, 244–250. https://doi.org/10.1016/j.psychres.2018.08.020.

Kahan, D. (2015). Adult physical inactivity prevalence in the Muslim world: Analysis of 38 countries. *Preventive Medicine Reports*, *2*, 71–75. https://doi.org/10.1016/j.pmedr.2014.12.007.

Kalmbach, D. A., Pillai, V., Roth, T., & Drake, C. L. (2014). The interplay between daily affect and sleep: A 2-week study of young women. *Journal of Sleep Research*, *23*(6), 636–645. https://doi.org/10.1111/jsr.12190.

Kang, S., & Xiang, X. (2017). Physical activity and health services utilization and costs among U.S. adults. *Preventive Medicine*, *96*, 101–105. https://doi.org/10.1016/j.ypmed.2016.12.043.

Karimi, M., & Brazier, J. (2016). Health, health-related quality of life, and quality of life: What is the difference? *Pharmaco Economics*, *34*(7), 645–649. https://doi.org/10.1007/s40273-016-0389-9.

Knaus, M. C., Lechner, M., & Reimers, A. K. (2018). *For better or worse?—The effects of physical education on child development*. IZA Institute of Labor Economics. Discussion paper series no. 11268 https://papers.ssrn.com/sol3/papers.cfm?abstract_id=3111142.

Kohl, H. W., Craig, C. L., Lambert, E. V., Inoue, S., Alkandrari, J. R., Leetongin, G., … Kahlmeier, S. (2012). The pandemic of physical inactivity: Global action for public health. *Lancet*, *380*(9838), 294–305. https://doi.org/10.1016/S0140-6736(12)60898-8.

Kristjansson, A. L., James, J. E., Allegrante, J. P., Sigfusdottir, I. D., & Helgason, A. R. (2010). Adolescent substance use, parental monitoring, and leisure-time activities: 12-year outcomes of primary prevention in Iceland. *Preventive Medicine*, *51*(2), 168–171. https://doi.org/10.1016/j.ypmed.2010.05.001.

Kuntsche, E., & Kuntsche, S. (2019). Parental drinking and characteristics of family life as predictors of preschoolers' alcohol-related knowledge and norms. *Addictive Behaviors*, *88*, 92–98. https://doi.org/10.1016/j.addbeh.2018.08.024.

Lee, I., Shiroma, E. J., Lobelo, F., Puska, P., Blair, S. N., & Katzmarzyk, P. T. (2012). Effect of physical inactivity on major non-communicable diseases worldwide: An analysis of burden of disease and life expectancy. *The Lancet*, *380*(9838), 219–229. https://doi.org/10.1016/s0140-6736(12)61031-9.

Loveland-Cherry, C. (2011). Role of families in health promotion. In M. Craft-Rosenberg & S. R. Pehler (Eds.), *Encyclopedia of family health* (pp. 912–913). London: Sage. https://doi.org/10.4135/9781412994071.n297.

Mannelli, P., & Wu, L. (2016). Family substance use screening: Less to hide, more to gain. *Indian Journal of Medical Research*, *143*(6), 682–684. https://doi.org/10.4103/0971-5916.191920.

Martire, L. M., Schulz, R., Helgeson, V. S., Small, B. J., & Saghafi, E. M. (2010). Review and meta-analysis of couple-oriented interventions for chronic illness. *Annals of Behavioral Medicine*, *40*(3), 325–342. https://doi.org/10.1007/s12160-010-9216-2.

Meyler, D., Stimpson, J. P., & Peek, M. K. (2007). Health concordance within couples: A systematic review. *Social Science and Medicine*, *64*, 2297–2310. https://doi.org/10.1016/j.socscimed.2007.02.007.

Mireku, M. O., Barker, M. M., Mutz, J., Dumontheil, I., Thomas, M. S., Röösli, M., … Toledano, M. B. (2019). Night-time screen-based media device use and adolescents sleep and health-related quality of life. *Environment International*, *124*, 66–78. https://doi.org/10.1016/j.envint.2018.11.069.

Morton, K. L., Atkin, A. J., Corder, K., Suhrcke, M., & Sluijs, E. M. (2015). The school environment and adolescent physical activity and sedentary behaviour: A mixed-studies systematic review. *Obesity Reviews*, *17*(2), 142–158. https://doi.org/10.1111/obr.12352.

Persson, C., & Benzein, E. (2014). Family health conversations: How do they support health? *Nursing Research and Practice*, *2014*, 1–11. https://doi.org/10.1155/2014/547160.

Prakash, P., & Singh, A. R. (2018). The role of group therapy to improvement quality of life in individuals with alcohol dependence. *Indian Journal of Health and Well-Being*, *9*(2), 205–210.

Pyper, E., Harrington, D., & Manson, H. (2017). Do parents' support behaviours predict whether or not their children get sufficient sleep? A cross-sectional study. *BMC Public Health*, *17*(1), 432. https://doi.org/10.1186/s12889-017-4334-4.

Rebar, A. L., Stanton, R., Geard, D., Short, C., Duncan, M. J., & Vandelanotte, C. (2015). A meta-meta-analysis of the effect of physical activity on depression and anxiety in non-clinical adult populations. *Health Psychology Review*, *9*(3), 366–378. https://doi.org/10.1080/17437199.2015.1022901.

Rebar, A. L., & Taylor, A. (2017). Physical activity and mental health; it is more than just a prescription. *Mental Health and Physical Activity*, *13*, 77–82. https://doi.org/10.1016/j.mhpa.2017.10.004.

Robles, T. F. (2014). Marital quality and health: Implications for marriage in the 21st century. *Current Directions in Psychological Science*, *23*(6), 427–432. https://doi.org/10.1177/0963721414549043.

Robles, T. F., & Kiecolt-Glaser, J. K. (2003). The physiology of marriage: Pathways to health. *Physiology and Health*, *79*, 409–416. https://doi.org/10.1106/S0031-9383(03)00160-4.

Robles, T., Slatcher, R., Trombello, J., & McGinn, M. (2014). Marital quality and health: A meta-analytic review. *Psychological Bulletin, 140*(1), 140–187. https://doi.org/10.1037/a0031859.

Sacco, P., Cagle, J. G., Moreland, M. L., & Camlin, E. A. (2017). Screening and assessment of substance use in hospice care: Examining content from a national sample of psychosocial assessments. *Journal of Palliative Medicine, 20*(8), 850–856. https://doi.org/10.1089/jpm.2016.0538.

Sajjadi, H., Shushtari, Z. J., Mahboubi, S., Rafiey, H., & Salimi, Y. (2018). Effect of socioeconomic status, family smoking and mental health through social network on the substance use potential in adolescents: A mediation analysis. *Public Health, 157*, 14–19. https://doi.org/10.1016/j.puhe.2018.01.004.

Schuler, M. S., Tucker, J. S., Pedersen, E. R., & Damico, E. J. (2019). Relative influence of perceived peer and family substance use on adolescent alcohol, cigarette, and marijuana use across middle and high school. *Addictive Behaviors, 88*, 99–105. https://doi.org/10.1016/j.addbeh.2018.08.025.

Seidell, J. C., & Halberstadt, J. (2015). The global burden of obesity and the challenges of prevention. *Annals of Nutrition and Metabolism, 66*(2), 7–12. https://doi.org/10.1159/000375143.

Sher, T., Braun, L., Domas, A., Bellg, A., Baucom, D. H., & Huole, T. T. (2014). The partners for life program: A couples based approach to cardiac risk reduction. *Family Process, 53*(10), 131–149. https://doi.org/10.1111/famp.12061.

Simor, P., Krietsch, K. N., Köteles, F., & Mccrae, C. S. (2015). Day-to-day variation of subjective sleep quality and emotional states among healthy university students—A 1-week prospective study. *International Journal of Behavioral Medicine, 22*(5), 625–634. https://doi.org/10.1007/s12529-015-9464-4.

Sobal, J., & Hanson, K. (2011). Family meals and body weight in US adults. *Public Health Nutrition, 14*(9), 1555–1562. https://doi.org/10.1017/S1368980011000127.

Stubbs, B., Koyanagi, A., Hallgren, M., Firth, J., Richards, J., Schuch, F., … Vancampfort, D. (2017). Physical activity and anxiety: A perspective from the World Health Survey. *Journal of Affective Disorders, 208*, 545–552. https://doi.org/10.1016/j.jad.2016.10.028.

Tanaka, C., Tanaka, S., Inoue, S., Miyachi, M., Suzuki, K., Abe, T., & Reilly, J. J. (2019). Results from the Japan 2018 report card on physical activity for children and youth. *Journal of Exercise Science & Fitness, 17*(1), 20–25. https://doi.org/10.1016/j.jesf.2018.10.001.

United Nations Development Program. (2017). *UNDP support to the implementation of sustainable development goal 3: Ensure healthy lives and well-being for all at all ages*. https://www.undp.org/content/undp/en/home/sustainable-development-goals/goal-3-good-health-and-well-being/targets.html.

United States Department of Health and Human Services. (2015). *National survey on drug use and health*. (p.2012). https://www.icpsr.umich.edu/icpsrweb/NAHDAP/studies/34933.

Utter, J., Larson, N., Berge, J. M., Eisenberg, M. E., Fulkerson, J. A., & Neumark-Sztainer, D. (2018). Family meals among parents: Associations with nutritional, social and emotional wellbeing. *Preventive Medicine, 113*, 7–12. https://doi.org/10.1016/j.ypmed.2018.05.006.

Valle, G., Weeks, J., Taylor, M., & Eberstein, I. (2013). Mental and physical health consequences of spousal health shocks among older adults. *Journal of Aging and Health, 25*(7), 1121–1142. https://doi.org/10.1177/0898264313494800.

Wagner, D. (2015). Mental health benefits of exercise in children. *Child Adolescent Psychiatry, 32*(1), 37–50. https://www.psychiatrictimes.com/child-adolescent-psychiatry/mental-health-benefits-exercise-children.

WHO. (1948). *Constitution of the World Health Organization*. The Constitution was adopted by the International Health Conference held in New York from 19 June to 22 July 1946, signed on 22 July 1946 by the representatives of 61 States (Official Records of WHO, no. 2, p. 100), and entered into force on 7 April 1948.

World Health Organization. (2003). *Which are the most effective and cost-effective interventions for tobacco control?*. http://www.euro.who.int/__data/assets/pdf_file/0004/74722/E82993.pdf.
World Health Organization. (2015). *Physical activity and adults*. https://www.who.int/dietphysicalactivity/factsheet_adults/en/.
World Health Organization. (2016). *Global health observatory: Prevalence of tobacco smoking* https://www.who.int/gho/tobacco/use/en.
World Health Organization. (2019). *Global action plan on physical activity 2018–2030: More active people for a healthier world*. Geneva:World Health Organization Press. https://www.who.int/ncds/prevention/physical-activity/global-action-plan-2018-2030/en/.

CHAPTER 12

The interface between family well-being and government policy in Australia

Janet Stanley
Sustainable Society Institute, School of Design, University of Melbourne, Melbourne, VIC, Australia

The problem of growing inequality and poverty

Over the past four decades, there has been a growing trend toward increasing inequality between families in Australia (Barrett & Whelan, 2018), which parallels similar trends in most high-income countries (Organization of Economic Cooperation and Development, 2018). This is particularly manifesting itself in inequality of wealth (as distinct from income) (Davidson, Saunders, Bradbury, & Wong, 2018; Swan et al., 2016). There are also pockets of absolute poverty in Australia, which reflect having insufficient income to meet basic daily needs like paying for food and accommodation. It is estimated that around 6% of the total Australian population are in absolute poverty, among which long-term unemployed, homeless people; those supporting people in the household with a severe disability; and some Indigenous Australian families are greatly overrepresented (Stanley, Stanley, & Hansen, 2017). However, it is common to view poverty in terms of relative poverty, or income disparity, defined by the Organization of Economic Cooperation and Development (OECD) as the percentage of household incomes below half the median household income, sometimes described as the *poverty line* (OECD, 2019). This chapter is focused on the relative poverty or inequality in Australia. It presents the large city of Melbourne as a case example that explores some explanations as to why, despite Australia being a very wealthy country, a culture of poverty remains associated with some families.

Despite a growing Australian economy over the past 27 years, a record of continuous economic growth unparalleled in Australia's history, most of the gains have occurred in the upper income brackets. Over the past 20 years, real incomes grew for those in the top 20% more than five times as much

Cross-Cultural Family Research and Practice
https://doi.org/10.1016/B978-0-12-815493-9.00012-0

as the incomes of the bottom 20% (Barrett & Whelan, 2018). This results in an estimated 2.9 million people or 13.3% of the Australian population living below the poverty line, measured (as is usual in Australia) as 50% of the median household income, less housing costs (Davidson et al., 2018).

In the rich countries of the OECD, inequality has been increasing across almost all countries, including Australia (OECD, 2018). Australia ranks as having the 14th highest rate of relative poverty of 32 countries in the OECD (2019). These relative poverty rates are more than double those of the lowest ranked rates of relative poverty (Denmark and Finland each at about 6%), and somewhat lower than the highest-ranked countries (Russia and the United States at 17%). In other words, Australia falls above the middle of the ranking of OECD countries for rates of relative poverty. The United Kingdom is quite similar in relative poverty rates to Australia, ranking 16th, just below Australia, with a rate of relative poverty of about 11%.

In Australia, disturbingly, those in relative poverty include 739,000 children under the age of 15 (17.3% of all children—more than 1 in 6), and 410,000 youth between the ages of 15 and 24 (13.9%) (Davidson et al., 2018). Other people at risk of experiencing disadvantage include those on a low income, those with poor health or disability, some older people, and people from culturally and linguistically diverse backgrounds. Most people living below the poverty line (53%) rely on social security payments as their main source of income (Davidson et al., 2018). Due to the low levels of income support, people need to supplement their social security payments with private income or move out of the social security system altogether, but clearly over half are not able to raise their income sufficiently to come off the welfare. Of particular note is the freezing of Newstart Allowance (welfare benefits paid to unemployed persons) since 1994, resulting in a progressive deepening of poverty for people in households relying mainly on that payment. Thus, there is a deep divide between those with adequate income and those getting left behind. These families are commonly located in long-standing areas of disadvantage in major cities and regional areas, where there are less opportunities and services. This risks the creation of entrenched disadvantage, some families likely to experience an adverse impact on longer-term functioning and intergenerational disadvantage (Davidson et al., 2018; Skattebol & Redmond, 2019).

This chapter includes a critical analysis of a common explanation for disadvantage in families that poverty reflects personal inadequacy of the poor person. The chapter looks beyond this traditional explanation, and argues that the lack of adequate infrastructure and service provision, and poor welfare provisions for some groups of people, is partly creating and maintaining the culture of poverty. An "umbrella" approach is taken, crossing disciplines, offering evidence from the literature and from research on this issue in Melbourne. It argues that government policies and expenditure, and also the lack of them, may be facilitating a culture of poverty for some families that risks becoming entrenched such that is very difficult to change and achieve social mobility, even though a country may be regarded as wealthy. While the chapter's focus is on Melbourne, Australia, the causes and effects on inequality in this example city illustrate issues common in many cities in high-wealth countries.

The influences on family functioning

The conditions that influence good outcomes for families can be viewed as falling in four major areas (Stanley et al., 2017). The first of these are societal trends, often on an international scale, such as climate change, population growth, the world economy, international conflict, technological change, dominant ideological stances, and environmental loss. Probably the most critical, climate change, will have an increasingly adverse impact on families, especially those experiencing disadvantage. As outlined above, such families have fewer resources to provide protection against climate change impacts and extreme events, while at the same time they are likely to be more exposed due to factors such as living in poorer quality housing that may also be built on land at risk of flooding but is more affordable (Brotherhood of St. Laurence, 2007; Stanley, 2014). The adverse impact of climate change and associated extreme events, especially on families experiencing disadvantage, is likely to increase overtime in the light of Australian and international policy inaction and failures (Stanley, 2014).

The second area within which to understand impacts on family functioning encompasses government policy, and how government views or understands social policy and responsibility for the well-being of the population. This includes resources and services such as employment, income support, housing, transport, health, education, emergency services, land use, and community services. It also encompasses the opportunity cost of spending on other sectors of the economy, such as on fuel rebates and other subsidies for big business, on defense spending, and also on tax rebates for middle-class earners, such as Australian government subsidies for private health insurance, costing $AUD 6 billion annually (Duckett, 2019). Equity and social justice are an important concern of social policy: who gets what under what circumstance (McLelland, 2014). In contemporary society in Australia, economic policy is given overwhelming priority, while social policy is given much less consideration (Alexander & Floyd, 2018).

The third area impacting families comprises community dynamics, encompassing issues such as attachment to community, social capital, sense of place, and belonging and social cohesion. The fourth area refers to features of families and individuals, including age, gender, personality and attitudes, and physical and mental health, as well as cultural beliefs and practices. Entrenched disadvantage can lead to a position that could be described as characterized by low self-esteem, expectation of failure, and feelings of helplessness. In some situations, this is reenforced by negative judgments and blame from others in their community and wider, including the media and politicians. It may be characterized by language such a "dole bludgers," which is Australian slang for a belief that people have no intention of working but prefer to live on government handouts.

Poor outcomes in one or more of the major areas, outlined above, can also adversely impact on broader society, leading to environmental, social, and economic costs in areas such as health, productivity, antisocial behavior, and disrupted social cohesion. Equity is a value judgment and while everyone will never be equal in all aspects, Sen's work argues for fairness such that people should be provided with the minimum capability or opportunities, supported by a society that provides conditions to make this possible (Sen, 1992). This should apply to current living generations and future generations, a position particularly poignant in the light of climate change and the current levels of environmental destruction.

The specific drivers of poverty

This section outlines particular aspects of the areas described above that interact, leading to a risk of entrenched poverty for some families, even in a wealthy country. Research undertaken in Victoria (a state of Australia) examined and measured the particular attributes or conditions that drive social exclusion and well-being (see, e.g., Stanley, Stanley, & Hensher, 2012). Social exclusion is defined as the inability of a person or household to participate fully in society due largely to government policy failures. The measurable definition of social exclusion was adapted from the work of Burchardt, LeGrand, and Piachaud (2002), while well-being was measured using two prominent instruments in the psychological literature, subjective well-being and psychological well-being (Ryan & Deci, 2001). A model based on those associations that were statistically significant ($p < .01$) identified the major drivers of risk of social exclusion as: low income (embedding education and work); poor accessibility (transport or mobility); low social capital, especially bridging social capital that links people with broader community members, and connections with the community (personal relationships and connections); and low self-esteem, confidence, and poor control over your personal environment (such as capabilities to make choices, problem solve). In turn, social inclusion was highly associated with well-being, with model fit to the data suggesting that the direction of impact was such that improving social inclusion improved personal well-being.

The drivers of impacts do not sit neatly in one of the four areas that impact on families and individuals. There is interconnection between these four major areas of impact, with a heavy reliance on the provision of capabilities by government policy (see, e.g., Brain, Stanley, & Stanley, 2019; Vella-Brodrick & Stanley, 2013). Each area has a cluster of specific attributes that may not carry equal weight on impact and may also have a co-related impact across the areas. What is really being shown here is a system of intersecting impacts.

By way of example, opportunities for education and work are related to the ability to generate an adequate income. While there may be local educational services provided, this can be difficult to utilize where the location is inaccessible and there is no public transport, where the family has insecure housing necessitating frequent moves, and where the costs associated with schooling are unable to be met. Interpersonal relationships and connections with the broader community are very important, especially those associations that offer the greatest access to societal opportunities—bridging social capital (Stanley, Stanley, Vella-Brodrick, & Currie, 2010). However, this again needs the ability to be mobile to reach these opportunities. Actualization, or achieving confidence, self-esteem, the ability to be creative, a belief in personal control and thinking clearly, are more likely to be achieved when a person has their basic needs met and adequate levels of inclusion and well-being are achieved (Maslow, 1954; Vella-Brodrick & Stanley, 2013).

It is argued in this chapter that it is the combined impact of the failure to meet basic needs, as outlined by Maslow back in 1954, exacerbated through government policy failures that promotes a culture of poverty and disadvantage in many families in wealthy countries, with the risk of entrenched disadvantage that is passed from one family to neighbors and future families. Sociological research has shown the negative effect on community members of degraded environments that have features such as vandalized and neglected buildings, and heavy polluting freeways intersecting living locations and schools (see, e.g., Shields & Wooden, 2003). These basic needs are those reflected in the research findings reported above, as well as work by Inglehart (2000): physiological needs (food, water, shelter); safety; personal interactional needs (love and belonging); personal esteem and confidence; that leads to self-actualization; and a term frequently used at present, resilience, or the ability to cope in adversity.

The culture of adversity

There is not a lot of interest in the culture of adversity or disadvantage, nor the depth of impact that inequality may be having on individuals, families and indeed, society in general. Pioneers on societal impacts across countries, Wilkinson and Pickett (2009), undertook a study of 23 of the world's richest nations and individual states in the United States. They found that the societies that do best for their citizens are those with the narrowest income differentials, such as Japan, as distinct from those countries that are most unequal, such as the United States. This inequality is represented in 11 different health and social problems that were examined by Wilkinson and Pickett, including child well-being, social mobility, trust, and community life. The extent of this inequality varies between social issues and locations, but proves to be statistically important in many situations (Wilkinson & Pickett, 2009).

The psychological impact of social exclusion has been empirically investigated (Abrams, Hogg, & Marques, 2005; Haslam, Cruwys, Haslam, & Jetten, 2015). It was found that the impact of social exclusion is almost always negative, as the need to belong is fundamental to human life, as reported by Maslow (1954). Impacts of social exclusion can include lowered self-esteem, anger, frustration, emotional denial, and cognitive impairment, resulting in behaviors that may include attempts to find different sources of validation and defensive reactions. In the author's experience with child protection, the diminution of trust and the believe of a lack of control over their environment, leading to a form of coping that aligns with helplessness, probably related to low self-esteem that can occur in excluded families, is both damaging and very difficult to change. Much of this pattern is established in highly disadvantaged families with very young children. Self-defeating behaviors, procrastination, and poor performance on analytical reasoning tasks may occur (Twenge & Baumeister, 2004). Thus, there would seem to be some evidence for the presence of a culture of disadvantage and a theoretical basis for work on inequality and the poverty cycle (Duncan, Dockery, & Cassells, 2014).

Policy deficits and the creation of social exclusion

Recent work has also linked the culture of poverty with structural deficits (Prieur, 2018). By structure, the author is referring to socioeconomic and cultural conditions in geographical areas of disadvantage. Prieur notes that an earlier theorist (Young, 2011) found that poor living conditions and associated stigmatization can lead to unhelpful emotional responses related to the experience of inequality. She states that economic deprivation, possibly supplemented with discrimination and stigmatization, is a source of a series of emotions, such as humiliation or frustration, although her major interest is in criminal behavior. New Australian research, referred to below, explores further how government policy in areas such as financial distribution, land use, urban planning, and the provision of infrastructure and services, can impact on people, discussed in "Infrastructure" section.

Income distribution

Current Australian Government national policy is that certain groups of people without income, or with a very low income, can be offered a pension or allowance, the amount varying according to a judgment about "worthiness" and "shared responsibility." Thus, disadvantage may be viewed as being due to a personal or family deficit, where the cause of disadvantage could be present as a result of personal and/or family inadequacy, such as laziness, "bludging off welfare" and "leaners not lifters" (both terms are insults suggesting people prefer to be economically dependent on the welfare from the government) (Whiteford, 2019). Thus, there is a tacit requirement that people make every effort to get off welfare, such as find employment as soon as possible, whatever or where ever, it is.

The Australian government also provides funding to selected nongovernment organizations to assist individuals or families to improve their well-being through a range of interventions, such as "case management," or offering personal guidance as to how to change their adverse circumstances. Some nongovernment organizations are also expected to raise money independently and, until the last couple of decades, organize their own approach to assisting people. However, there is now a strong tendency for state and national governments to pass operational duties to welfare organizations and recipients, combined with closely defined programs that government wishes to be implemented, and usually require defined eligibility to access these programs.

Such a disbursement of decision-making can be seen in the National Disability Insurance Scheme (NDIS), being rolled out across Australia since 2016. This is a major government initiative that seeks to place the person with a disability as having the choice and control over what services they receive (Australian Department of Human Services, 2018). According to the government website, eligibility for the NDIS is if a person has a "permanent and significant disability" and supports are provided where this is "reasonable and necessary," such as where the service offers "value for money" and likely to be "effective and beneficial to the participant" (Australian Department of Human Services, 2018). However, while this approach may result in short-term relief for some families, there has been little longer-term amelioration of entrenched disadvantage experienced by many families that support one or more members with a disability (see, e.g., Dickinson, 2019).

Indeed, tax relief for the wealthy is common in Australia. A recent report from the welfare agency, Anglicare, found that the impact of Australian government tax concessions, negative gearing of investment loans (being able to write off the cost of loans to purchase investments like housing to rent to others against other income for the purpose of assessing income tax), capital gains tax concessions, the use of discretionary trusts (the ability to spread income and capital among members of a family), the exemption from the Goods and Services Tax of private health insurance and education, and the exemption from capital gains tax of people's place of residence, resulted in $AUD 68.55 billion each year going to the wealthiest 20% of households, compared with $AUD 6.1 billion that goes to the poorest 20% (Dawson & Smith, 2018).

Work from the United Kingdom highlights the problems of social policy emulating the market economy, with emphasis on consumption, bank credit, as well as structures like defined contracts, business models in government, and depersonalized interactions (Blond, 2010; Jordan, 2010). A similar situation is present in Melbourne. The unemployment benefit system:

> *operates more like a system of penalties than an employment service, requiring participants to apply for 20 jobs a month or go on Work for the Dole programs rather preparing them for work.*
>
> ***(Whiteford, 2019)***

The Federal Australian Government has introduced programs such as the *robo-debt* system, a program that computer matches data from the tax office with income reported to Centrelink, the agency responsible for federal welfare payments. Such a scheme was introduced as the government had the impression that there was widespread rorting of welfare entitlements. Repayment demands were sent to 200,000 welfare recipients demanding back payments and imposing sanctions (Whiteford, 2019). A review of this program found that:

> *the 'online compliance system' lacked procedural fairness at every stage and had put thousands of innocent Australians through the trauma of having to prove they do not owe the money the welfare agency demanded.*
>
> ***(Towell, 2017)***

Such an approach to welfare recipients overlooks the value of interpersonal interactions, collective action, networks and community, and trust, leaving people feeling isolated and helpless, with little sense of belonging and low self-esteem. These are words that seem to be reflected by the social psychologists who have examined the personal and family impact of social exclusion, and similar to the findings the empirical study in Victoria on social exclusion and well-being, reported above.

Infrastructure

A further area of government policy that compounds disadvantage relates to the provision of infrastructure. Blond (2010, p.55), in association with the "Big Society" movement in the United Kingdom, identifies this as an area that has failed in the United Kingdom: "British infrastructure is in a terrible state." He particularly mentions transport, communication, and education, noting that too much has also been spent on the wrong things, a problem that is also recognized in Melbourne with a massive freeway building program (see Jacks, 2018; Preiss & Towell, 2018).

In Australia, areas of poverty in new suburbs, where there is a rapidly growing population, have fallen behind in Australia in the provision of infrastructure (Brain et al., 2019). This reflects a failure in land use planning, where urban areas are growing beyond the capacity to meet basic infrastructure needs. New research found that residents in the six fastest growing outer suburbs of Greater Melbourne, growing at 2% and over per annum, being Cardinia, Casey, Hume, Melton, Whittlesea, and Wyndham, went backwards relative to the state as a whole over the 1992–2017 period, in terms of an ability to generate income, as well as in many social indicators. The lack of public transport in outer suburbs and traffic congestion across Melbourne was particularly revealing, leading to poor accessibility to work, services, and recreation.

Indicators between local government areas show quite strong spatial associations, particularly with respect to differences in travel time to central Melbourne (which is highly correlated with distance to Melbourne). For example, as travel times from a local government area to central Melbourne increases, population and job densities decrease, median house prices decline, and open space per resident increases. However, there are numerous negative effects (Brain et al., 2019). Economic capital per person declines, the proportion of higher educated people declines, the proportion of jobs that are high-tech declines, and local government productivity declines. Trust in others declines (defined as the proportion of local government residents who think that most people in general can be trusted). The proportion of people living near public transport declines, and public transport use for the journey to work also declines. Car use increases for the journey to work and the proportion of commutes that are longer than 2 hours increases. Finally, reports of heart disease and obesity increase.

The extent of these disadvantages can be traced to a calculation of infrastructure deficit spending in the six outer edge local governments of Melbourne, described above. Analysis by the National Institute of Economic and Industry Research concludes that there was accumulated underinvestment (in relation to the average investment in Victoria) of around $AUD 125 billion as at 2015–16, building at an annual rate of approximately $AUD 5 billion over the 25-year period from 1992 to 2017. This deficit is in the areas of transport infrastructure, commercial capital (defined as direct private sector investment in commercial infrastructure that will be undertaken because the public sector expenditures will create the profitable opportunities to justify the expenditures), community capital (such as hospitals and schools), skills development, and knowledge service industries (innovation and digital opportunities).

Early child development standards provide a good indicator of current and future disadvantage. The Australian Early Development Census (Australian Government, 2016) has examined the percentage of children on school entry who have the required developmental milestones. The Victorian average is 19.9% of children with one or more developmental delays[a] on reaching school age. Children in disadvantaged suburbs, particularly in outer Melbourne, show a lower developmental average. This

[a]Five developmental milestones were measured: physical health and well-being; social competence; emotional maturity; language and cognitive skills; and communication skills and general knowledge.

pattern of outer suburban disadvantage is reflected in measures of school retention and literacy and numeracy standards, all showing standards below the average Victorian measures. It is also reflected in high levels of youth unemployment. In all, 15 local governments in Melbourne have youth unemployment rates between around 15% and 20%, and a further 15 being between 10% and 14.5%. These figures do not include the levels of underemployment, nor disengagement from education and searching for work. In February 2017, the underemployment rate for youth aged 15–24 years of age sat at about 18% (Vandenbroek, 2017). Thus, poor early child development is a matter of serious concern to both the individual and family, as well as the Australian community, as it reduces many opportunities available over a lifetime.

How this can be changed

This chapter argues that entrenched disadvantage can lead to what could be described as a culture of poverty in a family, characterized by low self-esteem, expectation of failure, and feelings of helplessness, thus making it very difficult for cultural change and social mobility to take place. In some situations, this culture is reenforced by negative judgments and blame from the community and politicians. The reasons for poverty, social exclusion, and disadvantage have been traditionally seen as personal and family failures, with "solutions" based on the family intervention and case management. However, inequality remains and risks growing in wealthy countries such as in Australia and Europe.

The major Australian Research Council funded research project on the drivers of social exclusion, reported above, gives important clues as to how disadvantage can be addressed. Other research has shown that entrenched poverty in some local government areas in Melbourne have a significant level of undercapitalization by government, especially around transport and accessibility, thus offering few opportunities for those experiencing poverty to change their circumstances, thus perpetuating the culture of poverty. This situation has also been shown to be replicated in other Australian capital cities (Brain, 2019 unpublished research).

This imbalance between need and structural provisions reflects a failure in land use planning, where urban areas are growing beyond the capacity to meet basic infrastructure needs and basic services. An important part of the solution is the introduction of 20-minute neighborhoods, where most people can reach most services most of the time within a 20-minute trip by active transport or public transport. Such an approach not only improves accessibility but also improves the acquisition of social capital and sense of community, facilitating improved inclusion and family well-being and also improving health outcomes and employment opportunities, thus in turn promoting capabilities and reducing exclusion and poverty.

The value of accessibility, being able to get to places to meet needs, has been clearly shown in research in relation to improvements to well-being, in general, and particularly in areas of competence, positive relations with others, self-acceptance, and autonomy, all measures that should assist to counteract the negative experiences of social exclusion, outlined above. Thus, improving land use and transport planning, especially around public transport should be a high priority, also offering greater affordability for government.

The pursuit of more compact settlement development patterns will capture external benefits of density and improved productivity, particularly benefits from the clustering of activities (Stanley et al., 2017). Increasing density of population and jobs in the middle suburbs will also reduce the cost of infrastructure when compared with new housing developments on the fringe. Increases in density in low-rise buildings, especially in areas close to public transport, are already occurring in Melbourne. For example, the City of Hume has experienced net density growth (excluding open space) from 11 to 12 dwellings per hectare to 17 or 18 dwellings per hectare (personal communication, Hume Council planning department). This needs to increase further in order to manage infrastructure costs and achieve neighborhoods organized on the concept of a 20-minute city (Stanley et al., 2017).

This chapter argues that greater consideration should be given to how government policy can impact disadvantage, both in terms of provision of infrastructure and improved land use and urban planning, as well as how the punitive provision of welfare services may be increasing the negative impact of social exclusion and emotional stress. Revision of the administrative approach for welfare recipients, from one based on compliance, is required, as insights from psychology suggests that a punitive process can compound the problem that the welfare system is wanting to address—that is, moving people toward financial independence. A review of some welfare payments is urgently needed, especially the unemployment benefits, ironically called "Newstart," which, as noted has remained unchanged for 25 years and has widened the pool of recipients to include single parents (Davidson et al., 2018). Swan et al. (2016) report evidence that the egalitarian impact of the social security system has reduced overtime, partly due to its effect of continuing income tax cuts, noted above, which have reduced the overall tax take.

The culture of poverty related to a growing inequality of opportunities in Australia is, in the author's view, an unacceptable position in a wealthy country. This is especially so in relation to the poverty gap between indigenous and nonindigenous Australians, a topic of considerable complexity, thus not discussed in this chapter, but not meaning to be overlooked. Inequality can be recognized in growing rates of unemployment, especially long-term unemployment, which has doubled in the past decade, slow growth in real wages, growing housing unaffordability, some rising health-care costs, and inadequate education funding (Cerasa, 2019; Swan et al., 2016). It is a position that will be highly likely to deteriorate overtime, particularly in the context of climate change. While addressing disadvantage will involve considerable costs, it is argued here that the benefits of social externalities are rarely included in decisions about expenditure on infrastructure and welfare payments. Thus, there is also an urgent need to make these benefits transparent, so the value to society as a whole can be better understood.

The ultimate goal should be to improve the quality of the social environment in our societies with economic outcomes viewed as a subsystem of social well-being, both encompassed within the goal of a sustainable natural environment. Such approaches are being promoted in other international cities. For example, in Vancouver, Canada, government policies address housing affordability, such as inclusionary zoning near public transport and density bonuses. London's congestion pricing, where those who can afford to drive in central London pay a high price for this privilege, reduces congestion and allows for a high-quality public transport system that is available for all people (Metro Vancouver, 2015; Transport for London, 2019).

The 20-minute city idea is proposed in *Plan Melbourne Refresh*, the document that guides planning for Melbourne to 2050, although there is little evidence of implementation of the scheme (Victorian Government, 2017). The policy of forming 20-minute cities is well developed in much of Scandinavia and Eastern Europe, in the context of increased density of urban areas. This may be part of the reason for lower levels of inequality in these countries than that found in Australia. While closer scrutiny and further research is needed to understand the political and policy differences between countries that lead to inequality and disadvantage, it is hoped that this chapter offers a broader perspective on possible reasons for an ongoing culture of poverty in wealthy countries.

References

Abrams, D., Hogg, M., & Marques, J. A. (2005). Social psychological framework for understanding social inclusion and exclusion. In D. Abrams, M. Hogg, & J. Marques (Eds.), *The social psychology of inclusion and exclusion* (pp. 1–24). United Kingdon: Taylor & Francis Group.

Alexander, S., & Floyd, J. (2018). *Carbon civilisation and the energy future: Life beyond this brief anomaly*. The Simplicity Institute. http://simplicityinstitute.org.

Australian Department of Human Services. (2018). *What are reasonable and necessary supports?* https://www.ndis.gov.au/participants/reasonable-and-necessary-supports.

Australian Government. (2016). *Australian early development census 2015*. https://www.aedc.gov.au.

Barrett, G., & Whelan, S. (2018). *The fair go is a fading dream, but don't write if off*. Sydney Democracy Network. October 24. http://sydneydemocracynetwork.org/article-the-fair-go-is-a-fading-dream-but-dont-write-it-off/.

Blond, P. (2010). *Red tory: How the left and right have broken Britain and how we can fix it*. London: Faber and Faber.

Brain, P., Stanley, J. R., & Stanley, J. K. (2019). *Melbourne: How big, how fast and at what cost?* Melbourne: Melbourne Sustainable Society Research Paper, University of Melbourne.

Brotherhood of St. Laurence. (2007). *Equity in response to climate change roundtable*. Melbourne, Australia: Brotherhood of St. Laurence.

Burchardt, T., LeGrand, J., & Piachaud, D. (2002). Degrees of exclusion: Developing a dynamic, multidimensional measure. In J. Hills, J. LeGrand, & D. Piachaud (Eds.), *Understanding social exclusion* (pp. 30–43). Oxford: Oxford University Press.

Cerasa, D. (2019). The continually shifting narrative on Newstart. *The Age*. 2 August. 18.

Davidson, P., Saunders, P., Bradbury, B., & Wong, M. (2018). *Poverty in Australia, 2018*. Australian Council of Social Service/University of New South Wales Poverty and Inequality Partnership Report No. 2, Sydney: Australian Council of Social Service.

Dawson, E., & Smith, W. (2018). *The cost of privilege*. http://www.anglicare.asn.au/docs/default-source/default-document-library/cost-of-privilege-report.pdf.

Dickinson, H. (2019). Here's what needs to happen to get the NDIS back on track. *The Conversation*, May 30. https://theconversation.com/heres-what-needs-to-happen-to-get-the-ndis-back-on-track-117835.

Duckett, S. (2019). Is it time to ditch the private health insurance rebate? It's a question Labor can't ignore. *The Conversation*, February 28. https://theconversation.com/is-it-time-to-ditch-the-private-health-insurance-rebate-its-a-question-labor-cant-ignore-111171.

Duncan, A., Dockery, A. M., & Cassells, R. (2014). *Falling through the cracks: Poverty and disadvantage in Australia*. Focus on the States Report Series, No. 1. Perth, Australia: Bankwest Curtin Economics Centre.

Haslam, C., Cruwys, T., Haslam, S., & Jetten, J. (2015). Social connectedness and health. In N. Pachana (Ed.), *Encyclopedia of geropsychology* (pp. 1–10). Singapore: Springer. https://doi.org/10.1007/978-981-287-082-7_46.

Inglehart, R. (2000). Globalization and postmodern values. *The Washington Quarterly*, Winter. https://www.uv.es/genomica/spa/inves/Inglehart.pdf.

Jacks, T. (2018). 1.4 million Melbournians can't reach public transport. *The Age*, October 26. 13.

Jordan, B. (2010). *What's wrong with social policy and how to fix it*. Cambridge: Polity Press.

Maslow, A. (1954). *Motivation and personality* (3rd ed.). New York: Harper and Row Publishers.

McLelland, A. (2014). Values, concepts and social policy design. In A. McClelland & P. Smyth (Eds.), *Social policy in Australia: Understanding for action* (3rd ed., pp. 16–31). Melbourne, Australia: Oxford University Press.

Metro Vancouver. (2015). *Regional affordable housing strategy: Services and solutions for a liveable region*. Update Draft, August. http://www.metrovancouver.org/services/regional-planning/PlanningPublications/RegionalAffordableHousingStrategyUpdateAug19-2015.pdf.

Organization for Economic Cooperation and Development (OECD). (2019). *Income inequality, OECD data*. https://data.oecd.org/inequality/income-inequality.htm#indicator-chart.

Organization for Economic Cooperation and Development. (OECD). (2018). *An overview of growing income inequalities in OECD countries: Main findings*. http://www.oecd.org/social/inequality-and-poverty.htm.

Preiss, B., & Towell, N. (2018). Libs go all in on East West Link reboot. *The Age*, June 29, 8.

Prieur, A. (2018). Towards a criminology of structurally conditioned emotions: Combining Bourdieu's field theory and cultural criminology. *European Journal of Criminology*, *15*(3), 344–363. https://doi.org/10.1177/1477370817737242.

Ryan, R., & Deci, E. (2001). On happiness and human potentials: A review of research on hedonic and eudaimonic well-being. *Annual Review of Psychology*, *52*, 141–166.

Sen, A. (1992). Capability and wellbeing. In M. Nussbaum & A. Sen (Eds.), *The quality of life* (pp. 30–53). Oxford: Clarendon Press.

Shields, M., & Wooden, M. (2003). Investigating the role of neighbourhood characteristics in determining life satisfaction. *Melbourne Institute Working Paper No. 24/03, September*. Victoria, Australia: University of Melbourne.

Skattebol, J., & Redmond, G. (2019). Troubled kids? Locational disadvantage, opportunity structures and social exclusion. *Children's Geographies*, *17*(1), 76–89. https://doi.org/10.1080/14733285.2018.1487031.

Stanley, J. (2014). Climate change: A new challenge for social policy. In A. McClelland & P. Smyth (Eds.), *Social policy in Australia: Understanding for action* (pp. 230–241). Melbourne, Australia: Oxford University Press.

Stanley, J. K., Stanley, J. R., & Hansen, R. (2017). *How great cities happen: Integrating people, land use and transport*. Cheltenham, United Kingdom: Edward Elgar.

Stanley, J. K., Stanley, J. R., & Hensher, D. (2012). Mobility, social capital and sense of community: What value? *Urban Studies*, *49*, 3595–3609. https://doi.org/10.1177/0042098012447002.

Stanley, J., Stanley, J., Vella-Brodrick, D., & Currie, G. (2010). The place of transport in facilitating social inclusion via the mediating influence of social capital. *Research in Transportation Economics*, *29*, 280–286. https://doi.org/10.1016/j.retrec.2010.07.035.

Swan, W., Cooney, M., Clyne, C., Hetherington, D., Oliver, D., Whiteford, P., et al. (2016). *Inequality: The facts and the future*. Canberra, Australia: Chifley Research Centre Inclusive Prosperity Commission.

Towell, N. (2017). Parliamentary committee finds Centrelink robo-debt system has had 'profoundly negative impact'. June 22. *Sydney Morning Herald*.

Transport for London. (2019). *Congestion charge*. https://tfl.gov.uk/modes/driving/congestion-charge.

Twenge, J., & Baumeister, R. (2004). Social exclusion increases aggression and self-defeating behavior while reducing intelligent thought and prosocial behavior. In D. Abrams, M. Hogg, & J. Marques (Eds.), *Social psychology of inclusion and exclusion* (pp. 27–46). New York: Psychology Press. https://doi.org/10.4324/9780203496176.

Vandenbroek. (2017). *Underemployment statistics: A quick guide*. Research Paper Series 2016–2017, Parliamentary Library, Parliament of Australia.

Vella-Brodrick, D., & Stanley, J. (2013). The significance of transport mobility in predicting well-being. *Transport Policy*, *29*, 236–242. https://doi.org/10.1016/j.tranpol.2013.06.005.

Victorian Government. (2017). *Plan Melbourne refresh 2017–2050: Summary document*. https://www.planmelbourne.vic.gov.au/__data/assets/pdf_file/0009/377127/Plan_Melbourne_2017-2050_Summary.pdf.

Whiteford, P. (2019). Future budgets are going to have to spend more on welfare, which is fine. It's spending on us. *The Conversation*, March 7. https://theconversation.com/future-budgets-are-going-to-have-to-spend-more-on-welfare-which-is-fine-its-spending-on-us-111498.

Wilkinson, R., & Pickett, K. (2009). *The spirit level: Why equality is better for everyone*. London: Allen Lane.

Young, J. (2011). *The criminological imagination*. Cambridge: Polity Press.

CHAPTER 13

Family migration: Demographics and sociopsychological issues

David L. Sam[a] and Mary B. Setrana[b,c]
[a]Department of Psychosocial Science and Center for International Health, University of Bergen, Bergen, Norway
[b]Centre for Migration Studies, University of Ghana, Legon-Accra, Ghana
[c]Department of Sociology, University of Johannesburg, Johannesburg, South Africa

Not only is migration one of the major challenges the world faces today (United Nations Population Fund (UNFPA), n.d.), but it is also one that is growing in scope, complexity, and diversity. Moreover, migration is also becoming unsafe and unmanageable, as exemplified by activities on the Southern border of the United States, and the Trans-Mediterranean crossing from North Africa to Southern Europe. Migration has important ramifications for both the sending and receiving countries, the individuals and groups that are migrating, and those who are left behind in the countries of origin. Indeed, the United Nation's (2018) Sustainable Development Goals (SDG) for 2030 recognized in several parts of its agenda about the importance of migration. In all, 10 or 11 out of the 17 SDGs contain targets and indicators that are relevant to migration or mobility underscoring the importance of migration for global development. The European Commission puts it thus: "*If it is to be fit for the 21st century, development thinking must fully integrate the role of migration and mobility as development enablers*" (European Commission, 2018). Brunson McKinley (2006), the former Director General of the International Organization for Migration (IOM), aptly puts it as migration is "*an essential, inevitable and potentially beneficial component of the economic and social life of every country and region*" and that "*well-informed choices by migrants, governments, home and host communities, civil societies, and the private sector can help realize the positive potential of immigration in social, economic and political terms.*"

The increased attention devoted to international immigration parallels the rising number of international migrants. It is estimated that if the

☆This chapter was prepared when the first author was a short-term J1-scholar at the Chicago School of Professional Psychology, and also a Visiting Scholar at the University of Ghana, Legon-Accra, Ghana.

https://doi.org/10.1016/B978-0-12-815493-9.00013-2

immigrant population grows at a similar rate as in the past two decades, it is likely to reach 405 million, a doubling in one generation by 2050, compared to the 258 million as of 2017 (European Commission, 2015; International Organization for Migration, 2018). Embedded within these escalating numbers are different groups of migrants, ranging from those who have migrated voluntarily (such as *immigrants*) and those who do so involuntarily (such as *refugees*). Further, the nature of their status as migrants may be permanent, as in the case of refugees or temporary such as the situation of *asylum-seekers*. The 2017 United Nations High Commission for Refugee (UNHCR) Report indicated that in 2016, as many as 65.6 million people had been forcibly displaced worldwide as a result of persecution, conflict, violence, or human rights violations. The majority of these people end up as refugees or asylum-seekers and are collectively referred to as forced migrants.

Migration, whether voluntary or involuntary, has an impact on the family. Statistics on family migration are, however, scarce and sparse, even though family migration constitutes a major part of international migration. This chapter highlights the sociopsychological aspects of family migration. More specifically, we look at some individual migrants (such as children) within the family and their adaptation in a new cultural context; as transformed gender relations and its impact on men in the face of migration, fostering, and family separation together with transnational parenting. The chapter, however, begins with the demographics of family migration after a brief discussion of who or what constitutes family migration.

Family migration: A definition

As a term, family migration refers to people who migrate due to family ties. While this definition is simple and appears to be straightforward, it is much more complex than the definition suggests. The term is bogged down by several definitional issues and challenges and comes with a number of policy issue implications and human rights ramifications. In spite of its simplicity, the term, according to the International Migration Outlook (Organization for Economic Cooperation and Development (OECD), 2017), encompasses four categories of migrants: *family formation* (i.e., when a resident national or foreigner marries a foreigner or sponsors, respectively, for admission or status change); *accompanying family* (i.e., family members are admitted together with the principal migrant); *family reunification* (i.e., family members migrate after the arrival of a principal migrant who sponsors their admission). Normally, the family ties should predate the arrival of the principal migrant; and *international adoption* (i.e., a resident national or foreigner adopts a child of foreign nationality resident abroad).

Having four different categories under a single definition is itself an indication of how complex family migration is and underscores the heterogeneity of the people captured under the term. Family migration includes people accompanying migrant workers or refugees; people reuniting with family members who have migrated previously or people forming new family units with nationals of the destination countries. Family migrants also come with diverse types of family links with their counterparts in both the countries of destination and origin, including spouses, children, parents, and siblings, highlighting the transnational nature of this group. The complexity of the term originates from the lack of consensus in the definition of family, specifically, who qualifies as a family member.

Murdock (1949) conceptualized the family as a social group that is characterized by common residence, economic cooperation, and reproduction. The social group is also conceptualized as including adults of both sexes, of which at least two of them maintain a socially approved sexual relationship, and one or more children (i.e., own or adopted by the sexually cohabiting adults). The last couple of generations have seen societies undergoing major changes following migration and globalization, and these changes have made Murdock's definition less tenable in contemporary societies. The definition of the family has also become contentious when one tries to conceptualize it from Western and non-Western societal perspectives. Murdock's conceptualization of the family, referred to the nuclear family, is the type of family that is at the heart of the family migration. The nuclear family is conceptualized as spouse/partner and minor children with a common residential address. However, from a sociocultural perspective, the focus on the nuclear family is not defensible because this conceptualization excludes a large group of extended family members who may desire to be recognized under the umbrella of family migration.

To understand who is or may be covered under family migration regulation, it is important to recognize that spouse and minor children (biological or adopted) constitute the primary focus (or the nucleus) of family migration. Parents (i.e., of the migrant and spouse in the nucleus) and adult children, and other people constitute secondary focus. The relations to members of the nuclear family are typically considered a person's closest family ties (direct relations). Children of migrants are direct descendants and therefore direct beneficiaries of family migration. However, whether these children are eligible under the family migration umbrella is dependent on the age of the child and national policies (OECD, 2017). A person's parents and siblings are direct ascendants and are at the same genealogic level. Since spouses are direct relations, the legal implication is that spouses of children may be treated the same way as directly related adult children. But this may not be necessarily so.

When deciding on who qualifies, or can be beneficiary of family migration rights, a distinction is often made between *dependents* and *nondependents*, as well as the age of the potential beneficiary. On legal grounds, children 18 years or older are considered as adults and therefore are seen as ineligible for family migration consideration. Whereas parents and grandparents are de-facto ineligible for family migration consideration because they are adults, they may be eligible as *dependents* due to their advanced age, disability, or poor health (OECD, 2017). Married children are oftentimes not considered part of the original nuclear family, as they represent a separate (nuclear) family of their own.

Notions of nuclear families get much complex when trying to define who a spouse is. This complexity will arise, when a person is not married to a partner but linked through some form of registered partnership or long-term marriage-like cohabitation. Such a person may or not be recognized as a spouse and could, therefore, be disqualified as family migration beneficiary. When the partner is a parent of common children, this may count positively and be relevant when deciding on one's status as a spouse. Other complicating issues that may arise are spouses in a polygamous marriage, orphaned grand-children, and adopted children.

The extended family that includes cousins, nieces/nephews, and uncles are oftentimes not eligible for family migration consideration. Countries differ in their recognition of family membership, and whether individual members of a family qualify for family migration status. Of interest to our discussion, is, for instance, the notion that an individual family member, who entered into a country as a refugee may be able to sponsor member(s) of his or her nuclear family under a reunification program. Different national governments have different qualifying criteria when such individuals apply for residence under the family reunification program.

Demographics of family migration

Statistics on family migration is generally poor, but more so in low-income countries, due to a lack of capacity or political will to collect data (Migration Data Portal, 2018). Moreover, many low-income countries are collectivist and see the family as extended, and may complicate the numbers involved in this migration. Statistics on family migration in OECD countries, however, are much more elaborate, and perhaps rightly so because family migration accounts for about 40% of all international migration to OECD countries. Indeed, family migration has in recent years become the main form of migration to the OECD area. For instance, in 2015, more than 1.6 million family migrants received a residence permit in the OECD area in Europe, and about a third of intra-European movements are associated with family reasons (i.e., around 400,000 people in 2015 alone; OECD, 2017). Considering the sheer size and importance of family migration to OECD countries, the OECD in 2017 painted a detailed portrait of family migration. In this portrait, they indicate, among others that family formation is a major driver of all family migration, and in 2015, over 25% of all family migrants were young children, 15 years or young. Furthermore, 60% of all family migration were women. See OECD (2017) for more of the portrait of family migration.

Sociopsychological aspects of family migration

Women and children form the largest group when it comes to family migration, and in this section of the chapter, we want to highlight sociopsychological issues that impact these two groups of migrants. More specifically, the focus is first on (im)migrant children or children in immigrant families followed by women and the transnational fostering. As part of the section on children in immigrant families, we also direct attention to two groups of children that have come to dominate the US-America migration debate in recent years. The two groups are commonly known by their acronyms as the DREAMers and DACA. DREAMers stands for Development, Relief, and Education for Alien Minors Act, and DACA refers to Deferred Action for Childhood Arrivals. Whereas adopted children fall under the category of family migration, this group is not focused on in this chapter. This is because many of these children are adopted into the family of nonmigrants in the society of destination. Similarly, the chapter does not focus on unaccompanied minors, because they migrate without their families. Moreover, they are also not in the position to sponsor their family members, even though it may be possible in later years after they have established themselves.

Children in immigrant families

Immigrant children constitute a sizable proportion of the national population of many Western countries (Hernandez, 2012), and they are also one of the fastest expanding groups of the population. For instance, in the United States, a 47% increase in the number of immigrants was observed from 1990 to 1997, which was compared with only a 7% increase in the number of children of native-born parents (Hernandez & Charney, 1998). From 1994 to 2017, the percentage of immigrant children increased by 51%, from 18% to 27% (Camarota & Zeigler, 2016). In line with UN's SDGs core principle of "*leaving no one behind*," including immigrants, it is important to direct attention to this large sector of the population in many countries. The main focus of SDG 1 is to *end poverty in all its forms everywhere* by 2030. Although immigrants are not directly identified by name, they constitute a sizable proportion of the population in many countries. For instance, nearly 4 out of 10 immigrant children in Norway live in poverty (Statistics Norway, 2018). This is despite the fact that Norway is ranked first on the UN's Human Development Index (HDI, 2018). It is reasonable to argue that this proportion may be larger in many other countries.

Children in immigrant families include both first- and second-generation immigrants as well as those referred to as 1.5-generation[a] immigrants. Strictly speaking, second- and higher-generation immigrant children do not directly come under family migration. Research on children in immigrant families quite often does not distinguish between children who migrated on the basis of family ties. Moreover, even if second-generation immigrant children fall outside this definition, they grow up in the homes of immigrant families and may experience similar sociopsychological factors as their first- and 1.5-generation immigrant counterparts. As such, we do not make any distinction in our discussion.

Sociopsychological issues relating to children in immigrant families

From childhood to adulthood, human beings go through major developmental changes including changes in biological, cognitive, and psychosocial domains. These changes are affected, both directly and indirectly, by important contexts, including family, school, and peers (Suarez-Orozco, Motti-Stefanidi, Marks, & Katsiaficas, 2018). For immigrant children, these processes are influenced by the varied cultural frameworks in which they grow up (Motti-Stefanidi, Berry, Chryssochoou, Sam, & Phinney, 2012). An important developmental task that young people face during the transition from childhood to adulthood is identity formation that deals with how they see themselves as an individual, conceptualize who they are, what they are, and who they hope to be in the future. For migrant children, the two important aspects of their identity formation are as a member of their

[a]The term 1.5 generation refers to individuals who immigrate to a new country before or during their early teens.

own ethnic group (Phinney, 1990) and, to varying degrees, as a member of the larger society (Phinney & Devich-Navarro, 1997). The extent to which immigrant children develop a preference for either the ethnic or national group or combine them into a bicultural identity has been implicated for their psychological adjustment (Motti-Stefanidi et al., 2012). For instance, research has found that migrant children who strongly identify both their ethnic heritage and the larger society to which they have settled, referred to as biculturals, have better adaptation outcome (Chen, Benet-Martínez, Wu, Lam, & Bond, 2013).

By means of socialization within the family and the community, adolescents learn the values, traditions, and practices of their culture. In many immigrant families (particularly those from non-Western societies), these values typically emphasize interdependence among family members (Fuligni, 1998; Kwak, 2003), including the expectation of mutual support within the family. This is in contrast to the family values of independence that prevails in Western societies. The relationship formation is closely linked to independence-interdependence values and their associated values of autonomy vs relatedness (Greenfield, Keller, Fuligni, & Maynard, 2003). With increased age, there are countervailing pressures for greater autonomy; peers from other groups provide models of alternative ways of interacting with parents and others. A strong sense of obligation to the family may benefit immigrant youth by providing clear roles and a sense of direction (Fuligni, Tseng, & Lam, 1999), but it may also create conflicts, as adolescents strive for more autonomy in their own lives. Concomitantly, immigrant parents may be less willing to grant autonomy to their children who are striving for resulting in intergenerational discrepancies in value orientations (Costigan & Dokis, 2006; Santisteban & Mitrani, 2003). As they get older, adolescents are more oriented toward peers generally and particularly toward members of the opposite sex. Immigrant parents are often stricter than other parents in Western societies in regulating social behaviors like dating (Kagitcibasi, 2003). As immigrant youth make friends across ethnic group lines, conflicting pressures from parents and peers regarding social contacts can pose a challenge for these adolescents (Berry, Phinney, Sam, & Vedder, 2006).

Schools are one of the most important developmental contexts for all young people, as these contexts may be instrumental in helping children acquire the knowledge and the thinking skills, as well as the behaviors and values that are considered important for their welfare (Horenczyk & Tartar, 2012). A successful trajectory through school is an asset for the future employment opportunities and choices of most youth (Masten & Motti-Stefanidi, 2009) and is considered by many immigrant students as the avenue for upward social mobility and for a better life than their parents had (Fuligni, 1997, 1998). Immigrant parents may have very high expectations for their children's academic achievements right from the beginning as part of their desire to see their children succeed in the new country (Hill & Torres, 2010). These efforts by immigrant parents are a result of the increasing struggles that they encounter relative to their peer natives; however, Cathles, Ou, Sasso, Setrana, and van Veen (2018) argue that for migrant children, the quality of education that they receive is an important asset to combat the challenges that they face.

In each of these situations, immigrant children are faced with decisions as to the extent to which they retain the expectations, values, and behaviors of their family and community and adopt those of the larger society. Because of the varying influences and opportunities, they face in making choices on these issues, and immigrant children's development may follow different trajectories, some of which may be positive, and others may be negative (Oppedal & Toppelberg, 2016). Stated in a different way, these choices may conflict with each other and result in making the migrant child vulnerable to adaptation problems (Motti-Stefanidi, 2018).

The adaptation outcome of children in immigrant families

Immigrant children's development may depend on different trajectories, and it is not surprising that research findings on immigrant children's adaptation are contradictory and inconsistent. Whereas some reviews, particularly older ones, point to both vulnerabilities and maladaptation (Beiser et al., 2011), others identify resilience and very good adaptation (Motti-Stefanidi, 2018). These discrepancies have probably also contributed to the notion of the immigrant adaptation paradox (Sam, Vedder, Liebkind, Neto, & Virta, 2008), which asserts that first-generation immigrants show better adaptation than their later generation counterparts (Marks, Ejesi, & García Coll, 2014).

Research suggests that psychological problems are not exclusive to immigrant children, and that immigrant children do not necessarily manifest the greater incidence of disorders than their native peers (Berry et al., 2006), although the experience of migration, and conflicting cultural values may complicate what otherwise should be normal developmental transition result in adaptation problems.

Intergenerational conflicts in family values have been suggested to be the basis of maladaptation among immigrant children (Dinh, Sarason, & Sarason, 1994; Li Chiu, Feldman, & Rosenthal, 1992; Rosenthal, 1996).

However, the psychosocial situation that underlies intergenerational conflicts or acculturation gap has been questioned as exclusive to immigrant families and therefore a plausible basis for the maladaptation (Kwak, 2003; Lim, Yeh, Liang, Lau, & McCabe, 2008). Findings from the 13-country comparative study among ethnocultural youth (i.e., ICSEY study), Berry et al. (2006) suggest that some form of discrepancies in intergenerational family values exists in both immigrant and national groups. These discrepancies are neither uniquely an immigrant family phenomenon, nor do they inevitably lead to adaptation problems. Some groups of immigrant children may have increased intergenerational conflicts, and these may affect their adaptation and in others, albeit large intergenerational conflicts, may not result in adaptation problems.

Studies examining how immigrant youth define and relate to their own ethnicity and that of the national group abound (Manuh, 1999; Phinney & Devich-Navarro, 1997; Rosenthal & Feldman, 1992). How an immigrant defines his or her ethnicity, and how this relates to his or her adaptation, has often been examined against the status assigned to the immigrant in the society of residence (Liebkind, Mähönen, Varjonen, & Jasinskaja-Lahti, 2016). It has often been suggested that if the group to which the immigrant belongs is a disparaged one, adaptation problems such as low self-esteem, may abound. However, various reviews have found conflicting results (Branscombe, Schmitt, & Harvey, 1999; Jasinskaja-Lahti, Liebkind, & Solheim, 2009). In the ICSEY study, Berry et al. (2006) found that immigrant children generally had a high ethnic identity, and this was consistently higher than their national identity (i.e., the identity with the society of settlement). In addition, the study found positive relationships between both ethnic and national identities and self-esteem across ethnic groups and across countries. This was irrespective of the country of residence. However, the overall adaptation was moderated by the degree of perceived discrimination.

School adjustment and academic achievement issues relating to children in immigrant families

The centrality of schooling in young people's well-being in general and the acculturation of immigrant children and youth in schools is very well documented in the acculturation literature (Reiser, 2009). An important aspect of migrant children's adaptation is their school adjustment and academic achievement as their success in these two areas may be good indicators of their overall development and their acculturation (Sam, 2014; Vedder &

Motti-Stefanidi, 2016). Positive adaptation in the school context may also be a broad predictor of positive adult outcomes in terms of employment, social and interpersonal relationships, and psychological well-being, and is associated with lower levels of societal problems, such as welfare dependency and criminal behavior.

The intrinsic value of education is the satisfaction that comes with being cognitively and socially resourceful and having easy access to a school career that is conducive to a person's further thriving (Robeyns, 2006; Vedder & Motti-Stefanidi, 2016). This function comes together with an instrumental value: schools support the acquisition of competencies that increase and eventually determine individuals' employability as well as their productivity; hence, this function is important for persons' material well-being (Dustmann & Theodoropoulosy, 2010).

For immigrant parents, school adjustment of their children and their academic achievement may be the indirect crowning of their own ambition. Quite often, immigrant parents begin at the lower end of the occupational ladder (Hayfron, 2006; Liebig & Widmaier, 2009) and may be unable to rise very high. Thus, supporting their children to succeed in school may be the pathway toward prosperous future in the new country. As such, immigrant parents have high regard for education and seek good education for their children (Okagaki, Helling, & Bingham, 2009; Vedder, Horenczyk, & Liebkind, 2006), something these parents directly or indirectly communicate to their children (Portes & Fernández-Kelly, 2008).

In spite of immigrant parents' concern about their children's school adjustment and academic achievements, they oftentimes appear to be less involved in the actual school activities of their children (Kiselica, Changizi, Cureton, & Gridley, 1995). This is because of their limited understanding of the school system of the new society and partly because of their own beliefs regarding a parental role in schooling (Atkinson, 1986). Immigrant parents' high aspirations quite often are not in tandem with the resources needed to successfully support their children's learning (Spera, 2005). On the one hand, they may encourage their children to do well and behave properly in school (Strand, 2011), and on the other hand they fail to organize sufficient linguistic and cognitive resources that can furnish their children with the scaffold for high achievement (Dustmann, Frattini, & Lanzara, 2012). Similarly, they remain on the sideline when it comes to getting involved in what takes place at school. Yao (1988) points out that Asian parents, for instance, hold school teachers and their authority in such high esteem that they hardly question their work with the belief that parent's main responsibility toward the child is at home, and the teachers are fully responsible with what takes place at school. In short, immigrant parents fail to see the relevance of close collaboration between school and home and between parents and teachers regarding the children. Accordingly, while refraining from "interfering" with what happens at school, they nevertheless put pressure on their children to work hard without achieving the expected outcome. An OECD comparative study (OECD, 2006), for instance, showed that although immigrant students in the participating countries generally were more motivated to achieve in school and held more positive attitudes than their native counterparts, the achievements of immigrant students in mathematics, reading, and science lagged behind their national peers.

DREAMers and DACA: Migrant children in the shadows

Whereas this group of migrant children is not recognized under the definition of family migration, they are of particular interest in the United States and warrant attention under family migration because their presence in the United States is linked to their parents' (i.e., family's) undocumented status. Of the approximately 11 million undocumented immigrants in the United States, more than a million are undocumented young people who were brought to the United States as minors and have spent most of their lives living in the United States. They are collectively known as the DREAMers, a term that is an acronym for Development, Relief, and Education for Alien Minors Act (DREAM). Because of their undocumented status, these young people risked deportation, which is why they tend to live anonymously and in the shadows, hiding from the authorities (Suárez-Orozco, Yoshikawa, Teranishi, & Suárez-Orozco, 2011). In 2012, the Obama Administration signed the Deferred Action for Childhood Arrivals (DACA) program to help some DREAMers to find a pathway toward citizenship (Human Rights Campaign, 2018).

DREAM was originally introduced in 2001 by two Senators (Dick Durbin—a Democrat and Orrin Hatch—a Republican) as a bill that would have granted legal status to undocumented immigrants who were brought to the United States as children and had received obligatory school in the United States. The bill could not get the needed support in the two houses of the US congress when it was first introduced. Since several versions of the bill have been introduced over the years, the latest being in 2017, but it has never passed both houses of the US Congress. While the term DREAMer originally took its name from the bill in Congress, it presently has a double meaning about the undocumented youth who have big hopes and dreams for a better future.

Because of their undocumented status, many of these young people lived in the shadows and in fear of being deported (Suárez-Orozco et al., 2015; Teranishi, Suárez-Orozco, & Suárez-Orozco, 2015). As pointed out, the Obama administration sought to offer some relief to the DREAMers, with the introduction of the DACA, which is also an acronym for Deferred Action for Childhood Arrivals. DACA was a 2-year deferment from deportation actions and provided eligibility for a work permit upon satisfying a number of requirements such as lack of criminal record and continuous residency in the United States. The introduction of the DACA resulted in some 800,000 undocumented migrant youth (i.e., DREAMers) to register and be on the pathway for legal status in the United States (FileRight, 2017). However, the Trump administration tried to abolish it in 2017 and was put on hold by a number of Federal Courts when immigration activists filed cases against the move. For now, DACA is on hold with the hope that a comprehensive immigration law that can be signed by the Trump administration will straighten out the status of this group of migrants. Unfortunately, at the time of writing this chapter, no one knows what is ahead for DREAMEers and DACA youth. There is no bill ready to be signed by the Trump government, and there appears to be an impasse between two aisles of the US Congress. On the one hand, DACA to become legitimate citizens of the country is viewed as "rewarding" an "illegal" behavior of parents. On the other hand, it is necessary to bring undocumented immigrants out of the shadows and contribute economically to the nation that they regard as home. The case has passed through the courts and at the time of writing, it is in the hands of the Supreme Court of the United States.

Whether they are DREAMers or DACA youth, they share an undocumented immigrant background, and they face a host of challenges that can undermine their psychological well-being. Having grown up in the United States, only to discover in their adolescence that they risk deportation and lack many basic rights, many DREAMers and DACA have experienced hopelessness, shame and self-blame, anxiety, fear of deportation, and concerns about blocked social mobility. There have even been reports of suicide among undocumented young people who felt that they could not overcome the barriers imposed by their irregular status (Benuto, Casas, Cummings, & Newlands, 2018; Venkataramani, Shah, O'Brien, Kawachi, & Tsai, 2017).

Gendered dynamics of family migration

The simple definition of family migration oversimplifies the complexities associated with the vulnerable groups (such as women and children) within households (Sam, 2006). The definition (and categorization) overlooks the increasing involvement of women in international migration, thereby creating the impression that women are still associational migrants to the principal migrant who is usually the male. For decades, international migration studies, and in particular those dealing with labor issues, either explicitly focused only on men or implicitly assumed that most migrants were males (Pedraza, 1991; Zlotnik, 1990). Women's participation in international migration was regarded as negligible (Kofman, 2000), even though several Western countries were observed to have large numbers of female migrants (Pedraza, 1991). Prior to the 1970s, international migration was rarely classified by gender, and because many women remained in the private sphere of domestic life, they were largely excluded from public discourse such as policy formation and research agendas.

Until recently, immigration laws regarding women were based on coverture law, i.e., the status of a woman acquires upon marriage under common law. This means that a wife normally has no legal identity of her own other than that of her husband. In terms of international migration, women's immigration status is often a derivative of her husband's status (Clifford & Pearce, 2004). One form of coverture law in international migration is the family reunification program which allows the spouse and children to join migrants who are already in a country legally.

Thus, while migration of women is not a new phenomenon, women for a very long period were simply overlooked. Furthermore, the fact that until recently, and in many parts of the world, they migrated short distances and engaged in circular migration (Awumbila, Deshingkar, Kandilige, Teye, & Setrana, 2018). However, the "novelty" as described by Marchetti (2019) lies with the fact that women are increasingly embarking on international migration (referred to as the feminization of migration) for diverse purposes. In quantitative terms, there has been an increase of female international migration from 46.6% to 48.8% between 1960 and 2000 and recently, females makeup about 48.4% of all international migrants, although this figure is slightly less than the 49.3% in 2000 (UNDESA, 2017). Whichever way we look at it, there is still an increasing participation of women in international migration compared to the 1960s when women were not recognized in migration surveys and studies.

Despite the increasing trends, there are geographical differences and these are due to factors such as aging female migrants who had migrated several decades and increasing need of care for these developed countries where the dependent population is higher (IOM, 2018). According to IOM

(2018) between 2000 and 2017, Europe showed the highest proportion of female migrants (from 51.6% in 2000 to 52.0% in 2017). Northern America, in the second place, showed a 1% increase during the same period (i.e., from 50.5% in 2000 to 51.5% in 2017). Asia bucks the trend, with a decrease from 46.2% in 2000 to 42.4% in 2017.

The 19th and early 20th centuries experienced more male migration than females. During the 1980s and early 1990s, the different experiences between men and women migration were acknowledged and criticized (Marchetti, 2019). Studies on gender and migration were initiated by a group of social historians and labor sociologists whose ideas were influenced by Marx in the 1970s on the role of migration in the capitalist expansion (Marchetti, 2019). This was linked to the 1984 special issue of the International Migration Review edited by Mirjana Morokvasic.

The second wave was started by a group of feminist and black scholars on migration, reproductive work, and multiple oppressions experienced by minority women such as black women. There were several studies on how domestic work (either paid or unpaid) was unequally distributed among white and black women in the history of the United States (Rollins, 1985). During this same period, intersectionality approaches were developed to give prominence to migrant women as being vulnerable groups who were often ignored in various fields of academic research (Dwyer, 1999).

The third wave was a combination of factors such as the UN World Women Conference in Beijing in 1995 that contributed to the debate on gender issues in a global governance perspective (Meyer & Prügl, 1999, cited in Marchetti, 2017). In 1998, the United Nations Population Division first released data on women's migration. There was also the collapse of the Soviet Union which had a massive effect on large-scale migration of women who were mostly employed domestic, sex, and care workers from East to West Europe (Marchetti, 2017; Vianello, 2014). Following the 1990s awakening, there were new publications on women's participation in domestic work. These developments also resonated well in developing continents such as Africa, with growing interest among scholars on women international migration (Adepoju, 2010; Manuh, 1999; Tonah & Setrana, 2017), although the focus of the analysis was quite different from what was happening in Western countries. African women were not necessarily traveling for domestic work but for education, trading among others (Adepoju, 2010). Contemporary studies on migration have expanded based on these initial debates to include all kinds of comprehensive qualitative and quantitative discussions on gender and migration.

In essence, these developments have contributed to the changing patterns and trends of migration, moving it from a more male-dominated experience to include other members of the household such as women. Apart from Ravenstein (1885), all the other theories of migration developed afterward did not acknowledge the inclusion of women in the migration experience. However, Ravenstein criticized for classifying women as short-distance migrants, which is better rated than the others such as the NELM which focusses on economic reasons for migration. The NELM tires migration decision-making to the household; however, it should also be noted that within the households, women were mostly not consulted when it came to issues of migration and breeding winning. For example, studies in some parts of Africa show that even up until now, some women who want freedom or break away from patriarchal norms, leave run away from their communities and households without informing husbands and children or fathers and brothers. Others are also forced into all kinds of migration against their will (Awumbila et al., 2018).

Gendered experiences of family migration

Psychologists such as Maccoby (1988, 2000) and Gilligan (1982) recognized the shift toward the need to include gender in migration studies in the fields of psychology. Both male and female movements have had an impact on the fundamental and traditional family systems of many countries. Migration has empowered and emancipated women by giving them the opportunity to play an active part in negotiating their roles, rights, and responsibilities across borders as well as assimilating them into their respective destination societies (Adepoju & van der Wiel, 2010; De Haas, 2007 cited in Tonah & Setrana, 2017).

The traditional notion of the male as a "bread winner" has undergone some fundamental changes; remittances transferred by female migrants are used to support and improve the well-being of the household. Trends in remittance flows indicate that male migrants remit higher amounts than their female counterparts (Adepoju & van der Wiel, 2010, cited in Tonah & Setrana, 2017). Analysis indicates that in 2011, for instance, males remitted the US $1446 compared to US $878 sent by female migrants (Mohapatra & Ratha, 2011).

As a result of female migration, gender division of labor and its associated roles have been renegotiated to fill the gaps established from the increasing absence of women in the home. Some scholars (such as Sluzki, 1979) emphasize the conflict created among families due to migration.

Such scholars explain how the change of power relations and the empowerment of women outside the home generate family conflicts (Hondagneu-Sotelo, 1992). While international migration offers women opportunities for women who migrate to work, at the same time it disempowers men who have to give off some of their so-called "productive time" to what is so-called "reproductive work" which is nonpaid. In some context at the origin, the left-behind men enforce their patriarchal rights by handing over the productive work to another female of his or the spouse's family. At the destination, men's desire to be rigid about his traditional expectations of their spouses has led to conflict and even domestic violence and divorce (see, e.g., Espin, 1999). Women on the other hand, double-up roles as reproductive workers and breadwinners at the same time. Manuh (1999), in her studies among Ghanaian immigrants in Toronto, pointed out that the conducive and favorable policies of women empower the Ghanaian wife, and this contradicts the home origin values and expectations of a wife. These contradictions create tensions among couples, thereby breeding distrust.

On the other hand, Charsley (2005), in his study among Pakistan male spouses, also showed contradictory findings. The Pakistani men in this study had limited power because it was the wives who supported their migration to the United Kingdom. They lose their patriarchal rights as husbands, and they no longer take household decisions to a great extent because they no longer have the breadwinning position and are mostly given the house-keeping jobs especially for women in Pakistan. These issues were a source of discomfort to the Pakistani immigrants. Frustrations experienced by the Pakistani men help explain instances where such marriages have ended in the husband's violence, desertion, or taking a second wife.

Family migration results in double acculturation for the couple, in that the couple as individuals have to adapt to a new culture, and the marriage itself also has to adapt to the new culture. The couple has to learn to accommodate to a new cultural context (Ataca & Berry, 2002). For women joining their husbands as part of family reunification, there is the danger of the couple had grown apart during the transition between the time the husband migrates and when the wife joins him. This may hamper the adaptation process when effort has to be directed toward salvaging the marriage.

Family migration also adds new tasks to the traditional unpaid tasks a woman has. These new tasks include ensuring the survival of the culture of the family and that of the heritage culture (Sam, 2006; Shaheed, 1999) in a possibly nonsupportive social environment; socializing a child in multiple cultures (Maiter & George, 2003); maintaining cordial social networks of kin and networks of obligations including economic remittances (Curran & Saguy, 2001; Salaff, 2002). In addition to their own efforts to adjust to two cultural systems (as part of their acculturation), immigrant women, and in particular mothers, have the added responsibility of helping their children deal with adjusting to two or more cultures, including the values of their own ethnic group, and those of the society of settlement. For instance, mothers try to support the learning of the children's heritage language as well as training them in the national language. Furthermore, as culture bearers, women bear the responsibility of ensuring the survival of their traditional culture in the society of settlement. Using Hong Kong Chinese women in Canada as examples, Salaff (2002) describes how these women struggle to sustain cordial kinship relationships when they are abroad. It is the responsibility of the Chinese woman to build social support networks of a wide variety when they at home in China or overseas. While trying to maintain a good social network support system, the woman may nevertheless be isolated at home (Ataca & Berry, 2002; Mirdal, 1984).

The participation of women in international migration has been strictly categorized by some scholars as either having a positive or a negative effect (Awumbila & Torvikey, 2017). For those who describe it as a positive outcome, international migration provides opportunities for women escaping rigid and cruel traditional practices, gain economic independence, and empowerment. On the contrary, the international migration of females has also been a source of vulnerability. Some women are exploited and abused in the hands of their foreign employers, others fall in the hands of smugglers and traffickers in their attempt to seek greener pastures.

Transnational parenting

Child raising is an important duty for parents and it creates a bond between parents and their children and this is an important component of the family structure of many traditions. However, there are challenges couples face at their destination with family reunification applications, which makes it difficult for them to move their children from countries of origin to the new country. Rigid immigration policies are resulting in an increasing number of "left-behind children." Migrant parents who have left behind sons and daughters in their home country often aspire to bring these children to the destination country (Dankyi, 2011).

Migrant parents who have left their children behind in their country of origin outsource their parenting duties to schools, families, and organizations that provide care for their children in their absence (Abutima, 2017;

Chen, Qiuqiong, Rozelle, Shi, & Zhang, 2014). Migrant parents support their children back home with remittances to pay for their fees, medical care, daily upkeeping among others. The absence of parents has several effects on the "left-behind children." For instance, some of the children perform poorly at schools while trying to adjust to parental absence (Abutima, 2017). Children who live with both parents and have their parents making a full commitment in their upbringing tend to have high educational attainment, social competence, self-esteem, and life satisfaction. On the other hand, children who do not have both parents due to international migration face challenges such as poor performance and attendance in school and emotional instability among others (Abutima, 2017; Dankyi, 2011).

Abutima (2017) releases two divergent views with regard to the effect of transnational parents on children's education: the first group of researchers argue the positive impact of parental migration on the left-behind children's rate of attending school (e.g., Hanson & Woodruff, 2003) and the other researchers have also argued that parental migration has a negative impact on school attendance due to the receipt of remittances and the role that it plays in the provision of home and educational supplies. These different views highlight the significant effect of contextual factors within the home community of the "left-behind children" as well as the contextual factors of the migrant parent(s). These contextual factors include adequate supervision of children when out of school, the emotional issues of a parent's presence and parents' immigration status in the host country among other economic activities. Children living without one or both parents in the home country have generally been portrayed to have worse educational performance compared to other children living with both parents in the home country (cited in Abutima, 2017). Parental migration, on the other hand, has increased school enrolment, retention, and improved educational outcomes. In other studies, parental migration did not have an adverse impact on the academic performance of children from migrant households because of the high level of support that extended kin in the Ghanaian context provided to such children (Dankyi, 2011, pp. 133–162). Unlike studies in Ecuador where left-behind children were considered as children of unpatriotic citizens and treated with disdain (Cortes, 2008). Increasing female migration accounted for the negative impact of parental migration on the educational outcomes of left-behind children (Cortes, 2015; Jampaklay, 2006).

These gendered experiences highlight the nuances within the nuclear family setting. While in the home country, fosterage is either outsourced to neutral persons and or family members, mothers and fathers in some cases also travel to occupy similar positions of caring for others in order to raise money to take care of the children. While migration has opportunities in terms of the remittances sent home for the upkeep of the family, it also has negative consequences on children when the outsourced caregiving is not in adequate hands, or the parents do not have the right documentation and job to remit for the upkeep of the child or in situations where there are communication barriers between parents and left-behind children. Missing in the whole area of migrant children's academic attainment is a systematic examination of school attainment of migrant children in the new society and left-behind children. The question is that do children left behind by their migrant parents fare worse than those who live with their parents in the new country, or the opposite, where migrant children do poorly.

Policy implications and conclusions

Family migration has an effect not only on the migrant and family at the destination but also on the left-behind families. In situations, where the traditional fosterage is in place, the family support system aids parenting while away, but in cases where such are not available migrants outsource their parental roles and in each case, the children are either affected positively by the remittances sent to them or negatively by remittances not being sent at all which also has a negative influence on school attendance, enrollment, and performance. It is worth noting that all these experiences are due to factors such as immigration policies at the destination which may not allow migrants to come in with their children or even in cases where they could, childcare is expensive to the extent that the little wages earned cannot cater for the children adequately.

Women migration empowers them depending on the conditions at both ends of the migration. In some cases, they may be empowered due to better immigration policies while in other cases, they are exposed to abuses. Sometimes women who are empowered at the destination, find themselves turn back to the old disempowerment when they eventually return because the home country does not allow them to practice the learned values. In some instances, these negative circumstances also happen to men who are disempowered at the destination and gain back their power upon return.

Considering the size of family migration, and the challenges individuals and families involved face, policies that can ease family migration, and support remittances sent to the country of origin may go along way in reaching a sustainable development in both sending countries and host countries.

References

Abutima, T. K. (2017). Comparative analysis of educational performance of children. In S. Tonah, B. M. Setrana, & J. A. Arthur (Eds.), *Migration and development in Africa: Trends, challenges, and policy implications* (pp. 89–105). Maryland, USA: Lexington Books.

Adepoju, A. (Ed.). (2010). *International migration within, to and from Africa in a globalised world.* Accra: Sub-Saharan Publishers.

Adepoju, A., & van der Wiel, A. (2010). *Seeking greener pastures abroad: A migration profile of Nigeria.* Ibadan: Safari Books Limited Sub-Saharan Publishers.

Ataca, B., & Berry, J. W. (2002). Psychological, sociocultural, and marital adaptation of Turkish immigrant couples in Canada. *International Journal of Psychology, 37*(1), 13–26. https://doi.org/10.1080/00207590143000135.

Atkinson, D. R. (1986). Cultural issues in school and family. In L. B. Golden & D. Capuzzi (Eds.), *Helping families help children* (pp. 237–244). Springfield, IL: Charles C. Thomas.

Awumbila, M., & Torvikeh, G. D. (2017). Women on the move: An historical analysis of female migration in Ghana. In M. Awumbila, M. Badasu, & J. K. Teye (Eds.), *Migration in a globalizing world: Perspectives from Ghana* (pp. 125–139). Migration Reader Series, University of Ghana.

Awumbila, M., Deshingkar, P., Kandilige, L., Teye, J. K., & Setrana, M. (2018). Please, thank you and sorry—Brokering migration and constructing identities for domestic work in Ghana. *Journal of Ethnic and Migration Studies*, 1–17. https://doi.org/10.1080/1369183X.2018.1528097.

Beiser, M., Zilber, N., Simich, L., Youngmann, R., Zohar, A. H., Taa, B., & Hou, F. (2011). Regional effects on the mental health of immigrant children: Results from the New Canadian Children and Youth Study (NCCYS). *Health & Place, 17*(3), 822–829.

Benuto, L. T., Casas, J. B., Cummings, C., & Newlands, R. (2018). Undocumented, to DACAmented, to DACAlimited: Narratives of Latino students with DACA status. *Hispanic Journal of Behavioral Sciences, 40*(3), 259–278. https://doi.org/10.1177/0739986318776941.

Berry, J. W., Phinney, J. S., Sam, D. L., & Vedder, P. (2006). *Immigrant youth in cultural transition: Acculturation, identity, and adaptation across national contexts.* Lawrence Erlbaum.

Branscombe, N. R., Schmitt, M. T., & Harvey, R. D. (1999). Perceiving pervasive discrimination among African Americans: Implications for group identification and well-being. *Journal of Personality and Social Psychology*, 77(1), 135–149. https://doi.org/10.1037/0022-3514.77.1.135.

Camarota, S. A., & Zeigler, K. (2016). *Immigrants in the United States: A profile of the foreign-born using 2014 and 2015 Census Bureau data.* Washington, DC: Center for Immigration Studies.

Cathles, A., Ou, D., Sasso, S., Setrana, M., & van Veen, T. (2018). *Where do you come from, wheredo you go? Assessing skills gaps and labour market outcomes of young adults with different immigration backgrounds.* CESifo Working Papers, 7157, July. https://papers.ssrn.com/sol3/papers.cfm?abstract_id=3250748.

Charsley, K. (2005). Unhappy husbands: Masculinity and migration in transnational Pakistani marriages. *Journal of the Royal Anthropological Institute, 11*, 85–105.

Chen, S. X., Benet-Martínez, V., Wu, W. C., Lam, B. C., & Bond, M. H. (2013). The role of dialectical self and bicultural identity integration in psychological adjustment. *Journal of Personality, 81*(1), 61–75. https://doi.org/10.1111/j.1467-9655.2005.0027.x.

Chen, X., Qiuqiong, H., Rozelle, S., Shi, Y., & Zhang, L. (2014). Effect of migration on children's educational performance in rural China. In J. C. Brada, P. Wachtel, & D. T. Yang (Eds.), *China's economic development* (pp. 323–343). London: Palgrave Macmillan. https://doi.org/10.1057/9781137469960_11.

Clifford, E. J., & Pearce, S. C. (2004). *Women and current US immigration practices.* http://newmedia.colorado.edu/~socwomen/socactivism/womenimm.pdf.

Cortes, P. (2015). The feminization of international migration and its effects on the children left behind: Evidence from the Philippines. *World Development, 65*, 62–78. https://doi.org/10.1016/j.worlddev.2013.10.021.

Cortes, R. (2008). *Children and women left behind in labour sending countries: An appraisal of social risks*. New York: United Nations Children's Fund (UNICEF), Policy, Advocacy and Knowledge Management Section, Division of Policy and Practice. https://pdfs.semanticscholar.org/892a/62e7366e785997135d68d310e8c656abf93e.pdf.

Costigan, C. L., & Dokis, D. P. (2006). Relations between parent–child acculturation differences and adjustment within immigrant Chinese families. *Child Development*, 77(5), 1252–1267. https://doi.org/10.1111/j.1467-8624.2006.00932.x.

Curran, S. R., & Saguy, A. C. (2001). Migration and cultural change: A role for gender and social networks? *Journal of International Women's Studies, 2*, 54–77. https://vc.bridgew.edu/jiws/vol2/iss3/4/.

Dankyi, E. (2011). Growing up in a transnational household: A study of children of international migrants in Accra. *Ghana Studies, 14*, 133–162.

De Haas, H. (2007). *Remittances, migration and social development: A conceptual review of literature*. Social Policy and Development Programme Papers No. 34. Geneva: United Nations Research Institute for Social Development.

Dinh, K. T., Sarason, B. R., & Sarason, I. G. (1994). Parent-child relationships in Vietnamese immigrant families. *Journal of Family Psychology, 8*(4), 471.

Dustmann, C., Frattini, T., & Lanzara, G. (2012). Educational achievement of second-generation immigrants: An international comparison. *Economic Policy, 27*(69), 143–185. https://doi.org/10.1111/j.1468-0327.2011.00275.x.

Dustmann, C., & Theodoropoulosy, N. (2010). Ethnic minority immigrants and their children in Britain. *Oxford Economic Papers, 62*, 209–233. https://doi.org/10.1093/oep/gpq004.

Dwyer, C. (1999). Contradictions of community: Questions of identity for young British Muslim women. *Enviornment and Planning, 3*, 53–68. https://doi.org/10.1068/a310053.

Espin, O. M. (1999). *Women crossing boundaries: A psychology of immigration and transformations of sexuality*. New York: Routledge.

European Commission. (2015). *Legal migration in the EU*. European Political Strategy Centre. https://ec.europa.eu/epsc/sites/epsc/files/strategic_note_issue_2.pdf.

European Commission. (2018). *International cooperation and development* Migration and asylum. https://ec.europa.eu/europeaid/node/672_es.

FileRight. (2017). *The difference between DACA and the DREAM Act*. https://www.fileright.com/blog/the-difference-between-daca-and-the-dream-act.

Fuligni, A. J. (1997). The academic achievement of adolescents from immigrant families: The role of family background, attitudes, and behavior. *Child Development, 68*(2), 351–363. https://doi.org/10.1111/j.1467-8624.1997.tb01944.x.

Fuligni, A. J. (1998). The adjustment of children from immigrant families. *Current Directions in Psychological Science*, 7(4), 99–103. https://doi.org/10.1111/1467-8721.ep10774731.

Fuligni, A. J., Tseng, V., & Lam, M. (1999). Attitudes toward family obligations among American adolescents with Asian, Latin American, and European backgrounds. *Child Development, 70*(4), 1030–1044. https://doi.org/10.1111/1467-8624.00075.

Gilligan, C. (1982). *A different voice*. Cambridge, MA: Harvard University Press.

Greenfield, P. M., Keller, H., Fuligni, A., & Maynard, A. (2003). Cultural pathways through universal development. *Annual Review of Psychology, 54*(1), 461–490. https://doi.org/10.1146/annurev.psych.54.101601.145221.

Hanson, G. H., & Woodruff, C. (2003). *Emigration and educational attainment in Mexico*. Mimeo., University of California at San Diego, CA, USA.

Hayfron, J. E. (2006). Immigrants in the labour market. In D. L. Sam & J. W. Berry (Eds.), *The Cambridge handbook of acculturation psychology* (pp. 439–468). Cambridge: Cambridge University Press.

Hernandez, D. J. (2012). Resources, strengths, and challenges for children in immigrant families in eight affluent countries. In A. Masten, K. Liebkind, & D. J. Hernandez (Eds.), *Realizing the potential of immigrant youth* (pp. 17–40). Cambridge: Cambridge University Press.

Hernandez, D. L., & Charney, E. (1998). *From generation to generation: The health and well-being of children in immigrant families.* Washington, DC: National Academy Press.

Hill, N. E., & Torres, K. (2010). Negotiating the American dream: The paradox of aspirations and achievement among Latino students and engagement between their families and schools. *Journal of Social Issues, 66*(1), 95–112. https://doi.org/10.1111/j.1540-4560.2009.01635.x.

Hondagneu-Sotelo, P. (1992). Overcoming patriarchal constraints: The reconstruction of gender relations among Mexican immigrant women and men. *Gender & Society, 6*(3), 393–415. https://doi.org/10.1111/j.1540-4560.2009.01635.x.

Horenczyk, G., & Tartar, M. (2012). Conceptualizing the school acculturative context: School, classroom and the immigrant student. In A. S. Masten, K. Liebkind, & D. J. Hernandez (Eds.), *Realizing the potential of immigrant youth* (pp. 359–375). Cambridge: Cambridge University Press.

Human Development Index (HDI). (2018). *United Nation human development reports.* http://hdr.undp.org/en/2018-update.

Human Rights. (2018). *Dream Act.* https://www.hrc.org/resources/dream-act.

International Migration Organization. (2018). *World migration report, 2018.* Geneva: International Migration Organization.

Jampaklay, A. (2006). Parental absence and children's school enrollment: Evidence from a longitudinal study in Kanchanaburi, Thailand. *Asian Population Studies, 2*(1), 93–110. https://doi.org/10.1080/17441730600700598.

Jasinskaja-Lahti, I., Liebkind, K., & Solheim, E. (2009). To identify or not to identify? National disidentification as an alternative reaction to perceived ethnic discrimination. *Applied Psychology An International Review, 58*(1), 105–128. https://doi.org/10.1111/j.1464-0597.2008.00384.x.

Kagitcibasi, C. (2003). Autonomy, embeddedness and adaptability in immigration contexts. *Human Development, 46*(2–3), 145–150. https://doi.org/10.1159/000068584.

Kiselica, M. S., Changizi, J. C., Cureton, V. L. L., & Gridley, B. E. (1995). Counselling children and adolescents in schools. Salient multicultural issues. In J. G. Ponteroto, J. M. Casas, L. A. Suzuki, & C. M. Alexander (Eds.), *Handbook of multicultural counselling* (pp. 516–532). Thousand Oaks, CA: Sage Publications.

Kofman, E. (2000). The invisibility of skilled female migrants and gender relations in studies of skilled migration in Europe. *International Journal of Population Studies, 6*, 45–59. https://doi.org/10.1002/(SICI)1099-1220(200001/02)6:1<45::AID-IJPG169>3.0.CO;2-B.

Kwak, K. (2003). Adolescents and their parents: A review of intergenerational family relations for immigrant and non-immigrant families. *Human Development, 46*(2–3), 115–136. https://doi.org/10.1159/000068581.

Li Chiu, M., Feldman, S. S., & Rosenthal, D. A. (1992). The influence of immigration on parental behavior and adolescent distress in Chinese families residing in two Western nations. *Journal of Research on Adolescence, 2*(3), 205–239.

Liebig, T., & Widmaier, S. (2009). *Children of immigrants in the labour markets of EU and OECD countries: An overview.* Organization for Economic Cooperation and Development (OECD), Directorate for Employment, Labour and Social Affairs. https://doi.org/10.1787/9789264086395-2-en.

Liebkind, K., Mähönen, T.-A.E., Varjonen, S. A., & Jasinskaja-Lahti, I. (2016). Acculturation and identity. In D. L. Sam & J.W. Berry (Eds.), *Cambridge handbook of acculturation psychology* (2nd ed., pp. 30–49). Cambridge: Cambridge University Press.

Lim, S. L., Yeh, M., Liang, J., Lau, A. S., & McCabe, K. (2008). Acculturation gap, intergenerational conflict, parenting style, and youth distress in immigrant Chinese American families. *Marriage & Family Review, 45*(1), 84–106. https://doi.org/10.1080/01494920802537530.

Maccoby, E. E. (1988). Gender as a social category. *Developmental Psychology*, *24*(6), 755–765.
Maccoby, E. E. (2000). Perspectives on gender development. *International Journal of Behavioral Development*, *24*(4), 398–406.
Maiter, S., & George, U. (2003). Understanding context and culture in the parenting approaches of immigrant South Asian mothers. *Affilia*, *18*(4), 411–428. https://doi.org/10.1177/0886109903257589.
Manuh, T. (1999). 'This place is not Ghana': Gender and rights discourse among Ghanaian men and women in Toronto. *Ghana Studies*, *2*, 77–96.
Marchetti, R. (Ed.). (2017). *Debating migration to Europe: Welfare vs identity*. Routledge.
Marchetti, S. (2019). Gender, migration and globalisation: An overview of the debates. In A. Triandafyllidiou (Ed.), *Handbook of migration and globalisation* (pp. 444–456). Edward Elgar Publishing.
Marks, A. K., Ejesi, K., & García Coll, C. (2014). Understanding the US immigrant paradox in childhood and adolescence. *Child Development Perspectives*, *8*(2), 59–64. https://doi.org/10.1111/cdep.12071.
Masten, A. S., & Motti-Stefanidi, F. (2009). Understanding and promoting resilience in children: Promotive and protective processes in schools. In C. R. Reynolds & T. B. Gutkin (Eds.), *The handbook of school psychology* (4th ed., pp. 721–738). New York: Wiley.
McKinley, B. (2006). *18 December; International migrants day*. http://un.org.ua/en/information-centre/news/383-2006-12-18-11-20-13.
Meyer, M. K., & Prügl, E. (1999). *Gender politics in global governance*. Lanham, MD: Roman and Littlefield.
Migration Data Portal. (2018). *Family migration*. https://migrationdataportal.org/themes/family-migration.
Mirdal, G. M. (1984). Stress and distress in migration: Problems and resources of Turkish women in Denmark. *International Migration Review*, *18*(4), 984–1003. https://doi.org/10.1177/019791838401800405.
Mohapatra, S., & Ratha, D. (2011). Remittance markets in Africa. *World Development*, *3*, 59.
Motti-Stefanidi, F. (2018). Immigrant and refugee youth positive adaptation and development. In J. E. Lansford & P. Banati (Eds.), *Handbook of adolescent development research and its impact on global policy* (pp. 217–237). Oxford: Oxford University Press.
Motti-Stefanidi, F., Berry, J. W., Chryssochoou, S., Sam, D. L., & Phinney, J. S. (2012). Positive immigrant youth adaptation in context. In A. Masten, K. Liebkind, & D. J. Hernandez (Eds.), *Realizing the potential of immigrant youth* (pp. 117–158). Cambridge: Cambridge University Press.
Murdock, G. P. (1949). *Social structure*. New York: Macmillan.
Okagaki, L., Helling, M. K., & Bingham, G. E. (2009). American Indian college students' ethnic identity and beliefs about education. *Journal of College Student Development*, *50*, 157–176. https://doi.org/10.1353/csd.0.0060.
Oppedal, B., & Toppelberg, C. (2016). Culture competence: A developmental task of acculturation. In J. W. Berry & D. L. Sam (Eds.), *The Cambridge handbook of acculturation psychology revised* (pp. 71–92). Cambridge: Cambridge University Press.
Organization for Economic Cooperation and Development (OECD). (2006). *Where immigrant students succeed: A comparative review of performance and engagement in PISA 2003*. Paris: OECD.
Organization for Economic Cooperation and Development (OECD). (2017). *International migration outlook 2017*. Paris: OECD Publishing.https://doi.org/10.1787/migr_outlook-2017-en.
Pedraza, S. (1991). Women and migration: The social consequences of gender. *Annual Review of Sociology*, *17*(1), 303–325. https://doi.org/10.1146/annurev.so.17.080191.001511.
Phinney, J. S. (1990). Ethnic identity in adolescents and adults: Review of research. *Psychological Bulletin*, *108*(3), 499–514.

Phinney, J. S., & Devich-Navarro, M. (1997). Variations in bicultural identification among African American and Mexican American adolescents. *Journal of Research on Adolescence*, 7(1), 3–32.

Portes, A., & Fernández-Kelly, P. (2008). No margin for error: Educational and occupational achievement among disadvantaged children of immigrants. *The Annals of the American Academy of Political and Social Science*, *620*, 12–36. https://doi.org/10.1177/0002716208322577.

Ravenstein, E. G. (1885). The laws of migration. *Journal of the Statistical Society of London*, *48*(2), 167–235.

Reiser, L. (2009). Immigrant children and education. In F. Chang-Muy & E. P. Congress (Eds.), *Social work with immigrants and refugees: Legal issues, clinical skills and advocacy* (pp. 209–234). New York: Springer.

Robeyns, I. (2006). Three models of education: Rights, capabilities and human capital. *Theory and Research in Education*, *4*, 69–84. https://doi.org/10.1177/1477878506060683.

Rollins, J. (1985). *Between women: Domestics and their employers*. Philadelphia, PA: Temple University Press.

Rosenthal, D. (1996). Vietnamese adolescents in Australia: Relationships between perceptions of self and parental values, intergenerational conflict, and gender dissatisfaction. *International Journal of Psychology*, *31*(2), 81–92.

Rosenthal, D. A., & Feldman, S. S. (1992). The nature and stability of ethnic identity in Chinese youth: Effects of length of residence in two cultural contexts. *Journal of Cross-Cultural Psychology*, *23*(2), 214–227. https://doi.org/10.1177/0022022192232006.

Salaff, J. W. (2002). Women's work in international migration. In E. Chow (Ed.), *Transforming gender and development in East Asia* (pp. 217–238). London: Routledge.

Sam, D. L. (2006). Acculturation of immigrant children and women. In D. L. Sam & J. W. Berry (Eds.), *The Cambridge handbook of acculturation psychology* (1st ed., pp. 403–418). Cambridge: Cambridge University Press.

Sam, D. L. (2014). Immigrant families in a global context: Challenges and future directions. In D. L. Sam (Ed.), *Global perspectives on well-being in immigrant families* (pp. 311–320). New York: Springer.

Sam, D. L., Vedder, P., Liebkind, K., Neto, F., & Virta, E. (2008). Immigration, acculturation and the paradox of adaptation in Europe. *European Journal of Developmental Psychology*, *5*(2), 138–158. https://doi.org/10.1080/17405620701563348.

Santisteban, D. A., & Mitrani, V. B. (2003). The influence of acculturation processes on the family. In K. M. Chun, P. Balls Organista, & G. Marín (Eds.), *Acculturation: Advances in theory, measurement, and applied research* (pp. 121–135). Washington, DC: American Psychological Association. https://doi.org/10.1037/10472-009.

Shaheed, F. (1999). Constructing identities: Culture, women's agency and the Muslim world. *International Social Science Journal*, *51*(159), 61–73. https://doi.org/10.1111/1468-2451.00177.

Sluzki, C. (1979). Migration and family conflict. *Family Process*, *18*(4), 379–390.

Spera, C. (2005). A review of the relationship among parenting practices, parenting styles, and adolescent school achievement. *Educational Psychology Review*, *17*, 120–146. https://doi.org/10.1111/j.1545-5300.1979.00379.x.

Statistics Norway. (2018). *Flere barn i husholdninger med vedvarende lavinntekt* [Child poverty and children in low-income families; Concepts and research] https://www.ssb.no/inntekt-og-forbruk/artikler-og-publikasjoner/flere-barn-i-husholdninger-med-vedvarende-lavinntekt.

Strand, S. (2011). The limits of social class in explaining ethnic gaps in educational attainment. *British Educational Research Journal*, *37*(2), 197–229. https://doi.org/10.1080/01411920903540664.

Suárez-Orozco, C., Katsiaficas, D., Birchall, O., Alcantar, C. M., Hernandez, E., Garcia, Y., … Teranishi, R. T. (2015). Undocumented undergraduates on college campuses: Understanding their challenges and assets and what it takes to make an undocufriendly campus. *Harvard Educational Review*, *85*(3), 427–463. https://doi.org/10.17763/0017-8055.85.3.427.

Suarez-Orozco, C., Motti-Stefanidi, F., Marks, A., & Katsiaficas, D. (2018). An integrative risk and resilience model for understanding the adaptation of immigrant origin children and youth. *American Psychologist, 73*(6), 781–796. https://doi.org/10.1037/amp0000265.

Suárez-Orozco, C., Yoshikawa, Y., Teranishi, R., & Suárez-Orozco, M. (2011). Living in the shadows: The developmental implications of unauthorized status. *Harvard Education Review, 81*(3), 438–472. https://doi.org/10.17763/haer.81.3.g23x203763783m75.

Teranishi, R. T., Suárez-Orozco, C., & Suárez-Orozco, M. (2015). *In the shadows of the ivory tower: Undocumented undergraduates and the liminal state of immigration reform.* http://hdl.handle.net/10919/83640.

Tonah, S., & Setrana, B. M. (2017). Introduction. In S. Tonah, B. M. Setrana, & J. A. Arthur (Eds.), *Migration and development in Africa: Trends, challenges, and policy implications* (pp. 1–39). Maryland: Lexington Books.

United Nations. (2018). *Sustainable development goals.* https://sustainabledevelopment.un.org/sdgs.

United Nations, Department of Economic and Social Affairs, Population Division. (2017). *Trends in international migrant stock: The 2017 revision.* https://www.un.org/en/development/desa/population/migration/data/estimates2/docs/MigrationStockDocumentation_2017.pdf.

United Nations Population Fund (UNFPA). (n.d.) Meeting the challenges of migration. Progress since the ICPD. United Nations Population Fund. https://www.unfpa.org/sites/default/files/pub-pdf/migration_icpd.pdf.

Vianello, F. A. (2014). Ukrainian migrant workers in Italy: Coping with and reacting to downward mobility. *Central and Eastern European Migration Review, 3*(1), 85–98.

Vedder, P., Horenczyk, G., & Liebkind, K. (2006). Problems in ethno-cultural diverse educational settings and strategies to cope with these challenges. *Educational Research Review, 1*(2), 157–168.

Vedder, P., & Motti-Stefanidi, F. (2016). Children, families and school. In D. L. Sam & J.W. Berry (Eds.), *The Cambridge handbook of acculturation psychology* (2nd ed., pp. 464–482). Cambridge: Cambridge University Press.

Venkataramani, A. S., Shah, S. J., O'Brien, R., Kawachi, I., & Tsai, A. C. (2017). Health consequences of the US Deferred Action for Childhood Arrivals (DACA) immigration programme: A quasi-experimental study. *The Lancet Public Health, 2*(4), e175–e181. https://doi.org/10.1016/S2468-2667(17)30047-6.

Yao, E. L. (1988). Working effectively with Asian immigrant parents. *Phi Delta Kappan, 70*, 223–225.

Zlotnik, H. (1990). International migration policies and the status of female migrants. *International Migration Review, 24*, 372–381. https://doi.org/10.1177/019791839002400208.

CHAPTER 14

The antiimmigrant sentiment and its impact on immigrant families

Selcuk R. Sirin, Esther J. Sin, Clare Clingain, and Elysia Choi
Department of Applied Psychology, New York University, New York, NY, United States

The rate of human migration is consistently rising. According to the United Nations (UN) Migration Agency, over the past four-and-a-half decades, the number of international migrants has increased from 153 million in 1990 to 244 million in 2015. There are now more than 60 million people who are forced out of their country by violence, famine, or other catastrophes. The United States, Germany, and Russia have been the top three destinations for migrants and India, Mexico, and Russia are the top three places migrants are coming from (United Nations Migration Agency, 2018). International migration, especially in the United States and Europe, has been publicized due to the political aspect. However, contrary to what the mass media portrays, migration is happening everywhere. In Africa, within country migration has been on a rise where it exceeds the number of international migration (United Nations Migration Agency, 2018). An economic report of Africa by the United Nations has also reported that migration within Africa is promoting economic benefits (United Nations Conference on Trade and Development, 2018). Also, within country migration usually happens from rural areas to big cities in Asian countries, such as India and China, where poor migrants "return temporarily to their villages when agricultural labor demands are high and permanently in old age" (Castles & Miller, 2019). Although within country migration in places like Africa has brought beneficial changes, due to the increasing difficulty of international migration to places like the United States and Europe, within country migration is the only option left (Mcauliffe & Kitimbo, 2019). Even when one is successful at international migration, they receive mixed messages from their new countries.

The rise in migrations around the globe has led to a divided world. In the United States, 75% of people believe that immigration is a good thing and 83% of Americans favor children who have not gained legal approval

Cross-Cultural Family Research and Practice
https://doi.org/10.1016/B978-0-12-815493-9.00014-4

to remain in the United States being given a chance to become US citizens (Brenan, 2018). However, 22% of Americans believe that immigration is the nation's most important problem and this number is the highest it has ever been (Newport, 2018). At the border of Ceuta, Spain, there have also been reports of many African migrants trying to cross the fence over to Europe (Nordland, 2018). People show mixed emotions as a result of these migrants crossing borders: "The people who don't support them say they come to invade us – that they want to steal, take our jobs and take advantage of our health system. But they just want to work and send money back to their families" (CBS NEWS, 2019). The YouGov-Cambridge Globalism Project polls have shown that people believe the "costs of immigration outweigh the benefits for their country, ranging from 31% in the United States to 37% in the United Kingdom, 40% in Germany and 50% in Italy." This antiimmigrant sentiment is also happening in countries like Australia, where 46% of people agreed that immigration should be reduced (Rogers de Waal, 2019).

However, when taking a closer look at the data, people also do not believe that immigration should decrease by "a lot" and all of the countries had positive thoughts about "qualified professionals coming here with a job offer" (Rogers de Waal, 2019). Thus, migration is a prominent topic for many countries; the experiences of refugees, asylum seekers, interregional migrants, and immigrants all differ with one another and with the increasing rates of migration, we encourage researchers to delve deeper into the lives of migrants everywhere. Throughout the rest of this chapter, we will focus on immigrants and acculturation within the United States and Europe in order to better portray their specific experiences. This focus reflects, in part, that over half of all international immigrants move to one of 10 high-income countries in Western Europe or North America (International Organization for Migration, 2015). We will draw from research conducted in the United States and Europe to highlight the unique contexts in which immigrants' lives are embedded in, the challenges and stressors they encounter, and the social support that may buffer their experiences. We conclude with recommendations for future directions min research and social policy.

Current macro-context in the United States and Europe

Immigration can be a stressful transition for families as they negotiate multiple social and cultural forces between their home and host cultures. This process of acculturation has become decisively more stressful in the recent years due to the global political shift. With the recent political realignment in the United States and Europe, there is now a growing antiimmigrant climate, which creates a toxic environment to raise children for immigrants. In the United States, with the 2016 election of President Trump, immigration is at the center of political debate to a degree that American's rating immigration a problem is now the highest it has ever been in history (Brenan, 2018). Similarly, in Europe, every major election in the recent years has involved a divisive debate about the role of migrants (Crabtree & Kluch, 2017).

As reviewed in the Migration Policy Institute report by Pierce, Bolter, and Selee (2018), looking deeper into the antiimmigration climate in the United States, we can see that the Trump administration is unique in that no other American administration in the past has focused so much on immigration policies. To name a few, the administration has "mandated interviews for all visa applicants, increased scrutiny applied to temporary visa issuance, and drastically narrowed the eligibility criteria for asylum" (Pierce et al., 2018, p. 1). The number of refugees allowed to enter the United States has dropped to the same levels as when the US resettlement program began in 1980, and the Temporary Protected Status given to people who have experienced natural disasters and violent conflicts have been terminated for people from certain countries. Some policies (e.g., zero-tolerance policy) have taken the media by storm due to their controversial nature.

Some countries in Europe have also started implementing stricter migration policies. Similar to Trump administration's intensive policies on immigration enforcement, the Italian government passed a bill "making it easier to deport migrants and strip them of the Italian citizenship if they are guilty of serious crimes" (British Broadcasting Commission, 2018). In Europe, the populist radical right parties, which believe in exclusive nationalism and antiimmigrant policies, have been around for decades but have had little electoral success until very recently (Akkerman, 2018; Schain, 2018). However, the radical right parties have benefitted from any perceived loss of the current government's control over migration because these parties are seen as the face of antiimmigration with constant statements about their disapproval (Akkerman, 2018). The mainstream parties are forced to respond to the radical right parties to regain voters. For example, in Austria, Denmark, Italy, the Netherlands, Norway, and Switzerland, radical right parties have increasingly joined governing coalitions (Schain, 2018). Nordic countries have also tightened policies "reducing family reunification, restricting access to refugee and other protected statuses, and limiting access to public assistance benefits" (Widfeldt, 2018, p. 1).

With Trump's election, policy changes and executive orders have led to a climate of fear and stress for the immigrant communities (Capps, Muzaffar, Gelatt, Bolter, & Ruiz Soto, 2018). This fear is evident in the changes in behavior of the people. Crime reports are dropping in immigrant neighborhoods and utilization of public benefits, such as health care, have decreased (Pierce et al., 2018). Although not all of the proposed antiimmigration changes have been implemented, the sharp rise in strict immigration enforcement continues to stir up more antiimmigration sentiment, and the US Federal Bureau of Investigation has reported more hate crimes against Muslims (Kishi, 2017; McHugh, 2018). Hate crimes have also spiked in other countries like Germany, where 3598 hate crimes in 2016 more than doubled to 7913 in 2017 with the biggest reason for hate crimes being racism and xenophobia (Organization for Security and Co-operation in Europe Office for Democratic Institutions and Human Rights, 2019). In the United States, there has been a decrease in the number of unauthorized immigrants crossing the border and in Europe, immigration levels have fallen to pre-2015 levels (Lopez, Kristen, & Radford, 2018; Silver, 2018). While the immigrants face financial stress and negative impacts on youth development in all domains, the recent enforcement changes are not particularly benefitting the US citizens (Gelatt, Koball, Bernstein, Runes, & Pratt, 2017).

Divisive rhetoric and policy debates have real consequences. Any closer look into immigrant communities in the West will show us signs of a stressful environment. Stillman (2018) in her article on deportation tells the story of Laura. Laura was on her way home from her waitressing job when a police officer stopped her for driving between the lines. Once the officer realized that Laura was unauthorized, he had called Border Control. That one phone call turned her life upside down. She was sent back to Reynosa in Mexico, a city known for its violence and the home of Laura's abusive exboyfriend. After a few weeks, she was found dead and her exboyfriend admitted to strangling her to death. Bell (2018) reports an interview with an Austrian citizen, Friedrich, who voiced his opinion that the migrants in Austria "don't want to integrate." Many citizens in Austria have similar opinions to Friedrich. Meanwhile, there are people like Saleha Yaghubi from Afghanistan that moved to Austria in 2016 with her son and are pending the grant to stay. Saleha mentioned that she "can't sleep at night. I wake up and pray that all these problems will go away." Due to the rise in stricter laws enforcing immigrants and forcing unauthorized immigrants out of the country, the immigrants living in these antiimmigrant sentiment countries are undoubtedly suffering.

Although mixed messages from the host countries have led to fear and anxiety in immigrants, there is also increased support for immigrants from some members of the host country and from fellow immigrant community members. For example, while the Trump administration is pushing for stricter immigration enforcement, the divided nature of the US government is working in favor of immigrants. Several Trump decisions are now challenged or stopped by the courts. The funds required to be provided by Congress to commence these policy changes have made the Congress resist some of these policy changes (Pierce et al., 2018). Some state governments are supporting and others resisting Federal US antiimmigrant policies. For example, Texas has made it a criminal offense for the local police, government department heads, or elected officials to refuse to assist federal immigration enforcement; whereas California lawmakers have made the Trump administration's deportation efforts more difficult (McHugh, 2018). There have also been many protestors and demonstrators demanding an end to detention centers holding migrant children (McLaughlin & Chavez, 2019).

The conflicted nature of the debate about immigration is also visible in European countries. Although there is evidence of public debate on migrants in Germany, in small towns like Altena, with a 10% drop in their population, most of the neighborhood was welcoming of refugees (Shubert & Schmidt, 2017). Although some were fearful, people like Bernadette Koopman who acted as a mentor for the new refugees believed in their rights to ask for help (Shubert & Schmidt, 2017). Even undocumented migrants in Paris have stormed the Pantheon for the rights to stay in Paris (BBC News, 2019). Thus, although there is a sense of fear and hatred for migrants in the United States and Europe, the social support from lawmakers, citizens of the host countries, and fellow immigrant community members bolster resilience within the immigrant community.

The integrative risk and resilience model

These case studies highlight the ways in which policy changes and antiimmigrant sentiment can affect immigrants' lives. With a closer look, we can see how such changes affect youth. Maria and Joel's children are growing up without a father figure, while Saleha's son watches his mother's daily fear, likely taking on some of his mother's worry. A comprehensive view of the effects of immigration policy and antiimmigration sentiment on immigrant youth development was elucidated by Suárez-Orozco, Motti-Stefanidi, Marks, and Katsiaficas (2018) integrative risk and resilience model. Borrowing from Bronfenbrenner and Morris's (2006) ecological framework, García Coll et al.'s (1996) integrative development model of minority youth, and Motti-Stefanidi, Berry, Chryssochoou, Sam, and Phinney's (2012) integrative resilience model of immigrant youth development, the integrative risk and resilience model utilizes a multilevel approach to understand the development and adaptation of immigrant youth. The primary factor that distinguishes the current model from its predecessors is its inclusion and exploration of the global context and global forces. The unique ways in which immigrant youth develop and adapt to their environments are inextricable linked to the global forces around them. Relevant contextual factors for refugee and undocumented youth are also embedded in the model, providing a robust framework through which to understand the nuances and complexities of immigrant youth development. For the purpose of this chapter, we will discuss specific layers of this model as they relate to raising children in the current antiimmigrant sentiment in each of the developmental contexts (see Fig. 14.1).

The most distal ecological context of the integrative framework is global forces, which are defined as economic, political, and social realities

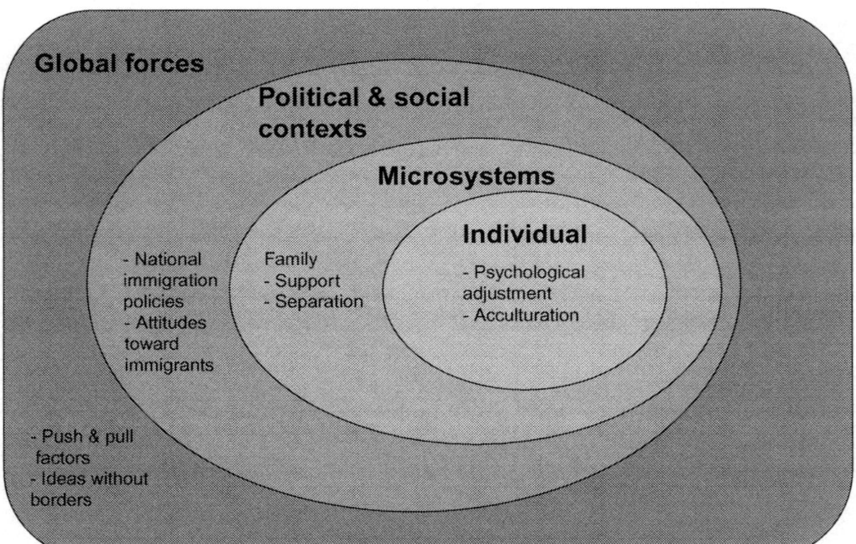

Fig. 14.1 A simplified version of the integrative risk and resilience model by Suárez-Orozco et al. (2018).

that have unduly influence on immigrant youth's adaptation compared to the adaptation of their nonimmigrant counterparts (Suárez-Orozco et al., 2018). In the US context, the economic realities of the *American dream*—that one can create a stable and viable living for oneself from hard work—serve as "pull factors" for mass immigration to the country. The economic opportunities and protections offered by the European Union also draw waves of immigrants from postcolonial countries. Paired with the economic possibilities of the United States and Europe is the relative economic underdevelopment of the nations from which these immigrants depart; thus, the economic realities consist of both "pull" and "push" factors (Suárez-Orozco et al., 2018). Political and social realities often result in the forced migration of people, including the recent Syrian refugee crisis, which has caused both internal and external displacement of more than 11 million Syrian people (Sirin & Rogers-Sirin, 2015). Moreover, hateful ideologies such as Islamophobia and xenophobia—"ideas without borders"—continue to intensify in some Western countries (Suárez-Orozco et al., 2018, p. 786). Together, these various global forces generate the environment in which immigrant youth will develop. However, these environments are not identical across Western countries. For example, both Canada and Italy are wealthy countries who have long histories of immigration, yet Canada doesn't grapple with the same push of antiimmigrant ideologies as does Italy.

Embedded within global forces are political and social contexts regarding perceptions of immigrants (Suárez-Orozco et al., 2018). The United States and Europe are experiencing new waves of antiimmigrant sentiment, which manifest into restrictive national immigration policies. A rise in nationalism and assimilationist ideologies limits the definition of what it means to be a member of a given society, particularly for immigrants of color. Such changes in political and social attitudes toward immigrants directly impact the next level of the model: immigrant families. Families serve as a vital source of social support for immigrant youth. The rise in deportation and family separation has lasting psychological consequences for immigrant youth, with prolonged separation linked to negative psychological well-being (Suárez-Orozco et al., 2018). The "deportation nation" attitude of the current US administration threatens sources of familial support for immigrant youth.

Political and social attitudes also impact another microsystem for youth: schools. Outside of the family, schools are the environment in which youth spend most of their time. Youth are exposed to new ideas, languages, and cultures through their interactions with peers, teachers, and the education system of their respective location. How political and social attitudes affect schools varies largely across and within contexts. For example, New York City and Berlin boast public schools designed to help youth assimilate to American or German culture while maintaining their ethnic identities and cultures. Despite their successes, the decentralized education systems of the United States and Germany face pushback from teachers and policy makers who claim immigrant youth overstretch the limits of teachers who aren't trained to handle such diverse backgrounds, particularly when it comes to language (Public Radio International, 2017), resulting in the tracking of migrant youth into nonregular or remedial education streams and ostracization from the larger student body.

At the center of the integrative model proposed by Suárez-Orozco et al. (2018) is the individual. The unique task which immigrant origin youth face at this level are acculturation—process of adapting to a new country that involves heritage-culture retention and receiving-culture acquisition (Berry, 1997)—and acculturative stress—the turmoil that arises from the adjustments. As we will discuss in the following section, acculturation and acculturative stress have implications for immigrant families, immigrant youths, and their mental health and well-being. Balancing both the home culture and the host culture are crucial to immigrant youth's adaptation and well-being.

Acculturative stress and immigrant families

The global and local contexts in which the individuals acculturate are important factors to consider because families' acculturation experiences unfold within the local community of resettlement (Schnittker, 2002), and their acculturation processes are affected by the immigrant group's experience in the larger society (Gibson, 2001). As stated above, acculturation is broadly defined as the process of adapting to a new country and involves heritage-culture retention and receiving-culture acquisition (Berry, 1997). It involves complex changes on multiple levels including individual, family, and cultural (Berry, 1997, 2002) as a result of contact with different cultures and social systems (Berry, 2002; Berry, Trimble, & Olmedo, 1986; Redfield, Linton, & Herskovits, 1936). Immigrant families with children are the most vulnerable group that are directly impacted by the antiimmigrant sentiment, which shape the group's acculturation experiences.

Acculturation can be a stressful experience for a variety of reasons. The unique stressors that immigrants experience during the acculturation process are referred to as *acculturative stress* (Berry, 2006). Immigrant families experience these acculturative stressors as they navigate and negotiate different cultures and environments (Staudenmeyer, Macciomei, Del Cid, & Patel, 2016). Sources of acculturative stress can come from a number of changes, such as the experience of taking on a minority status for the first time, learning a new language, navigating through novel institutional systems (e.g., schools), and creating new social supports and networks (Kia-Keating, Capous, Juang, & Bacio, 2016). Further, the antiimmigrant sentiment manifested in threats of deportation, separation from families, and documentation status also contribute to acculturative stress (Bekteshi & Kang, 2018; Suarez-Orozco & Suarez-Orozco, 2001). Acculturative stress can also be experienced through negative attitudes, stereotypes, and discrimination (Rogers-Sirin, Ryce, & Sirin, 2014). Thus, immigrant families experience unique challenges and acculturative stress as their acculturation experiences unfold in the host country.

Acculturative stress implications for immigrant families

A key contributor to immigrants and their families' acculturative stress is forced separation, which has negative implications for children's development and family functioning. Specifically, with 16.7 million people having at least one unauthorized immigrant family member (foreign-born noncitizens who are not legal residents) living in the same house, and more than 5.9 million citizen children living with at least one family member who is unauthorized in the United States (Mathema, 2017), fear of separation is a prevalent anxiety experienced by families. The act of separation is a traumatic experience for the whole family. Once a family separates, children report emotional problems and mental health issues such as post-traumatic stress disorder (PTSD) symptoms (Rojas-Flores, Clements, Koo, & London, 2017; Zayas, Aguilar-Gaxiola, Yoon, & Rey, 2015). Further, separation affects the family structure. Being separated from the family leads to financial stress and poor living conditions (Chaudry et al., 2010). If a father is deported, the mother is forced to provide for the family by working longer hours, which leads the older children to take on the role of primary caregiver for their younger siblings (Chaudry et al., 2010; Dreby, 2012), while some family members may be forced to live with other relatives and friends (Chaudry et al., 2010). Children in these circumstances manifest changes in psychosocial behaviors such as crying more, and these behaviors affect short- and-long-term development

(Chaudry et al., 2010). Not only does the actual separation lead to trauma, the common exposure to politics of deportation leads immigrant families to be fearful for their future and for deportation to their home countries (Dreby, 2012). The fear that is instilled in immigrant families affects family functioning and children's developmental outcomes (Lorenzo-Blanco et al., 2017). For example, when the family is affected by stress and fear, reports of family functioning go down and youth become more likely to take part in health-risk behaviors (Lorenzo-Blanco et al., 2017). These negative effects are expected to increase in the current environment as the level of the deportation threat has escalated in the recent years.

Although the family members involved in a deportation of an unauthorized immigrant experience the most amount of stress and trauma, so do their friends and the whole immigrant community. Unauthorized immigrants report higher levels of immigration challenges, but groups report similar levels of fear of deportation (Arbona et al., 2010). Traumatic events, such as deportation or detainment, produce a less severe harm for the larger number of Latinos who are not directly experiencing deportation (Dreby, 2012). Children with a deported parent suffer the most but children involved in the immigrant community are affected by immigration enforcement and are suffering from mental health issues (Gulbas et al., 2015). There is a spillover effect where even if one does not feel like they are directly at risk, the fear and concern influence the family and community members (Quiroga, Medina, & Glick, 2014).

Aside from the separation and fear of deportation, immigrants also struggle with hostile attitudes, discrimination, stereotypes, and prejudice (Deaux, 2006; Rogers-Sirin et al., 2014). Theoretical underpinning and research have long emphasized the negative effects that discrimination and stigma have on psychological well-being (Allport, 1954, 1979; Erikson, 1956; Spencer, 1999). The impact of racism and discrimination may be particularly salient during adolescence, as this is an important developmental period of identity and self-concept (Erikson, 1968; Harter, 1998). Families are also negatively affected by perceived discrimination. For example, immigrant groups may internalize the negative perceptions about their ethnic group, which in turn may lead to negative perceptions of the family (Schwartz et al., 2013).

Intergenerational conflict as a result of varying degrees of acculturation between parents and children is another contributing factor of acculturative stress. Portes and Rumbaut (2001) conducted a longitudinal study of second-generation immigrants that demonstrated that dissonant acculturation (when compared to parents children master the language and cultural norms of the host country and disconnect from ethnic culture faster than their parents) diminishes parents' ability to act as protectors and authorities in their children's lives, and this in turn is linked to increased mental health risk. The selective pattern, or when both parents and children maintain healthy connections to both cultures, has the most positive outcomes, in part because immigrant families in this pattern were better able to cope with discrimination by maintaining support within their families and ethnic communities. These findings demonstrate the importance of examining dynamics between immigrant parents and children.

A number of studies have found consistent evidence of the negative impact of acculturative stressors on mental health outcomes of immigrant youth (e.g., Fine & Sirin, 2007; García Coll & Marks, 2009; Sirin et al., 2015; Sirin, Ryce, Gupta, & Rogers-Sirin, 2013). These findings have been consistent across ethnic groups, such as Asian immigrants (Cho & Haslam, 2010; Mui & Kang, 2006), Arab immigrants (Abdulrahim & Baker, 2009), Pakistani immigrants (Jibeen & Khalid, 2011), and Latino immigrants (Crockett et al., 2007; Hovey, 2000; Mejía & McCarthy, 2010). Further, longitudinal studies that have examined the relation between trajectories of acculturative stress and mental health among immigrant youth have found that greater exposure to acculturative stress predicted significantly more withdrawn/depressed, somatic, and anxious/depressed symptoms overtime (e.g., Sirin et al., 2013). Similarly, Smokowski, Rose, and Bacallao (2010) examined the effects of acculturative stress overtime, and their study found that while there was an overall decrease in internalizing mental health symptoms overtime, acculturative stress was associated with higher levels of internalizing symptoms and a more stable trajectory of symptoms, as opposed to improvement, overtime. In sum, understanding the acculturation process and the impact of acculturative stressors in the lives of immigrant youth and their families are critical as they affect their overall psychological health and development.

Family support as a source of resilience

Despite the challenges immigrant families face in the acculturation process, family support emerges as protective factors that play an important role in positive outcomes for immigrant families and their youth. Resilience is defined as the process of positive adaptation in the context of adversity (Luthar, Cicchetti, & Becker, 2000). According to the literature, resilience involves two major conditions: (1) the individual encounters significant threat or adversity and (2) despite the adversity, the individual achieves positive adaptation during development (e.g., Luthar & Zigler, 1991; Masten, Best, & Garmezy, 1990; Rutter, 1990). As discussed above, immigrant youth and immigrant families experience a wide range of stressful situations before, during, and after they immigrate to their host countries. These challenges shape children's and parents' psychological health and well-being (Rogers-Sirin et al., 2014). Despite the acculturative stress experienced during the acculturation process, relationships within the family provide a protective factor against the negative impacts of acculturative stress (Arends-Toth & van de Vijver, 2006). Thus, learning about family support and its buffering role against the negative impacts of acculturation are important to understand the successful adaptation of immigrants and their families.

Studies have been unequivocal in demonstrating family support as a particularly robust protective factor against mental health outcomes. For example, in a study of immigrant youth, parental support buffered the negative effects of high acculturative stress on anxiety and depressive symptoms (Crockett et al., 2007). Similarly, Leong, Park, and Kalibatseva (2013) found that among Latino and Asian immigrant families, family cohesion serves as a protective factor for depression. Leidy, Guerra, and Toro (2010) found that acculturative differences, challenges faced in child's education, and discrimination are some of the factors that affect immigrants' parents to parent and foster family cohesion. They suggest that that it is important to improve positive parenting and healthy family functioning through strategies that help immigrant parents navigate these significant acculturation gaps. Most recently, a metaanalysis examining various forms of social support, such as teachers, peers, and family members, found that in contrast to other sources of support, the importance of family support was prevalent across all developmental stages (Rueger, Malecki, Pyun, Aycock, & Coyle, 2016) and was negatively associated with depression.

Further, a number of strategies to strengthen family relationship have been found for specific immigrant populations. For example, among a sample of recent Chinese immigrants, strengths and resilience are drawn from family support, social support, communication, and spiritual well-being (Xia, Zhou, & Xie, 2005). Usita and Blieszner (2002) found that Japanese immigrant adolescents used strategies, such as receiving help, asking for clarification from parents, and using humor as ways to navigate the acculturation challenges. Latino values, including familismo (strong familial ties), personalismo (interpersonal relationships), and respeto (respect) have been found to promote family connectedness and solidarity (Cardoso & Thompson, 2010).

Families that demonstrate strong Latino cultural orientation tend to have a sound sense of familismo, which protect against negative mental health problems (Ayón, Marsiglia, & Bermudez Parsai, 2010). Among Asian immigrants, access to social support from relatives and families partially mediated the relationship between acculturative stress and depressive symptoms; specifically, immigrants who reported lower levels of acculturative stress reported higher social support and, in turn, experienced fewer depressive symptoms (Xu & Chi, 2013).

Family support serves an important role in immigrant families, especially since the acculturation processes unfold within a family context (Ward & Geeraert, 2016). These findings collectively highlight the need to take into consideration the strong evidence of the protective effects of family social support on immigrant outcomes. However, it is equally important to acknowledge that the changes in global forces put these protective factors at risk as families are separated under harsh deportation policies and discrimination is heightened. In the following section, we provide recommendations policy makers, practitioners, and researchers to help navigate through the challenges faced by immigrants and their families.

Future directions: Call for action

The devastating effects of increased antiimmigrant climate for families are real and here to stay for the foreseeable future. Despite the challenges, there are roles that policy makers, practitioners, and researchers can execute to help steer the challenges and shape the lives of millions of children and their families who currently live in the United States and Europe as immigrants. Below, we provide recommendations policy makers, practitioners, and researchers.

Recommendations for policy makers

There is now a consensus among scholars and policy experts about the influence of the antiimmigrant sentiment on immigrant families but how to respond to this challenge with policy at different levels of local, regional, and national is still debated. As family separation is the most prominent in the United States, the first policy suggestion is to crucially and heavily consider family unity (Gubernskaya & Dreby, 2017; Langhout et al., 2018; Mathema, 2017). Multiple studies mentioned throughout this chapter exhibit how important family unity is and how inimical family separation can be. Even a few weeks without parents could lead to lifelong trauma for everyone in the family. Separation not only affects the immediate family, but also it affects the whole community by infusing fear (Dreby, 2012; Langhout et al., 2018). A heavier emphasis on family unity can be initiated through establishing a temporary visa status for family members who are trying to receive lawful residency or expanding the eligibility criteria for programs like Deferred Action for Childhood Arrivals, which allows individuals without lawful presence in the United States, and who were brought to the United States in childhood, to receive a renewable 2-year period of deferred action from deportation and become eligible for a work permit (Gubernskaya & Dreby, 2017; Hainmueller et al., 2017; Langhout et al., 2018). Related to the first suggestion on family unity, policy makers should also discourage deportation. The discouragement of deportation would lift the anxiety and fear that immigrant communities face but also lift the "burden on state-run welfare and foster care systems" (Gubernskaya & Dreby, 2017, p. 425). The third recommendation for policy makers is to consider how the antiimmigrant sentiment affects the mental health of immigrant families. One of the most concerning consequences of immigrant enforcement is the mental health problems induced in children and family members within the immigrant community. Providing safe havens through the community, the state, and local jurisdictions or better promotion of health care to legal immigrants could provide a sense of protection and lead to increased trust (Langhout et al., 2018; Raymond-Flesch, Siemons, Pourat, Jacobs, & Brindis, 2014).

All in all, the scientific evidence and policy reports suggest the need to promote social equality, and for the immigrants and the host countries to try to accommodate for one another (Cobb, Meca, Xie, Schwartz, & Moise, 2017; Segal & Mayadas, 2005). These policy recommendations require public support and political will. Specifically, focusing on the contributions of immigrants to their relative host societies, and highlighting the social and economic costs of current antiimmigrants policies, would be important starting points. It is also important to note that current polling shows that younger generations are more supportive of these more humane immigrant policy suggestions, which gives hope for a more positive political climate in the future.

Recommendations for practitioners

Antiimmigrant sentiment manifests itself in the lives of immigrant families and youth through acculturative stress, which is known to have adverse impacts on minority immigrants' mental health. Mental health practitioners have a responsibility to serve immigrant communities to ensure the safety and well-being of children and families in all of the settings they serve, such as schools, hospitals, and community centers. First, mental health practitioners can use family interventions for immigrant clients even at those times when the issues involve only one member of the family. Maintaining and strengthening family social support is crucial for all but it is clear that for immigrant families, this protective factor is critical for mental health since research has shown the protective qualities of social support on psychological distress (Crockett et al., 2007). Second, mental health practitioners should adapt interventions and clinical practices to their clients' home cultures and norms. Lack of culturally competent practices can alienate immigrants from accessing and maintaining services and may result in the microaggressions committed by the therapist (Rogers-Sirin, Melendez, Refano, & Zegarra, 2015). Third, mental health practitioners should pay attention to intergenerational differences between parents and children in terms of acculturation and acculturative stress. How the magnitude of the numerous global forces and political and social contexts affects immigrants may differ from parents to children, from one generation to the next, which may result in intergenerational family conflicts. Thus, practitioners should be mindful of the specific manifestations of acculturative stress in both parents and children.

Recommendations for researchers

The current research field needs more rigorous and systematic empirical data to understand how the current antiimmigrant sentiment affects immigrant families. For example, we do not know how the current zero-tolerance policies of the Trump administration will affect children who are separated from their parents. Similarly, the services provided to refugees throughout Europe deserve more scientific scrutiny as many of the interventions currently in effect are not evidence based. In order to answer these and many other questions that involve immigrant families, we would like to make a number of suggestions. First, given that the developmental effects of the current policies will take some time to observe, there is an acute need to launch longitudinal studies to examine the relation between acculturative stress and developmental outcomes among individuals and varying aspects of family dynamics as the majority of the research to date has been cross sectional (for a review, see Bekteshi & Kang, 2018).

Second, mixed-methods studies could provide a more nuanced and deeper investigation of the actual acculturation processes experienced by immigrants and their families since we have limited knowledge base about the issues facing immigrant families. Third, acculturative stressors may affect individuals and families differently depending on the source of stressors (e.g., structural conditions and cultural factors). While there have been some efforts to understand these factors (Bostean & Gillespie, 2018; Oropesa & Landale, 2004; Padilla & Borrero, 2006), a consistent methodological approach to examining these factors has remained largely overlooked in research on immigration and family relationships. Fourth, as a result of nonrandom sampling, convenience sampling, data collection at a limited number of sites, and the homogeneity of the sample, most of the research findings may not be generalizable to broader populations with different characteristics (Bekteshi & Kang, 2018). Further, studies consistently identify their small sample size as the most frequently reported bias.

Future research should seek to address the above-described limitations in research designs, which would contribute in advancing the field's collective knowledge on the implications of demographic factors (e.g., ethnicity, gender, social class, and immigration status) on immigrant youth and family development. Fifth, the impact that immigration and acculturation-related stressors have in deteriorating family relationships are important for immigrant health researchers (Rook, 1984; Rook, Luong, Sorkin, Newsom, & Krause, 2012). It is important for future research to consider the extent to which immigration-related family relationship changes affect immigrant health outcomes and changes overtime. Finally, given the vulnerable populations that are under major surveillance, researchers should be more ethically and sensitively aware from data collection to publication without putting the population at further risk.

References

Abdulrahim, S., & Baker, W. (2009). Differences in self-rated health by immigrant status and language preference among Arab Americans in the Detroit metropolitan area. *Social Science & Medicine, 68*(12), 2097–2103. https://doi.org/10.1016/j.socscimed.2009.04.017.

Akkerman, T. (2018). *The impact of populist radical-right parties on immigration policy agendas: A look at the Netherlands.* Washington, DC: Migration Policy Institute.

Allport, G. W. (1954). *The nature of prejudice.* Cambridge, MA: Addison-Wesley Publishing Company.

Allport, G. W. (1979). *The nature of prejudice: 25th anniversary edition.* New York, NY: Perseus Books Group.

Arbona, C., Olvera, N., Rodriguez, N., Hagan, J., Linares, A., & Wiesner, M. (2010). Acculturative stress among documented and undocumented Latino immigrants in the United States. *Hispanic Journal of Behavioral Sciences, 32*(3), 362–384. https://doi.org/10.1177/0739986310373210.

Arends-Toth, J., & van de Vijver, F. (2006). Issues in the conceptualization and assessment of acculturation. In M. Bornstein & L. Cote (Eds.), *Acculturation and parent-child relationships* (pp. 33–62). Mahwah, NJ: Lawrence Erlbaum.

Ayón, C., Marsiglia, F. F., & Bermudez Parsai, M. (2010). Latino family mental health: Exploring the role of discrimination and familismo. *Journal of Community Psychology, 38*(6), 742–756. https://doi.org/10.1002/jcop.20392.

BBC News. (2018). *Italy migrant crisis: Government passes tough bill.* Retrieved from https://www.bbc.com/news/world-europe-45625833?intlink_from_url=https://www.bbc.com/news/topics/cnx753je2q4t/europe-migrant-crisis&link_location=live-reporting-story.

BCC News. (2019). *Black vest' protesters storm Panthéon in Paris.* Retrieved from https://www.bbc.com/news/world-europe-48969438.

Bekteshi, V., & Kang, S. W. (2018). Contextualizing acculturative stress among Latino immigrants in the United States: A systematic review. *Ethnicity & Health*, 1–18. https://doi.org/10.1080/13557858.2018.1469733.

Bell, B. (2018). *Salzburg summit: Austria leads EU's anti-migrant drive.* Retrieved from https://www.bbc.com/news/world-europe-45558510.

Berry, J. W. (1997). Immigration, acculturation, and adaptation. *Applied Psychology An International Review, 46*, 5–34. https://doi.org/10.1080/026999497378467.

Berry, J. W. (2002). Acculturation: Living successfully in two cultures. *International Journal of Intercultural Relations, 29*, 697–712. https://doi.org/10.1016/j.ijintrel.2005.07.013.

Berry, J. W. (2006). Acculturative stress. In P. T. Wong & L. J. C. Wong (Eds.), *Handbook of multicultural perspectives on stress and coping* (pp. 287–298). New York: Springer.

Berry, J. W., Trimble, J. E., & Olmedo, E. L. (1986). Assessment of acculturation. In W. J. Lonner & J. W. Berry (Eds.), *Cross-cultural research and methodology series*: Vol. 8. Field methods in cross-cultural research (pp. 291–324). Thousand Oaks, CA: Sage Publications, Inc.

Bostean, G., & Gillespie, B. J. (2018). Acculturation, acculturative stressors, and family relationships among Latina/o immigrants. *Cultural Diversity and Ethnic Minority Psychology, 24*(1), 126. https://doi.org/10.1037/cdp0000169.

Brenan/Gallup, Inc. (2018). *Record-high 75% of Americans say immigration is good thing.* Retrieved from https://news.gallup.com/poll/235793/record-high-americans-say-immigration-good-thing.aspx?version=print.

Bronfenbrenner, U., & Morris, P. A. (2006). The bioecological model of human development. In R. M. Lerner & W. Damon (Eds.), *Handbook of child psychology: Theoretical models of human development* (pp. 793–828). Hoboken, NJ: John Wiley & Sons.

Capps, R., Muzaffar, C., Gelatt, J., Bolter, J., & Ruiz Soto, A. G. (2018). *Revving up the deportation machinery: Enforcement and pushback under Trump.* Washington, DC: Migration Policy Institute.

Cardoso, J. B., & Thompson, S. J. (2010). Common themes of resilience among Latino immigrant families: A systematic review of the literature. *Families in Society, 91*(3), 257–265. https://doi.org/10.1606/1044-3894.4003.

Castles, S., & Miller, M. J. (2019). *Migration in the Asia-Pacific region.* Migration Policy Institute. Retrieved from https://www.migrationpolicy.org/article/migration-asia-pacific-region.

CBS News. (2019). *Fortified Morocco-Spain border fence is under pressure from African migrants.* CBS News.1–20. Retrieved from https://www.cbsnews.com/news/fortified-morocco-spain-border-fence-is-under-pressure-from-african-migrants/.

Chaudry, A., Capps, R., Pedroza, J. M., Castaneda, R. M., Santos, R., & Scott, M. M. (2010). *Facing our future: Children in the aftermath of immigration enforcement.* PsycEXTRA Dataset. https://doi.org/10.1037/e726272011-001.

Cho, Y.-B., & Haslam, N. (2010). Suicidal ideation and distress among immigrant adolescents: The role of acculturation, life stress, and social support. *Journal of Youth and Adolescence, 39*(4), 370–379. https://doi.org/10.1007/s10964-009-9415-y.

Cobb, C. L., Meca, A., Xie, D., Schwartz, S. J., & Moise, R. K. (2017). Perceptions of legal status: Associations with psychosocial experiences among undocumented Latino/a immigrants. *Journal of Counseling Psychology, 64*(2), 167–178. https://doi.org/10.1037/cou0000189.

Crabtree, S., & Kluch, S. (2017). *Terrorism, migration trouble many in Europe.* Retrieved from https://news.gallup.com/poll/212405/terrorism-migration-trouble-europe.aspx?version=print.

Crockett, L., Iturbide, M., Torres Stone, R., McGinley, M., Raffaelli, M., & Carlo, G. (2007). Acculturative stress, social support, and coping: Relations to psychological adjustment among Mexican American college students. *Cultural Diversity and Ethnic Minority Psychology*, *13*(4), 347–355. https://doi.org/10.1037/1099-9809.13.4.347.

Deaux, K. (2006). *To be an immigrant*. New York: Sage.

Dreby, J. (2012). The burden of deportation on children in Mexican immigrant families. *Journal of Marriage and Family*, *74*, 829–845. https://doi.org/10.1111/j.1741-3737.2012.00989.x.

Erikson, E. (1968). *Identity: Youth and crisis*. New York: Norton and Co.

Erikson, E. H. (1956). The problem of ego identity. *Journal of the American Psychoanalytic Association*, *4*(1), 56–121. https://doi.org/10.1177/000306515600400104.

Fine, M., & Sirin, S. R. (2007). Theorizing hyphenated lives: Researching marginalized youth in times of historical and political conflict. *Social and Personality Psychology Compass*, *1*, 16–38. https://doi.org/10.1111/j.1751-9004.2007.00032.x.

García Coll, C., Lamberty, G., Jenkins, R., McAdoo, H. P., Crnic, K., Wasik, B. H., & Vázquez García, H. (1996). An integrative model for the study of developmental competencies in minority children. *Child Development*, *67*, 1891–1914. https://doi.org/10.2307/1131600.

García Coll, C., & Marks, A. K. (2009). *Immigrant stories: Ethnicity and academics in middle childhood*. New York: Oxford University Press.

Gelatt, J., Koball, H., Bernstein, H., Runes, C., & Pratt, E. (2017). *State immigration enforcement policies (Rep.)*. Washington, DC: Urban Institute.

Gibson, M.A. (2001). Immigrant adaptation and patterns of acculturation. *Human Development*, *44*(1), 19–23. https://doi.org/10.1159/000057037.

Gubernskaya, Z., & Dreby, J. (2017). US immigration policy and the case for family unity. *Journal on Human Migration and Human Security*, *5*(2), 417–430. https://doi.org/10.1177/233150241700500210.

Gulbas, L. E., Zayas, L. H., Yoon, H., Szlyk, H., Aguilar-Gaxiola, S., & Natera, G. (2015). Deportation experiences and depression among U.S. citizen-children with undocumented Mexican parents. *Child: Care, Health and Development*, *42*(2), 220–230. https://doi.org/10.1111/cch.12307.

Hainmueller, J., Lawrence, D., Martén, L., Black, B., Figueroa, L., Hotard, M., … Laitin, D. D. (2017). Protecting unauthorized immigrant mothers improves their children's mental health. *Science*, *357*(6355), 1041–1044. https://doi.org/10.1126/science.aan5893.

Harter, S. (1998). The development of self-representations. In W. Damon & N. Eisenberg (Eds.), *Handbook of child psychology: Social, emotional, and personality development* (pp. 553–617). Hoboken, NJ: John Wiley & Sons.

Hovey, J. D. (2000). Acculturative stress, depression, and suicidal ideation in Mexican immigrants. *Cultural Diversity & Ethnic Minority Psychology*, *6*(2), 134–151. https://doi.org/10.1037/1099-9809.6.2.134.

International Organization for Migration. (2015). *World migration report 2015*. Geneva: International Organization for Migration.

Jibeen, T., & Khalid, R. (2011). Moderators of acculturative stress in Pakistani immigrants: The role of personal and social resources. *International Journal of Intercultural Relations*, *35*(5), 523–533. https://doi.org/10.1016/j.ijintrel.2011.04.002.

Kia-Keating, M., Capous, D., Juang, L., & Bacio, G. (2016). Family factors: Immigrant families and intergenerational considerations. In S. Patel & D. Reicherter (Eds.), *Psychotherapy for immigrant youth* (pp. 49–70). Cham: Springer.

Kishi, K. (2017). *Assaults against Muslims in U.S. surpass 2001 level*. Pew Research Center, Fact Tank. November 15, 2017 www.pewresearch.org/fact-tank/2017/11/15/assaults-against-muslims-in-u-s-surpass-2001-level/.

Langhout, R. D., Buckingham, S. L., Oberoi, A. K., Chavez, N. R., Rusch, D., Esposito, F., & Suarez-Balcazar, Y. (2018). Statement on the effects of deportation and forced separation on immigrants, their families, and communities. *American Journal of Community Psychology*, *62*(1–2), 3–12. https://doi.org/10.1002/ajcp.12256.

Leidy, M. S., Guerra, N. G., & Toro, R. I. (2010). Positive parenting, family cohesion, and child social competence among immigrant Latino families. *Journal of Family Psychology*, *24*(3), 252. https://doi.org/10.1037/a0019407.

Leong, F., Park, Y. S., & Kalibatseva, Z. (2013). Disentangling immigrant status in mental health: Psychological protective and risk factors among Latino and Asian American immigrants. *American Journal of Orthopsychiatry*, *83*(2), 361–371. https://doi.org/10.1111/ajop.12020.

Lopez, G., Kristen, B., & Radford, J. (2018). *Key findings about U.S. immigrants*. Retrieved from http://www.pewresearch.org/fact-tank/2018/09/14/key-findings-about-u-s-immigrants/.

Lorenzo-Blanco, E. I., Meca, A., Unger, J. B., Romero, A., Szapocznik, J., Piña-Watson, B., … Schwartz, S. J. (2017). Longitudinal effects of Latino parent cultural stress, depressive symptoms, and family functioning on youth emotional well-being and health risk behaviors. *Family Process*, *56*(4), 981–996. https://doi.org/10.1111/famp.12258.

Luthar, S. S., Cicchetti, D., & Becker, B. (2000). The construct of resilience: A critical evaluation and guidelines for future work. *Child Development*, *71*(3), 543–562. https://doi.org/10.1111/1467-8624.00164.

Luthar, S. S., & Zigler, E. (1991). Vulnerability and competence: A review of research on resilience in childhood. *American Journal of Orthopsychiatry*, *61*, 6–22. https://doi.org/10.1037/h0079218.

Masten, A., Best, K., & Garmezy, N. (1990). Resilience and development: Contributions from the study of children who overcome adversity. *Development and Psychopathology*, *2*, 425–444. https://doi.org/10.1017/S0954579400005812.

Mathema, S. (2017). *Keeping families together*. Retrieved from https://www.americanprogress.org/issues/immigration/reports/2017/03/16/428335/keeping-families-together/.

Mcauliffe, M., & Kitimbo, A. (2019). *African migration: What the numbers really tell us* (pp. 1–7). World Economic Forum https://www.weforum.org/agenda/2018/06/heres-the-truth-about-african-migration/.

McHugh, M. (2018). *In the age of Trump: Populist backlash and progressive resistance create divergent state immigrant integration contexts*. Washington, DC: Migration Policy Institute.

McLaughlin, E. C., & Chavez, N. (2019). *'Close the camps': Protesters across country demand end to migrant detention centers*. Retrieved from https://www.cnn.com/2019/07/02/us/detention-facilities-protests-migrants-close-camps/index.html.

Mejía, O. L., & McCarthy, C. J. (2010). Acculturative stress, depression, and anxiety in migrant farm work college students of Mexican heritage. *International Journal of Stress Management*, *17*(1), 1–20. https://doi.org/10.1037/a0018119.

Motti-Stefanidi, F., Berry, J., Chryssochoou, X., Sam, D. L., & Phinney, J. (2012). Positive immigrant youth adaptation in context: Developmental, acculturation, and social-psychological perspectives. In A. S. Masten, K. Liebkind, & D. J. Hernandez (Eds.), *The Jacobs Foundation series on adolescence. Realizing the potential of immigrant youth* (pp. 117–158). New York, NY: Cambridge University Press. https://doi.org/10.1017/CBO9781139094696.008.

Mui, A. C., & Kang, S.Y. (2006). Acculturation stress and depression among Asian immigrant elders. *Social Work*, *51*(3), 243–255. https://doi.org/10.1093/sw/51.3.243.

Newport, F. (2018). *Immigration urge to top of most important problem list* Gallup. 1–10. Retrieved from https://news.gallup.com/poll/237389/immigration-surges-top-important-problem-list.aspx?version=print.

Nordland, R. (2018). 'All of Africa is here': Where Europe's southern border is just a fence. *The New York Times*, 1–9. Retrieved from https://www.nytimes.com/2018/08/19/world/africa/ceuta-morocco-spain-migration-crisis.html.

Oropesa, R. S., & Landale, N. S. (2004). The future of marriage and Hispanics. *Journal of Marriage and Family*, *66*, 901–920. https://doi.org/10.1111/j.0022-2445.2004.00061.x.

Organization for Security and Co-operation in Europe (OSCE) Office for Democratic Institutions and Human Rights (ODIHR). (2019). *Hate crime reporting Germany*. Retrieved from http://hatecrime.osce.org/germany.

Padilla, A. M., & Borrero, N. E. (2006). The effects of acculturative stress on the Hispanic family. In P. T. P. Wong & L. C. J. Wong (Eds.), *Handbook of multicultural perspectives on stress and coping*. (pp. 299–317). Dallas, TX: Spring Publications. https://doi.org/10.1007/0-387-26238-5_13.

Pierce, S., Bolter, J., & Selee, A. (2018). *U.S. immigration policy under Trump: Deep changes and lasting impacts*. Washington, DC: Migration Policy Institute.

Portes, A., & Rumbaut, R. G. (2001). *Legacies: The story of the immigrant second generation*. New York: Sage.

Public Radio International. (2017). *There are more than 400,000 refugee kids in Germany: Educating them is a 'national task'* Retrieved from: https://www.pri.org/stories/2017-10-12/there-are-about-400000-refugee-kids-germany-educating-them-n.

Quiroga, S. S., Medina, D. M., & Glick, J. (2014). In the belly of the beast. *American Behavioral Scientist, 58*(13), 1723–1742. https://doi.org/10.1177/0002764214537270.

Raymond-Flesch, M., Siemons, R., Pourat, N., Jacobs, K., & Brindis, C. D. (2014). "There is no help out there and if there is, it's really hard to find": A qualitative study of the health-care access and concerns of Latino "dreamers". *Journal of Adolescent Health, 55*, 323–328. https://doi.org/10.1016/j.jadohealth.2013.10.051.

Redfield, R., Linton, R., & Herskovits, M. J. (1936). Memorandum on the study of acculturation. *American Anthropologist, 56*, 149–152.

Rogers de Waal, J. (2019). *There is no mass public revolt against globalisation*. YouGov-Cambridge Centre. Retrieved from https://yougov.co.uk/topics/international/articles-reports/2019/05/03/there-no-mass-public-revolt-against-globalisation.

Rogers-Sirin, L., Melendez, F., Refano, C., & Zegarra, Y. (2015). Immigrant perceptions of therapists' cultural competence: A qualitative investigation. *Professional Psychology: Research and Practice, 46*(4), 258–269. https://doi.org/10.1037/pro0000033.

Rogers-Sirin, L., Ryce, P., & Sirin, S. R. (2014). Acculturation, acculturative stress, and cultural mismatch and their influences on immigrant children and adolescents' well-being. In R. Dimitrova, M. Bender, & F. D. R. van de Vijver (Eds.), *Global perspectives on well-being in immigrant families* (pp. 11–30). New York, NY: Springer.

Rojas-Flores, L., Clements, M. L., Koo, J. H., & London, J. (2017). Trauma and psychological distress in Latino citizen children following parental detention and deportation. *Psychological Trauma: Theory, Research, Practice, and Policy, 9*(3), 352–361. https://doi.org/10.1037/tra0000177.

Rook, K. S. (1984). The negative side of social interaction: Impact on psychological well-being. *Journal of Personality and Social Psychology, 46*, 1097–1108. https://doi.org/10.1037/0022-3514.46.5.1097.

Rook, K. S., Luong, G., Sorkin, D. H., Newsom, J. T., & Krause, N. (2012). Ambivalent versus problematic social ties: Implications for psychological health, functional health, and interpersonal coping. *Psychology and Aging, 27*, 912–923. https://doi.org/10.1037/a0029246.

Rueger, S.Y., Malecki, C. K., Pyun, Y., Aycock, C., & Coyle, S. (2016). A meta-analytic review of the association between perceived social support and depression in childhood and adolescence. *Psychological Bulletin, 142*(10), 1017–1067. https://doi.org/10.1037/bul0000058.

Rutter, M. (1990). Psychosocial resilience and protective mechanisms. In J. Rolf, A. S. Masten, D. Cicchetti, K. H. Nuechterlein, & S. Weintraub (Eds.), *Risk and protective factors in the development of psychopathology* (pp. 181–214). New York: Cambridge University Press.

Schain, M. A. (2018). *Shifting tides: Radical-right populism and immigration policy in Europe and the United States*. Washington, DC: Migration Policy Institute.

Schnittker, J. (2002). Acculturation in context: The self-esteem of Chinese immigrants. *Social Psychology Quarterly*, 56–76. https://doi.org/10.2307/3090168.

Schwartz, S. J., Waterman, A. S., Umaña-Taylor, A. J., Lee, R. M., Kim, S. Y., Vazsonyi, A. T., … Williams, M. K. (2013). Acculturation and well-being among college students from immigrant families. *Journal of Clinical Psychology, 69*, 298–318. https://doi.org/10.1002/jclp.21847.

Segal, U. A., & Mayadas, N. S. (2005). Assessment of issues facing immigrant and refugee families. *Child Welfare*, *84*, 563–583.

Shubert, A., & Schmidt, N. (2017). *Through Germany's open door: What life is really like for refugees.* Retrieved from https://www.cnn.com/2017/09/04/europe/germany-refugees/index.html.

Silver, L. (2018). *Immigration concerns fall in Western Europe, but most see need for newcomers to integrate into society* Retrieved from http://www.pewresearch.org/fact-tank/2018/10/22/immigration-concerns-fall-in-westerneuropebut-most-see-need-for-newcomers-to-integrate-into-society/.

Sirin, S. R., Rogers-Sirin, L., Cressen, J., Gupta, T., Ahmed, S. F., & Novoa, A. D. (2015). Discrimination-related stress effects on the development of internalizing symptoms among Latino adolescents. *Child Development*, *86*(3), 709–725. https://doi.org/10.1111/cdev.12343.

Sirin, S. R., & Rogers-Sirin, L. (2015). *The educational and mental health needs of Syrian refugee children (p. 13).* Washington, DC: Migration Policy Institute.

Sirin, S. R., Ryce, P., Gupta, T., & Rogers-Sirin, L. (2013). The role of acculturative stress on mental health symptoms for immigrant adolescents: A longitudinal investigation. *Developmental Psychology*, *49*(4), 736. https://doi.org/10.1037/a0028398.

Smokowski, P. R., Rose, R. A., & Bacallao, M. (2010). Influence of risk factors and cultural assets on Latino adolescents' trajectories of self-esteem and internalizing symptoms. *Child Psychiatry & Human Development*, *41*(2), 133–155.

Spencer, M. B. (1999). Social and cultural influences on school adjustment: The application of an identity-focused. *Educational Psychologist*, *34*(1), 43–57. https://doi.org/10.1207/s15326985ep3401_4.

Staudenmeyer, A., Macciomei, E., Del Cid, M., & Patel, S. G. (2016). Immigrant youth life stressors. In S. Patel & D. Reicherter (Eds.), *Psychotherapy for immigrant youth* (pp. 3–24). Cham: Springer.

Stillman, S. (2018). *When deportation is a death sentence* Retrieved from https://www.newyorker.com/magazine/2018/01/15/when-deportation-is-a-death-sentence.

Suárez-Orozco, C., Motti-Stefanidi, F., Marks, A., & Katsiaficas, D. (2018). An integrative risk and resilience model for understanding the adaptation of immigrant-origin children and youth. *American Psychologist*, *73*(6), 781–796. https://doi.org/10.1037/amp0000265.

Suarez-Orozco, C., & Suarez-Orozco, M. M. (2001). *Children of immigration.* Cambridge, MA: Harvard University Press.

United Nations Conference on Trade and Development. (2018). *Economic development in Africa report 2018.* United Nations Conference on Trade and Development. Retrieved from http://unctad.org/en/PublicationsLibrary/aldcafrica2018_en.pdf.

United Nations Migration Agency. (2018). *World migration report 2018.* https://doi.org/10.18356/f45862f3-en.

Usita, P. M., & Blieszner, R. (2002). Immigrant family strengths: Meeting communication challenges. *Journal of Family Issues*, *23*(2), 266–286. https://doi.org/10.1177/0192513X02023002005.

Ward, C., & Geeraert, N. (2016). Advancing acculturation theory and research: The acculturation process in its ecological context. *Current Opinion in Psychology*, *8*, 98–104. https://doi.org/10.1016/j.copsyc.2015.09.021.

Widfeldt, A. (2018). *The growth of the radical right in Nordic countries: Observations from the past 20 years.* Washington, DC: Migration Policy Institute.

Xia, Y. R., Zhou, Z. G., & Xie, Z. (2005). Strengths and resilience in Chinese immigrant families: An initial effort of inquiry. In V. L. Bengtson, A. C. Acock, K. R. Allen, P. Dilworth-Anderson, & D. M. Klein (Eds.), *Sourcebook of family theory and research* (pp. 108–111). Thousand Oaks, CA: Sage.

Xu, L., & Chi, I. (2013). Acculturative stress and depressive symptoms among Asian immigrants in the United States: The roles of social support and negative interaction. *Asian American Journal of Psychology*, *4*(3), 217. https://doi.org/10.1037/a0030167.

Zayas, L. H., Aguilar-Gaxiola, S., Yoon, H., & Rey, G. N. (2015). The distress of citizen-children with detained and deported parents. *Journal of Child and Family Studies*, *24*(11), 3213–3223. https://doi.org/10.1007/s10826-015-0124-8.

CHAPTER 15

Family interventions and armed conflict

Joop T.V.M. de Jong
Amsterdam Institute for Social Science Research, University of Amsterdam, Amsterdam, Netherlands
Amsterdam University Medical Center, Amsterdam, Netherlands
Boston University School of Medicine, Boston, MA, USA

Dealing with war

The perennial horrors of war and the plight of families

For most of human history, war was the natural pastime of governments, as Pinker (2018) mentions in his book *Enlightenment now*. At the dawn of the modern era, the great powers were pretty much always at war. As major wars around the globe became shorter and less frequent, the wars that did take place became more lethal. But overall, after the second world war, the rates of genocide decreased (Melander, Pettersson, & Themnér, 2016). Mass killings or other war-related deaths still took place in Indonesia (1965–66: 700,000 deaths), the Chinese Cultural Revolution (1966–75: 600,000), Burundi (1965–73: 140,000), the Bangladesh War of Independence (1971: 1.7 million), north versus south Sudan (1956–72: 500,000), Amin's regime in Uganda (1972–79: 500,000), Ethiopia (1974–91: 750,000), Angola (1975–2002, 1 million), Mozambique (1981–92: 500,000) Pol Pot in Cambodia (1975–79: 2.5 million), killing of political enemies in Vietnam (1965–75: 500,000), killings in Bosnia (1992–95: 225,000), Rwanda (1994: 700,000), Darfur (2003–8: 375,000), and Syria (250,000 battle deaths as of 2016). This incomplete list of figures reflects unfathomable human fear, sadness and grief, physical and mental disabilities, huge loss of property and productivity, and gigantic flows of internally displaced people and refugees.

Throughout history, the brunt of the burden of war fell on the family. The family tried to identify its dead, bury them, and perform the mourning rituals customary in its culture. It had to prevent starvation to death when crops were destroyed by wartime politics of the scorched earth. Families had to rebuild their houses or move to other areas when the war made their

Cross-Cultural Family Research and Practice
https://doi.org/10.1016/B978-0-12-815493-9.00015-6

homeland unsafe. Families had to come to terms with loyalties and with perpetrators or collaborators in their communities.

The pathways by which the family as a unit, rather than only the individuals within it, comes to terms with war, are affected by several major challenges such as: (1) loss and grief; (2) disruption of family relations and social support networks due to breakdown of social fabric including public health and educational services; (3) parental stress affecting child-rearing practices; (4) economic damage; (5) reconciliation and dealing with stigmatization or other negative attitudes as a result of association with perpetrators; (6) chaos and uncertainty associated with everyday living; (7) finding meaning in the horrors of war—if there was any meaning; and (8) if survivors are from collectivistic cultures, address the attack on collective identity.

In addition to the problems that armed conflict has on a family, trauma-related problems of a family member have adverse consequences for the entire family. Commonly, aggressiveness, outbursts, detachment, or irritability (Monson, Taft, & Friedman, 2009) isolates trauma victims and is associated with marital and family difficulties. Family members are also affected by the victim's transmission of trauma-related experiences, behaviors, emotions, and attitudes. This phenomenon, known as "secondary trauma," involves the vicarious experience of similar feelings, thoughts, and actions through empathic identification with the trauma victim (Matsakis, 2004). As a result, families of trauma victims may suffer from impaired marital and family relations, reduced family cohesion, social dysfunction, and heightened family conflict (Henry et al., 2011; Nelson-Goff, Crow, Reisbig, & Hamilton, 2007; Slobodin & De Jong, 2015).

The impact of the interaction between armed conflict and culture results in a variety of sociopsychological manifestations and of terminologies used to express these. Psychological trauma or stress are concepts that did not exist, or had no indigenous linguistic equivalent, in conflict affected areas throughout history. Even today, most European languages have no synonym for stress and use the English word. Around the world, many use the Greek word for wound, i.e., trauma, both in medicine and psychology. The lingua franca of many low- and middle-income countries (LMIC) listed above with their death toll has no words that are synonymous with today's language of psychotraumatology. If that is the case, what do we know about the expression of traumatic distress around the globe?

The universality of posttraumatic stress across time

If adverse and horrible events were common throughout the history of mankind, did people suffer from (post)traumatic stress, or is Post-Traumatic

Stress Disorder (PTSD) a recent western invention? One of the proponents of the latter point of view is Young (1995), who claimed that PTSD is a culture derived diagnosis that only existed in the late 20th century. The historicity debate addresses the question whether PTSD—like other manifestations of psychopathology—is a valid diagnosis from a transhistorical perspective. Since the initial 19th century descriptions in the scientific literature of conditions that resemble posttraumatic stress, there has been controversy about PTSD's validity as a psychiatric condition. Shay (1991) suggested that elements of the posttraumatic stress disorder could be identified in Homer's Iliad. Ben-Ezra (2003) asserts that the posttrauma symptoms of nightmares, sleep disturbances, and increased anxiety have not changed in 4000 years. Dean (1997) identified symptoms of PTSD in the accounts of veterans of the American Civil War. De Jong and Hinton (2018) summarize the historicity debate by saying that the symptoms of trauma-related disorder change over time. A given historical era expresses itself, to some extent, in an idiosyncratic way in the presentation of individual suffering. Depending on the conceptualization of mind, body and vulnerabilities, and the effect of trauma in a certain historical period, one would expect variations in the salience of PTSD and other trauma-related symptoms. One would expect PTSD symptoms to be present, but not necessarily emphasized in experience and presentation. Several authors argue that the term "posttraumatic stress syndrome"—syndrome in the sense of a collection of signs and symptoms—better conveys the sense of the lived, contextualized experience of trauma than does the narrower term PTSD; it better describes the lived experience of illness than does the disease concept of PTSD (De Jong & Hinton, 2018; Eisenberg, 1977; Hinton & Good, 2016). The historicity and universality debate parallels the debate on PTSD and culture. Further on in this chapter, I argue that although PTSD may be a valid universal diagnosis, there are other expressions of psychological distress that are more relevant or salient.

In what circumstances are people hurt by adverse events?

As we just saw, (post)traumatic stress may be an affliction of all times. One may then wonder in what circumstances people are troubled or hurt by these ubiquitous and perennial adverse events. There are several approaches to assessing the burden of traumatic stressors and its effect on survivors. Benight and Bandura (2004) mention four key features contributing to disabling forms of stress: perilousness, unpredictability, uncontrollability, and unpreventability. They regard unpreventability of the traumatic events as a key feature of many natural disasters. Similarly, an international consensus meeting tried to address the question when people get hurt, by focusing on mass violence, war, and disasters (Hobfoll et al., 2007). The participants concluded that there are several ways in which stressful events may reach traumatic proportions for individuals, families, and communities. First, the sheer physical, social, and psychological demands of situations involving mass casualty may be overwhelming—either directly (by the extent of pain, injury, destruction, or devastation) or because of their grotesque and incongruous elements (e.g., the massacres in Cambodia's killing fields); or by their symbolic salience (beheading as practiced by ISIS, or imposed killing of one's own family members by child soldiers in Mozambique, Sierra Leone, or Uganda); or people may be abhorred by the personal relevance (e.g., assuming that an act of terror could reach one's own family at any moment, as happens in ongoing wars in the Arab world or in Nigeria in areas affected by Boko Haram, or as happened in areas afflicted by the Lord's Resistance Army in Uganda).

Second, the devastation of resources can impoverish the capacity of individuals, families, and communities to cope with a traumatic situation and recover from its consequences, especially where individuals or communities already have depleted psychosocial and economic resources due to prior trauma, a history of psychiatric disorder, socioeconomic disenfranchisement, or cycles of violence like in the Great Lake area in Africa or in the Middle East. The loss, or threatened loss, of attachment bonds articulates with the previous elements of witnessing horror and direct personal threat. Many traumatic events involve a reactivation of attachment systems and ensuing agony and distress (such as looking for missing relatives or searching casualty lists).

Third is the loss of territory, or safety within a territory. This loss can be either direct via temporary or permanent relocation, or indirect as people's previously secure base is infiltrated by threat and horror. In many instances of mass casualty, the ongoing violence, the aftershocks, and the secondary losses and adversities that follow the initial phase mean that there may be no clearly demarcated period that can be termed posttrauma. Finally, many trauma survivors struggle with challenges to a sense of meaning and justice in the face of shattered assumptions. This may be due to the way in which they were exposed to traumatic events (e.g., being involved in a war they perceive as senseless, or being an innocent victim), or the way they were treated during the post-traumatic aftermath (e.g., via discriminatory distribution of resources, or discrimination after relocation or migration to safer areas) (Hobfoll & de Jong, 2014).

Geographically, there is an inverse relationship between the risk of getting involved in an armed conflict and the availability of mental health providers or services. There is a broad area of the globe—stretching from Central America to sub-Saharan West Africa, over East Africa to the Middle East, and fanning out to several Asian countries—where armed conflict is rampant. These countries are characterized by weak economies, weak nation-states, and faulty democracies, and these same countries have the lowest density of psychologists, psychiatrists, psychiatric nurses, and other mental health professionals. For example, African conflict-affected countries such as Angola, Sierra Leone, Liberia, Guinea-Bissau, Rwanda, Burundi, Eritrea, Mozambique, and Sudan have one or a few psychiatrists and very few clinical psychologists in each country. Violence may cause an exodus of intellectuals (cf. Syria, Iran, Iraq, and Afghanistan) or they may have been killed in a genocide (cf. Cambodia, Rwanda) (De Jong, 2010a, 2010b, 2011). Therefore, one may wonder what resources people used to come to terms with their psychological pain, either in the past, or until today in underserved areas of the world. In the next paragraph, I will summarize various rituals, ceremonies as well as dissociation and psychogenic illness as generic human resources to process horrendous events.

Rituals and healing

In *Les rites de passage*, Van Gennep (1909) mentioned that rituals have three phases: separation, liminality, and incorporation. For example, when people marry, they first *separate* from their environment, have a *liminal* phase such as a wedding ceremony or a honeymoon, and upon return *incorporate* in society. Turner (1969) distinguished two types of liminality, *rituals of status elevation*, in which the subject or novice is conveyed from a lower to a higher position, and *rituals of status reversal*, in which subjects with low status exercise ritual authority over their superiors. Critical moments of transition that societies ritualize are the important times of childbirth, initiation, puberty, betrothal, marriage, retirement, and death. Rituals and ceremonies are fundamental to human growth, development, and socialization in families, communities, or cultural subgroups. But they also disturb the social order, involving a rearrangement of relationships, and require adaptation by families and communities (Goffman, 1971). In other words, ritual and ceremony serve to acknowledge and adapt to the change without threatening the overall social order. The ceremony sanctions a transformation and is often led by a representative of the established order such as a priest, a healer, a mayor, a professor, a general, or a senior student. Communities' capacity to heal depends on their connectedness with one another and with their family and cultural histories. During times of mass distress, the psychological sense of connection between past, present, and future can easily become disrupted. Being able to draw on the rituals, strengths, stories, scripts, and themes of past generations helps people reconnect their transitional pathways. This enables families to reunite their communities, enhancing their collective resilience (Boszormenyi-Nagy & Spark, 1973; Landau, 2007; Weine et al., 2008). In addition to the function of marking and transforming life transitions and helping people to move from one stage of life to another, through a liminal or threshold stage, there are several other functions of rituals and ceremonies that are relevant for this chapter.

First, ceremonies are symbolic events that have much in common with theater as Scheff (1979) describes in his book *Catharsis in Healing, Ritual, and Drama.* Meineck (2012) suggests that Athenian tragedy offered a form of performance-based collective "catharsis" or "cultural therapy" by providing a place where the traumatic experiences faced by the spectators were reflected upon the gaze of the masked characters performing before them. He quotes Socrates stating that soldiers had to be violent in battle but humane at home. Rituals and the enactment in tragedies served the violence-transformation-restoration transition, similar to the modern era transition from a "combat mode" to a "civic mode" among war veterans and former combatants. The same motif is found in the study of the Native American Ritual and the way it contributes to staged and enacted ceremonies among former warriors and combatants. Similarly, therapeutic ceremonies designed by mental health professionals to facilitate the integration of a traumatized family member by the family may be of significant value. For example, Obenchain and Silver (1992) designed a "welcome home" ceremony in the treatment of American war veterans to address "sanctuary trauma, social isolation, and alienation." Like in an ancient Greek tragedy, the reattachment to family and the reintegration in society are fundamental aspects of recovery of war veterans, freedom fighters, rebels, and child soldiers alike.

In addition to catharsis and reintegration in the family and society, ritual has a second function, which one may describe as the balancing of emotions. Scheff (1979) introduced the concept of *aesthetic distance* to indicate an intermediate level of arousal that is optimal for adaptation, catharsis, and integration. A person may be underdistanced when he or she is overwhelmed by the emotion, and overdistanced when he or she is not at all aroused. In a ritual, it is essential that the danger or the psychological pain be represented but not experienced as real. The value of a ceremony thus is enhanced if the participants are at an aesthetic distance, being both emotionally engaged and yet aware of its symbolic nature. Psychotherapy can be seen as a ritualistic encounter (Frank & Frank, 1991). Many psychotherapists use rituals and other distancing techniques, either to mitigate or to increase emotional involvement in the therapeutic process, for example, when a (co-)therapist increases engagement by formulating an emotion-loaded working hypothesis via a one-way screen or telephone in family therapy; or when a therapist stimulates the reliving or sharing of an experience in psychodrama, Al Peso therapy, or group therapy. On the other hand, the therapist may aim at mitigating emotions as happens in guided imagery in Ericksonian hypnotherapy, or when a therapist uses the metaphor of a TV screen in exposure-based therapies (Al-Krenawi, 1999; Rossi, 1980; Yalom, 1980). Many of these psychotherapeutic techniques are not unique and can be found in local healing traditions around the world (De Jong, 1986, 1987, 2001).

In addition to balancing of emotions, rituals share other important characteristics with healing practices. They may be perceived as culturally more congruent or appropriate than psychotherapy, especially in conflict zones with unavailable, inaccessible, or unaffordable mental health professionals, as I mentioned before. When as the Nongovernment Organization Transcultural Psychosocial Organisation (TPO) we developed our public mental health programs in over 20 postconflict zones in Africa and Asia (De Jong, 2002, 2011), we wondered which psychotherapies we should add to the higher tier of our public mental health pyramid—a paradigm that is similar to the more intense and specialized interventions offered in the advanced steps of collaborative care. When discussing and practicing CBT (Cognitive Behavioral Therapy), EMDR (Eye Movement Desensitization and Reprocessing), TFT (Thought Field Therapy, similar to EFT), and NET (Narrative Exposure Therapy), the participating psychologists and psychiatrists from war affected countries were all in favor of TFT—due to its ritual tapping of fingers on different body parts, which they felt would be more appealing to the inhabitants of their home countries. They also preferred NET over Trauma-Focused CBT, because NET uses a rope as a time line for negative and positive life events, and because of its narrative character. Modifying individual self-cognitions in Trauma-Focused CBT or EMDR was perceived as a senseless exercise, for example, when we ask a patient whether she can modify her self-perception of *being a loser* into *I am a winner*. However, modifying the way a patient was perceived by their *co-villagers* as a way of "group" cognitive restructuring seemed to be a good alternative in their interdependent societies.

In addition to the question whether a ritualistic or "collectivistic" treatment approach may be culturally more appropriate in low- or middle-income countries (LMIC), or among refugees or immigrants, healing rituals and ceremonies are widely used around the world, as shown further on. But first, it seems wise to address the question whether PTSD is the most important expression of major distress and trauma, and should thus get top priority in family interventions in conflict areas.

The universality of PTSD versus other expressions of distress

As mentioned previously, De Jong and Hinton (2018) argue that PTSD may be regarded as a universally *valid* diagnosis. However, they also assert that the diagnosis lacks ecological *utility* for various reasons. First, the PTSD diagnosis is highly comorbid: a metaanalysis among adult refugees in high-income countries (HIC) showed that 71% of those diagnosed with depression also had PTSD, and that 44% of those diagnosed with PTSD also had depression (Fazel, Wheeler, & Danesh, 2005). In other words, the diagnosis per se will not guide a therapist who has to choose an evidence-based treatment protocol. The latter is further compounded by the fact that the response to trauma includes not just PTSD and depression, but also psychological and psychosomatic symptoms, bereavement, grief-related disorders, anxiety, panic attacks, acting out and substance abuse, and crisis of existential meaning.

PTSD further lacks ecological utility because it co-occurs with, and does not incorporate, idioms of distress (IOD) or cultural syndromes. In LMIC, these other expressions of distress are locally more relevant and at times more prevalent. For example, Hassan et al. (2015) describe that Syrian Arabic has a range of idioms to indicate general distress ("heaviness in the heart, cramps in the guts" and other complaints); one idiom is *habit qalbi*, "failing or crumbling of the heart," corresponding to the somatic reaction of sudden fear; and *atlan ham*, referring to anticipated anxiety and worry. The syndrome *halat ikti'ab* resembles depression with concepts such as brooding, darkening of mood, aches, and a gloomy outlook, and may be accompanied by a variety of medically unexplained somatic symptoms and fatigue, as well as signs of social isolation and not talking much. IODs and cultural syndromes are a challenge for mental health professionals, because they may either occur concurrently with other expressions of psychopathology, or they may overlap with internationally known ICD or DSM diagnostic constructs, or they may be more or less unique locally salient symptom clusters. In other words, an idiom can be indicative of a psychopathological state, but not necessarily so. This may lead to diagnostic errors. Either a clinician can miss a diagnosis such as PTSD or depression because associated features of idioms and cultural syndromes are most prominent, or the associated features can be overlooked because of the presence of the diagnosis of, for example, PTSD (De Jong & Hinton, 2018). Moreover, IODs and cultural syndromes are not only often socially more salient and prominent, they may also play a dominant role in help-seeking pathways.

The challenge of IODs becomes predominantly manifest when dealing with dissociation, which refers to a lack of integration of personal memory or identity (Spiegel et al., 2011). Dissociation is interpreted in different ways across disciplines. Mental health professionals tend to regard dissociative disorders as real, spontaneous alterations in brain states that reflect neurobiological phenomena, or alternatively as fragments of one's self, "unconscious" conflicts, subpersonalities, or "alters." Dissociation may manifest itself as behavior that could be symptomatic of disorganized attachment associated with child maltreatment and family violence, followed by poorer outcomes in many areas of one's life course (Green & Goldwyn, 2002). In contrast, anthropologists prefer to interpret dissociation as imaginary, socially constructed role performances governed by interpersonal expectations, power dynamics and cultural script, or as discursive processes of attributing action and experience to agencies other than the self. Anthropologists often refer to dissociation as an IOD (De Jong & Reis, 2013; Seligman & Kirmayer, 2008). Around the globe, the use of trance and dissociation expressing distress and traumatic events is omnipresent. Hundreds of millions of people in the Americas and Africa use dissociation on a regular basis in Pentecostal or apostolic churches, expressing a range of hassles going from worries, to "thinking too much," to traumatic stress. In Balaji, the largest center of religious and ritual healing in India, thousands of visitors per year go into a trance during their healing process. Most of the patients visit the shrine with their families. Reinterpreting the complaint of the "identified patient" during the healing ritual as spirits haunting the family is highly reminiscent of relabeling in systemic family therapy. Moreover, there are hundreds of such ritual healing shrines and temples around India, Nepal, or Sri Lanka (Basu, 2009; Dwyer, 2003; Kakar, 1983; Pakaslahti, 2009; Satijam, Singhm, Nathawat, & Sharmam, 1981; Sax & Weinhold, 2010; Thirthalli et al., 2016).

Worldwide, healers—estimated at 1 per 300–500 inhabitants—often use dissociation on a daily base during their diagnostic divination of patients in distress (De Jong, 2001). The importance of dissociation in various (post) conflict settings is accentuated by the advent of disturbing and at times capricious and hard to domesticate "new" spirits, difficult to treat by healers and humanitarian interventionists alike. For example, the *gamba* spirits in Mozambique, the *cen* in Uganda, *jok jok* in Sudan, *ihahamuka* with group fainting at the anniversary of the genocide in Rwanda, *ibisazi* in Burundi, or similar dissociative mass psychogenic illness phenomena in Nepal, Sri Lanka, or in garment factories in Cambodia (Allen, 1991; Baines, 2007; Geffray, 1990; Gobodo-Madikizela & van der Merwe, 2009; Hagengimana & Hinton, 2009; Honwana, 2003; Igreja et al., 2010; Perera, 2001; Reis, 2016; Tol, Jordans, Reis, & De Jong, 2009; Van Duijl, Cardeña, & de Jong, 2005; Van Ommeren et al., 2001; Ventevogel, 2016). De Jong and Reis (2013) coined the term "*collective trauma processing*" to describe mass possession trance in Guinea-Bissau as an indigenous way to process the horrendous traumatic stress in an African postwar context where mental health professionals were virtually absent.

Epidemic expressions of distress and anxiety are a global phenomenon. "Mass hysteria," mass psychogenic illness, or medically unexplained illness tends to occur in connection with distress caused by exams or other stressors in schools, environmental pollution, work conditions, exposure to toxins, exposure to nuclear fallout, or the stressors of war and other disasters. A well-known example among western veterans is the Gulf War Syndrome (for a description of veterans' syndromes, see http://veterans.iom.edu). Another example is the *yao yan* or "salt hysteria" when large numbers of the Chinese bought salt to protect themselves against the fallout after the Fukushima disaster in Japan. One may regard these often relatively brief epidemics as another form of processing traumatic stress or common distress (Bartholomew & Sirois, 2000). The characteristics of these epidemics are that: (1) there is a high level of stress in a population; (2) culture-specific explanations provide credibility to an interpretation of threat; (3) complaints manifest themselves among individuals with a vulnerability to angst or with other vulnerability traits, or among persons who share characteristics with other manifest cases; and (4) complaints have a relatively short duration and do not develop into chronic afflictions[a] (cf. De Jong, 2010a, 2010b).

Apart from issues related to the universalism of diagnostic categories versus the relativism of considering other expressions of distress, our psychological and psychiatric profession tends to believe that evidence from

[a]Veterans' syndromes are an exception. Over more than hundred years, there is a debate about the problems of soldiers returning home after a war, and about the possible relation between their complaints and secondary benefits such as allowances.

epidemiological or outcome studies plays an important role in deciding what the priorities are in humanitarian emergency settings. In reality, however, there are local and international dynamics—which I refer to as the *traumascape*—that guide the focus, the size, and the nature of assistance for survivors (cf. De Jong, 2007). The international community often determines the main concern of humanitarian aid, for example, whether it is migration of refugees, terrorism, human rights or child rights, governance, gender-based or domestic violence, former combatants or child soldiers.

Child soldiers

Over the past two decades, child soldiers became a chosen and important vulnerable group for interventionists and researchers alike. Psychodynamically, this interest in child soldiers is an intriguing phenomenon with its questions about good and evil, and about innocence versus agency of a child. There are many child soldier studies in LMIC, but virtually no studies on adult former combatants in the third world. On the other hand, there is an abundance of veteran studies in high-income countries that overall fall outside the scope of this chapter. A variety of former child soldier programs aim at reintegration, traditional healing, family interventions, and multisectorial interventions. Disarmament, demobilization, and reintegration (DDR) programs for child soldiers typically comprise economic and education support packages, such as vocational training, apprenticeships, microfinance loans, and formal or informal schooling. Kohrt, Sauharda, and Maharjan (2015) provide an excellent overview of child soldier interventions. They mention that DDR programs should involve caregivers and families in tangible ways to promote the psychosocial well-being of reintegrated child soldiers (Borisova, Betancourt, & Willett, 2013). In Sierra Leone, the inclusion of family-based elements resulted in the greater impact of an intervention (Betancourt, Newnham, McBain, & Brennan, 2013). A study of multisectorial reintegration assistance in Burundi found that former child soldiers who were socially integrated within families and communities had high work and employment rates, they showed literacy rates above the national average, and had no differences in present functioning and mental health compared to never-recruited peers (Jordans et al., 2012). While coaching, on-the-job, and vocational training, and provision of material kits were associated with improved socioeconomic reintegration trajectories, a sense of satisfaction with received services is most strongly and recurrently associated with better outcomes. Community connection and a sense of future were also identified as protective factors among Colombian former child soldiers (Cortes & Buchanan, 2007). Among Colombian and Nepali child soldiers, feeling connected to the community, and hope for the future were strong predictors of functioning in school, work, and social activities among children exposed to political violence.

Given the high burden of stigma and psychosocial problems among former child soldiers, using traditional healing rituals is an important trend in humanitarian interventions, particularly as combined with Western mental health services. In a study in Mozambique, traditional ceremonies and cleansing rituals played a vital role in rebuilding community trust and cohesion, as these ceremonies were believed to clean the child soldiers from their "contaminated" atrocities of war, and thus helped to promote family and community acceptance (Boothby, 2006; Boothby & Thomson, 2013). Traditional healers in Mozambique facilitated reintegrating of ex-combatants by "taking the war out of the people" (Nordstrom, 1997). In Sierra Leone, communities and families employed rituals of welcome to incorporate returned girl soldiers into acceptable social positions (McKay & Mazurana, 2004). Honwana (2006) describes traditional healers in Angola conducting communal cleansing rituals for child soldiers to purify them of their exposure to killing. In our work among Sudanese refugees in northern Uganda, the elderly used cleansing rituals for women and girls who were raped by rebel groups, such as a ritual asking them to walk over eggshells before reentering their homesteads.

Reunification rituals or rituals of reintegration, however, may come at a cost to former child soldiers. For example, child soldiers are threatening to adults because they do not fall into socially expected roles of submission to adult authority (Boyden, 2003). Rituals may restore the expected socially submissive role of children, but this may represent a lost opportunity for children to gain a greater voice in social processes. Speaking of girl soldiers, traditional rituals often reinforce gender discrimination by promoting the status of (older) males and threatening the well-being of women and girls (Denov, 2007). In Nepal, girl soldiers refused to participate in rituals during reintegration because they were perceived as a form of subjugation and antithetical to the reason they originally joined armed groups (Kohrt et al., 2015). In Guinea-Bissau, the Kiyang-yang possession movement among the largest ethnic group was perceived as a threat by the state, but the power of its female leaders and of youngsters taking over healing functions was equally perceived as a threat by the male gerontocracy of that ethnic group (De Jong, 1987; De Jong & Reis, 2010, 2013). Worldwide and throughout history, shifts in responsibilities between men and women take place in war zones when men go to battle or leave abroad. Upon their return, family dynamics have to readjust, which may be detrimental to the emancipation of women who are often expected to take up their previous subjugated role.

Socio-ecological approaches and the role of family therapy

The past 20 years accelerated research and clinical progress regarding the understanding and treatment of traumatic stress reactions. New treatments have been developed, and demonstrated some of the highest levels of effectiveness of the treatment of mental health disorders (Powers, Halpern, Ferenschak, Gillihan, & Foa, 2010). However, because we have extrapolated from clinical populations, we have overemphasized individual-cognition-brain aspects of this process. Unfortunately, these do not accurately apply to nonclinical populations, and furthermore we have made the mistake of thinking that because a model works in a clinical population in a given culture, it is the final solution for those who attempt to recover and find a new balance after extreme stress. Clinicians should try to get more insight into passageways, systemic and socio-ecological factors influencing the impact of distress in order to create synergy between external factors and community actors, and their own clinical work (Hobfoll & de Jong, 2014).

Currently, a popular conceptual framework for psychosocial interventions in humanitarian settings is that of ecological resilience (Tol, Song, & Jordans, 2013). In the context of children and war, ecological resilience has been defined as the assets and processes on all socio-ecological levels that have been shown to be associated with good developmental outcomes after exposure to situations of armed conflict (Tol et al., 2009). Ecological resilience refers to a process whereby families and individuals attain desirable outcomes despite significant risks to their adaptation and development. These processes are thought to involve a dynamic relationship between risk, protective, and promotive factors at different levels of the social ecology (e.g., individual, family, school, neighborhood levels) (Betancourt, Meyers-Ohki, Charrow, & Tol, 2013; Ungar, Ghazinour, & Richter, 2013). As mentioned at the beginning of the current chapter, families experiencing war suffer multiple secondary stressors related to the fighting, including family disruption, homelessness, financial instability, impaired access to education and health services, and the chaos and uncertainty associated with everyday living (Pynoos, Steinberg, & Piacentini, 1999, Shaw, 2003). Longitudinal studies show that secondary adversities accumulate (Rutter, 1985). There is a strong relationship between the number of risk factors to which children are exposed and adaptive functioning (e.g., Gutman, Sameroff, & Eccles, 2002; Masten, 2001). For example, Neuner, Schauer, Catani, Ruf, and Elbert (2006) have shown a significant increase in children's risk of psychiatric disorder associated with accumulating adverse life events. Catani, Schauer, and Neuner (2008) found a correlation between exposure to war and domestic violence in Afghanistan and Sri Lanka, likely caused by parents' exposure to war events. Silva et al. (2000), reviewing data on children, reported that witnessing domestic violence as a risk factor was more strongly associated with subsequent PTSD than was the abuse by itself.

In times of armed conflict, families are the most important source of resilience to come to terms with distress, adverse events, and secondary adversities. When homes and communities are destroyed, the family is often regarded as the most important remaining social institution. Families play a central role in helping family members to cope with refugee trauma and displacement. The interconnectedness of family members means that traumatic experiences have the potential to impact a large circle of family members beyond the original victim and that trauma results in both individual and family-level mental health problems. On the other hand, the family serves as a protective factor, in which the family itself becomes an important context for processing the trauma and providing social support (Boehnlein et al., 1995; Nickerson, Bryant, Silove, & Steel, 2011; Weine et al., 2008).

Punamäki, Diab, Isosävi, Kuittinen, and Qouta (2018) provide an excellent overview of the family systems approach to war trauma and mental health. In the systems approach, families are composed of multiple dynamically interacting relational subsystems typically involving marital, parent-children, and sibling relationships (Coyne, Downey, & Boergers, 1992; Minuchin, 1974). Families are conceptualized according to their structure (e.g., boundaries between subsystems, dominance hierarchies, and communication transparency) and relational context (e.g., autonomy vs intimacy; harshness vs warmth) (Blass & Blatt, 1992; Kerig, 2005; Olson, 2000). Research has delineated, for instance, enmeshed families, characterized by too thin boundaries with easily evoked emotional spillover between parents and children, and disengaged families, characterized by too thick boundaries and a lack of support and emotional closeness among members.

Attachment theory is highly informative for understanding family relations in cases of war and other traumatic conditions. A sense of security is a core motivator for young children, and the goodness of fit among the emotional, cognitive, and behavioral responses of parents and children is essential to survival and mental health (Bowlby, 1969/1982). War experiences can affect attachment and family relations, and research on the transgenerational transmission of trauma has documented negative phenomena, such as scapegoating, enmeshed and withdrawn parenting, role reversals, and silenced and biased communication (Dalgaard, Todd, Daniel, & Montgomery, 2016). Cohesive families, in turn, enjoy a balance between autonomy and intimacy, meaning that members have access to both emotional support and individuality and privacy (Cecchin, Lane, & Ray, 1993; Kerig, 2005; Lindblom et al., 2017; Minuchin, 1974). As an example, research reveals highly supportive and resilience-enhancing parent-child relationships among war-traumatized families (Fossion et al., 2015).

Even though it is logical to intervene at the family level, systematic reviews of evidence show that this type of intervention is heavily underrepresented, at least when targeting children (Jordans, Pigott, & Tol, 2016; Jordans, Tol, Komproe, & de Jong, 2009). Slobodin and De Jong (2015) conducted a systematic review of intervention studies for traumatized immigrants and refugees, specifically looking at family interventions. Only six experimental studies met the inclusion criteria. Four of them describe school-based interventions, and two present multifamily support groups. Whereas families seem to be a logical way to come to terms with war and mass disaster, there appears to be an amazing lack of research on family therapy in (post)conflict areas.

Explaining the paucity of literature on family therapy in (post) conflict areas

I would like to share a few thoughts or hypotheses that may explain the paucity of research, which in my opinion is caused by various therapeutic and methodological hurdles that warrant additional research efforts and which will be described in the next section. A first explanation may be that family therapy is an extreme example of the 10/90 divide, meaning that 90% of publications in the world are related to 10% of the world's population. The systems field is likely even more skewed, because it mainly addresses, like a major part of psychology, WEIRD populations (Western Educated Industrialized Rich Democratic). There are a few interesting exceptions, such as Multifamily therapy (MFT), which is a combination of psychodynamic group therapy with systemic therapy that aims to elicit behavioral changes through the restructuring of interactional patterns in families. MFT is applied in settings with refugees and veterans (McFarlane, 2002; Van Ee, 2018). Weine et al. (2004, 2008) did multifamily group work with

Bosnian and Kosovar refugees in Chicago to change the knowledge, beliefs, and attitudes of family members, and to increase access to mental health services. Rather than a systems approach, their low intensity family support intervention used family problem-solving and family resilience, which in (post)war conditions may be more transferable and scalable. Other studies focusing on groups of disadvantaged war survivors—instead of WEIRD study participants—will be discussed later on.

A second explanation that illustrates the methodological complexity of family research is the enormous variety of family structures and dynamics around the globe. Each component or building block of system therapy raises a myriad of questions and doubts across cultures that have hardly been addressed in basic research or dealt with in the cultural adaptation of interventions. A fundamental problem in this regard is our lack of insight into what a normal family is in different cultural contexts (see also Chapters 2 and 18). For example, autonomy is hardly relevant in interdependent cultures or may even be regarded as synonymous to selfishness or egocentrism. When family relationships are close, and interpersonal reliance and interdependency are highly valued in collectivist cultures, concepts such as lacking boundaries, dependent or avoidant personality, or enmeshment may be misapplied. Boundaries between parents and children are often a product of cultural scripts and dominant display rules related to respect, and do not allow an assessment in terms of enmeshed or disengaged families. When mental health professionals emphasize the necessity of separation from parents as a developmental goal, this can undermine relatedness in interdependent cultures. The relationship between parents and children may also be challenged when the familial hierarchy changes in a new environment and adults become dependent on their children (cf. Slobodin & De Jong, 2015). Using terms such as enmeshed or rigid parenting across cultures may easily result in ethnocentrism or stereotyping. Matrilinearity and patrilinearity, matrifocality, virilocality, matrifocal concubinate, or polygyny complicate the assessment of normal versus deviant boundaries between subsystems, or the evaluation of dominance hierarchies. A paradoxical intervention such as the prescription of "no-change" presupposes a Western symmetric relation between therapist and family, which may have unexpected outcomes in cultural contexts where a therapist has ascribed authority instead of acquired authority as in Western countries. Circular interviewing, the distinction between content and metalevel, or the hybridization or creolization of family cultures after leaving the country of origin, may pose a conundrum even to an experienced family therapist (De Jong, 1991). Further on, I give an example of family dynamics related to honor, and question which criteria are appropriate to assess whether family dynamics in such a situation are normal or deviant.

A third explanation for the paucity of research that holds for all forms of psychotherapy is the disparity between a psychological and a magic-spiritual-religious paradigm. Personalized illness explanations, such as sorcery, witchcraft, the evil eye, spirits of ancestors or previous reincarnations, are dominant around the world. What are the implications for family therapists when a majority of war-affected citizens have a firm belief in these illness attributions? De Jong and Reis (2013) compared research on traditional healing resources with the approach of clinical psychology and psychiatry. They proposed a comprehensive model of five ontological dimensions involved in suffering that are addressed in any form of healing: (1) intrahuman (mind-body); (2) interhuman (social interactions, e.g., on the micro, *meso*, and community level; ethnic group); (3) suprahuman (e.g., ancestors, god, embodied entities and forces); (4) extrahuman (ecology, nature, cosmos, animals); and (5) time, which is a different dimension reflecting the relationship between the past, present, and future (e.g., refugee family members' experiences are framed by developmental time, family life cycle, time since exposure to trauma, and time of resettlement, with service needs likely to change over time) (Weine et al., 2008). Psychology and psychiatry focus primarily on the intrahuman dimension. Family therapists prefer a wider scope, when they embrace the interhuman dimension including the family and the community. However, most affected citizens in a war zone or conflict area interpret their suffering on a suprahuman (ancestors, spirits, God, Allah) or extrahuman level. For example, in Cambodia and Sri Lanka, the horrors of the Khmer Rouge and the war were often attributed to bad karma from a previous reincarnation. In my opinion, spirit possession and dissociation acquire their importance and popularity in various places, including the expansion of apostolic churches and its religious practices, by allowing individuals and families to straddle the intra/interhuman dimensions and the suprahuman.[b]

A fourth explanation for our lack of research is that the five ontological dimensions challenge the worldview of family therapists and other interventionists. Academic humanitarian interventionists wanting to train local

[b]The supreme art of systems therapy is a therapist who is well versed in the local spirit cosmology and other aspects of the suprahuman dimension, and is able on a metalevel to interpret and use spirits or other forces as a multidimensional explanatory framework and a vehicle for family interventions.

health workers or paraprofessionals in providing trauma treatments live in another existential dimension than their trainees. Many interventionists feel that trainees need protoprofessionalization (De Swaan, 2004), when trainees get acquainted with the basics of psychology and human behavior, to understand that there is more under the sky than magic-religious machinations, or spirits and other supranatural beings who need to be placated with a ceremony or a sacrifice. In other words, the interpretation of human suffering and its causes is different for system therapists and the local survivors of political violence. This implies that there often is a tremendous misfit between what Western science can offer and what is locally seen as relevant for the solution. Fine-tuning, synthesizing, and bridging the ontological dimensions by family therapists and survivors is necessary to forge a relationship that enables training and agreement on both training and treatment goals. Therefore, developing expertise in negotiating and synthesizing explanatory models transcending these different dimensions are part and parcel of culturally competent psychotherapy.

A fifth explanation for the lack of fruition of family therapy in non-Western settings is related to the biopsychosociospiritual parameters related to the so-called debate on universality and relativism regarding normalcy, deviancy, and human rights. In addition, there is the issue of which value system is dominant for a therapist. Is that a universalist human rights based or Hippocratic value system, or does she respect or allow values that may clash with her academic or religious persuasion[c]? Let me provide the example of honor revenge to illustrate this point. Honor revenge may occur when a family is convinced that a family member, for example, a flirting teenage girl or adult woman, has affected the honor of the family and needs appropriate punishment. The punishment may consist of physical violence,

[c]A personal anecdote may illustrate my point. As a 26-year-old doctor working for the liberation movement of Guinea-Bissau in a hospital in south Senegal, I had to amputate a leg of a peasant-refugee with a gangrenous foot. In the daily clinical meeting with the nurses, I mentioned that I was going to discuss the intervention with the patient. The 25 nurses looked at me with disbelief. Why tell him? Did I think he would sleep well if I told him that he was going to be operated? And did I not teach them some time ago that someone has to be in a good physical and mental condition before an operation? And what was the point of telling him? If we would not operate he would die, so is there a choice? And who is that gentleman Hippocrates and that oath I mentioned? So he lived in Greece 500 years before Christ, but doctor, this is West Africa in 1974. After a very long African-Socratic discussion I gave in. I did not inform the patient. Laid awake myself. The next afternoon I visited the patient in the ward and the white bandage where the leg used to be lighted up from the darkness. The patient spread out his arms with tears in his eyes. He hugged me and thanked me for saving his life.

immolation, or killing, whereas in the same family fathers or husbands may have several wives, or brothers can have premarital sex, as may happen among refugees or inhabitants from the Middle East (cf. De Jong & Hinton, 2018). If honor revenge occurs, does it indicate a cohesive family, with the caveat that men have more autonomy whereas women protect the family honor, *namus*, by being chaste? Are men, but not women, allowed premarital or extramarital sexual intimacy for the higher sake of family honor? Is the family honor or chastity a higher order cultural dimension a therapist has to respect because it overrules differences in preferential gender roles, or the harshness of the chastising?

The example illustrates differences in norms and values, but it also illustrates the complexity of system thinking. From a universal human rights perspective, or from a gender or family viewpoint, one may regard the implicated norms and behaviors as intolerable.[d] But if one chooses a wider historical-societal system level to analyze this phenomenon, one may draw a different conclusion. Patriarchal, tribal Bedouins in the Middle East or North Africa mainly lived outside the reach of weak states with flawed justice systems, which is a phenomenon that is present even today in parts of Syria and most of its neighboring countries. Within such a context, the concept of honor had a social function by promoting endogamy, chasteness, and peaceful coexistence of families and clans, with possible serious sequelae for women, as we just saw. This complexity may result in a puzzling family dynamic for system

[d]This case is less eccentric than one may think. Several times I was involved in forensic cases of honor revenge killing in the Netherlands, with an estimated death rate of about 50–100 persons/year among immigrants from the Middle East and north Africa. In a systemic therapy with a Middle Eastern family that seemed to function reasonably well, a colleague and I were suddenly confronted with a half cutoff ear of the daughter, wounds on her head, and the wall of the dining room smeared with blood. She had been promised to a cousin in the Middle East, but she had fallen in love with a boy from her school. (Migrant families back home often pay substantial amounts of money to arrange a family marriage with a cousin or niece in the "West".) On another occasion, during a consultancy in Pakistan on a large project on family violence, one of the reports estimated that of the 5–7000 annual honor revenge killings in the world, approximately 4–5000 take place in Pakistan. Pakistani papers frequently mention, for example, "stove burnings," i.e., housewives who open their stove and burn their face, in reality often a discontent family member who puts her on fire, e.g., after pouring kerosene. These are examples of a wide range of similar women-unfriendly practices that are perpetuated in that specific political-religious-cultural context (cf. Patel, 2003). When the inherent values have been interiorized by women for generations, and reinforced by persistent questionable interpretations of the Koran, a family therapist has to find a delicate balance. The fate of the Middle Eastern refugee girl is not uncommon—apart from the amount of violence—among immigrant families, with a majority of the teenage girls thinking that there is no way to escape these norms, even if they are perceived as imposed.

therapists in assessing possible deviancy. Do these therapists attribute salience to the century-old, culturally ingrained, culture-specific, and collectivistic construct of family honor and its implications? Or do they emphasize an emancipated, universalist gender and human rights discourse, and regard such a family as dysfunctional or pathological from their academic point of view? And, importantly, what do these considerations entail in terms of family interventions or the formulation of hypotheses to be addressed in the sessions?

Friedman (1982) argued that families in all cultures select behavioral repertoires that suit them. He distinguished *cultural costume* and *cultural camouflage*. *Cultural costume* is protective because it prescribes a series of behavioral repertoires in a society; an example would be when honor in a Middle Eastern family promotes chastity and peaceful coexistence. *Cultural camouflage* occurs when local culture is used as a smokescreen to hide reality or interaction rules. For example, this can happen when in such a family an arranged marriage of a family member is a smokescreen to obtain material wealth by transplanting a marriage partner to a high-income country. Similarly, in a family session in HIC, disclosure of private information about family matters is encouraged, whereas in interdependent or collectivistic (sub)cultures this may lead to family conflicts and increase anxiety among participants, for example, in a context of family violence or honor revenge (Ganesan, 2006; Hendriks et al., 2019). In other words, the unraveling of biopsychosociospiritual elements is not only related to normalcy or deviancy, but also to cultural sensitivity. Cultural adaptation of interventions may lead to increased engagement (Lau, 2006), commitment (Castro, Barrera, & Martinez, 2004), and reduce dropout (Hwang, 2006). In contrast, interventions that are culturally *in*sensitive may not only be less effective, but may even be harmful (Christopher, Wendt, Marecek, & Goodman, 2014). In a similar vein, although culturally adapted psychotherapeutic interventions may be more effective, the questions raised in this part of the chapter show the complexity of cultural adaptation (Chowdhary et al., 2014; Hinton & Jalal, 2014; La Roche & Lustig, 2010; Smith, Domenech-Rodrıguez, & Bernal, 2010).[e]

[e]The last authors describe the process of cultural adaptation. Though cultural adaptation is beyond the focus of this chapter, there are several other models that can be used in an eclectic way for culturally diverse populations. For example, the *Cultural Accommodation Model* (CAM) (Leong & Lee, 2006) developed in the United States, helping therapists to recognize cultural factors that differentiate patients from each other and adapt their treatment. The *assimilation model* attempts to add crucial elements of the patient's culture based on a thorough assessment of the cultural background of a patient and of relevant stressors. Another approach is the universal therapist variables described by Frank and Frank (1991) in their book *Persuasion and healing*.

A sixth and final explanation is that family and systems therapy is somewhat lagging behind the development of other mental health and psychosocial services (MHPSS) in disaster-affected areas. Family therapy has not yet succeeded in developing and testing the effectiveness and efficacy of family interventions in evidence-based research, and in manualizing and disseminating scalable family interventions in mass disaster settings, even though it is obvious that there are many methodological, pragmatic, and ethical obstacles to conduct randomized controlled trials (RCTs) in conflict-affected populations and disseminate interventions to all those in need (Slobodin & De Jong, 2015). The aforementioned Multifamily Therapy is promising for refugees and veterans, but due to its generic design it could be improved substantially with clear guidelines for the intervention. The guidelines should focus on the active ingredients of MFT, such as working with other families, recognition and identification, understanding between family members, mentalization, unraveling behavioral sequences, and improved parenting (Van Ee, 2018).

Other forms of MHPSS show a proliferation of psychosocial and psychiatric interventions in humanitarian settings over the last decades. Although major knowledge gaps remain, collaboration between scholars, practitioners, policy makers, Nongovernment Organizations, and the United Nations (UN) resulted in a wide range of assessment instruments and evidence-based psychosocial interventions in humanitarian crises. Effective, scalable, and at times sustainable interventions have been developed by partnerships between these actors and humanitarian practitioners. Training of trainer approaches and task sharing with less specialized local health workers, teachers, social workers or women's organizations can cover the needs of large groups of survivors in a cost-effective way. For example, research assisted in prioritizing humanitarian programming by understanding the most critical mental health and psychosocial needs and by unpacking the risk, protective, and promotive factors linked to MHPSS concerns; improving interventions by testing assumptions in MHPSS program theories of change; evaluating whether and how both locally and externally developed psychosocial activities meet their aims; examining how proven interventions may most effectively be culturally adapted, disseminated, and implemented; how to strengthen needs assessments and program monitoring through the development and testing of measurement tools; and how to understand barriers and facilitators to implementing MHPSS activities (Lee et al., 2019). Examples of guidelines covering a psychosocial continuum from well-being to moderate emotional and severe mental disorders are the Interagency Standing Committee Guidelines[f] (IASC, 2007), the mhGAP-HIG Guidelines, the Problem Management plus guidelines, or the Building Back Better concept in disaster situations. All these interventions are freely available on the websites of the World Health Organization, the United Nations High Commission on Refugees, Nongovernment Organizations, and academic institutions.[g]

To conclude, family and systems therapy has not yet evolved into transferable and translatable interventions in humanitarian settings specifically, and in low- and middle-income countries more generally. These are the very settings that are disproportionally affected in terms of psychological morbidity or mortality by political violence, armed conflict, and mass disaster. The relative strength of the generic systems approach raises a myriad of questions when applying the fundamentals of its intervention philosophy in settings that require a public mental health approach due to the relative absence of mental health practitioners. Over the past years, family therapists realized that the strength of working with the family should be combined with other therapeutic interventions. The most important modalities that gained popularity were parenting and school-based interventions aimed at children and youth. I try to summarize this literature in the remaining part of this chapter, while also mentioning some promising developments in multifamily therapy.

A holding environment in times of war and disaster

The paradigm shift from resilience as a personal trait to the concept of ecological resilience went along with a shift from intrapsychic psychological processes to a socio-ecological and public mental health approach in

[f]Some of these current humanitarian assessment methods show striking similarities to assessment methods in systems approaches. Thus, Landau (2007) and Landau and Saul (2004) describe the transitional field map as a schematic representation of a community's members, problems, resources, events, themes, and histories that exist within every level of the network, including biological, cultural, and individual psychosocial systems; natural and artificial support systems; and ecosystems. Further, the transitional field map underscores that each level within a system (family and/or community) affects the others.

[g]Recent studies in the MHPSS field focus on research questions such as transdiagnostic group interventions with Syrian adolescents in Jordan; phone and web application for Syrian refugees in Lebanon and Europe; child psychosocial interventions in Lebanon; multicomponent behavioral intervention delivered by lay helpers with conflict-affected adults in Pakistan; adapting and evaluating guided self-help intervention with South Sudanese refugees in northern Uganda; and efficacy research on psychological first aid in Sierra Leone. Again, family interventions are lacking (Tol, Reis, Susanty, & de Jong, 2010).

settings of armed conflict and mass disaster. We saw that ecological resilience refers to a process whereby families and individuals attain desirable outcomes despite significant risks to their adaptation and development. Ecological resilience and its concomitant processes involve a dynamic relationship between risk, protective, and promotive factors at different levels of the social ecology (e.g., individual, family, school, neighborhood levels). There is by now robust evidence that armed conflict affects children's mental health and psychosocial well-being via multiple pathways (Jordans et al., 2016; Miller & Jordans, 2016). These include direct exposure to the violence as well as the effects of the aforementioned secondary adversities.

Critical among the direct and indirect pathways are a subset of daily stressors related to family functioning. Armed conflict increases the risk of interparental violence, child maltreatment, and impaired parenting related to parental psychological distress that may be influenced by substance and alcohol abuse as seen in many refugee settings (cf. Catani et al., 2008; Sriskandarajah, Neuner, & Catani, 2015). This illustrates the crucial importance of interventions that reduce family violence and support parents' own mental health and psychosocial well-being. More broadly, the salience of daily stressors at all levels of children's social ecology underscores the need for multilevel interventions that target ongoing environmental stressors. Addressing environmental stressors has to go hand in hand with helping children to heal from or become more resilient to the effects of toxic stressors such as harsh parenting and the use of violence by teachers, whether directly or indirectly related to armed conflict. However, despite growing calls for preventive and socio-ecological interventions, few evidence-based interventions transcend the narrow focus on direct work with children. Although child-focused interventions, particularly when school- and group-based, may strengthen social ties among children and between children and teachers, they do not address ongoing threats to children's well-being at home or in the community. While the challenges to altering stressful social conditions are formidable, recent studies have documented that toxic stressors can be successfully reduced. Jordans et al. (2016) do not suggest that such interventions should supplant direct work with children, but that they should form essential components of multilevel approaches to supporting children in conflict and postconflict settings. More recently, Jordans et al. (2018) describe their multilevel ecological approach among refugees and postconflict populations. Let us have a look at parenting and school- or group-based interventions.

Parenting interventions

Parenting interventions in a stable, high-income society consist broadly of group-based programs, home-based programs, and video-feedback programs (Mourik, 2019). After military deployment among veterans in HIC, the family is affected via components of posttraumatic stress, such as irritability combined with social withdrawal and numbing. From a clinical perspective, it appears that providing explanations of and strategies to deal with emotional numbing and withdrawal may benefit the soldiers' families. Involving the veteran's family and the children is less stigmatizing than treatment seeking by veterans who are identified as patients (McFarlane, 2009; Tanielian & Jaycox, 2008). Several veteran studies emphasize the importance of parenting practices and the role of mothers for the regulation of children's adjustment in the aftermath of trauma exposure (Gewirtz, Forgatch, & Wieling, 2008; Gewirtz, Pinna, Hanson, & Brockberg, 2014). In view of the vast literature on Western veterans that largely falls outside the scope of this chapter, what do we know about parenting in LMIC?

As mentioned before, trauma researchers increasingly recognize the limitations of individual approaches and advocate for the development of family-based interventions. The rationale in LMIC is that the family is the most important remaining social institution; the family is the most proximal social environment in which children learn patterns of adjustment; and the highly trained professionals needed for individual trauma treatments are hardly available in conflict-affected areas. Working with trauma using a lens focused on social interactional relationships, however, is a relatively new field of study with few evidence-based interventions (Gewirtz et al., 2008, 2014).

Both in HIC and LMIC, one of the leading models for understanding the development of child behavior problems is Patterson's coercive family process model rooted in social learning theory. Patterson's model focuses on child-parent behavioral interactions in which children's noncompliance is reinforced by escape and/or parents' escalating punishments are reinforced by compliance. A more recent trend is to lean less on coercion, but rather on prevention. One prevention framework draws on Patterson's five effective positive parenting practices that promote healthy child development: skill encouragement, limit setting, child monitoring, interpersonal problem-solving, and positive involvement. The twin goals of a parenting-based intervention following mass trauma would be to support parents to provide: (i) a social environment with structure, security, and emotional warmth; and (ii) a social environment that addresses the traumatic event. Longitudinal studies show that effective parenting practices provide a protective environment surrounding youngsters. When parents are able to monitor, set limits, encourage skill development, problem-solve, and be positively involved, their children are more likely to show resilience in the face of traumatic events ranging from war and natural disasters to community and intrafamilial violence and abuse (Gewirtz et al., 2008). Dubow et al. (2012) confirmed that intact parental mental health and positive parenting practices protected children's mental health from military violence in Israeli and Palestinian families. In sum, interventionists increasingly realize the importance of combining a dual focus on parenting skills and parenting stress.

Another ecological model situates parent's expectations and childcare practices within a theoretical framework that relates higher-level cultural factors with child-rearing practices (Super & Harkness, 1986, 2002; Worthman, 2010a, 2010b). It draws upon the concept of the developmental niche as a framework to study how higher-order influences structure the environment for child development. The developmental niche is described as the composite of three subsystems: (1) the physical and social settings in which the child lives (the people they spend time with, and the social roles—e.g., work vs play—children fill throughout the day); (2) childcare customs and practices (referring to sequences of behavior that are commonly used and accepted when interacting with children in given situations or stages of development); and (3) parents' psychology (beliefs that commonly accompany childcare customs including beliefs concerning the nature and needs of children, parental and community goals for child-rearing, and caretaker beliefs about effective rearing techniques). Burkey et al. (2016) used this developmental niche to describe "maturing" over time in Nepal. Their study is a good example of balancing respect and incorporation of local customs and beliefs in the design of evidence-based interventions. Parents, teachers, and other respondents in their study spoke about rearing strategies to promote change in children's behavior and shared ideas about when to best address behavioral problems. The study concluded that interventions building on existing concepts of verbal reminders and guidance contained within the *samjhaune* concept (a preference for verbal reminders over physical punishment) have the potential to be more culturally compelling and could utilize existing local knowledge and expertise.

There is a widespread consensus that experimental designs with multicomponent, multimodal and multisectorial interventions targeting children, parents, the family system, and the larger community are needed in order to provide a deeper understanding of the relational impact and dynamics between trauma, and child/family functioning. What about the existing evidence in postwar communities? In a RCT, Sim, Annan, Puffer, Salhi, and Betancourt (2014) significantly reduced the use of harsh parenting practices among displaced Burmese families in Thailand, working solely with parents. They used a methodology they successfully replicated in postconflict Liberia. In another RCT in Liberia, Puffer et al. (2015) evaluated the impact of a brief parenting intervention, *Parents Make the Difference* (PMD). The program led to a 55.5% reduction in caregiver-reported use of harsh punishment practices. The program also increased the use of positive behavior management strategies and improved caregiver-child interactions. The average caregiver in the treatment group reported a 4.4% increase in positive interactions, while the average child of a caregiver assigned to the treatment group reported a 17.5% increase. The program did not have a measurable impact on child well-being, cognitive skills, or household adoption of malaria prevention behaviors.

On a wider ecological level, using an intersectoral approach, Devries et al. (2015) did an RCT to evaluate an intervention aimed at reducing the use of violent disciplinary practices by teachers in Uganda. The intervention achieved a significant reduction, with 17% fewer students in the intervention group than the control group reporting past-week incidents of violence by teachers.

Mutamba et al. (2018) evaluated group interpersonal psychotherapy (IPT) delivered within a low-resource health system in postconflict Northern Uganda in a nonrandomized trial design. Working with caregivers of children with nodding disease, they found that group IPT delivered by lay community health workers is an effective psychological treatment for common mental health problems in caregivers and has psychological benefits for the children as well.

Jordans, Tol, Ndayisaba, and Komproe (2013) conducted a controlled evaluation of a brief parenting psychoeducation intervention in Burundi. The two-session psychoeducation intervention was offered to groups of parents of children who had been screened for elevated psychosocial distress. The intervention had a beneficial effect on reducing conduct problems compared to the control condition ($d = 0.60$), especially among boys. Parents evaluated the intervention positively, with increased awareness of positive parenting strategies and appropriate disciplinary techniques reported as the most common learning points.

Miller and Jordans (2016) suggest that direct work with children is best conceptualized as one element of a comprehensive approach that also targets ongoing stressors and increases the availability of supportive resources in the family and community. Such an approach is illustrated in an RCT by O'Callaghan, McMullen, Shannon, Rafferty, and Black (2013). They achieved significant and lasting improvements in the mental health of sexually exploited Congolese girls by combining trauma-focused cognitive behavioral therapy with livelihood support and the active engagement of the girls' caregivers in the intervention. They found unusually large treatment effect sizes that were still present at 3-month follow-up.

A feasibility study in Northern Uganda based on Parent Management Training (PMT) added sessions on the intergenerational transmission of violence, substance abuse, and other risk-taking behaviors (Wieling et al., 2017). Mothers reported remarkable changes in their relationships with children after only a few sessions focused on positive involvement and were ready to challenge their long-standing views regarding harsh punishment. However, the research team also saw that without more intensive support and practice regarding limit setting, mothers readily abandoned the alternative discipline strategies they learned.

The Multifamily Approach (MFA) is a family-oriented method provided in a group setting, derived from Multifamily Therapy (Asen & Scholz, 2010). Mentalization is an important aspect of these groups. To increase mentalizing capacities and parental emotional availability, the group can be used for mirroring purposes and be encouraged to offer feedback. If, for example, the group is composed of parents with children, the parents may be encouraged in coaching each other when interacting with their child. In some activities, parents are invited to play with children other than their own—thereby increasing their critical observation and curiosity. A qualitative feasibility study and a manual showed that MFA is feasible for families with a child with a disability in the West Bank (Mooren et al., 2018; WTF, 2015). Another feasibility study has been conducted among survivors of sexual trauma in Congo.

Adhikari et al. (2018) did a study that evaluated the feasibility, acceptability, and outcomes of a combined school- and family-based intervention, delivered by psychosocial counselors, for children with behavior problems in rural Nepal. Significant behavioral change was shown in child outcome measures in this uncontrolled evaluation, alongside qualitative findings suggesting feasibility and acceptability, to support moving toward a controlled trial to determine effectiveness.

Betancourt et al. (2011) and Chaudhury et al. (2016) did several studies in Rwanda on family-based prevention of mental health problems in children in families affected by HIV, as well as studies on a family-based intervention to reduce intimate partner violence (IPV). In their mixed methods approach, they collected qualitative data for cultural adaptation of both their measures to assess local mental health constructs, and to adapt their family-based preventive and curative interventions. After the qualitative phase, they collected quantitative data for a prevalence study of mental health problems; and they conducted a feasibility study of family-strengthening interventions that should finally result in an RCT. The research group also seriously attempted to elicit local sources of resilience and incorporate these in their culturally attuned or adapted interventions. They used free listing exercises, clinician key informant interviews, and key informant interviews with adults and children (ages 10–17) to investigate strengths and sources of resilience in Rwandan children and families at risk for psychosocial difficulties due to HIV/AIDS. Five forms of protective resources emerged through this research: perseverance (*kwihangana*); self-esteem/self-confidence (*kwigirira ikizere*); family unity/trust (*kwizerana*); good parenting (*kurera neza*); and collective/communal support (*ubufasha abaturage batanga*). Operating within individual, family, and collective/community systems, these resources support children at multiple ecological levels. In their family-based intervention to reduce intimate partner violence (IPV), they addressed family conflict and problems related to alcohol use to promote child mental health and family functioning within HIV-affected families. A family home-visiting, evidence-based intervention designed to identify and enhance resilience and communication in families to promote mental health in children was adapted and developed through a series of pilot study phases prior to being tested in open and RCTs in families affected by caregiver HIV. They found reductions in alcohol use and IPV among caregivers, and improved family functioning, lower levels of violence, and problem drinking, as well as improved child mental health, among the intervention group. In addition, they assessed the feasibility and acceptability of the Family Strengthening Intervention (FSI), a preventive intervention to reduce mental health problems and bolster resilience among children living in households affected by caregiver HIV. Families reported high satisfaction with the FSI. Caregiver-reported improvements in family connectedness, good parenting, social support, and children's pro-social behavior were sustained and strengthened from postintervention to 6-month follow-up. Additional significant improvements in caregiver-reported child perseverance/self-esteem, depression, anxiety, and irritability were seen at follow-up (Betancourt et al., 2014).

The studies by Betancourt and colleagues are a good example of the current state of the art in balancing the cultural adaptation of instruments and interventions while retaining essential intervention components and treatment fidelity. Similar mixed methods strategies were used in Indonesia and are described elsewhere (De Jong, 2002; De Jong & Van Ommeren, 2002; Tol et al., 2009). It is obvious that disaster, whether man-made or natural, may complicate existing parenting programs for other types of mental and neurological disorders. Patel et al. (2018) mention several rehabilitation programs for children with developmental, behavioral, and emotional problems such as autism, spectrum disorder, and ADHD.

For this review, I included all studies that are relevant for the theme of this chapter, though obviously a systematic review is needed to synthesize all the literature on the topic. In my opinion, it is good to realize that although these studies have shown promise, so far none of these interventions have been scaled. Also, questions remain about the training needed for lay health workers to deliver such interventions among conflict-affected populations, outside the context of supervision and training by highly trained clinicians working on the research trials.

Conclusion

This chapter described the negative impact of war on families. It showed that the expression of suffering is determined by history and culture. It is noteworthy that poor LMIC with a high prevalence of armed conflict suffer from an enormous shortage of mental health professionals. Inhabitants of those regions may express their plight in psychological disorders that are recognizable for academics. Although PTSD may be a valid universal diagnosis, it lacks ecological utility around the globe because other expressions of psychological trauma and distress are either more prevalent or have more salience in the local context.

When mass violence, war, and disasters occur, there are several ways in which stressful events may reach traumatic proportions for individuals, families, and communities. But if there are hardly any mental health professionals, how do people get to terms with their distress? Throughout history, mankind used rituals, ceremonies, dissociation, and psychogenic illness as resources to process horrendous events. After wars, rituals have helped former combatants to reintegrate. Moreover, rituals share important characteristics with healing practices around the world and may be perceived as culturally more congruent or appropriate than psychotherapy. The diagnostic phase before psychotherapy may be challenged by cultural syndromes and idioms of distress that may be hard to recognize and hard to disentangle from formal ICD or DSM diagnostic constructs. This also applies to the phenomena of trance, dissociation, or mass psychogenic illness.

One of the long-standing target groups for humanitarianism are child soldiers. A variety of demobilization and reintegration programs for former child soldiers are mentioned. The chapter also mentions traditional healing including rituals and ceremonies, family interventions, and multisectorial interventions. Interventions in the past years reflect a trend toward socio-ecological approaches, shifting from individual resilience toward ecological resilience. The socio-ecological approach also led to a revival of interest in family therapy, especially in wartimes when families are often the last resort for help. Whereas, on the one hand, families are a logical way to come to terms with war and mass disaster, on the other hand, there are methodological and therapeutic hurdles that have resulted in an amazing lack of research on family therapy in (post)conflict areas. The chapter provides six possible reasons that alone or in combination explain this paucity of research on families that as yet may be carried out.

We have seen that LMIC are the very settings that are disproportionally affected in terms of psychological morbidity or mortality by political violence, armed conflict, and mass disaster. The last decade has seen an increase in tested psychological treatments, which in general are individual or group-based. This may have led to the realization that we need more evidence on family interventions. Moreover, family interventionists realized that the strength of working with the family should be combined with other therapeutic interventions. The most important modalities that gained popularity were parenting and school-based interventions aimed at children and youth. A dozen studies are mentioned in a wide range of (post)conflict areas that use a family or parenting approach, or a school-based approach, often working with local village health workers or other interventionists. These studies show a revival of family-related interventions and seem to be a promising future trend in postwar and postconflict settings. However, the focus needs to turn to exploring holistic, scalable, and sustainable programs that consider and target the multiple risk factors affecting families in these settings.

Acknowledgment

I would like to thank Felicity Brown, Mark Jordans, and Stevan Weine for their helpful and constructive comments.

References

Adhikari, R. P., Upadhaya, N., Satinsky, E. N., Burkey, M. D., Kohrt, B. A., & Jordans, M. J. D. (2018). Feasibility study of a family- and school-based intervention for child behavior problems in Nepal. *Child and Adolescent Psychiatry and Mental Health*, *12*, 20. https://doi.org/10.1186/s13034-018-0226-3.

Al-Krenawi, A. (1999). An overview of rituals in Western therapies and intervention: Argument for their use in cross-cultural therapy. *International Journal for the Advancement of Counselling*, *21*, 3–17.

Allen, T. (1991). Understanding Alice: Uganda's Holy Spirit movement in context. *Africa*, *61*, 370–400.

Asen, E., & Scholz, M. (2010). *Multifamily therapy: Concepts and techniques*. London: Routledge.

Baines, E. K. (2007). The haunting of Alice: Local approaches to justice and reconciliation in northern Uganda. *The International Journal of Transitional Justice*, *1*, 91–114.

Bartholomew, R., & Sirois, F. (2000). Occupational mass psychogenic illness: A transcultural perspective. *Transcultural Psychiatry*, *37*, 495–524. https://doi.org/10.1177/136346150003700402.

Basu, H. (2009). Contested practices of control: Psychiatric and religious mental health care in India. *Curare*, *32*(1+2), 28–39.

Ben-Ezra, M. (2003). Flashbacks and PTSD. *The British Journal of Psychiatry*, *183*, 75.

Benight, C. C., & Bandura, A. (2004). Social cognitive theory of posttraumatic recovery: The role of perceived self-efficacy. *Behaviour Research and Therapy*, *42*, 1129–1148. https://doi.org/10.1016/j.brat.2003.08.00.

Betancourt, T. S., Meyers-Ohki, S., Stulac, S. N., Barrera, A. B., Mushashi, C., & Beardslee, W. R. (2011). Nothing can defeat combined hands (Abashize hamwe ntakibananira): Protective processes and resilience in Rwandan children and families affected by HIV/AIDS. *Social Science & Medicine*, *73*, 693–701. https://doi.org/10.1016/j.socscimed.2011.06.053.

Betancourt, T. S., Meyers-Ohki, S. E., Charrow, A. P., & Tol, W. A. (2013). Interventions for children affected by war: An ecological perspective on psychosocial support and mental health care. *Harvard Review of Psychiatry*, *21*, 70–91. https://doi.org/10.1097/HRP.0b013e318283bf8f.

Betancourt, T. S., Newnham, E. A., McBain, R., & Brennan, R. T. (2013). Post-traumatic stress symptoms among former child soldiers in Sierra Leone: Follow-up study. *British Journal of Psychiatry*, *203*, 196–202. https://doi.org/10.1192/bjp.bp.112.113514.

Betancourt, T. S., Ng, L. C., Kirk, C. M., Munyanah, M., Mushashi, C., Ingabire, C., … Sezibera, V. (2014). Family-based prevention of mental health problems in children affected by HIV and AIDS: An open trial. *AIDS*, *28*, S359–S368. https://doi.org/10.1097/QAD.0000000000000336.

Blass, R. B., & Blatt, S. J. (1992). Attachment and separateness. A theoretical context for the integration of object relations theory with self psychology. *The Psychoanalytic Study of the Child*, *47*, 189–203. https://doi.org/10.1080/00797308.1992.11822671.

Boehnlein, J. K., Tran, H. D., Riley, C., Vu, K. C., Tan, S., & Leung, P. K. (1995). A comparative study of family functioning among Vietnamese and Cambodian refugees. *Journal of Nervous and Mental Disease*, *183*, 768–773. https://doi.org/10.1097/00005053-199512000-00007.

Boothby, N. (2006). What happens when child soldiers grow up? The Mozambique case study. *Intervention: International Journal of Mental Health, Psychosocial Work and Counselling in Areas of Armed Conflict*, *4*(3), 244–259. https://doi.org/10.1080/10926771.2013.814742.

Boothby, N., & Thomson, B. (2013). Child soldiers as adults: The Mozambique case study. *Journal of Aggression, Maltreatment & Trauma*, *22*(7), 735–756. https://doi.org/10.1080/10926771.2013.814742.

Borisova, I. I., Betancourt, T. S., & Willett, J. B. (2013). Reintegration of former child soldiers in Sierra Leone: The role of caregivers and their awareness of the violence adolescents experienced during the war. *Journal of Aggression, Maltreatment & Trauma, 22*(8), 803–828. https://doi.org/10.1080/10926771.2013.824059.

Boszormenyi-Nagy, I., & Spark, G. (1973). *Invisible loyalties: Reciprocity in intergenerational family therapy*. New York: Harper and Row.

Bowlby, J. (1969/1982). *Attachment and loss: Attachment. Vol. 1*. New York: Basic Books.

Boyden, J. (2003). The moral development of child soldiers: What do adults have to fear? *Peace and Conflict: Journal of Peace Psychology, 9*(4), 343–362. https://doi.org/10.1207/s15327949pac0904_6.

Burkey, M. D., Ghimire, L., Adhikari, R. P., Wissow, L. S., Jordans, M. J. D., & Kohrt, B. A. (2016). The ecocultural context and child behavior problems: A qualitative analysis in rural Nepal. *Social Science & Medicine, 159*, 73–82. https://doi.org/10.1016/j.socscimed.2016.04.020.

Castro, F. G., Barrera, M., & Martinez, C. R. (2004). The cultural adaptation of prevention interventions: Resolving tensions between fidelity and fit. *Prevention Science, 5*(1), 41–45.

Catani, C., Schauer, E., & Neuner, F. (2008). Beyond individual war trauma: Domestic violence against children in Afghanistan and Sri Lanka. *Journal of Marital and Family Therapy, 34*(2), 165–176. https://doi.org/10.1111/j.1752-0606.2008.00062.x.

Cecchin, G., Lane, G., & Ray, W. (1993). *Irreverence: A strategy for therapists of survival*. London: Karnac Books.

Chaudhury, S., Brown, F. L., Kirk, C. M., Mukunzi, S., Nyirandagijimana, B., Mukandanga, J., … Betancourt, T. S. (2016). Exploring the potential of a family based prevention intervention to reduce alcohol use and violence within HIV-affected families in Rwanda. *AIDS Care, 28*(Suppl. 2), 118–129. https://doi.org/10.1080/09540121.2016.1176686.

Chowdhary, N., Jotheeswaran, A. T., Nadkarni, A., Hollon, S. D., King, M., Jordans, M. J. D., … Patel, V. (2014). The methods and outcomes of cultural adaptations of psychological treatments for depressive disorders: A systematic review. *Psychological Medicine, 44*, 1131–1146. https://doi.org/10.1017/S0033291713001785.

Christopher, J. C., Wendt, D. C., Marecek, J., & Goodman, D. M. (2014). Critical cultural awareness: Contributions to a globalizing psychology. *American Psychologist, 69*(7), 645. https://doi.org/10.1037/a0036851.

Cortes, L., & Buchanan, M. J. (2007). The experience of Columbian child soldiers from a resilience perspective. *International Journal for the Advancement of Counselling, 29*, 43–55.

Coyne, J. C., Downey, G., & Boergers, J. (1992). Depression in families: A systems perspective. In D. Cicchetti (Ed.), *Developmental perspectives on depression: Rochester symposium on developmental psychopathology* (pp. 211–249). Rochester, NY: University of Rochester.

Dalgaard, N. T., Todd, B. K., Daniel, S. I. F., & Montgomery, E. (2016). The transmission of trauma in refugee families: Associations between intrafamily trauma communication style, children's attachment security and psychosocial adjustment. *Attachment & Human Development, 18*(1), 69–89. https://doi.org/10.1080/14616734.2015.1113305.

Dean, E. T. (1997). *Shook over Hell: Post-traumatic stress, Vietnam, and the Civil War* (pp. 101–114). Cambridge, MA: Harvard University Press.

De Jong, J., & Hinton, D. (2018). Traumascape: An ecological–cultural–historical model for extreme stress. In D. Bhugra & K. Bhui (Eds.), *Textbook of cultural psychiatry* (pp. 347–363). Cambridge: Cambridge University Press.

De Jong, J.T.V.M. (1986). Palmwijn in nieuwe zakken, oftewel enkele gedachten naar aanleiding van een Ndöp-ritueel [Palm wine in new bags, some thoughts on a Ndöp ritual]. *Ned Tijdschrift voor Psychiatrie, 28*(1), 15–29.

De Jong, J.T.V.M. (1987). *A descent into African psychiatry*. Amsterdam: Royal Tropical Institute.

De Jong, J.T.V.M. (1991). Vreemde gezinnen? Systeemtherapie met Marokkaanse, Surinaamse en Turkse gezinnen [Strange families? Systemic therapy with Moroccan, Surinamese and Turkish families]. *Maandblad voor Geestelijke Volksgezondheid, 9*, 924–938.

De Jong, J.T.V.M. (2001). Remnants of the colonial past: The difference in outcome of mental disorders in high- and low-income countries. In D. Bhugra & R. Littlewood (Eds.), *Colonialism and mental health* (pp. 131–167). New Delhi: Oxford University Press.

De Jong, J.T.V.M. (Ed.), (2002). *Trauma, war and violence: Public mental health in socio-cultural context*. New York: Plenum-Kluwer.

De Jong, J.T.V.M. (2007). Traumascape: An ecological-cultural-historical model for extreme stress. In D. Bhugra & K. Bhui (Eds.), *Textbook of cultural psychiatry* (pp. 347–364). Cambridge: Cambridge University Press.

De Jong, J.T.V.M. (2010a). A public health framework to translate risk factors related to political violence and war into multilevel preventive interventions. *Social Science & Medicine, 70*, 71–79. https://doi.org/10.1016/j.socscimed.2009.09.044.

De Jong, J.T.V.M. (2010b). Psychopathologie [Psychopathology]. In J.T.V.M. de Jong & S. Colijn (Eds.), *Handboek culturele psychiatrie en psychotherapie*. [Handbook of cultural psychiatry and psychotherapy]Utrecht: de Tijdstroom.

De Jong, J.T.V.M. (2011). (Disaster) Public mental health. In D. J. Stein, M. J. Friedman, & C. Blanco (Eds.), *Trauma and mental health: Resilience and posttraumatic disorders* (pp. 217–262). London: Wiley-Blackwell.

De Jong, J.T.V.M., & Reis, R. (2010). Kiyang-yang, a West-African post-war idiom of distress. *Culture, Medicine and Psychiatry, 34*(2), 301–321.

De Jong, J.T.V.M., & Reis, R. (2013). Collective trauma processing: Dissociation as a way of processing postwar traumatic stress in Guinea Bissau. *Transcultural Psychiatry, 50*(5), 644–661. https://doi.org/10.1177/1363461513500517.

De Jong, J.T.V.M., & Van Ommeren, M. H. (2002). Toward a culture-informed epidemiology: Combining qualitative and quantitative research in transcultural contexts. *Transcultural Psychiatry, 39*, 422–433. https://doi.org/10.1177/136346150203900402.

De Swaan, A. (2004). *Zorg en de Staat* [Care and the State] (6th ed.). Amsterdam: Bert Bakker.

Denov, M. (2007). Is culture always right? The dangers of reproducing gender stereotypes and inequalities in psychosocial interventions for war-affected children. In L. Dowdney (Ed.), *Psychosocial web page* (pp. e3–e6). London: Coalition to Stop the Use of Child Soldiers.

Devries, K. M., Knight, L., Child, J. C., Mirembe, A., Nakuti, J., Jones, R., … Watts, C. (2015). The good school toolkit for reducing physical violence from school staff to primary school students: A cluster-randomised controlled trial in Uganda. *The Lancet Global Health, 385*, e378–e386. https://doi.org/10.1016/S2214-109X(15)00060-1.

Dubow, E. F., Boxer, P., Huesmann, L. R., Landau, S., Dvir, S., Shikaki, K., & Ginges, J. (2012). Cumulative effects of exposure to violence on posttraumatic stress in Palestinian and Israeli youth. *Journal of Clinical Child & Adolescent Psychology, 41*(6), 837–844. https://doi.org/10.1080/15374416.2012.675571.

Dwyer, G. (2003). *The divine and the demonic. Supernatural affliction and its treatment in North India*. London: Routledge Curzon.

Eisenberg, L. (1977). Disease and illness: Distinctions between professional and popular ideas of sickness. *Culture, Medicine and Psychiatry, 1*, 9–23.

Fazel, M., Wheeler, J., & Danesh, J. (2005). Prevalence of serious mental disorder in 7000 refugees resettled in western countries: A systematic review. *The Lancet, 365*, 1309–1314. https://doi.org/10.1016/S0140-6736(05)61027-6.

Fossion, P., Leys, C., Vandeleur, C., Kempenaers, C., Braun, S., Verbanck, P., & Linkowski, P. (2015). Transgenerational transmission of trauma in families of Holocaust survivors: The consequences of extreme family functioning on resilience, sense of coherence, anxiety and depression. *Journal of Affective Disorders, 171*, 48–53. https://doi.org/10.1016/j.jad.2014.08.054.

Frank, J. D., & Frank, J. B. (1991). *Persuasion and healing: A comparative study of psychotherapy* (3rd ed.). Baltimore, MD: The Johns Hopkins University Press.

Friedman, E. H. (1982). The myth of the shiksa. In M. McGoldrick (Ed.), *Ethnicity and family therapy*. New York: Guildford.

Ganesan, M. (2006). Psychosocial response to disasters—Some concerns. *International Review of Psychiatry, 18*(3), 241–247. https://doi.org/10.1080/09540260600656035.

Geffray, C. (1990). *La cause des armes au Mozambique. Anthropologie d'une guerre civile.* Paris: Karthala.

Gewirtz, A., Forgatch, M., & Wieling, E. (2008). Parenting practices as potential mechanisms for child adjustment following mass trauma. *Journal of Marital and Family Therapy, 34*, 177–192. https://doi.org/10.1111/j.1752-0606.2008.00063.x.

Gewirtz, A. H., Pinna, K. L. M., Hanson, S. K., & Brockberg, D. (2014). Promoting parenting to support reintegrating military families: After deployment, adaptive parenting tools. *Psychological Services, 11*(1), 31–40. https://doi.org/10.1037/a0034134.

Gobodo-Madikizela, P., & van der Merwe, C. (2009). *Memory, narrative, and forgiveness: Perspectives on the unfinished journeys of the past.* Newcastle, UK: Cambridge Scholars.

Goffman, E. (1971). *The presentation of self in everyday life*. London: Pelican Books.

Green, J., & Goldwyn, R. (2002). Annotation: Attachment disorganization and psychopathology: New findings in attachment research and their potential implications for developmental psychopathology in childhood. *Journal of Child Psychology and Psychiatry, 43*, 835–846. https://doi.org/10.1111/1469-7610.00102.

Gutman, L. M., Sameroff, A. J., & Eccles, J. S. (2002). The academic achievement of African American students during early adolescence: An examination of multiple risk, promotive, and protective factors. *American Journal of Community Psychology, 30*, 367–399.

Hagengimana, A., & Hinton, D. E. (2009). Ihahamuka, a Rwandan syndrome of response to the genocide: Blocked flow, spirit assault, and shortness of breath. In Hinton, D. E., & Good, B. J. (Eds.), *Culture and panic disorder* (pp. 205–229). Stanford, CA: Stanford University Press.

Hassan, G., Kirmayer, L. J., Mekki-Berrada, A., Quosh, C., el Chammay, R., Deville-Stoetzel, J. B., … Ventevogel, P. (2015). *Culture, context and the mental health and psychosocial well-being of Syrians: A review for mental health and psychosocial support staff working with Syrians.* Geneva: United Nations High Commission on Refugees.

Hendriks, T., Schotanus-Dijkstra, M., Hassankhan, A., Sardjo, W., Graafsma, T., Bohlmeijer, E., & de Jong, J.V.M.T. (2019). Resilience and well-being in the Caribbean: Findings from a randomized controlled trial of a culturally adapted multi-component positive psychology intervention. *Journal of Positive Psychology*, https://doi.org/10.1080/17439760.2019.1590624.

Henry, S. B., Smith, D. B., Archuleta, K. L., Sanders-Hahs, E., Nelson Goff, B. S., Reisbig, A. M. J., … Scheer, T. (2011). Trauma and couples: Mechanisms in dyadic functioning. *Journal of Marital and Family Therapy, 37*, 311–332. https://doi.org/10.1111/j.1752-0606.2010.00203.x.

Hinton, D. E. & Good, B. J. (Eds.), (2016). *Culture and PTSD: Trauma in historical and global perspective*. Pennsylvania: University of Pennsylvania Press.

Hinton, D. E., & Jalal, B. (2014). Guidelines for the implementation of culturally sensitive CBT among refugees and in global contexts. *Intervention, 12*(1), 78–93. https://doi.org/10.1097/WTF.0000000000000069.

Hobfoll, S. E., & de Jong, J.T.V.M. (2014). The course and limitations of natural recovery from trauma: The centrality of threats to attachment and safety and their reinstatement. In L. A. Zoellner & N. C. Feeny (Eds.), *Facilitating resilience and recovery following traumatic events* (pp. 69–89). New York: Guilford.

Hobfoll, S. E., Watson, P., Bell, C. C., Bryant, R. A., Brymer, M. J., Friedman, M. J., … Ursano, R. J. (2007). Five essential elements of immediate and mid-term mass trauma intervention: Empirical evidence. *Psychiatry, 70*(4), 283–315. https://doi.org/10.1521/psyc.2007.70.4.283.

Honwana, A. (2003). Undying past: Spirit possession and the memory of war in southern Mozambique. In Meyer, B., & Pels, P., (Eds.), *Magic and modernity: Interfaces of revelation and concealment* (pp. 60–80). Stanford, CA: Stanford University Press.

Honwana, A. (2006). *Child soldiers in Africa*. Baltimore, MD: University of Pennsylvania Press.

Hwang, W.-C. (2006). The psychotherapy adaptation and modification framework: Application to Asian Americans. *American Psychologist, 61*(7), 702. https://doi.org/10.1037/0003-066X.61.7.702.

Igreja, V., Dias-Lambranca, B., Hershey, D. A., Racin, L., Richters, A., & Reis, R. (2010). The epidemiology of spirit possession in the aftermath of mass political violence in Mozambique. *Social Science and Medicine, 71*, 592–599.

Inter-Agency Standing Committee (IASC). (2007). *IASC guidelines on mental health and psychosocial support in emergency settings*. Geneva: IASC, World Health Organization.

Jordans, M. J., Tol, W. A., Ndayisaba, A., & Komproe, I. H. (2013). A controlled evaluation of a brief parenting psychoeducation intervention in Burundi. *Social Psychiatry and Psychiatric Epidemiology, 48*(11), 1851–1859. https://doi.org/10.1007/s00127-012-0630-6.

Jordans, M. J. D., Komproe, I. H., Tol, W. A., Ndayisaba, A., Nisabwe, T., & Kohrt, B. A. (2012). Reintegration of child soldiers in Burundi: A tracer study. *BMC Public Health, 12*, 905.

Jordans, M. J. D., Pigott, H., & Tol, W. A. (2016). Interventions for children affected by armed conflict: A systematic review of mental health and psychosocial support in low- and middle-income countries. *Current Psychiatry Reports, 18*(9), https://doi.org/10.1007/s11920-015-0648-z.

Jordans, M. J. D., Tol, W. A., Komproe, I. H., & de Jong, J.T.V.M. (2009). Systematic review of evidence and treatment approaches: Psychosocial and mental health care for children in war. *Child and Adolescent Mental Health, 14*, 2–14. https://doi.org/10.1111/j.1475-3588.2008.00515.x.

Jordans, M. J. D., van den Broek, M., Brown, F. L., Coetzee, A., Ellermeijer, R. E. C., Hartog, K., … Miller, K. E. (2018). Supporting children affected by war. In N. Morina & A. Nickerson (Eds.), *Mental health in refugee and post-conflict populations* (pp. 261–281). Cham: Springer.

Kakar, S. (1983). *Shamans, mystics and healers. A psychological enquiry into India and its healing traditions*. Oxford: Oxford University Press.

Kerig, P. K. (2005). Revisiting the construct of boundary dissolution. *Journal of Emotional Abuse, 5*, 5–42. https://doi.org/10.1300/J135v05n02_02.

Kohrt, B. A., Sauharda, R., & Maharjan, S. M. (2015). Child soldiers. In J. Wright (Ed.), *Vol. 3. International encyclopedia of the social & behavioral sciences* (2nd ed., pp. 430–438). https://doi.org/10.1016/B978-0-08-097086-8.64113-2.

La Roche, M. J., & Lustig, K. (2010). Cultural adaptations: Unpacking the meaning of culture. *Scientific Review of Mental Health Practice*, 7(2), 26–30.

Landau, J. (2007). Enhancing resilience: Families and communities as agents for change. *Family Process, 46*, 351–365. https://doi.org/10.1111/j.1545-5300.2007.00216.x.

Landau, J., & Saul, J. (2004). Facilitating family and community resilience in response to major disaster. In F. Walsh & M. McGoldrick (Eds.), *Living beyond loss* (pp. 285–309). New York: Norton.

Lau, A. S. (2006). Making the case for selective and directed cultural adaptations of evidence-based treatments: Examples from parent training. *Clinical Psychology: Science and Practice, 13*(4), 295–310. https://doi.org/10.1111/j.1468-2850.2006.00042.x.

Lee, C., Nguyen, A. J., Haroz, E., Tol, W., Aules, Y., & Bolton, P. (2019). Identifying research priorities for psychosocial support programs in humanitarian settings. *Global Mental Health, 6*, https://doi.org/10.1017/gmh.2019.19.

Leong, F. T. L., & Lee, S.-H. (2006). A cultural accommodation model for cross-cultural psychotherapy: Illustrated with the case of Asian Americans. *Psychotherapy: Theory, Research, Practice, Training, 43*, 410–423.

Lindblom, J., Peltola, M. J., Vänskä, M., Heitanen, J. K., Laasko, A., … Punamäki, R. (2017). Early family system types predict children's emotional attention biases at school age. *International Journal of Behavioral Development, 41*(2), 245–256. https://doi.org/10.1177/0165025415620856.

Masten, A. S. (2001). Ordinary magic: Resilience processes in development. *American Psychologist, 56*, 227–238.

Matsakis, A. (2004). Trauma and its impact on families. In D. R. Catherall (Ed.), *Handbook of stress, trauma, and the family* (pp. 12–26). New York, NY: Brunner-Routledge.

McFarlane, W. (2002). *Multiple-family groups in the treatment of severe psychiatric disorders*. New York: Guilford Press.https://doi.org/10.1097/YCO.0b013e32832c9064.

McFarlane, A. C. (2009). Military deployment: The impact on children and family adjustment and the need for care. *Current Opinion in Psychiatry, 22*, 369–373.

McKay, S., & Mazurana, D. (2004). *Where are the girls? Girls in fighting forces in Northern Uganda, Sierra Leone, and Mozambique: Their lives during and after war*. Montreal: Rights and Democracy.

Meineck, P. (2012). Combat trauma and the tragic stage: "Restoration" by cultural catharsis. *Intertexts, 16*(1), 7–24. https://doi.org/10.1353/itx.2012.0008.

Melander, E., Pettersson, T., & Themnér, L. (2016). Organized violence, 1989–2015. *Journal of Peace Research*. https://doi-org.proxy.uba.uva.nl:2443/10.1177/0022343316663032.

Miller, K. E., & Jordans, M. J. D. (2016). Determinants of Children's mental health in war-torn settings: Translating research into action. *Current Psychiatry Reports, 18*, 58. https://doi.org/10.1007/s11920-016-0692-3.

Minuchin, S. (1974). *Families and family therapy*. Cambridge, MA: Harvard University Press.

Monson, C., Taft, C., & Friedman, S. (2009). Military-related PTSD and intimate relationships: From description to theory-driven research and intervention development. *Clinical Psychology Review, 29*, 707–714. https://doi.org/10.1016/j.cpr.2009.09.002.

Mooren, T., Rabaia, Y., Mitwalli, S., Reiffers, R., Koenen, R., & de Man, M. (2018). The Multi-family approach to facilitate a family support network for Palestinian parents of children with a disability: A descriptive study. *The Lancet, 391*(Suppl. 2), S49. https://doi.org/10.1016/S0140-6736(18)30415-X.

Mourik, K. (2019). Personal communication.

Mutamba, B. B., Kane, J. C., de Jong, J.T.V.M., Okello, J., Musisi, S., & Kohrt, B. A. (2018). Psychological treatments delivered by community health workers in low-resource government health systems: Effectiveness of group interpersonal psychotherapy for caregivers of children affected by nodding syndrome in Uganda. *Psychological Medicine, 48*(15), 2573–2583. https://doi.org/10.1017/S0033291718000193.

Nelson-Goff, B. S., Crow, J. R., Reisbig, A. M. J., & Hamilton, S. (2007). The impact of individual trauma symptoms of deployed soldiers on relationship satisfaction. *Journal of Family Psychology, 21*, 344–353. https://doi.org/10.1037/0893-3200.21.3.344.

Neuner, F., Schauer, E., Catani, C., Ruf, M., & Elbert, T. (2006). Post-tsunami stress: A study of posttraumatic stress disorder in children living in three severely affected regions in Sri Lanka. *Journal of Traumatic Stress, 19*, 339–347. https://doi.org/10.1002/jts.20121.

Nickerson, A., Bryant, R. A., Silove, D., & Steel, Z. (2011). A critical review of psychological treatments of posttraumatic stress disorder in refugees. *Clinical Psychology Review, 31*, 399–417. https://doi.org/10.1016/j.cpr.2010.10.004.

Nordstrom, C. (1997). *A different kind of war story*. Philadelphia, PA: University of Pennsylvania Press.

O'Callaghan, P. S., McMullen, J., Shannon, C., Rafferty, H., & Black, A. (2013). A randomized controlled trial of trauma-focused cognitive behavioral therapy for sexually exploited, war-affected Congolese girls. *Journal of the American Academy of Child & Adolescent Psychiatry, 52*(4), 359–369. https://doi.org/10.1016/j.jaac.2013.01.013.

Obenchain, J. V., & Silver, S. M. (1992). Symbolic recognition: Ceremony in a treatment of post-traumatic stress disorder. *Journal of Traumatic Stress*, *5*(1), 37–43. https://doi.org/10.1002/jts.2490050105.

Olson, D. (2000). Circumplex model of marital and family systems. *Journal of Family Therapy*, *22*, 144–167. https://doi.org/10.1111/1467-6427.00144.

Pakaslahti, A. (2009). Health-seeking behaviour for psychiatric disorders in North India. In Incayawar, M., Wintrob, R., & Bouchard, L. (Eds.), *Psychiatrists and traditional healers: Unwitting partners in global mental health* (pp. 149–166). Oxford: Wiley-Blackwell.

Patel, R. (2003). *Woman versus man. Socio-legal gender inequality in Pakistan*. Karachi: Oxford University Pres.

Patel, V., Saxena, S., Lund, C., Thornicroft, G., Baingana, F., Bolton, P., … UnÜtzer, J. (2018). The Lancet Commission on global mental health and sustainable development. *The Lancet*, *392*, 1553–1598. https://doi.org/10.1016/S0140-6736(18)31612-X.

Perera, S. (2001). Spirit possessions and avenging ghosts: Stories of supernatural activity as narratives of terror and mechanisms of coping and remembering. In Das, V., Kleinman, A., Lock, M., Ramphele, M., & Reynolds, P. (Eds.), *Remaking a world: Violence, social suffering and recovery* (pp. 157–200). Berkeley: University of California Press.

Pinker, S. (2018). *Enlightenment now: The case for reason, science, humanism and progress*. London: Penguin.

Powers, M. B., Halpern, J. M., Ferenschak, M. P., Gillihan, S. J., & Foa, E. B. (2010). A meta-analytic review of prolonged exposure for posttraumatic stress disorder. *Clinical Psychology Review*, *30*, 635–641. https://doi.org/10.1016/j.cpr.2010.04.007.

Puffer, E. S., Green, E. P., Chase, R. M., Sim, A. L., Zayzay, J., Friis, E., … Boone, L. (2015). Parents make the difference: A randomized-controlled trial of a parenting intervention in Liberia. *Global Mental Health*, *2*, e15. https://doi.org/10.1017/gmh.2015.12.

Punamäki, R.-L., Diab, S. Y., Isosävi, S., Kuittinen, S., & Qouta, S. R. (2018). Maternal pre- and postnatal mental health and infant development in war conditions: The Gaza Infant Study. *Psychological Trauma Theory Research Practice and Policy*, *10*(2), 144–153. https://doi.org/10.1037/tra0000275.

Pynoos, R. S., Steinberg, A. M., & Piacentini, J. C. (1999). A developmental psychopathology model of childhood traumatic stress and intersection with anxiety disorders. *Biological Psychiatry*, *46*, 1542–1554. https://doi.org/10.1016/S0006-3223(99)00262-0.

Reis, R. (2016). Children's idioms of distress. In L. Manderson, E. Cartwright, & A. Hardon (Eds.), *The Routledge handbook of medical anthropology* (pp. 36–42). London: Routledge, Case Study 2.1.

Rossi, E. L. (Ed.), (1980). *(Vols. I–IV). The nature of hypnosis and suggestion. The collected papers of Milton H. Erickson on hypnosis*. New York: Irvington Publishers.

Rutter, M. (1985). Resilience in the face of adversity. Protective factors and resistance to psychiatric disorder. *The British Journal of Psychiatry*, *147*, 598–611. https://doi.org/10.1192/bjp.147.6.598.

Satijam, D., Singhm, C. D., Nathawat, S. S., & Sharmam, V. (1981). A psychiatric study of patients attending Mehandipur Balaji temple. *Indian Journal of Psychiatry*, *23*(3), 247–250.

Sax, W. S., & Weinhold, J. (2010). Rituals of possession. In Brosius, C. & Husken, U. (Eds.), *Ritual matters: The dynamics of change and stability* (pp. 236–252). Delhi: Routledge.

Scheff, T. J. (1979). *Catharsis in healing, ritual, and drama*. Los Angeles, CA: University of California Press.

Seligman, R., & Kirmayer, L. J. (2008). Dissociative experience and cultural neuroscience: Narrative, metaphor and mechanism. *Culture, Medicine and Psychiatry*, *32*, 31–64.

Shaw, J. A. (2003). Children exposed to war/terrorism. *Clinical Child and Family Psychology Review*, *6*, 237–246. https://doi.org/10.1023/B:CCFP.0000006291.10180.bd.

Shay, J. (1991). Learning about combat stress from Homer's Iliad. *Journal of Traumatic Stress*, *4*, 561–579.

Silva, R., Alpert, M., Munoz, D. M., Singh, S., Matzner, F., & Dummit, S. (2000). Stress and vulnerability to posttraumatic stress disorder in children and adolescents. *American Journal of Psychiatry, 157*, 1229–1235. https://doi.org/10.1176/appi.ajp.157.8.1229.

Sim, A., Annan, J., Puffer, E., Salhi, C., & Betancourt, T. (2014). *Building happy families: Impact evaluation of a parenting and family skills intervention for migrant and displaced Burmese families in Thailand*. New York: International Rescue Committee.

Slobodin, O., & De Jong, J.T.V.M. (2015). Family interventions in traumatized immigrants and refugees: A systematic review. *Transcultural Psychiatry, 52*(6), 723–742. https://doi.org/10.1177/1363461515588855.

Smith, T., Domenech-Rodrıguez, M. M., & Bernal, G. (2010). Culture. *Journal of Clinical Psychology, 67*, 166–175. https://doi.org/10.1002/jclp.20757.

Spiegel, D., Loewenstein, R. J., Lewis-Fernandez, R., Sar, V., Simeon, D., Vermetten, E., … Dell, P. F. (2011). Dissociative disorders in DSM-5. *Depression and Anxiety, 28*(12), e17–e45. https://doi.org/10.1002/da.20923.

Sriskandarajah, V., Neuner, F., & Catani, C. (2015). Parental care protects traumatized Sri Lankan children from internalizing behavior problems. *BMC Psychiatry, 15*, 203. https://doi.org/10.1186/s12888-015-0583-x.

Super, C. M., & Harkness, S. (1986). The developmental niche: A conceptualization at the interface of child and culture. *International Journal of Behavioral Development, 9*(4), e545–e569. https://doi.org/10.1177/016502548600900409.

Super, C. M., & Harkness, S. (2002). Culture structures the environment for development. *Human Development, 45*(4), e270–e274. https://doi.org/10.1159/000064988.

Tanielian, T. L., & Jaycox, L. H. (2008). *Invisible wounds of war: Psychological and cognitive injuries, their consequences, and services to assist recovery*. Santa Monica, CA: Rand.

Thirthalli, J., Zhou, L., Kumar, K., Gao, J., Vaid, H., Liu, H., et al. (2016). Traditional, complementary, and alternative medicine approaches to mental health care and psychological well-being in India and China. *The Lancet Psychiatry, 3*(7), 660–672. https://doi.org/10.1016/S2215-0366(16)30025-6.

Tol, W. A., Jordans, M. J. D., Reis, R., & De Jong, J.T.V.M. (2009). Ecological resilience: Working with child related psychosocial resources in war-affected communities. In D. Brom, R. Pat-Horenczyk, & J. Ford (Eds.), *Treating traumatized children: Risk, resilience, and recovery* (pp. 164–189). London: Routledge.

Tol, W. A., Reis, R., Susanty, D., & de Jong, J.T.V.M. (2010). Communal violence and child psychosocial wellbeing: Qualitative findings from Poso, Indonesia. *Transcultural Psychiatry, 47*(1), 112–135. https://doi.org/10.1177/1363461510364573.

Tol, W. A., Song, S., & Jordans, M. J. (2013). Annual research review: Resilience and mental health in children and adolescents living in areas of armed conflict—A systematic review of findings in low- and middle-income countries. *Journal of Child Psychology and Psychiatry, 54*, 445–460. https://doi.org/10.1111/jcpp.12053.

Turner, V. W. (1969). *The ritual process*. London: Routledge & Kegan.

Ungar, M., Ghazinour, M., & Richter, J. (2013). Annual research review: What is resilience within the social ecology of human development? *Journal of Child Psychology and Psychiatry, 54*, 348–366. https://doi.org/10.1111/jcpp.12025.

Van Duijl, M., Cardeña, E., & de Jong, J.T.V.M. (2005). The validity of DSM-IV dissociative disorders categories in SW Uganda. *Transcultural Psychiatry, 42*(2), 219–241. https://doi.org/10.1177/1363461505052666.

Van Ee, E. (2018). Multi-family therapy for veteran and refugee families: A Delphi study. *Military Medical Research, 5*, 25. https://doi.org/10.1186/s40779-018-0170-9.

Van Gennep, A. (1909). *Les rites de passage* [The rites of passage]. Paris: Émile Nourry.

Van Ommeren, M., Sharma, B., Komproe, I., Poudyal, B., Sharma, G. K., Cardeña, E., & de Jong, J.T.V.M. (2001). Trauma and loss as determinants of medically unexplained epidemic illness in a Bhutanese refugee camp. *Psychological Medicine, 31*(7), 1259–1267. https://doi.org/10.1017/S0033291701004470.

Ventevogel, P. (2016). Borderlands of Mental Health: Explorations in Medical Anthropology. Psychiatric Epidemiology and Health Systems Research in Afghanistan and Burundi (Ph.D. thesis). Amsterdam: University of Amsterdam.

War Trauma Foundation. (2015). *The multi-family approach in humanitarian settings.* Amsterdam: War Trauma Foundation. www.wartrauma.nl www.interventionjournal.com.

Weine, S., Kulauzovic, Y., Klebic, A., Besic, S., Mujagic, A., Muzurovic, J., et al. (2008). Evaluating a multiple-family group access intervention for refugees with PTSD. *Journal of Marital and Family Therapy*, *34*(2), 149–164.

Weine, S., Muzurovic, J., Kulauzovic, Y., Besic, S., Lezic, A., Mujagic, A., … Pavkovic, I. (2004). Family consequences of refugee trauma. *Family Process*, *43*, 147–160. https://doi.org/10.1111/j.1545-5300.2004.04302002.x.

Wieling, E., Mehus, C., Yumbul, C., Moellerherm, J., Ertl, V., Achan, L., … Catani, C. (2017). Preparing the field for feasibility testing of a parenting intervention for war-affected mothers in northern Uganda. *Family Process*, *56*, 376–392. https://doi.org/10.1111/famp.12189.

Worthman, C. M. (2010a). *Formative experiences: The interaction of caregiving, culture, and developmental psychobiology.* Cambridge: Cambridge University Press.

Worthman, C. M. (2010b). The ecology of human development: Evolving models for cultural psychology. *Journal of Cross Cultural Psychology*, *41*(4), 546–562. https://doi.org/10.1177/0022022110362627.

Yalom, I. D. (1980). *Existential psychotherapy.* New York: Basic Books.

Young, A. (1995). *The harmony of illusions: Inventing post-traumatic stress disorder.* Princeton, NJ: Princeton University Press.

PART 4

Family interventions

CHAPTER 16

Couple and family therapy across the globe: Cultural adaptations

Norman B. Epstein[a], Mariana K. Falconier[a], and Frank M. Dattilio[b]
[a]Department of Family Science, University of Maryland, College Park, MD, United States
[b]Department of Psychiatry, Harvard Medical School, Boston, MA, United States

Couple and family therapy[a] (CFT) is a relatively young mental health discipline, but it has developed rapidly over the past few decades, in terms of conceptual models, systematic training of clinicians, and widespread dissemination around the world. Major strides have been made in some countries to formalize CFT as a distinct profession with a unique body of knowledge, theoretical concepts, and treatment methods that should be regulated via certification or licensure. However, members of other established mental health professions (e.g., psychiatry, psychology, social work, professional counseling) commonly have argued that CFT is a subspecialization of their disciplines, and many clinicians in the United States and other countries practice couple and family therapy within the scope of their academic degrees and licenses in those other professions. As CFT has developed and spread across the globe, its status as a distinct profession or as a clinical specialization has varied from one country to another, but it is safe to say that there is consensus that providing competent and effective treatment for members of distressed relationships requires significant conceptual knowledge and clinical skills that differ from those used in forms of individual psychotherapy. This consensus is reflected in the vast number of national level CFT associations around the world and the growth of the International Family Therapy Association (IFTA).

This chapter describes the roots of CFT and its remarkable spread around the world in a fairly brief period, and we focus on ways in which

[a]We use the title couple and family therapy rather than the traditional "marriage and family therapy" in order to reflect the more inclusive nature of the field, which addresses committed couple relationships, between members of various sexual orientations, that may or may not involve legal marriage.

Cross-Cultural Family Research and Practice
https://doi.org/10.1016/B978-0-12-815493-9.00016-8

its broad dissemination has created a need for clinicians to consider cultural fit when applying theories and techniques that were developed in countries with different cultural values and belief systems. Although there may be some universal characteristics of couple and family relationships across cultures, sociocultural variations require that clinicians be prepared to adapt their application of therapy models to meet the needs of their clients. Cultural aspects of therapy models are likely to be salient whenever models that were developed in one culture are applied with clients from another culture. Furthermore, changes in the sociopolitical context within a country over time (e.g., the opening of a society that had long been constrained by an oppressive political regime) produce cultural shifts that therapists must take into account. This chapter focuses on situations in which a model that was developed within one culture is introduced via any of a number of possible pathways to a country with a different culture. Other contexts in which cultural factors should be considered include when a therapist within a country conducts therapy with clients from different regions within that country (e.g., with rural vs urban families) with subcultural differences, when a therapist treats cultural intermarriages, when a therapist treats first-generation immigrant couples and families, and when working with relationships of citizens/residents of second, third or greater generational immigrant backgrounds who are bicultural. In all of these situations, clinicians must either adapt traditional CFT models to be appropriate for the different context (the process through which the model is conducted will be different) or develop an indigenous CFT model designed specifically for the needs and cultural characteristics of the local population. In this chapter, we discuss factors that need to be addressed either when adapting traditional CFT models for diverse cultural contexts or constructing indigenous models. Working with differences among the members of intercultural relationships is beyond the scope of the present discussion.

Cultural sensitivity is an important concept in the CFT field, as writers (e.g., McGoldrick, Pearce, & Giordano, 2007) have stressed its importance and outlined some differences among cultural groups that clinicians need to take into account. However, the enthusiasm with which CFT has spread internationally appears to have contributed to a degree of dissemination of models and methods that may not take cultural factors into account sufficiently. The pitfalls of underestimating cultural factors are especially relevant when popular models have been applied in culturally diverse settings without any empirical investigations addressing what works well and what may not, a circumstance that is most often the case. As we describe in this chapter, cultural factors potentially account for a significant percentage of the variance in effects that CFT interventions have on clients and should be given a prominent place in the practice and empirical evaluation of CFT.

In the sections that follow, we briefly trace the roots of CFT as a significant paradigm shift in the understanding and treatment of relational and individual problems, "migration" pathways and patterns through which CFT models have spread around the world, forms of training and practice of CFT around the world, key cultural considerations in the practice of CFT, and some recommendations for culturally sensitive CFT, including suggestions for research.

Development of couple and family therapy as a mental health field

Consideration of the role that culture plays in conceptual models of CFT and their application in clinical practice requires an understanding of the forces that contributed to the development of the field. CFT initially was established as a counterpoint to the medical and psychodynamic model of individual psychological functioning that focused on intrapsychic processes in dysfunctional personality development and was dominated by interventions intended to counteract pathological internal functioning. The rise of CFT occurred to a large extent from growing recognition that significant interpersonal relationships play a major role in individuals' mental health. Clinical observations and research on small group dynamics applied to families (Winter & Ferreira, 1969), interpersonal models of psychiatry (e.g., Sullivan, 1953), the child guidance movement (that included parents in treatment of their children, albeit via separate parental sessions), the social work framework of intervening with the individual's social context, and studies that explored the possible role of family communication processes in the etiology and maintenance of schizophrenia (e.g., Bateson, Jackson, Haley, & Weakland, 1956; Lidz, Cornelison, Fleck, & Terry, 1957; Wynne, Ryckoff, Day, & Hirsch, 1958) paved the way for the emergence of family therapy in the mid-1950s (Nichols & Schwartz, 2004).

The impetus for the practice of family therapy was shaped significantly by the pioneering work of Don Jackson, Nathan Ackerman, and Murray Bowen. Based on Bateson and colleagues' research on communication in families of individuals diagnosed with schizophrenia, Jackson's approach to family therapy involved an integration of conceptual developments in the fields of systems theory, cybernetics, biology, and communication. Jackson

(1967) viewed the family as a homeostatic system that resisted change and an individual member's symptoms as maintaining family homeostasis. During the 1960s, the communications model developed at the Mental Research Institute in Palo Alto by Bateson, Jackson, Haley and Weakland had a major influence on the CFT field. Ackerman, a psychoanalytic-oriented child psychiatrist by training, at first followed the child guidance model in which the child was treated by a psychiatrist and the mother by a social worker. However, he shifted to having the same therapist see the mother and the child individually, and eventually held conjoint treatment sessions in which the family was the unit for treatment and diagnosis (Nichols & Schwartz, 2004). Bowen, a psychiatrist with psychodynamic training, who specialized in schizophrenia, initiated a project at the National Institute of Mental Health in the United States in which he hospitalized individuals diagnosed with schizophrenia and their whole families. He began holding multifamily meetings that eventually evolved into family therapy that assisted family members in increasing self-differentiation and decreasing their emotional reactivity to each other. Jackson, Ackerman, and Bowen, whose approaches all were based on the belief that human problems could be understood and treated in the context of family interactions, learned about each other's ideas in a 1955 meeting of the American Orthopsychiatric Association (Nichols & Schwartz, 2004). Subsequently, their combined influences began to establish family therapy as not only a treatment modality but rather as a distinct mental health field conceptually different from others regarding the etiology, diagnosis, and treatment of psychological problems.

Other major influences on the development of the field came from experiential therapists, in particular Virginia Satir (1964, 1972) and Carl Whitaker (Napier & Whitaker, 1978), who believed that family problems arise when rigid interaction patterns block members from experiencing and expressing their needs and associated emotions. Through eliciting intense emotional encounters among the family members during sessions by means of activities such as role playing and sculpting, the therapist encourages the members not only to express distress, anger and individual needs, but also caring for one another. Satir emphasized how family communication patterns limit members' abilities to be in touch with their needs and feelings and convey those to others, ultimately affecting the individual's self-esteem and degree of self-actualization. She identified rigid roles that family members took on with each other that constrained expression of their individuality. Her therapy sessions used experiential interventions such as sculpting to free members to express themselves. Whitaker eschewed theory for the value of the therapist interjecting oneself into the family and disrupting their rigid negative patterns, sometimes through unorthodox behavior. These experiential concepts and methods were strong counterpoints to the traditional medical model and added momentum to the growing CFT field.

In the 1970s, the structural family therapy model developed by Salvador Minuchin (Minuchin, 1974) gained prominence. Subsequently, in the mid-to-late 1970s and the 1980s strategic family therapy blossomed through the work of the brief therapy group that included Paul Watzlawick, John Weakland, and Richard Fisch (1974), the work of Jay Haley (Haley, 1976) and Cloe Madanes (Madanes, 1981), and the Milan group of Mara Selvini-Palazzoli, Luigi Boscolo, Gianfranco Cecchin, and Juliana Pratta (Selvini-Palazzoli, Boscolo, Cecchin, & Prata, 1978). Several other therapists (e.g., Ivan Boszormenyi-Nagy, Carlos Sluzki) were pioneers during the early development of the family therapy field, and this brief review cannot adequately acknowledge all of the individuals who made contributions.

As family therapists strove to differentiate their field from the established mental health professions of psychiatry and psychology, they focused mostly on family interpersonal processes as the problems in need of change, and even actively avoided attention to intrapsychic processes (Haley, 1976). The overarching theoretical framework was general systems theory (von Bertalanffy, 1968), as applied to family relationships, although various models that were developed (e.g., structural family therapy, variations of strategic therapy, intergenerational family therapy) interpreted and applied systems principles in quite different ways (Steinglass, 1987).

It is understandable that the early focus in the field on family system dynamics, sometimes to the exclusion of intrapsychic processes of thought and emotion, served a function of distinguishing family therapy from the established traditional mental health professions of psychiatry and psychology. However, the conceptual pendulum in the field has swung back to a moderate position in more recent years, with a significant shift toward examining links between couple and family processes and the intrapersonal functioning (cognitions and emotional responses) of the individual members. One of the conceptual shifts was the application of constructivist concepts in the 1980s to understand and treat family problems. Central to constructivism is the concept that individuals actively and subjectively process their experiences in daily life, selectively perceiving events, categorizing experiences, and attaching meaning to perceptions. Nichols and Schwartz (2004, p. 98) state that constructivism was key in "alerting us to the importance of cognition in family life." Prime examples of constructivism in major CFT models are reframing in strategic and structural therapies, in which the therapist introduces clients to new meanings of family events that they had interpreted negatively, and the central concept in solution-focused therapy that clients have the capacity to behave more effectively but have been hampered by negative cognitions about their problems (Berg, 1994; de Shazer, 1985). The common element across the models is modifying family members' negative cognitions about their relationships (Nichols & Schwartz, 2004). Nichols and Schwartz (2004) noted the value of focusing on intrapsychic experiences but argued that it should be integrated with attention to systemic family behavioral patterns.

Another shift in the CFT field toward returning some attention to family members' intrapsychic cognitions that began in the late 1960s and developed further through the 1980s and 1990s was the application of the "cognitive revolution" in psychology to the CFT field. Initial behavioral models focused on basic learning processes (classical conditioning, operant conditioning, modeling) and primarily assessed and worked toward modifying negative couple and family behavioral patterns (e.g., Jacobson & Margolin, 1979; Liberman, 1970; Patterson, 1971; Stuart, 1969). However, in the 1980s and 1990s, an integrative cognitive-behavioral couple and family therapy (CBCFT) model was developed that examined mutual influences between family members' behaviors toward each other and their cognitions about those behaviors (Baucom & Epstein, 1990; Beck, 1988; Dattilio & Padesky, 1990; Ellis, Sichel, Yeager, DiMattia, & DiGiuseppe, 1989; Epstein, 1982; Epstein, Schlesinger, & Dryden, 1988). Although CBCFT clinicians tend to embrace some degree of an expert role in guiding clients in examining the appropriateness of their cognitions, to a large extent they apply a constructivist framework emphasizing the subjectivity of individuals' appraisals of life events and focus on the advantages and disadvantages of individuals' cognitions such as their personal standards for the characteristics of a "good" relationship.

Given how powerful intrapsychic emotional processes are in couple and family life, it is striking that it took a while before CFT models were developed that more systematically assess and intervene with members' emotional responses to each other. CBCFT has integrated attention to the mutual influences among family members' cognitions, emotions, and behavioral interactions (Epstein & Baucom, 2002). Similarly, Gottman's (1999) approach to couple therapy, which draws heavily on cognitive-behavioral concepts and methods, emphasizes the impact that strong negative emotional responses has on partners' behavioral interactions and feelings of intimacy. Furthermore, emotionally focused therapy (EFT; Greenberg & Johnson, 1988; Johnson, 1996) places primary emphasis on emotions that shape members' behaviors and cognitions about each other, and focuses on ways in which partners regulate each other's emotions (Greenberg & Goldman, 2008).

In addition to the influence of constructivism on structural and strategic approaches, another major epistemological shift took place in the CFT field toward the end of the 1980s. It began with the feminist critique of the strategic and structural family therapy models for their inattention to issues of privilege and power in families and the broader society, and their reinforcement of stereotyped sex roles in the family. Family therapists such as Rachel Hare-Mustin (1978) and Virginia Goldner (1987) described the sociopolitical and cultural contexts that shape family therapy theories and interventions. This development was parallel to the emergence of postmodernist epistemologies such as social constructionism that view theories and knowledge as socially constructed and associate families' and individuals' mental health struggles with the marginalizing power of society's dominant discourses. The result of these influences in CFT can be seen in the emergence in the 1990s of the narrative therapy (White & Epston, 1990) and the collaborative approach (e.g., Andersen, 1991; Anderson & Goolishian, 1988) that emphasizes an egalitarian relationship between therapist and clients, which reached international audiences of therapists. Both the feminist and social constructionist CFT approaches have contributed to the incorporation of social justice principles and a focus on issues of inclusion and diversity in the field. All of these conceptual developments were tied to cultural factors influencing the thinking of the model developers and in turn have implications for the relevance of their application in diverse cultures. For example, families in some countries with hierarchical social structures such as China tend to view therapists as respected teachers who provide knowledge and guidance (Epstein et al., 2012; Epstein, Curtis, Edwards, Young, & Zheng, 2014), so they may be uncomfortable at least initially with collaborative approaches.

Finally, another development in the CFT field that can have implications for cultural adaptations of models has been the application of couple and family interventions with a wide range of individuals' psychological and physical health problems, such as depression, anxiety, substance abuse, eating disorders, and coping with severe physical health risks such as cancer (Epstein, Baucom, Kirby, & LaTaillade, 2019; Fischer & Baucom, in press; Fischer, Baucom, & Cohen, 2016). This approach maximizes the potential for treatment effectiveness by capturing both internal psychological processes of individual family members and interpersonal processes among family members (and places responsibility on clinicians to be well versed in assessment and treatment methods in both domains). As we describe in this chapter, such systemic interventions for problems that traditionally are viewed as residing within individuals may vary in the degree to which they are consistent with clients' and therapists' culturally based conceptions of the causes of individuals' health problems. Within cultures in which mental and physical health problems are attributed primarily or entirely to intrapersonal psychological or physical processes, CFT clinicians must be sensitive to how they introduce systemic conceptualizations and interventions for presenting problems.

A key trend in the history of the CFT field that has implications for use of models in diverse cultures was the significant degree to which the early conceptual development and research base for the systemic paradigm occurred primarily in the United States and some other English-speaking Western countries. Overall, the initial impetus for systems-based CFT approaches came primarily from the United States, even though some prominent pioneers (e.g., Ivan Boszormenyi-Nagy, Salvador Minuchin, Cloe Madanes, Carlos Sluzki) immigrated to the United States from other countries and lent their diverse cultural perspectives to the predominant Western framework, and clinicians in some other countries developed new approaches or variations on existing systemic models. For example, the Milan Group in Italy (Selvini-Palazzoli et al., 1978) developed a variation of strategic family therapy, and Michael White from Australia collaborated with David Epston from New Zealand (White & Epston, 1990) in developing narrative therapy. Similarly, Norwegian Tom Andersen (Andersen, 1991) developed the practice of using reflective teams to build collaborative relationships between therapists and clients in which family members observe clinicians' conversations about their impressions of the family's interactions during sessions, which provides new perspectives for the family. In addition, Greenberg and Johnson (1988) in Canada developed EFT for couples. Overall, CFT rapidly became international, but there were few instances in which therapy models developed initially in non-Western countries and were shaped by diverse cultural values and traditions.

The "migration" of CFT models to other countries no doubt introduced cultural variation in their use, but those developments were for the large part implicit (i.e., therapists who adopted a model that was developed within another culture implemented it in ways that made it consistent with their own cultures and thus palatable for their clients), and those cultural variations of imported models were untested empirically. We have described some examples of those cultural adaptations in this chapter, and their implications for achieving some consistency around the world in the education, training, and competence evaluation of therapists.

"Migration" of couple and family therapy throughout the world

CFT has been welcomed in numerous countries around the world during the past few decades, due to the recognition of the strong influence that family functioning has on individuals' well-being and the need for clinical services that focus on family processes. For example, in Russia increases in the divorce rate, family violence, alcoholism, and child abandonments, and a decrease in the average age of children making suicide attempts have alerted policy makers and mental health agencies to the need for family therapy services (Bebtschuk, Smirnova, & Khayretdinov, 2012). Similar examples of societal changes that have alerted policy makers and mental health organizations to increase mental health services for children and their families include mainland China's rapidly rising divorce rate and increases in depression and anxiety among school children (Epstein et al., 2012) and negative effects that rapid economic development and traditions of long work hours are having on work-family conflict and individual health in Asian countries including Japan, Taiwan, China, and South Korea (Kim, Cho, Lee, Marion, & Kim, 2005; Luo & Cooper, 2015).

As we have noted, CFT primarily originated in the United States beginning in the 1950s and spread around the world, with a number of significant innovations in theory and methods coming from clinicians in some other countries. There have been three major processes through which knowledge and expertise in CFT have "migrated" around the world. First, pioneers who developed therapy models commonly wrote books describing their model's theoretical concepts, as well as clinical assessment and intervention methods, often with detailed case examples that gave distant readers a sense of the value of these groundbreaking systemic approaches. Because the United States had a long history of academic publishing in many fields, books became the main method through which these models initially were disseminated, to the extent that book publishers sought international markets. Developers of CFT models also published brief descriptions of their approaches in professional journals, again mostly in English and mostly disseminated to other countries to the degree that the journals had international readerships. Consequently, initially knowledge about CFT was disseminated quickly among Western English-speaking countries, due to the common language in which its developers published their work, and it made its way into other countries more slowly due to language barriers.

Second, professionals in many non-English–speaking countries who were aware of CFT developments abroad traveled to the United States to receive training either short-term through a private institute (e.g., the Ackerman Institute in New York, the Mental Research Institute in Palo Alto, the Philadelphia Child Guidance Clinic) or more intensive training in one of the programs that had been developed in academic institutions. These professionals returned to their countries with knowledge and skills in systemic CFT models and often would offer trainings locally through their own private institutes or via courses/seminars at universities (Roberts et al., 2014). They also returned with books written by the model developers, which commonly were translated into the languages of their home countries and facilitated wider dissemination. A significant example we have already noted that led to the development of an innovative therapy model in another country was Italian psychoanalyst Mara Selvini-Palazzoli, who made a major professional shift to a family systems theoretical framework and visited the United States to observe the work of Ivan Boszormenyi-Nagy, James Framo, and Salvador Minuchin in Philadelphia in 1967. She returned to Milan and with colleagues established the Center for the Study of the Family and of Group Techniques, where the focus increasingly was on a systemic approach to family therapy. In 1971, she and colleagues Luigi Boscolo, Gianfranco Cecchin, and Giuliana Prata founded their Milan Group approach that expanded strategic therapy concepts developed in the United States, focusing on dysfunctional communication processes in families of individuals with psychiatric disorders (Ng, 2005).

A similar process occurred as Asian scholars visited Western family therapy training centers and returned to develop groundbreaking family therapy programs at home. For example, Professor Wai Yung Lee grew up in Hong Kong, completed graduate studies abroad in Canada and the United States, worked as a family therapist in the United States for many years primarily at the Minuchin Center for the Family in New York City (where she is still a faculty member), then returned to Hong Kong and established the Hong Kong University (HKU) Family Institute. She began the first master's program in marriage and family therapy in Asia, as well as founded the Asian

Academy of Family Therapy that has grown to include family scholars, therapists, and policy makers from several Asian countries (Semmelhack & Lan, 2018). Similarly, Professor Xiaoyi Fang has developed a family therapy program for masters' and doctoral students in the Institute for Developmental Psychology at Beijing Normal University after receiving training in CFT at the University of Maryland, Baltimore between 1994 and 1996 (Deng, Lin, Lan, & Fang, 2013). Subsequently, Dr. Fang was a Fulbright Scholar for a year in the Department of Family Science at the University of Maryland, College Park, and focused on the department's AAMFT accredited couple and family therapy program. He also organized three international family therapy conferences in China (2012, 2014, and 2016), bringing together Western and Chinese clinicians, researchers, and graduate students.

Third, the spread of CFT into non-English–speaking countries also occurred when many of the pioneers and later developers of therapy models (e.g., Minuchin, Satir, Watzlawick) traveled to other countries (e.g., Taiwan, Mainland China, Mexico) to make invited presentations of their approaches, commonly with local professionals providing translation (Chao & Huang, 2013; Deng et al., 2013; Roberts et al., 2014). The non-English–speaking countries had further access to the models developed in English-speaking Western countries to the extent that English-speaking professionals in the host countries translated the visitors' books and created local training programs. In many cases, this type of input was widely accepted within host countries, but sometimes professional traditions and territorial conflicts interfered. For example, Le Goff (1999) described how the predominant psychoanalytic orientation in France, in which theoretical assumptions led therapists to avoid including family members in an individual's treatment, for many years posed a major obstacle to clinicians' openness to CFT approaches. Le Goff (1999) noted that it was not until Selvini-Palazzoli and her colleagues' *Paradox and Counterparadox* text was translated into French in 1978 that French therapists became interested in systemic family therapy.

Sometimes the timing of opportunities for originators of family therapy models to visit and conduct training was determined by sociopolitical events. For example, the major civil unrest and occupation of French universities in May 1968 led to the cancellation of a visit by Ivan Boszormenyi-Nagy, and it was not until 1988 that he was re-invited to present his work (Le Goff, 1999). As another notable example, 1989 marked the end of a political regime in Romania that had suppressed the profession of psychology and related disciplines, and in 1993 Zoltan and Agnes Konya, a pediatrician and a family physician who were among the initiators of family therapy in

Romania, met two family therapists from Ireland and invited them to present training in systemic therapy in Romania (Konya & Konya, 2013). Konya and Konya both shifted their careers toward therapy and were the founding members of the Cluj Systemic Family Therapy Association in Romania. Similarly, beginning in 1976, at the end of the Cultural Revolution that severely cut off China's society from the outside world beginning in 1966, the Chinese government's new "open door" policy led German family therapists Margarete Haass-Wiesegart and Ann-Katrin Scheerer to visit China and introduce Chinese psychiatrists and psychologists to systemic family therapy models (Deng et al., 2013; Sim & Hu, 2009). The German therapists tended to present the Western-developed models in their original form, so Chinese therapists adapted the models to their own culture in an ad hoc manner (Epstein et al., 2012). Since that time, developers and practitioners of major CFT models such as structural, MRI brief therapy, experiential (particularly Satir's model), multigenerational, and cognitive-behavioral have visited China, and those models are taught in programs such as the MFT program at Beijing Normal University (Deng et al., 2013). Another example is Argentina, in which the first family therapy conference was organized by Carlos Sluzki in 1970, but the military government that ruled the country between 1976 and 1983 closed academic and public mental health centers, and family therapy pioneers such as Minuchin, Falicov, and Sluzki moved to the United States to continue their work. However, family systemic thinking persisted in the private sector, within agencies such as the Center for Studies of Families and Couples and the Private Center for Psychotherapies and the publication of the first Spanish language journal Terapia Familiar from 1978 to 1993 (Herscovici, Gonzalez, Label, Vega, & Garcia, 2013).

Similarly, after the 1973 military coup in Chile, many mental health professionals left the country, including some who obtained training in family therapy abroad (Herscovici et al., 2013). They and other psychologists, psychiatrists, and social workers who learned about systemic approaches to family therapy in the early 1980s founded family therapy training programs at public and private institutes within Chile.

Thus, CFT has spread to many countries via direct interaction with the originators of models or indirectly via secondary paths involving "second-generation" therapists who were trained in the originators' models visiting other countries and passing along their expertise. Sometimes that process was facilitated by the free flow of information and professional exchanges among countries, but sometimes it was influenced by political forces.

Currently, due to widespread dissemination of therapy models through all three modalities, as well as through YouTube videos and other Internet media, plus the establishment of training centers in many countries, the worldwide "migration" pathways for CFT have become more complex. Consequently, forms of CFT practice currently exist in a multitude of countries, offering valuable interventions for problems in relational and individual functioning. What is less clear is (a) the degree to which the models that are practiced in diverse cultures conform to the core concepts and methods of their originators' work, (b) the degree to which models that were developed in one cultural context have been applied sensitively and appropriately in other countries with different cultures, and (c) the degree to which the quality of training and practice of CFT are monitored and regulated across cultures.

CFT professional associations, training, certification, and licensure around the world

The spread of CFT around the world is reflected in the presence of professional CFT associations in numerous countries, with many countries having multiple associations representing CFT clinicians. It is beyond the scope of this chapter to catalogue all of the CFT-related professional associations, but in this section we describe examples of their roles in defining CFT clinical competencies, establishing a code of ethics for clinical practice, providing educational programs (e.g., annual conferences, workshops) and offering professional certification in the field.

As we have described, the roots of the CFT field were largely in the United States and some other English-speaking countries with Western cultural traditions. The American Association for Marriage and Family Therapy (AAMFT) was founded in 1942 as the American Association of Marriage Counselors and has developed into a vast organization that accredits education/training programs, certifies approved supervisors, and has membership requirements involving core educational course content and supervised clinical practice. Approximately 50,000 AAMFT members from across the United States, Canada, and abroad have access to a variety of continuing education sources, including an annual conference, institutes, various publications, and offerings of local affiliated organizations. Currently, all 50 US states and the District of Columbia, as well as two Canadian Provinces (Ontario and Quebec) regulate the CFT profession.

The Canadian Association for Marriage and Family Therapy (CAMFT) was founded in 1995 and provides Registered Marriage and Family

Therapist (RMFT) credentialing that distinguishes family therapists from other mental health practitioners in terms of education, clinical training, and experience, with RMFT qualifications (with associated ethics code) very similar to those for clinical membership in AAMFT. CAMFT also certifies Approved Supervisors in a manner similar to AAMFT. CAMFT's stated mission is to promote couple/marriage and family therapy as a distinct profession in Canada, although to date only two provinces regulate it.

Within the United Kingdom, the Association for Family Therapy and Systemic Practice (AFT) is the overarching CFT association that provides membership based on accredited education and clinical training, tied to a code of ethics, as well as a variety of training resources including conferences, training courses, workshops, professional publications, and access to research findings. There are three levels of training, with the highest level based on prior training in another relevant profession, followed by a 4-year graduate program and registration with the United Kingdom Council for Psychotherapy. The Family Therapy Association of Ireland (FTAI) is a registering body with ethics code for members who have completed a 4-year FTAI training program at one of two nationally accredited clinical sites. FTAI also registers family therapy supervisors. FTAI is registered with the European Association of Psychotherapy and the Irish Council for Psychotherapy.

Also among the English-speaking Western-oriented nations, the Australian Association of Family Therapy (AAFT) has an established family and systemic psychotherapy training program curriculum and accreditation standards (coursework and supervised practice), a list of accredited training centers (some within universities and others in institutes and therapy centers), and a Clinical Family Therapist membership category. AAFT sponsors a variety of continuing education programs, including workshops and an annual conference, and the *Australia and New Zealand Journal of Family Therapy*. By 1979, all of the Australian states had their own family therapy associations, but the most developed association in Victoria became AAFT in 2012, with the other state associations disbanding and joining the new national organization. Currently, CFT is not regulated as a distinct mental health profession in Australia.

General membership in the New Zealand Association of Family Therapy (NZAFT) is open to anyone with an interest in a systemic approach to family therapy. There are two levels of clinical practitioner membership, focused either on education and supervised training specifically in family therapy or on working with families or systems. NZAFT also certifies approved family therapy supervisors. Currently, CFT is not regulated as a distinct mental health profession in New Zealand.

Across Europe, many countries have family therapy associations that promote systemic work, help professionals network, and conduct conferences. In general, family therapy training in Europe is at the postgraduate (master's) level, through institutes commonly attached to universities, after individuals have completed bachelor's degree programs that have qualified them to practice professionally in fields such as psychology. For example, the German Society of Systemic Family Therapy, Counseling, and Family Therapy (DGSF) was formed in 2000 due to the merger of two family therapy associations. The membership comprises individuals from a variety of professions (e.g., psychologists, medical doctors, social workers, educators). DGSF certifies types of education and clinical training in systemic practice (e.g., systemic therapy and counseling; systemic child and youth therapy), as well as expertise in systemic supervision. In February 2019, systemic therapy with adults was added to the list of treatments supported by public health insurance in Germany.

As we have noted previously, there initially was resistance to family therapy in France, due to the strong psychoanalytic tradition there. However, Le Goff (1999) noted that systemic, analytic, and intergenerational/contextual family therapy models currently are the most popular therapy models. They are conducted in public psychiatric practices (without specific training qualifications) in the private sector (with no insurance coverage and no inclusion of family therapy as a formal profession), and on an ad hoc basis in some small local clinical associations. There is little training in universities and no nationally recognized degrees (Le Goff, 1999). The oldest association is the Association of Systemic and Family Therapists (ATSF) that provides loose links among local associations, primarily through national conferences. However, the French Society of Family Therapy (SFTF) that was established in 1995 does require specific education and supervised training for membership, and it provides continuing education via seminars, conferences, and a journal.

In Spain, family therapy is a recognized discipline, represented by the Spanish Federation of Family Therapy (FEATF) that was founded in 1991 and is linked to the European Family Therapy Association (EFTA) (Roberts et al., 2014). Spanish universities increasingly included coursework in concepts and methods of systemic family therapy during the 1980s and 1990s, and beginning in 1997 the Spanish Association for Research and Development in Family Therapy has promoted research that supports family therapy practice. Nevertheless, Roberts et al. noted the need for accreditation of family therapists in Spain.

Despite the high prevalence of relational and individual psychological problems in Russia, the development of the CFT field has been limited. There is a strong stigma regarding mental illness and seeking any form of psychotherapy, in addition to negative residue from the prior period of "punitive psychiatry" in the USSR, in which political dissidents were given psychiatric diagnoses that justified treatment, including incarceration (Bebtschuk et al., 2012; Simakhodskaya, 2013). There are few options for education and clinical training, including the 2-year full-time systemic family therapy master's degree program conducted within the Faculty of Social Sciences at the Moscow campus of the National Research University Higher School of Economics and a variety of training offerings (e.g., lectures, seminars, workshops, observation of experts' sessions, live supervision) at the Center of Systemic Family Therapy, a postgraduate institute in Moscow (Bebtschuk et al., 2012; Simakhodskaya, 2013). Otherwise, professionals tend to seek out their own CFT training on an ad hoc basis. Currently, there is no licensing of family therapists or national association.

The Family and Marital Therapies Association of Turkey (AETD) was established in 1997 and encompasses a wide variety of members, such as psychiatrists, psychologists, psychological counselors, social workers, and family medical practitioners. AETD provides continuing education opportunities such as national conferences and has hosted the International Family Therapy Association conference.

Family therapy has strong roots in Italy, beginning with Selvini-Palazzoli and colleagues' founding of their Milan school and the subsequent founding in Rome of the Research Center for Family and Relational Therapy in 1972, by Luigi Cancrini and colleagues who emphasized strategic and structural concepts and procedures from Haley and Minuchin's work and founding in 1984 of the Italian Association for Relational Psychology and Psychotherapy (SIPPR), which represents Italy within the European Family Therapy Association (EFTA) (Manfrida, Giachi, & Eisenberg, 2013). Of the many Italian training institutions that have developed over time that address family relationships and therapy, most include the term "relational therapy" rather than family therapy in their titles, because in Italy family therapy is viewed as one of various contexts for intervening with significant interpersonal relationships influencing individuals' lives. Families are viewed not only as resources for their members but also as potential threats to individuals' abilities to overcome severe national economic conditions, unemployment, lack of affordable housing for the younger generation, etc. A variety of family therapy models are taught in numerous systemic-relational schools, which commonly have a psychology foundation. Registration of therapists with regional medical or psychological associations that set ethical codes and professional qualification standards is required by law. Nevertheless, Manfrida et al. (2013) noted that family or relational therapy is sought more by individuals than couples and families, that adverse economic conditions in Italy have reduced the accessibility of therapy, and that there is a lack of unity among the many competing systemic-relational schools.

Family therapy was introduced in the Netherlands during the 1970s. It grew during the next decade, but declined during the 1990s due to the lack of a clear theoretical base and empirical support regarding its effectiveness (Wagenaar & Baars, 2012). The Dutch Association for Relational and Family Therapy (NVRG), the national association for family therapists, provides networking and educational resources through resources such as annual conferences, master classes, and a professional journal. It also certifies family therapy trainers who must meet professional standards and conduct training with therapists (mostly clinical psychologists and psychotherapists) with university-level education. Required training that results in registered membership in NVRG includes education regarding theory and research, supervised clinical practice, and undergoing one's own therapy (Wagenaar & Baars, 2012). Initially traditional structural, strategic, and cybernetic models were most popular in the Netherlands, but narrative therapy, functional family therapy, and emotionally focused therapy are currently major models in practice. However, Wagenaar and Baars noted that overall family therapy is not considered the treatment of choice in the mental health field, due to its cost, continued limited empirical support, and an emphasis on individualism over family ties in the Netherlands.

Family therapy associations also are very common across Scandinavia. For example, the Finnish Family Therapy Association was founded in 1984, the Swedish Society for Family Therapy (SFFT) was formed in 1987 and comprised nine local associations and the Danish Association for Systemic and Narrative Therapy and Consultation (STOK) was founded in 1988 and brings together local family therapy groups across Denmark.

The Israeli Association for Marital and Family Therapy and Family Life Education is an interdisciplinary organization (e.g., social workers, psychologists, educational counselors, family physicians, and art therapists who focus on CFT) founded in 1977. It provides education and training through conferences, seminars, workshops, online publications, and a professional journal. Its Committee for Certification of Marital and Family Therapists provides the only certification of CFTs in Israel, based on academic education and clinical training, and there is no national professional licensure. The association has an ethics code and an ethics committee that oversees members' adherence to the code.

Thus, CFT models have migrated extensively across Europe, but their adoption among mental health professionals has been limited by traditional popularity of individual psychotherapy models practiced by members of dominant mental health professions, as well as competition among schools promoting systemic approaches. The lack of consistent training and practice guidelines across as well as within countries continues to pose barriers to the development of the field. However, the growing influence of international CFT organizations such as the European Family Therapy Association (EFTA) and the International Family Therapy Association (IFTA) that we describe later is drawing attention to the importance of developing cross-cultural standards for education and training.

In Asia, a number of countries have well-developed CFT professions, with varying degrees of professional associations, formal educational programs, training, certification, and governmental licensure. For example, the Korean Association of Family Therapy (KAFT) (established in 1988) offers two levels of certification as a Couple and Family Counseling Professional in South Korea. Individuals who have been registered as members of KAFT for a year can apply to be evaluated for certification at one of the two levels based on their academic degree related to family therapy from a formal educational institution, at least a bachelor's degree for the lower level certificate and at least a master's degree for the higher level certificate, coursework (e.g., family relations, family therapy, human development, counseling ethics) hours of supervised clinical practice, clinical case presentations, and a written exam. An additional supervisor grade certification requires more extensive academic and clinical preparation and specific supervision training. KAFT offers a variety of training programs, including an annual conference and additional workshops during the year. A second association, the Korean Association of Family Relations, also conducts certification of family counseling, based on requirements for academic degree and specific coursework, as well as clinical training. Nevertheless, there currently is no licensure in couple and family therapy in South Korea.

In Taiwan, there are two professional organizations focused on CFT—the Association of Couple and Family Therapy in Taiwan (ACFT) (founded in 2006) and the Taiwan Association for Marriage and Family Counseling

(TAMFC) (founded in 2003). Both associations provide family therapy training courses and tiered certification for family therapists with different levels of coursework and supervised clinical training. Community therapy centers also offer family therapy training. Chao and Huang (2013) noted that family therapy is practiced primarily by psychiatrists, psychologists, and social workers in Taiwan, all with their own national licensure, and there is no licensure for CFT as a distinct profession.

In Mainland China, a small number of CFT training programs with structured coursework were established in academic settings (e.g., the Institute for Developmental Psychology at Beijing Normal University, the Institute of Psychology in the Chinese Academy of Sciences), and in 2007 the title of Marriage and Family Counselor was recognized as an occupation by the Ministry of Human Resources and Social Security, with educational and training criteria established in 2008 (Deng et al., 2013). However, in spite of the *occupational* recognition, CFT is not considered a separate mental health *field*, and there still is no CFT licensure in China, with most practicing clinicians having degrees in more traditional established mental health fields such as psychology. There is no national CFT association, but the Chinese Association of Social Psychology is in the process of establishing a marriage and family psychology specialization. As we have noted previously, Chinese clinicians have actively imported Western CFT models, but they currently are advocating for the development of adapting those models more to be compatible with Chinese culture (e.g., collectivist, parental authority and protectiveness roles, filial piety, marriage as a link between two families, upholding the honor of one's family) as well as developing an indigenous Chinese family therapy (Deng et al., 2013).

The Japan Association of Family Therapy (JAFT) (founded in 1984) includes members from various mental health professions (e.g., psychiatrists, social workers, clinical psychologists, court mediators), with approximately a third being medical doctors. An early impetus for the development of family therapy in Japan was a high incidence of school refusal among youth, which has been conceptualized as based on family boundary issues (Kameguchi & Murphy-Shigematsu, 2001). There is licensing of psychiatrists and social workers in Japan, but not psychologists or family therapists. Enns and McRae (2007) noted that the percentage of intercultural intimate relationships has increased rapidly, and that clinicians are faced with helping couples negotiate conflicting cultural norms, such as those associated with communication styles, expressions of intimacy, childrearing practices, gender roles, values regarding education, work styles, and extended family relations.

The CFT field in India has been dominated by social workers, psychiatrists, and psychologists (Rastogi, 2010). Initially, mental health professionals in New Delhi organized family therapy training workshops that were taught by instructors from England, and the Indian Association for Family Therapy (IAFT) was founded in 1991, offering workshops, other trainings, and annual conferences. In spite of growing interest in the field and the presence of faculty with CFT interests at several universities, there are no social science-based CFT degree programs, and Rastogi (2010) noted that the only training program at a university is medically focused. To date, there is no CFT licensure in India.

Thus, a variety of Western-derived CFT models have been imported in many Asian countries, although training programs and settings for clinical practice still are relatively limited. Some attention has been paid to potential barriers to the use of traditional CFT models within Asian cultures, based on cultural differences regarding factors such as family structure, roles, communication styles, and family expectations for the therapist's role. However, future growth of the field, including the development of indigenous CFT models will require further conceptual work.

In Central and South American countries, psychoanalysis has long been the predominant model, with its practice dominated by psychiatrists and psychologists (Roberts et al., 2014). However, innovators such as David Liberman focused on neurotic processes in family relationships and applied the communication framework that had developed in North America, eventually leading to the establishment of systemic therapy associations such as Sociedad Argentina de Terapia Familiar (SAFT) in Argentina in 1978. In the 1970s SAFT emphasized the MRI structural approaches, but today there is more integration of systemic therapy with cognitive-behavioral, mindfulness, and EMDR models (Herscovici et al., 2013). In addition, institutes such as INTERFAS offer training in postmodern approaches such as narrative and collaborative models, sometimes through postgraduate certificate programs at universities. The prominent University of Buenos Aires emphasizes a psychoanalytic approach, with an elective course in systems therapy, with most systemic training provided in private universities and institutes. Before the 1990s, family therapy training took place primarily in postgraduate private institutes, but beginning in the 1990s accredited university undergraduate psychology programs (that qualify graduates to engage in clinical practice as psychologists) began to require education in family psychology, albeit without clinical training that still occurs most often in postgraduate programs at private institutes (Herscovici et al., 2013).

In Chile, family therapy is taught widely in undergraduate psychology programs, and private institutes that previously provided postgraduate family therapy clinical training are now affiliated with universities in order to provide master's degrees. The practice of all forms of psychotherapy is restricted to psychologists and psychiatrists, but Herscovici et al. (2013) noted the important role that social workers play as well. Initially, strategic, structural, intergenerational, and systemic models were prominent, whereas solution-focused, narrative, and dialogical approaches also have become popular, and other approaches involving aspects such as attachment, cognition, and mindfulness have been introduced (Herscovici et al., 2013). Currently, there is no consensus among the family therapy training institutions regarding the process of certifying couple or family therapists.

In Peru, the overall limited offerings of therapy within social service agencies and the psychoanalytic tradition of focusing on intrapsychic causes of life problems have led family therapists to create opportunities to provide treatment by networking with a variety of other helping professionals such as priests, psychologists, lawyers, and doctors (Roberts et al., 2014). Family therapists work in public hospitals, government institutions, and private clinics, but hospitals and health centers providing family therapy are few in number overall. In terms of training, the Catholic University of Lima offers a family therapy postgraduate certificate program, the Peruvian Center for Family and Couple Therapy offers a masters in family therapy program with emphasis on gestalt therapy in conjunction with the Autonomous University of Barcelona, and the Institute for Training and Family Development focuses on training in brief and solution-focused family therapy models. Despite the limited number of training programs, many clinicians without training provide treatment to families (Roberts et al., 2014).

Family therapy has been welcomed by mental health professionals in Mexico due to the consistent valuing of the family as the foundation of all levels of society, and therapists have addressed the negative effects of societal stressors including poverty, social inequality, discrimination toward indigenous groups, drug trafficking, and violence (Roberts et al., 2014). Family therapists commonly apply postmodern, social constructionist, and brief therapy models, commonly collaborating with government agencies to address social problems such as domestic violence. Family therapists also are part of health-care teams in hospitals, and the Department of Family Medical Therapy at Angeles Lomas Hospital in Mexico City has a Diploma in Medical Family Therapy program (Roberts et al., 2014). Overall, the

Mexican Association of Family Therapy, which was founded in 1981 and has its own ethics code and scientific journal, recognizes seven training programs in universities and institutes, although formal national government certification of family therapists has yet to be achieved.

Although CFT models have been imported by many countries in Central and South America, based on the common emphasis on family relationships, there is considerable diversity in the models that have become popular, influenced in part by sociopolitical conditions in each country and the degree of influence from traditional psychoanalytic practice and its focus on intrapsychic processes. Family therapists often have limited professional recognition, and CFT models tend to be practiced by members of dominant mental health professions such as psychologists and psychiatrists who received postgraduate training. The extent to which Western-derived CFT models are adapted explicitly for local cultural factors often is unclear.

The development of family therapy within Africa has been quite diverse, due to the vast differences in cultures and social contexts across the continent. It has tended to follow a pattern similar to many other regions, with mental health professionals becoming aware of family models via professional literature and considering their relevance for addressing the needs of local families, experimenting with family interventions in their clinical practice, inviting experts from the United States and other countries to make presentations at conferences in Africa, and traveling to the experts' training sites (Mason & Shuda, 1999). However, the concepts and methods from Western models commonly have been integrated significantly with indigenous African regional approaches to address individual and family problems, including long-standing traditional healing involving divining, counseling, and herbalism that are applied within a strong restorative context of communalism, peace, empathy, and acceptance (Ng, 2005). African family therapists often have been critical of traditional Western family therapy models that are seen as perpetuating colonial Western negative views of African people and cultures, as well as ignoring the strengths of their cultures (e.g., long established healing models that have a strong core spiritual base and utilize methods such as divining, storytelling, and trances that are not considered valid within Western models of medicine), the long history of oppression and abuse from apartheid and other aspects of colonial rule, and devastating effects of AIDS on families (Bakker & Snyders, 1999; Kasiram & Oliphant, 2007; Ng, 2005; Nwoye, 2001; Roberts et al., 2014). In addition, Western observers commonly have overlooked the high level of diversity of tribes with their associated religions, languages, and cultural traditions across Africa (Roberts et al., 2014). The primary professional organization, the South African Association of Marital and Family Therapy, was founded in 1981 and has held biennial conferences. Those conferences have reflected the development of family therapy in Africa, as the initial meetings focused on presentations by international experts, but increasingly the presenters were from African countries and described more indigenous family therapy approaches based on local cultural traditions and practices (Mason & Shuda, 1999). Overall, there still is limited mental health care, including family therapy, available in countries such as Nigeria, and the populous tends to rely on traditional healers for assistance with psychological problems that they attribute to spiritual problems (Adkeson, 2003). Thus, even though African family therapists have imported CFT models that largely were developed within Western cultures, they have needed to adapt the models greatly, including the characteristics of the clinician-therapist relationship, to address the widespread history of discrimination and trauma experienced by families in many regions of Africa, cultural values regarding family roles, and the major role of traditional healers.

"Superordinate" family therapy associations

In attempts to increase coherence in the field in terms of defining the CFT professional identity and competencies, standardizing education and training requirements, and sharing knowledge broadly among the many countries that have local family therapy communities, broader "superordinate" CFT associations have been formed that encompass associations from several countries, from a region of the world, or even around the globe. Examples are the European Family Therapy Association (EFTA), the Asian Academy of Family Therapy (AAFT), and the International Family Therapy Association (IFTA).

The European Family Therapy Association (EFTA) developed during the 1980s as a result of family therapists in various European countries seeking cross-cultural communication and training, and those in countries that lacked family therapy professional associations seeking a professional base (www.efta-nfto.com). The EFTA website currently lists 28 national association links, as well as regional association links. In addition, membership is available to individuals and institutes. The website describes significant variability in the CFT training opportunities, standards, and regulation from one country to another, so a goal of EFTA has been to increase high-quality training and practice across Europe. One major means toward that goal is conferences that bring together CFT clinicians, trainers, and researchers. In addition, EFTA's Training Institutes Chamber plays an advisory role regarding criteria for high-quality systemic family therapy training and practice. It sets standards for qualifications of individual therapists, supervisors, and training programs.

The Asian Academy of Family Therapy (AAFT) was established in 2012 by Dr. Wai Yung Lee and her colleagues who previously had established the Hong Kong University Family Institute (Tamura, Lee, & Cheng, 2017). It is based in Hong Kong but includes members from several Asian countries, including Japan, Korea, Taiwan, Mainland China, Singapore, and Malaysia. AAFT not only addresses cultural diversity across Asia but also focuses on shared factors such as collectivist cultural values, influences of Buddhism and Confucianism, extended family and kinship relationships, and filial piety. Tamura et al. (2017) noted that because the mental health systems in Asian countries tend to largely provide individually oriented interventions, a mission of AAFT is to promote systems thinking and services. AAFT's conferences rotate from country to country each year, and a current goal is to provide credentialing for Asian family therapists, taking into account not only some shared aspects of Asian culture but also models for training and practice that are tailored to each country or region.

The International Family Therapy Association (IFTA) was founded in 1987 and comprises educators/trainers, therapists, researchers, and other professionals from around the world (North America, Central and South America, Europe, Africa, Asia, Australia) who directly work with families or prepare others to do so. Both individual therapists and CFT associations hold affiliations with IFTA. IFTA holds conferences to develop the quality of family therapy practice and promotes continuing education, develops standards for clinical training and practice, and publishes the *Journal of Family Psychotherapy*. Through its International Accreditation Commission for Systemic Therapy Education, it provides two levels of accreditation for CFT education and training programs (with considerable overlap with AAMFT's Commission on Accreditation standards). Furthermore, IFTA certifies family therapists and systemic supervisors. Given the diverse backgrounds of the international clinicians and training programs that are evaluated by its Accreditation Commission, IFTA and similar "superordinate" associations serve not only important standardization challenges but also the crucial task of building in cultural sensitivity in its standards for education, training, licensure, and program accreditation.

Forms of training and practice of couple and family therapy around the world

As we have described, in some countries, commonly following the model of AAMFT in the United States, CFT has been established formally as a mental health field distinct from disciplines such as psychiatry, psychology, and social work, with separate professional associations, accreditation standards for training programs, licensing boards, etc. AAMFT developed core competencies for therapists that are achieved through extensive education and supervised clinical practice. Its Commission on Accreditation for Marriage and Family Therapy Education (COAMFTE) has established standards for training programs, involving their systemically oriented mission, qualifications of program faculty members and approved clinical supervisors, program goals and associated measurable student learning outcomes, resources for achieving program goals, and core course content (www.coamfte.org). However, as we have noted, clinicians from other mental health professions can practice CFT within the scope of their own licenses, at times failing to take into account their ethical codes that preclude practicing outside the scope of one's professional expertise. Members of other mental health professions commonly argue that CFT is an area of specialization within their disciplines.

In most countries, family therapy is practiced by individuals from diverse disciplines, certification standards from professional associations for CFT education and supervised clinical practice vary widely, and there very commonly is no legal recognition of the CFT profession. In some countries, professionals receive CFT training within a postgraduate degree program, but many countries (e.g., Argentina, Romania) have a different educational structure that does not require mental health professionals to hold a master's or doctoral degree. In spite of the proliferation of CFT-oriented professional associations, many countries still have no national association that sets the standards for adequate education and clinical training, and there is no required license. Consequently, CFT training programs vary widely internationally in their scope (number and content of courses, including therapy models taught), length, and quality. In addition, whereas in some countries there are academic graduate level CFT programs, training in many countries is offered by private agencies that are not subject to university educational policies and standards. Most of these institutes offer training in one particular model of therapy, which limits the breadth of therapists' training. Furthermore, the degree to which training of therapists in diverse cultures includes explicit attention to adapting standard Western-derived CFT models to be compatible with local cultural beliefs and traditions is unclear. Development of indigenous CFT models is crucial for the field, but sharing of information about such efforts from one country to another has been limited.

An additional issue stemming from CFT being taught in many countries outside the university context as a postgraduate certificate is that there has been limited, if any, opportunity for research on the efficacy and cultural relevance of the treatments. CFT models have been disseminated internationally mostly by practitioners for practitioners, without consideration of empirical evaluation of the quality of clinicians' implementation of the models or the relevance of the conceptual frameworks for clients in diverse cultures. Although it is likely that therapists commonly implicitly adapted the models appropriately to be consistent with their local cultural traditions, this ad hoc process has precluded research on the effectiveness of "imported" CFT models.

As we have noted, some regional and global CFT associations have attempted to address the potential problem of unsystematic adaptation and application of traditional CFT models by creating accreditation standards for training programs. This has ensured that particular areas of content are covered in coursework and a minimum number of hours of supervised clinical practice are accrued by trainees (a requirement that does not exist for licensure as a mental health professional in some countries, e.g., Argentina, Russia). We believe that it is advantageous to have a central national professional CFT organization to advance the identity of the profession and set standards for training and clinical practice. Ideally, global CFT associations can facilitate the existence of comparable training and practice standards around the world, but this also creates a challenge of embedding standards for evaluating culture-specific adaptations of models that can be assessed empirically. Even though the practice of CFT across cultures must take into account particular factors that may be prominent in some cultures but not others (e.g., the central role of healing traditions in African cultures; the emphasis on relational harmony in Asian cultures), there appear to be a number of key relatively universal cultural factors that should be considered. The following section outlines these factors.

Summary of international approaches to professional organization, training, program accreditation, and licensure

In sum, CFT is now practiced in many parts of the world by professionals who are either certified or licensed as such, or who conduct relationship therapies within the scope of their credentials within other mental health professions (e.g., psychology, psychiatry, social work, professional counseling). The depth and breadth of their CFT training varies widely, depending on the availability of basic and/or advanced training through academic programs and institutes, central organizations, or institutions that may be accredited by national professional associations that connect, represent, and provide continuing education programs for clinicians across the country. In some countries such as the United States and Canada, there are long-standing centralized professional CFT associations that set standards for training and practice of the profession, accredit CFT training programs, offer continuing education programs and conferences, and disseminate CFT theories, research, and methods through the publication of journals. Those central national organizations often are linked to local chapters, with whom they have developed local (e.g., state, provincial) licensing boards. The specific training, program accreditation, and licensure have contributed to strengthening the identity of CFT as a distinct mental health field in those countries to some extent, although as we have noted members of other mental health professions commonly practice CFT as specialties within their disciplines. Some other English-speaking countries with Western cultures such as the United Kingdom, Ireland, and Australia also have large central CFT professional organizations with similar roles, but despite their advanced CFT training and accredited programs, CFT is not regulated as a distinct mental health profession. Furthermore, in many Western European and Latin American countries, CFT has been practiced for decades but without a national accreditation system for training programs or a central organization or institution that represents the profession and sets standards for its practice. In many of these countries, the development of CFT professionally and institutionally has been affected by a strong psychoanalytic tradition that focuses on individual therapy. Most of those countries have several CFT organizations that offer their own trainings, certifications, conferences, and publication outlets and may compete with each other. Then, in Eastern European, Asian, and African countries CFT has developed more recently, so it is unsurprising that there is a lack of strong central institutions that promote common standards for training, legal regulation, and practice of the profession. Overall, the global pattern is that CFT is still marginally distinguished as a mental health profession even in countries with certification or licensure. However, it is important to note that in the absence of a single national professional organization, many professional organizations in different countries increasingly have joined superordinate regional family therapy organizations (e.g., EFTA, AAFT) to address cultural variation while seeking common standards for the training and practice of CFT.

Key cultural and sociopolitical factors that influence family relationships and the process of conducting couple and family therapy

The CFT models that have been more prominent in a country or culture, either in their original form or adapted, or indigenous new models that have been developed, have been influenced by the local cultural traditions and sociopolitical contexts (Borcsa, Hanks, & Vetere, 2013). For example, we noted that family therapy in France has been greatly influenced by the long-standing prominent status of psychoanalysis. Eastern European countries opened their doors to family therapy after the lifting of the iron curtain and increased acceptance of models that value independence of individuals as well as family connections. Andersen's (1991) reflecting team approach that emphasizes egalitarian relations between clinicians and family members developed in Norway's social-democratic state model (Borcsa et al., 2013). Structural family therapy has been relevant for dealing with common conflicts in three-generation Chinese families in which grandparents have played major roles in caring for grandchildren while their adult children are focused on their jobs (Epstein et al., 2012). Narrative therapy has grown in popularity across Africa due to its relevance for helping families develop self-enhancing narratives for their lives to counteract societal level as well as individual family histories of oppression and victimization (Bakker & Snyders, 1999; Nwoye, 2001).

In this section, we describe a number of characteristics of families that can be considered relatively universal across countries and cultures but still require therapists to conduct assessment and treatment that is tailored to the local culture. We also note ways in which the *process* through which therapy is conducted is shaped by the cultural context.

Cultural views about causes and treatment of mental health problems

It is natural for people to develop personal assumptions about causes of physical and mental health problems, and those views commonly are shaped by the narratives that are dominant in the local culture. Consequently, members of a culture are likely to consider CFT models and how they are implemented relevant for their lives and palatable to the extent that the models' implicit and explicit assumptions about causes and treatment of mental health problems are compatible with local cultural beliefs and traditions. For example, common core beliefs in African cultures that psychological problems such as depression and PTSD are caused by individuals' spiritual crises account for families turning to traditional healers rather than mental health professionals. However, Nwoye (2006) applies an adapted version of narrative therapy to assist African families experiencing problems with child psychological functioning and family relational problems by identifying and modifying narratives so they more accurately reflect actual causes of problems. Nwoye illustrated this approach with a case description regarding a female adolescent with psychosomatic symptoms. The therapist initially interviewed the girl individually to uncover her personal narrative about the causes of her symptoms, using forms of listening that combined empathy, interest in hearing the details of her views, deconstruction of her story to identify any gaps and ambiguities that suggest limitations to her narrative, "unpacking" of contradictions or unwarranted inferences, and identification of meanings that the client and family members attached to the factors they assumed were causes of the presenting problems. In this case, the girl inferred that her parents *intentionally* mistreated her and did so because they *devalued* her, but the parents saw no connection between family interactions and their child's emotional distress symptoms. The girl's disclosures to the therapist rather than her parents reflected the common condition of African parents not hearing about their children's emotions. The therapist subsequently avoided conjoint family sessions that would convey family responsibility for the girl's problems and would violate the cultural prohibition against children complaining about their parents publicly. Instead, the therapist interviewed the parents about the factors that had influenced their decisions regarding their daughter (e.g., why they sent her to live with a grandmother for 3 years) that greatly distressed her. When the therapist shared these alternative explanations with the daughter, it shifted her negative narrative to a much more benign one, created family healing in which the parents were motivated to express their caring for their daughter more directly, and reduced the girl's expression of somatic symptoms.

Furthermore, we have described how many cultures attach stigma to mental health and family relational problems, viewing them as signs of moral failure, as well as to seeking help for them. Consequently, CFT practitioners need to develop ways to frame the implementation of their models as useful resources for competent people, i.e., shifting to a strength-based framework. Thus, Epstein et al. (2012) described how families in China commonly avoid stigmatized mental health treatments. However, parents can be encouraged to participate in joint therapy sessions with their symptomatic children if the therapist avoids suggesting that the parents' behavior contributed to the problems and emphasizes how they can fulfill their societally prescribed role of raising competent and successful offspring by using the resources and suggestions provided by the professional. Thus, the more therapists apply CFT models that are compatible with clients' cultural views about causes and appropriate treatment of problems, *and* are willing to adapt their ways of implementing the models to make them palatable (e.g., avoiding discussions that can elicit loss of face), the more the clients will become engaged in treatment.

As already discussed, in many cultures indigenous systems for obtaining help for life problems and psychological distress, such as individuals identified as healers, are people's default option, and formal mental health services are avoided (and often not trusted). Religious, spiritual, and philosophical perspectives influence individuals' conceptions of fate versus control over their environment. For example, the Yoruba in Nigeria do not view mental health problems such as depression and PTSD as disorders that can be improved through therapy, but rather as spiritual attacks (Adkeson, 2003), so traditional family therapy does not work well, and it is crucial to integrate dream interpretation, folk tales, prayer, and invoking the gods and ancestors. Consequently, family therapists who use traditional CFT models must make it clear to potential clients that treatment will address highly valued aspects of their life experiences, such as spirituality, perhaps in collaboration with indigenous healers.

Collectivist versus individualistic cultural values, family boundaries, and family roles

Western-derived CFT models commonly include assumptions and values that emphasize autonomy and self-sufficiency as positive characteristics and admirable goals. When such models are applied in more collectivist cultures that emphasize loyalty to one's family and broader systems over self-actualization, therapists need to avoid attributing pathology to close family relationships. In terms of therapy process, they should focus on engaging family members' motivation to contribute to the group as a basis for taking some new actions that simultaneously can benefit individual members. Thus, parents from an Asian culture that strongly values education as a primary path to success in life and views a child's academic accomplishment as a positive reflection on the entire family may monitor their child's activities very closely and restrict leisure activities so the child can focus on schoolwork, with the unintended result that the child has become depressed (Epstein et al., 2012). Rather than encouraging the child to express his or her emotional distress to the parents (violating the principle of filial piety and risking the parents' loss of face), a family therapist not only can join with the parents' positive goal of raising a competent and productive child and show respect for their position in the family hierarchy but also present psychoeducational information suggesting that students who continuously need to concentrate on challenging academic material often are at risk of anxiety and depression. The therapist can explain that this danger can be alleviated if the child receives emotional support from parents and opportunities for leisure breaks to "recharge" their energy. Thus, the therapist supports the parents' societally sanctioned role but modifies their conception somewhat to best guide their child to be successful and healthy (Epstein et al., 2014). The modified conceptualization of parental role behavior influences the emotional quality of the parent-child relationship without challenging the traditional family hierarchy.

Furthermore, therapists need to be sensitive to implications of cultural differences in the foundations of couple/marital relationships. Some cultures emphasize romantic attachment, whereas others focus on contractual marriage that serves important functions such as economic viability, raising children, and the union of the partners' families of origin (Falconier, Randall, & Bodenmann, 2016). Therapists whose interventions fail to take those core family functions into account are unlikely to establish solid therapeutic alliances with their clients.

Gender roles and women's and children's rights

The application of CFT models also must take into account cultural variation regarding gender roles and the rights of women and children. For example, in countries with traditional patriarchal family traditions, therapists may need to deal with their own self-of-therapist issues such as discomfort when working with families in which males dominate females and in which children are not allowed to express their opinions and feelings. Thus, in traditional Chinese families, women marry into their husband's family and have less power than their spouse and mother-in-law, and the principle of filial piety discourages children from challenging their parents. Clinicians who apply models such as structural family therapy or cognitive-behavioral couple and family therapy that are not only respectful of traditional hierarchies but also promote egalitarian relationships are faced with intervening in a culturally sensitive manner. The challenge involves creating opportunities for disempowered family members to have a voice without more dominant members feeling overly threatened and losing face.

When practicing in diverse cultures, family therapists need to be aware of the current socioeconomic and political context that influences gender roles and parent-child relationships. For example, the traditional patriarchal relationships we have previously described within families in mainland China have been shifting significantly as rapid economic growth has led to vast numbers of women in higher education and the workforce, with associated economic power (Epstein et al., 2012). Social advances in children's rights also have altered acceptable parenting practices, such as reducing traditional use of physical punishment. Nevertheless, it is crucial to probe carefully for family members' beliefs and personal family traditions regarding the roles of adults and children, as well as indications of how those role enactments have influenced the members' mental and physical health. It also is important for the clinician to be knowledgeable about local support services (or lack thereof) for family members whose well-being has been compromised by family patterns, as those services may be quite different from those to which the clinician is accustomed.

Indigenous and professional social support systems

It is important for couple and family therapists to take an ecological approach to assessing and intervening with the levels of support that are available to families, as well as any barriers that interfere with family members making use of available resources. Even though CFT has spread around the world, varying levels of stigma still exist across cultures for seeking help for individual and relational problems. Even when crises motivate families to contact professionals, the members may be very reluctant to disclose personal information to outsiders and to each other. Within cultures that emphasize strong boundaries around a family and privacy regarding family issues, conveying the value of CFT while respecting traditional boundaries is essential. Nwoye's (2006) case example that we described previously illustrates how a CFT model can be implemented in a manner that reduces uncomfortable direct communication among family members who are accustomed to either consult indigenous healers or keep their problems to themselves. It also is important for family therapists to become familiar with levels of resources within the local culture (e.g., kinship networks, religious groups, social service agencies) and be prepared to facilitate families' connections with them. Cultural variations in a family's links with their extended family and who is considered kin also must be understood in order to devise relevant systemic treatment plans. Building relationships with indigenous sources of support creates opportunities for CFT clinicians to provide holistic treatment.

In countries with a history of political oppression, therapists need to be sensitive to family members' unease about self-disclosure.

Cultural conceptions of the role of the therapist

The role of the therapist vis-à-vis the family varies across CFT models (Nichols & Schwartz, 2004). For example, in models such as cognitive-behavioral, structural, and strategic, the therapist tends to be relatively directive, whereas in constructivist and social constructionist models such as solution-focused and narrative therapies, the clinician emphasizes the clients as experts on their own lives and as the initiators of change. However, in applying models, clinicians need to consider local cultural expectations regarding the role of helpers such as therapists. For example, members of some cultures (e.g., China) hold professionals such as teachers and therapists in high regard as experts who can provide them direction in solving problems, whereas families in many Western cultures prefer to experience self-efficacy and are more likely to view therapists as consultants (Epstein et al., 2012, 2014). Members of some collectivist cultures (e.g., those in many Central and South American countries) are likely to experience unease with therapists who maintain professional distance, and they may make efforts to establish more personal relationships with clinicians, e.g., by bringing food to sessions to share with the therapist. Ethical codes of CFT associations may portray accepting food and other gifts from clients as inappropriate boundary violations, but cultural sensitivity may require more flexibility in this area.

In addition, the models of individual therapy that were prominent when family therapy was introduced to a culture likely influenced which CFT models were embraced by mental health professionals as well as by the public. As we have noted previously, the strong psychoanalytic tradition we described in France and in Central and South American countries sometimes has posed a barrier to the development of systemic CFT models. CFT practitioners must be sensitive to the professional traditions in the local region and forge collaborative alliances with established practitioners.

Communication patterns and styles

Family communication processes are among the core concepts of systemic models of couple and family therapy, and therapists need to be aware of and honor cultural variations in the ways members of families communicate with each other and outsiders. A key difference exists between *low-context and high-context communication* (Hall, 1976). Low-context communication emphasizes explicit verbal content to convey meanings and tends to be more common in Western individualistic cultures, whereas in high-context communication, meanings are influenced significantly by contextual factors such as the type of relationship between the speaker and the listener (e.g., a power differential) and nonverbal behavior that may reinforce or qualify the verbal content (Falconier et al., 2016). For example, therapy models that emphasize overt verbal expressions of love and forms of physical affection (hugs, kisses) as central to family members' communication of caring for each other can be inappropriate within cultures in which caring is conveyed contextually. For example, a parent's strict monitoring and control of a child's activities may appear authoritarian to an outsider but be understood by both parent and child as based on the parent's love and desire to keep the child safe. Therapists who themselves were socialized within one mode of communication are at risk of misinterpreting messages exchanged by family members from a different culture, and may attempt to impose a form of communication that is uncomfortable to family members. Consequently, therapists need to identify the core assumptions about family communication that underlie their preferred therapy models and the degree of fit between them and culturally normative communication styles in the local population. However, it also is important to avoid applying stereotyped concepts about a culture's communication patterns to individual cases, as there is significant within-group variation, and each family has its own culture that involves traditions, values, and interaction patterns.

Professional and ethical standards

We have described how many CFT professional associations around the world have established ethical codes to guide clinical practice, but there is no universal set of principles; neither can therapists rely on governmental statutes to guide them consistently across cultures. For example, in some countries therapists have mandated reporting duties (e.g., regarding child abuse and neglect, duty to warn potential victims of violence). Cultures have different relationship standards for acceptable parenting practices, partner aggression, sexual relationships outside marriage, etc. Consequently, when working cross-culturally, a therapist whose training included adherence to particular professional and ethical codes must become familiar with cultural variations in standards of conduct for therapists and their clients. They are faced with deciding when local standards violate their personal codes sufficiently to warrant explicit discussions with clients versus when they can accept cultural relativity regarding their clients' behavior.

Research on effectiveness of couple and family therapy

It is beyond the scope of this chapter to provide a detailed review of the existing empirical support for couple and family therapy models, especially on an international scale. However, in this section we present some encouraging evidence from treatment outcome studies that supports the use of CFT approaches for problems in individual and relationship functioning. Because conducting methodologically rigorous treatment outcome studies is very time consuming and expensive, and grant funding is very limited for clinical trials on CFT approaches that are not focused on effects on DSM disorders, the efforts of clinical researchers have been impressive, although preliminary in several areas. Sprenkle (2012) provided an evaluative overview of a special issue of the *Journal of Marital and Family Therapy* devoted to reviews of CFT outcome research, targeting several major presenting problems and identifying degrees to which studies have examined efficacy (effects in controlled laboratory settings), effectiveness (effects in "real-world" clinical settings), and implementation in diverse countries and cultures. He concluded that research has established the validity of CFT as a mental health field, but noted that evidence is still quite limited for many popular CFT models, as well as their application in diverse populations within community clinical settings. In this section, we briefly summarize the current state of empirical support for CFT models, including their application across cultures, also referring to chapters in the present volume that review outcome research on CFT interventions.

One of the major presenting problems for which research has examined the effects of CFT interventions is child and adolescent conduct disorders and delinquency. Overall, the findings have been very encouraging, especially for Multisystemic Therapy (MST) that includes structural, strategic, social learning, and cognitive-behavioral principles and has demonstrated efficacy, effectiveness, and applicability across cultures (Henggeler & Sheidow, 2012). Solid empirical support also has been found for family-based interventions for substance abusing adolescents, notably Multidimensional Family Therapy (MDFT; Liddle, Rowe, Dakof, Henderson, & Greenbaum, 2009) that integrates interventions to improve parenting and the family environment, interventions with extrafamilial systems such as schools, and individual therapy and drug counseling for adolescents. Other family-based treatments for child and adolescent internalizing and externalizing presenting problems (e.g., depression, anxiety, attention deficit disorder, conduct disorder) for which there has been good empirical support are Family Cognitive-Behavioral Therapy for child anxiety disorders (Kendall, Hudson,

Gosch, Flannery-Schroeder, & Suveg, 2008), Parent-Child Interaction Therapy (Eyberg, Boggs, & Algina, 1995), the Triple-P Positive Parenting Programme (Sanders, 1999), and Attachment-Based Family Therapy (Diamond, Siqueland, & Diamond, 2003). Further description of parenting interventions, especially for parenting infants, is provided in Chapter 18 and the Triple P Positive Parenting Programme and its adaptations in diverse cultural contexts are provided in Chapter 19. To date, there has been limited research overall on cultural adaptations of other family treatments for child and adolescent problems.

Regarding CFT interventions for adult problems, literature reviews (Baucom, Shoham, Mueser, Daiuto, & Stickle, 1998; Fischer et al., 2016; Lebow, Chambers, Christensen, & Johnson, 2012; Snyder & Halford, 2012; Chapter 17) indicate that the most consistently empirically supported couple therapy models for reducing overall relationship distress have been cognitive-behavioral couple therapy (CBCT; Baucom & Epstein, 1990; Epstein & Baucom, 2002), integrative behavioral couple therapy (IBCT; Christensen, Dimidjian, & Martell, 2015), and emotionally focused couple therapy (EFCT; Greenberg & Goldman, 2008; Johnson, 1996). Furthermore, some couple education programs that were developed in university research settings and provide education about risk factors for relationship distress and strategies for counteracting them, as well as training in communication and conflict management skills, have demonstrated good empirical support for helping couples develop strong relationships, with positive effects decreasing over time (Markman & Rhoades, 2012; Chapter 17). Unfortunately, there has been limited research on applicability of these programs across diverse cultures internationally, and most other popular relationship education programs remain virtually untested.

There has been rapid growth in research on common factors (characteristics of therapists, clients, and the therapist-client therapeutic alliance) that contribute to findings that various CFT models have comparable positive effects on presenting problems of couples and families (Sprenkle, Davis, & Lebow, 2009). However, the research showing minimal differences among models has been conducted primarily with Western samples of couples and families, and little is known about the extent to which differences between particular models may exist in diverse cultures. This is an important direction for future research on common factors in CFT.

In addition, a growing body of empirical evidence that couple-based interventions (especially cognitive-behavioral and emotionally focused) improve a variety of individual adult psychological problems, such as depression, anxiety disorders, posttraumatic stress disorder, substance abuse, eating disorders, and coping with severe physical health problems such as cancer and heart disease is accumulating (Epstein et al., 2019). For example, studies have found support for behavioral couples therapy (BCT; Birchler, Fals-Stewart, & O'Farrell, 2008) for adults with substance abuse problems. Unfortunately, Sprenkle (2012) noted that those evidence-based CFT treatments for substance abuse have not been adopted widely in clinical settings in the United States, and international application has been limited.

Overall, studies investigating the efficacy and effectiveness of couple interventions for relationship quality and for individual psychopathology have been conducted primarily with white heterosexual European and North American samples (Sprenkle, 2012). Sprenkle also noted that effects of many popular couple therapy models (e.g., Bowen therapy, narrative therapy) have for the most part remained untested. As with CFT approaches for child and adolescent problems, the most research has been conducted on treatment efficacy, with limited studies on effectiveness in "real-world" clinical settings, and very little research on cross-cultural adaptations.

Research on family psychoeducation (FPE) models for major mental disorders (especially schizophrenia and bipolar disorder) that primarily include cognitive-behavioral interventions have demonstrated positive effects in clinical trials in many countries (e.g., the United States, China, Australia, Japan, Italy, India, Pakistan, Denmark); at present this appears to be the most widely disseminated CFT approach internationally (Lucksted, McFarlane, Downing, Dixon, & Adams, 2012; Chapter 20). Chapter 20 examines ways in which those CFT approaches can be adapted cross-culturally, and the widespread appeal of the programs to families in diverse contexts is encouraging. However, it is not consistently explicit in the literature how the standard FPE components such as psychoeducation (symptoms of a disorder, known causes, effective treatments, prognosis, and constructive relations between identified patients and other family members), training in communication and problem solving skills, and managing relapses are adapted from one culture to another.

In summary, research on CFT models has focused primarily on structured programs (consistent with the need to control the content of interventions in clinical trials) that commonly involve integration of elements from a number of traditional CFT models. There has been very little research on efficacy, effectiveness, and outcomes of cross-cultural adaptations of traditional CFT models (e.g., structural, strategic, narrative) that have proliferated around the globe. Although existing findings are valuable in identifying interventions that address particular presenting problems effectively, they do not clarify how core CFT models can be adapted to match the cultural and sociopolitical factors that we have covered in this chapter that influence families and the process of conducting therapy. It is our hope that clinicians and researchers will collaborate increasingly to develop and evaluate culturally sensitive treatments for families around the world.

Recommendations

The rapid rise and dissemination of CFT to address the major influences of intimate relationships on the well-being of children and adults has been a significant advance in the mental health field. As a relatively new paradigm, it has been characterized by not only considerable enthusiasm among those who practice it, but also a large degree of variation in conceptual models, forms of training, regulation of programs offering education and supervised clinical training, and public access to adequately trained professionals. In addition, we have noted that the body of empirical research testing the effectiveness of CFT treatments has been encouraging but still is quite limited, especially regarding effects when models that were developed in one culture have been imported into other contexts. Treatment outcome studies are complex and expensive, but nonetheless it is crucial that more research be conducted across cultures in order to ensure that families are receiving the help they need. Large-scale efforts such as government-supported clinical trials, including CFT interventions that are being conducted in the United Kingdom for the treatment of depression (Baucom et al., 2018), are a major step in the right direction, although the resources for such projects likely are lacking in some countries. Government grants for CFT studies continue to be scarce even in wealthy countries such as the United States, in spite of substantial research evidence regarding the role that intimate relationships play in physical and mental health. Efforts to impress these findings on legislators and make a strong case for cross-cultural research on CFT approaches should be pursued further.

The growth of CFT professional associations in many countries has been very encouraging. However, we believe that "superordinate" associations that create links among practitioners from many cultures and sociopolitical contexts are essential to balance acceptance of diversity with the need for standards for CFT education, supervised training, and clinical practice. Those larger associations are the wave of the future, and they may be instrumental in creating opportunities for cross-cultural therapy outcome research.

At the level of the individual clinician, it is important that therapists examine the cultural frame of reference of their therapy models. Cultural sensitivity is a self-of-therapist factor, and minimizing blind spots in one's work with diverse clients requires immersing oneself in clients' cultures and encouraging feedback from the clients, and ideally from peer supervisors, regarding the manner in which one conducts therapy.

References

Adkeson, M. (2003). Indigenous family work in Nigeria: The Yoruba experience. In K. S. Ng (Ed.), *The family therapy and counseling series. Global perspectives in family therapy: Development, practice and trends* (pp. 147–160). New York, NY: Brunner/Routledge.

Andersen, T. (1991). *The reflecting team*. New York, NY: Norton.

Anderson, H., & Goolishian, H. (1988). Human systems as linguistic systems: Preliminary and evolving ideas about the implications for clinical theory. *Family Process*, *27*, 371–393. https://doi.org/10.1111/j.1545-5300.1988.00371.

Bakker, T. M., & Snyders, F. J. A. (1999). The (hi)stories we live by: Power/knowledge and family therapy in Africa. *Contemporary Family Therapy*, *21*, 133–154. https://doi.org/10.1023/A:1021639324279.

Bateson, G., Jackson, D. D., Haley, J., & Weakland, J. (1956). Toward a theory of schizophrenia. *Behavioral Science*, *1*, 251–264. https://doi.org/10.1002/bs.3830010402.

Baucom, D. H., & Epstein, N. (1990). *Cognitive-behavioral marital therapy*. New York: Brunner/Mazel.

Baucom, D. H., Fischer, M. S., Worrell, M., Corrie, S., Belus, J. M., Molyva, E., & Boeding, S. E. (2018). Couple-based intervention for depression: An effectiveness study in the National Health Service in England. *Family Process*, *57*(2), 275–292. https://doi.org/10.1111/famp.12332.

Baucom, D. H., Shoham, V., Mueser, K. T., Daiuto, A. D., & Stickle, T. R. (1998). Empirically supported couples and family therapies for adult problems. *Journal of Consulting and Clinical Psychology*, *66*, 53–88. https://doi.org/10.1037/0022-006X.66.1.53.

Bebtschuk, M., Smirnova, D., & Khayretdinov, O. (2012). Family and family therapy in Russia. *International Review of Psychiatry*, *24*, 121–127. https://doi.org/10.3109/09540261.2012.656305.

Beck, A. T. (1988). *Love is never enough*. New York, NY: Harper and Row.

Berg, I. K. (1994). *Family-based services: A solution-focused approach*. New York, NY: Norton.

Birchler, G. R., Fals-Stewart, W., & O'Farrell, T. J. (2008). Couple therapy for alcoholism and drug abuse. In A. S. Gurman (Ed.), *Clinical handbook of couple therapy* (4th ed., pp. 523–544). New York: Guilford Press.

Borcsa, M., Hanks, H., & Vetere, A. (2013). The development of family therapy and systemic practice in Europe: Some reflections and concerns. *Contemporary Family Therapy*, *35*, 342–348. https://doi.org/10.1007/s10591-013-9266-2.

Chao, W., & Huang, W. (2013). The development of family therapy practice and training in Taiwan: A review and prospects for the future. *Contemporary Family Therapy*, *35*, 319–328. https://doi.org/10.1007/s10591-013-9261-7.

Christensen, A., Dimidjian, S., & Martell, C. R. (2015). Integrative behavioral couple therapy. In A. S. Gurman, J. L. Lebow, & D. K. Snyder (Eds.), *Clinical handbook of couple therapy* (5th ed., pp. 61–94). New York: Guilford Press.

Dattilio, F. M., & Padesky, C. A. (1990). *Cognitive therapy with couples*. Sarasota, FL: Professional Resource Exchange.

de Shazer, S. (1985). *Keys to solutions in brief therapy*. New York, NY: Norton.

Deng, L., Lin, X., Lan, J., & Fang, X. (2013). Family therapy in China. *Contemporary Family Therapy*, *35*, 420–436. https://doi.org/10.1007/s10591-013-9273-3.

Diamond, G. S., Siqueland, L., & Diamond, G. M. (2003). Attachment-based family therapy for depressed adolescents: Programmatic treatment development. *Clinical Child and Family Psychology Review*, *6*, 107–127. https://doi.org/10.1023/A:1023782510786.

Ellis, A., Sichel, J. L., Yeager, R. J., DiMattia, D. J., & DiGiuseppe, R. (1989). *Rational-emotive couples therapy*. New York: Pergamon.

Enns, C. Z., & McRae, J. (2007). International mental health professionals in Japan: Challenges and opportunities. *Psychology International*, *18*(3), 11–12. Retrieved from: https://www.apa.org/international/pi/2007/05/japan.

Epstein, N. (1982). Cognitive therapy with couples. *American Journal of Family Therapy*, *10*(1), 5–16. https://doi.org/10.1080/01926188208250432.

Epstein, N., Schlesinger, S. E., & Dryden, W. (Eds.), (1988). *Cognitive-behavioral therapy with families*. New York: Brunner/Mazel.

Epstein, N. B., & Baucom, D. H. (2002). *Enhanced cognitive-behavioral therapy for couples: A contextual approach*. Washington, DC: American Psychological Association.

Epstein, N. B., Baucom, D. H., Kirby, J. S., & LaTaillade, J. J. (2019). Cognitive-behavioral couple therapy. In K. S. Dobson & D. J. A. Dozois (Eds.), *Handbook of cognitive-behavioral therapies*. (4th ed.)(pp. 433–463). New York, NY: Guilford Press.

Epstein, N. B., Berger, A. T., Fang, J. J., Messina, L., Smith, J. R., Lloyd, T. D., … Liu, Q. X. (2012). Applying Western-developed family therapy models in China. *Journal of Family Psychotherapy*, *23*, 217–237. https://doi.org/10.1080/08975353.2012.705661.

Epstein, N. B., Curtis, D. S., Edwards, E., Young, J. L., & Zheng, L. (2014). Therapy with families in China: Common factors influencing the therapeutic alliance. *Contemporary Family Therapy*, *36*, 201–212. https://doi.org/10.1007/s10591-014-9302-x.

Eyberg, S., Boggs, S., & Algina, J. (1995). Parent-child interaction therapy: A psychosocial model for the treatment of young children with conduct problem behavior and their families. *Psychopharmacology Bulletin*, *31*, 83–91.

Falconier, M. K., Randall, A. K., & Bodenmann, G. (2016). Cultural considerations in understanding dyadic coping across cultures. In M. K. Falconier, A. K. Randall, & G. Bodenmann (Eds.), *Couples coping with stress: A cross-cultural perspective* (pp. 23–35). New York, NY: Routledge.

Fischer, M. S., & Baucom, D. H. Couple-based interventions for relationship distress and psychopathology. In J. N. Butcher, J. Hooley, & P. C. Kendall (Eds.), Psychopathology: Understanding, assessing and treating adult mental disorders, (in press), Washington, DC: American Psychological Association.

Fischer, M. S., Baucom, D. H., & Cohen, M. J. (2016). Cognitive-behavioral couple therapies: Review of the evidence for the treatment of relationship distress, psychopathology, and chronic health conditions. *Family Process*, *55*, 423–442. https://doi.org/10.1111/famp.12227.

Goldner, V. (1987). Instrumentalism, feminism and the limits of family therapy. *Journal of Family Psychology*, *1*, 109–116. https://doi.org/10.1037/h0080445.

Gottman, J. M. (1999). *The marriage clinic: A scientifically based marital therapy*. New York, NY: W.W. Norton.

Greenberg, L. S., & Goldman, R. N. (2008). *Emotion-focused couples therapy: The dynamics of emotion, love, and power*. Washington, DC: American Psychological Association.

Greenberg, L. S., & Johnson, S. M. (1988). *Emotionally-focused therapy for couples*. New York, NY: Guilford Press.

Haley, J. (1976). *Problem solving therapy*. San Francisco, CA: Jossey-Bass.

Hall, E. T. (1976). *Beyond culture*. New York, NY: Doubleday.

Hare-Mustin, R. (1978). A feminist approach to family therapy. *Family Process*, *17*, 181–194. https://doi.org/10.1111/j.1545-5300.1978.00181.x.

Henggeler, S. W., & Sheidow, A. J. (2012). Empirically supported family based treatments for conduct disorder and delinquency in adolescents. *Journal of Marital and Family Therapy, 38*, 30–58. https://doi.org/10.1111/j.1752-0606.2011.00244.x.

Herscovici, C. R., Gonzalez, S. C., Label, H., Vega, R. Z., & Garcia, N. C. (2013). Development and practice of the systems paradigm in Argentina, Chile and Peru. *Contemporary Family Therapy, 35*, 200–211. https://doi.org/10.1007/s10591-013-9241-y.

Jackson, D. D. (1967). *Therapy, communication and change*. Palo Alto, CA: Science and Behavior Books.

Jacobson, N. S., & Margolin, G. (1979). *Marital therapy: Strategies based on social learning and behavior exchange principles*. New York, NY: Brunner/Mazel.

Johnson, S. M. (1996). *The practice of emotionally focused marital therapy*. New York, NY: Brunner/Mazel.

Kameguchi, K., & Murphy-Shigematsu, S. (2001). Family psychology and family therapy in Japan. *American Psychologist, 56*, 65–70. https://doi.org/10.1037/0003-066X.56.1.65.

Kasiram, M., & Oliphant, E. (2007). Indigenizing family therapy for South African use: Challenges and changes. *Journal of Family Psychotherapy, 18*, 71–82. https://doi.org/10.1300/J085v18n02_06.

Kendall, P. C., Hudson, J. L., Gosch, E., Flannery-Schroeder, E., & Suveg, C. (2008). Cognitive-behavioral therapy for anxiety disordered youth: A randomized clinical trial evaluating child and family modalities. *Journal of Consulting and Clinical Psychology, 76*, 282–297. https://doi.org/10.1037/0022-006X.76.2.282.

Kim, G. S., Cho, W. J., Lee, C. Y., Marion, L. N., & Kim, M. J. (2005). The relationship of work stress and family stress to the self-rated health of women employed in the individual sector in Korea. *Public Health Nursing, 22*, 389–397. https://doi.org/10.1111/j.0737-1209.2005.220503.x.

Konya, Z., & Konya, A. (2013). Family therapy in Romania. *Contemporary Family Therapy, 35*, 275–287. https://doi.org/10.1007/s10591-013-9253-7.

Le Goff, J. F. (1999). Family therapy in France: A brief overview. *Contemporary Family Therapy, 21*, 77–88. https://doi.org/10.1023/A:1021914721496.

Lebow, J. L., Chambers, A. L., Christensen, A., & Johnson, S. M. (2012). Research on the treatment of couple distress. *Journal of Marital and Family Therapy, 38*, 145–168. https://doi.org/10.1111/j.1752-0606.2011.00249.x.

Liberman, R. P. (1970). Behavioral approaches to family and couple therapy. *American Journal of Orthopsychiatry, 40*, 106–118. https://doi.org/10.1111/j.1939-0025.1970.tb00683.x.

Liddle, H., Rowe, C., Dakof, G., Henderson, C., & Greenbaum, P. (2009). Multidimensional family therapy for early adolescent substance abusers: Twelve month outcomes of a randomized controlled trial. *Journal of Consulting and Clinical Psychology, 77*, 12–25. https://doi.org/10.1037/a0014160.

Lidz, T., Cornelison, A., Fleck, S., & Terry, D. (1957). Intrafamilial environment of the schizophrenic patient. II: Marital schism and marital skew. *The American Journal of Psychiatry, 114*, 241–248. https://doi.org/10.1176/ajp.114.3.241.

Lucksted, A., McFarlane, W., Downing, D., Dixon, L., & Adams, C. (2012). Recent developments in family psychoeducation as an evidence-based practice. *Journal of Marital and Family Therapy, 38*, 101–121. https://doi.org/10.1111/j.1752-0606.2011.00256.x.

Luo, L., & Cooper, C. L. (2015). *Handbook of research on work-life balance in Asia*. Cheltenham: Edward Elgar Publishing.

Madanes, C. (1981). *Strategic family therapy*. San Francisco, CA: Jossey-Bass.

Manfrida, G., Giachi, E., & Eisenberg, E. (2013). Steps to a therapy of human relationships: The evolution of family therapy in Italy. *Contemporary Family Therapy, 35*, 376–387. https://doi.org/10.1007/s10591-013-9269-z.

Markman, H. J., & Rhoades, G. K. (2012). Relationship education research: Current status and future directions. *Journal of Marital and Family Therapy, 38*, 169–200. https://doi.org/10.1111/j.1752-0606.2011.00247.x.

Mason, J., & Shuda, S. (1999). The history of family therapy in South Africa. *Contemporary Family Therapy, 21*, 155–172. https://doi.org/10.1023/A:1021691308349.

McGoldrick, M., Pearce, J., & Giordano, J. (2007). *Ethnicity and family therapy* (3rd ed.). New York, NY: Guilford Press.

Minuchin, S. (1974). *Families and family therapy*. Cambridge, MA: Harvard University Press.

Napier, A.Y., & Whitaker, C. A. (1978). *The family crucible*. New York, NY: Harper & Row.

Ng, K. S. (2005). The development of family therapy around the world. *The Family Journal: Counseling and Therapy for Couples and Families, 13*, 35–42. https://doi.org/10.1177/1066480704270264.

Nichols, M. P., & Schwartz, R. C. (2004). *Family therapy: Concepts and methods* (6th ed.). Boston, MA: Pearson.

Nwoye, A. (2001). History of family therapy: The African perspective. *Journal of Family Psychotherapy, 12*, 61–77. https://doi.org/10.1300/J085v12n04_05.

Nwoye, A. (2006). A narrative approach to child and family therapy in Africa. *Contemporary Family Therapy, 28*, 1–23. https://doi.org/10.1007/s10591-006-9691-6.

Patterson, G. R. (1971). *Families*. Champaign, IL: Research Press.

Rastogi, M. (2010). Family therapy in India: An overview. *The International Connection (newsletter of the International Family Therapy Association), 23*(1), 3.

Roberts, J., Abu-Baker, K., Fernandez, C. D., Garcia, N. C., Fredman, G., Kayma, H., … Vega, R. Z. (2014). Up close: Family therapy challenges and innovations around the world. *Family Process, 53*, 544–576. https://doi.org/10.1111/famp.12093.

Sanders, M. R. (1999). Triple P-Positive Parenting Program: Towards an empirically validated multilevel parenting and family support strategy for the prevention of behavior and emotional problems in children. *Clinical Child and Family Psychology Review, 2*, 71–90. https://doi.org/10.1023/A:1021843613840.

Satir, V. M. (1964). *Conjoint family therapy*. Palo Alto, CA: Science and Behavior Books.

Satir, V. M. (1972). *Peoplemaking*. Palo Alto, CA: Science and Behavior Books.

Selvini-Palazzoli, M., Boscolo, L., Cecchin, G., & Prata, G. (1978). *Paradox and counterparadox*. New York, NY: Jason Aronson.

Semmelhack, D., & Lan, J. (2018). Lee, Wai Yung. In J. L. Lebow, et al. (Eds.), *Encyclopedia of couple and family therapy*. https://doi.org/10.1007/978-3-319-15877-8_753-1.

Sim, T., & Hu, C. (2009). Family therapy in the Forbidden City: A review of Chinese journals from 1978 to 2006. *Family Process, 48*, 559–583. https://doi.org/10.1111/j.1545-5300.2009.01302.x.

Simakhodskaya, Z. (2013). Bringing emotionally focused couples therapy to Russia. *Psychology International, 24*(3), 6–8. Retrieved from: https://www.apa.org/international/pi/2013/10/russia.

Snyder, D. K., & Halford, W. K. (2012). Evidence-based couple therapy: Current status and future directions. *Journal of Family Therapy, 34*, 229–249. https://doi.org/10.1111/j.1467-6427.2012.00599.x.

Sprenkle, D. H. (2012). Intervention research in couple and family therapy: A methodological and substantive review and an introduction to the special issue. *Journal of Marital and Family Therapy, 38*, 3–29. https://doi.org/10.1111/j.1752-0606.2011.00271.x.

Sprenkle, D. H., Davis, S. D., & Lebow, J. L. (Eds.), (2009). *Common factors in couple and family therapy: The overlooked foundation for effective practice*. New York, NY: Guilford Press.

Steinglass, P. (1987). A systems view of family interaction. In T. Jacob (Ed.), *Family interaction and psychopathology: Theories, methods, and findings* (pp. 25–65). New York, NY: Plenum.

Stuart, R. B. (1969). An operant-interpersonal treatment for marital discord. *Journal of Consulting and Clinical Psychology, 33*, 675–682. https://doi.org/10.1037/h0028475.

Sullivan, H. S. (1953). *The interpersonal theory of psychiatry*. New York, NY: W.W. Norton.

Tamura, T., Lee, W. Y., & Cheng, V. (2017). Asian Academy of Family Therapy. In J. L. Lebow, et al. (Eds.), *Encyclopedia of couple and family therapy*. https://doi.org/10.1007/978-3-319-15877-8_940-1.

von Bertalanffy, L. (1968). *General systems theory*. New York, NY: Braziller.
Wagenaar, K., & Baars, J. (2012). Family and family therapy in the Netherlands. *International Review of Psychiatry*, *24*, 144–148. https://doi.org/10.3109/09540261.2012.656301.
Watzlawick, P., Weakland, J., & Fisch, R. (1974). *Change: Principles of problem formation and problem resolution*. New York, NY: Norton.
White, M., & Epston, D. (1990). *Narrative means to therapeutic ends*. New York, NY: Norton.
Winter, W. D., & Ferreira, A. J. (1969). *Research in family interaction*. Palo Alto, CA: Science and Behavior Books.
Wynne, L. C., Ryckoff, I., Day, J., & Hirsch, S. I. (1958). Pseudomutuality in the family relationships of schizophrenics. *Psychiatry*, *21*, 205–220. https://doi.org/10.1080/00332747.1958.11023128.

CHAPTER 17

Culture, couple relationship standards, and couple relationship education and therapy

W. Kim Halford[a] and Fons van de Vijver[b]
[a]School of Psychology, University of Queensland, St. Lucia, QLD, Australia
[b]Department of Culture Studies, Tilburg University, Tilburg, Netherlands

Committed couple relationships of marriage and marriage-like partnerships exist across almost all known cultures and likely reflect shared evolutionary imperatives to form such partnerships to enhance the chance of raising viable offspring (Buss, 2016). At the same time, there is considerable variation between cultures in the form and functions of couple relationships. For example, in some cultural contexts marriages are arranged by extended kinship networks, and the function of marriage has a focus on meeting the collective needs of the kinship groups (Coontz, 2005). In these cultural contexts divorce and separation often are strongly socially disapproved of, or even banned entirely. In other cultural contexts, couple partnerships are initiated by the partners, most often by cohabitation, where the focus primarily is on the individual needs of those partners (Coontz, 2005). In these cultural contexts, separation often is socially acceptable and occurs at high rates (Coontz, 2005; Chapter 3).

The first half of the current chapter reviews the forms and functions of couple relationships cross-culturally, with a particular focus on how standards (beliefs about what should be in good relationships) are similar and different across diverse cultures. The second half of the chapter describes interventions to enhance couple functioning, focusing on couple relationship education (RE) and couple therapy. The chapter includes an analysis of how these couple interventions can appropriately address issues of culture.

Cross-Cultural Family Research and Practice
https://doi.org/10.1016/B978-0-12-815493-9.00017-X

Couple relationships, evolution, and culture

Evolution and couple relationships

Committed couple relationships developed as an evolutionary adaptation to engage both biological parents, and extended family, in raising human children (Fletcher, Simpson, Campbell, & Overall, 2015). As Bjorklund and Myers (Chapter 2) note, humans are an extreme variant of the primate evolutionary trend toward a large brain, extended social groups, and an extended period of immaturity among offspring. The developmental period of offsprings' reliance on adult caregivers is much longer in humans than other species. To address the survival challenges of caring for dependent offspring, humans evolved to became cooperative breeders with parents engaged in long-term, committed pair bondings (Fletcher et al., 2015; Lewis, Al-Shawaf, Conroy-Beam, Asao, & Buss, 2017).

The attachment system involves a set of hormonal and behavioral responses to another that is associated with emotional bonding and caregiving (Bowlby, 1969). The attachment system originally evolved as a means of strengthening the bond between mothers and offspring, which had an evolutionary advantage in enhancing maternal caregiving (Bowlby, 1969). However, during the last 2–3 million years the attachment system evolved in human ancestors to create romantic love and pair bonding, which served to commit partners to each other in an ongoing, predominantly monogamous relationship (Buss, 2016; Fletcher et al., 2015). Development of committed couple relationships had evolutionary advantages for both parents and offspring. For men, women's relationship commitment increased paternal certainty, that is the father could be more confident that a child born of their partner was biologically their child (Buss, 2016). For women, men's relationship commitment was associated with a male investment in offspring, both through direct caregiving and through the provision of resources (e.g., food and protection) (Buss, 2016). Evolutionary psychologists have shown that men and women both seek indicators of relationship commitment from potential long-term partners, as such commitment is associated with participating in the care of offspring (Lewis et al., 2017). Commitment to the couple relationship (i.e., remaining in the relationship) benefits offspring, as it is associated with greater survival rates in offspring across a wide range of cultures (Geary, 2000).

Culture and couple relationships

One limitation of evolution as a means to adapt to changing environments is that its effects occur relatively slowly. For example, there was an evolutionary life history shift that occurred in human ancestors to move from ape-like ancestors living in polygamous groups, to the predominantly monogamous pair bonding evident in *Homo sapiens*. These evolutionary changes occurred across approximately 6 million years (Kaplan, Hooper, & Gurven, 2009). In contrast, major human societal changes in the last 10,000 years, like the development of agriculture, industrialization, and the emergence of information technologies have occurred too rapidly for evolution to keep pace and select adaptations to adjust to these changing aspects of the environment.

Darwin believed human adaptation to the environment involved more than just evolution, as he stated in *The Descent of Man*: "Now if someone man in tribe, more sagacious than others, invented a new snare, or other means of attack or defense, the plainest self-interest, without the assistance of much reasoning power, would prompt the other members of the social group to imitate him, and all would thus profit" (Darwin, 1877/2004, p. 155). Humans do not inherit through evolution a set of fixed action patterns, but rather evolution provides complex psychology that allows behavioral flexibility in response to variable environments (Tooby & Cosmides, 1992). Culture develops from the extraordinarily highly developed capacity of humans to learn from each other, to accumulate knowledge across time and generations, and to codify values and beliefs that guide that behavioral flexibility to an adaptive solution to environmental challenges (Boyd & Richerson, 2009). It is quite likely that human capacity to develop culture evolved in response to rapidly changing environments, as were evident in major fluctuations in global mean temperatures in the mid-Pleistocene era (Richerson & Boyd, 2005).

Somewhat different cultures develop in different social groups of humans in response to the environments in which humans are living, and these differences can be quite stable across time (Richerson & Boyd, 2005). When one culture is better adapted to the current environment, it often takes over another culture. The process of cultural competition can involve conflict where one cultural group conquers another, or by one culture attracting members of another cultural group to join it.

Social changes like the development of agriculture, industrialization, increasing control by women over their fertility, and increasing economic independence of women have modified the cultural values attached to the couple relationship. The accumulated social changes have made marriage less critical to addressing survival needs for the majority of people in developed high-income countries (Coontz, 2005). For example, married women's increasing engagement in the paid workforce in Western countries across the second half of the 20th century has given more women sufficient economic independence from their husbands to allow those women to leave distressed marriages (Sayer & Bianchi, 2000). This is not to say that parental separation does not have some negative consequences, there are still some advantages in life outcomes (e.g., educational attainment and mental health) of children of intact parental relationships relative to offspring of separated parents (Amato, 2010; Waite & Gallagher, 2000). However, in high-income countries the negative consequences of parental separation for children seem to be concentrated in a small proportion of separated families, particularly those with severe ongoing parental conflict (Amato, 2010). In contrast, an intact couple relationship remains crucial for survival in low-income countries where shared resources within the extended family are essential for the survival of all family members (Geary, 2000).

Across the last few hundred years in developed countries, the concept of couple relationship quality shifted to focus less exclusively on survival needs, and include the idea of the partners meeting each other's psychological needs (Coontz, 2005). In that context, when psychologists working in economically developed countries began to evaluate relationship quality it was in terms of the partners' satisfaction with aspects of the couple's interactions, such as feeling loved, experiencing psychological intimacy, and managing conflict effectively (e.g., Locke & Wallace, 1959; Spanier, 1976).

Recent changes in human culture have further modified the aspirations for couple relationships, at least within high-income countries (Agnew, 2014). Globalization and developments in information technology have facilitated the expansion of travel, social media platforms, and online dating applications that vastly increase access to potential mates from different backgrounds (Sanri & Goodwin, 2014). Exposure to expanded alternatives potentially alters the courtship process and the expectations of what is sufficient in couple relationship quality. In Western countries, young adults now delay marriage more than previous generations, on average have more relationships and sexual partners before marriage, are more likely to cohabit as an alternative or prelude to marriage and are more likely to engage in casual sexual experiences than earlier generations (Petersen & Hyde, 2010; Sassler, 2010). Finkel, Hui, Carswell, and Larson (2014) suggest that availability of a wider suite of potential mates, the idealized portrayals of mates and relationships, and the changing cultural context emphasizing self-fulfillment, create increasingly higher expectations of marriage, such as the idea that marriage should help partners achieve higher psychological functioning and self-actualization.

The changing functions of marrige (and similar committed couple relationships) are evident in changing reasons for separation. In hunter-gatherer societies, separation of partners from a couple relationship was rare, as such a separation could threaten the survival of the partners, their offspring, and the extended family (Lewis et al., 2017). In developed countries, in the middle of the 20th century, the stated reasons for divorces were predominantly severe conflict, violence, or adultery (de Graaf & Kalmijn, 2006), which all are conditions that threaten the survival advantages for offspring associated with marriage. However, by the early 2000s, many people describe psychological motives ("I felt we had grown apart," "We were not communicating") as the reasons for their divorce (de Graaf & Kalmijn, 2006), and severe marital problems now predict only about half of the divorces in a high-income country like the United States (Amato & Rogers, 1997).

Since the late 20th century, as documented by Lesthaege (Chapter 4), cohabitation rather than formal marriage has become more and more common. In high-income countries, the majority of couples cohabit, either as a prelude or an alternative to marriage (Chapter 4). Rates of break up among cohabiting couples are much higher than for married couples and can reflect that one partner aspires to a better relationship (Chapter 3), even if the current relationship has no major shortcomings (Sanri & Goodwin, 2014). Thus, the threshold of what makes a couple relationship acceptable to continue—at least in high-income countries—has increased from meeting survival needs, through meeting basic psychological needs for safety and intimacy, to developing psychological well-being and growth in the partners. In many ways, this couple relationship progression parallels Maslow's (1943) hierarchy of individual needs in which prioritize first meeting the most basic needs (e.g., food, water, sex, safety). When those needs are met, people then seek to meet psychological needs (e.g., a sense of accomplishment and respect from others). Finally, the needs at the top of the hierarchy involve self-actualization needs, reflected in achieving one's full potential, creative activities, personal growth, and self-expression.

Relationship standards

Concept and assessment of relationship standards

Epstein and Baucom (2002, p. 72) defined couple relationship standards as "… personal beliefs about the characteristics an intimate relationship … should have." We add to this definition that such personal beliefs usually reflect the cultural values of the society in which individuals are raised.

We focus on the construct of relationship standards as a way of thinking about culture and its association with couple relationships, as standards embody what is believed to be important in relationships. Further on in the chapter, we discuss the striking cultural differences (and similarities) of relationship standards across diverse cultures.

There are five published measures of couple relationship standards. Researchers developed four of these measures based on research and clinical experience within Western cultural contexts: three in the United States (Baucom, Epstein, Rankin, & Burnett, 1996; Eidelson & Epstein, 1982; Vangelisti & Daly, 1997), and one in New Zealand (Fletcher, Simpson, Thomas, & Giles, 1999). A limitation of each of these four measures was that they focus, understandably, on relationship standards reflecting values in the society in which they were developed. Specifically, all of the assessed standards reflected aspects of romantic love (e.g., attraction, expressions of love and caring, and psychological intimacy and closeness) as central to couple relationships, which reflects Western ideals but not necessarily ideals of other cultures (Wong & Goodwin, 2009).

The fifth published measure is the Cross-Cultural Couple Relationship Standards Scale (CCCRSS) (Hiew, Halford, van de Vijver, & Liu, 2015a; Iqbal, Ayub, van de Vijver, & Halford, 2019), which was designed to assess a broader range of relationship standards likely to be salient across a diverse range of cultures (Halford, Leung, et al., 2018a, 2018b; Iqbal et al., 2019). We describe the content of this measure as a way of explaining the scope of relationship standards that might be salient in different cultures. The CCCRSS assesses 10 standards, 5 of which reflect Western relationship standards: demonstration of love (e.g., "Express their love for each other in words every day"), demonstration of caring (e.g., "Regularly do work for each other"), intimacy expression (e.g., "Tell each other when they are feeling positive emotions"), intimacy responsiveness (e.g., "Ask each other about their thoughts"), and relationship effort ("Persist in trying to make things better when the relationship has problems") (Halford, Leung, et al., 2018a; Hiew, Halford, van de Vijver, & Liu, 2015b; Iqbal et al., 2019). Across diverse cultures, these five scales form a superordinate scale called Couple Bond (Halford, Leung, et al., 2018a; Iqbal et al., 2019).

The next four standards in the CCCRSS reflect values seen as important in many collectivistic cultures (e.g., in much of Asia and Africa), but which have not been included in Western conceptualization or measurement of relationship standards (Halford, Leung, et al., 2018a; Hiew et al., 2015a; Iqbal et al., 2019). These standards are relations with the extended family (e.g., "Do not disagree with family elders"), face (e.g., "Avoid doing things that might lower others' opinions of their partner or the couple"), relational harmony (e.g., "Do not speak about things that may lead to conflict"), and traditional gender roles (e.g., "The man financially supports his partner and children"). In traditional Chinese culture, and indeed in many collectivistic cultures, fulfilling one's social role, and maintaining harmony—particularly within the family—is seen as of paramount importance (Shi & Wang, 2009). Relatedly, Chinese value protecting one's own and others' face (social image and credibility) during interaction (Kam & Bond, 2008). Finally, gender role standards have been more traditional among Chinese than Western couples (Marshall, 2008) although traditional gender roles are gradually being displaced by more egalitarian roles in some parts of China (Tang, Chua, & Jiaqing, 2010). Across Chinese, Western, Pakistani and Thai cultures, these four scales form a superordinate scale called Family Responsibility (Halford, Leung, et al., 2018a; Hiew et al., 2015a; Iqbal et al., 2019).

The 10th and final standard is the Importance of Religion (e.g., "The relationship is blessed by God" and "Follow religious teaching in how they lead their life together"). This was included because in developing the scale we sought to assess standards that might influence relationship quality and people who are more religious tend to have more stable, satisfying marriages compared to those who are less religious, including Christians living in the United States (Mahoney, Pargament, Tarakeshwar, & Swank, 2008) and Muslims living in majority Muslim countries (Yeganeh & Shaikhmahmoodi, 2013). A number of religion-related relationship standards might link religiosity and relationship satisfaction. One potential standard is the belief that one's marriage should be blessed by a divine being (marital sanctification); believing that one's marriage is sanctified is associated with marital satisfaction (Kusner, Mahoney, Pargament, & DeMaris, 2014). Second, partners' sharing of religious practices (e.g., praying together and raising children in a religious faith) is associated with relationship satisfaction (Mahoney et al., 2008). Across diverse cultural and religious groups, the Importance of Religion couple relationship standards was found to have a single internally consistent structure (Skellern et al., 2019).

Cultural differences in relationship standards

In examining cultural differences in relationship standards, our research group has tried to examine standards across cultures that are different in important ways from Western cultures where much of the extant research on couple relationships has been done. Therefore we have examined standards in cultures seen as more collectivistic (e.g., Chinese, Thai and Pakistani), or more religious and traditional (e.g., Pakistani and Thai) than Western cultures.

Westerners endorse a little more strongly relationship standards based on individualistic ideals of romantic love and psychological intimacy (Couple Bond) than do Chinese (Epstein, Chen, & Beyder-Kamjou, 2007; Halford, Leung et al., 2018a; Hiew et al., 2015b), or Pakistanis (Iqbal et al., 2019). Chinese and Pakistanis much more strongly endorse relationship standards (medium to large effect sizes) based on collectivistic ideals regarding extended family relations, face, relationship harmony, and traditional gender roles (Family Responsibility) than do Westerners (Halford, Leung et al., 2018a, 2018b; Hiew et al., 2015a, 2015b; Iqbal et al., 2019). At the same time, the mean endorsement of Couple Bond standards is higher than the mean endorsement of Family Responsibility standards by both Chinese and Westerners (Hiew et al., 2015a, 2015b), suggesting that Couple Bond standards are cross-culturally seen as important. In fact, endorsement of Couple Bond by Chinese was high and similar to Westerners in a young, predominantly single sample of participants living in Australia (Hiew et al., 2015a), and were endorsed only slightly less, a small effect size of $d=0.20$ standard deviations, in married Chinese than married Westerners living in Australia (Hiew et al., 2015b). In contrast, there are large effect size differences between cultures in the endorsement of Family Responsibility standards (Halford, Leung et al., 2018a, 2018b; Hiew et al., 2015a, 2015b; Iqbal et al., 2019).

There also are large and potentially important cultural differences in the importance attached to religion in couple relationships (Skellern et al., 2019). Specifically, as might be expected, those without a religious faith living in Western countries (United States and Australia) rate religion as much less important to a couple relationship than Christians living in Western countries, and the Non-religious Westerners also rated religion as less important than Thai Buddhists or Pakistani Muslims (Skellern et al., 2019). Interestingly, Pakistani Muslims rated religion as more important in couple relationships than did Thai Buddhists or Western Christians.

In summary, the strength of endorsement of Family Responsibility and Importance of Religion standards vary substantially cross-culturally. In contrast, Couple Bond standards are rated as of high importance across all cultures examined, and vary with only small effect sizes. As noted previously in the current chapter, there is considerable evidence that romantic attachment evolved in humans as a means to promote shared caregiving of offspring

(Fletcher et al., 2015), suggesting that strong endorsement of Couple Bond standards might be pervasive across cultures and reflect the evolutionary imperative of promoting strong couple bonds to enhance offspring survival. This then raises the question of what accounts for the cultural variations in relationship standards.

Culture, adaptation, and relationship standards

Relationship standards that vary greatly between culture, particularly Family Responsibility standards, seem to relate to broader dimensions of cultural variation. One more widely used typology of culture is that of Inglehart and Baker (2000), who suggest cultures vary along two major dimensions of values: (1) Survival vs Self-expression, and (2) Traditional vs Secular/Rational. Survival values prioritize collective behaviors that promote economic and physical security and are endorsed most strongly in low-income countries (Dobewell & Strack, 2014; Kashima, Bain, & Perfors, 2019), notably in African, Islamic countries, and countries strongly influenced by the Orthodox Russian religion (World Values Survey, 2016). Self-expression values prioritize personal freedom and quality of life and are most strongly endorsed in the high-income countries (Kashima et al., 2019), notably the majority English speaking countries, and predominantly protestant areas of Europe (World Values Survey, 2016).

People living in countries with high survival values (and low self-expressive values), such as Pakistan (Iqbal et al., 2019), strongly endorse Family Responsibility standards. For example, endorsing the importance of maintaining harmony in the couple relationship is valued in high survival value context, which might reflect the consequences for survival if a couple separate in that context. People living in countries with high self-expressive values (and low survival values), like Australia and the United States (Hiew et al., 2015a; Skellern et al., 2019), rate Family Responsibility standards as relatively unimportant. For example, maintaining harmony is rated as unimportant when self-expression is a core value of the culture.

Traditional values emphasize the importance of God and religion, that development of obedience and conformity to social rule is important; whereas Secular-rational values prioritize the importance of science and rational thought, developing independence and self-determination in life (Inglehart & Baker, 2000). A related concept is cultural tightness, which refers to a culture having strong social norms and low tolerance of deviance (Gelfand et al., 2011). Tight cultures tend to occur when the environment has distal ecological and historical threats (e.g., high population density, resource scarcity, a history of territorial conflict, and disease and environmental threats), and tight cultures are argued to be a means under harsh survival threats to ensure adherence to required social behaviors (Gelfand et al., 2011). Tight cultures and high religiosity covary (Gelfand et al., 2011), and it might be that religion is a cultural adaptation that promotes adherence to tight social norms. For example, many religions have strong prohibition of sex outside marriage, and strong disincentives to divorce, which serve to promote and sustain a marriage, which has particular adaptive significance in harsh environments with strong survival pressures. The importance of religion in relationships is very highly endorsed in Pakistan (Iqbal et al., 2019), which also has strong endorsement of traditional values (World Values Survey, 2016), and much less strongly in Australia (Skellern et al., 2019), which strongly endorses Secular-Rational values (World Values Survey, 2016).

Almost all of the research on relationship standards has been with different-sex couples. However, 1 recent study assessed relationship standards in 414 same-sex attracted adults living in Australia using the Rainbow Couple Relationship Standards Scale (RCRSS), an adapted form of CCCRSS (Baker & Halford, 2019). RCRSS used the existing scales of the CCCRSS and added three new scales argued to be of particular relevance to same-sex couples: Relationship Outness (public disclosure of the relationship), Sexual Openness (acceptance of open sexual relationship with extra-dyadic sexual partners), and Dyadic Coping with Homophobic discrimination.

For people living in a Western country, same-sex attracted people's standards were similar for men and women, and for singles vs those in a relationship (Baker & Halford, 2019). Same-sex attracted people's standards were very similar in the endorsement of Couple Bond, Family Responsibility, and Relationship Effort standards as those of heterosexuals. As the evolutionary selection effects on same-sex and different-sex attracted adults are very similar (Galperin et al., 2013), the high rated importance of Couple Bond likely reflects that shared evolutionary imperative to form committed couple relationships.

Religion was rated as somewhat less important to the couple relationship by same-sex attracted people relative to heterosexuals. This might well reflect that many religions often have been hostile to same-sex attracted individuals, and religion-based beliefs are a commonly reported source of familial rejection, and lack of positive gay identity among sexual minorities (Anderton, Pender, & Asner-Self, 2011). The often negative views of same-sex attraction by organized religions, might account for low rates of religiosity in same-sex attracted adults compared with heterosexuals (Australian Bureau of Statistics, 2013).

The Relationship Outness and Dyadic Coping with Homophobia scales were rated as of high importance and assessed potentially important standards that reflect some distinctive challenges for same-sex couple relationships (Baker & Halford, 2019). Interestingly both Relationship Outness and Dyadic Coping with Homophobia scales loaded onto to the superordinate Couple Bond scale (Baker & Halford, 2019). In other words, the importance attached to expressing things like love and caring, and intimacy were associated with the importance of supporting each other to cope with homophobia and being open about the same-sex relationship to others. Other research finds that having a positive gay identity is associated with greater willingness to be open about a same-sex relationship, and higher relationship satisfaction (Pepping, Cronin, Halford, & Lyons, 2019). Together these findings suggest mutual associations between being in a high-quality couple relationship, having a positive gay identity, mutual support to cope with homophobia, and comfort with being open about the same-sex relationship.

Sexual openness received the low mean endorsement of importance by both male and female same-sex couples. This low rating likely reflects that monogamy is still normative within same-sex attracted individuals (Lyons & Hosking, 2014). Also, there is some evidence that endorsement of monogamy by gay men has increased in Western countries over recent decades with factors such as the decriminalization of homosexuality facilitating the pursuit of stable, monogamous relationships (Peplau & Fingerhut, 2007). However, prevalence and importance are not the same thing. While there might be low mean importance placed on Sexual Openness, there still might be a substantial prevalence of nonmonogamous relationships, and Sexual openness might be important to a small proportion of same-sex attracted adults.

Association of standards with relationship satisfaction

Researchers have suggested three ways in which relationship standards can influence relationship satisfaction. First, satisfaction might be a function of the extent to which the relationship meets the partners' relationship standards, and there does seem to be an association between these constructs (e.g., Baucom et al., 1996). However, it is questionable if the construct of relationship satisfaction is conceptually distinct from the extent to which the relationships meets your standards. Predicting satisfaction from the extent to which your relationship meets your relationship standards seems circular in that satisfaction can be seen as just another way of saying the relationship meets your standards.

A second way in which relationship standards have been suggested to influence relationship satisfaction is that some standards are adaptive and others maladaptive for couple relationships. For example, standards such as shared decision-making and frequent demonstration of caring have been suggested to enhance relationship satisfaction (Baucom et al., 1996), whereas believing that any disagreement is destructive and that partners should be able to sense each other's needs without overt communication, have been proposed to adversely affect relationship satisfaction (e.g., Eidelson & Epstein, 1982). Some modest associations exist between endorsement of some of these standards and relationship satisfaction in Western couples (e.g., Baucom et al., 1996). However, the country and its culture moderate the association between endorsement of specific standards and satisfaction (Goodwin & Gaines, 2004). This suggests that at least some standards may not be universally adaptive, but rather different standards may be appropriate for different cultural contexts.

Third, the similarity between partners' standards might predict relationship satisfaction (Hiew et al., 2015b). This prediction is based on the assumption that if both partners believe similar behaviors are important, then it is more likely the relationship can meet each partner's standards.

In a series of studies using the CCCRSS, a strong endorsement of Couple Bond standards was associated with high relationship satisfaction in Western and Chinese couples (Halford, Leung, et al., 2018b; Hiew et al., 2015b). The level of endorsement of family responsibility standards did not predict relationship satisfaction, but an agreement between the partners on the importance of family responsibility standards did predict high satisfaction in both Western and Chinese couples (Halford, Leung, et al., 2018b; Hiew et al., 2015b). Relationship Effort (also known as relationship self-regulation), which refers to the importance of putting effort into the relationship in order to sustain satisfaction (Wilson, Charker, Lizzio, Halford, & Kimlin, 2005) predicted relationship satisfaction in Western couples, but the association with satisfaction in other cultures is untested. In recently collected, as yet unpublished, data endorsement of the importance of religion in marriage was associated with high relationship satisfaction in Malay Muslim couples (Nordin, Halford, Barlow, & Mastor, 2019). It is likely that, within religious communities, agreement between partners on the importance of religion might be associated with relationship satisfaction, but this hypothesis has not yet been tested.

It is striking that there are some substantial cross-cultural differences in strength of endorsement of relationships standards, particularly on Family Responsibility and Importance of Religion. Yet the association of most standards with satisfaction has been similar across diverse cultures. Specifically, high endorsement of Couple Bond predicts satisfaction, reflecting the universality of that set of standards as central to a satisfying relationship. In contrast, it is agreement between partners on Family Responsibility standards that predicts high satisfaction, suggesting that is not a universally endorsed standard and what is required is for partners to agree about the standard's importance.

It also is striking that, across cultures, there are few significant and only small effect size gender differences in relationship standards. This is true in heterosexual adults among Westerners, and Chinese living in Australia (Hiew et al., 2015a, 2015b); Chinese living in Hong Kong (Halford, Leung et al., 2018a, 2018b); Pakistani Muslims (Iqbal et al., 2019); and Malay Muslims (Nordin et al., 2019). Moreover, this is also true of same-sex attracted adults living in Western countries (Baker & Halford, 2019). There also are few significant differences between standards of people who are single (when standards refer to what they want in a future relationship) and partnered people (whose standards most likely reflect aspirations in the current relationship) (Baker & Halford, 2019; Halford, Leung et al., 2018a; Hiew et al., 2015a; Iqbal et al., 2019; Skellern et al., 2019). Moreover, endorsement of standards does not covary with age (Halford, Leung et al., 2018a). The findings of little to no differences in standards by gender, sexual orientation, age, or relationship status; but some large differences between cultures; suggest relationship standards reflect broad cultural values that might well be developed prior to adulthood.

Consistent with the proposition that broad cultural values prior to adulthood are a key influence on couple relationship standards, when people change the culture they live in there are only some small effect size shifts in standards, at least in the first generation, and these shifts are not uniformly toward the new culture (Halford, Leung et al., 2018b). For example, Chinese women who migrate to Australia endorse Couple Bond standards and Family Responsibility standards more strongly than Hong Kong Chinese women. The higher endorsement of Couple Bond is a shift toward the majority Western culture, but the higher endorsement of Family Responsibility is a shift away from the majority Western culture's typical endorsement of standards. We analyze the effects of migration and partnering with someone of a different culture on relationship standards, and the association of standards with satisfaction, in the section that follows.

Migration and intercultural couples

Prevalence across countries

Rates of international migration have increased markedly in the 21st century and cultural diversity is increasing in many Western countries (International Organization for Migration, 2015). In many parts of the world intercultural relationships are common, 1 in 3 marriages in Australia, Hawaii, Singapore, and Taiwan, and more than 1 in 10 marriages in France, South Korea, and mainland United States, and 5% of Canadian marriages are intercultural (Australian Bureau of Statistics, 2012; Lee & Edmonston, 2005; Statistics Canada, 2012). As noted previously, people from different cultures can have important differences in their standards for what constitutes a good couple relationship (Halford, Leung et al., 2018b; Hiew et al., 2015a, 2015b) and similarity on some standards (notably Family Responsibility standards) are associated with couples' relationship satisfaction—including in intercultural couples (Hiew et al., 2015b).

Selection, acculturation, and convergence

There is a well-replicated finding that migrants to Western countries endorse many of the values of the majority culture somewhat more strongly than do nonmigrants (e.g., Feldman, Mont-Reynaud, & Rosenthal, 1992; Tharp, Meadow, Lennhoff, & Satterfield, 1968). The greater endorsement of Western values by migrants might reflect selection effects, as those who hold Western values might be more likely to migrate to Western countries. In addition to any selection effect, Western values might reflect acculturative adjustment, which refers to a process of adopting the majority values by the migrant. There is greater Westernization in values in second and subsequent generations of migrants relative to first- generation migrants (Feldman et al., 1992), which likely reflects acculturative adjustment although also might reflect cohort effects (i.e., social changes between generations).

There are variations across cultural domains in the extent to which migrants endorse Western values. Feldman et al. (1992) found some Westernization of all values of Chinese Americans relative to Hong Kong Chinese, but substantial differences across value domains. The largest similarity to Western values in Chinese migrants (relative to Hong Kong Chinese) was found for tradition (e.g., perceived importance of rites and rituals). The largest dissimilarity from Western values was for the importance of extended family living as a residential unit. Their study suggests that value differences between migrants and nonmigrants living in their culture of origin are domain-specific and that certain core values persist even in second-generation migrants.

Individual differences in the alignment of migrants with the majority culture across value domains might be a function of domain-specific processes of socialization. In studies of Turkish and Moroccan migrant families in Germany and the Netherlands, parents were found to transmit collectivistic values to adolescents, but not individualistic values (Phalet & Schönpflug, 2001). It is quite likely that individualistic values were transmitted by socialization agents from mainstream groups, such as peers and teachers. So, different transmission mechanisms could lead to differences in the extent of Westernization of different cultural value domains (Güngör, Fleischmann, & Phalet, 2011).

There is a tacit assumption in many studies that newly arrived migrants have preferences close to their ancestral culture and then move partially or entirely to the new culture. However, as noted previously, differences between migrants' and nonmigrants' values might be due at least partially to self-selection in the migrant group and be evident at the time of migration. Once people migrate, it has been suggested that sometimes migrants overshoot their adjustment to the new group, presumably to emphasize their link with the mainstream group (Triandis, Kashima, Shimada, & Villareal, 1986). Conversely, when confronted with rejection and discrimination, migrants can retain or even strengthen their cultural identity; for example by retaining or even strengthening their original religious identity in the diaspora (Verkuyten, 2016).

There are likely to be diverse influences on the development of couple relationship standards in migrants beyond the majority culture of origin and destination. One noteworthy influence is being in an intercultural relationship, as being in a relationship with a person of majority culture is a strong acculturation influence on migrants. It is a well-replicated finding that intercultural couples (with at least one native partner) acculturate more to majority culture than couples in which both partners are nonnative (Dribe & Lundh, 2008; Kaya, Halford, Hiew, Sheffield, & van de Vijver, 2019). At the same time, the migrant spouse also seems to influence the native's cultural identification. For example, both partners in intercultural Chinese-Western couples identify strongly with both Western and Chinese cultures (Kaya et al., 2019), and to endorse relationship standards midway between the cultures of origin of the two partners (Hiew et al., 2015b). These patterns likely reflect a mixture of three influences: selection effects, as Chinese and Westerners who value each other's cultural standards might be more likely to form an intercultural relationship; accommodation effects, where partners move toward each other's standards; and acculturation effects reflecting movement toward to the majority cultural values by the migrant spouse.

We summarize the key points about culture and relationship standards covered so far, before turning to the implications of these finding for couple interventions. Couple Bond relationship standards seem cross-culturally highly valued, a strong endorsement of those standards predicts couple relationship satisfaction. The universality of Couple Bond standards seems to reflect evolutionary imperatives shared across cultures. Family Responsibility and Importance of Religion standards vary a lot cross-culturally, and agreement between partners on these standards predicts relationship satisfaction cross-culturally. Family Responsibility and Importance of Religion standards reflect broader cultural values of Survival vs Self-expression and Traditional vs Secular/rational cultural values. Relationship standards are remarkably similar across genders and sexual orientation within cultures and seem to be learned early in life. However, migrants to a different culture often have different relationship standards to their culture of origin, which reflects a mixture of selection and acculturation effects. Moreover, being in an intercultural relationship with someone of the majority culture intensifies this difference.

Couple interventions

There are two general types of interventions to enhance couple relationships: couple RE and couple therapy. Couple RE seeks to enrich couples' relationships and help couples to sustain healthy, mutually satisfying, and stable relationships. Evidence-based RE draws on the substantial research on what influences couple relationship satisfaction and stability and uses an approach to RE that has been evaluated in well-designed research trials (Halford, Markman, & Stanley, 2008). Evidence-based RE usually is brief, typically consisting of 10–12 hours curriculum that introduces key relationship knowledge (e.g., the importance of commitment and developing shared and realistic relationship expectations) and skills (e.g., couple communication, problem-solving, coping). RE usually works with couples who are currently satisfied in their relationship and are committed to that relationship. Here RE builds on the high level of positive emotion typical of currently satisfied couples and has a strong emphasis on building the positive foundations for a great life together.

Couple therapy is somewhat distinct from RE, couple therapy is usually addressed to couples who are distressed in their relationship. Couple therapy often has to manage the high levels of negative effect in the relationship and address the ambivalence many distressed couples feel about whether they wish the relationship to continue (Halford & Snyder, 2012). Evidence-based couple therapy is often extensive in duration, with evidence-based approaches often involving 20 or more sessions of therapy (Halford & Snyder, 2012).

Couple relationship education

Couple RE began in Western countries in the early to mid-20th century as brief counsel offered by religious organizations to marrying couples with the aim of promoting stable, healthy couple relationships (Hunt, Hof, & DeMaria, 1998). In the 1970s psychologists began building on these pioneering efforts by applying knowledge from psychological science to inform the design and content of RE and to evaluate the effects of such programs in a range of secular settings (Halford, Markman, Kline, & Stanley, 2003). RE has developed to include a number of evidence-informed programs, offered to couples across many phases of the couple relationship (e.g., marriage preparation, transition to parenthood, forming a stepfamily, balancing work and family), and that can be accessed in diverse formats (e.g., face-to-face, online, or video conference) (Halford, 2011).

Although RE is conceptually distinct from couple therapy, there is a substantial proportion of couples presenting for RE who have at least mild relationship distress (De Maria, 2005). For these distressed couples, RE might have similar benefits as for couple therapy in producing an increase in relationship satisfaction immediately after RE. However, universal and selective offerings of RE typically attract a majority of couples who are currently highly satisfied, and there might be little immediate scope for increases in satisfaction. In universal and selective offerings of RE, the potential benefit of RE are in helping couples to sustain such high satisfaction.

Approaches to relationship education

There are two general approaches to RE that are evidence-based and widely adopted: assessment with feedback and curriculum-based education (Halford, 2011). Assessment with feedback is based on the assumption that providing couples with key understandings of the strengths and challenges in their relationship will enable them to make any required changes to strengthen their relationship. Curriculum-based education is based on the assumption that teaching couples certain crucial relationship skills and knowledge will empower them to sustain a mutually satisfying relationship.

Assessment with feedback

Assessment with feedback involves conducting a structured assessment of the couple relationship, most commonly using a self-report inventory completed by both partners. Feedback involves an educator-led semistructured discussion on the assessment results with the couple, which sometimes is supplemented with skill training (Halford, 2011). There are several important strengths common to the widely used inventories of PREmarital Preparation And Relationship Enhancement (PREPARE) (Olson, Fournier, & Druckman, 1996), Facilitating Open Couple Communication Understanding and Study (FOCCUS: Premarital Assessment and feedback intervention) (Markey & Micheletto, 1997), and RELATionship Evaluation (RELATE) (Busby, Holman, & Taniguchi, 2001). First, each has adequate reliability, content, and construct validity (Larson, Newell, Topham, & Nichols, 2002). Second, each inventory predicts the trajectory of relationship satisfaction in the early years of marriage (Fowers & Olson, 1986; Holman, 2001; Larsen & Olson, 1989; Williams & Jurich, 1995). Thus, each inventory assesses factors relevant to relationship outcomes. Third, there is structured training on the use of these inventories (e.g., Olson, Dyer, & Dyer, 1997). The ease of access, structured approach to use of the inventory, and the training available, likely at least partly explains the widespread adoption of inventory-based relationship education. A significant limitation of all the inventories is that they were developed within the Western cultural context, focus almost exclusively on the Couple Bond related constructs (Larson et al., 2002), and their appropriateness to non-Western couples is untested.

There have been a number of studies evaluating the effects of assessment with feedback on relationship satisfaction both for couples seeking relationship education (e.g., Halford et al., 2010, 2012; Knutson & Olson, 2003; Larson, Vatter, Galbraith, Holman, & Stahmann, 2007); and distressed couples seeking therapy (Cordova et al., 2005, 2014). However, only one study has shown in a randomized controlled trial that there is a **sustained** benefit to couple relationships from assessment and feedback alone (Cordova et al., 2014), and that intervention included a booster intervention after 12 months. Given the lack of replication of long-term benefits from assessment and feedback, I do not consider this approach further even though it is quite widely used in many Western countries.

Curriculum-based relationship education

Curriculum-based RE has a focus on active training of key relationship skills, although curriculum-based approaches usually also include a significant emphasis on building awareness and cognitive change (Halford et al., 2008). There are three curriculum-based RE programs that have been evaluated in two or more randomized controlled trials, that included follow-up of more than 12 months, and found long-term benefit of the programs: the *Prevention and Relationship Enhancement Program*, which sometimes has been referred to as the *Prevention and Relationship Education Program* (PREP; Markman, Stanley, & Blumberg, 2010), *Couple Commitment And Relationship Enhancement* (Couple CARE; Halford et al., 2006), and *Couples Coping Enhancement Training* (CCET; Bodenmann & Shantinath, 2004). These programs have a number of content areas in common. For example, skills training in positive communication and conflict management are included in PREP, Couple CARE, and CCET. There also are significant variations in content across time within programs and between programs. For example, in PREP there are multiple foci of intervention. In the earliest version of PREP, the most time was devoted to prevention of destructive conflict as that was argued to be central to the prevention of relationship problems, but more recent versions have increased the emphasis on shared positive activities, building commitment, and other positive aspects of the relationship (Markman et al., 2010). Individual and conjoint coping with daily hassles as well as life stress is a major component of CCET that receives little attention in most other programs although Couple CARE does include some content on this topic. In Couple CARE the development of each partner's capacity to identify and implement self-change to enhance the relationship (relationship self-regulation or effort) is a core focus that was not in the other programs although recent versions of PREP include some focus on this.

A major strength of the evidence-based approaches to RE is that training is focused on potentially modifiable variables that predict relationship outcomes. In particular, teaching couple communication is a key focus of all curriculum-based RE. Observed effective communication is reliably correlated with relationship satisfaction cross-culturally (Woodin, 2011; Chapter 7) predicts the trajectory of couple relationship satisfaction (e.g., Hanzel & Segrin, 2009; McNulty & Russell, 2010) and is endorsed by couples as important in what makes for a good relationship (Halford, Leung et al., 2018b; Hiew et al., 2015b; Iqbal et al., 2019). Similarly, couples' dyadic coping (Bodenmann, Meuwly, & Kayser, 2011) and couple relationship self-regulation (Halford, Lizzio, Wilson, & Occhipinti, 2007) predict couple relationship satisfaction, which are the focus of the CCET and Couple CARE programs, respectively. However, all the evidence-based approaches to RE were developed and have been evaluated in Western cultural contexts.

The latest version of Couple CARE was influenced by some of the research on cross-cultural differences in couple relationship standards and includes some attention to family responsibility relationships standards (Halford, 2017). However, the appropriateness and effects of RE in non-Western cultural contexts remain untested.

Couple therapy

Several approaches to couple therapy produce moderate effect size and often clinically significant reductions in couple relationship distress. Randomized clinical trials have been conducted on six different couple-based treatments for couple distress: traditional behavioral, cognitive behavioral, integrative behavioral, emotionally focused, integrative systemic, and insight-oriented couple therapy (Snyder, Castellani, & Whisman, 2006). These approaches can usefully be conceptualized as varying along a continuum anchored by a focus on behavior change (behavior exchange and skills-building interventions) on one end, to interventions emphasizing intrapersonal dynamics and vulnerabilities on the other end. However, of these approaches, only two, cognitive-behavioral and emotion-focused couple therapy have had replications of their effectiveness (Halford & Pepping, 2019), and the current chapter focused on those two approaches.

Cognitive-behavioral couple therapy

Cognitive-behavioral couple therapy (CBCT) began as behavioral couple therapy (BCT). BCT began, and largely has remained, as conjoint therapy that targets change in the couple's interactions. Based on the premise that a relationship is defined by the exchange of behaviors between partners, BCT began as the application of behavioral contracting to the treatment of relationship problems. The details of the procedures have been refined, particularly focusing on encouraging each partner to take responsibility for making unilateral positive change, but an emphasis on promoting more positive and less negative relationship behaviors remains an important element of BCT.

In the early 1970s, a second element was added to BCT: communication and problem- solving skills training. Based on early observational studies that distressed couples showed very negative communication (Gottman, Markman, & Notarius, 1977; Wills, Weiss, & Patterson, 1974), couples were trained to use new communication behaviors. These skills were conceptualized as providing couples with the means to enhance intimate communication and resolve their current and future sources of conflict (Jacobson & Margolin, 1979; Weiss, Hops, & Patterson, 1973). In most applications of

BCT the communication skills targeted in training were identified by the therapist, based on contrasting the couple's current communication with a model of adaptive marital communication.

As some of the earliest research on couples showed, observable behavior accounts for only a modest proportion of the variance in relationship satisfaction (Wills et al., 1974). Although BCT included recognition of the importance of internal mediators of external experience from its earliest writings (Weiss, 1984), research in the 1980s lead to a focus on altering cognitions of individual partners. Many of the strategies incorporated in BCT were standard cognitive therapy procedures applied to relationship problems. For example, self-instructional strategies were used to modify negative attributions (Baucom & Lester, 1986) or control anger (Schindler & Vollmer, 1984) and guided discovery, Socratic dialogue, and behavioral experiments were used to challenge irrational relationship beliefs (Baucom & Epstein, 1990). BCT in its 1990s form had become CBCT and consisted of the three components of behavior exchange, communication training, and cognitive restructuring.

CBCT including behavior change, communication training, and relationship cognition change has an exclusive focus on the Couple Bond. That focus seems likely to be salient to many couples cross-culturally, given the seemingly universal aspiration to having a close Couple Bond. At the same time, CBCT concepts are largely silent on how to address issues of Family Responsibilities and cultural differences in what might be appropriate couple communication, or on how to address the role of religion in couple relationships.

In the last two decades, the concept of acceptance has become a guiding principle in a "third wave" of cognitive behavior therapy. In individual treatments (e.g., Hayes, Strosahl, & Wilson, 2012; Neacsiu & Linehan, 2014) acceptance techniques encourage clients to tolerate and even mindfully embrace negative feelings (e.g., sadness and anxiety) and the situations that create those negative feelings. Couple relationship acceptance has been defined as "an adaptive and constructive affective-cognitive reaction to an event, usually a behavior by the partner" (Doss & Christensen, 2006, p. 289).

Interestingly, the emphasis on acceptance has been developed largely by applying ideas drawn from Eastern (Hindu and Buddhist) practices of mindfulness to couple interventions (Pepping & Halford, 2015). This is one of the first attempts to include non-Western concepts into evidence-based couple therapy. The idea of acceptance does seem consistent with relationship standards like maintaining harmony, as increasing acceptance of minor negative behaviors by a partner should reduce conflict over that behavior, and preserving face through compassion and acceptance of the partner.

The combination of acceptance and cognitive-behavioral approaches has been labeled Integrative Behavioral Couple Therapy (IBCT; Christensen & Doss, 2017). IBCT includes two primary acceptance interventions: unified detachment and empathic joining. In unified detachment, the therapist works to help the couple reduce emotional reactivity to a problem by developing an intellectual understanding of that problem. Essentially, this is an interactive form of psychoeducation for couples, which is similar to psychoeducation used in many forms of individual therapy. For example, new parents might be engaged in discussion with the therapist about the challenges of the transition to parenthood, where information can be provided on common challenges for new parents, such as how parents share the many hours of work that having an infant adds to the couple household; the effects of sleep deprivation on new parents; and managing infant sleeping, crying, and feeding problems (Halford, Petch, & Creedy, 2015). It is easy to imagine that this process could be extended to include discussion of cultural values related to standards for relationships.

Empathic joining involves discussions of a core couple problem so it becomes a vehicle for emotional intimacy. In these therapist-guided discussions, partners are encouraged to reveal previously unshared and emotionally vulnerable information about the behaviors the partner views as problematic. A major focus in such discussion is encouraging the expression of emotion and vulnerability.

Emotion-focused couple therapy

Emotion-focused couple therapy (EFCT) combines an experiential, intrapsychic focus on inner emotional experience with an emphasis on cyclical, self-reinforcing interactions (Johnson, Hunsley, Greenberg, & Schindler, 1999). More specifically, EFCT is based on the assumption that attachment is the underlying basis for romantic relationships and that couple relationship problems develop because of insecure attachment bonds between the spouses (Johnson, 2019). Consequently, the goal of EFCT is to enhance romantic attachment bonds between partners. This is achieved in a three-step process. The first step is to de-escalate negative cycles of couple interaction, such as anger, critical demanding, and emotional distancing. These negative cycles are argued to arise from attachment insecurity (Greenman, Johnson, & Wiebe, 2018). The therapist promotes de-escalation by slowing down the interaction and promoting emotional awareness (e.g., through therapist empathy with each partner.) The second step involves promoting positive bonding interactions that make explicit spouse's attachment needs and encourage the expression of emotional vulnerabilities. The third and final step involves the consolidation of gains, which involves the promotion of new cycles of couple interaction that promote the attachment between the spouses.

A key element of EFCT is the assumption that therapy needs to focus on helping spouses identify and focus on emotion. In particular, it is argued that couple therapists need to help spouses identify emotions associated with negative couple interaction that might not be immediately apparent (Greenman et al., 2018). It is further argued that some emotions are obvious but are secondary responses to primary emotions (Greenberg, 2012). Secondary emotions include anger and frustration that often are overtly expressed and tend to mask primary emotions such as sadness or fear, which are argued to relate to attachment needs (Johnson, 2019). For example, one partner might respond with anger (secondary emotion) to their spouse being late home from work. This might show as the first partner raising their voice and criticizing the spouse for being selfish and inconsiderate. EFCT would argue that anger is a secondary emotion to an attachment-related primary emotion. For example, the spouse's lateness elicits a primary emotion of feeling unimportant or unloved in the first partner, and sadness at feeling unloved or unimportant elicits the secondary emotion of anger. The EFCT therapists first identify and empathize with the anger, and then seek to elicit the primary emotion of sadness.

EFCT is strongly focused on couple attachment, which like CBT would seem to address issues of the Couple Bond behaviors rated as important to most couples across diverse cultures. At the same time, there seems to be considerable cultural variation in the acceptability of emotional expressiveness. For example, Chinese couples are less likely than Western couples to overtly express positive affection (Hiew, Halford, van de Vijver, & Liu, 2015c; Chapter 7). People from cultures with lower acceptability of emotional expressiveness might struggle to accept the EFCT focus on emotion.

Couple therapy models and culture

Both CBCT and EFCT conceptualizations of therapy focus on the couple relationship, with attention primarily focused on aspects of the Couple Bond. For example, communication skills training in CBCT is intended, in part, to enhance emotional intimacy between the partners. Similarly, drawing out attachment needs in EFCT is intended to achieve a similar increase in experienced intimacy.

None of the evidence-based approaches to couple therapy pay specific attention to aspects of the couple relationship like the role of Family Responsibility or the Importance of Religion, which are of great importance in some non-Western cultures. That is not to say that these aspects of couple relationship would not receive any attention from therapists using a CBCT or EFCT approach. For example, in Epstein and Baucom's (2002) book on CBCT, they do describe analyzing the context of couple relationships, which includes culture. But the Western cultural origins of the evidence-based couple therapies is reflected in the major focus of their therapeutic approaches.

Effects of couple relationship education and therapy

Effects of couple relationship education

There are now over 150 trials of curriculum-based RE that have been summarized in literature reviews and metaanalyses (Cowan & Cowan, 2014; Halford & Bodenmann, 2013; Hawkins, Blanchard, Baldwin, & Fawcett, 2008). The majority of these studies have been done in the United States, with a small number of studies in Western Europe and Australia. No rigorous controlled studies have been done in non-Western countries. A widely cited, comprehensive metaanalysis by Hawkins et al. (2008) showed RE produced small to moderate immediate improvements in couple communication, $d=0.44$, and relationship satisfaction, $d=0.36$. However, almost all of the studies available at the time of the review only had follow-ups of 6 months or less, and the long-term effects of RE were unclear. Metaanalyses of RE for couples having their first child (Pinquart & Teubert, 2010) and for couples forming stepfamilies (Lucier-Greer & Adler-Baeder, 2012) also showed that RE produced small to moderate immediate effects on couple relationship.

Halford and Bodenmann (2013) reviewed 17 randomized controlled trials of RE that included follow-ups of at least 12 months (range 1–5 years), almost all of which were published after the Hawkins et al. (2008) metaanalysis. Halford and Bodenmann (2013) found 14 of 17 studies reported RE helped maintain relationship satisfaction. Effect size improvements in relationship satisfaction were highly variable. There was a null overall effect at 3-year follow-up for the very large-scale Building Strong Families project with 6212 low-income unmarried couples with child(ren) (Wood, Moore, Clarkwest, & Killewald, 2014), but in that study, only a minority of the participants attended even half of RE session they were offered. There were overall very small effects for the similarly large scale Supporting Healthy Marriage study with 6298 couples at 3-year follow-up (Lundquist et al., 2014). However, there are multiple examples of medium to large size effects in smaller scale trials at 1-year (Ledermann, Bodenmann, & Cina, 2007) and 5-year follow-up (Schulz, Cowan, & Cowan, 2006).

In summary, there is a substantial evidence based on the effects of RE in Western countries, but the effects of RE across trials are inconsistent. Metaanalyses of the available studies report small-to-medium effect size immediate improvement in relationship skills and relationship satisfaction, but there are a number of studies reporting null findings. At follow-ups of 1–5 years again the findings on RE effects are mixed, with small to moderate positive effects in most studies, but null or very small effects found in some studies (Halford & Bodenmann, 2013). Importantly, two particularly large-scale, well-funded trials of RE with low-income couples reported overall null (Wood et al., 2014) or very small effects (Lundquist et al., 2014). This has led some scholars to argue that RE is at best a weak intervention (Bradbury & Lavner, 2012), and that widespread use of RE is not justified (Johnson, 2012). In contrast, others have argued that RE might have selective benefits for particular couples (Halford & Bodenmann, 2013) and that this might explain the inconsistencies in findings on RE effects. The appropriateness of current RE approaches to non-Western countries is as yet untested.

Effects of couple therapy

Numerous reviews show that several approaches to couple therapy produce statistically and clinically significant reductions in relationship distress (e.g., Lebow, Chambers, Christensen, & Johnson, 2012; Snyder & Halford, 2012). In a review of previously published metaanalytic studies, Shadish and Baldwin (2003) reported a large mean effect size of $d=0.84$ for couple therapy, which they noted was similar to the effect sizes of the most effective psychological and pharmacological treatments available for individual psychological disorders. There is little evidence of differential effectiveness across different theoretical orientations to couple therapy, particularly once other covariates (e.g., the reactivity of measures) are controlled (Shadish & Baldwin, 2003). Moreover, data from follow-up at 6 months or longer showed that the positive effects of couple therapy were generally maintained though there was evidence that treatment effects gradually attenuated across time. However, the vast majority of these studies have been done in the United States, with a small number of studies in Western Europe and Australia, but no studies have been done in non-Western countries.

In addition to BCT, the other approach to couple therapy replicated in multiple trials to be effective is EFCT. In four randomized trials, EFCT was superior to a wait-list control in reducing relationship distress, with a weighted mean effect size of $d = 1.31$ (Johnson, 2002). The mean effect size is notably larger than reported for other therapies. However, couples in EFCT studies tend to be less distressed than couples in most other studies, and the larger effect size seems to reflect the greater responsiveness of moderately distressed couples to therapy, rather than EFCT being a more effective form of couple therapy (Shadish & Baldwin, 2003).

In summary, two forms of couple therapy—BCT and EFCT—have been demonstrated to be effective in multiple randomized controlled trials. There is no replicated evidence of differential efficacy between approaches to couple therapy. Across all evaluated approaches to couple therapy, there is a noteworthy minority of 25%–30% of couples who show no improvement from these therapies. In addition, all couple therapies show some attenuation of gains at follow-up. Specifically, assessment at 2 years or longer after termination indicates significant deterioration in up to 45% of successfully treated couples (Christensen, Atkins, Baucom, & Yi, 2010; Jacobson, Schmaling, & Holtzworth-Munroe, 1987; Snyder, Wills, & Grady-Fletcher, 1991).

Treatment of adult problems

The strong association of couple relationship distress with individual psychological disorder provides a rationale for applying a couple therapy to treat those individual issues. Moreover, even when couple relationship distress is not present, couple-based interventions can help partners collaborate in the treatment and management of an individual psychological disorder in one spouse (Mueser & Brunette, 2003).

Baucom, Whisman, and Paprocki (2012) differentiate couple-based interventions that address a psychological disorder in one of the partners from couple therapy for relationship distress. Couple interventions for individual psychological disorders sometimes do focus on the treatment of relationship distress, based on the premise that relationship distress is a major stressor that contributes to the development or maintenance of an individual psychological disorder. However, often the couple intervention is adapted to address the specific challenges of the particular disorder. There are two forms of adaptations of couple intervention for the treatment of individual disorders. First, disorder-specific interventions focus on changing couple interactions that influence either the individual disorder or important components of its treatment. For example, BCT to address alcohol abuse specifically targets having the couple modify alcohol availability in their home, increasing couple and family activities not associated with drinking, spouse support of the drinking partner's attempts to modify his or her drinking, and reducing conflict about the drinking that ironically often precipitates drinking (O'Farrell & Fals-Stewart, 2006). Second, there are partner-assisted interventions in which the spouse serves as a support and coach in assisting the other partner with individual problems. For example, Fischer and Baucom (2018) provide descriptions of couple-based interventions for obsessive-compulsive disorder in which the partner assists in the conduct of exposure and response prevention.

A metaanalysis (Powers, Vedel, & Emmelkamp, 2008) and a more recent comprehensive review (O'Farrell & Clements, 2012) of disorder-specific couple therapy for the treatment of alcohol abuse found it was more effective than individual behavioral treatment of alcohol abuse in reducing alcohol problems. Moreover, only couple therapy, and not individual therapy, enhanced relationship satisfaction. Thus, when there is coexisting couple relationship distress and/or alcohol abuse in a partner, couple treatment is the treatment of choice as it is overall more effective than individual treatments. Moreover, when the partner of a problem drinker seeks help, there has been some success in working with this partner to prompt the problem drinker to present for treatment (O'Farrell & Clements, 2012).

A metaanalysis of couple treatment of depression based on 14 randomized controlled trials found it was equally effective in reducing depression as individual therapy, but only couple therapy enhanced relationship satisfaction (Barbato, D'Avanzo, & Parabiaghi, 2018). When depression is treated with individual evidence-based treatments, relationship distress predicts poor treatment outcome (Denton et al., 2010; Whisman, 2001). Even when individual treatment successfully treats depression, this typically does not resolve relationship distress (Snyder & Whisman, 2003), and ongoing relationship distress predicts relapse in those people successfully treated (Denton et al., 2010). Thus, when there is coexisting couple relationship distress and depression in a partner, couple treatment is the treatment of choice as it is equally effective as an individual treatment of the depression and provides the added value of enhancing the couple relationship at the same time. The combined effect of couple treatment on both depression and relationship satisfaction is likely to be of importance in the maintenance of treatment effects, as ongoing relationship distress predicts relapse after individual treatment of depression (Whisman, Johnson, Be, & Li, 2012).

One caveat on the general proposition that couple-based treatment is the treatment of choice for alcohol and depression is that there has been no research on how acceptable couple-based treatment is to clients presenting with alcohol or depression as their key individual concern. Clearly, a couple's approach to therapy has to be acceptable to both the client with the individual problems and their spouse and the effectiveness research has been limited to couples in which couple intervention is acceptable to both partners.

As was the case for couple therapy for the treatment of relationship distress, couple treatments for individual adult problems have been developed and evaluated in Western cultures. The appropriateness, acceptability, and effectiveness of these treatments in non-Western cultures are untested.

Cultural adaptation of couple interventions

There has been some research on the effectiveness of couple interventions with different cultural groups within Western countries. For example, there is now a sizeable research base on the effects of relationship education with low-income, predominantly African-American, and Hispanic couples within the United States (see Hawkins & Ericson, 2015). However, there is much more speculation than evidence about how couple interventions might need to be adapted to be effective in non-Western cultural contexts.

Culture and presenting couple concerns

Couple interventions likely require some tailoring and adaptation to be appropriate for specific cultural groups (Huang, 2005). As documented in the current chapter, there are some cultural commonalities in relationship standards (high importance is attached to the Couple Bond) and some differences in relationship standards, particularly with respect to Family Responsibilities, and the Importance of Religion. Perhaps reflecting these different standards, the most common areas of relationship concern of couples show some commonalities and some differences cross-culturally (Su et al., 2015). Specifically, Taiwanese men and women rated their most important concern as raising children, whereas that was the least most important concern as rated by US men and women (Su et al., 2015). This suggests that couple therapy might need to have somewhat different foci in different cultures.

US couples rated communication as the most important relationship concern, and Taiwanese couples rated it as the second most important concern (Su et al., 2015), suggesting it is widely perceived as an important area of concern. Given this commonality, interventions that focus attention on couple communication seem to be addressing issues salient to couples across many cultures. As described previously the evidence-based couple interventions (cognitive-behavioral couple therapy, emotion-focused therapy, and the evidence-based relationship education programs), all have focus on couple communication and promoting couple intimacy. It seems likely that the domain of communication is salient to many couples across diverse cultures, but what constitutes adaptive couple communication might vary cross-culturally (Halford, Lee, Hiew, & van de Vijver, 2018; Chapter 7).

It also is interesting to note that Chinese couples in Taiwan report less overall relationship concerns across six areas of common concern than do US couples (Su et al., 2015). This does not reflect that Chinese couples are more satisfied than Western couples, most research shows that Chinese couples rate their mean satisfaction as lower than do Western couples (e.g., Halford, Lee, et al., 2018; Hiew et al., 2015b). However, the areas of concern that couples rated might have omitted areas of concern to Chinese that might be less important to Westerners. For example, areas like relations with extended family, maintaining face, and religion were not included in the areas of concern couples rated.

Addressing diverse relationship standards

Western-trained couple therapists may not always be aware of important relationship standards held by members of other cultural groups. For example, the high importance attached in some cultures to religion in couple relationships might be ignored. This suggests that couple assessment measures that have been developed with Western couples might need cultural adaptation to include assessment of a range of potential areas of couple concern that are not necessarily emphasized by Western couples or therapists.

There is one example of evaluation of a culturally tailored couple intervention to address religious standards: a relationship education program for African American couples living in the southern states of the United States (Beach et al., 2011). Given that religion is important to many African American couples, the intervention included shared prayer sessions, and participants valued this component of the program although the research design did not allow a specific analysis of whether incorporating shared prayer enhanced couple relationship outcomes.

One potentially important difference between couple therapy and education as it has been developed and practiced with Western couples, and what might be appropriate for other cultures, is the role of extended family.

In Chinese culture and many other collectivistic cultures, the couple relationship is viewed as part of the extended family, whereas in most Western cultures there is a more clearly demarked differentiation of the couple relationship from extended family (Hiew et al., 2015a). Given that difference, RE and therapy might need to engage with extended family in collectivistic cultures. This could involve talking about extended family issues in couple only sessions, or even bringing in the extended family to some therapy sessions.

Cultural diversity within countries warrants attention by couple educators and therapists. As described previously, there have been very high rates of migration to some Western countries, and there are many couples drawn from distinctively different cultures from the majority culture. There also are quite high rates of intercultural relationships. For example, Australia, Singapore, and some states of the United States have 30% or more of all marriages being intercultural marriages (Hiew, Halford, & Liu, 2014). Differing partner relationship standards can occur in any couples but is more common in intercultural couples (Halford, Leung, et al., 2018b; Hiew et al., 2015b). Anecdotally it has been reported that partners in intercultural couples rarely attribute differences in relationship standards to cultural differences (Hiew et al., 2014), and helping partners to understand the cultural influences that have shaped their relationship standards could promote the acceptance described by Doss and Christensen (2006) as important in Integrative Behavioral Couple Therapy.

There is some evidence that attention to cultural differences within countries influences access and acceptability of couple interventions. For example, an online survey of same-sex attracted adults found that respondents felt therapists' lack of knowledge about Gay culture was a barrier to respondents seeking out couple therapy (Pepping, Lyons, Halford, Cronin, & Pachankis, 2017). Respondents also believed that therapists needed to pay attention to distinctive challenges same-sex couples face, such as homonegativity. That need not mean developing whole new models of couple therapy, but adapting existing evidence-based approaches to include attention to these issues. For example, Rainbow Couple CARE is a relationship education program to address the needs of same-sex couples that adapts Couple CARE designed for heterosexual couples (Pepping, Cronin, Halford, & Lyons, in press).

Couple interventions for adult disorders

The extension of couple interventions to treatment of adult disorders has been an important development, which has the potential to significantly enhance psychological treatments for a range of adult disorders (Fischer &

Baucom, 2018). For example, there has been some use of couple interventions for substance abuse in US services for military veterans (Wesley, 2016), and the national health service in the United Kingdom has been expanding use of couple interventions in the treatment of depression (Baucom et al., 2018).

One reason for the focus on couple relationships in the treatment of disorders like depression, substance abuse, and anxiety disorders is the centrality of the couple relationship to most adults' lives (Fischer & Baucom, 2018; Halford, 2011). However, almost all the research examining how couple relationships interact with individual psychopathology and how couple interventions can enhance individual as well as couple functioning has done in Western countries, most notably the United States and to a lesser extent in Canada, Australia, and Germany. It has been argued that the couple relationship is of less importance in non-Western collectivistic societies, as intimacy and support needs are shared across the broader family network (Lewinsohn & Werner, 1997). Moreover, the mutual absorption and disregard of others' views that characterize romantic love in individualistic societies sometimes are regarded as dysfunctional in collectivistic societies, in which group needs are prioritized over personal desires (Dion & Dion, 1993). Hence it is crucial to test, rather than presume, that the promising results of couple interventions for adult disorders in Western countries are generalizable to other cultures.

Conclusions

The form and function of couple relationships cross-culturally show some striking consistencies and some equally striking variations. Committed couple relationships (marriage and similar committed relationships) seem universal across cultures. From an evolutionary perspective, the primary function of the couple relationship is to enhance survival of offspring, and the universality of couple relationships likely reflects that evolutionary imperative. In environments where survival is challenging, marriage remains crucial to the social networks in which the couple reside as the pooling of resources and cooperative parenting benefits all. In these contexts, marriages are arranged and supported by extended family and social networks, and divorce is rare. In environments where survival pressures are less demanding, the couple relationship is less critical to collective survival, relationship formation and dissolution tend to be seen as a matter for the individuals in the relationship, and rates of separation are high.

Social changes have modified the cultural context in which relationships exist, and influenced standards for the couple relationship, which typically expand to include psychological intimacy and enhancement of individual partner well-being. Couple Bond standards are important across diverse cultures, which likely reflects the evolutionary selection of those who form a strong attachment to a romantic partner. The importance of Family Responsibility standards and the Importance of Religion show considerable cross-cultural variation, which likely reflects the broader cultural values within the societies in which couples are forming relationships. Family Responsibility standards and the Importance of Religion standards are less strongly endorsed in Western than non-Western cultures. Much of the psychology of close relationships theory that has developed in Western countries pays limited attention to these aspects of couple relationships.

Couple RE and couple therapy have been developed in Western countries. All evidence-based approaches focus predominantly on the Couple Bond, which has been shown to be effective in enhancing the relationships of Western couples. On the one hand, this focus is likely to be salient across diverse cultures, given the high endorsement of the Couple Bond across diverse cultures. On the other hand, the lack of attention to relationship standards of importance in non-Western cultures raises serious questions about the acceptability and appropriateness of existing approaches to RE and couple therapy for couples in non-Western countries, or even for cultural minority groups living in Western countries. Ultimately the effectiveness of existing couple interventions for culturally diverse couples is an empirical question, which is as yet untested. So it is unclear if we have a robust psychological science of how to enhance couple relationships or a culturally specific set of procedures.

References

Agnew, C. R. (Ed.), (2014). *Social influences on romantic relationships*. Cambridge: Cambridge University Press.

Amato, P. R. (2010). Research on divorce: Continuing trends and new developments. *Journal of Marriage and Family*, *72*(3), 650–666. https://doi.org/10.1111/j.1741-3737.2010.00723.x.

Amato, P. R., & Rogers, S. J. (1997). A longitudinal study of marital problems and subsequent divorce. *Journal of Marriage and the Family*, *59*(3), 612–624. https://doi.org/10.2307/353949.

Anderton, C., Pender, D., & Asner-Self, K. (2011). A review of the religious identity/sexual orientation identity conflict literature: Revisiting Festinger's cognitive dissonance theory. *Journal of LGBT Issues in Counseling*, *5*(3–4), 259–281. https://doi.org/10.1080/15538605.2011.632745.

Australian Bureau of Statistics. (2012). *Cultural diversity in Australia. Reflecting a nation: Stories from the 2011 Census, 2012–2013.* (Issue 2071.0). Retrieved from http://www.abs.gov.au.

Australian Bureau of Statistics. (2013). *Australian social trends: Same-sex couples.* (cat. no. 4102.0). Retrieved from http://www.abs.gov.au.

Baker, N. A., & Halford, W. K. (2019). Assessment of couple relationship standards in same sex attracted adults. *Family Process*, https://doi.org/10.1111/famp.12447. Advanced online publication.

Barbato, A., D'Avanzo, B., & Parabiaghi, A. (2018). Couple therapy for depression. *Cochrane Database of Systematic Reviews*, https://doi.org/10.1002/14651858.CD004188.pub3.

Baucom, D. H., & Epstein, N. (1990). *Cognitive behavioral marital therapy*. New York: Brunner/Mazel.

Baucom, D. H., Epstein, N., Rankin, L. A., & Burnett, C. K. (1996). Assessing relationship standards: The inventory of specific relationship standards. *Journal of Family Psychology*, *10*, 72–88. https://doi.org/10.1037/0893-3200.10.1.72.

Baucom, D. H., Fisher, M. S., Worrell, M., Corrie, S., Belus, J. M., Molyva, E., & Boeding, S. E. (2018). Couple-based intervention for depression: An effectiveness study in the National Health Service in England. *Family Process*, *57*(2), 275–292. https://doi.org/10.1111/famp.12332.

Baucom, D. H., & Lester, G. W. (1986). The usefulness of cognitive restructuring as an adjunct to behavioral marital therapy. *Behavior Therapy*, *17*(4), 385–403. https://doi.org/10.1016/S0005-7894(86)80070-3.

Baucom, D. H., Whisman, M. A., & Paprocki, C. (2012). Couple-based interventions for psychopathology. *Journal of Family Therapy*, *34*, 250–270.

Beach, S. R. H., Hurt, T. R., Fincham, F. D., Franklin, K. J., McNair, L. M., & Stanley, S. M. (2011). Enhancing marital enrichment through spirituality: Efficacy data for prayer focused relationship enhancement. *Psychology of Religion and Spirituality*, *3*(3), 201–216. https://doi.org/10.1037/a0022207.

Bodenmann, G., Meuwly, N., & Kayser, K. (2011). Two conceptualizations of dyadic coping and their potential for predicting relationship quality and individual well-being: A comparison. *European Psychologist*, *16*, 255–266. https://doi.org/10.1027/1016-9040/a000067.

Bodenmann, G., & Shantinath, S. D. (2004). The Couples Coping Enhancement Training (CCET): A new approach to prevention of marital distress based upon stress and coping. *Family Relations*, *53*(5), 477–484. https://doi.org/10.1111/j.0197-6664.2004.00056.x.

Bowlby, J. (1969). *Attachment and loss*. New York: Basic Books.

Boyd, R., & Richerson, P. J. (2009). Culture and the evolution of human cooperation. *Philosophical Transactions of the Royal Society*, *364*, 3281–3288. https://doi.org/10.1098/rstb.2009.0134.

Bradbury, T. N., & Lavner, J. A. (2012). How can we improve preventive and educational interventions for intimate relationships? *Behavior Therapy*, *43*, 113–122. https://doi.org/10.1016/j.beth.2011.02.008.

Busby, D. M., Holman, T. B., & Taniguchi, N. (2001). RELATE: Relationship evaluation of the individual, family, cultural and couple contexts. *Family Relations*, *50*, 308–317. https://doi.org/10.1111/j.1741-3729.2001.00308.x.

Buss, D. M. (2016). *The evolution of desire: Strategies of human mating* (4th ed.). New York: Basic Books.

Christensen, A., Atkins, D. C., Baucom, B., & Yi, J. (2010). Marital status and satisfaction five years following a randomized clinical trial comparing traditional versus integrative behavioral couple therapy. *Journal of Consulting and Clinical Psychology*, *78*(2), 225–235. https://doi.org/10.1037/a0018132.

Christensen, A., & Doss, B. D. (2017). Integrative behavioural couple therapy. *Current Opinion in Psychology*, *13*, 111–114. https://doi.org/10.1016/j.copsyc.2016.04.022.

Coontz, S. (2005). *Marriage, a history: From obedience to intimacy, or how love conquered marriage.* New York: Viking Press.

Cordova, J. V., Fleming, C. J., Morrill, M. I., Hawrilenko, M., Sollenberger, J. W., Harp, A. G., … Wachs, A. E. (2014). The marriage checkup: A randomized controlled trial of annual relationship checkups. *Journal of Consulting and Clinical Psychology*, *83*(5), 592–604. https://doi.org/10.1037/a0037097.

Cordova, J. V., Scott, R. L., Dorian, M., Mirgain, S., Yaeger, D., & Groot, A. (2005). The marriage checkup: An indicated preventive intervention for treatment-avoidant couples at risk for marital deterioration. *Behavior Therapy*, *36*(4), 301–309. https://doi.org/10.1016/S0005-7894(05)80112-1.

Cowan, P. A., & Cowan, C. P. (2014). Controversies in couple relationship education (CRE): Overlooked evidence and implications for research and policy. *Psychology, Public Policy, and Law*, *20*(4), 361–383. https://doi.org/10.1037/law0000025.

Darwin, C. (1877/2004). *The descent of man.* London: Penguin Classics.

de Graaf, P. M., & Kalmijn, M. (2006). Divorce motives in a period of rising divorce: Evidence from a Dutch life-history survey. *Journal of Family Issues*, *27*, 483–505. https://doi.org/10.1177/0192513X05283982.

De Maria, R. M. (2005). Distressed couples and marriage education. *Family Relations*, *54*, 242–253. https://doi.org/10.1111/j.0197-6664.2005.00019.x.

Denton, W. H., Carmody, T. J., Rush, A. J., Thase, M. E., Trivedi, M. H., Arnow, B. A., & Keller, M. B. (2010). Dyadic discord at baseline is associated with lack of remission in the acute treatment of chronic depression. *Psychological Medicine*, *40*, 415–424. https://doi.org/10.1017/S0033291709990535.

Dion, K. K., & Dion, K. L. (1993). Individualistic and collectivistic perspectives on gender and the cultural context of love and intimacy. *Journal of Social Issues*, *49*, 53–69. https://doi.org/10.1111/j.1540-4560.1993.tb01168.x.

Dobewell, H., & Strack, M. (2014). Relationship of Inglehart's and Schwartz's value dimensions revisited. *International Journal of Psychology*, *49*(4), 240–248. https://doi.org/10.1002/ijop.12004.

Doss, B. D., & Christensen, A. (2006). Acceptance in romantic relationships: The frequency and acceptability of partner behavior inventory. *Psychological Assessment*, *18*(3), 289–302. https://doi.org/10.1037/1040-3590.18.3.289.

Dribe, M., & Lundh, C. (2008). Intermarriage and immigrant integration in Sweden: An exploratory analysis. *Acta Sociologica*, *51*, 329–354. https://doi.org/10.1177/0001699308097377.

Eidelson, R. J., & Epstein, N. (1982). Cognition and relationship maladjustment: Development of a measure of dysfunctional relationship beliefs. *Journal of Consulting and Clinical Psychology*, *50*, 715–720.

Epstein, N., & Baucom, D. H. (2002). *Enhanced cognitive-behavioural therapy for couples: A contextual approach.* Washington, DC: American Psychological Association.

Epstein, N. B., Chen, F., & Beyder-Kamjou, I. (2007). Relationship standards and marital satisfaction in Chinese and American couples. *Journal of Marital and Family Therapy*, *31*, 59–74. https://doi.org/10.1111/j.1752-0606.2005.tb01543.x.

Feldman, S. S., Mont-Reynaud, R., & Rosenthal, D. A. (1992). When East moves West: The acculturation of values of Chinese adolescents in the US and Australia. *Journal of Research on Adolescence*, *2*, 147–173. https://doi.org/10.1207/s15327795jra0202_3.

Finkel, E. J., Hui, C. M., Carswell, K. L., & Larson, G. M. (2014). The suffocation of marriage: Climbing Mount Maslow without enough oxygen. *Psychological Inquiry*, *25*(1), 1–41. https://doi.org/10.1080/1047840X.2014.863723.

Fischer, M. S., & Baucom, D. H. (2018). Couple-based interventions for relationship distress and psychopathology. In J. N. Butcher, J. Hooley, & P. C. Kendall (Eds.), *Psychopathology: Understanding, assessing and treating adult mental disorders* (pp. 661–686). Washington, DC: American Psychological Association.

Fletcher, G. J., Simpson, J. A., Thomas, G., & Giles, L. (1999). Ideals in intimate relationships. *Journal of Personality and Social Psychology, 76*, 72–89. https://doi.org/10.1037/0022-3514.76.1.72.

Fletcher, G. O., Simpson, J. O., Campbell, L., & Overall, N. C. (2015). Pair-bonding, romantic love, and evolution: The curious case of *Homo sapiens*. *Perspectives on Psychological Science, 10*, 20–36. https://doi.org/10.1177/1745691614561683.

Fowers, B. J., & Olson, D. H. (1986). Predicting marital success with PREPARE: A predictive validity study. *Journal of Marital and Family Therapy, 12*, 403–413. https://doi.org/10.1111/j.1752-0606.1986.tb00673.x.

Galperin, A., Haselton, M. G., Fredrick, D. A., Poore, J., von Hippel, W., Buss, D., & Gonzaga, G. C. (2013). Sexual regret: Evidence for evolved sex differences. *Archives of Sexual Behavior, 42*, 1145–1161. https://doi.org/10.1007/s10508-012-0019-3.

Geary, D. C. (2000). Evolution and proximate expression of human paternal investment. *Psychological Bulletin, 126*, 55–77. https://doi.org/10.1037/0033-2909.126.1.55.

Gelfand, M. J., Raver, J. L., Nishii, L., Leslie, L. M., Lun, J., Lim, B. C., … Yamaguchi, S. (2011). Differences between tight and loose cultures: A 33-nation study. *Science, 332*, 1100–1104. https://doi.org/10.1126/science.1197754.

Goodwin, R., & Gaines, S. O. (2004). Relationship beliefs and relationship quality across cultures: Country as a moderator of dysfunctional beliefs and relationship quality in three former Communist societies. *Personal Relationships, 11*, 267–279. https://doi.org/10.1111/j.1475-6811.2004.00082.x.

Gottman, J., Markman, H., & Notarius, C. (1977). The topography of marital conflict: A sequential analysis of verbal and nonverbal behavior. *Journal of Marriage and the Family*, 461–477. https://doi.org/10.2307/350902.

Greenberg, L. S. (2012). Emotions, the great captains of our lives: Their role in the process of change in psychotherapy. *American Psychologist, 67*, 697–707. https://doi.org/10.1037/a0029858.

Greenman, P. S., Johnson, S. M., & Wiebe, S. (2018). Emotionally focused therapy for couples: At the heart of science and practice. In B. H. Fiese, M. Celano, K. Deater-Deckard, E. N. Jouriles, & M. A. Whisman (Eds.), *APA handbook of contemporary family psychology: Family therapy and training* (pp. 291–305). Washington, DC: American Psychological Association.

Güngör, D., Fleischmann, F., & Phalet, K. (2011). Religious identification, beliefs, and practices among Turkish Belgian and Moroccan Belgian Muslims. Intergenerational continuity and acculturative change. *Journal of Cross-Cultural Psychology, 42*, 1356–1374. https://doi.org/10.1177/0022022111412342.

Halford, W. K. (2017). *CoupleCARE: Couple commitment and relationship enhancement* (2nd ed.). Brisbane: Australian Academic Press.

Halford, W. K. (2011). *Marriage and relationship education: What works and how to provide it*. New York, NY: Guilford Press.

Halford, W. K., & Bodenmann, G. (2013). Effects of relationship education on maintenance of couple relationship satisfaction. *Clinical Psychology Review, 33*, 512–525. https://doi.org/10.1016/j.cpr.2013.02.001.

Halford, W. K., Chen, R., Wilson, K. L., Larson, J., Bussby, D., & Holman, T. (2012). Does therapist guidance enhance assessment-based feedback as couple relationship education? *Behavior Change, 29*(4), 199–212. https://doi.org/10.1017/bec.2012.20.

Halford, W. K., Lee, S., Hiew, D. N., & van de Vijver, F. J. R. (2018). Indirect couple communication and relationship satisfaction in Chinese, Western, and Chinese-Western intercultural couples. *Couple and Family Psychology: Research and Practice*, 7, 183–200. https://doi.org/10.1037/cfp0000109.

Halford, W. K., Leung, P. W. L., Hyng-Chueng, C., Chau-Wan, L., Hiew, D., & Van de Vijver, F. J. R. (2018a). Couple relationship standards and migration: Comparing Hong

Kong Chinese with Australian Chinese. *Family Process, 57*(4), 996–1011. https://doi.org/10.1111/famp.12337.

Halford, W. K., Leung, P. W. L., Hyng-Chueng, C., Chau-Wan, L., Hiew, D., & Van de Vijver, F. J. R. (2018b). Relationship standards and relationship satisfaction in Chinese, Western, and intercultural couples living in Australia and Hong Kong. *Couple and Family Psychology: Research and Practice*, 7, 127–142. https://doi.org/10.1037/cfp0000104.

Halford, W. K., Lizzio, A., Wilson, K. L., & Occhipinti, S. (2007). Does working at your marriage help? Couple relationship self-regulation and satisfaction in the first 4 years of marriage. *Journal of Family Psychology, 21*, 185–194. https://doi.org/10.1037/0893-3200.21.2.185.

Halford, W. K., Markman, H. J., Kline, G. H., & Stanley, S. M. (2003). Best practice in couple relationship education. *Journal of Marital and Family Therapy, 29*(3), 385–406. https://doi.org/10.1111/j.1752-0606.2003.tb01214.x.

Halford, W. K., Markman, H. J., & Stanley, S. M. (2008). Strengthening couple relationships with education: Social policy and public health perspectives. *Journal of Family Psychology, 22*, 497–505. https://doi.org/10.1037/a0012789.

Halford, W. K., Moore, E. M., Wilson, K., Dyer, C., Farrugia, C., & Judge, K. (2006). *Couple commitment and relationship enhancement.* Brisbane: Australian Academic Press.

Halford, W. K., & Pepping, C. (2019). What every therapist needs to know about couple therapy. *Behavior Change*, https://doi.org/10.1017/bec.2019.12.

Halford, W. K., Petch, J., & Creedy, D. (2015). *Clinical guide to helping new parents.* New York, NY: Springer.

Halford, W. K., & Snyder, D. K. (2012). Universal processes and common factors in couple therapy and relationship education. *Behavior Therapy, 43*(1), 1–12. https://doi.org/10.1016/j.beth.2011.01.007.

Halford, W. K., Wilson, K., Watson, B., Verner, T., Larson, J., Busby, D., & Holman, T. (2010). Couple relationship education at home: Does skill training enhance relationship assessment and feedback? *Journal of Family Psychology, 24*(2), 188. https://doi.org/10.1037/a0018786.

Hanzel, A., & Segrin, C. (2009). The role of conflict resolution styles in mediating the relationship between enduring vulnerabilities and marital quality. *Journal of Family Communication, 9*(3), 150–169. https://doi.org/10.1080/15267430902945612.

Hawkins, A. J., Blanchard, V. L., Baldwin, S. A., & Fawcett, E. B. (2008). Does marriage and relationship education work? A meta-analytic study. *Journal of Consulting and Clinical Psychology, 76*(5), 723–734. https://doi.org/10.1037/a0012584.

Hawkins, A. J., & Ericson, S. E. (2015). Is couple and relationship education effective for low income participants? A meta-analytic study. *Journal of Family Psychology, 29*(1), 59–68. https://doi.org/10.1037/fam0000045.

Hayes, S. C., Strosahl, K. D., & Wilson, K. G. (2012). *Acceptance and commitment therapy: The process and practice of mindful change.* New York, NY: Guilford Press.

Hiew, D., Halford, W. K., & Lui, S. (2014). Loving diversity: Living in intercultural relationships. In A. Abela & J. Walker (Eds.), *Contemporary issues in family studies: Global perspectives on partnerships, parenting and support in a changing world* (pp. 87–99). Oxford: Wiley-Blackwell.

Hiew, D. N., Halford, W. K., van de Vijver, F. J. R., & Liu, S. (2015a). The Chinese-Western intercultural couple standards scale. *Psychological Assessment, 27*(3), 816–826. https://doi.org/10.1037/pas0000090.

Hiew, D. N., Halford, W. K., van de Vijver, F. J. R., & Liu, S. (2015b). Relationship standards and satisfaction in Chinese, Western, and intercultural Chinese-Western couples in Australia. *Journal of Cross-Cultural Psychology, 46*(5), 684–701. https://doi.org/10.1177/0022022115579936.

Hiew, D. N., Halford, W. K., van de Vijver, F. J. R., & Liu, S. (2015c). Communication and relationship satisfaction in Chinese, Western, and intercultural Chinese-Western couples. *Journal of Family Psychology, 30*(20), 193–202. https://doi.org/10.1037/fam0000144.

Holman, T. B. (2001). *Premarital prediction of marital quality or break up: Research, theory and practice*. New York: Kluwer.

Huang, W.-J. (2005). An Asian perspective on relationship and marriage education. *Family Process*, *44*(2), 161–173. https://doi.org/10.1111/j.1545-5300.2005.00051.x.

Hunt, R. A., Hof, L., & DeMaria, R. (1998). *Marriage enrichment: Preparation, mentoring, and outreach*. Philadelphia, PA: Brunner/Mazel.

Inglehart, R., & Baker, W. E. (2000). Modernization, cultural change, and the persistence of traditional values. *American Sociological Review*, *65*(1), 19–51. https://doi.org/10.2307/2657288.

International Organization for Migration. (2015). *World migration report 2015*. Geneva: International Organization for Migration.

Iqbal, S., Ayub, N., van de Vijver, F., & Halford, W. K. (2019). Couple relationship standards in Pakistan. *Couple and Family Psychology: Research and Practice*, https://doi.org/10.1037/cfp0000124. Advance online publication.

Jacobson, N. S., & Margolin, G. (1979). *Marital therapy: Strategies based on social learning and behavior exchange principles*. New York, NY: Brunner/Mazel.

Jacobson, N. S., Schmaling, K. B., & Holtzworth-Munroe, A. (1987). Component analysis of behavioral marital therapy: Two year follow-up and prediction of relapse. *Journal of Marital and Family Therapy*, *13*, 187–195.

Johnson, M. D. (2012). Healthy marriage initiatives: On the need for empiricism in policy implementation. *American Psychologist*, *67*, 296–308. https://doi.org/10.1037/a0027743.

Johnson, S. (2019). Attachment in action—Changing the face of 21st century couple therapy. *Current Opinion in Psychology*, *25*, 101–104. https://doi.org/10.1016/j.copsyc.2018.03.007.

Johnson, S. M. (2002). Marital problems. In D. H. Sprenkle (Ed.), *Effectiveness research in marriage and family therapy* (pp. 163–190). Alexandria, VA: American Association for Marriage and Family Therapy.

Johnson, S. M., Hunsley, J., Greenberg, L., & Schindler, D. (1999). Emotionally focused couples therapy: Status and challenges. *Clinical Psychology: Science and Practice*, *6*(1), 67–79. https://doi.org/10.1093/clipsy.6.1.67.

Kam, C. C. S., & Bond, M. H. (2008). Role of emotions and behavioural responses in mediating the impact of face loss on relationship deterioration: Are Chinese more face-sensitive than Americans? *Asian Journal of Social Psychology*, *11*, 175–184. https://doi.org/10.1111/j.1467-839X.2008.00254.x.

Kaplan, H. S., Hooper, P. L., & Gurven, M. (2009). The evolutionary and ecological roots of human social organization. *Philosophical Transactions of the Royal Society B*, *364*, 3289–3299. https://doi.org/10.1098/rstb.2009.0115.

Kashima, Y., Bain, P. G., & Perfors, A. (2019). The psychology of cultural dynamics: What is it, what do we know, and what is yet to be known? *Annual Review of Psychology*, *70*, 499–529. https://doi.org/10.1146/annurev-psych-010418-103112.

Kaya, M., Halford, W. K., Hiew, D. N., Sheffield, J., & van de Vijver, F. J. R. (2019). Ethnic identification and relationship satisfaction in Chinese, Western and Chinese-Western intercultural couples. *Couple and Family Psychology: Research and Practice*, *8*(3), 12–136. https://doi.org/10.1037/pcf0000120.

Knutson, L., & Olson, D. H. (2003). Effectiveness of PREPARE program with premarital couples in community settings. *Marriage & Family: A Christian Journal*, *6*(4), 529–546.

Kusner, K. G., Mahoney, A., Pargament, K. I., & DeMaris, A. (2014). Sanctification of marriage and spiritual intimacy predicting observed marital interactions across the transition to parenthood. *Journal of Family Psychology*, *28*(5), 604–614. https://doi.org/10.1037/a0036989.

Larsen, A. S., & Olson, D. H. (1989). Predicting marital satisfaction using PREPARE: A replication study. *Journal of Marital and Family Therapy*, *15*, 311–322. https://doi.org/10.1111/j.1752-0606.1989.tb00812.x.

Larson, J. H., Newell, K., Topham, G., & Nichols, S. (2002). A review of three comprehensive premarital assessment questionnaires. *Journal of Marital and Family Therapy*, *28*, 233–239. https://doi.org/10.1111/j.1752-0606.2002.tb00360.x.

Larson, J. H., Vatter, R. S., Galbraith, R. C., Holman, T. B., & Stahmann, R. F. (2007). The RELATionship Evaluation (RELATE) with therapist-assisted interpretation: Short-term effects on premarital relationships. *Journal of Marital and Family Therapy*, *33*(3), 364–374. https://doi.org/10.1111/j.1752-0606.2007.00036.x.

Lebow, J. L., Chambers, A. L., Christensen, A., & Johnson, S. M. (2012). Research on the treatment of couple distress. *Journal of Marital and Family Therapy*, *38*(1), 145–168. https://doi.org/10.1111/j.1752-0606.2011.00249.x.

Ledermann, T., Bodenmann, G., & Cina, A. (2007). The efficacy of the couples coping enhancement training (CCET) in improving relationship quality. *Journal of Social and Clinical Psychology*, *26*, 940–959. https://doi.org/10.1521/jscp.2007.26.8.940.

Lee, S. M., & Edmonston, B. (2005). New marriages, new families: U.S. racial and Hispanic intermarriage. *Population Bulletin*, *60*, 3–36. Retrieved from http://www.prb.org/Publications/Reports/2005/NewMarriagesNewFamiliesUSRacialandHispanicIntermarriagePDF593KB.aspx.

Lewinsohn, M. A., & Werner, P. D. (1997). Factors in Chinese marital process: Relationship to marital adjustment. *Family Process*, *36*, 43–61. https://doi.org/10.1111/j.1545-5300.1997.00043.x.

Lewis, D. M. G., Al-Shawaf, L., Conroy-Beam, D., Asao, K., & Buss, D. M. (2017). Evolutionary psychology: A how to guide. *American Psychologist*, *72*(4), 353–373. https://doi.org/10.1037/a0040409.

Locke, H. J., & Wallace, K. M. (1959). Short marital adjustment and prediction tests: Their reliability and validity. *Marriage and Family Living*, *21*(3), 251–255. https://doi.org/10.2307/348022.

Lucier-Greer, M., & Adler-Baeder, F. (2012). Does couple and relationship education work for individuals in stepfamilies? A meta-analytic study. *Family Relations*, *61*(5), 756–769. https://doi.org/10.1111/j.1741-3729.2012.00728.x.

Lundquist, E., Hsueh, J., Lowenstein, A. E., Faucetta, K., Gubits, D., Michalopoulos, C., & Knox, V. (2014). *A family strengthening program for low-income families: Final impacts from the supporting healthy marriage evaluation.* Washington, DC: United States Department of Health and Human Services.

Lyons, A., & Hosking, W. (2014). Prevalence and correlates of sexual partner concurrency among Australian gay men aged 18–39. *AIDS and Behavior*, *18*, 801–809. https://doi.org/10.1007/s10461-013-0613-y.

Mahoney, A., Pargament, K. I., Tarakeshwar, N., & Swank, A. B. (2008). Religion in the home in the 1980s and 1990s: A meta-analytic review and conceptual analysis of links between religion, marriage, and parenting. *Psychology of Religion and Spirituality*, *S*(1), 63–101. https://doi.org/10.1037/1941-1022.S.1.63.

Markey, B., & Micheletto, M. (1997). *Instructor manual for FOCCUS.* Omaha, NE: Archdiocese of Omaha.

Markman, H. J., Stanley, S. M., & Blumberg, S. L. (2010). *Fighting for your marriage* (3rd ed.). San Francisco, CA: Jossey-Bass, Inc.

Marshall, T. C. (2008). Cultural differences in intimacy: The influence of gender-role ideology and individualism—Collectivism. *Journal of Social and Personal Relationships*, *25*, 143–168. https://doi.org/10.1177/0265407507086810.

Maslow, A. H. (1943). A theory of human motivation. *Psychological Review*, *50*(4), 370–396. https://doi.org/10.1037/h0054346.

McNulty, J. K., & Russell, V. M. (2010). When "negative" behaviors are positive: A contextual analysis of the long-term effects of problem-solving behaviors on changes in relationship satisfaction. *Journal of Personality and Social Psychology*, *98*(4), 587–604. https://doi.org/10.1037/a0017479.

Mueser, K. T., & Brunette, M. F. (2003). Schizophrenia spectrum disorders. In D. K. Snyder & M. A. Whisman (Eds.), *Treating difficult couples: Helping clients with coexisting mental and relationship disorders* (pp. 137–158). New York: Guilford Press.

Neacsiu, A. D., & Linehan, M. M. (2014). Dialectical behavior therapy for borderline personality disorder. In D. H. Barlow (Ed.), *Clinical handbook of psychological disorders* (5th ed., pp. 394–461). New York: Guilford Press.

Nordin, N., Halford, W. K., Barlow, F., & Mastor, K. A. (2019). Relationship standards and marital satisfaction in Malay Muslims. (Paper submitted for publication).

O'Farrell, T. J., & Clements, K. (2012). Review of outcome research on marital and family therapy in treatment for alcoholism. *Journal of Marital and Family Therapy, 38*(1), 122–144. https://doi.org/10.1111/j.1752-0606.2011.00242.x.

O'Farrell, T. J., & Fals-Stewart, W. (2006). *Behavioral couples therapy for alcoholism and drug abuse.* New York: Guilford.

Olson, D. H., Dyer, P., & Dyer, G. (1997). *Growing together: Leaders manual. A group program for couples.* Minneapolis, MN: Life Innovations.

Olson, D. H., Fournier, D. G., & Druckman, J. M. (1996). *PREPARE.* Minneapolis, MN: Life Innovations.

Peplau, L. A., & Fingerhut, A. W. (2007). The close relationships of lesbians and gay men. *Annual Review of Psychology, 58*, 405–424. https://doi.org/10.1146/annurev.psych.58.110405.085701.

Pepping, C., Cronin, T. J., Halford, W. K., & Lyons, A. (2019). Minority stress and same-sex relationship satisfaction: The role of concealment motivation. *Family Process, 58*(2), 494–508. https://doi.org/10.1111/famp.12365.

Pepping C., Cronin T.J., Halford W.K. and Lyons A., Couple relationship education for same-sex couples: A preliminary evaluation of Rainbow Couple CARE, Journal of Couple and Relationship Therapy, in press.

Pepping, C., & Halford, W. K. (2015). Mindfulness and couple relationships. In E. Shohin, W. van Gordon, & M. D. Griffiths (Eds.), *Mindfulness and Buddhist-derived approaches in mental health and addiction* (pp. 391–412). Heidelberg: Springer.

Pepping, C. A., Lyons, A., Halford, W. K., Cronin, T. J., & Pachankis, J. E. (2017). Couple interventions for same sex couples: A consumer survey. *Couple and Family Psychology: Research and Practice, 6*, 258–273. https://doi.org/10.1037/cfp0000092.

Petersen, J. L., & Hyde, J. S. (2010). A meta-analytic review of research on gender differences in sexuality, 1993 - 2007. *Psychological Bulletin, 136*(1), 21–38. https://doi.org/10.1037/a0017504.

Phalet, K., & Schönpflug, U. (2001). Intergenerational transmission of collectivism and achievement values in two acculturation contexts: The case of Turkish families in Germany and Turkish and Moroccan families in the Netherlands. *Journal of Cross-Cultural Psychology, 32*, 186–201. https://doi.org/10.1177/0022022101032002006.

Pinquart, M., & Teubert, D. (2010). A meta-analytic study of couple interventions during the transition to parenthood. *Family Relations, 59*(3), 221–231. https://doi.org/10.1111/j.1741-3729.2010.00597.x.

Powers, M. B., Vedel, E., & Emmelkamp, P. M. G. (2008). Behavioral couples therapy (BCT) for alcohol and drug use disorders: A meta-analysis. *Clinical Psychology Review, 28*(6), 952–962. https://doi.org/10.1016/j.cpr.2008.02.002.

Richerson, P. J., & Boyd, R. (2005). *Not by genes alone: How culture transformed human evolution.* Chicago, IL: University of Chicago Press.

Sanri, C., & Goodwin, R. (2014). The influence of globalization and technological development on intimate relationships. In C. R. Agnew (Ed.), *Social influences on romantic relationships* (pp. 11–32). Cambridge: Cambridge University Press.

Sassler, S. (2010). Partnering across the life course: Sex, relationships, and mate selection. *Journal of Marriage and Family, 72*(3), 557–575. https://doi.org/10.1111/j.1741-3737.2010.00718.x.

Sayer, L. C., & Bianchi, S. M. (2000). Women's economic independence and the probability of divorce: A review and re-examination. *Journal of Family Issues*, *21*(7), 906–943. https://doi.org/10.1177/019251300021007005.

Schindler, L., & Vollmer, M. (1984). Cognitive perspectives in behavioral marital therapy: Some proposals for bridging theory, research, and practice. In K. Hahlweg & N. S. Jacobson (Eds.), *Marital interaction: Analysis and modification* (pp. 309–324). New York: Guilford Press.

Schulz, M. S., Cowan, C. P., & Cowan, P. A. (2006). Promoting healthy beginnings: A randomized controlled trial of a preventive intervention to preserve marital quality during the transition to parenthood. *Journal of Consulting and Clinical Psychology*, *74*(1), 20–31. https://doi.org/10.1037/0022-006X.74.1.20.

Shadish, W. R., & Baldwin, S. A. (2003). Meta-analysis of MFT interventions. *Journal of Marital and Family Therapy*, *29*(4), 547–570. https://doi.org/10.1111/j.1752-0606.2003.tb01694.x.

Shi, L., & Wang, L. (2009). A multilevel contextual model for couples from Mainland China. In M. Rastogi & V. Thomas (Eds.), *Multicultural couple therapy* (pp. 297–316). Thousand Oaks, CA: Sage. https://doi.org/10.4135/9781452275000.n15.

Skellern, S. K., Sanri, C., Iqbal, S., Ayub, N., Jarukasemthawee, S., Pitsitunkaragan, K., … & Halford, W. K. (2019) Assessment of the perceived importance of religion in couple relationships in Christians, Muslims and Buddhist and the non-Religious. (Manuscript submitted for publication).

Snyder, D. K., Castellani, A. M., & Whisman, M. (2006). Current status and future directions in couple therapy. *Annual Review of Psychology*, *57*, 317–344. https://doi.org/10.1146/annurev.psych.56.091103.070154.

Snyder, D. K., & Halford, W. K. (2012). Evidence based couple therapy: Current status and future directions. *Journal of Family Therapy*, *34*(3), 229–249. https://doi.org/10.1111/j.1467-6427.2012.00599.x.

Snyder, D. K., & Whisman, M. A. (2003). *Treating difficult couples: Helping clients with coexisting mental and relationship disorders*. New York: Guilford Press.

Snyder, D. K., Wills, R. M., & Grady-Fletcher, A. (1991). Long-term effectiveness of behavioral versus insight-oriented marital therapy: A 4-year follow-up study. *Journal of Consulting and Clinical Psychology*, *59*(1), 138–141. https://doi.org/10.1037/0022-006X.59.1.138.

Spanier, G. B. (1976). Measuring dyadic adjustment: New scales for assessing the quality of marriage and similar dyads. *Journal of Marriage and the Family*, *38*(1), 15–28. https://doi.org/10.2307/350547.

Statistics Canada. (2012). *Mixed unions in Canada*. Retrieved from https://www12.statcan.gc.ca/nhs-enm/2011/as-sa/99-010-x/99-010-x2011003_3-eng.cfm.

Su, L. P., Miller, R. B., Canlas, J. M., Tsui-Chan, L., Hsiao, Y.-L., & Willoughby, B. J. (2015). A cross-cultural study of perceived marital problems in Taiwan and the United States. *Contemporary Family Therapy*, *37*, 165–175. https://doi.org/10.1007/s10591-015-9337-7.

Tang, C. S. K., Chua, Z., & Jiaqing, O. (2010). A gender perspective on Chinese social relationships and behavior. In M. H. Bond (Ed.), *The Oxford handbook of Chinese psychology: Oxford library of psychology*. (pp. 533–553). New York, NY: Oxford University Press. https://doi.org/10.1093/oxfordhb/9780199541850.013.0032.

Tharp, R. G., Meadow, A., Lennhoff, S. G., & Satterfield, D. (1968). Changes in marriage roles accompanying the acculturation of the Mexican-American wife. *Journal of Marriage and the Family*, *30*, 404–412. https://doi.org/10.2307/349908.

Tooby, J., & Cosmides, L. (1992). The psychological foundations of culture. In J. H. Barlow, L. Cosmides, & J. Tooby (Eds.), *The adapted mind: Evolutionary psychology and the generation of cultures* (pp. 19–136). New York: Oxford University Press.

Triandis, H. C., Kashima, Y., Shimada, E., & Villareal, M. (1986). Acculturation indices as a means of confirming cultural differences. *International Journal of Psychology*, *21*, 43–70. https://doi.org/10.1080/00207598608247575.

Vangelisti, A. L., & Daly, J. A. (1997). Gender differences in standards for romantic relationships. *Personal Relationships*, *4*, 203–219. https://doi.org/10.1111/j.1475-6811.1997.tb00140.x.

Verkuyten, M. (2016). The integration paradox empiric evidence from the Netherlands. *American Behavioral Scientist*, *60*, 583–596. https://doi.org/10.1177/0002764216632838.

Waite, L. J., & Gallagher, M. (2000). *The case for marriage: Why married people are happier, healthier, and better off financially*. New York: Doubleday.

Weiss, R. L. (1984). Cognitive and behavioral measures of marital interaction. In K. Hahlweg & N. S. Jacobson (Eds.), *Marital interaction: Analysis and modification* (pp. 337–355). New York: Guilford Press.

Weiss, R. L., Hops, H., & Patterson, G. R. (1973). A framework for conceptualizing marital conflict: A technology for offering it; some data for evaluating it. In L. A. Hamerlynck, L. C. Handy, & E. J. Mash (Eds.), *Behavior change: Methodology, concepts and practice* (pp. 309–342). Champaign, IL: Research Press.

Wesley, K. C. C. (2016). The use of behavioural couple therapy and couple based interventions in the treatment of substance abuse disorders. *Addiction Research and Theory*, *24*(2), 88–92. https://doi.org/10.3109/16066359.2015.1022160.

Whisman, M. A. (2001). Marital adjustment and outcome following treatments for depression. *Journal of Consulting and Clinical Psychology*, *69*(1), 125–129. https://doi.org/10.1037/0022-006X.69.1.125.

Whisman, M. A., Johnson, D. P., Be, D., & Li, A. (2012). Couple-based interventions for depression. *Couple and Family Psychology: Research and Practice*, *1*(3), 185–198. https://doi.org/10.1037/a0029960.

Williams, L., & Jurich, J. (1995). Predicting marital success after five years: Assessing the predictive validity of FOCCUS. *Journal of Marital and Family Therapy*, *21*, 141–153. https://doi.org/10.1111/j.1752-0606.1995.tb00149.x.

Wills, T. A., Weiss, R. L., & Patterson, G. R. (1974). A behavioral analysis of the determinants of marital satisfaction. *Journal of Consulting and Clinical Psychology*, *42*(6), 802–811. https://doi.org/10.1037/h0037524.

Wilson, K. L., Charker, J., Lizzio, A., Halford, W. K., & Kimlin, S. (2005). Assessing how much couples work at their relationship: The behavioral self-regulation for effective relationships scale. *Journal of Family Psychology*, *19*, 385–393. https://doi.org/10.1037/0893-3200.19.3.385.

Wong, S., & Goodwin, R. (2009). Experiencing marital satisfaction across three cultures: A qualitative study. *Journal of Social and Personal Relationships*, *26*, 1011–1028. https://doi.org/10.1177/0265407509347938.

Wood, R. G., Moore, Q., Clarkwest, A., & Killewald, A. (2014). The long-term effects of building strong families: A program for unmarried parents. *Journal of Marriage and the Family*, *76*, 446–463. https://doi.org/10.1111/jomf.12094.

Woodin, E. M. (2011). A two-dimensional approach to relationship conflict: Meta-analytic findings. *Journal of Family Psychology*, *25*(3), 325. https://doi.org/10.1037/a0023791.

World Values Survey. (2016). *Live culture map (1981–2015)*. Retrieved from http://www.worldvaluessurvey.org/WVSContents.jsp.

Yeganeh, T., & Shaikhmahmoodi, H. (2013). Role of religious orientation in predicting marital adjustment and psychological well-being. *Sociology Mind*, *3*, 131–136. https://doi.org/10.4236/sm.2013.32020.

CHAPTER 18

Parenting

Jennifer E. Lansford[a] and Marc H. Bornstein[b,c,d]
[a]Sanford School of Public Policy, Duke University, Durham, NC, United States
[b]*Eunice Kennedy Shriver* National Institute of Child Health and Human Development, Bethesda, MD, United States
[c]UNICEF, New York, NY, United States
[d]Institute for Fiscal Studies, London, United Kingdom

This chapter provides a review of parenting in diverse cultural contexts. We begin by describing the key responsibilities of parents, how parenting differs across cultural contexts, and how culture may affect links between particular forms of parenting and child outcomes. We next describe how the international agenda reflected in the United Nations Convention on the Rights of the Child and the Sustainable Development Goals, in addition to national policies and laws related to parenting, set the stage for parent-child relationships. We then describe parenting programs that aim to change parenting as a way of improving child outcomes. We draw on standards for parenting programs developed by UNICEF (2017) and illustrate how those standards work in practice by providing examples from specific programs in a number of low-, middle-, and high-income countries. We also consider how to adapt parenting programs for particular cultural contexts and conclude with a series of recommendations.

Key responsibilities of parents

Parents in all cultures are tasked with responsibilities for providing for their children's physical health and safety, cognitive stimulation, affection, and behavioral socialization so that children develop the characteristics needed to be well-adjusted members of their cultural group (Bornstein, 2015). These parenting responsibilities may fall to biological parents, stepparents, adoptive parents, other family members such as siblings and grandparents, or nonfamily caregivers. Cultural groups differ in who is most likely to provide different types of care. We focus in this chapter on parents but recognize the importance of nonparental caregivers as well.

Cross-Cultural Family Research and Practice
https://doi.org/10.1016/B978-0-12-815493-9.00018-1

Physical health and safety

Infants depend on their parents for their very survival, and parents are responsible for promoting physical health and safety throughout childhood and adolescence. Feeding and nutrition are important aspects of promoting physical health. The World Health Organization (2018) recommends exclusive breastfeeding for 6 months and supplementing breastfeeding with solid foods for 2 years to optimize nutrition, immune function (as antibodies are passed from mothers to infants through breastmilk), and mother-child bonding. In most countries, breastfeeding duration falls short of these goals (World Health Organization, 2018). Beyond breastfeeding, feeding and mealtimes remain central parts of parent-child interactions. In samples from the United States, children and adolescents who eat at least three meals per week with their parents are more likely to be of a healthy weight and less likely to engage in disordered eating (Hammons & Fiese, 2011). The benefits of family meals extend beyond nutrition, as eating together as a family on a regular basis is related to fewer behavior problems, less substance use, less depression, and better school performance in children, even controlling for other factors (Child Trends, 2013).

Parents are also important role models with respect to other aspects of physical health and safety. For example, the probability of adolescents' eating a low-fat diet is five times higher if their mothers also eat a low-fat diet (Rossow & Rise, 1994), a relation that does not weaken as adolescents age. Likewise, children whose parents smoke are themselves more likely to smoke (Green, Macintyre, West, & Ecob, 1991). In part links between parents' and their children's health-related behaviors are found because, for example, parents often are responsible for preparing meals for their children and shopping for groceries the entire family consumes, and through observing their parents' behaviors, children also develop beliefs about what are acceptable and expected behaviors for themselves.

Cognitive stimulation

Cognitive stimulation is important for children's brain development, language acquisition, school readiness, and eventual academic achievement. Examples of parenting behaviors that involve cognitive stimulation include language interaction (e.g., singing, talking, reading), provision of learning materials and exposure to learning opportunities (e.g., books, magazines), and physical interaction (e.g., sports, playing games). Even before birth, fetuses already can distinguish the sound of their own mother's voice from the voices of other women (Kisilevsky et al., 2003), and parents are an important source of linguistic input throughout childhood. Infants' vocalizations are highly contingent on their mothers' vocalizations to them (and, reciprocally, mothers' vocalizations are contingent on their infants' vocalizations to them), a finding that has been replicated in 11 countries (Bornstein, Putnick, Cote, Haynes, & Suwalsky, 2015). Children's vocabulary is correlated with the amount of language to which they have been exposed (Hurtado, Marchman, & Fernald, 2008) as well as to the quality of the linguistic input, including nonverbal cues that can help children understand the meaning of new words (Cartmill et al., 2013). The most effective way for children to acquire language is through dyadic, contingent exchanges with their parents, rather than through hearing language passively as adults speak to each other or other children or via media (Topping, Dekhinet, & Zeedyk, 2013). In one study, mothers were asked to teach their 2-year-old child two new words, one at a time. During one of the two teaching periods, a cell phone call placed by the experimenter disrupted the mother-child interaction. Children successfully learned the new word during the teaching period that was not interrupted but did not learn the new word during the teaching period that was interrupted by the cell phone call, even though parents said the word the same number of times in both conditions (Reed, Hirsh-Pasek, & Golinkoff, 2017). These findings illustrate the central role of parent-child interactions in language acquisition. Children with more advanced vocabularies and language skills are better prepared for school and have an advantage as they progress through education systems (Duncan et al., 2007).

Cognitive stimulation can come from parents' and children's everyday conversations as well as through more scripted exchanges, such as when parents read to children, tell them stories, or sing songs (Bornstein, 2006). In a study of different forms of cognitive stimulation provided by parents in more than 127,000 families with children younger than 5 years in 28 countries, the range was large (Bornstein & Putnick, 2012). For example, across all countries, 25% of caregivers read books to their child in the last 3 days, but caregivers in countries high on the Human Development Index (which encompasses average life expectancy at birth, gross national income per capita, and mean number of years of education completed; United Nations Development Programme, 2018) were at least twice as likely to have read to their children than the average effect across all countries. To illustrate the range in a study focused specifically on infants, almost no caregivers in Burkina Faso read to their infants in the last 3 days, whereas more than half of caregivers in Trinidad and Tobago had read to their infants in the last

3 days (Bornstein, Putnick, Lansford, Deater-Deckard, & Bradley, 2015). Of course, access to books depends on having material resources that are more likely present in high- than low-income countries, but differences in cognitive aspects of caregiving were found even using indicators that did not rely on material resources. For example, 35% of caregivers across all countries had told their child stories in the last 3 days, but caregivers in countries low on the Human Development Index were less than half as likely to have told their children stories as the average effect across all countries (Bornstein & Putnick, 2012). Large differences across countries were also found for naming, counting, and drawing; playing; singing; and taking the infant or child outside.

One of the most important ways that parents of older children and adolescents provide cognitive stimulation is through their involvement in their child's education. At the most basic level, parents' involvement in education includes whether they enable their children to attend school, which is a particular challenge in low-income countries when financial and logistical burdens such as prohibitive distances to schools are barriers to attendance (Lelei & Weidman, 2012). Parental involvement in education can occur in home-based activities (e.g., helping with homework, advising on which courses to take) or at school (e.g., volunteering in the classroom, talking with teachers; Park & Holloway, 2013). Children and adolescents whose parents are more involved in their education generally perform better in school and have higher educational aspirations (Hill et al., 2004).

Provision of affection

Affection, support, and responsiveness, with a foundation in early bonding, are expressed through social and emotional relationships, building trust and attachment, and behavioral interactions such as hugging, holding, and loving physical contact (Rohner & Smith, 2019). Responsive parenting includes prompt response to a child's behavior that is appropriate to the child's needs and developmental phase. Structure is associated with discipline, supervision, and protection of the child from harm, abuse, and neglect. These parenting interactions are expressed through positive disciplinary practices and a safe, secure, and consistent environment, all of which contribute to children's perception that they are loved and accepted.

According to Interpersonal Acceptance-Rejection Theory (Rohner & Lansford, 2017), children universally need to feel loved and accepted by their parents. Parents can express acceptance in physical, verbal, or symbolic forms, such as through hugging, offering verbal praise, or engaging in a culturally specific ritual such as giving a gift on a birthday (Rohner & Smith, 2019). Children interpret parents' messages about acceptance or rejection within cultural frames of reference that help shape the meaning that specific parental behaviors will convey to children.

If children do not perceive themselves as being loved and accepted, their well-being suffers. In metaanalyses testing the major postulates of Interpersonal Acceptance-Rejection Theory across 31 countries, maternal and paternal acceptance predicted better psychological adjustment for boys and girls, and men's and women's recollections of their parents' acceptance when they were children were related to better adjustment in adulthood (e.g., Khaleque, 2013; Khaleque & Ali, 2017). At the other end of the spectrum, these metaanalyses also demonstrate that feeling rejected by parents is universally related to children's maladaptation.

Behavioral socialization

Socialization relates to parenting that promotes the development of values, attitudes toward life, and identity; it is often an expression of cultural, social, and religious morals and expectations (Grusec & Davidov, 2010). Grusec and Davidov (2010) described a model of parents' socialization of children that focuses on five domains. First, in the protection domain, parents comfort children when they are distressed, so children learn that they will be kept safe and that they can eventually cope with distress on their own. Second, in the reciprocity domain, an egalitarian parent-child relationship is established in which each responds to the reasonable requests of the other. Third, in the control domain, the parent-child relationship is hierarchical, with parents correcting children's misbehaviors through rewards, punishments, and reasoning so that ultimately children will be able to control their own behavior. Fourth, in the guided learning domain, parents teach their children at a level appropriate to particular developmental stages so that children eventually internalize parents' values and ways of thinking. Finally, in the group participation domain, children learn what is acceptable behavior for a social group by watching others and engaging in everyday activities with the group so that ultimately they will be well socialized to be part of the group. Parent-child relationships function more smoothly when they operate rarely in the control domain and instead rely on the other domains for promoting reciprocal and mutually beneficial exchanges that induce children to accept their parents' socialization attempts willingly.

Theories of parenting have distinguished between parenting styles (the overall emotional climate of the parent-child relationship) and parenting practices (the specific behaviors in which parents engage; for a review see Rudolph, Lansford, & Rodkin, 2017). Children are more likely to be receptive to their parents' socialization attempts using particular parenting practices if the overall parenting style is characterized by love and warmth (Darling & Steinberg, 1993). In the context of loving and warm relationships, children are more likely to regard their parents as having legitimate authority to try to direct their behavior, more willing to comply with their parents' requests, and more likely to internalize the messages parents are trying to convey (Darling, Cumsille, & Peña-Alampay, 2005). Developmentally, in diverse countries, including Chile, the Philippines, and the United States, older adolescents are less likely than younger adolescents to believe that parents have legitimate authority over their lives and less likely to believe that they must obey parents if they disagree with them.

Culture, parenting, and child development

In ecological theories of development (e.g., Bronfenbrenner & Morris, 2006), parent-child relationships are situated not only in broader family systems but also in cultural contexts that encompass beliefs and norms that set the stage for expectations of parenting. Culture is reproduced over time through parent-child interactions, and the ways in which parents and children come to understand how to behave toward one another are highly dependent on how they observe others in their communities behaving (Rogoff, 2003). Culture can affect parents' behaviors as well as the meaning those behaviors convey, which can lead to differences between cultural groups in the beliefs parents hold and behaviors they enact (Bornstein, 1995). Culture-level factors, such as the normativeness of particular parenting behaviors, also can affect links between parenting behaviors and child outcomes.

Form vs function

Specific behaviors in which parents engage have been described as the "form" of parenting, and the underlying meaning of the specific behaviors has been described as their "function." A particular form of parenting may have the same or different functions across different cultural contexts. Forms and functions can be conceptualized as a two (forms) by two (functions) matrix with four possible cells (Bornstein, 1995). First, if the form and function are the same across cultural contexts, this describes a universal aspect of parenting. For example, mothers have been found to respond contingently to infants' vocalizations in consistent ways in a number of cultural groups, including Argentina, Belgium, Brazil, Cameroon, France, Israel, Italy, Japan, Kenya, South Korea, and the United States; mothers' contingent vocalizations increase the likelihood that infants will in turn continue to vocalize (Bornstein, Putnick, Cote, et al., 2015).

Second, if the form and function are different across cultural contexts, this describes cultural specificity. For example, mother-infant interactions in Japan are more characterized by mothers orienting infants to themselves, whereas mother-infant interactions in the United States are more characterized by mothers orienting infants to objects in the environment (Bornstein, Cote, Haynes, Suwalsky, & Bakeman, 2012). The former type of mother-infant interaction functions to promote closeness and dependency between the mother and infant, whereas the latter form of mother-infant interaction functions to promote the infant's autonomy, individuation, and exploration (Rothbaum, Weisz, Pott, Miyake, & Morelli, 2000).

Third, if the form is the same across cultural contexts, but the function is different, this circumstance describes plasticity. For example, direct eye contact between parents and children (the form of parenting) may serve different functions in different cultural groups. In European American families, direct eye contact serves to establish that the child and parent are attending to one another so they can have an undistracted conversational exchange (Trevarthen, 1998), whereas in some Native American families, direct eye contact is a sign of disrespect and aggression, so parents would not encourage children to establish eye contact before communicating (Mitchum, 1989).

Fourth, if the form is different across cultural contexts, but the function is the same, this situation describes context specificity. For example, in European American families, physical affection and saying, "I love you" serve the function of conveying parents' love to their children (Cheah, Li, Zhou, Yamamoto, & Leung, 2015). In contrast, in Chinese American immigrant families, parents are less likely to use physical affection and verbal expressions of love but are instead more likely to convey love to their children by providing guidance and educational opportunities (Cheah et al., 2015). An important aspect of distinguishing the form of parenting from its function is that the meaning conveyed by a particular behavior is often more important than the behavior itself. Thus, more important than *how* parents demonstrate warmth, love, and affection or rejection, hostility, or neglect is whether children perceive themselves as being loved or rejected.

Mean-level differences in parenting

Many studies related to a large number of parenting domains have demonstrated mean-level differences in different types of parenting across cultural groups. For example, in terms of physical caregiving, there are wide cultural differences in the amounts of physical contact infants have with their parents. During the first year of life, hunter-gatherer !Kung infants spend 90% of their time in skin-to-skin contact with their mother or another caregiver, being carried in slings during waking hours and co-sleeping at night. This circumstance stands in stark contrast to infants in many Western cultures, who are more likely to be placed in devices such as strollers and infant seats, that are not attached to parents, and who sleep not only in separate beds but also often in separate rooms (Diamond, 2012). In terms of socioemotional interactions, cultural differences characterize both the likelihood that parents will respond to infant crying and the length of time it takes parents to respond. For example, although practices may have changed in more recent times, early studies showed that German parents ignored infant crying one-third of the time and often waited for 10–30 minutes when they did respond (Diamond, 2012). In contrast, Efe Pygmy mothers and other caregivers respond to virtually all instances of infant crying within 10 seconds (Diamond, 2012). In terms of play, parents in different cultural groups vary in both their beliefs about play and their engagement in play with their children. Parents in technologically advanced societies are more likely to endorse and engage in play as a way of promoting their children's cognitive and social development, and parents in more agrarian or hunter-gatherer societies are more likely to regard play as incidental to child development (Roopnarine, 2011). As children develop and spend less time in direct contact with parents, forms of distal supervision and monitoring become more salient. The mean level of supervision differs across cultural groups, with different expectations about how much control parents will have over their adolescents' activities and decisions (Qin, Pomerantz, & Wang, 2009).

Mean-level differences in particular kinds of parenting are related to mean-level differences in corresponding child behaviors. For example, there have been naturalistic observations of dyads of mothers and 5-month-old infants in Argentina, Cameroon, Italy, Kenya, South Korea, and the United States. The mothers in these different countries differed widely in the opportunities they provided for their infants to sit independently, to sit with the support of adults or furniture, and to sit on different surfaces (Karasik, Tamis-LeMonda, Adolph, & Bornstein, 2015). Infants in Cameroon and Kenya experienced the most opportunities to sit independently and were the most proficient sitters.

Understanding parents' beliefs underlying different behaviors offers insights into why mean-level differences in the behaviors exist across cultures. For example, in a comparison of mother–infant naturalistic interactions in Argentina, Belgium, Israel, Italy, and the United States, mothers differed in the emphasis they placed on different forms of caregiving, accounting for mean-level cultural differences in nurturing, physical, social, didactic, material, and language domains (Bornstein, Putnick, Park, Suwalsky, & Haynes, 2017). Differences in maternal behavior were related to infants' behaviors in corresponding domains, such that infants whose mothers encouraged physical development were more physically developed, whereas infants whose mothers emphasized social engagement paid more attention to their mothers, for instance.

Differences in physical environments also can account for mean-level differences in parents' behaviors. For example, in cultural contexts with high infant mortality rates, caregivers prioritize infants' physical health and safety in an effort to promote survival, often by keeping infants in direct physical contact (e.g., with infants strapped to mothers' bodies as they engage in daily tasks and responding quickly to signals of infants' distress; LeVine et al., 1996). By contrast, in cultural contexts with low infant mortality rates, caregivers focus less on infants' survival and focus more on other aspects of development, such as cognitive stimulation (LeVine et al., 1996). Other aspects of physical environments also influence parenting. For example, in Nigeria, there are distinct rainy and dry seasons that result in fluctuations in the availability of food, which lead Yoruba parents to socialize their children in particular ways related to food (Babatunde & Setiloane, 2014). Parents teach children not to visit other families at mealtimes, to wait patiently for food, and to eat rare and valuable meat and fish at the end of meals, as ways to socialize children to be thrifty, to be able to delay gratification, and to show proper etiquette. In these ways, mean-level differences in parenting can result from differences in belief systems as well as differences in environmental features of different cultural groups.

Culture as a moderator of links between parenting and child outcomes

Just as when a given form of parenting has different functions in different cultural contexts, culture may moderate the association between a given form of parenting and child outcomes. One aspect of culture that has been investigated as a potential moderator is cultural normativeness, or how accepted and frequent a particular belief or behavior is within a cultural context. The idea of normativeness as a moderator was initially proposed as a way to understand links between corporal punishment and child outcomes (Deater-Deckard & Dodge, 1997). In a study of mother-child dyads in six countries (China, India, Italy, Kenya, the Philippines, and Thailand), cultural normativeness (defined both in terms of how frequently mothers reported using corporal punishment and the extent to which mothers believed other parents in their community used corporal punishment) moderated the link between mothers' own use of corporal punishment and their children's aggression and anxiety, as reported by mothers and children (Lansford et al., 2005). In all six countries, frequent corporal punishment was related to more child aggression and anxiety. However, the association between corporal punishment and child aggression and anxiety was weaker in countries in which corporal punishment was more normative, with the weakest link in Kenya (where corporal punishment was most normative) and the strongest link in Thailand (where corporal punishment was least normative).

One reason that normativeness might moderate the link between a particular form of parenting and child outcomes is that the meaning conveyed to the child by a parent's behavior might be moderated by the cultural normativeness of the behavior. Specifically, a behavior that many parents in a given culture use is likely different from the meaning conveyed by the same behavior occurring in a cultural context where the behavior is relatively unusual. This hypothesis has received some empirical support. For example, children's perceptions of maternal hostility mediate the link between mothers' use of corporal punishment and child aggression and anxiety, with children in cultural contexts where corporal punishment is more normative less likely to perceive mothers who use corporal punishment as being hostile (Lansford et al., 2010). Nevertheless, these findings should not be interpreted as meaning that corporal punishment is not detrimental so long as it is culturally accepted. At an individual level, corporal punishment in a context in which it is culturally accepted may not convey to children that their parents are personally rejecting of them and may have a weaker link to that individual child's adjustment. However, at the group level cultures in which corporal punishment is accepted also have higher levels of endorsement of other forms of aggression and higher rates overall of violence (Lansford & Dodge, 2008). Corporal punishment is more likely to be used in "cultures of violence" in which spousal abuse, homicides, and other forms of violence also are more common (Ember & Ember, 1994). The cultural spillover theory has been used to explain how attitudes and acceptance of violence in one domain are likely to spill over into attitudes and acceptance of violence in other domains (Baron & Straus, 1989).

Extending normativeness theory beyond corporal punishment, the normativeness of some other forms of discipline also moderates links between those forms of discipline and child outcomes (Gershoff et al., 2010). Specifically, mothers' perceptions of the normativeness of expressing disappointment toward children for their misbehavior moderate the association between the frequency of mothers' expressing disappointment and their children's anxiety, with children reporting more anxiety symptoms if their mothers frequently express disappointment in cultural contexts in which expressions of disappointment are not normative. Similar results were found for children's perceptions of the normativeness of mothers' yelling. Despite the weaker links between expressing disappointment and yelling with child anxiety in contexts of greater normativeness, the means of child anxiety were higher in contexts where mothers frequently expressed disappointment and yelled, consistent with findings regarding corporal punishment in broader cultures of violence.

Discipline aside, cultural normativeness also moderates links between other forms of parenting and child outcomes. For example, in nine countries (China, Colombia, Italy, Jordan, Kenya, Philippines, Sweden, Thailand, and the United States), five aspects of parenting (expectations regarding family obligations, monitoring, psychological control, behavioral control, warmth/affection) when children were 10 years old were examined in relation to five aspects of youth adjustment at age 12 (social competence, prosocial behavior, academic achievement, externalizing behavior, internalizing behavior) (Lansford et al., 2018). In seven of the eight analyses in which cultural normativeness moderated the association between parenting and youth adjustment, a particular parenting behavior was more likely to be related to better youth adjustment if the parenting behavior was more normative. To illustrate, mothers' expectations regarding children's family obligations were more strongly related to children's social competence in cultures in which family obligation expectations were more normative. The findings suggested that youth adjustment was better when their parents' beliefs and behaviors were aligned with the beliefs and behaviors of other parents in the cultural group.

Extended family and community support in parenting

Historically, mothers have received more research attention than fathers, and mothers have more often been characterized as primary caregivers and fathers have been characterized as playmates (Lamb, 2010). However, in many countries, gender roles have become more equitable over time, and research now suggests that fathers play many important roles in child development (Kuhns & Cabrera, 2018). Mother-child and father-child relationships are, of course, situated in extended families and community contexts that offer both supports and challenges. In some cultural contexts, older siblings provide much of the day-to-day care (Kramer & Hamilton, 2019), and in others, grandmothers, aunts, and other family members often share caregiving responsibilities, or even take primary responsibility for caregiving (Keller & Chaudhary, 2017). In many cultural groups, families live in extended family households in which parenting functions are performed not only between parents and their biological children but also by many family members (Sear, 2015). Extended family support is even more crucial when parents migrate either to urban areas within a country or to other countries to work, leaving their children behind in the care of other relatives, a phenomenon that has become common in many parts of the world (e.g., Peng & Wong, 2015). Recognizing the importance of nonparental caregiving in many cultural contexts, scholars have argued for the need to expand an understanding of attachment beyond parent-child dyads to encompass children's attachments within a network of relationships surrounding the child (Keller & Chaudhary, 2017).

Expectations regarding family obligations, for both parents and children, vary within and between cultures. In a comparison of different ethnic groups within the United States, European American adolescents believed they were less obligated to help and respect extended family members than Asian American or Latin American adolescents (Fuligni, Tseng, & Lam, 1999). More authoritarian attitudes as well as warmer, more controlling, and less neglectful behaviors are related to mothers' and fathers' holding more expectations for children's family obligations in China, Colombia, Italy, Jordan, Kenya, the Philippines, Sweden, Thailand, and the United States (Lansford et al., 2016).

Policy contexts of parenting

Resources available through government agencies, such as Ministries of Health and Education, often support parenting. The international development agenda and national policies also shape parenting. Indeed, policies are a key way in which macro-level contexts can affect parenting. Cultural beliefs about the acceptability and advisability of particular behaviors can be concretized in the form of policies.

International agenda for parenting

Developmental science and cultural psychology recognize the importance of culture in shaping parenting behaviors, but a human rights perspective suggests that taking an extreme position on cultural relativism is not always in children's best interests. For example, child marriage and female genital mutilation are still common practices in many parts of the world, endorsed by parents for economic, religious, and cultural reasons. The international community increasingly has recognized that child marriage and female genital mutilation are harmful to children's well-being and should not be tolerated regardless of their cultural normativeness (e.g., Coleman, 1998; Sundaram, Travers, & Branson, 2018). The United Nations Convention on the Rights of the Child was a landmark step in the international agendum related to parenting and child development, and the Sustainable Development Goals are the current organizing framework for international efforts that have many implications for parenting.

Convention on the Rights of the Child

The Convention on the Rights of the Child (CRC) is the most widely ratified convention in the history of the United Nations, having been ratified by all countries except the United States. The CRC outlines the international human rights agenda for children, including children's right to protection from abuse and exploitation and right to participate in decisions affecting their lives, among others (United Nations, 1989). Upon ratifying the CRC, countries are subject to periodic reviews by international bodies to evaluate how well countries are faring with respect to protecting children's rights as outlined in the CRC and to recommend areas in need of improvement. The CRC can be credited with much of the international progress in changing laws to protect children and ensure that national systems are in place to provide for children's best interests (Seymour, 2009). Parts of the CRC have been understood as applying specifically to parenting. For example, not only are countries tasked with ensuring children's right to a childhood and right to protection from abuse and exploitation, but, by extension as children's primary caregivers, so are parents.

Sustainable Development Goals

The Sustainable Development Goals adopted by the United Nations General Assembly in 2015 are meant to guide the international agenda through 2030 (United Nations, 2015). Seventeen goals cover a wide range of issues, such as poverty, climate action, gender equality, and health and well-being.

To facilitate monitoring and evaluation of progress, each goal is operationalized into a series of specific targets with criteria to determine whether each target is being met. For example, Goal 16 involves "peace, justice, and strong institutions." Target 16.2 is to: "End abuse, exploitation, trafficking, and all forms of violence against and torture of children" (Global Initiative to End All Corporal Punishment of Children, 2018). The criterion for determining progress is operationalized as the percentage of children aged 1–17 who experienced any physical punishment and/or psychological aggression by caregivers in the past month (Danish Institute for Human Rights, 2018).

National laws related to parenting

Many different types of laws have the potential to influence parent-child relationships. These can range from laws that have implications for families' financial status, as poverty is a major influence on parenting (e.g., Morris et al., 2017), to laws that directly prohibit particular types of parenting behaviors. There are also laws that have implications for how much time parents can spend with children (e.g., child custody laws in the case of parental divorce, as described elsewhere in this volume). Here, we focus on three types of laws: (1) laws related to parental leave policies as an example of how laws that affect the time parents spend with children can affect parent-child relationships and child well-being; (2) legal bans on corporal punishment as an example of how laws have been used to prohibit particular parenting behaviors; and (3) laws related to education as an example of how national laws can affect the development of children and adolescents at the intersection of contexts within and outside the family.

Parental leave policies set the stage for mother-child and father-child relationships

Laws are a public instantiation of a country's stance on the importance of engaging in or avoiding particular behaviors (Linos, 2013). In the case of parental leave policies, countries differ widely in whether parents are legally entitled to time away from work following the birth or adoption of a child and, if so, whether the time is paid or unpaid and whether only mothers or both mothers and fathers are entitled to parental leave (Bartel, Rossin-Slater, Ruhm, Stearns, & Waldfogel, 2018). The Nordic countries (Denmark, Finland, Iceland, Norway, and Sweden) enjoy the most generous parental leave policies (Eydal & Rostgaard, 2016). In Sweden, for example, every couple is guaranteed 480 days of paid parental leave, 90 days of which must be used by the father or forfeited (International Labor Organization, 2014).

Fewer than 10 countries have no paid parental leave policy; the United States is the only high-income country without a national paid parental leave policy. In the United States, parental leave laws are applied only to large organizations rather than to all employers, and even large organizations guarantee unpaid time away from work only for a limited period (Bartel et al., 2018).

Whether leave policies can be used by fathers as well as mothers has important implications for gender roles within families and at a societal level (Eydal & Rostgaard, 2016). In some countries, the idea of mothers as primary caregivers continues to abide, and only mothers are eligible for parental leave. However, in other countries explicit attempts are made to engage fathers and mothers equally in caregiving, such as in social policies providing for paternal as well as maternal leave following the birth of a child (Shim, 2016). When fathers take parental leave and have time devoted to caring for their children, they are more likely to perceive themselves as active co-parents, and children relate to both parents more equitably (McHale & Sirotkin, 2019).

Parental leave policies are important both for parents and children. For example, paid parental leave policies increase the likelihood that mothers will breastfeed (Hamad, Modrek, & White, 2019), with all of the attendant benefits of breastfeeding for mothers and infants (World Health Organization, 2018). In a comparison of infant mortality rates from 1969 to 2010 across 19 Organization for Economic Cooperation and Development (OECD) countries, job-protected paid leave reduced infant mortality, whereas unpaid leave or leave that did not protect the parent's job did not reduce infant mortality (Shim, 2016). These effects were evident even after controlling for gross domestic product per capita, health expenditures, child health indicators, and social expenditure measures. More generous parental leave policies also have the potential to promote better outcomes later in childhood, given links between parents' employment characteristics and children's cognitive, socioemotional, and behavioral outcomes (Han, Hetzner, & Brooks-Gunn, 2019).

Corporal punishment outlawed in 59 countries

Corporal punishment has been outlawed in all settings, including by parents in their homes, in 59 countries as of February 2020 (Global Initiative to End All Corporal Punishment of Children, 2020). In 1979, Sweden became the first country to outlaw corporal punishment, followed by Finland (1983), Norway (1987), and Austria (1989). More recently, the number of countries passing bans has accelerated as part of a global movement to end corporal punishment, in large part in response to international monitoring related to protecting children from all forms of abuse as outlined in the Convention on the Rights of the Child.

In some countries, attitudes about corporal punishment became less favorable prior to the ban, facilitating the change in the law. For example, in Sweden, in the years leading up to the ban, national surveys demonstrated that parents' endorsement of the statement that corporal punishment is sometimes necessary fell from 53% in 1965 to 35% in 1971 (Durrant, 1996). Over this same time period parents increasingly agreed with the statement that children should be brought up without corporal punishment (from 35% to 60%; Durrant, 1996). Endorsement of corporal punishment continued to decline after the legal ban. In other countries, bans were passed to change attitudes and behaviors (sometimes in response to international pressure) rather than because attitudes and behaviors were changing already. For example, even 3 years after corporal punishment was outlawed in Togo, 77% of parents in a nationally representative sample reported that their children had experienced corporal punishment in the last month, and 37% reported that using corporal punishment is necessary to rear a child properly (Lansford et al., 2017).

Outlawing corporal punishment is not always sufficient to change parents' attitudes and behaviors. Unless the change in law is publicized, parents may not be aware that corporal punishment is no longer legal. For example, 1 year after the legal ban in Germany, 70% of German parents were not aware of the ban (Bussmann, 2004). In a study of nationally representative samples of 56,371 parents in 8 countries using data from UNICEF's Multiple Indicator Cluster Survey, parents' use of corporal punishment and belief that corporal punishment is necessary to rear a child properly were not systematically related to the legal status of corporal punishment (Lansford et al., 2017). For example, Togo outlawed corporal punishment in 2007, yet more parents reported believing it was necessary to use corporal punishment after than before the ban (37% in 2010 compared to 33% in 2006). However, in other countries, the percentage of parents who reported believing it is necessary to use corporal punishment declined over time (e.g., from 14% in 2005 to 10% in 2012 in Ukraine, which outlawed corporal punishment in 2004; from 57% in 2005–6 to 43% in 2009–10 in Sierra Leone, which has not outlawed corporal punishment as of 2020).

Beliefs and behaviors related to corporal punishment change most when legal bans are accompanied by publicity and educational campaigns (Bussmann, Erthal, & Schroth, 2011). Publicity sometimes involves public awareness campaigns to ensure that parents are aware of the change in the law (Durrant & Smith, 2011). Educational campaigns generally involve teaching parents about alternate forms of discipline. For example, the Positive Discipline in Everyday Parenting Program, which has been adapted for use in many countries, tries to reduce parents' use of punishment by increasing their awareness of different problem-solving strategies during parent-child conflicts, helping parents take a more child-centered approach to parenting, and building emotionally close parent-child relationships (Durrant et al., 2017).

Efforts to change beliefs and behaviors might have the most potential for success when they are adapted to address existing beliefs held by particular cultural groups. For example, conservative Christians use corporal punishment more frequently than other religious groups in the United States, likely because of a "spare the rod spoil the child" interpretation of biblical teachings (Mahoney & Boyatzis, 2019). At a Christian college, students were randomly assigned to an intervention that presented only scientific findings on the detrimental effects of corporal punishment, an intervention that presented scientific findings in addition to a biblical reinterpretation of the passages most often cited as endorsements of corporal punishment, or a control group. Attitudes changed the most from before to after the intervention in the biblical reinterpretation group (Perrin, Miller-Perrin, & Song, 2017). The findings suggest that educational efforts to change beliefs about corporal punishment and decrease its use should attend to why parents believe corporal punishment is necessary.

Laws related to education

Sustainable Development Goal 4 aims to achieve universal completion of primary and secondary education by 2030, but 1/6 of lower secondary school age adolescents, and 1/3 of upper secondary school age adolescents are not in school (United Nations Educational, Scientific and Cultural Organization (UNESCO) Institute of Statistics and Global Education Monitoring Report, 2016). Countries vary widely in their education systems, including compulsory ages of schooling, whether parents are expected to pay for schooling-related expenses, and supports and barriers that might affect parents' ability to foster their children's education and eventual transition to the work force (Putnick & Bornstein, 2015; Singh & Mukherjee, 2018). Large within-country variation is a function of families' socioeconomic status, urban vs rural residence, and other demographic or regional differences (Crivello & van der Gaag, 2016).

Some countries, such as China (Ministry of Education, 2018), have national laws that standardize the curriculum, textbooks, and pedagogy to reduce disparities in education across the country. In other countries, such as the United States (United States Department of Education, National Center for Education Statistics, 2018), the education system is highly localized in terms of content and structure. In the United States, schools depend on local property taxes for funding, meaning that parents have a larger role to play in selecting neighborhoods (and thereby schools), and students' educational experiences vary depending on their parents' involvement (Smith, Wohlstetter, Kuzin, & De Pedro, 2011). Nevertheless, even in countries with relatively standardized school systems, parents can still affect educational outcomes through providing extracurricular opportunities, such as cram schools, to prepare for national exams, and by conveying their values and motivations regarding education.

Parenting programs in diverse cultural contexts

Parenting programs are developed to improve parents' knowledge, attitudes, and behaviors, often with the ultimate goals of improving children's physical, cognitive, social, and emotional development (Lansford & Bornstein, 2007). Parenting programs should carefully consider what the desired outcomes are and tailor the content to match the desired outcomes. Desired outcomes often involve one of three categories: (1) improving child outcomes (e.g., physical health; cognitive, social, and emotional development), (2) changing parent outcomes (e.g., knowledge, attitudes, and practices related to caregiving), and (3) both child and parent outcomes. The content of parenting programs might differ if the goal is to change cognitive development vs socioemotional outcomes in children or to change parents' knowledge vs behavior. The desired outcome will also influence many other decisions about the program covered in the recommendations that follow regarding the mode of delivery, who should deliver the program, and so forth.

In general, programs offer more opportunities for behavior change with higher frequency, intensity, and duration (Maher, Marcynyszyn, Corwin, & Hodnett, 2011). Together, these three factors determine the "dose" of the program that families receive—the overall amount of the program that is delivered. Frequency refers to how often the program is delivered. Intensity is strength of an intervention or how much of the intervention is delivered within each session. Duration is the length of the program from start to finish.

Parenting programs generally are more effective when they are delivered face-to-face using active rather than passive learning (Centers for Disease Control and Prevention, 2009). Lectures, brochures, and videos might improve parents' knowledge without changing their parenting practices or behaviors. Delivery will be more effective if it is interactive and involves practice in role playing situations or in actual interactions with children rather than just reading pamphlets or listening to lectures. The most effective delivery modes involve parents and children together so leaders can model how to interact with children, problem solve in discipline situations, and provide other hands-on experiences in real-life situations. Strategies associated with improving both child and parent outcomes include training parents how to interact positively with their children on a daily basis in real-life everyday scenarios, modeling newly learned skills for parents with their own child, and giving parents the opportunity to role-play with a peer or trainer (Child Welfare Information Gateway, 2013). Demonstrating (e.g., how to position infants to decrease risk of sudden infant death syndrome) and offering opportunities to practice can help change parents' behaviors.

Programs are delivered in a variety of settings, most often in participants' homes, schools, health clinics, and community centers (Lansford & Bornstein, 2007). Home visiting has been found to be particularly effective, perhaps because these programs tend to be delivered to individual families rather than to groups, so messages can be targeted to meet the specific needs of individual families (Grantham-McGregor & Walker, 2015). Programs delivered in participants' homes also have the advantage of being able to identify safety hazards and resources in homes that can be pointed out to parents to enhance safety or make use of resources (Dodge et al., 2014). If programs are delivered in school, health care, or community settings instead of at participants' homes, providing transportation, refreshments, and child care for children not participating in the program are important for improving participation rates (Mytton, Ingram, Manns, & Thomas, 2014).

Intensive home visiting programs in which program providers work with individual families in their homes are often the most effective, but these may not be feasible because they are more expensive and resource-intensive than other modalities. If less intensive modalities are needed for logistical reasons, outcomes can be improved by using more than one modality. For example, if a program is primarily delivered in a group setting, offering individual counseling as a way to address individual parents' unique concerns and questions might make the program more effective. As an example of how different modes of delivery can be combined to improve outcomes, consider a program in Iran that aimed to improve oral health outcomes in young children (Mohebbi, Virtanen, Vahid-Golpayegani, & Vehkalahti, 2009). The strategy included either giving parents pamphlets with information on feeding habits, sugar intake, bacteria transmission, and oral hygiene ("pamphlet group"), and/or three bimonthly phone calls delivered during a 6-month period reminding parents of the oral health instructions delivered by clinic staff ("pamphlet + reminder"). The pamphlet + reminder group had fewer children with cavities than the control group, but the pamphlet-only group did not differ from the control group, meaning that handing out pamphlets alone was not sufficient to improve oral health outcomes.

Integrating parenting programs with existing programs and policies can maximize the success of the parenting program (Jeria, 2017). Parenting programs need a sustainable source of funding, which is more feasible when such programs are integrated into existing programs. In addition, programs that receive community support and that are multisectoral (drawing resources from the health sector, education sector, and others) will be more likely to gain support from policymakers.

Particularly if the goal is to change not just parents' knowledge but parents' behavior or children's outcomes, it is important to involve more than one family member. Often multiple family members care for children including mothers, fathers, grandparents, aunts, uncles, siblings, and others (Keller & Chaudhary, 2017). Changing the knowledge, attitudes, or even behaviors of one of these caregivers may not be enough to change the knowledge, attitudes, or behaviors of the other caregivers, or the development of children because children's developmental outcomes reflect treatment by all of their caregivers. Thus, it is important whenever possible for programs to treat families as systems that include multiple members rather than targeting just the mother (the most common approach in parenting programs).

UNICEF recommendations for parenting programs

UNICEF (2017) produced a set of recommendations for parenting programs through nine standards covering the content of parenting programs, target populations, contexts in which programs are administered, and integration of parenting programs in larger systems. The specific standards include: (1) support nurturing care to promote holistic child development; (2) build on a theory of change leading to desired results; (3) tailor content to the child's developmental stage; (4) serve vulnerable children and their families; (5) involve all parents and key caregivers; (6) adapt to culture and context and build on positive parenting practices; (7) integrate into existing delivery platforms; (8) engage trained workforce and service providers; and (9) improve through systematic monitoring and evaluation.

To illustrate how these standards can be implemented in practice, consider Standard 3, regarding tailoring content to the child's developmental stage. For some outcomes, the program should begin even before a child's birth. For example, programs designed to promote breastfeeding should ideally start with women while they are still pregnant and then continue to provide support after the birth of the child. For other types of outcomes, there may be different critical developmental periods during which interventions might be targeted (e.g., transition to school). The Mexican Oportunidades conditional cash transfer program illustrates how the timing of the program is important. In analyses examining the long-term effects of participation in Oportunidades at 5- and 10-year follow-up, dose-response analyses comparing early (families enrolled immediately into the program) and late (families enrolled 18 months later) enrollment favored early enrollment across physical, cognitive, social, and emotional developmental outcomes (Fernald, Gertler, & Neufeld, 2008, 2009). That is, beginning the program 18 months earlier in the child's development was related to more improvement across many domains of child development. Not only is implementation of the program earlier in the child's lifetime generally more effective than later implementation, it is also important to be sure the content of the program is appropriate given the child's developmental stage. For example, a program in South Africa employed a trained physiotherapist to deliver the program to low-income families with children younger than 2.5 years, infected with HIV (Potterton, Stewart, Cooper, & Becker, 2010). Over a 12-month period, the home stimulation curriculum was updated every 3 months to match with the child's developmental level. Compared to a control group, the mental development of the children in the program improved more over time.

Next, consider Standard 5 of involving all key caregivers. When it is possible to include fathers, programs should not frame fathers' role as just helping mothers. Instead, programs should make it clear that fathers' positive involvement directly benefits children. Findings from program evaluations demonstrate stronger effects on both child and parenting behaviors when fathers are involved in the program (Lundahl, Tollefson, Risser, & Lovejoy, 2008). Fathers and other family members can be incorporated in parenting programs in a variety of ways. For example, a community-based program targeted fathers through fathers' clubs in Haiti (Sloand, Astone, & Gebrian, 2010). The fathers' clubs consisted of regular meetings among fathers to discuss their involvement in childcare. Fathers also attended health education sessions with input from the nurse or village health agent where several practices (e.g., exclusive breastfeeding for 6 months, immunizing children, providing adequate fluids, and seeking appropriate health care when the child is sick) were discussed. Fathers' knowledge, skills, and performance in early childhood care were targeted. The presence of a fathers' club was associated with increased vaccination status, growth monitoring, and vitamin A supplementation. However, child mortality rates and child weight did not differ before and after the initiation of fathers' clubs in the villages.

Illustrative examples of parenting programs

At least 40 national parenting programs have been implemented in at least 33 countries (Lansford & Bornstein, 2007). These programs range from focusing primarily on physical caregiving to focusing on cognitive stimulation and socioemotional aspects of caregiving. Most parenting programs focus on young children. For example, in a mapping of parenting programs conducted by UNICEF (2014), 72% of programs focused on birth through the age of 3 years, 12% on ages 3–6, and 6% on children in the earliest grades of school through age 8, and only 10% of programs cut across these age ranges. Even fewer parenting programs focus on adolescents. Most of the parenting programs in the UNICEF review focused on child health outcomes, particularly in low-income countries. In middle- and upper-middle-income countries, psychosocial stimulation was more likely to be included, but health promotion still constituted a larger focus. To provide a sense of what approaches have been taken and what effects have been found in relation to parent and child outcomes, we review briefly examples of programs that have been implemented in Chile, India, Jamaica, Jordan, Nepal, Turkey, and the United States.

Crece Contigo in Chile

Crece Contigo began as a multisectoral program targeting pregnant women who were then followed until their child was 4 years old; the program has since expanded to follow children until the age of 9 years (Jeria, 2017). The program takes a holistic approach by providing prenatal care to mothers to improve children's health at birth and well-child checks to monitor children's physical health, by promoting skin-to-skin contact and breastfeeding both for their physical and psychosocial benefits, and by targeting early childhood education to promote cognitive growth. Evaluations of the program have revealed positive effects on child outcomes and that the effects are stronger when parents take more advantage of the benefits offered through the program and when the program has been active in the community for longer (Torres et al., 2018). One of the important features of Crece Contigo is that it is integrated with the Ministries of Health, Education, and Social Development, which embeds the program in existing national structures that make it sustainable in the long term. An important lesson from Crece Contigo is that it is not sufficient just to expand programming to reach more families; it is necessary to attend carefully to the quality of the services offered, because negative effects of poor quality programming have been found (Berlinski & Schady, 2015).

Integrated Child Development Services in India

A breastfeeding promotion program in India was intensive, with local village workers belonging to the Integrated Child Development Services scheme, the largest multisectoral maternal and child health and nutrition program in the world (Kapil, 2002). Auxiliary nurse midwives who directed immunization clinics and other healthcare providers visited mothers with low literacy levels monthly, for 1 year; the visits were supplemented with monthly neighborhood meetings led by community representatives to reinforce the breastfeeding messages (Bhandari et al., 2003). This program illustrates the importance of attending to the issue of who delivers the program. Parenting programs are usually delivered by professionals, such as doctors, nurses, and educators, or paraprofessionals who do not have the same credentials as professionals but who are trained to deliver the program. Programs sometimes are more effective if they are delivered by professionals because professionals tend to be respected authority figures who are trusted by the community as experts in the health and well-being of children. However, using professionals to deliver programs can increase program costs and may not be feasible if there are not enough professionals available to meet the needs of the program. Paraprofessionals can be recruited from local populations, such as women in a given community with relatively high levels of education who can be trained to deliver the program, providing a cost-effective solution to delivering programs to parents (Cooper et al., 2009). Regardless of who delivers the program, rapport between the parents and program providers is important. Training and supervision are also important regardless of who delivers the program. In training the trainer models, it is important to ensure that everyone receives adequate training at all levels. Even after training has concluded, program staff benefit from peer support, mentoring, coaching, and ongoing training. Taking these steps can help programs maintain fidelity over time by keeping staff in tune with program goals, reminding them about effective ways of meeting goals, and handling any problems that occur along the way in delivering the program.

Early Childhood Home Visiting Intervention in Jamaica

The Early Childhood Home Visiting Intervention in Jamaica was revised over time through a series of pilot studies. The pilot studies tested whether factors such as the use of doctors and nurses vs paraprofessionals to deliver the intervention, and the use of purchased toys vs home-made toys, made a difference to the parenting and child development outcomes (Grantham-McGregor & Walker, 2015). This example highlights the importance of monitoring and evaluation in the development and implementation of an intervention. In its current instantiation, paraprofessionals work with mothers and their children in their homes from birth to age 4 years. Mothers are encouraged to be responsive to their child's emotions, behaviors, and vocalizations; to provide cognitive stimulation (e.g., by labeling and describing things in the environment); and to show the child love and positive feedback. The trainer scaffolds the activities for mothers, and trainers teach the mothers to scaffold interactions for their child. Randomized controlled trials showed that a 24-month version of the program with weekly home visits had positive effects on children's cognitive development. The program is also important in illustrating that holistic approaches generally are more effective at improving outcomes than approaches targeting a single domain. A nutrition supplementation program in Jamaica was either combined with a psychosocial stimulation program as part of the Early Childhood Home Visiting Intervention or not (Gardner et al., 2005; Walker, Chang, Powell, & Grantham-McGregor, 2005). Nutritional supplementation alone was ineffective in improving cognitive development. Nutrition supplementation was effective in improving children's cognitive outcomes only in combination with psychosocial stimulation.

Better Parenting Program in Jordan

The Better Parenting Program has been implemented nationally in over 200 locations in Jordan as a way to help parents of children ages 0–8 years to provide a loving, stimulating, and safe home environment to promote children's physical, cognitive, and psychosocial development (Al-Hassan & Lansford, 2014). The program consists of 16 hours of lessons that focus on particular parenting knowledge, attitudes, and behaviors. Centralized trainers teach social workers, health workers, kindergarten teachers, and paraprofessionals how to deliver the lessons. A randomized evaluation of the Better Parenting Program found that, compared to a control group that did not receive the intervention, parents in the intervention group gained parenting knowledge, spent more time playing and reading books with their children, used more explanations during the course of disciplining their child, and more accurately perceived behaviors indicative of child neglect (Al-Hassan & Lansford, 2011). For some programs and in some settings, thinking creatively about who is best suited to deliver the program can improve effectiveness. For example, the Better Parenting Program in Jordan wanted to target fathers but was having a difficult time reaching them. Imams were recruited to deliver the program to fathers during Friday prayer services, which proved an effective means of reaching fathers both because the men were in attendance already and because the Imams were respected by the fathers.

BabyWASH in Nepal

The BabyWASH (Water, Sanitation, and Health) initiative aims to prevent stunting, which affects 38% of children in South Asia from birth to 59 months, by improving nutrition and sanitation in ways that will have direct benefits for maternal and child health (Aguayo & Menon, 2016). In Nepal, the Suaahara Program is a holistic 5-year program that provides services to the most vulnerable children and parents; the program has resulted in improvements in nutrition and household health and has reduced stunting of children younger than 2 years (Aguayo & Menon, 2016). The BabyWASH approach recognizes that, in general, the most vulnerable populations benefit the most from parenting programs. If resources are limited and a universal program is not an option, programs can do the most good by targeting the most at-risk families (by virtue of their poverty, lack of education, or other risk factors). Younger age groups also tend to benefit more than older age groups. Targeting prevention programs to children as young as possible is likely to yield the largest benefits. Preventing a problem before it occurs is better than intervention to solve a problem after it has happened already. Children who start out behind have a harder time catching up the longer they have been deprived of nutrition, cognitive stimulation, emotional security, or other resources.

Early Enrichment Project in Turkey

The Turkish Early Enrichment Project was implemented as a training program to teach mothers educational activities to engage in with their

3–5-year-old children (Kagitcibasi, 1997). Mothers participated in group meetings and intensive guided discussions about how to provide cognitive stimulation for their child. The program also included home visits, and provision of worksheets and books for mothers to use with their children to promote preliteracy and prenumeracy skills. The entire program was delivered over the course of 60 weeks spanning 2 years. The program was evaluated by comparing preprogram scores with immediate postprogram scores and scores from 7 years to 19 years following the program. Even 19 years following the program, individuals whose mothers had participated had better educational and occupational outcomes than individuals whose mothers had not participated in the program (Kagitcibasi, Sunar, Bekman, Baydar, & Cemalcilar, 2009).

Family Connects in the United States

Family Connects provides one to three nurse home visits to all mothers who give birth in communities that have adopted the program, which is currently available in 20 communities in 12 states. During the home visits, which occur in the first 3 weeks of the infants' life, nurses assess newborns' and their families' risks and strengths and then act as liaisons to connect families with services in their communities. By having this early contact with families, nurses are able to make referrals to more intensive parenting programs or services depending on the families' level of need. In Durham, North Carolina, the first community to adopt the program, randomized controlled trials demonstrated that by the time the infant was 6 months, families participating in the program were more likely to be connected to community resources and using higher quality childcare (as measured by the state's rating system) (Dodge, Goodman, Murphy, O'Donnell, & Sato, 2013). Mothers were 28% less likely to report clinical levels of anxiety, and infants used 34% fewer emergency medical services, with effects sustained through the age of 2 years (Dodge et al., 2013). In addition, observers who were blind to the control or intervention status of the family observed mothers in the intervention to be more sensitive to and accepting of the infant, as well as having homes that were characterized by better safety and availability of books, toys, and learning materials (Dodge et al., 2014). An important aspect of the evaluation of the Family Connects program is that cost-benefit analyses demonstrate that for every $1 spent to implement the program $3.02 was saved in emergency health care costs, suggesting that, at the level of an entire community, investment in the program could result in long-term health savings.

Making parenting programs culturally appropriate

A metaanalysis of 76 studies found that mental health interventions were 4 times more effective if they were targeted at a particular cultural group rather than mixing together participants from a variety of cultural backgrounds (Griner & Smith, 2006). Making parenting programs culturally appropriate is likewise important. The two major ways of making parenting programs culturally appropriate are to adapt an existing program to make both the content and delivery mechanisms acceptable for a new setting or to develop a new program that is specific to a particular cultural group. Special considerations may be needed if participants are immigrant or refugee parents to take into consideration perspectives from their countries of origin and destination (Bornstein, 2017).

Adapting existing programs

Often, rather than developing a new parenting program, it is cost-effective to adapt an existing program that has demonstrated effectiveness in other settings. Using an existing program has several advantages, including being able to capitalize on a curriculum that has been developed already, having access to training materials for program providers and educational materials for families, and offering reassurance that the program has the potential to deliver intended results, to name a few. These advantages can result in savings in both time and expenses that would be associated with developing an entirely new program. Despite these advantages, using an existing program in new settings also has some potential disadvantages. In particular, the existing program may not meet the needs of parents in the new setting, or the mode of delivery or other logistics may not work well in the new setting. To take advantage of existing programs while at the same time being sensitive to the needs of families in new settings, an alternative option is to adapt the program by retaining some features of the original program and changing other features to tailor it for the new setting.

In Mexico, one program that used a variety of approaches, including live demonstrations and individualized counseling, was based on both an established program (The Injury Prevention Program) and a locally developed program (*Pal Kai*, or "healthy child"). The program was delivered in private clinics that were subsidized at different rates for upper-, middle-, and lower-socioeconomic status (SES) groups. The program consisted of elements of injury prevention counseling delivered by nurses and trained health-care workers as part of overall health promotion. Counseling was upgraded as part of this intervention. For upper- and middle-SES groups, the upgrading involved lectures and demonstrations lasting 6 hours and utilizing audio-visual materials from several sources. In addition, the middle-SES group received clinic-based counseling that lasted 15–20 minutes, whereas the lower-SES group received 30-minute home visits by trained nurses and health promoters. All three SES groups demonstrated improvements from before to after the intervention in safety practices related to using caution, although results were mixed for other outcomes (Mock et al., 2003).

In adapting parenting programs to new settings, it is important to pay attention to the "demand-side" of parenting programs. For example, cultural preferences should be integrated, and the strengths parents bring to the program should be recognized. Making cultural adaptations requires knowledge of the local population's customs, beliefs, preferences, and prohibitions. For example, a program targeting behaviors associated with hygiene and hand washing practices was evaluated in Bangladesh (Luby et al., 2010). Field workers introduced soap or sanitizer and instructed mothers about when and how to wash their hands. The Bangladesh program took into account local customs (e.g., sanitizer used did not contain alcohol because many Muslims in Bangladesh are reluctant to use products that contain alcohol). The evaluation showed that waterless hand sanitizer was readily adopted by the community and reduced hand contamination as much as soap.

Developing new programs to address a cultural need

Even when a new program is developed, delivery must sometimes balance standardization to ensure that all families have access to effective elements of the program with flexibility to ensure that individual families receive services that may be of particular relevance to them. For example, in a food insecure region of India, locally trained counselors provided monthly nutrition education to caregivers with children between 5 and 11 months of age (Kilaru, Griffiths, Ganapathy, & Ghosh, 2005). The counseling messages focused on the preparation and use of developmentally appropriate local foods as well as feeding advice. The counselors were trained to be sensitive to household constraints (e.g., limited financial resources, available household food, decision-making capacity, and privilege within the family structure). In this way, the program was tailored for individual families within the broader program structure. The program resulted in greater dietary diversity and more rapid weight gain for children who were otherwise at risk of malnutrition.

Special considerations with immigrant or refugee parents

Regardless of the target population, parenting programs are more successful when they recognize strengths that parents already have as a starting point and avoid bringing a deficit perspective that implies the interventionist will address the shortcomings of the parent (Green, McAllister, & Tarte, 2004). Working with immigrant and refugee parents can introduce the need for particular caution in understanding parents' beliefs and behaviors within the framework of both their country of origin and their country of destination (Bornstein, 2017; Cote, Kwak, Putnick, Chung, & Bornstein, 2015). When immigrant parents behave in a way that is acceptable, or even desired, in their country of origin that conflicts with expectations in the country of destination, immigrant parents may be unduly criticized. For example, several traditional healing practices, such as coining and cupping, that are used in the United States primarily by immigrant families leave marks that have been taken as signs of abuse and reported to Child Protective Services (Killion, 2017). Thus, special care is needed when working with immigrant and refugee parents to ensure that their perspectives are understood in a culturally relevant way.

Conclusions and recommendations

Parents in all cultural contexts are tasked with the important responsibilities of providing for their children's physical health and safety, cognitive stimulation, affection, and behavioral socialization. The particular behaviors parents use to fulfill those responsibilities may differ across cultures, and the meaning delivered by specific parenting behaviors also may differ. Cultural differences in parenting may be characterized by mean-level differences in beliefs and behaviors, and culture may moderate links between parenting and child development outcomes. Policies related to parenting are featured in international development agenda, both in the Convention on the Rights of the Child and in the Sustainable Development Goals guiding international development through 2030. National laws have implications for organizing parents' time with children (e.g., parental leave), behaviors that parents can and cannot engage in with children (e.g., legal bans on corporal punishment), and the intersection of contexts for child development within and outside the family (e.g., national laws related to education). Parenting programs have been implemented in many countries in an effort to address both international and national goals related to child outcomes.

Evaluations of parenting programs in a wide range of cultural contexts led UNICEF (2017) to recommend nine standards to consider when implementing programs. Recommendations focus on promoting holistic child development, using theory to guide the development of interventions to ensure that the expected pathways of influence are clear, tailoring content to children's developmental stage, serving vulnerable families, involving all key caregivers, adapting programs to particular cultural contexts, integrating programs into existing delivery platforms, engaging trained service providers, and engaging in ongoing monitoring and evaluation. A systematic review and metaanalysis evaluated 17 trials of 4 parenting programs developed in the United States and Australia, which were adapted for use in 10 countries in 5 regions of the world. The programs produced benefits to children and parents that were equal to, or in some cases exceeded, the benefits of the programs in their original countries (Gardner, Montgomery, & Knerr, 2016). Thus, it is possible to transport parenting interventions to culturally distinct contexts.

Parenting programs that address the health and safety of young children are far more prevalent than parenting programs addressing cognitive stimulation, affection, and behavioral socialization or programs targeting parents of older children or adolescents. In part this focus makes sense, given the documented benefits of investment in early childhood and the dire consequences in terms of morbidity and mortality if health and safety of young children are not preserved (Black et al., 2017). Nevertheless, even in settings with few resources to support children and families, most children survive. Helping parents to provide cognitive stimulation, affection, and behavioral socialization in ways that will optimize children's development is an important direction for future research, programming, and policy. In addition, because adolescents are often neglected, relative to children, in parenting programs, especially in low- and middle-income countries, tailoring future parenting programs to issues that are salient during adolescence, such as substance use, sexual behavior, academic achievement, and career planning, has the potential to help parents and adolescents navigate what can be a difficult developmental period. Situating parenting within a cultural framework will enhance both basic research aimed at understanding parent-child relationships as well as implementing culturally sensitive policies and programs that can support parents in their responsibilities to promote children's physical, cognitive, and socioemotional well-being.

Acknowledgments

Preparation of this chapter was supported by the *Eunice Kennedy Shriver* National Institute of Child Health and Human Development grant RO1-HD054805 and the Intramural Research Program of the NIH/NICHD, United States, and an International Research Fellowship at the Institute for Fiscal Studies (IFS), London, United Kingdom, funded by the European Research Council (ERC) under the Horizon 2020 research and innovation program (grant agreement No 695300-HKADeC-ERC-2015-AdG).

References

Aguayo, V. M., & Menon, P. (2016). Stop stunting: Improving child feeding, women's nutrition and household sanitation in South Asia. *Maternal & Child Nutrition, 12*, 3–11. https://doi.org/10.1111/mcn.12283.

Al-Hassan, S. M., & Lansford, J. E. (2011). Evaluation of the Better Parenting Program in Jordan. *Early Child Development and Care, 181*, 587–598. https://doi.org/10.1080/03004431003654925.

Al-Hassan, S. M., & Lansford, J. E. (2014). A randomised evaluation of the Better Parenting Programme in Jordan. *Early Childhood Matters, 122*, 7–11.

Babatunde, E. D., & Setiloane, K. (2014). Changing patterns of Yoruba parenting in Nigeria. In H. Selin (Ed.), *Parenting across cultures: Childrearing, motherhood and fatherhood in non-western cultures* (pp. 241–252). New York, NY: Springer.

Baron, L., & Straus, M. A. (1989). *Four theories of rape in American society. A state-level analysis*. New Haven, CT: Yale University Press.

Bartel, A. P., Rossin-Slater, M., Ruhm, C. J., Stearns, J., & Waldfogel, J. (2018). Paid family leave, fathers' leave-taking, and leave-sharing in dual-earner households. *Journal of Policy Analysis and Management, 37*, 10–37. https://doi.org/10.1002/pam.22030.

Berlinski, S. & Schady, N. (Eds.), (2015). *The early years: Child wellbeing and the role of public policy*. New York: Palgrave Macmillan.

Bhandari, N., Bahl, R., Mazumdar, S., Martines, J., Black, R. E., & Bhan, M. K. (2003). Effect of community-based promotion of exclusive breastfeeding on diarrhoeal illness and growth: A cluster randomised controlled trial. *The Lancet, 361*, 1418–1423. https://doi.org/10.1016/S0140-6736(03)13134-0.

Black, M. M., Walker, S. P., Fernald, L. C. H., Andersen, C. T., DiGirolamo, A. M., Lu, C., … Lancet Early Childhood Development Series Steering Committee. (2017). Early childhood development coming of age: Science through the life course. *The Lancet, 389*, 77–90. https://doi.org/10.1016/S0140-6736(16)31389-7.

Bornstein, M., Putnick, D., Park, Y., Suwalsky, J., & Haynes, O. (2017). Human infancy and parenting in global perspective: Specificity. *Proceedings of the Royal Society B, 284*, 20172168. https://doi.org/10.1098/rspb.2017.2168.

Bornstein, M. H. (1995). Form and function: Implications for studies of culture and human development. *Culture & Psychology, 1*, 123–137. https://doi.org/10.1177/1354067X9511009.

Bornstein, M. H. (2006). Parenting science and practice. In W. Damon, K. A. Renninger, & I. E. Sigel (Eds.), *Child psychology in practice: Vol. 4. Handbook of child psychology* (6th ed., pp. 893–949). New York, NY: Wiley.

Bornstein, M. H. (2015). Children's parents. In R. M. Lerner (Series Ed.) & M. H. Bornstein & T. Leventhal (Vol. Eds.), *Ecological settings and processes in developmental systems: Vol. 4. Handbook of child psychology and developmental science* (7th ed., pp. 55–132). Hoboken, NJ: Wiley.

Bornstein, M. H. (2017). The specificity principle in acculturation science. *Perspectives on Psychological Science, 12*, 3–45. https://doi.org/10.1177/1745691616655997.

Bornstein, M. H., Cote, L. R., Haynes, O. M., Suwalsky, J. T. D., & Bakeman, R. (2012). Modalities of infant-mother interaction in Japanese, Japanese American immigrant, and European American dyads. *Child Development, 83*, 2073–2088. https://doi.org/10.1111/j.1467-8624.2012.01822.x.

Bornstein, M. H., & Putnick, D. L. (2012). Cognitive and socioemotional caregiving in developing countries. *Child Development, 83*, 46–61. https://doi.org/10.1111/j.1467-8624.2011.01673.x.

Bornstein, M. H., Putnick, D. L., Cote, L. R., Haynes, O. M., & Suwalsky, J. T. D. (2015). Mother-infant contingent vocalizations in 11 countries. *Psychological Science, 26*, 1272–1284. https://doi.org/10.1177/0956797615586796.

Bornstein, M. H., Putnick, D. L., Lansford, J. E., Deater-Deckard, K., & Bradley, R. H. (2015). A developmental analysis of caregiving modalities across infancy in 38 low- and middle-income countries. *Child Development, 86*, 1571–1587. https://doi.org/10.1111/cdev.12402.

Bronfenbrenner, U., & Morris, P. A. (2006). The bioecological model of human development. In R. M. Lerner & W. Damon (Eds.), *Handbook of child psychology: Theoretical models of human development* (pp. 793–828). Hoboken, NJ: Wiley.

Bussmann, K.-D. (2004). Evaluating the subtle impact of a ban on corporal punishment of children in Germany. *Child Abuse Review, 13*, 292–311. https://doi.org/10.1002/car.866.

Bussmann, K.-D., Erthal, C., & Schroth, A. (2011). Effects of banning corporal punishment in Europe: A five-nation comparison. In J. E. Durrant & A. B. Smith (Eds.), *Global pathways to abolishing physical punishment: Realizing children's rights* (pp. 299–322). New York, NY: Routledge.

Cartmill, E. A., Armstrong, B. F., III, Gleitman, L. R., Goldin-Meadow, S., Medina, T. N., & Trueswell, J. C. (2013). Quality of early parent input predicts child vocabulary 3 years later. *Proceedings of the National Academy of Sciences, 110*, 11278–11283. https://doi.org/10.1073/pnas.1309518110.

Centers for Disease Control and Prevention. (2009). *Parent training programs: Insight for practitioners*. Atlanta, GA: Centers for Disease Control and Prevention.

Cheah, C. S. L., Li, J., Zhou, N., Yamamoto, Y., & Leung, C. Y. Y. (2015). Understanding Chinese immigrant and European American mothers' expressions of warmth. *Developmental Psychology, 51*, 1802–1811. https://doi.org/10.1037/a0039855.

Child Trends. (2013). *Family meals*. Available from: https://www.childtrends.org/wp-content/uploads/2012/09/96_Family_Meals.pdf.

Child Welfare Information Gateway. (2013). *Parent education to strengthen families and reduce the risk of maltreatment*. Washington, DC: U.S. Department of Health and Human Services, Children's Bureau.

Coleman, D. L. (1998). The Seattle compromise: Multicultural sensitivity and Americanization. *Duke Law Journal, 47*, 717–783.

Cooper, P., Tomlinson, M., Swartz, L., Landman, M., Molteno, C., Stein, A., … Murray, L. (2009). Improving quality of mother-infant relationship and infant attachment in socioeconomically deprived community in South Africa: Randomised controlled trial. *British Medical Journal, 338*, b974. https://doi.org/10.1136/bmj.b974.

Cote, L. R., Kwak, K., Putnick, D. L., Chung, H. J., & Bornstein, M. H. (2015). The acculturation of parenting cognitions: A comparison of South Korean, Korean immigrant, and European American mothers. *Journal of Cross-Cultural Psychology, 46*, 1115–1130. https://doi.org/10.1177/0022022115600259.

Crivello, G., & van der Gaag, N. (2016). *Between hope and a hard place: Boys and young men negotiating gender, poverty and social worth in Ethiopia*. Young Lives Working Paper 160 University of Oxford.

Danish Institute for Human Rights. (2018). *The human rights guide to the Sustainable Development Goals*. Available from: http://sdg.humanrights.dk/.

Darling, N., Cumsille, P., & Peña-Alampay, L. (2005). Rules, legitimacy of parental authority, and obligation to obey in Chile, the Philippines, and the United States. In J. Smetana (Ed.), *New directions for child and adolescent development. Changing boundaries of parental authority during adolescence* (pp. 47–60). San Francisco, CA: Jossey-Bass.

Darling, N., & Steinberg, L. (1993). Parenting style as context: An integrative model. *Psychological Bulletin, 113*, 487–496.

Deater-Deckard, K., & Dodge, K. A. (1997). Externalizing behavior problems and discipline revisited: Nonlinear effects and variation by culture, context, and gender. *Psychological Inquiry, 8*, 161–175. https://doi.org/10.1207/s15327965pli0803_1.

Diamond, J. (2012). *The world until yesterday: What can we learn from traditional societies?* New York, NY: Penguin Books.

Dodge, K. A., Goodman, W. B., Murphy, R. A., O'Donnell, K., & Sato, J. (2013). Randomized controlled trial of universal postnatal nurse home visiting: Impact on emergency care. *Pediatrics, 132*, S140–S146.

Dodge, K. A., Goodman, W. B., Murphy, R. A., O'Donnell, K., Sato, J., & Guptill, S. (2014). Implementation and randomized controlled trial evaluation of universal postnatal nurse home visiting. *American Journal of Public Health, 104*, S136–S143. https://doi.org/10.2105/AJPH.2013.301361.

Duncan, G. J., Dowsett, C. J., Claessens, A., Magnuson, K., Huston, A. C., Klebanov, P., ... Japel, C. (2007). School readiness and later achievement. *Developmental Psychology, 43*, 1428–1446.

Durrant, J., Plateau, D. P., Ateah, C. A., Holden, G. W., Barker, L. A., Stewart-Tufescu, A., ... Ahmed, R. (2017). Parents' views of the relevance of a violence prevention program in high, medium, and low human development contexts. *International Journal of Behavioral Development, 41*, 523–531. https://doi.org/10.1177/0165025416687415.

Durrant, J. E. (1996). The Swedish ban on corporal punishment: Its history and effects. In D. Frehsee, W. Horn, & K.-D. Bussmann (Eds.), *Family violence against children: A challenge for society* (pp. 19–25). Berlin: Walter de Gruyter.

Durrant, J. E. & Smith, A. B. (Eds.), (2011). *Global pathways to abolishing physical punishment: Realizing children's rights*. New York, NY: Routledge.

Ember, C. R., & Ember, M. (1994). War, socialization, and interpersonal violence: A cross-cultural study. *Journal of Conflict Resolution, 38*, 620–646. https://doi.org/10.1177/0022002794038004002.

Eydal, G. B. & Rostgaard, T. (Eds.), (2016). *Fatherhood in the Nordic welfare states: Comparing care policies and practice*. Bristol: Policy Press.

Fernald, L. C. H., Gertler, P. J., & Neufeld, L. M. (2008). Role of cash in conditional cash transfer programmes for child health, growth, and development: An analysis of Mexico's Oportunidades. *The Lancet, 371*, 828–837. https://doi.org/10.1016/S0140-6736(08)60382-7.

Fernald, L. C. H., Gertler, P. J., & Neufeld, L. M. (2009). 10-year effect of Oportunidades, Mexico's conditional cash transfer programme, on child growth, cognition, language, and behavior: A longitudinal follow-up study. *The Lancet, 374*, 1997–2005. https://doi.org/10.1016/S0140-6736(09)61676-7.

Fuligni, A. J., Tseng, V., & Lam, M. (1999). Attitudes toward family obligations among American adolescents with Asian, Latin American, and European backgrounds. *Child Development, 70*, 1030–1044. https://doi.org/10.1111/1467-8624.00075.

Gardner, F., Montgomery, P., & Knerr, W. (2016). Transporting evidence-based parenting programs for child problem behavior (age 3–10) between countries: Systematic review and meta-analysis. *Journal of Clinical Child and Adolescent Psychology, 45*, 749–762. https://doi.org/10.1080/15374416.2015.1015134.

Gardner, J. M. M., Powell, C. A., Baker-Henningham, H., Walker, S. P., Cole, T. J., & Grantham-McGregor, S. M. (2005). Zinc supplementation and psychosocial stimulation: Effects on the development of undernourished Jamaican children. *The American Journal of Clinical Nutrition, 82*, 399–405. https://doi.org/10.1093/ajcn/82.2.399.

Gershoff, E. T., Grogan-Kaylor, A., Lansford, J. E., Chang, L., Zelli, A., Deater-Deckard, K., & Dodge, K. A. (2010). Parent discipline practices in an international sample: Associations with child behaviors and moderation by perceived normativeness. *Child Development, 81*, 487–502.

Global Initiative to End All Corporal Punishment of Children. (2018). *Countdown to universal prohibition*. Available from: https://endcorporalpunishment.org/countdown/.

Grantham-McGregor, S., & Walker, S. (2015). The Jamaican early childhood home visiting intervention. *Early Childhood Matters, 2015*, 28–34. https://doi.org/10.1111/j.1467-8624.2009.01409.x.

Green, B. L., McAllister, C. I., & Tarte, J. M. (2004). The strengths-based practices inventory: A tool for measuring strengths-based service delivery in early childhood and family support programs. *Families in Society, 85*, 327–334. https://doi.org/10.1177/104438940408500310.

Green, G., Macintyre, S., West, P., & Ecob, R. (1991). Like parent like child? Associations between drinking and smoking behaviour of parents and their children. *British Journal of Addiction, 86*, 745–758. https://doi.org/10.1111/j.1360-0443.1991.tb03100.x.

Griner, D., & Smith, T. B. (2006). Culturally adapted mental health intervention: A meta-analytic review. *Psychotherapy, 43*, 531–548. http://psycnet.apa.org/journals/pst/43/4/.

Grusec, J. E., & Davidov, M. (2010). Integrating different perspectives on socialization theory and research: A domain-specific approach. *Child Development, 81*, 687–709. https://doi.org/10.1111/j.1467-8624.2010.01426.x.

Hamad, R., Modrek, S., & White, J. S. (2019). Paid family leave effects on breastfeeding: A quasi-experimental study of US policies. *American Journal of Public Health, 109*, 164–166. https://doi.org/10.2105/AJPH.2018.304693.

Hammons, A. J., & Fiese, B. H. (2011). Is frequency of shared family meals related to the nutritional health of children and adolescents? *Pediatrics, 127*, e1565–e1574. https://doi.org/10.1542/peds.2010-1440.

Han, W.-J., Hetzner, N. P., & Brooks-Gunn, J. (2019). Employment and parenting. In M. H. Bornstein (Ed.), *Handbook of parenting* (3rd ed.). New York, NY: Taylor and Francis.

Hill, N. E., Castellino, D. R., Lansford, J. E., Nowlin, P., Dodge, K. A., Bates, J. E., & Pettit, G. S. (2004). Parent academic involvement as related to school behavior, achievement, and aspirations: Demographic variations across adolescence. *Child Development, 75*, 1491–1509. https://doi.org/10.1111/j.1467-8624.2004.00753.x.

Hurtado, N., Marchman, V. A., & Fernald, A. (2008). Does input influence uptake? Links between maternal talk, processing speed and vocabulary size in Spanish-learning children. *Developmental Science, 11*, F31–F39. https://doi.org/10.1111/j.1467-7687.2008.00768.x.

International Labor Organization. (2014). *Maternity and paternity at work: Law and practice across the world*. Geneva: International Labor Organization.

Jeria, M. B. (2017). Chile Crece Contigo: 10 years on. *Early Childhood Matters, 2017*, 10–14.

Kagitcibasi, C. (1997). The Turkish early enrichment project and the mother-child education program. *Journal of Adolescent and Adult Literacy, 41*, 70–72. https://www.jstor.org/stable/40026302.

Kagitcibasi, C., Sunar, D., Bekman, S., Baydar, N., & Cemalcilar, Z. (2009). Continuing effects of early enrichment in adult life: The Turkish Early Enrichment Project 22 years later. *Journal of Applied Developmental Psychology, 30*, 764–779. https://doi.org/10.1016/j.appdev.2009.05.003.

Kapil, U. (2002). Integrated child development services (ICDS) scheme: A programme for holistic development of children in India. *Indian Journal of Pediatrics, 69*, 597–601. https://doi.org/10.1007/BF02722688.

Karasik, L. B., Tamis-LeMonda, C. S., Adolph, K. E., & Bornstein, M. H. (2015). Places and posture: A cross-cultural comparison of sitting in 5-month-olds. *Journal of Cross-Cultural Psychology*, *46*, 1023–1038. https://doi.org/10.1177/0022022115593803.

Keller, H., & Chaudhary, N. (2017). Is the mother essential for attachment? Models of care in different cultures. In H. Keller & K. A. Bard (Eds.), *The cultural nature of attachment: Contextualizing relationships and development* (pp. 109–138). Cambridge, MA: MIT Press.

Khaleque, A. (2013). Perceived parental warmth, and children's psychological adjustment, and personality dispositions: A meta-analysis. *Journal of Child and Family Studies*, *22*, 297–306. https://doi.org/10.1007/s10826-012-9579-z.

Khaleque, A., & Ali, S. (2017). A systematic review of meta-analyses of research on interpersonal acceptance-rejection theory: Constructs and measures. *Journal of Family Theory and Review*, *9*, 441–458. https://doi.org/10.1111/jftr.12228.

Kilaru, A., Griffiths, P. L., Ganapathy, S., & Ghosh, S. (2005). Community-based nutrition education for improving infant growth in rural Karnataka. *Indian Pediatrics*, *42*, 425–432. https://indianpediatrics.net/may2005/may-425-432.htm.

Killion, C. M. (2017). Cultural healing practices that mimic child abuse. *Annals of Forensic Research and Analysis*, *4*(2), 1042.

Kisilevsky, B. S., Hains, S. M., Lee, K., Xie, X., Huang, H., Ye, H. H., … Wang, Z. (2003). Effects of experience on fetal voice recognition. *Psychological Science*, *14*, 220–224. https://doi.org/10.1111/1467-9280.02435.

Kramer, L., & Hamilton, T. (2019). Sibling caregiving. In M. H. Bornstein (Ed.), *Handbook of parenting* (3rd ed.). New York, NY: Taylor and Francis.

Kuhns, C., & Cabrera, N. (2018). Fathering. In M. H. Bornstein (Ed.) & M. Arterberry, K. Fingerman, & J. E. Lansford (Assoc. Eds.), *The SAGE encyclopedia of lifespan human development* (pp. 858–860). Thousand Oaks, CA: Sage.

Lamb, M. E. (2010). How do fathers influence child development? Let me count the ways. In M. E. Lamb (Ed.), *The role of the father in child development* (5th ed., pp. 1–26). Hoboken, NJ: Wiley.

Lansford, J. E., & Bornstein, M. H. (2007). *Review of parenting programs in developing countries*. New York, NY: UNICEF.

Lansford, J. E., Cappa, C., Putnick, D. L., Bornstein, M. H., Deater-Deckard, K., & Bradley, R. H. (2017). Change over time in parents' beliefs about and reported use of corporal punishment in eight countries with and without legal bans. *Child Abuse & Neglect*, *71*, 44–55. https://doi.org/10.1016/j.chiabu.2016.10.016.

Lansford, J. E., Chang, L., Dodge, K. A., Malone, P. S., Oburu, P., Palmérus, K., … Quinn, N. (2005). Physical discipline and children's adjustment: Cultural normativeness as a moderator. *Child Development*, *76*, 1234–1246. https://doi.org/10.1111/j.1467-8624.2005.00847.x.

Lansford, J. E., & Dodge, K. A. (2008). Cultural norms for adult corporal punishment of children and societal rates of endorsement and use of violence. *Parenting: Science and Practice*, *8*, 257–270. https://doi.org/10.1080/15295190802204843.

Lansford, J. E., Godwin, J., Alampay, L. P., Uribe Tirado, L. M., Zelli, A., Al-Hassan, S. M., … Tapanya, S. (2016). Mothers', fathers', and children's perceptions of parents' expectations about children's family obligations in nine countries. *International Journal of Psychology*, *51*, 366–374. https://doi.org/10.1002/ijop.12185.

Lansford, J. E., Godwin, J., Zelli, A., Al-Hassan, S. M., Bacchini, D., Bombi, A. S., … Uribe Tirado, L. M. (2018). Longitudinal associations between parenting and youth adjustment in twelve cultural groups: Cultural normativeness of parenting as a moderator. *Developmental Psychology*, *54*, 362–377. https://doi.org/10.1037/dev0000416.

Lansford, J. E., Malone, P. S., Dodge, K. A., Chang, L., Chaudhary, N., Tapanya, S., … Deater-Deckard, K. (2010). Children's perceptions of maternal hostility as a mediator of the link between discipline and children's adjustment in four countries. *International Journal of Behavioral Development*, *34*, 452–461. https://doi.org/10.1177/0165025409354933.

Lelei, M., & Weidman, J. C. (2012). Education development in Kenya: Enhancing access and quality. In C. Acedo, D. Adams, & S. Popa (Eds.), *Quality and qualities: Tensions in education reforms* (pp. 143–162). Rotterdam: Sense Publishers.

LeVine, R. A., LeVine, S., Dixon, S., Richman, A., Leiderman, P. H., Keefer, C. H., & Brazelton, T. B. (1996). *Child care and culture: Lessons from Africa.* New York, NY: Cambridge University Press.

Linos, K. (2013). *The democratic foundations of policy diffusion: How health, family, and employment laws spread across countries.* New York, NY: Oxford University Press.

Luby, S. P., Kadir, M. A., Sharker, Y., Yeasmin, F., Unicomb, L., & Islam, M. S. (2010). A community randomized controlled trial promoting waterless hand sanitizer and handwashing with soap, Dhaka, Bangladesh. *Tropical Medicine and International Health, 15*, 1508–1516. https://doi.org/10.1111/j.1365-3156.2010.02648.x.

Lundahl, B. W., Tollefson, D., Risser, H., & Lovejoy, M. C. (2008). A meta-analysis of father involvement in parent training. *Research on Social Work Practice, 18*, 97–106. https://doi.org/10.1177/1049731507309828.

Maher, E. J., Marcynyszyn, L. A., Corwin, T. W., & Hodnett, R. (2011). Dosage matters: The relationship between participation in the Nurturing Parenting Program for infants, toddlers, and preschoolers and subsequent child maltreatment. *Children and Youth Services Review, 33*, 1426–1434. https://doi.org/10.1016/j.childyouth.2011.04.014. https://www.sciencedirect.com/science/journal/01907409/33/8.

Mahoney, A., & Boyatzis, C. J. (2019). Parenting, religion, and spirituality. In M. H. Bornstein (Ed.), *Handbook of parenting* (3rd ed.). New York, NY: Taylor and Francis.

McHale, J. P., & Sirotkin, Y. (2019). Coparenting in diverse family systems. In M. H. Bornstein (Ed.), *Handbook of parenting* (3rd ed.). New York, NY: Taylor and Francis.

Ministry of Education. (2018, January 5). *Review of China's education reform in 2017.* Retrieved from: http://en.moe.gov.cn/News/Top_News/201801/t20180130_326023.html.

Mitchum, N. T. (1989). Increasing self-esteem in Native-American children. *Elementary School Guidance and Counseling, 23*, 266–271. https://www.jstor.org/stable/42873996.

Mock, C., Arreola-Risa, C., Trevino-Perez, R., Almazan-Saavedra, V., Zozaya-Paz, J. E., Gonzalez-Solis, R., … Hernandez-Torre, M. H. (2003). Injury prevention counselling to improve safety practices by parents in Mexico. *Bulletin of the World Health Organization, 81*, 591–598.

Mohebbi, S. Z., Virtanen, J. I., Vahid-Golpayegani, M., & Vehkalahti, M. M. (2009). A cluster randomised trial of effectiveness of educational intervention in primary health care on early childhood caries. *Caries Research, 43*, 110–118. https://doi.org/10.1159/000209343.

Morris, A. S., Robinson, L. R., Hays-Grudo, J., Claussen, A. H., Hartwig, S. A., & Treat, A. E. (2017). Targeting parenting in early childhood: A public health approach to improve outcomes for children living in poverty. *Child Development, 88*, 388–397. https://doi.org/10.1111/cdev.12743.

Mytton, J., Ingram, J., Manns, S., & Thomas, J. (2014). Facilitators and barriers to engagement in parenting programs: A qualitative systematic review. *Health Education & Behavior, 41*, 127–137. https://doi.org/10.1177/1090198113485755.

Park, S., & Holloway, S. D. (2013). No parent left behind: Predicting parental involvement in adolescents' education within a sociodemographically diverse population. *Journal of Educational Research, 106*, 105–119. https://doi.org/10.1080/00220671.2012.667012.

Peng, Y., & Wong, O. M. H. (2015). Who takes care of my left-behind children? Migrant mothers and caregivers in transnational child care. *Journal of Family Issues, 37*, 2021–2044. https://doi.org/10.1177/0192513X15578006.

Perrin, R., Miller-Perrin, C., & Song, J. (2017). Changing attitudes about spanking using alternative biblical interpretations. *International Journal of Behavioral Development, 41*, 514–522. https://doi.org/10.1177/0165025416673295.

Potterton, J., Stewart, A., Cooper, P., & Becker, P. (2010). The effect of a basic home stimulation programme on the development of young children infected with HIV. *Developmental Medicine and Child Neurology, 52*, 547–551. https://doi.org/10.1111/j.1469-8749.2009.03534.x.

Putnick, D. L., & Bornstein, M. H. (2015). Is child labor a barrier to school enrollment in low- and middle-income countries? *International Journal of Educational Development, 41*, 112–120. https://doi.org/10.1016/j.ijedudev.2015.02.001.

Qin, L., Pomerantz, E. M., & Wang, Q. (2009). Are gains in decision-making autonomy during early adolescence beneficial for emotional functioning? The case of the United States and China. *Child Development, 80*, 1705–1721. https://doi.org/10.1111/j.1467-8624.2009.01363.x.

Reed, J., Hirsh-Pasek, K., & Golinkoff, R. M. (2017). Learning on hold: Cell phones sidetrack parent-child interactions. *Developmental Psychology, 53*, 1428–1436. https://doi.org/10.1037/dev0000292.

Rogoff, B. (2003). *The cultural nature of human development*. New York, NY: Oxford University Press.

Rohner, R. P., & Lansford, J. E. (2017). Deep structure of the human affectional system: Introduction to Interpersonal Acceptance-Rejection Theory. *Journal of Family Theory & Review, 9*, 426–440. https://doi.org/10.1111/jftr.12219.

Rohner, R. P., & Smith, R. (2019). Parental acceptance/rejection. In M. H. Bornstein (Ed.), *Handbook of parenting* (3rd ed.). New York, NY: Taylor and Francis.

Roopnarine, J. L. (2011). Cultural variations in beliefs about play, parent–child play, and children's play: Meaning for childhood development. In A. D. Pellegrini (Ed.), *The Oxford handbook of the development of play* (pp. 19–27). New York, NY: Oxford University Press.

Rossow, I., & Rise, J. (1994). Concordance of parental and adolescent health behaviors. *Social Science & Medicine, 38*, 1299–1305. https://doi.org/10.1016/0277-9536(94)90193-7.

Rothbaum, F., Weisz, J., Pott, M., Miyake, K., & Morelli, G. (2000). Attachment and culture: Security in the United States and Japan. *American Psychologist, 55*, 1093–1104. https://doi.org/10.1037/0003-066X.55.10.1093.

Rudolph, K. D., Lansford, J. E., & Rodkin, P. (2017). Interpersonal theories of developmental psychopathology. In D. Cicchetti (Ed.), *Developmental psychopathology* (pp. 243–311). New York, NY: Wiley.

Sear, R. (2015). Beyond the nuclear family: An evolutionary perspective on parenting. *Current Opinion in Psychology*, 7, 98–103. https://doi.org/10.1016/j.copsyc.2015.08.013.

Seymour, D. (2009). *1989-2009: Convention brings progress on child rights, but challenges remain*. UNICEF. Available from: https://www.unicef.org/rightsite/237_241.htm.

Shim, J. (2016). Family leave policy and child mortality: Evidence from 19 OECD countries from 1969 to 2010. *International Journal of Social Welfare, 25*, 215–221. https://doi.org/10.1111/ijsw.12186.

Singh, R., & Mukherjee, P. (2018). Push out, pull out, or opting out? Reasons cited by adolescents for discontinuing education in four low- and middle-income countries. In J. E. Lansford & P. Banati (Eds.), *Handbook of adolescent development research and its impact on global policy* (pp. 238–259). New York, NY: Oxford University Press.

Sloand, E., Astone, N. M., & Gebrian, B. (2010). The impact of fathers' clubs on child health in rural Haiti. *Field Action Report, 100*, 201–204. https://doi.org/10.2105/AJPH.2008.152439.

Smith, J., Wohlstetter, P., Kuzin, C. A., & De Pedro, K. (2011). Parent involvement in urban charter schools: New strategies for increasing participation. *School Community Journal, 21*, 71–94.

Sundaram, L., Travers, E., & Branson, M. (2018). How to end child marriage around the world: Strategies, progress and action needed. In J. E. Lansford & P. Banati (Eds.), *Handbook of adolescent development research and its impact on global policy* (pp. 173–195). New York, NY: Oxford University Press.

Topping, K., Dekhinet, R., & Zeedyk, S. (2013). Parent–infant interaction and children's language development. *Educational Psychology*, *33*, 391–426. https://doi.org/10.1080/01443410.2012.744159.

Torres, A., Lopez, B., Parra, V., Vazquez, C., Segura-Pérez, S., Cetin, Z., & Pérez-Escamilla, R. (2018). Chile Crece Contigo: Implementation, results, and scaling-up lessons. *Child: Care, Health and Development*, *44*, 4–11. https://doi.org/10.1111/cch.12519.

Trevarthen, C. (1998). The concept and foundations of infant intersubjectivity. In S. Braten (Ed.), *Intersubjective communication and emotion in early ontogeny* (pp. 15–46). New York, NY: Cambridge University Press.

United Nations. (1989). *Convention on the Rights of the Child*. Available from: https://www.ohchr.org/en/professionalinterest/pages/crc.aspx.

United Nations. (2015). *Transforming our world: The 2030 agenda for sustainable development*. Available from: https://sustainabledevelopment.un.org/post2015/transformingourworld/publication.

United Nations Development Programme. (2018). *Human Development Index*. Available from: http://hdr.undp.org/en/content/human-development-index-hdi.

United Nations Educational, Scientific and Cultural Organization (UNESCO) Institute of Statistics and Global Education Monitoring Report. (2016). *Education for people and planet: Creating sustainable futures for all*. UNESCO. Available from: http://unesdoc.unesco.org/images/0024/002457/245752e.pdf.

United Nations International Children's Emergency Fund (UNICEF). (2014). *Mapping of Early Childhood Development parenting programs in low and middle income countries*. New York, NY: UNICEF.

United Nations International Children's Emergency Fund (UNICEF). (2017). *Standards for Early Childhood Development parenting programmes in low and middle income countries*. New York, NY: UNICEF.

United States Department of Education, National Center for Education Statistics. (2018). *The condition of education 2018*. [NCES 2018-144].

Walker, S. P., Chang, S. M., Powell, C. A., & Grantham-McGregor, S. M. (2005). Effects of early childhood psychosocial stimulation and nutritional supplementation on cognition and education in growth-stunted Jamaican children: Prospective cohort study. *The Lancet*, *366*, 1804–1807. https://doi.org/10.1016/S0140-6736(05)67574-5.

World Health Organization. (2018). *Breastfeeding*. Available from: http://www.who.int/topics/breastfeeding/en/.

CHAPTER 19

Evidence-based parenting support across cultures: The Triple P—Positive Parenting Program experience

Karen M.T. Turner, Meghna Singhal, Cari McIlduff, Saumya Singh, and Matthew R. Sanders
Parenting and Family Support Centre, The University of Queensland, Brisbane, QLD, Australia

Of all the modifiable influences affecting children's health, mental health, general well-being, and life course trajectories, none is more important than the parenting they receive (Sanders & Mazzucchelli, 2018). Parenting impacts many aspects of child development, including cognition, emotion regulation, language, social skills, personal values, and academic achievement, to name a few (Center on the Developing Child at Harvard University, 2016). Research from diverse disciplines, including behavioral and social sciences, molecular biology, neuroscience, and epigenetics, has provided evidence on essential ingredients of children's healthy development (Beach, Brody, Barton, & Philibert, 2016). These ingredients include physical care and nurturance (meeting children's survival needs, such as food, warmth, shelter, clothing, safety); emotional care (love, affection, developing secure attachments, creating a caring environment that helps children learn emotion regulation); social care (helping children develop the social competencies they require to thrive in their social networks and broader community); and disciplining and socialization (providing a role model for desirable behavior, setting limits in a developmentally and culturally appropriate manner, and teaching children the skills to function independently). However, exposure to toxic early family experiences, including poverty, family conflict and violence, harsh or inconsistent parenting, or child maltreatment can adversely impact children's normal development, including their physical and mental health (Felitti et al., 1998; Moffitt & Klaus-Grawe Think Tank, 2013).

Cross-Cultural Family Research and Practice
https://doi.org/10.1016/B978-0-12-815493-9.00019-3

Combined, this knowledge makes parenting a key target for supporting optimal child outcomes across the globe.

Parenting and culture are indivisible and the cultural context in which families live can greatly influence parenting practices. Beliefs, values, norms, and customs within a culture are shaped by ecology and history, creating complex systems of cognition, emotion, and behavior that inevitably affect the way families raise their children (Bhawuk, 2015). Social group affiliation involves identification with, and belongingness to, an ethnic/racial/cultural group (Fung, Wong, & Park, 2018) and this influences individuals' beliefs about the kind of parenting practices that will enable them to achieve their parenting goals, their aspirations for their child, as well as family routines and rituals. For example, individualistic cultures value individuation, independence, self-reliance, personal needs and rights; whereas collectivistic cultures downplay independence and instead promote interdependence, conformity, obligations, duty, and respect (Greenfield, Keller, Fuligni, & Maynard, 2003). These cultural dimensions impact all aspects of life including attributions, emotions, cognitions, social behaviors, language and communication, group processes, and leadership (Weng, 2015), all of which affect the parenting journey.

Culturally based parenting values and practices are diverse. They may differ between parents from different cultural backgrounds, but may also vary between parents with the same cultural background, even within the same family, and between generations. They are also dynamic, not static. For example, when families such as migrant or refugee families have increasing contact with a new culture, acculturation may lead to changes in traditionally held parenting beliefs and approaches, both between and within generations (Bornstein & Bohr, 2011). Culture and parenting are significantly influenced by the contexts and systems within which they exist, not least of these being poverty, which affects many nondominant cultures globally (e.g., in developing countries and minority cultures within developed countries).

Living in poverty is one factor that affects both children and parents and has a pervasive impact on parents' capacity to undertake their parenting responsibilities. Parents living in low-resource environments are more likely to experience stressors such as poorer physical health, food, and housing insecurity, limited access to quality health and family support services, exposure to family and neighborhood violence, chronic unemployment, substance abuse, mental illness, and child maltreatment. Poverty adversely affects child well-being (Bornstein & Bradley, 2014), and is a risk factor for physical, emotional, and behavioral problems in children (Najman, Clavarino, et al., 2010; Najman, Hayatbakhsh, et al., 2010). The interaction of childhood poverty and race has been shown to impact adult health outcomes (Nikulina & Widom, 2014). The mediators for this association between poverty and child difficulties have been found to be lower quality physical environment (including frequent physical punishment and exposure to community violence), maternal emotional unresponsiveness, family stress, and fewer stimulating experiences (Eamon, 2000). Such factors can be buffered or exacerbated by cultural practices.

Children in low- and middle-income countries (LMICs), due to the effects of poverty and deprivation, have been found to be at high risk for a range of developmental and health-related difficulties (Engle et al., 2007), which in turn make parenting more difficult (Ward, Sanders, Gardner, Mikton, & Dawes, 2016). Though country income categories may hide a great deal of within-country variation (prime examples being India and China), countries labeled low- or middle-income by the World Health Organization share key characteristics, of which poverty and weak health, mental health, and social welfare systems are fundamental (Mounier-Jack, Mayhew, & Mays, 2017; Saraceno et al., 2007). These factors can place communities, families, and children at risk for adverse outcomes, pointing to the need for effective parenting interventions to be widely accessible in these settings (Engle et al., 2011; World Health Organization & WHO Collaborating Centre for Violence Prevention, 2010).

Across different cultural models of parenting in high-income countries (HICs) and LMICs, parenting practices (in particular, parental responsiveness and exposure to stimulating materials and experiences, and to some extent discipline practices) have been identified as key factors related to child well-being and cognitive and social-emotional competence (Bradley & Corwyn, 2005). By targeting parenting practices, it is possible to weaken the link between poverty and poor child development outcomes, moving toward achieving United Nations Sustainable Development Goals such as ensuring good health well-being for all, ensuring inclusive and equitable quality education, and reducing inequality (United Nations, 2015).

Numerous parenting programs designed to provide support and skills training to parents have been disseminated the world over and have been shown to be effective across diverse contexts. Trials of parenting programs related to child health and development targets conducted in LMICs have shown promising outcomes for parents and children in culturally diverse and low-resource communities in countries such as South Africa (Cooper et al., 2009), Ethiopia (Klein & Rye, 2004), China (Jin et al., 2007), and Iran (Oveisi et al., 2010). However, research often lacks methodological rigor due to resource constraints, leading to an agenda for transporting programs that have already undergone evaluation elsewhere, and testing contextual fit and sustainability. Ideally, such programs should have an evidence base that they reduce risk factors (e.g., limited emotional responsivity and poor parent-child interaction, coercive and inconsistent parenting, and parental distress such as stress, depression, and partner conflict) and poor outcomes, be cost effective, and feasible to take to scale (Ward et al., 2016).

Evidence-based parenting support (EBPS) is a term used to refer to parenting programs that have undergone years of rigorous empirical scrutiny, including randomized controlled clinical efficacy trials to prove favorable child and family outcomes, real-world effectiveness trials to prove clinical utility, and broader dissemination trials to prove generalizability and sustainability. That is, when several rigorous trials with different samples support its effectiveness, a program is considered to be evidence based (Kumpfer & United Nations Office on Drugs and Crime, 2009). There is now an international movement toward investment in EBPS prevention and early intervention programs which have the potential to reduce inequalities perpetuated by poverty and have the high benefit-to-cost ratio important for public spending accountability, particularly in low-resource communities (Engle et al., 2011). However, the well-established literature on parenting support program effectiveness comes predominantly from HICs, rather than LMICs where community structure (e.g., collectivist vs individualist societies), cultural differences (e.g., expectations and values relating to child behavior and development), and disproportionate poverty must be considered.

Parenting support across cultures

EBPS programs that are considered to be robust have been subjected to controlled evaluations across sites, investigators, family characteristics, cultures, and countries. However, transporting an EBPS program to another culture or country is not necessarily a straightforward process. It may require cultural adaptation and accommodation of the original program, which is a time-consuming process, requiring careful assessment of the local political, religious, and economic context as well as the social context of family practices and cultural norms (Kumpfer, Pinyuchon, Teixeira de Melo, & Whiteside, 2008). There has been limited documentation of how or if these EBPS programs have met the specific ongoing needs in each situation. This is particularly important and relevant for low-resource contexts, such as LMICs, where the prevalence rate of child maltreatment and psychological disorders in children is equivalent to or higher than their higher-income counterparts (Mejia, Calam, & Sanders, 2012).

Since a country, or even a large geographical area such as a state or a district within a state, may have large variability in terms of languages spoken and child-rearing practices, it is important to determine the extent to which an EBPS program needs to and can be altered to fit the needs, cultural and social context, and administrative mandates of the local setting without violating fidelity to the model. Gardner, Montgomery, and Knerr (2016) conducted a systematic review to examine the extent to which EBPS programs for reducing child behavioral problems are effective when transported to new countries. They found that EBPS interventions (with only minor adaptations) were effective in new cultural contexts, despite huge cultural, economic, and policy differences between transported trials, and between origin and transported-to countries.

Ward et al. (2016) noted that concerns about culture, and therefore need for adaptation, assume that programs are fixed and nondynamic and that they involve dictating strategies to parents. However, many EBPS programs are culturally flexible, with collaborative values built into their training and delivery, such as self-determined goals in line with personal values, tailored delivery according to individual learning styles (e.g., preference for role play and storytelling over didactic teaching). They also draw attention to child-rearing practices which may themselves be in a state of flux and cite the example of physical punishment, which may be culturally normative in certain countries while at the same time be prohibited by the new policies of that country. Furthermore, cultural practices may shift from one generation to another within the same community, with younger parents having more access to communication technologies, and thus, ideas about raising children may be readily transported from one cultural context to another. In view of these findings, it is important that programs are based on flexible, respectful principles that allow tailoring to parents' unique circumstances, and where parents learn to problem solve to meet their own family goals in ways that are both acceptable and effective for them (Ward et al., 2016).

Our experience with the Triple P—Positive Parenting Program

What is Triple P?

The Triple P—Positive Parenting Program is one of the most extensively evaluated systems of parenting support in the world, with over 320 evaluation studies and 630 conceptual papers by researchers across 36 countries over the last 40 years. It is a preventively oriented multilevel system of parenting support designed as a population-based strategy to enhance the knowledge, skills, and confidence of parents in the task of raising their children (Sanders, 2012). It draws on social learning theory (Bandura, 1977; Patterson, 1982), applied behavior analysis (Baer, Wolf, & Risley, 1968), research on child development and developmental psychopathology (Hart & Risley, 1995; Rutter, 1985a, 1985b), social information processing models (Dodge, 1986), secure attachment, healthy relationships and nurturing environments (Biglan, Flay, Embry, & Sandler, 2012), and public health principles (Farquhar et al., 1985).

Triple P has many distinguishing features in its flexibility, varied delivery modalities, multidisciplinary approach, and focus on self-regulation and generalization of parenting skills. The primary aims of Triple P are to promote children's development and family relationships and prevent child maltreatment and behavioral, emotional, and developmental problems in children. Triple P teaches parents strategies to encourage their child's social and language skills, emotional self-regulation, independence, and problem-solving ability. Attainment of these skills promotes family harmony, reduces parent-child conflict and risk of child maltreatment, fosters successful peer relationships, and prepares children for successful experiences at school and later in life. Five core principles of positive parenting form the basis of the program: (1) creating a safe and engaging environment for children; (2) having a positive learning environment; (3) using consistent, nonviolent discipline; (4) having reasonable expectations; and (5) taking care of oneself as a parent (Sanders, 2012).

Parents are taught a variety of child management skills including building relationships through spending quality time, talking and affection; providing brief contingent praise and attention for desirable behavior; arranging engaging activities in high-risk situations (such as when parents have time pressure, competing demands, or are unable to be responsive; or when children have little to do or are exposed to physical danger); establishing limits and rules; monitoring problem behavior; using directed discussion and planned ignoring for minor problem behavior; giving clear, calm instructions; and backing up instructions with logical consequences, quiet time (nonexclusionary time-out) and time-out (i.e., brief removal from an activity where a problem is occurring, to a safe space to calm down and avoid escalation of the problem). These strategies promote emotion self-regulation and self-control (such as focusing attention, following rules and limits, and controlling impulsive behavior). Parents learn to apply these skills both at home and in the community and to generalize and maintain parenting skills across settings and over time. While all principles and strategies are routinely introduced, content is individually tailored as families develop their own goals and select strategies to form their own personalized parenting plans.

The Triple P system involves a unique blending of universal and targeted interventions on a continuum of increasing intensity and narrowing population reach for parents of children from birth to age 16. The multilevel approach utilizes the guiding principle of selecting the "minimally sufficient" effective intervention to meet the needs of parents in order to maximize efficiency, contain costs, avoid overservicing, and ensure that support becomes widely available to parents in the community. This tiered continuum aligns with the principle of proportionate universalism (Marmot & Bell, 2012) where a combination of universal and targeted services is made available according to need. The levels of Triple P are described briefly here.

Level 1. *Universal Triple P* uses health promotion and social marketing strategies to increase favorable community attitudes toward parents and children; promote the use of positive parenting practices; destigmatize and normalize the process of seeking help for parenting issues; increase parents' receptivity toward participating in a parenting program; and increase the visibility and reach of EBPS programs.

Level 2. *Selected Triple P/Brief Primary Care Triple P* is delivered through primary care or community services as brief 10–20 minute individual sessions on a specific concern (e.g., disobedience, fighting, and aggression) or a 90-minute group seminar series covering information about causes, and strategies for the prevention and management of challenging child behavior.

Level 3. *Primary Care Triple P/Triple P Discussion Groups* comprise a more intensive (e.g., 3–4 half-hour individual sessions or 2-hour discussion groups), selective prevention strategy targeting parents who have mild and relatively discrete concerns about their child's behavior or development. This intervention level incorporates active skills training and the selective use of parenting tip sheets or workbooks covering common developmental and behavioral issues.

Level 4. *Standard Triple P/Group Triple P/Self-Directed Triple P/Triple P Online* are indicated prevention/early intervention programs (e.g., 10 individual 60-minute sessions or eight interactive online modules) targeting families of higher risk children with detectable subclinical problems, or who meet diagnostic criteria, with the aim of preventing the progression of problem behavior. Group (e.g., five 2-hour groups plus three brief telephone consultations) and self-directed (a 10-session workbook) variants at this level of intervention can also be offered as an indicated prevention approach targeting an entire population to improve parenting capacity and reach individual children at risk.

Level 5. *Enhanced Triple P* is an indicated level of intervention for families with additional risk factors. It adds modules that focus on areas such as partner support and communication, mood management and stress coping skills, and anger management skills for parents. Usually, at this level of intervention, children have behavior problems that are complicated by additional family adversity factors such as parental mental health or relationship issues. Families typically complete a Level 3 or 4 intervention prior to Level 5, but practitioners may run Level 5 sessions concurrently with, or even prior to, parenting sessions based on their case formulation and family need.

What is the evidence for Triple P across cultures?

Variants of Triple P have been deployed in 39 countries to date, in many different cultural contexts, including culturally and linguistically diverse (CALD) populations in Australasia (Australia, New Zealand), United Kingdom (England, Scotland, Wales) and Jersey, North America (Canada, United States), Europe (Austria, Belgium, Denmark, France, Germany, Greece, Ireland, Luxembourg, Portugal, Sweden, Switzerland, The Netherlands); non-Western cultures in high-income countries including Asia (Hong Kong, Japan, Singapore) and South America (Argentina, Chile, Curaçao); and LMICs in Europe (Romania), the Middle East (Iran, Turkey), Asia (China, India, Indonesia, Malaysia, Vietnam), South America (Brazil, Columbia, Costa Rica, Mexico, Panama), Africa (Kenya, South Africa); and with Indigenous and First Nations families in Australia, Canada, New Zealand, and the United States.

Example research evaluations of program acceptability and outcomes for First Nations, non-Western/European, and Culturally and Linguistically Diverse (e.g., migrant and refugee) families are detailed in Table 19.1. There has been consistently high acceptability of the program and significant positive child and parent outcomes reported, with varying levels of adaptation required—from surface level (e.g., language translation, altered session length, and structure, locally relevant teaching examples, culturally connected facilitators) to deep-level cultural adaptation, such as new or adjunct program resources that explicitly include cultural values (e.g., teaching aids developed for Māori parenting groups in New Zealand with graphics linking the five Triple P positive parenting principles to eight Māori tikanga or

Table 19.1 Triple P evaluation in different cultural contexts.

Author (year)	Country	Language	Design and sample size	Population	Age	Program and adaptation	Assessment times	Program attrition	Main findings[a]
First Nations									
1. Houlding, Schmidt, Stern, Jamieson, and Borg (2012)	Canada	English	Retrospective structured interviews (11 parents; 8 providers)	Inuit, Metis and First Nations parents and local service providers (treatment)	3–12	Indigenous Group Triple P (L4) Further surface adaptation	na	na	Improved parenting (different to family of origin) and child behavior; high acceptability of program, but requested Canadian Aboriginal DVD.
2. Keown et al. (2018)	New Zealand	English, Māori	RCT (41 Int; 29 WL)	Rural and urban Māori families (targeted)	3–7	2 Primary Care Discussion Groups (L3) Deep/structural adaptation	Pre, Post, 6-month FU	4.9%	Improved parenting (overreactivity), confidence, and child behavior; reduced conflict over parenting; high acceptability.
3. Andersson et al. (2019)	Australia	English	Service evaluation 277 Int	Urban, rural, and remote Indigenous families involved in child protection (targeted)	2–12	Indigenous Group Triple P (L4) Deep/structural adaptation	Pre, Post	9.2%	Improved parenting (reduced coercion), parenting confidence, and child behavioral and emotional adjustment; high acceptability.

Continued

Table 19.1 Triple P evaluation in different cultural contexts—cont'd

Author (year)	Country	Language	Design and sample size	Population	Age	Program and adaptation	Assessment times	Program attrition	Main findings[a]
4. Turner et al. (2007)	Australia	English	RCT (26 Int; 25 WL)	Urban Australian Indigenous families (targeted)	1–13	Indigenous Group Triple P (L4) Deep/structural adaptation	Pre, Post, 6-month FU	23.1%	Improved parenting (laxness) and child behavior; high acceptability.
Asia									
5. Au et al. (2014)	Hong Kong	Chinese	RCT (pilot) 8 Int; 9 WL	Chinese parents of children with ADHD (treatment)	3–7	Group Triple P (L4) Translation, surface adaptation	Pre, Post, 6-month FU	Nil	Improved parenting efficacy and child behavior; high acceptability.
6. Chan, Leung, and Sanders (2016)	Hong Kong	Chinese	RCT (27 Int; 31 comparison program control; 31 WL)	Chinese parents (universal)	2–6	Group Triple P (L4) Non-directive program comparison condition Translation, surface adaptation	Pre, Post	10.3% (L4) 19.4% (comp)	Improved child behavior in Triple P group only; no difference on parenting stress.
7. Crisante and Ng (2003)	Australia	Chinese	Feasibility pilot (45 Int)	Chinese migrant parents living in Australia (targeted)	3–10	Group Triple P (L4) Translation, surface adaptation	Pre, Post	5%	Improved prosocial behavior, no change on behavior problems; high satisfaction. Parenting not measured.

8. Fujiwara, Kato, and Sanders (2011)	Japan	Japanese	Service evaluation 91 Int; 24 Matched control	Japanese mothers in Tokyo (targeted)	2–8	Group Triple P (L4) Translation, surface adaptation	Pre, Post	Not reported	Improved parenting (laxness, overreactivity, verbosity), parenting confidence, and child behavior; reduced parental depression. Acceptability not measured.
9. Guo, Morawska, and Sanders (2016)	China	Chinese	RCT 40 Int; 41 WL	Chinese parents in Shanghai (universal)	6–10	Group Triple P (L4) Translation, surface adaptation	Pre, Post, 6-month FU	7.5%	Improved parenting, parenting confidence, parental adjustment, child behavior at home and school, and significant increase in parents' satisfaction with children's academic achievement.

Continued

Table 19.1 Triple P evaluation in different cultural contexts—cont'd

Author (year)	Country	Language	Design and sample size	Population	Age	Program and adaptation	Assessment times	Program attrition	Main findings[a]
10. Leung, Fan, and Sanders (2013)	Hong Kong	Chinese	RCT 42 Int; 39 WL	Chinese parents with a child with a developmental disability (targeted)	3–5	Group Triple P (L4) Translation, surface adaptation	Pre, Post, 6-month FU	7.1%	Improved parenting (laxness, overreactivity) and child behavior; reduced parental stress and conflict over parenting; high acceptability.
11. Leung, Sanders, Ip, and Lau (2006)	Hong Kong	Chinese	Service evaluation 661 Int	Chinese parents (targeted)	2–12	Group Triple P (L4) Translation, surface adaptation	Pre, Post	12%	Improved parenting confidence and child behavior; reduced parental stress and anxiety. Parenting style and acceptability not measured. Greater change in child behavior related to low SES, immigrant status and higher parenting stress.

12. Leung, Sanders, Leung, Mak, and Lau (2003)	Hong Kong	Chinese	RCT 46 Int; 45 WL	Chinese parents (targeted)	3–7	Group Triple P (L4) Translation, surface adaptation	Pre, Post	6.5%	Improved parenting (laxness, overreactivity, verbosity) parenting efficacy and satisfaction, and child behavior and social and emotional wellbeing; reduced conflict over parenting; high acceptability.
13. Matsumoto, Sofronoff, and Sanders (2007)	Australia	Japanese	RCT 25 Int; 25 WL	Japanese parents in Australia (universal)	2–10	Group Triple P (L4) Translation, surface adaptation	Pre, Post, 3 month FU	Nil	Improved parenting (laxness, overreactivity), confidence and child behavior; reduced parental stress and conflict over parenting, high acceptability, and usefulness of skills.

Continued

Table 19.1 Triple P evaluation in different cultural contexts—cont'd

Author (year)	Country	Language	Design and sample size	Population	Age	Program and adaptation	Assessment times	Program attrition	Main findings[a]
14. Matsumoto, Sofronoff, and Sanders (2010)	Japan	Japanese	RCT 28 Int; 26 WL	Japanese parents in Tokyo (universal)	2–10	Group Triple P (L4) Translation, surface adaptation	Pre, Post	10.7%	Improved parenting (laxness, overreactivity), confidence, and child behavior and emotional wellbeing; high acceptability.
15. Sawada, Ono, Miyajima, Tsukamoto, and Kato (2016)	Japan	Japanese	Service evaluation 8 Int	Japanese parents in Tokyo accessing psychiatric outpatient services (treatment)	1–18	Group Triple P (L4) Translation, surface adaptation	Pre, Post	Nil	Improved parenting (laxness); reduced parental anxiety.
16. Singhal and Turner (2018)	India	English	Quasiexperimental pilot 75 Int; 27 WL	Indian corporate employee parents (universal)	2–12	Triple P Seminar (L2) Surface adaptation	Pre, 2 week FU	NA	Improved parenting confidence; high acceptability.
17. Sumargi, Sofronoff, and Morawska (2014)	Australia	Indonesian	Acceptability pilot 27 Int	Indonesian parents in Australia (universal)	2–12	Triple P Seminar (L2) Translation, surface adaptation	Pre, Post, 3 month FU	Nil	Improved parenting (laxness) and child behavior; high acceptability.

18. Sumargi, Sofronoff, and Morawska (2015)	Indonesia	Indonesian	RCT (72 Int; 71 WL)	Indonesian (Javanese and Chinese) parents in Surabaya	2–12	Triple P Seminar Series—3 seminars (L2) Translation, surface adaptation	Pre, Post, 6 month FU	8.3%	Improved parenting (laxness, overreactivity), confidence and child behavior; reduced parental stress; high acceptability.
19. Zhou et al. (2017)	Singapore	English	Service evaluation (64 L4; 43 L4+5) Allocated to inter-vention intensity according child mal-treatment risk	Chinese, Malay and Indian parents involved in the child protection system (treatment)	2–12	Standard/Group Triple P (L4) Pathways Triple P (L5) No adaptation reported	Pre, Post	17.2% (L4) 38.7% (L4+5)	Improved parenting (laxness, overreactivity), confidence and child behavior; reduced parental depression and stress; no additive impact of L5 on attributions or anger. Acceptability not measured.[a]
Middle East									
20. Aghebati, Gharraee, Hakim Shoshtari, and Gohari (2014)	Iran	Persian	RCT 14 Int; 13 WL	Mothers of children with ADHD in Tehran (treatment)	6–10	Group Triple P (L4) Translation, surface adaptation	Pre, Post	6.6%	Improved parenting, mother-child relationship and child behavior; reduced maternal depression, anxiety, and stress high acceptability.

Continued

Table 19.1 Triple P evaluation in different cultural contexts—cont'd

Author (year)	Country	Language	Design and sample size	Population	Age	Program and adaptation	Assessment times	Program attrition	Main findings[a]
21. Arkan (2014)	Turkey	Turkish	RCT 38 Int; 33 WL	Parents in Izmir (treatment)	12–18	Group Teen Triple P (L4) Translation, surface adaptation	Pre, Post, 3 month FU	2.6%	Reduced child emotional and behavioral and peer problems, hyperactivity, and parent-adolescent conflict; improved parent mental health; high acceptability.
22. Guvenir et al. (2014)	Turkey	Turkish	Population roll out 1526 Int (L2) 68 Int (L4)	Parents in Balçova, Izmir (universal)	1–19	Universal Triple P communications campaign (L1), Selected Triple P Seminars (L2), Group Triple P (L4)	Pre, Post	Not reported	L2: Reduced child hyperactivity, peer relations; improved prosocial behavior; L4: Reduced child emotional and behavioral problems and hyperactivity, improved peer relations; high acceptability.

23.	Kheirie, Shaeiri, Azad Fallah, and Rasulzade Tabatabaei (2009)	Iran	Persian	RCT 25 Int; 21 WL	Parents of children with ODD in Tehran (treatment)	6–7	Group Triple P (L4) Translation, surface adaptation	Pre, Post	Not reported	Significant decrease in ODD symptoms. Parenting and acceptability not measured.
24.	Moharreri, Shahrivar, Tehrani-Doost, and Mahmoudi-Gharaei (2008)	Iran	Persian	RCT 30 Int; 30 WL	Parents of children with ADHD in Tehran (treatment)	6–12	Group Triple P (L4) Translation, surface adaptation	Pre, Post	Nil	Improved parenting (laxness, overreactivity, verbosity); reduced conflict over parenting, parental depression, anxiety and stress; high acceptability. Child outcomes not measured.
25.	Ozyurt, Gencer, Ozturk, and Ozbek (2016)	Turkey	Turkish	RCT 28 Int; 29 WL	Parents of children with anxiety disorder in Izmir (treatment)	7–12	Group Triple P (L4) Translation, surface adaptation	Pre, 4 month FU	7.1%	Reduced child general anxiety level and anxiety disorder severity; improved parental anxiety and general wellbeing. Parenting and acceptability not measured.

Continued

Table 19.1 Triple P evaluation in different cultural contexts—cont'd

Author (year)	Country	Language	Design and sample size	Population	Age	Program and adaptation	Assessment times	Program attrition	Main findings[a]
26. Öztürk, Özyurt, and Akay (2018)	Turkey	Turkish	RCT 25 Int; 30 WL	Parents of children with ADHD in Izmir (treatment)	7–12	Group Triple P (L4) Translation, surface adaptation	Pre, Post	8%	Improved parenting (democratic attitude) and child behavior (ADHD symptoms, emotional wellbeing). Acceptability not measured.
27. Pouretemad, Khooshabi, Roshanbin, and Jadidi (2009)	Iran	Persian	Time series 8 Int	Mothers of children with ADHD in Tehran (treatment)	4–10	Group Triple P (L4) Translation, surface adaptation	Baseline, Pre, Mid, Post, 2 month FU	Not reported	Improved parent adjustment (stress); high acceptability. Child outcomes not measured.
28. Pourmohammadreza-Tajrishi, A'shouri, Afrouz, Arjmand-nia, and Ghobari-Bonab (2015)	Iran	Persian	Random clustering 20 Int; 20 Control	Mothers of children with intellectual disability in Varamin and Pakdasht provinces (targeted)	NA	Group Triple P (L4) Translation, surface adaptation	Pre, Post	Not reported	Improved mother-child relationship (in relation to conflict, closeness, dependency, and positive relationship). Child outcomes and acceptability not measured.

29. Sotoudeh Navroodi, Nicknam, Ahmadi, Roodbarde, and Azami (2018)	Iran	Persian	RCT 14 Int; 13 WL	Mothers of children with autism in Rasht (treatment)	6–5	Group Triple P (L4) Translation, surface adaptation	Pre, Post	Not reported	Improved maternal hope. Parenting and child outcomes, and acceptability not measured.
30. Tehrani-Doost, Shahrivar, Mahmoudi-Gharaie, and Alaghband-Rad (2009)	Iran	Persian	Service evaluation 120 Int	Mothers working at Tehran University (universal)	NA	Group Triple P (L4) Translation, surface adaptation	Pre, Post	Not reported	Improved parenting, parenting confidence, and child behavior; reduced conflict over parenting, maternal depression, anxiety and stress; high acceptability. Child outcomes not measured.
Latin America									
31. Barroilhet et al. (2014)	Chile	Spanish	Quasiexperimental 2 group design (Int vs WL) *N* not reported	Parents in low income areas in Santiago (targeted)	NA	Selected Triple P Seminars (L2) Translation, surface adaptation	Pre, Post, 6 month FU	23%	Improved parenting (coercive parenting); no group difference on child and family outcomes (both groups improved); high acceptability

Continued

Table 19.1 Triple P evaluation in different cultural contexts—cont'd

Author (year)	Country	Language	Design and sample size	Population	Age	Program and adaptation	Assessment times	Program attrition	Main findings[a]
32. Becerra Maciel (2016)	United States	Spanish	Retrospective service evaluation	Indigenous Mexican migrant parents (Mixtec, Zapotec, Purepecha) (targeted)	NA	Group Triple P (L4) Translation, deep/structural adaptation (discussion groups about the impact of violence on the mental health of the Indigenous community)	Post	Not reported	Literacy barriers (no written form of Indigenous languages) led to lay Indigenous co-facilitators being employed to reduce stigma and other barriers, and interpret and use experiential ways to convey program messages.
33. Errázuriz, Cerfogli, Moreno, and Soto (2016)	Chile	Spanish	Focus groups 34 Int	Parents in Santiago de Chile (targeted)	2–11	Group Triple P (L4) Translation, surface adaptation	Post	NA	Improved parenting, parent-child relationships, communication among caregivers, and child behavior; decreased parental stress.

34. Mejia, Calam, and Sanders (2015b)	Panama	Spanish	RCT 54 Int; 54 WL	Low income families in Panama City (targeted)	3–12	Primary Care Discussion Group (L3) Translation, surface adaptation	Pre, Post, 3 month FU, 6 month FU	14.8%	Improved parenting (hostility) and child behavior; reduced parental depression, anxiety and stress; high acceptability.
Africa									
35. Haslam et al. (2014)	Kenya	English	Single group design 30 Int	Parents living in Nairobi urban slums (targeted)	2–12	Group Triple P (L4) surface adaptation (tailored examples, additional conversation)	Pre, Post, 6 month FU	Not reported	Improved parenting (overreactivity), confidence and child behavior and emotional well-being; reduced parental depression, anxiety and stress; high acceptability.
36. Haslam et al. (2014)	Kenya	English with Kiswahili inter-preter support	Qualitative focus groups 51 Int	Parents living in Nairobi urban slums (targeted)	2–12	Strategies presented via video demon-stration with no adaptation	Post	NA	High acceptability and intention to use parenting strategies.
37. Haslam, Penman, and Sanders (2018)	Kenya	English	RCT 43 Int: 43 WL	Parents living in Embulbul (targeted)		Primary Care Discussion Group (L3) Deep adaptation (explicitly discussed culture/ beliefs)	Pre, Post	Not reported	No group difference on parenting and child outcomes (both groups improved); high acceptability

Continued

Table 19.1 Triple P evaluation in different cultural contexts—cont'd

Author (year)	Country	Language	Design and sample size	Population	Age	Program and adaptation	Assessment times	Program attrition	Main findings[a]
38. Wessels and Ward (2016)	South Africa	English	Qualitative focus groups 32 Int	Mothers living in battered women's shelters in Cape town	3–8	Strategies presented via video demonstration and tip sheets with no adaptation	Post	NA	High acceptability and intention to use parenting strategies. Some barriers related to written resources.
Refugee, migrant, and Culturally and Linguistically Diverse (CALD) families									
39. Afrif (2018)	New Zealand	English, Arabic with interpreters	Structured interviews First 7 volunteers from 500+ Int	Former refugee mothers from Africa, Sri Lanka, Burma, and Afghanistan (targeted)	1–21	Triple P Discussion Group Series (L3) Translation, surface adaptation, culturally linked practitioners	Post	NA	Improved parenting skills, confidence, child behavior; group format useful; strategies adapted to fit individual families; high acceptability.
40. Morawska et al. (2011)	Australia	English	Qualitative focus groups 137 Int	CALD parents living in Australia	2–10	Strategies presented via video demonstration and example tip sheet with no adaptation	Post	NA	High acceptability and intention to use parenting strategies.

41. Saeid (2018)	New Zealand	English, Arabic with inter-preters	Pilot evaluation 10 Int	Afghani women (targeted)	NA	Triple P Discussion Group Series (L3) Translation, surface adaptation, delivered how and when community requested	Post	Nil	High acceptability of program but lack of willingness to complete assessments (fear, low literacy); need for brief pictorial measures.

Note: *Int*, intervention; *WL*, wait list control.

[a]Results are presented for the longest follow-up assessment reported.

"ways of doing things"), and tailored professional training (e.g., two-way learning including an Indigenous and non-Indigenous trainer, and experiential learning processes) and posttraining support (e.g., mentoring by an experienced practitioner from the same cultural background). These latter deep adaptations have occurred in First Nations communities with a history of Western colonization.

What supports contextual fit of Triple P in different cultures?

Consumer informed program development

Triple P has been built on a consumer engagement approach where the experiences and opinions of parents are sought as potential recipients of programs being developed and also of practitioners who will be trained to deliver programs to families (Kirby & Sanders, 2012; Sanders & Kirby, 2014). In addition, it is designed for flexible tailoring according to parents' own values, priorities, and goals. Because of this, there is often no significant program adaptation required other than flexible delivery (discussed in more detail later). However, in some communities, minor delivery adaptation is insufficient to create a cultural fit and variation or addition of resources may be indicated.

Two processes have been developed to support culturally sensitive delivery of Triple P. Firstly, to preserve program integrity while enhancing accessibility in cultural contexts where there may be significant barriers to program adoption, we developed a process of accommodating cultural content into programs based on a *collaborative participation adaptation model* (CPAM) (Andersson et al., 2019). The CPAM process involves program developers in community consultation and partnership to adapt a program according to local enablers and barriers; acceptability of program content, resources, and delivery methods; and additional delivery support required to enhance program attendance and sustainability. The model involves consulting with elders and community leaders relating to cultural wisdom, as well as practitioners as end users, and parents as consumers. Key processes include sharing existing resources and gaining feedback relating to cultural and contextual fit (e.g., facilitated through structured meetings or focus groups, or independently through surveys and invited feedback), and piloting the adapted program before scaling up (including adapting training, ongoing evaluation, and implementation support). The process often identifies no need for major program adaptation (Morawska et al., 2011), but it has highlighted a need for program adaptations for indigenous communities in Australia (Turner, Richards, & Sanders, 2007) and New Zealand (Keown, Sanders, Franke, & Shepherd, 2018), which have shown positive outcomes.

The CPAM is the first step in exploring the fit of program principles and strategies with local cultural values and goals and developing culturally sensitive program resources and delivery methods to effectively engage families. We are currently developing and evaluating the next step—a *model of engaging communities collaboratively* (MECC) (McIlduff et al., 2020). To enhance community ownership and support for program adoption and long-term sustainment, the MECC processes seek to gain a deeper level of understanding of cultural and/or community dynamics that influence the implementation of programs such as Triple P. It involves a continuing cyclical feedback process to ensure appropriate adaptations to implementation occur. The MECC consists of extended periods of engagement within community: building relationships and establishing trust, mutual benefit and respect; establishing community-identified concerns and priorities; identifying relevant cultural traditions, values, beliefs, and norms; developing and integrating local solutions; appropriately evaluating program fit and outcomes; collaborative and cyclical adaptation; and seeking confirmation of interpretation of program outcomes prior to disseminating findings. This is done in a context of collaboration (e.g., with a local advisory group, organizations, leaders, and affected community members), capacity building and empowerment (e.g., training local practitioners and researchers), as well as addressing community enablers and barriers to program implementation (McIlduff, 2016). These processes are flexible to meet contextual needs, often occur simultaneously with others and can last from 1 week to several months dependent on contextual complexities. Model development has been informed by the knowledge, experience, and guidance of First Nations practitioners internationally, with the aim of developing a protocol that will inform culturally sensitive dissemination of EBPS and other programs, and optimal program implementation procedures for practitioners and agencies.

Focus on self-regulation

A central goal of Triple P is the development of parents' self-regulatory capacity (Sanders & Mazzucchelli, 2013), that is, the ability to monitor and regulate or change their own behavior (Karoly, 1993). Parents are supported to develop: *self-sufficiency* (the resilience, personal resources, knowledge, and skills they require to parent confidently and effectively with minimal additional support); *self-efficacy* (belief that change in their child's and their own behavior is possible and that they can resolve parenting problems); *self-management* (skills to change their parenting practices, such as personal goal setting, self-selection of parenting strategies, self-monitoring, self-evaluation, and further goal setting); a sense of *personal agency* (ownership of the change process and attributing positive change to their own or their child's efforts); and *problem solving* (defining a problem, formulating options, developing a parenting plan, executing it, evaluating the outcome and revising the plan as required) (Sanders & Turner, 2018). As the program offers a toolkit of parenting strategies for parents to develop their own approach according to their values and traditions, there is no imposition of prescriptive advice. Parents are introduced to parenting strategies via discussion, demonstration and role play so they can decide which approaches fit with their parenting goals and they develop a personalized parenting plan. This focus on enhancing parents' self-regulation has supported cross-cultural transportation of the program and relevance in extremely diverse contexts.

Flexible tailoring

Delivering a program with fidelity does not equate to rigid, formulaic delivery. Practitioners are encouraged to use their clinical judgment and ingenuity to flexibly tailor interventions to meet families' diverse needs, but in such a way that the integrity of essential program elements is not jeopardized, and the intervention is not moved beyond its evidence base (Mazzucchelli & Sanders, 2010). Within each level of intervention, considerable tailoring to parents' unique circumstances is possible to enable personally relevant risk and protective factors to be addressed. Key content, such as positive parenting principles and strategies, must be presented to parents. However, teaching examples and methods, homework tasks, and session lengths and numbers can all be varied to meet individual needs and goals. In this way, core concepts and procedures are preserved, but the idiosyncratic needs and preferences of particular parents, families, communities, or cultural groups can also be accommodated. As noted above, given the flexibility of Triple P in tailoring to parents' own values, priorities, and goals, little cultural adaptation may be required other than to find appropriate exemplars of strategies and work with families in a way that fits their learning style (e.g., storytelling or role play rather than relying on video or written resources) and cultural safety (e.g., including an elder or cultural advisor to create a safe place to learn and share).

Therapeutic relationship

Working collaboratively with parents allows responsivity to unique family and community contexts. Triple P promotes a guided participation model

(Sanders & Lawton, 1993) for sharing assessment results, developing a shared perspective of the family's goals, clarifying the nature and possible causes of the presenting problems, negotiating an intervention plan, and offering suggestions rather than prescriptions when parenting strategies are introduced. This process involves inviting parents to identify, from trigger material depicting social learning causes of problem behavior, which of the causes they consider relevant to their family. This encourages parents to take responsibility for understanding their child's motivations and actions, and how their own attributions and behavior may be contributing to a problem. Whether related to a parent's history, culture, or current living context, practitioners are prepared to confront and work through process issues such as disengagement, resistance, defensiveness, dependency, hostility, and conflict (Sanders & Burke, 2014).

Program resources

Program resources are designed to minimize potential barriers to engagement, such as an average sixth-grade reading level in parent resources to avoid difficulties with literacy, use of video and live practitioner modeling to demonstrate parenting strategies, and the inclusion of families from diverse ethnic and cultural backgrounds in the video demonstration and print resources. While positive parenting principles and strategies can cross cultures, what may vary according to culture are the goals and target behaviors, practical implementation of strategies, and ways of sharing information. Triple P resources have to date had 22 language translations from the original UK/US English (Arabic, Berber, Simplified Chinese, Traditional Chinese, Danish, Dutch, Flemish, French (universal, Canadian, European), German, Greek, Indonesian, Japanese, Malay, Māori, Papiamentu, Portuguese (European), Romanian, Spanish (Americas, Chile, United States), Swedish, Turkish, Urdu, Vietnamese), with surprisingly little need for other adaptation as acceptability and usefulness of the principles and strategies, seem almost universal (Morawska et al., 2011).

Criteria for making evidence-based parenting support deployable in culturally diverse and low-resource environments

Sociocultural context

Cultural nuances bring richness to family life and can be harnessed to create the most responsive and nurturing environment possible for children. However, for parenting programs to be successful across cultures, they need to be flexible and sensitive to the broader cultural milieu in which parenting takes place. Many liken culture to an iceberg, in that only a fraction is observable or recognizable, warning that it is those unseen aspects of culture that create risks of misunderstanding when working cross-culturally (Condon & LaBrack, 2015). Thus, flexibility and sensitivity should include consultation with locals who are knowledgeable about the historical and current context within which the population is parenting. This allows for EBPS to have contextual relevance both to traditional wisdom and contemporary ways of doing things. This is imperative as, even within cultures, there is a divergence of experiences, acculturation, values, and traditions. In-depth knowledge of these dynamics is not possible for an "outsider" unless significant time has been spent building relationships through a careful engagement process, and even then, some things will always remain a mystery to the "outsider." While it is ideal for parents to understand their own cultural values and beliefs and guide their children to develop desired qualities, there is often a lack of clarity about traditional values, beliefs, and parenting practices that were lost or ineffectively passed-down through generations due to the effects of acculturation, colonialism, and assimilation (Benzies, 2014). EBPS can, in fact, serve as a conduit to help reconnect families with traditional parenting values. Previous experiences with end-user engagement can help in guiding systematic adaptations of EBPS programs like Triple P and enhance contextual relevance, thereby increasing programs' cultural acceptability and feasibility in culturally diverse and low-resource environments (Mejia, Calam, & Sanders, 2015a).

In addition, macrosystem issues like political and social unrest, socioeconomic volatility, terrorism, inequitable gender norms, large populations, and limited health coverage are concerns that merit careful consideration (Mandal, Muralidharan, & Pappa, 2017; Parra-Cardona et al., 2018). An awareness of macrosystem issues is likely to help develop a better fit between service demand and supply. For example, a parent from a war-torn area in an LMIC may not be seeking the same support as a parent from an LMIC experiencing economic and political stability. Although low-resource environments can present a difficult sociopolitical and economic landscape, the opportunities for change emanate from the fact that LMICs have a young population dealing with multiple co-occurring issues, and intervening early may help in ensuring that people have long productive lives (Das Gupta et al., 2014). This is why investing in EBPS may be merited. The return on investment in parenting support (e.g., Sampaio et al., 2018) and early childhood education (e.g., Heckman, 2017) have been highlighted in HICs terms of improved education, health, social, and economic outcomes.

Resource context

EBPS strategies can never be a "one-size-fits-all" proposition as there is rarely a singular presentation of resource constraint. The World Bank classification of countries based on gross national income (GNI) per capita offers a helpful comparator between countries but does not capture the heterogeneity that exists within countries that are grouped together as low, middle, or high income. This should not, however, lead us to underestimate the relevance of country comparators as they can provide a good indication of public health expenditure needs (Kim, Vonneilich, Lüdecke, & von dem Knesebeck, 2017). For example, in many LMICs there is the simultaneous challenge of chronic infectious diseases (IDs) and a rise in noncommunicable diseases (NCDs) like coronary heart disease, diabetes, and cancer (Remais, Zeng, Li, Tian, & Engelgau, 2013), stretching limited systems for surveillance and disease control. This has led to hierarchical health service provision, where mental health and well-being are relegated to a secondary status in comparison to IDs and NCDs (Patel, 2007). In addition, health-care expenditure in low-resource environments tends to be out of pocket, burdening already resource-constrained people (Kumar et al., 2015), and many are forced to forego health care due to family financial priorities for basic survival needs (Kim et al., 2017).

Even if EBPS programs are planned, limited epidemiological data on child and adolescent health and well-being in low-resource environments may limit the identification of the best (i.e., highest risk) catchment area for intervention efforts (Patel, Flisher, Nikapota, & Malhotra, 2008). Additionally, the presence of a "social gradient in health," where a person with poorer access to resources also experiences greater health concerns than their well-off counterparts, may be more prevalent in low-resource environments necessitating a proportionate universalism approach (Marmot & Bell, 2012). This approach, as discussed earlier, proposes universal action for the reduction of the steepness in the social gradient of health, in which the scale and intensity of action proportionately match the experience of disadvantage. As an example, Triple P's public health approach with its tiered system of support, multiple entry points, and a largely preventive approach can help in providing support where it is needed the most (Sanders & Kirby, 2014).

Although studies in HICs have made a financial case for investing in EBPS, as an investment in EBPS programs like Triple P have a cost-benefit in developed parts of the world (Sampaio et al., 2018). It may be difficult to make a similar case in settings where it is limited or no proposed expenditure on EBPS. However, given the practical need to address multiple sustainable development goals (SDGs), an investment in EBPS may help in disseminating parenting interventions that could well be the glue connecting multiple development issues related to children by enhancing the agentic capacities of families and parents. For instance, intertwining an awareness-raising component on reducing diarrheal deaths in children with a traditional parenting intervention can address SDGs like good health and well-being, and clean water and sanitation.

Workforce context

A widely documented problem with working in low-resource environments, particularly in the area of mental health, is that of treatment gap (Kakuma et al., 2011; Patterson & Edwards, 2018). Put simply, it is a skewed equation between service providers (too few) and service users (too many) (Parra-Cardona et al., 2018). It is interesting to speculate how this may play out for EBPS, which is not necessarily a priority area in low-resource environments, and intervention and prevention services have yet to be established (Parra-Cardona et al., 2018). It is helpful to learn from extant strategies that mental health advocates have tested and employed to counter the trained human resource deficit in LMICs (Buttorff et al., 2012; Patel, Goel, & Desai, 2009). The task-shifting model (also referred to as "task sharing") where practitioners, lay health workers, and community members with lesser training, or training in other disciplines, are equipped with the skills to deliver interventions through educative and supervisory efforts (Day, Michelson, Thomson, Penney, & Draper, 2012; Singla et al., 2014) has growing advocacy and evidence (Healy, Kaiser, & Puffer, 2018; Rajaraman et al., 2012). Engagement with locally available human resources may also help in leveraging cultural wisdom in the delivery of EBPS, understanding the context, history, and learning needs of the service users. The balance to strike is in enlisting local community members (who may or may not have prior qualifications) to deliver EBPS without overburdening those with already heavy community responsibilities. They may implement the program informally through modeling and incidental teaching of strategies to other parents or may become accredited providers themselves if they have a formal parent support role. This is where co-facilitation and ongoing mentoring (Hodge, Turner, Sanders, & Forster, 2017), peer support, and leadership to reduce program burden (Hodge, Turner, Sanders, & Filus, 2017) are vital considerations for program sustainment.

Implications

There is significant evidence for EBPS programs in Western cultures where they are typically developed and researched, but there is a need for greater attention to be paid to issues of context and culture during the development, testing, and implementation of these interventions in LMICs and underserved and disenfranchised communities in HICs (Baumann et al., 2018). A number of myths and assumptions have slowed the progress of scaling EBPS programs, typically related to affordability. It is inappropriate to base costing estimates on program deployment in HICs, where labor costs are high and different formats of intervention might be used. For example, low cost and wide reach intervention options may be more appropriate in LMICs than the intensive multiweek individual programs often deployed in HICs. Some have advocated a "common elements" approach that involves the development of local programs based on general principles drawn from the parenting literature. However, such an approach requires ongoing evaluation and can have hidden costs (e.g., developing materials and resources for parents and practitioners, training and supporting a workforce to deliver programs with fidelity), and risks (e.g., the possibility of reduced or no program efficacy, difficulties sustaining implementation, inability to revise resources over time in response to evolving evidence). The major risk is the promulgation of programs that are not necessarily cheaper in the long run and that have no evidence of effectiveness.

Policy

A concerted effort has be made globally to decrease the rates of child mortality by investing in research and intervention on maternal and child health (Bishai et al., 2016). As child mortality rates decrease, the next step is to build the infrastructure to ensure that children thrive and not just survive. With a focus on major health system reforms, there is a unique opportunity available to tackle health and well-being challenges with novel approaches (Remais et al., 2013). To achieve sustainable development goals such as reducing poverty and inequality, and ensuring primary school completion for both girls and boys, governments must consider expanding high quality, cost-effective early child development programs (Engle et al., 2007) and EBPS offers a significant part of the solution. Coordinated public health spending to address surveillance, prevention, and treatment of disease and mental health issues is crucial to improve access to services, initially focusing on the most vulnerable populations, to begin to reduce population prevalence (Remais et al., 2013).

As cultural adaptation is not free of cost, dedicated funding is required to adapt and deliver evidence-based and culturally informed parenting programs to address the needs of diverse families (Sanders, Higgins, & Prinz, 2018). Initiatives that use EBPS programs that apply population-health principles should be funded as a public policy priority, including services appropriate to different cultural groups in the country. Long-term funding is needed to ensure proper planning, interagency engagement and collaboration, the development of evaluation frameworks, and implementation plans conforming to best-practice principles derived from implementation science. A culture of prevention (Parra-Cardona et al., 2018) needs to be strengthened and performance-based contracting and outcomes-focused government funding mechanisms developed for agencies (both not-for-profit and private) to implement EBPS. This requires agencies to incorporate manageable reporting on key clinical outcomes for each participating family. Funds are needed for rigorous tracking of outcomes through the linking of administrative data such as reports of child maltreatment, hospitalization due to child maltreatment-related injuries, and out-of-home care placements, where available.

Research

To maximize learnings from trials involving transported, culturally adapted programs, and implementation processes, their needs to be greater transparency in describing the adaptation process of efficacy, effectiveness, and scale up evaluations (van Mourik, Crone, de Wolff, & Reis, 2017). Because of the costs of funding randomized controlled trials (RCTs), findings from existing trials can be built upon using a well-constructed quasiexperimental design and longitudinal-observational studies, where RCT data is available in other contexts. New programs may require an incremental process of gathering evidence through varied methodologies to merit the employment of RCTs should funding become available. As administrative data sources are simply not available or are unreliable in many LMICs, it would be very useful to develop and validate a global parenting and family well-being survey to establish international comparators that would capture country-based population-level indicators of parenting capacity (knowledge, skills, behaviors) and family mental health and well-being. The availability of such tools would create momentum to invest in EBPS to ensure children's well-being is enhanced. Program evaluations could use such population-level indicators to link to other available administrative data to track change over time. Substantial implementation science research is needed in culturally diverse communities to promote the sustained deployment of culturally adapted evidence-based practices.

Practice

As noted earlier, much clinical nuance is required in working sensitively across cultures. While program content may not need to be drastically altered, practitioners need support in cultural sensitivity when working in communities that are not their own. Using a global lens to reach diverse families, particularly in low-resource settings where there is inadequate or no availability of professionals, the answer may be to find and support the capacity of community volunteers and lay personnel as EBPS practitioners or co-facilitators in delivering parenting programs. In this case, there must be attention paid to managing the burden of trauma and competing demands if they are local to complex, unstable communities. A possible solution to the potential barrier of practitioner training costs is to use new technologies to deliver professional training online. However, the evidence is yet required that satisfactory training outcomes can be achieved. This is an important area for future development and testing. There also need to be innovative program delivery models, such as media outreach, performance, text messaging, and low literacy resources. One creative example of the latter was a campaign using bread wrappers to distribute parenting information to adults looking after children in conflict zones in Syria (El-Khani, Cartwright, Redmond, & Calam, 2016).

Conclusion

There is evidence for the efficacy of many different approaches to disseminating parenting support in culturally diverse communities, from surface-level adaptations (e.g., language, graphics) to proven EBPS programs, to deep-level content adaptations addressing cultural values and traditions, to locally developed bespoke approaches (Mejia, Leijten, Lachman, & Parra-Cardona, 2017). However, given the time and cost involved in program development and the scientific evaluation required to reach evidence-based standard criteria, rather than creating new programs and training systems for each culturally diverse community, culturally adapting an existing EBPS program to match local needs is a cost-effective approach (Kumpfer, Magalhães, & Xie, 2017) that can be at least as effective as for the population for which the program was developed (Ortiz & Del Vecchio, 2013) and can significantly enhance family engagement (Kumpfer, Alvarado, Smith, & Bellamy, 2002). This stance is supported by our own experience with Triple P. While clinical judgment in balancing program fidelity and flexible delivery tailored to client needs is at the core of most EBPS programs if contextual or cultural fit is identified as a significant barrier to engagement, program developers can collaborate directly with communities to support locally tailored delivery, and resource adaptation and augmentation as required (Andersson et al., 2019).

It is concerning that so few parents worldwide participate in EBPS programs that can equip them with the skills and confidence to navigate their parenting journey well. To change this situation, there needs to be a policy-based investment to train and support a workforce that can deliver a variety of culturally informed EBPS programs relevant to the community and individual needs. Improved feedback loops between parenting and child development research, government policy and professional service delivery can only serve to enhance child, parent, family, and community outcomes. It is also incumbent on program developers to design programs that can be delivered flexibly in a culturally sensitive way, with appropriate implementation support to ensure fidelity to core program content. Now that the possibility of effectively transporting EBPS across cultures is proven, the field will benefit from further research into the factors that support effective program accommodation and adaptation, to increase program reach and access so that all families can be supported in their parenting role.

References

Afrif, A. (2018). *An exploration of former refugees' experiences of the Triple P Discussion Groups: A multi-informant approach*. Paper presented at the Helping Families Change Conference, Santa Rosa CA, USA.

Aghebati, A., Gharraee, B., Hakim Shoshtari, M., & Gohari, M. R. (2014). Triple P-Positive Parenting Program for mothers of ADHD children. *Iranian Journal of Psychiatry and Behavioral Sciences*, *8*(1), 59–65.

Andersson, E., McIlduff, C. D., Turner, K. M. T., Thomas, S., Davies, J., Elliott, E., … Jandu Yani, U. (2019). 'For All Families' Triple P—Positive Parenting Program in remote Australian Aboriginal communities: A study protocol for a community intervention trial. *BMJ Open*, *9*(10), 1–9. https://doi.org/10.1136/bmjopen-2019-032559.

Arkan, B. (2014). *Randomised controlled trial of group teen triple p for Turkish parents of teens with behavior problems*. Paper presented at the Helping Families Change Conference, Sydney, Australia.

Au, A., Lau, K., Wong, A. H., Lam, C., Leung, C., Lau, J., & Lee, Y. K. (2014). The efficacy of a Group Triple P (Positive Parenting Program) for Chinese parents with a child diagnosed with ADHD in Hong Kong: A pilot randomised controlled study. *Australian Psychologist*, *49*(3), 151–162. https://doi.org/10.1111/ap.12053.

Baer, D. M., Wolf, M. M., & Risley, T. R. (1968). Some current dimensions of applied behavior analysis. *Journal of Applied Behavior Analysis*, *1*(1), 91–97. https://doi.org/10.1901/jaba.1968.1-91.

Bandura, A. (1977). *Social learning theory*. Englewood Cliffs, NJ: Prentice-Hall.

Barroilhet, S., Haslam, D., Gaete, J., Barrueto, C., Puga, F., Fritsch, R., … Sanders, M. R. (2014). *Evaluating Triple P seminars in low resource Chilean schools*. Paper presented at the Helping Families Change Conference, Sydney, Australia.

Baumann, A. A., Mejia, A., Lachman, J. M., Parra-Cardona, R., López-Zerón, G., Amador Buenabad, N. G., … Domenech Rodríguez, M. M. (2018). Parenting programs for underserved populations in low- and middle-income countries: Issues of scientific integrity and social justice. *Global Social Welfare*, https://doi.org/10.1007/s40609-018-0121-0.

Beach, S. R., Brody, G. H., Barton, A. W., & Philibert, R. A. (2016). Exploring genetic moderators and epigenetic mediators of contextual and family effects: From gene x environment to epigenetics. *Developmental Psychopathology*, *28*(4pt2), 1333–1346. https://doi.org/10.1017/S0954579416000882.

Becerra Maciel, M. (2016). *Using Triple P with underserved Mixteco/indigenous populations in the child welfare system*. Paper presented at the Helping Families Change Conference, Banff, Canada.

Benzies, K. (2014). Parenting in Canadian Aboriginal cultures. In H. Selin (Ed.), *Parenting across cultures: Childrearing, motherhood and fatherhood in non-western cultures, science across cultures: The history of non-western science* (pp. 379–392). New York: Springer.

Bhawuk, D. P. S. (2015). Beliefs, values, norms, and customs (definitions). In J. M. Bennett (Ed.), *The SAGE encyclopedia of intercultural competence*. Thousand Oaks, CA: Sage.

Biglan, A., Flay, B. R., Embry, D. D., & Sandler, I. N. (2012). The critical role of nurturing environments for promoting human well-being. *American Psychologist*, *67*(4), 257–271. https://doi.org/10.1037/a0026796.

Bishai, D. M., Cohen, R., Alfonso, Y. N., Adam, T., Kuruvilla, S., & Schweitzer, J. (2016). Factors contributing to maternal and child mortality reductions in 146 low- and middle-income countries between 1990 and 2010. *PLoS One*, *11*(1), e0144908. https://doi.org/10.1371/journal.pone.0144908.

Bornstein, M. H., & Bohr, Y. (2011). Immigration, acculturation and parenting. In R. E. Tremblay, M. Boivin, & R. D. Peters (Eds.), *Encyclopedia on early childhood development*. Retrieved from http://www.child-encyclopedia.com/immigration/according-experts/immigration-acculturation-and-parenting.

Bornstein, M. H. & Bradley, R. H. (Eds.), (2014). *Socioeconomic status, parenting, and child development*. New York: Routledge.

Bradley, R., & Corwyn, R. (2005). Caring for children around the world: A view from HOME. *International Journal of Behavioral Development*, *29*(6), 468–478. https://doi.org/10.1080/01650250500146925.

Buttorff, C., Hock, R. S., Weiss, H. A., Naik, S., Araya, R., Kirkwood, B. R., … Patel, V. (2012). Economic evaluation of a task-shifting intervention for common mental disorders in India. *Bulletin of the World Health Organization*, *90*(11), 813–821. https://doi.org/10.2471/BLT.12.104133.

Center on the Developing Child at Harvard University. (2016). *From best practices to breakthrough impacts: A science-based approach to building a more promising future for young children and families* Retrieved from. https://developingchild.harvard.edu/resources/from-best-practices-to-breakthrough-impacts/.

Chan, S., Leung, C., & Sanders, M. (2016). A randomised controlled trial comparing the effects of directive and non-directive parenting programmes as a universal prevention programme. *Journal of Children's Services*, *11*(1), 38–53. https://doi.org/10.1108/Jcs-08-2014-0038.

Condon, J., & LaBrack, B. (2015). Culture, definition of. In J. M. Bennett (Ed.), *The SAGE encyclopedia of intercultural competence*. Thousand Oaks, CA: Sage.

Cooper, P. J., Tomlinson, M., Swartz, L., Landman, M., Molteno, C., Stein, A., … Murray, L. (2009). Improving quality of mother-infant relationship and infant attachment in socio-economically deprived community in South Africa: Randomised controlled trial. *British Medical Journal, 338*, b974. https://doi.org/10.1136/bmj.b974.

Crisante, L., & Ng, S. (2003). Implementation and process issues in using Group Triple P with Chinese parents: Preliminary findings. *Australian e-Journal for the Advancement of Mental Health, 2*(3). https://doi.org/10.5172/jamh.2.3.226.

Das Gupta, M., Engelman, R., Levy, J., Luchsinger, G., Merrick, T., & Rosen, J. E. (2014). *The power of 1.8 billion: Adolescents, youth and the transformation of the future*. Retrieved from https://www.unfpa.org/sites/default/files/pub-pdf/EN-SWOP14-Report_FINAL-web.pdf.

Day, C., Michelson, D., Thomson, S., Penney, C., & Draper, L. (2012). Evaluation of a peer led parenting intervention for disruptive behaviour problems in children: Community based randomised controlled trial. *British Medical Journal*, 344. https://doi.org/10.1136/bmj.e1107.

Dodge, K. (1986). A social information processing model of social competence in children. In M. Perlmutter (Ed.), *Minnesota symposia on child psychology*: Vol. 18. Cognitive perspectives on children's social and behavioral development (pp. 77–125). Hillsdale, NJ: Lawrence Erlbaum.

Eamon, M. K. (2000). Structural model of the effects of poverty on externalizing and internalizing behaviors of four- to five-year-old children. *Social Work Research, 24*(3), 143–154. https://doi.org/10.1093/swr/24.3.143.

El-Khani, A., Cartwright, K., Redmond, A., & Calam, R. (2016). Daily bread: A novel vehicle for dissemination and evaluation of psychological first aid for families exposed to armed conflict in Syria. *Global Mental Health, 3*, e15. https://doi.org/10.1017/gmh.2016.9.

Engle, P. L., Black, M. M., Behrman, J. R., Cabral de Mello, M., Gertler, P. J., Kapiriri, L., … the International Child Development Steering Group. (2007). Strategies to avoid the loss of developmental potential in more than 200 million children in the developing world. *Lancet, 369*(9557), 229–242. https://doi.org/10.1016/S0140-6736(07)60112-3.

Engle, P. L., Fernald, L. C. H., Alderman, H., Behrman, J., O'Gara, C., Yousafzai, A., … Iltus, S. (2011). Strategies for reducing inequalities and improving developmental outcomes for young children in low-income and middle-income countries. *The Lancet, 378*(9799), 1339–1353. https://doi.org/10.1016/S0140-6736(11)60889-1.

Errázuriz, P., Cerfogli, C., Moreno, G., & Soto, G. (2016). Perception of Chilean parents on the Triple P program for improving parenting practices. *Journal of Child and Family Studies, 25*(11), 3440–3449. https://doi.org/10.1007/s10826-016-0492-8.

Farquhar, J. W., Fortmann, S. P., Maccoby, N., Haskell, W. L., Williams, P. T., Flora, J. A., … Hulley, S. B. (1985). The Stanford Five-City Project: Design and methods. *American Journal of Epidemiology, 122*(2), 323–334. https://doi.org/10.1093/oxfordjournals.aje.a114104.

Felitti, V. J., Anda, R. F., Nordenberg, D., Williamson, D. F., Spitz, A. M., Edwards, V., … Marks, J. S. (1998). Relationship of childhood abuse and household dysfunction to many of the leading causes of death in adults: The Adverse Childhood Experiences (ACE) Study. *American Journal of Preventive Medicine, 14*(4), 245–258. https://doi.org/10.1016/S0749-3797(98)00017-8.

Fujiwara, T., Kato, N., & Sanders, M. R. (2011). Effectiveness of Group Positive Parenting Program (Triple P) in changing child behavior, parenting style, and parental adjustment: An intervention study in Japan. *Journal of Child and Family Studies, 20*(6), 804–813. https://doi.org/10.1007/s10826-011-9448-1.

Fung, J., Wong, M. S., & Park, H. (2018). Cultural background and religious beliefs. In M. R. Sanders & A. Morawska (Eds.), *Handbook of parenting and child development across the lifespan* (pp. 469–494). New York: Springer.

Gardner, F., Montgomery, P., & Knerr, W. (2016). Transporting evidence-based parenting programs for child problem behavior (age 3-10) between countries: Systematic review and meta-analysis. *Journal of Clinical Child and Adolescent Psychology*, *45*(6), 749–762. https://doi.org/10.1080/15374416.2015.1015134.

Greenfield, P. M., Keller, H., Fuligni, A., & Maynard, A. (2003). Cultural pathways through universal development. *Annual Review of Psychology*, *54*, 461–490. https://doi.org/10.1146/annurev.psych.54.101601.145221.

Guo, M., Morawska, A., & Sanders, M. R. (2016). A randomized controlled trial of Group Triple P with Chinese parents in mainland China. *Behavior Modification*, *40*(6), 825–851. https://doi.org/10.1177/0145445516644221.

Guvenir, T., Varol Tas, F., Ozbek, A., Gencer, O., Engin, G., Yaylaci, F., & Karaca, B. (2014). *Balcova's children will grow happier: First population roll out of Triple P in Turkey*. Paper presented at the Helping Families Change Conference, Sydney, Australia.

Hart, B. M., & Risley, T. R. (1995). *Meaningful differences in the everyday experience of young American children*. Baltimore, MD: Paul. H. Brooks.

Haslam, D., Gitonga, J., Okwach, J., Sanders, M., Johnson, C., & Brown, J. (2014). *Positive parenting for parents living in informal settlements in Kenya: Is Triple P relevant?* . Paper presented at the Helping Families Change Conference, Sydney, Australia.

Haslam, D., Penman, N., & Sanders, M. R. (2018). *A pilot RCT of Disobedience Discussion Groups in a very low resource setting* Paper presented at the Helping Families Change Conference, Santa Rosa CA, USA. http://helpingfamilieschange.org/wp-content/uploads/2018/04/C4.4-A-pilot-RCT-of-Disobedience-Discussion-Groups-in-a-very-low-resource-setting-Divna-Haslam.pdf.

Healy, E. A., Kaiser, B. N., & Puffer, E. S. (2018). Family-based youth mental health interventions delivered by nonspecialist providers in low- and middle-income countries: A systematic review. *Families, Systems & Health*, *36*(2), 182–197. https://doi.org/10.1037/fsh0000334.

Heckman, J. J. (2017). *There's more to gain by taking a comprehensive approach to early childhood development*. Retrieved from https://heckmanequation.org/www/assets/2017/01/F_Heckman_CBAOnePager_120516.pdf.

Hodge, L. M., Turner, K. M. T., Sanders, M. R., & Filus, A. (2017). Sustained Implementation Support Scale: Validation of a measure of program characteristics and workplace functioning for sustained program implementation. *The Journal of Behavioral Health Services & Research*, *44*(3), 442–464. https://doi.org/10.1007/s11414-016-9505-z.

Hodge, L. M., Turner, K. M. T., Sanders, M. R., & Forster, M. (2017). Factors that influence evidence-based program sustainment for family support providers in child protection services in disadvantaged communities. *Child Abuse and Neglect*, *70*, 134–145. https://doi.org/10.1016/j.chiabu.2017.05.017.

Houlding, C., Schmidt, F., Stern, S. B., Jamieson, J., & Borg, D. (2012). The perceived impact and acceptability of Group Triple P Positive Parenting Program for Aboriginal parents in Canada. *Children and Youth Services Review*, *34*(12), 2287–2294. https://doi.org/10.1016/j.childyouth.2012.08.001.

Jin, X., Sun, Y., Jiang, F., Ma, J., Morgan, C., & Shen, X. (2007). "Care for Development" intervention in rural China: A prospective follow-up study. *Journal of Developmental & Behavioral Pediatrics*, *28*(3), 213–218. https://doi.org/10.1097/dbp.0b013e31802d410b.

Kakuma, R., Minas, H., van Ginneken, N., Dal Poz, M. R., Desiraju, K., Morris, J. E., … Scheffler, R. M. (2011). Global Mental Health 5: Human resources for mental health care: Current situation and strategies for action. *Lancet*, *378*(9803), 1654–1663. https://doi.org/10.1016/S0140-6736(11)61093-3.

Karoly, P. (1993). Mechanisms of self-regulation: A systems view. *Annual Review of Psychology*, *44*, 23–52. https://doi.org/10.1146/annurev.ps.44.020193.000323.

Keown, L. J., Sanders, M. R., Franke, N., & Shepherd, M. (2018). Te Whanau Pou Toru: A randomized controlled trial (RCT) of a culturally adapted low-intensity variant of the Triple P-Positive Parenting Program for indigenous Maori families in New Zealand. *Prevention Science*, *19*(7), 954–965. https://doi.org/10.1007/s11121-018-0886-5.

Kheirie, M., Shaeiri, M. R., Azad Fallah, P., & Rasulzade Tabatabaei, K. (2009). Effect of the Triple P-Positive Parenting Program on children with oppositional defiant disorder. *Journal of Behavior Science*, *3*(1), 53–58.

Kim, T. J., Vonneilich, N., Lüdecke, D., & von dem Knesebeck, O. (2017). Income, financial barriers to health care and public health expenditure: A multilevel analysis of 28 countries. *Social Science & Medicine*, *176*, 158–165. https://doi.org/10.1016/j.socscimed.2017.01.044.

Kirby, J. N., & Sanders, M. R. (2012). Using consumer input to tailor evidence-based parenting interventions to the needs of grandparents. *Journal of Child and Family Studies*, *21*(4), 626–636. https://doi.org/10.1007/s10826-011-9514-8.

Klein, P. S., & Rye, H. (2004). Interaction-oriented early intervention in Ethiopia: The MISC approach. *Infants & Young Children*, *17*(4), 340–354.

Kumar, K., Singh, A., Kumar, S., Ram, F., Singh, A., Ram, U., ... Kowal, P. R. (2015). Socioeconomic differentials in impoverishment effects of out-of-pocket health expenditure in China and India: Evidence from WHO SAGE. *PLoS One*, *10*(8), e0135051. https://doi.org/10.1371/journal.pone.0135051.

Kumpfer, K. L., Alvarado, R., Smith, P., & Bellamy, N. (2002). Cultural sensitivity and adaptation in family-based prevention interventions. *Prevention Science*, *3*(3), 241–246. https://doi.org/10.1023/a:1019902902119.

Kumpfer, K. L., Magalhães, C., & Xie, J. (2017). Cultural adaptation and implementation of family evidence-based interventions with diverse populations. *Prevention Science*, *18*(6), 649–659. https://doi.org/10.1007/s11121-016-0719-3.

Kumpfer, K. L., Pinyuchon, M., Teixeira de Melo, A., & Whiteside, H. O. (2008). Cultural adaptation process for international dissemination of the strengthening families program. *Evaluation & the Health Professions*, *31*(2), 226–239. https://doi.org/10.1177/0163278708315926.

Kumpfer, K. L., United Nations Office on Drugs and Crime. (2009). *Guide to implementing family skills training programmes for drug abuse prevention*. New York: United Nations.

Leung, C., Fan, A., & Sanders, M. R. (2013). The effectiveness of a Group Triple P with Chinese parents who have a child with developmental disabilities: A randomized controlled trial. *Research in Developmental Disabilities*, *34*(3), 976–984. https://doi.org/10.1016/j.ridd.2012.11.023.

Leung, C., Sanders, M. R., Ip, F., & Lau, J. (2006). Implementation of Triple P-Positive Parenting Program in Hong Kong: Predictors of programme completion and clinical outcomes. *Journal of Children's Services*, *1*(2), 4–17. https://doi.org/10.1108/17466660200600010.

Leung, C., Sanders, M. R., Leung, S., Mak, R., & Lau, J. (2003). An outcome evaluation of the implementation of the Triple P-Positive Parenting Program in Hong Kong. *Family Process*, *42*(4), 531–544. https://doi.org/10.1111/j.1545-5300.2003.00531.x.

Mandal, M., Muralidharan, A., & Pappa, S. (2017). A review of measures of women's empowerment and related gender constructs in family planning and maternal health program evaluations in low- and middle-income countries. *BMC Pregnancy and Childbirth*, *17*(2), 119–127. https://doi.org/10.1186/s12884-017-1500-8.

Marmot, M., & Bell, R. (2012). Fair society, healthy lives. *Public Health*, *126*(Suppl. 1), S4–S10. https://doi.org/10.1016/j.puhe.2012.05.014.

Matsumoto, Y., Sofronoff, K., & Sanders, M. R. (2007). The efficacy and acceptability of the Triple P-Positive Parenting Program with Japanese parents. *Behaviour Change*, *24*(4), 205–218. https://doi.org/10.1375/bech.24.4.205.

Matsumoto, Y., Sofronoff, K., & Sanders, M. R. (2010). Investigation of the effectiveness and social validity of the Triple P Positive Parenting Program in Japanese society. *Journal of Family Psychology*, *24*(1), 87–91. https://doi.org/10.1037/a0018181.

Mazzucchelli, T. G., & Sanders, M. R. (2010). Facilitating practitioner flexibility within an empirically supported intervention: Lessons from a system of parenting support. *Clinical Psychology: Science and Practice*, *17*(3), 238–252.

McIlduff, C. (2016). *Engaging indigenous communities collaboratively towards positive change: Assessing the efficacy of the Model of Engaging Communities Collaboratively*. Paper presented at the Helping Families Change Conference, Banff, Canada.

McIlduff, C.D., Forster, M., Davies, J., Turner, K.M.T., Brown Wilson, C., & Sanders, M.R. (2020). *Model of Engaging Communities Collaboratively: Working towards an integration of implementation science, cultural adaptation and engagement* (Manuscript submitted for publication).

Mejia, A., Calam, R., & Sanders, M. R. (2012). A review of parenting programs in developing countries: Opportunities and challenges for preventing emotional and behavioral difficulties in children. *Clinical Child and Family Psychology Review*, *15*(2), 163–175. https://doi.org/10.1007/s10567-012-0116-9.

Mejia, A., Calam, R., & Sanders, M. R. (2015a). Examining delivery preferences and cultural relevance of an evidence-based parenting program in a low-resource setting of Central America: Approaching parents as consumers. *Journal of Child and Family Studies*, *24*(4), 1004–1015. https://doi.org/10.1007/s10826-014-9911-x.

Mejia, A., Calam, R., & Sanders, M. R. (2015b). A pilot randomized controlled trial of a brief parenting intervention in low-resource settings in Panama. *Prevention Science*, *16*(5), 707–717. https://doi.org/10.1007/s11121-015-0551-1.

Mejia, A., Leijten, P., Lachman, J. M., & Parra-Cardona, J. R. (2017). Different strokes for different folks? Contrasting approaches to cultural adaptation of parenting interventions. *Prevention Science*, *18*(6), 630–639. https://doi.org/10.1007/s11121-016-0671-2.

Moffitt, T. E., Klaus-Grawe Think Tank. (2013). Childhood exposure to violence and lifelong health: Clinical intervention science and stress-biology research join forces. *Development and Psychopathology*, *25*(4 Pt 2), 1619–1634. https://doi.org/10.1017/S0954579413000801.

Moharreri, F., Shahrivar, Z., Tehrani-Doost, M., & Mahmoudi-Gharaei, J. M. (2008). Efficacy of the Positive Parenting Program (Triple P) for a group of parents of children with attention deficit hyperactivity disorder (ADHD). *Iranian Journal of Psychiatry*, *3*(2), 59–63.

Morawska, A., Sanders, M., Goadby, E., Headley, C., Hodge, L., McAuliffe, C., … Anderson, E. (2011). Is the Triple P-Positive Parenting Program acceptable to parents from culturally diverse backgrounds? *Journal of Child and Family Studies*, *20*(5), 614–622. https://doi.org/10.1007/s10826-010-9436-x.

Mounier-Jack, S., Mayhew, S. H., & Mays, N. (2017). Integrated care: Learning between high-income, and low- and middle-income country health systems. *Health Policy and Planning*, *32*(Suppl. 4), iv6–iv12. https://doi.org/10.1093/heapol/czx039.

Najman, J. M., Clavarino, A., McGee, T. R., Bor, W., Williams, G. M., & Hayatbakhsh, M. R. (2010). Timing and chronicity of family poverty and development of unhealthy behaviors in children: A longitudinal study. *Journal of Adolescent Health*, *46*(6), 538–544. https://doi.org/10.1016/j.jadohealth.2009.12.001.

Najman, J. M., Hayatbakhsh, M. R., Clavarino, A., Bor, W., O'Callaghan, M. J., & Williams, G. M. (2010). Family poverty over the early life course and recurrent adolescent and young adult anxiety and depression: A longitudinal study. *American Journal of Public Health*, *100*(9), 1719–1723. https://doi.org/10.2105/AJPH.2009.180943.

Nikulina, V., & Widom, C. S. (2014). Do race, neglect, and childhood poverty predict physical health in adulthood? A multilevel prospective analysis. *Child Abuse & Neglect*, *38*(3), 414–424. https://doi.org/10.1016/j.chiabu.2013.09.007.

Ortiz, C., & Del Vecchio, T. (2013). Cultural diversity: Do we need a new wake-up call for parent training? *Behavior Therapy*, *44*(3), 443–458. https://doi.org/10.1016/j.beth.2013.03.009.

Oveisi, S., Ardabili, H. E., Dadds, M. R., Majdzadeh, R., Mohammadkhani, P., Rad, J. A., & Shahrivar, Z. (2010). Primary prevention of parent-child conflict and abuse in Iranian mothers: A randomized-controlled trial. *Child Abuse & Neglect*, *34*(3), 206–213. https://doi.org/10.1016/j.chiabu.2009.05.008.

Öztürk, Y., Özyurt, G., & Akay, A. P. (2018). The effects of the Triple P-Positive Parenting Programme on parenting, family functioning and symptoms of attention-deficit/hyperactivity disorder: A randomized controlled trial. *Psychiatry and Clinical Psychopharmacology*, 1–9. https://doi.org/10.1080/24750573.2018.1542189.

Ozyurt, G., Gencer, O., Ozturk, Y., & Ozbek, A. (2016). Is Triple P Positive Parenting Program effective on anxious children and their parents? 4th month follow up results. *Journal of Child and Family Studies*, *25*(5), 1646–1655. https://doi.org/10.1007/s1082.

Parra-Cardona, R., Leijten, P., Lachman, J. M., Mejía, A., Baumann, A. A., Amador Buenabad, N. G., … Domenech Rodríguez, M. M. (2018). Strengthening a culture of prevention in low- and middle-income countries: Balancing scientific expectations and contextual realities. *Prevention Science*, 1–11. https://doi.org/10.1007/s11121-018-0935-0.

Patel, V. (2007). Mental health in low- and middle-income countries. *British Medical Bulletin*, *81–82*(1), 81–96. https://doi.org/10.1093/bmb/ldm010.

Patel, V., Flisher, A. J., Nikapota, A., & Malhotra, S. (2008). Promoting child and adolescent mental health in low and middle income countries. *Journal of Child Psychology and Psychiatry*, *49*(3), 313–334. https://doi.org/10.1111/j.1469-7610.2007.01824.x.

Patel, V., Goel, D. S., & Desai, R. (2009). Scaling up services for mental and neurological disorders in low-resource settings. *International Health*, *1*(1), 37–44. https://doi.org/10.1016/j.inhe.2009.02.002.

Patterson, G. R. (1982). *A social learning approach to family intervention: III. Coercive family process*. Eugene, OR: Castalia.

Patterson, J. E., & Edwards, T. M. (2018). An introduction to global mental health. *Families, Systems & Health*, *36*(2), 137–143. https://doi.org/10.1037/fsh0000360.

Pouretemad, H. R., Khooshabi, K., Roshanbin, M., & Jadidi, M. (2009). The effectiveness of group positive parenting program on parental stress of mothers of children with attention-deficit/hyperactivity disorder. *Archives of Iranian Medicine*, *12*(1), 60–68.

Pourmohammadreza-Tajrishi, M. S., A'shouri, M., Afrouz, G. A., Arjmand-nia, A. A., & Ghobari-Bonab, B. (2015). The effectiveness of Positive Parenting Program (Triple-P) training on interaction of mother-child with intellectual disability. *Archives of Rehabilitation*, *16*(2), 128–137.

Rajaraman, R., Travasso, S., Chatterjee, A., Bhat, B., Andrew, G., Parab, S., & Patel, V. (2012). The acceptability, feasibility and impact of a lay health counsellor delivered health promoting schools programme in India: A case study evaluation. *BMC Health Services Research*, *12*(127), https://doi.org/10.1186/1472-6963-12-127.

Remais, J. V., Zeng, G., Li, G., Tian, L., & Engelgau, M. M. (2013). Convergence of non-communicable and infectious diseases in low- and middle-income countries. *International Journal of Epidemiology*, *42*(1), 221–227. https://doi.org/10.1093/ije/dys135.

Rutter, M. (1985a). Family and school influences on behavioural development. *Journal of Child Psychology and Psychiatry*, *26*(3), 349–368. https://doi.org/10.1111/j.1469-7610.1985.tb01938.x.

Rutter, M. (1985b). Resilience in the face of adversity. Protective factors and resistance to psychiatric disorder. *British Journal of Psychiatry*, *147*, 598–611. https://doi.org/10.1192/bjp.147.6.598.

Saeid, F. (2018). *Triple P for Afghan women: A pilot project for a mothers' and daughters' sewing group*. Paper presented at the Helping Families Change Conference, Santa Rosa CA, USA.

Sampaio, F., Barendregt, J. J., Feldman, I., Lee, Y. Y., Sawyer, M. G., Dadds, M. R., … Mihalopoulos, C. (2018). Population cost-effectiveness of the Triple P parenting programme for the treatment of conduct disorder: An economic modelling study. *European Child & Adolescent Psychiatry*, *27*(7), 933–944. https://doi.org/10.1007/s00787-017-1100-1.

Sanders, M. R. (2012). Development, evaluation, and multinational dissemination of the Triple P-Positive Parenting Program. *Annual Review of Clinical Psychology*, *8*, 345–379. https://doi.org/10.1146/annurev-clinpsy-032511-143104.

Sanders, M. R., & Burke, K. (2014). The "hidden" technology of effective parent consultation: A guided participation model for promoting change in families. *Journal of Child and Family Studies*, *23*(7), 1289–1297. https://doi.org/10.1007/s10826-013-9827-x.

Sanders, M. R., Higgins, D., & Prinz, R. (2018). A population approach to the prevention of child maltreatment: Rationale and implications for research, policy and practice. *Family Matters*, *100*, 62–70. Retrieved from https://aifs.gov.au/sites/default/files/8_a_population_approach_to_the_prevention_of_child_maltreatment.pdf.

Sanders, M. R., & Kirby, J. N. (2014). A public health approach to improving parenting and promoting children's well-being. *Child Development Perspectives*, *8*(4), 250–257. https://doi.org/10.1111/cdep.12086.

Sanders, M. R., & Lawton, J. M. (1993). Discussing assessment findings with families: A guided participation model of information transfer. *Child and Family Behavior Therapy*, *15*(2), 5–35. https://doi.org/10.1300/J019v15n02_02.

Sanders, M. R., & Mazzucchelli, T. G. (2013). The promotion of self-regulation through parenting interventions. *Clinical Child and Family Psychology Review*, *16*(1), 1–17. https://doi.org/10.1007/s10567-013-0129-z.

Sanders, M. R., & Mazzucchelli, T. G. (2018). How parenting influences the lives of children. In M. R. Sanders & T. G. Mazzucchelli (Eds.), *The power of positive parenting: Transforming the lives of children, parents, and communities using the Triple P system* (pp. 5–31). New York: Oxford University Press.

Sanders, M. R., & Turner, K. M. T. (2018). The Triple P system: Parenting support for every family. In B. H. Fiese (Ed.), Vol. 3. *American Psychological Association handbook of contemporary family psychology* (pp. 409–424). Washington, DC: APA Books.

Saraceno, B., van Ommeren, M., Batniji, R., Cohen, A., Gureje, O., Mahoney, J., … Underhill, C. (2007). Barriers to improvement of mental health services in low-income and middle-income countries. *The Lancet*, *370*(9593), 1164–1174. https://doi.org/10.1016/S0140-6736(07)61263-X.

Sawada, I., Ono, M., Miyajima, N., Tsukamoto, M., & Kato, N. (2016). *Evaluation study on Triple P trial for parents receiving psychiatric outpatient service in Japan*. Paper presented at the Helping Families Change Conference, Banff, Canada.

Singhal, M., & Turner, K. M. T. (2018). *Triple P in India: A feasibility study with corporate employees*. Paper presented at the Helping Families Change Conference, Santa Rosa CA, USA.

Singla, D. R., Weobong, B., Nadkarni, A., Chowdhary, N., Shinde, S., Anand, A., … Patel, V. (2014). Improving the scalability of psychological treatments in developing countries: An evaluation of peer-led therapy quality assessment in Goa, India. *Behaviour Research and Therapy*, *60*, 53–59. https://doi.org/10.1016/j.brat.2014.06.006.

Sotoudeh Navroodi, S. O., Nicknam, M., Ahmadi, A., Roodbarde, F. P., & Azami, S. (2018). Examining the effectiveness of group positive parenting training on increasing hope and life satisfaction in mothers of children with autism. *Iranian Journal of Psychiatry*, *13*(2), 128–134.

Sumargi, A., Sofronoff, K., & Morawska, A. (2014). Evaluation of a brief format of the Triple P-Positive Parenting Program: A pilot study with Indonesian parents residing in Australia. *Behaviour Change*, *31*(2), 144–158. https://doi.org/10.1017/bec.2014.7.

Sumargi, A., Sofronoff, K., & Morawska, A. (2015). A randomized-controlled trial of the Triple P-Positive Parenting Program Seminar Series with Indonesian parents. *Child Psychiatry and Human Development*, *46*(5), 749–761. https://doi.org/10.1007/s10578-014-0517-8.

Tehrani-Doost, M., Shahrivar, Z., Mahmoudi-Gharaie, J., & Alaghband-Rad, J. (2009). Efficacy of positive parenting on improving children's behavior problems and parenting styles. *Iranian Journal of Psychiatry and Clinical Psychology*, *14*(4), 371–379.

Turner, K. M. T., Richards, M., & Sanders, M. R. (2007). Randomised clinical trial of a group parent education programme for Australian indigenous families. *Journal of Paediatrics and Child Health*, *43*(6), 429–437. https://doi.org/10.1111/j.1440-1754.2007.01053.x.

Turner, K. M. T., Sanders, M. R., Keown, L. J., & Shepherd, M. (2018). A collaborative participation adaptation model. In M. R. Sanders & T. G. Mazzucchelli (Eds.), *The power of positive parenting: Transforming the lives of children, parents, and communities using the Triple P System* (pp. 310–320). New York: Oxford University Press.

United Nations. (2015). *Transforming our world: The 2030 agenda for sustainable development.* Retrieved from https://sustainabledevelopment.un.org/post2015/transformingourworld/publication.

van Mourik, K., Crone, M. R., de Wolff, M. S., & Reis, R. (2017). Parent training programs for ethnic minorities: A meta-analysis of adaptations and effect. *Prevention Science*, *18*(1), 95–105. https://doi.org/10.1007/s11121-016-0733-5.

Ward, C., Sanders, M. R., Gardner, F., Mikton, C., & Dawes, A. (2016). Preventing child maltreatment in low- and middle-income countries: Parent support programs have the potential to buffer the effects of poverty. *Child Abuse & Neglect*, *54*, 97–107. https://doi.org/10.1016/j.chiabu.2015.11.002.

Weng, L. (2015). Individualism and collectivism. In J. M. Bennett (Ed.), *The SAGE encyclopedia of intercultural competence*. Thousand Oaks, CA: Sage.

Wessels, I., & Ward, C. L. (2016). Battered women and parenting: Acceptability of an evidence-based parenting programme to women in shelters. *Journal of Child & Adolescent Mental Health*, *28*(1), 21–31. https://doi.org/10.2989/17280583.2015.1132425.

World Health Organization, & WHO Collaborating Centre for Violence Prevention (2010). *Violence prevention: The evidence.* Retrieved from http://www.who.int/iris/handle/10665/77936.

Zhou, Y. Q., Chew, Q. R. C., Lee, M., Zhou, J., Chong, D., Quah, S. H., … Tan, L. J. (2017). Evaluation of Positive Parenting Programme (Triple P) in Singapore: Improving parenting practices and preventing risks for recurrence of maltreatment. *Children and Youth Services Review*, *83*, 274–284. https://doi.org/10.1016/j.childyouth.2017.10.029.

CHAPTER 20

Culture and family-based intervention for schizophrenia, bipolar, and other psychotic-related spectrum disorders

Amy Weisman de Mamani, Olivia Altamirano, Merranda McLaughlin, and Daisy Lopez
Department of Psychology, University of Miami, Coral Gables, FL, United States

This chapter will focus on the role of culture and family-based psychological interventions for schizophrenia, bipolar, and related (SBR) psychotic spectrum disorders. Schizophrenia is a debilitating mental illness that affects approximately 1% of the population worldwide (Comer, 2015; Silverstein, Moghaddam, & Wykes, 2013). Psychosis, a condition defined by loss of contact with reality, is one of the defining features of this disorder. Examples of psychotic symptoms include hallucinations (seeing or hearing things that others do not), delusions (false beliefs), and incoherent speech (Gonzalez-Burgos, Fish, & Lewis, 2011; Potash et al., 2001; Waters, Blom, Jardri, Hugdahl, & Sommer, 2018). Most patients also experience negative symptoms or behavioral deficits including lack of motivation, flat affect, and poor hygiene (Lyne et al., 2018; Mairs, Lovell, & Keeley, 2013).

Bipolar disorder, also known as manic-depressive illness, also affects approximately 1% of the population worldwide (Fagiolini et al., 2013; Ferrari, Baxter, & Whiteford, 2011). This disorder is marked by unusual shifts in mood, energy, activity levels, and the inability to carry out day-to-day tasks (Oswald et al., 2017). For those afflicted, their moods can range from periods of being extremely "up," elated, and with energized behavior (known as manic episodes) to periods of being very sad, "down," or hopeless (known as depressive episodes) (Oswald et al., 2017). While schizophrenia and bipolar disorder are diagnoses traditionally used to organize these cluster of symptoms, we also include the term "related psychotic spectrum disorders" in this chapter. This is to acknowledge other psychotic disorders often included

Cross-Cultural Family Research and Practice
https://doi.org/10.1016/B978-0-12-815493-9.00020-X

within the literature (schizoaffective, schizophreniform, etc.) as well as the population of people who have similar symptoms and impairment to the groups above without meeting full criteria for either diagnosis (Ruhrmann, Schultze-Lutter, & Klosterkötter, 2010).

Physical health concerns, loneliness, and financial problems are among the highest-ranked challenges reported by people with psychotic disorders (Morgan et al., 2012). For many patients with SBR, the impact of the symptoms results in serious declines in personal, social, and occupational functioning (Comer, 2015). The risk of becoming homeless for people with SBR is approximately 10 times greater than the risk for the general population (Susser, Moore, & Link, 1993). For example, Copeland et al. (2009) found that 55% of patients with bipolar disorder reported lifetime homelessness and in that same study, 55% also reported a lifetime prevalence of incarceration.

Because the symptoms associated with SBR can be very debilitating, these illnesses often take a major toll, not only on those afflicted but also on their family members, who often become lifelong caregivers to their relatives (Ahlem et al., 2017; Bauer et al., 2011; Crowe & Brinkley, 2015). We begin this chapter with an overview of the importance of incorporating the family into interventions with SBR. We discuss variations in how SBR manifest and how family and societal reactions to the illness may differ across diverse cultures. Next, we review family-focused interventions that have been developed and tested on patients with schizophrenia and bipolar disorder and we discuss how specific interventions were tailored to address cultural differences in manifestations, illness conceptions, and treatment expectations. We conclude with a summary of the findings and outline specific take-home points that speak to the importance of utilizing family based, culturally informed, interventions that are consistent with the explanatory views and treatment expectations of diverse clients.

Why incorporate the family?

SBR is a family affair. Greater restrictions in insurance covered care, especially in the United States, have left families with more day-to-day responsibility for their ill loved ones, who often know little about mental illness (Mueser & Glynn, 1999; Weisman de Mamani, Dunham, Aldebot, Tuchman, & Wasserman, 2009). Only about 14% of patients with SBR achieve sustained recovery within the first 5 years of a psychotic episode (Insel, 2010), and only 10%–20% of patients are employed (Marwaha & Johnson, 2004).

Thus, family members often live with patients and assume a large portion of the caregiving responsibility (Pitschel-Walz, Leucht, Bäuml, Kissling, & Engel, 2001). Because family members tend to spend significant amounts of time with patients in the caregiving role, family relationships can constitute significant stressors (and sources of support) for both patients and caregivers. The US ethnic minorities are even more likely to assume care for an ill relative, than are mainstream Anglo Americans, and they often have fewer economic resources to assist them (Guarnaccia & Parra, 1996; Snowden, 2007). Thus, for many ethnic minorities in the United States, this responsibility appears even more pronounced. This is also true for individuals from many developing countries with limited resources available for the mentally ill (Chan, 2011).

Caregiver burden

As will be addressed extensively below, research strongly indicates that family treatments for SBR are effective in reducing patient psychiatric symptoms and improving patient functioning and quality of life. However, an equally important reason for family treatments for SBR is that caregivers themselves often report very high degrees of caregiver burden, depression, anxiety, and stress (Suro & Weisman de Mamani, 2013) and can benefit from professional care to address their own mental health needs. Accordingly, family therapy is a natural candidate for psychosocial interventions targeting SBR, as it addresses the familial stressors that affect both patient symptom trajectory and caregiver burden (Brown & Weisman de Mamani, 2018).

A meta-analysis by Shiraishi and Reilly (2018) examined all studies published between 1993 and 2017 that examined the impact of schizophrenia on family caregivers. Not surprisingly, results indicated that SBR can have devastating impacts on the family. Commonly reported negative outcomes included lack of sufficient personal and social resources to help care for the patient, family disruption, conflict in interpersonal relationships, and concerns around stigma and heredity. However, caregivers also frequently identified several ways that the patients' SBR positively affected their family system (2018). Commonly reported themes included increased family solidarity, greater affection and compassion for the ill family member, and increased personal growth of the patient and other family members because of coping with the SBR. While coping with SBR is indeed challenging, family treatments should aim to capitalize on helping these families recognize that coping with SBR can also lead to feelings of personal growth and resiliency, as well as greater solidarity among themselves and others coping with SBR.

An abundance of existing empirical studies has supported the benefit of including family members in the treatment of mental illness (discussed later in "Specific Family Interventions for SBR" section). Benefits include reductions in family member perceived burden, increased knowledge of illness, better treatment and medication adherence, better social functioning, increased quality of life, reduced medical costs, reduced psychiatric relapse, and less severe psychiatric symptoms (McFarlane, Dixon, Lukens, & Lucksted, 2003; Pitschel-Walz et al., 2001). Furthermore, treatment gains from family psychoeducation have been demonstrated in several cultures, including the United States (Goldstein & Miklowitz, 1995; McFarlane, 2016), China (Ling et al., 1999), Spain (Martínez & García, 2002), Scandinavia (Rund et al., 1994), and Britain (Barrowclough et al., 2001). Sadly, despite its demonstrated efficacy, very few families coping with serious mental illness receive any family psychoeducation or therapy, regardless of ethnicity (Dixon et al., 1999; Resnick, Rosenheck, Dixon, & Lehman, 2005). Moreover, ethnic minorities tend to utilize services less frequently than their Anglo counterparts do, and when they do, they appear to receive poorer quality of care (Satcher, 2001). As we will address further below, many ethnic minorities and individuals from developing cultures adhere to more collectivistic self-construal and are more family oriented (Weisman de Mamani et al., 2009; Weisman de Mamani, Kymalainen, Rosales, & Armesto, 2007). Thus, our view is that involving family members in treatment may improve this statistic.

Cross-cultural patterns in experiencing and responding to mental illness

In this section, we examine factors that impact the course of mental illness and discuss culturally relevant domains that should be addressed in family therapy for SBR. Although, the prevalence of schizophrenia and bipolar disorder appears to occur at similar rates around the world (as noted above), there is considerable variability regarding who is affected, how disorders are manifested, and how family members and others respond to SBR in a loved one. This information has clear implications for treating people afflicted with SBR and their caregivers.

A series of highly cited studies on this topic comes from the World Health Organization (Sartorius, Shapiro, & Jablensky, 1974; WHO, 1979, 1994). The WHO studies used standardized assessment procedures and were conducted at various sites around the world, over a period of several decades. Results of these studies consistently showed that patients with SBR who reside in the industrialized West had less favorable outcomes than their counterparts who reside in the developing countries. The resilience of these findings is striking and, despite a few critics (e.g., Edgerton & Cohen, 1994), most scholars accept the WHO's findings of cross-cultural differences in course of SBR. However, the reasons that may account for these cultural patterns are still hotly debated (e.g., Hopper, 2004). In this chapter, we argue that the finding of better illness course in traditional/developing cultures is likely due, in part, to increased family solidarity, greater reliance on religious and spiritual beliefs as a coping mechanism, and a worldview that lends itself to making less blameworthy attributions toward family members suffering from SBR.

It is important to note that in our view, antipsychotic pharmacotherapy is the first line of defense for patients in the psychotic spectrum (Hill et al., 2010; Lambert et al., 2010). However, even with pharmacotherapy, residual symptoms are common in patients with SBR. For example, 10%–60% of patients experience psychotic symptoms that are resistant to medication and medication adherence with antipsychotics is relatively poor (Lieberman et al., 2005; Lindenmayer, 2000; Nasrallah & Lasser, 2006). In the context of a large medication efficacy-effectiveness trial, Glick, Stekoll, and Hays (2011) present data indicating that having an available and supportive family member involved in patient care greatly improved patient outcomes and this was mediated in part by improving long-term adherence to medication. In group therapy, Maura and Weisman de Mamani (2018) found that adherence to therapy was greatly improved when a family member attended the group alongside the patient. In other words, patients were significantly less likely to drop out of treatment prematurely when a family member also attended therapy with them. Thus, family therapy may be effective for SBR, in part, because it increases adherence to medication and other aspects of a patients' professional care regimen.

Family cohesion

Ethnic minorities in the United States tend to have a more interdependent and interconnected view of themselves and their family members (Oyserman, Coon, & Kemmelmeier, 2002; Weisman de Mamani et al., 2007). Weisman, Rosales, Kymalainen, and Armesto (2005) found that greater perceptions of one's family as cohesive and supportive were associated with better general emotional well-being for both patients with schizophrenia and their family members (i.e., less reported stress, anxiety, and depression), and fewer specific psychiatric symptoms for patients, including fewer delusions, hallucinations, and disorganization in speech. In another study, family cohesion was found to serve as a buffer in the relationship between depression and suicidal ideation, with greater family support weakening the association between depression and suicidal ideation in children and young adults without SBR (Au, Lau, & Lee, 2009). Thus, attempting to improve perceptions of family cohesion within households coping with SBR should be an important treatment goal. Families tend to be more unified and cohesive in traditional (referring to societies that tend to put group goals above individual goals and have clearer norms for behaving) cultures and play a more central role in patients' lives than do their White counterparts (Lefley, 1990; Weisman de Mamani et al., 2009). This factor may contribute to the better course of illness that has been observed in traditional societies by the World Health Organization (WHO, 1979, 1994). Research conducted in the United States further indicates that Blacks and Latinos/Hispanics with SBR turn to family members for advice more often than do non-Hispanic Whites, yet Whites are more likely to seek help from the mental health professionals (Guarnaccia & Parra, 1996). Family treatments, therefore, may be more culturally syntonic for ethnic minorities making them more receptive to family therapy over individually oriented treatments.

Expressed emotion

Many family therapies for SBR are focused on addressing the emotional climate of the household and improving communication patterns. This is, in part, based on the mounting research in an area referred to as expressed emotion (EE). EE is a measure of the amount of criticism, hostility, and emotional overinvolvement (EOI), expressed from a family member toward a patient (Hooley, 2007). More specifically, there is now pervasive evidence that high-expressed emotion (e.g., high degrees of criticism, hostility, and EOI) is tied to greater relapse in individuals with schizophrenia (Bebbington & Kuipers, 1994; Brown & Rutter, 1966; Butzlaff & Hooley, 1998; Hooley & Parker, 2006), bipolar (Kim & Miklowitz, 2004), and a host of other disorders (e.g., depression (Hooley & Gotlib, 2000) and eating disorders (Medina-Pradas, Navarro, Lopez, Grau, & Obiols, 2011). In a comprehensive review of the literature, Kavanagh (1992) found that 48% of individuals living in a high-EE home relapsed as compared to 21% in low-EE homes. Ellis et al. (2014) further found that adolescents with bipolar disorder living in high- vs low-expressed emotion homes endorsed greater suicidal ideation. Furthermore, Wedig and Nock (2007) found that when parents were highly critical, a component of EE, adolescents with a range of mental health disorders were more likely to endorse self-injurious thoughts and behaviors.

Although the majority of cross-cultural studies have supported the predictive validity of EE, the construct is both a universal and a culturally bound paradigm. For example, several studies indicate that criticism is expressed less often in collective cultures and Hispanics and other minorities in the United States appear to have significantly lower rates of high EE and a better course of illness when compared to their White counterparts (Karno et al., 1987; Weisman de Mamani et al., 2007). Furthermore, what is considered critical, hostile, or emotionally overinvolved may vary across cultures (Jenkins & Karno, 1992). Kymalainen, Weisman, Rosales, and Armesto (2006) found that White family members in the United States tended to criticize undesirable, individual personality traits that violate norms of autonomy. On the other hand, Hispanic and Black family members were more likely to condemn behaviors that disrespect and shame other family members and defy collectivism and interdependence (Kymalainen et al., 2006). Family member warmth may be an even better predictor than high EE in this ethnic group (Kopelowicz et al., 2002; López et al., 2004; López, Nelson, Snyder, & Mintz, 1999). Family affection and harmony are highly prized among Latinos/Hispanics (Marín & Marín, 1991). Thus, high degrees of warmth from a loved one may be especially soothing to Hispanic patients with a mental illness.

Several studies have examined EE in Asian countries and most found that high EE, rated using traditional measures, appears to be associated with poor course of illness. Specifically, studies in China and Japan indicate that patients from high-EE homes relapse sooner and more frequently than those from low-EE homes (Mino et al., 1995; Ng, Mui, Cheung, & Leung, 2001). However, as with Hispanics, base rates of high EE appear to be lower in Asian countries. In India (Leff et al., 1987) and China (Ran, Leff, Hou, Xiang, & Ran, 2003), only about 25% of relatives were rated as high EE, compared to approximately 67% of non-Hispanic Whites in the United States (Vaughn, Snyder, Jones, Freeman, & Falloon, 1984). The low base rates of high EE may be due to the centrality of collectivism, conformity, modesty, and suppression of negative emotions among traditional cultures (Ran et al., 2003; Weisman de Mamani et al., 2009). Low base rates of high EE and these protective values may explain the WHO findings that patients with schizophrenia living in traditional societies have a better prognosis than do patients living in less traditional ones (Murphy & Raman, 1971; Saravanan et al., 2007; Wing, 1978).

African Americans are one of the only racial/cultural groups for which high EE has not been found to predict poorer prognosis for schizophrenia (Moline, Singh, Morris, & Meltzer, 1985; Tompson et al., 1995). In fact, a few studies suggest the opposite pattern. Rosenfarb, Bellack, and Aziz (2006) found that, for African Americans in the United States, more criticism and intrusive behaviors were associated with better patient outcomes. Specifically, they found that greater critical comments and intrusive behaviors were associated with lower levels of odd or unusual thinking in African-American patients (rated from the patient symptom profile) over a 2-year period (Rosenfarb et al., 2006). In line with Rosenfarb et al., Gurak and Weisman de Mamani (2017) found that having a low-EE caregiver was associated with greater symptom severity in African-American patients. Gurak and Weisman de Mamani, along with several other researchers (e.g., Davidson, 2001) have noted that African-American speech patterns tend to be more animated and expressive than their Anglo-American counterparts. Rogan and Hammer (1998) suggest that Blacks may even appreciate a more confrontational style, viewing this as a sign of honesty and sincerity. Thus, in African-American homes, behaviors designated as high-EE through traditional assessment methods may instead be interpreted by patients as indicating care and concern. Paradoxically, then, for African Americans, instead of exacerbating symptoms, "high-EE attitudes" may mitigate them.

Weisman de Mamani et al. (2007) further found that while White and Hispanic family members who expressed more criticism during the Camberwell Family Interview (CFI) were perceived as more critical by their ill family member, there was no association between EE criticisms and patient perceptions for African Americans. Tompson et al. (1995) also found that patient perceptions of their relatives' EE behaviors (based on scores from the Patient Interview for Assessing Patient Perceptions of Family Relationships) were completely congruent with Five Minute Speech Sample measures of EE. However, for African-American patients in the sample, patient perceived criticism was the sole predictor of symptom exacerbation. Thus, the traditional assessment measures of EE may have less predictive validity when compared with patient perceptions, especially in ethnic minority populations, as the meaning that patients ascribe to "criticism" may vary by ethnicity. On the other hand, patients' perceptions of criticism may be particularly important to address in psychotherapy with African-American families.

By targeting maladaptive communication patterns or perceived criticism for African Americans, family therapy can improve patient outcomes across many symptom clusters, well beyond the effects of antipsychotic medications and mood stabilizers alone (Falloon, Boyd, & McGill, 1984; Miklowitz & Chung, 2016). Family therapy can produce improvements in negative and disorganized symptoms by enhancing empathic communication between patients and family members. Improvements in communication, in turn, help patients and family members to gain coping mechanisms, reduce stress, and compensate for social deficits that influence patients' appraisal and response biases (Elis, Caponigro, & Kring, 2013).

In summary, while findings in the EE literature appear to be robust in some aspects, they are also complex. The literature overwhelmingly suggests that feeling criticized, unloved, or smothered by one's family members is likely to lead to poor course of illness for patients with SBR in most cultures. However, what is viewed as critical, hostile, or intrusive is in the eye of the beholder and is likely to be strongly influenced by the context and the societal values where one resides.

Causal attributions for SBR

Differences in the attributions that family members make for SBR symptoms also vary across cultures in ways that may account for some of the differences observed in base rates of high EE and in illness prognosis patterns observed by the WHO. In an analog study, Weisman and López (1997) found that after reading a vignette of a person described to have schizophrenia, Mexicans were more likely to perceive the negative behavioral symptoms as less controllable than were Whites. In turn, Whites expressed more intense unfavorable emotion (anger and frustration) toward the hypothetical person described to meet the criteria for schizophrenia. This study suggests that a less blaming view of schizophrenia and the associated symptoms may lead to lower levels of negative emotion and high EE (Weisman & López, 1997). There is an evidence to support this view in Whites and minorities alike. For example, in a Mexican-American sample (Weisman, López, Karno, & Jenkins, 1993) and in a non-Hispanic White sample (Weisman, Nuechterlein, Goldstein, & Snyder, 1998), Weisman and colleagues found that relatives designated as high EE were more likely to perceive the patient's illness and the associated symptoms as within the patient's personal control. In other words, they were less likely to buy into the notion that the patient had a legitimate illness and that their psychiatric symptoms were the unintended side effects. When family members attribute patients' symptoms to external (e.g., a biochemical/organic disease), rather than to internal controllable factors (e.g., laziness), they may be less likely to blame the person for the symptoms. Thus, educating patients and family members about the causes of SBR as well as fostering a more realistic understanding regarding the extent of a patient's ability to control his or her symptoms should be a major aim of family-focused psychoeducational programs for SBR.

These findings correspond to more general cultural psychology patterns where it has been shown that persons from individualistic countries tend to view behavior more as internally controlled, whereas people from collectivistic countries tend to attribute behavior more to external agents. Church and Katigbak (2012, 2017) have examined traitededness in cross-cultural studies. Overall, results indicate that behavior in more individualistic countries is more likely to be viewed as a consequence of one's traits (she gave this to him because she is kind), whereas behavior in collectivistic countries is often viewed as a consequence of norms and various external contingencies (she gave it to him because she was asked to give it).

Religion and spirituality

Religion and spirituality (R/S) is another critical component to consider when helping families to conceptualize and cope with mental illness (Weisman de Mamani, Tuchman, & Duarte, 2010). In general, there is a growing body of research that underscores the benefits of being religious and of addressing R/S in mental health treatments. A review article by Koenig, Larson, and Weaver (1998) summarized the data, which strongly indicated that religious individuals have lower rates of depression, alcohol, drug abuse, and anxiety disorders. A meta-analysis by McCullough, Hoyt, Larson, Koenig, and Thoresen (2000) also found a strong association between religious involvement and lower mortality rates. In other words, religious people appear to reap profound mental and physical health benefits.

These benefits also appear for those with SBR. For example, Mohr et al. (2012) reported in a multisite comparative study, including Canada, Switzerland, and the United States, that outpatients with schizophrenia were more religious than the general population and that greater private, religious involvement, was significantly associated with greater quality of life. Greater religious practices and support from the religious community were significantly associated with higher overall functioning and lower symptomology (negative, general, and overall symptoms). See Weisman de Mamani et al. (2010) for a critical review of the literature on incorporating religion into treatment for SBR.

There are likely multiple factors that may account for the associations between religiosity and mental health. For example, most religions endorse the importance of marriage and family and provide other networks that may bring psychosocial support that can promote mental health (Moreira-Almeida, Lotufo Neto, & Koenig, 2006). Religion can also provide health benefits through belief systems and meaning-making structures that help people to deal with stress, suffering, and life problems through frameworks that emphasize acceptance, endurance, and resiliency (Moreira-Almeida et al., 2006). Furthermore, religious practices such as meditation, prayer, and confession can help to maintain mental health by offering opportunities for meaningful engagement and soothing practices to cope with anxiety, anger, fear, and isolation (Moreira-Almeida et al., 2006).

The issue of addressing R/S in treatments for SBR has been controversial. Some researchers express concern that such a focus might lead to increased religious delusions among psychotic patients, which are already common (Getz, Fleck, & Strakowski, 2001; Torrey, 1995). There is some empirical evidence that suggests that for patients already exhibiting religious themes in their psychosis, incorporating R/S approaches in psychotherapy is counter-indicated as it may lead to greater conviction in delusional beliefs, greater symptom severity, and lower functioning levels (Siddle, Haddock, Tarrier, & Faragher, 2002; Weisman de Mamani et al., 2010). In general, however, others have argued that ignoring R/S issues in treatment for SBR provides a disservice to those patients who are religiously oriented, as the majority of studies indicate a positive response.

It is our view that R/S is highly compatible with psychotherapy. Adaptive uses of religion have some of the same functions as psychotherapy such as providing meaning to life, utilizing social support networks, and facilitating positive change (Stander, Piercy, Mackinnon, & Helmeke, 1994). Educating clients with SBR about the overlap between religion and psychotherapy and providing information regarding how to use faith to access positive therapeutic processes can be very effective in religious patients and their caregivers. Our view is that spirituality education is indicated for non-religious clients as well, as the main focus is on the development of empathy, forgiveness, appreciation, and a connectedness with others, which are all essential elements of the therapeutic processes (Sperry, 2001). An additional benefit of incorporating religion into treatment is that for many individuals in the United States, particularly ethnic minorities and people from traditional cultures, religion, and the religious community are often highly prominent and important aspects of their lives. In many minority cultures, churches and religious communities play a central role in their day-to-day lives. In addition, when access to mental health care is limited, as it is for many ethnic minorities of lower SES, individuals often turn to religious coping for strength and support (Lincoln & Mamiya, 1990; Pargament, 1997). Thus, ignoring religion is akin to ignoring the person.

Specific family interventions for SBR

In this section, we will describe several specific forms of family interventions that have been developed for SBR in the United States and elsewhere. Some treatments are offered to single families. Others are offered in a multifamily group format.

Family-focused therapy

Family-focused therapy (FFT; Miklowitz & Chung, 2016) is an evidence-based intervention for adults and children with bipolar disorder and their caregivers, usually given in conjunction with pharmacotherapy after an illness episode. Family-focused therapy emerged as a modification of behavioral family management (BFM; Falloon et al., 1984; Miklowitz & Goldstein, 1990), a 9-month family treatment for schizophrenia. BFM and FFT consist of sessions of psychoeducation, communication skills training, and problem-solving skills training for patients with SBR and their family members. Falloon et al. (1985) found that, for patients with schizophrenia who had just been discharged from the hospital, the combination of 9 months of BFM and neuroleptic medication was more effective in reducing rates of psychotic relapse and improving social functioning than 9 months of supportive individual therapy and neuroleptic medication. Miklowitz and Chung (2016) summarize over 30 years of research on FFT across eight randomized controlled trials with adults and adolescents with bipolar disorder. The authors conclude that FFT and mood-stabilizing medications hasten recovery from mood episodes, reduce recurrences, and levels of symptom severity compared to briefer forms of psychoeducation and medications with duration of over 1–2 years. Furthermore, the authors report that FFT improvement of symptoms is greater among patients living with high-expressed emotion relatives than patients living with low-expressed emotion relatives.

As far as we are aware, Ozerdem, Oguz, Miklowitz, and Cimilli (2009) conducted the first study to apply FFT outside of the United States. In a series of case studies, the authors explored the applicability of FFT in Turkey SBR and reported that FFT was easily applied to a Turkish sample with a few changes in format, and the participants greatly appreciated the psychoeducation module. The authors also found that the communication enhancement skills module was highly effective, however, quite challenging to implement. In the Turkish culture, positive feelings are more likely to be expressed via behavior such as cooking for the other, visiting, or doing a favor for that person. The problem-solving module, which required use of a worksheet in which families wrote down their definitions of specific problems, their intended solutions, and their solution and implementation plans, also presented some challenges. Although all families were able to hold weekly meetings, at home, in which they discussed and generated solutions, most resisted the formality of the problem-solving worksheets. The authors reported that very few homework assignments were returned completed and family members were reluctant to volunteer for the record keeping during sessions.

Friedman-Yakoobian, Mueser, Giuliano, Goff, and Seidman (2009, 2016) developed a family-directed cognitive adaptation (FCA) strategy to train family members in how to implement and provide the necessary environmental supports for patients with schizophrenia to compensate for their cognitive impairments. Patients and their relatives received 16 weekly individual-family sessions that included psychoeducation about schizophrenia and cognitive deficits related to the illness and collaborative help with identifying functional targets for the program. They received feedback about the patient's cognitive strengths and weaknesses and areas of functional difficulty. They were then taught a series of cognitive adaptation strategies to target specific areas of difficulty. Although this was a small pilot study with only 10 participants, the authors reported promising preliminary findings. They found that participants generally reported high levels of satisfaction with treatment and all families completed the 16-session intervention. Relatives reported reduced burden at termination and at follow-up. Furthermore, no participants were hospitalized during the treatment or follow-up period, and rates of work/school involvement increased from 30% at baseline to 50% at the end of treatment and follow-up. The authors purport that their pilot study provides preliminary support for the acceptability and feasibility of FCA.

Cai, Zhu, Zhang, Wang, and Zhang (2015) evaluated a comprehensive family therapy (CFT) approach aimed at improving cognition in Chinese patients with schizophrenia at six communities in Shanghai, People's Republic of China. The authors reported that CFT closely followed Liberman et al. (1998) "social and independent living skills" approach. As part of this approach, patients received social skills individualized training. This included five modules: medication management; symptom management; community reentry; recreation for leisure; and social independent living skills. There was also a family intervention piece that included psychoeducation provided to the patients' family members, including education regarding the causes, symptoms, treatments, diagnoses, and treatment of schizophrenia, as well as the common adverse effects of antipsychotics, and the prognosis and prevention of schizophrenia. The therapy was conducted weekly for 10 consecutive weeks. Participants were randomly assigned to receive either CFT or a usual daily care (UDC) group, which consisted of a monthly outpatient service and antipsychotic medication. The authors found that patients undergoing CFT showed greater improvement from baseline to 18-month follow-up assessments in cognitive functioning than those in the UDC group. More specifically, after 18 months of follow-up, the CFT group presented with better immediate memory, visuospatial skill, language, attention, and delayed memory compared with the UDC. The authors concluded that CFT is easily adaptable and may provide a promising approach to improving cognitive functioning in patients with schizophrenia. The authors noted that one of their rationales for adding family therapy is that it has been shown to be effective in treating people with schizophrenia within the People's Republic of China (Chien & Chan, 2013), where family members are more involved in patients' care. Interestingly, in this chapter, aside from a discussion of the prevalence of schizophrenia in China, there was no discussion of whether or how the treatment was tailored to adapt to Chinese culture.

Culturally informed therapy for schizophrenia

The first author of this chapter and her research team have developed and tested a family-focused treatment called culturally informed therapy for schizophrenia (CIT-S). Only one module of the intervention is specific to schizophrenia and this culturally informed treatment (CIT) could be easily tailored to any other SBR. CIT-S is a manualized approach that includes 5 modules, each lasting 3 sessions, for 15 weekly sessions. Every module is guided by a series of detailed handouts. The approach was developed in both English and Spanish and all materials are available in both languages. CIT is not tied to any specific ethnic or cultural group and is instead uniquely tailored to the beliefs, values, and traditions of the participants being served.

Module one, *family collectivism*, aims to foster or enhance family unity and helps family members to see one another as a team working toward shared goals. A few of the specific topics covered during this segment include family members' perspective of what the concept of family means to them, their views of their current family, and a discussion of how this compares to their ideal family. Gender and generational and other roles or hierarchies that exist within their family, and significant conflicts or alliances between members are also discussed. In this treatment segment, family members are praised for the decision to attend treatment, as this decision is indicative of a commitment to the patient and to the family unit. To further highlight family cohesion, therapists ask participants about their expectations for treatment. This is used as an opportunity to point out commonalities among family members, as therapists can almost always count on family members agreeing that they would all like to reduce tension in the household and that they would like to see improvement in the patients' symptoms and functioning. Cultural traditions that bring joy and connection to family members are discussed and encouraged throughout treatment. Increasing perceptions of family cohesion is a direct therapy goal because prior research indicates mental health benefits for patients with schizophrenia and their loved ones who view their families as more united (Weisman et al., 2005) and family-focused treatments for SBR have been found to be superior to individually oriented approaches (Pitschel-Walz et al., 2001).

Module two, *education*, is largely drawn from the work of Falloon et al. (1984) and Miklowitz and Goldstein (1997). This segment offers information regarding the symptoms of schizophrenia, the etiology of the illness and exacerbating factors, ways to create an environment that is protective against relapse, and relapse prevention drills should be the warning signs emerge. One specific goal of this module is to teach family members about the negative symptoms of the illness. Most are already aware that the behavioral excesses (e.g., hallucinations and delusions) reflect illness behavior because they are so unusual. However, many are unaware that the behavioral deficits (e.g., poor hygiene and social withdrawal) are also core illness symptoms. Thus, it is often the negative symptoms that are the target of family members' high-EE attitudes and blameworthy attributions (Weisman et al., 1998; Weisman & López, 1997). We explicitly aim to correct that during this module. The education module is especially well suited to ethnic minorities in the United States, who often expect therapy to provide more immediate information, advice, and problem-solving techniques (Tseng & Streltzer, 2001).

Module three, *spirituality*, begins with a discussion of the family's spiritual/religious history. Participants are asked to discuss their beliefs about God, morality, and the meaning of life. They are also asked to describe more behavioral components of religion, such as attendance of worship services, private prayer, and other activities such as reading religious scriptures. The aim in treatment is to help family members cultivate important spiritual practices, including emotional wisdom through forgiveness and empathy and connection with others and with a higher power. Outside of therapy, families are encouraged to tap into support systems inherent to many religious congregations to help cope with the illness. Participants are also directed to read spiritual/religious texts and find passages that may provide inspiration for more active ways to manage or come to terms with the illness, such as "God helps those who help themselves." They are asked to discuss religious practices and traditions that stem from their own or their family members' culture of origin, and they are encouraged to practice those they find soothing or adaptive.

Modules three and four, *spirituality* and *communication training*, were also drawn largely from the work of Falloon et al. (1984) and Miklowitz and Goldstein (1997), and have strong empirical support in the literature. Communication training focuses on enhancing the quality and effectiveness of the family's verbal interactions. Participants are encouraged to engage in communication skills including expressing positive feelings, active listening, making positive requests for change, and expressing negative feelings about specific behaviors. Role playing is the primary technique used to teach these skills. This method allows family members to practice new ways of communicating in a setting that is safe and nonjudgmental, with one important aim being to decrease the expression of high-EE attitudes.

The final module, *problem solving*, teaches concrete ways of discussing and resolving family problems. Techniques include brainstorming strategies, discussing the pros and cons of various ideas, agreeing on a solution, and implementing a plan of action. Family members work through several examples with the therapist and use homework to practice the techniques between sessions. Because this is the concluding module of CIT-S, families practice a "relapse drill," where they use problem-solving strategies to develop a detailed course of action in the event of a psychiatric relapse.

We have now published several papers that have evaluated different aspects of the intervention. Our findings indicate that, relative to a psychoeducation only treatment as usual condition, CIT-S is effective in reducing patient psychiatric symptoms (Weisman, Weintraub, Gurak, & Maura, 2014) and caregiver burden (Weisman de Mamani & Suro, 2016). It also reduces depression, anxiety, and stress in both patients and caregivers (Brown & Weisman de Mamani, 2018). In line with treatment goals, CIT-S increased family cohesion and this change appeared to mediate treatment effects (Brown & Weisman de Mamani, 2018). Contrary to expectations, we also found that patients that are more religious dropped out of treatment earlier and more often than did less religious participants (Gurak & Weisman de Mamani, 2017). This was surprising given that CIT-S directly targets religiosity during the third module of the intervention. However, results of a survival analysis indicated that nearly all participants who left the treatment prematurely did so before the religiosity segment began. In fact, no participants dropped out of treatment after completing the religiosity module. Thus, CIT-S may have been too compartmentalized and religious individuals may have perceived a disconnection between their beliefs and those of their mental health providers. We currently have a pilot study, funded by the John Templeton Foundation, underway, to develop a version of CIT-S that will better integrate religion throughout the intervention and offer it in a religious institution with an aim of improving retention of religious participants.

Multifamily group treatments

Multifamily group treatments for SBR are also showing promise. McFarlane developed a multifamily group (MFG; McFarlane, 2002) treatment that combines psychoeducation and skills training. MFG has been shown to lower relapse rates, improve social functioning, and decrease caregiver burden relative to customary care (Dyck et al., 2000; Hazel et al., 2004; McFarlane, Dushay, Stastny, Deakins, & Link, 1996; McFarlane et al., 1995). Kopelowicz et al. (2012) tested a cultural adaptation of MFG in Spanish-speaking Mexican-American patients with schizophrenia living in the United States, who had demonstrated a recent history of no adherence to their antipsychotic medication regimens. This is particularly salient given that individuals with schizophrenia generally do not comply with their antipsychotic prescriptions, with roughly half not fully adhering (Valenstein et al., 2004) and poor medication adherence being generally associated with a very poor course of illness.

Kopelowicz et al.'s (2012) approach to culture was a process-oriented one that recognizes the sociocultural context at the individual, patient, or family level. One major modification of MFG was to target improved adherence to medication using the principles of the theory of planned behavior. This is a social psychological theory (Ajzen, 1991) that has improved adherence to treatment in health-related interventions including treatment of coronary artery disease, hypertension, and human immunodeficiency virus infection (Kopelowicz et al., 2012). This study evaluated the benefits of adding either culturally modified MF adherence (MFG-A) or standard MFG (MFG-S) to treatment as usual (TAU) in a three-arm, randomized clinical trial conducted at two community mental health centers (CMHCs) in the Los Angeles area that also included TAU alone. At the end of the 1-year treatment, the authors found that their modified MFG-adherence approach was associated with higher medication adherence than the MFG standard or treatment as usual approach, and the MFG-adherence participants had a longer time to first hospitalization and were less likely to be hospitalized than those in MFG standard (and treatment as usual alone). The authors concluded that multifamily group therapy, specifically tailored to improve medication adherence through a focus on the beliefs and attitudes of the target population, is associated with improved outcome for Mexican-American adults with schizophrenia-spectrum disorders.

McDonell, Short, Hazel, Berry, and Dyck (2006) investigated the impact of multiple-family group treatment (MFGT) on outpatient and inpatient mental health service utilization in individuals with schizophrenia. The approach integrates psychoeducation and behavioral family therapy in a multiple-family group format (McFarlane, 2002). The treatment provides education about illness symptoms and management, and social support, with the goals of decreasing psychiatric symptoms, relapse, and hospitalization, and increasing social and vocational functioning. There are four specific phases: joining with the client and family, a psychoeducational workshop, relapse prevention, and vocational and social skills enhancement (McFarlane, 2002). After the joining and psychoeducational workshop phases, the final two phases use a problem-solving group format whereby consumers and families are taught to use a systematic strategy for overcoming illness and nonillness-related difficulties. Results of studies testing the efficacy of MFGT demonstrated that relative to standard care participants, the MFGT group had reduced community hospitalization during year 1 of the intervention and reduced state hospitalization at follow-up (McDonell et al., 2006). During the intervention period, MFGT participants demonstrated a significant increase in outpatient utilization as a direct consequence of the intervention. Previous studies on MFGT also demonstrated 1-year reductions in negative symptoms (Dyck et al., 2000) and 1- and 2-year reductions in caregiver distress (Hazel et al., 2004). McDonell et al. (2006) argue that family therapists who invest the time and effort to receive training in MFGT can expect reductions in community and state hospitalization in consumers receiving this treatment and in psychiatric symptoms and family caregiver distress.

Bradley et al. (2006) similarly tested multiple-family group treatment for schizophrenia in Australia for English-speaking families and explored the effectiveness of this approach for a newly arrived non-English–speaking migrant group of first-generation Vietnamese families. Participants were randomly assigned to a multiple-family group or a control group. The multiple-family group intervention included 26 sessions over 12 months and was delivered as an adjunct to case management services, which all consumers received. Outcomes were measured immediately after treatment and 18 months later. Relapse rates immediately after treatment were significantly lower for the multiple-family group than for the control group and they were also lower during the follow-up period. Symptom severity ratings from the Brief Psychiatric Rating Scale (Ventura et al., 1993) were also significantly lower for participants in the multiple-family group and vocational outcomes were improved. The authors found no differences in reductions in relapse and symptoms between the English- and Vietnamese-speaking family groups. The authors argue that multiple-family group treatment is an effective cognitive-behavioral intervention in the treatment of schizophrenia. The finding suggests continued application of and research on family interventions for non-English–speaking, migrant populations.

Chien and Chan (2013) evaluated the effectiveness of mutual support group intervention for Chinese families of people with schizophrenia in a randomized controlled trial with 24-month follow-up. As rationalization for the study, the authors argued that little attention has been paid to the effects on family members, particularly those in non-Western countries. Thus, they examined the effects of a 9-month family-led mutual support group for Chinese people with schizophrenia and compared this with a psychoeducation group and standard psychiatric care over a 24-month follow-up. The participants were randomly assigned into one of the three study groups. The mutual support and psychoeducation groups comprised 14 two-hour group sessions, with patients participating in at least 5 of the sessions and participants in standard care received routine psychiatric care only. Study findings indicated that the participants in the mutual support group showed significantly greater improvement in family and patient functioning, and in perceived social support. Results also indicated that this group showed significantly greater reductions in patients' symptom severity and length of rehospitalizations at 12- and 24-month follow-ups. The authors concluded that family-led mutual support group for schizophrenia produces longer-term benefits to both the patients' and families' functioning and relapse prevention for patients, compared with psychoeducation and standard care. Moreover, they reported that this support group program can be an effective family intervention for Chinese people coping with schizophrenia in the family.

Asmal, Mall, Emsley, Chiliza, and Swartz (2014) assessed the feasibility and acceptability of a family intervention for people with schizophrenia and their relatives in a socially deprived urban community in South Africa. The study also aimed to explore the contextual factors that could influence implementation of the intervention in this setting. The authors argue that the challenges of schizophrenia may be exacerbated in low-income contexts, such as sub-Saharan Africa and, despite a range of published studies on the pharmacological treatment of schizophrenia in this context, no studies had explored the efficacy of family treatments for schizophrenia in this region. To give some context, the authors describe that stigma and discrimination, low mental health literacy, and low priority of mental health may all contribute to the challenge of mental health-care delivery in sub-Saharan Africa. Cities are growing faster in Africa than elsewhere (United Nations-Habitat (UN-Habitat) & United Nations Human Settlements Programme, 2005). This change comes with its own set of challenges such as socioeconomic deprivation, economic inequality, substance abuse, and social fragmentation (United Nations–Habitat (UN-Habitat) & United Nations Human Settlements Programme, 2011) that can exacerbate mental illness. As a response to these challenges, the authors conducted a qualitative study to answer the question of how these challenges might affect the provision of a family support service in a real-world, low-income African context. They chose a multifamily group model that included patients and caregivers due to severe resource constraints, and to explore the acceptability of this method of treatment in the sub-Saharan African context.

Participants in the Asmal et al. (2014) study were divided into four groups based on the language preference (three in Afrikaans and one in English). Each group lasted for six sessions and included six to eight participants. The groups were semistructured and held once every 2 weeks, lasting for 90 minutes. The initial session focused on the biology of psychosis with subsequent sessions focused on medication and side effects; stress and expressed emotion; communication techniques; problem solving; and crisis planning. During these sessions, patients and families were invited to discuss their understanding of the topic of discussion, explore the factors that may influence adherence to a family intervention program, and offer input as to what format such a program should take.

Although the Asmal et al. (2014) study was primarily qualitative, the findings overall indicated that participants benefited from the treatment. Patients and their family members agreed that a psychoeducational frame helped to address gaps in their knowledge about the biology of schizophrenia and its treatment. Participants reported to find the content of each session relevant and accessible. Family members reported that the opportunity to share experiences with others who faced similar challenges was valuable and helped to decrease their sense of isolation, which many reported experiencing in their communities. However, participants also identified several additional needs that should be incorporated into programs to make them more amenable to the sub-Saharan African community. Patients described high levels of stigma within their communities. They also reported substance abuse and loneliness as major concerns that needed to be better addressed in treatment. Relatives identified poverty, physical illness, lack of support from other family members, and community violence as major contributors to the stress of being a carer and expressed that these areas would require greater attention in a sub-Saharan context. Interestingly, adherence to medication, which is frequently a concern for families in intervention studies elsewhere (McDonald, Garg, & Haynes, 2002; Zygmunt, Olfson, Boyer, & Mechanic, 2002), was not highlighted as a concern by participants of this study. The authors suggested that this is likely because in this region, all participants with schizophrenia were treated with depot antipsychotic medication, administered as a long-lasting injection, together with an assertive monitoring program. The authors concluded by saying that a family group model for addressing issues faced by people with schizophrenia and their families seems acceptable and feasible in this urban African context.

Summary and treatment recommendations

In this chapter, we reviewed the literature on family-focused interventions for SBR. Results across studies and cultures robustly indicate that incorporating the family into treatment generally reduces patient psychiatric symptoms and improves mental health of patients and their caregivers alike. The following basic elements of psychotherapy are likely helpful across cultures. For example, most families can benefit from education about SBR, and from learning communication and problem-solving techniques to help them cope and manage the illness. On the other hand, illness attributions, beliefs, and values vary from one culture to the next. Systematically checking in with patients and their family members about how they understand the patient's illness and what practices they view as useful (or detrimental) to them is necessary. Drawing from participants' unique cultural and family traditions can help them draw strength and gain solidarity in ways that are sanctioned and relevant within their communities. Offering care in settings that are familiar and valued (e.g., churches) and in close proximity (e.g., community centers) to where participants live may decrease barriers to treatment and reduce therapy attrition.

One limitation of this chapter and of the contemporary state of the field is that the large majority of the published randomized trials on family therapy for SBR have been conducted in the United States, with only a few small studies conducted elsewhere. Thus, while the cross-cultural research reviewed in this chapter does appear to indicate that certain general principles (e.g., positive communication and connectedness within families, increased knowledge of mental illness, and learning better strategies for solving problems around managing the illness) may universally help patients and their families manage mental illness, other aspects are likely to be culture specific. More representative research across the globe is needed to tease apart what elements of family therapy for SBR are helpful universally and which are culturally unique.

Interestingly, despite the fact that much of the research supporting family treatments for SBR is conducted in the United States, it noteworthy that in the United States only about 8% of patients with SBR have ever received any family therapy (Resnick et al., 2005). Factors that likely account for this include poor coordination of services, a lack of training among mental health practitioners on family-focused interventions, and the alarming fact that family care is not covered by many US-based insurance plans (e.g., Comer, 2015; Resnick et al., 2005). Given the empirical evidence in support of family therapy reviewed in this chapter, efforts to increase accessibility of affordable and integrated family care for patients with SBR is clearly needed.

In closing, based on our summary of the literature, we put forth the following specific recommendations for working with multicultural populations with SBR. (1) Incorporate the family into treatment whenever possible. (2) Tap into clients' religious and existential beliefs and practices and draw from those that may offer adaptive resources for coping with SBR in the family. (3) Elicit and discuss cultural differences in beliefs and values and recognize and respect those that differ from one's own. (4) Attempt to make treatments more accessible by offering them within the community and at religious institutions and other community settings (e.g., youth center), if possible.

References

Ahlem, H., Zgueb, Y., Ouali, U., Laajili, Y., Alaya, S. B., Jomli, R., & Nacef, F. (2017). Family burden in caregivers of schizophrenia patients. *European Psychiatry*, *41*(SS), S799–S800. https://doi.org/10.1016/j.eurpsy.2017.01.1541.

Ajzen, I. (1991). The theory of planned behavior. *Organizational Behavior and Human Decision Processes*, *50*(2), 179–211. https://doi.org/10.1016/0749-5978(91)90020-T.

Asmal, L., Mall, S., Emsley, R., Chiliza, B., & Swartz, L. (2014). Towards a treatment model for family therapy for schizophrenia in an urban African setting: Results from a qualitative study. *International Journal of Social Psychiatry*, https://doi.org/10.1177/0020764013488569.

Au, A. C.Y., Lau, S., & Lee, M. T.Y. (2009). Suicide ideation and depression: The moderation effects of family cohesion and social self-concept. *Adolescence*, *44*(176), 851–868.

Barrowclough, C., Haddock, G., Tarrier, N., Lewis, S. W., Moring, J., O'Brien, R., … McGovern, J. (2001). Randomized controlled trial of motivational interviewing, cognitive behavior therapy, and family intervention for patients with comorbid schizophrenia and substance use disorders. *American Journal of Psychiatry*, *158*(10), 1706–1713. https://doi.org/10.1176/appi.ajp.158.10.1706.

Bauer, R., Gottfriedsen, G., Binder, H., Dobmeier, M., Cording, C., Hajak, G., & Spiessl, H. (2011). Burden of caregivers of patients with bipolar affective disorders. *American Journal of Orthopsychiatry*, *81*(1), 139–148. https://doi.org/10.1111/j.1939-0025.2010.01081.x.

Bebbington, P., & Kuipers, L. (1994). The clinical utility of expressed emotion in schizophrenia. *Acta Psychiatrica Scandinavica*, *89*(s382), 46–53. https://doi.org/10.1111/j.1600-0447.1994.tb05865.x.

Bradley, G. M., Couchman, G. M., Perlesz, A., Nguyen, A. T., Singh, B., & Riess, C. (2006). Multiple-family group treatment for English- and Vietnamese-speaking families living with schizophrenia. *Psychiatric Services*, *57*(4), 521–530. https://doi.org/10.1176/ps.2006.57.4.521.

Brown, C. A., & Weisman de Mamani, A. (2018). The mediating effect of family cohesion in reducing patient symptoms and family distress in a culturally informed family therapy for schizophrenia: A parallel-process latent-growth model. *Journal of Consulting and Clinical Psychology*, *86*(1), 1–14. https://doi.org/10.1037/ccp0000257.

Brown, G. W., & Rutter, M. (1966). The measurement of family activities and relationships. *Human Relations*, *19*(3), 241–263. https://doi.org/10.1177/001872676601900301.

Butzlaff, R. L., & Hooley, J. M. (1998). Expressed emotion and psychiatric relapse. *Archives of General Psychiatry*, *55*(6), 547. https://doi.org/10.1001/archpsyc.55.6.547.

Cai, J., Zhu, Y., Zhang, W., Wang, Y., & Zhang, C. (2015). Comprehensive family therapy: An effective approach for cognitive rehabilitation in schizophrenia. *Neuropsychiatric Disease and Treatment*, https://doi.org/10.2147/NDT.S83569.

Chan, S. W. C. (2011). Global perspective of burden of family caregivers for persons with schizophrenia. *Archives of Psychiatric Nursing*, *25*(5), 339–349. https://doi.org/10.1016/j.apnu.2011.03.008.

Chien, W.T., & Chan, S.W. C. (2013). The effectiveness of mutual support group intervention for Chinese families of people with schizophrenia: A randomised controlled trial with 24-month follow-up. *International Journal of Nursing Studies*, *50*(10), 1326–1340. https://doi.org/10.1016/J.IJNURSTU.2013.01.004.

Church, A.T., Katigbak, M. S., & Chiu, C.-Y. (2012). Culture and personality. In M. J. Gelfand & Y.-Y. Hong (Eds.), Vol. 2. *Advances in culture and psychology* (pp. 139–204). Oxford: Oxford University Press.

Church, A. T., & Katigbak, M. S. (2017). Trait consistency and validity across cultures. In A. T. Church (Ed.), *The Praeger handbook of personality across cultures* (pp. 279–308). Santa Barbara, CA: ABC-CLIO/Praeger.

Comer, R. (2015). *Abnormal psychology*. New York, NY: Worth Publishers.

Copeland, L., Miller, A., Welsh, D. E., McCarthy, J. F., Zeber, J. E., & Kilbourne, A. M. (2009). Clinical and demographic factors associated with homelessness and incarceration among VA patients with bipolar disorder. *American Journal of Public Health*, *99*(5), 871–877. https://doi.org/10.2105/AJPH.2008.149989.

Crowe, A., & Brinkley, J. (2015). Distress in caregivers of a family member with serious mental illness. *Family Journal*, *23*(3), 286–294. https://doi.org/10.1177/1066480715572967.

Davidson, M. N. (2001). Know thine adversary: The impact of race on styles of dealing with conflict. *Sex Roles*, *45*(5/6), 259–276. https://doi.org/10.1023/A:1014301429426.

Dixon, L., Lyles, A., Scott, J., Lehman, A., Postrado, L., Goldman, H., & McGlynn, E. (1999). Services to families of adults with schizophrenia: From treatment recommendations to dissemination. *Psychiatric Services*, *50*(2), 233–238. https://doi.org/10.1176/ps.50.2.233.

Dyck, D., Short, R., Hendryx, M., Norell, D., Myers, M., Patterson, T., … McFarlane, W. (2000). Management of negative symptoms among patients with schizophrenia attending multiple-family groups. *Psychiatric Services*, *51*(4), 513–519. https://doi.org/10.1176/appi.ps.51.4.513.

Edgerton, R. B., & Cohen, A. (1994). Culture and schizophrenia: The DOSMD challenge. *British Journal of Psychiatry*, *164*(02), 222–231. https://doi.org/10.1192/bjp.164.2.222.

Elis, O., Caponigro, J. M., & Kring, A. M. (2013). Psychosocial treatments for negative symptoms in schizophrenia: Current practices and future directions. *Clinical Psychology Review*, *33*(8), 914–928. https://doi.org/10.1016/J.CPR.2013.07.001.

Ellis, A. J., Portnoff, L. C., Axelson, D. A., Kowatch, R. A., Walshaw, P., & Miklowitz, D. J. (2014). Parental expressed emotion and suicidal ideation in adolescents with bipolar disorder. *Psychiatry Research*, *216*(2), 213–216. https://doi.org/10.1016/j.psychres.2014.02.013.

Fagiolini, A., Forgione, R., Maccari, M., Cuomo, A., Morana, B., Dell'Osso, M., & Pellegrini, F. (2013). Prevalence, chronicity, burden and borders of bipolar disorder. *Journal of Affective Disorders*, *148*(2–3), 161–169. https://doi.org/10.1016/j.jad.2013.02.001.

Falloon, I. R. H., Boyd, J. L., & McGill, C. W. (1984). *Family care of schizophrenia: A problem-solving approach to the treatment of mental illness*. New York, NY: Guilford Press.

Falloon, I. R. H., Boyd, J. L., McGill, C. W., Williamson, M., Razani, J., Moss, H. B., … Simpson, G. M. (1985). Family management in the prevention of morbidity of schizophrenia. *Archives of General Psychiatry*, *42*(9), 887. https://doi.org/10.1001/archpsyc.1985.01790320059008.

Ferrari, A. J., Baxter, A. J., & Whiteford, H. A. (2011). A systematic review of the global distribution and availability of prevalence data for bipolar disorder. *Journal of Affective Disorders*, *134*(1–3), 1–13. https://doi.org/10.1016/j.jad.2010.11.007.

Friedman-Yakoobian, M. S., Mueser, K. T., Giuliano, A. J., Goff, D. C., & Seidman, L. J. (2016). Family-directed cognitive adaptation pilot: Teaching cognitive adaptation to families of individuals with schizophrenia. *American Journal of Psychiatric Rehabilitation*, *19*(1), 62–74. https://doi.org/10.1080/15487768.2015.1125401.

Friedman-Yakoobian, M. S. M., Mueser, K. T. K., Giuliano, A., Goff, D. C. D., & Seidman, L. J. L. (2009). Family-directed cognitive adaptation for schizophrenia. *Journal of Clinical Psychology*, *65*(8), 854–867. https://doi.org/10.1002/jclp.20611.

Getz, G. E., Fleck, D. E., & Strakowski, S. M. (2001). Frequency and severity of religious delusions in Christian patients with psychosis. *Psychiatry Research*, *103*(1), 87–91. https://doi.org/10.1016/S0165-1781(01)00262-1.

Glick, I. D., Stekoll, A. H., & Hays, S. (2011). The role of the family and improvement in treatment maintenance, adherence, and outcome for schizophrenia. *Journal of Clinical Psychopharmacology*, *31*(1), 82–85. https://doi.org/10.1097/JCP.0b013e31820597fa.

Goldstein, M. J., & Miklowitz, D. J. (1995). The effectiveness of psychoeducational family therapy in the treatment of schizophrenic disorders. *Journal of Marital and Family Therapy*, *21*(4), 361–376. https://doi.org/10.1111/j.1752-0606.1995.tb00171.x.

Gonzalez-Burgos, G., Fish, K. N., & Lewis, D. A. (2011). GABA neuron alterations, cortical circuit dysfunction and cognitive deficits in schizophrenia. *Neural Plasticity*, https://doi.org/10.1155/2011/723184.

Guarnaccia, P. J., & Parra, P. (1996). Ethnicity, social status, and families' experiences of caring for a mentally ill family member. *Community Mental Health Journal*, *32*(3), 243–260. https://doi.org/10.1007/BF02249426.

Gurak, K., & Weisman de Mamani, A. (2017). Caregiver expressed emotion and psychiatric symptoms in African-Americans with schizophrenia: An attempt to understand the paradoxical relationship. *Family Process*, *56*(2), 476–486. https://doi.org/10.1111/famp.12188.

Hazel, N. A., McDonell, M. G., Short, R. A., Berry, C. M., Voss, W. D., Rodgers, M. L., & Dyck, D. G. (2004). Impact of multiple-family groups for outpatients with schizophrenia on caregivers' distress and resources. *Psychiatric Services*, *55*(1), 35–41. https://doi.org/10.1176/appi.ps.55.1.35.

Hill, M., Crumlish, N., Whitty, P., Clarke, M., Browne, S., Kamali, M., & Ocallaghan, E. (2010). Nonadherence to medication four years after a first episode of psychosis and associated risk factors. *Psychiatric Services*, *61*(2), 189–192. https://doi.org/10.1176/ps.2010.61.2.189.

Hooley, J. M. (2007). Expressed emotion and relapse of psychopathology. *Annual Review of Clinical Psychology*, *3*(1), 329–352. https://doi.org/10.1146/annurev.clinpsy.2.022305.095236.

Hooley, J. M., & Gotlib, I. H. (2000). A diathesis-stress conceptualization of expressed emotion and clinical outcome. *Applied and Preventive Psychology*, *9*(3), 135–151. https://doi.org/10.1016/S0962-1849(05)80001-0.

Hooley, J. M., & Parker, H. A. (2006). Measuring expressed emotion: An evaluation of the shortcuts. *Journal of Family Psychology*, *20*(3), 386–396. https://doi.org/10.1037/0893-3200.20.3.386.

Hopper, K. (2004). Interrogating the meaning of "culture" in the WHO international studies of schizophrenia. In *Schizophrenia, culture, and subjectivity*. Cambridge University Press.

Insel, T. R. (2010). Rethinking schizophrenia. *Nature*, *468*(7321), 187. https://doi.org/10.1038/nature09552.

Jenkins, J. H., & Karno, M. (1992). Special article: The meaning of expressed emotion: Theoretical issues raised by cross-cultural research. *American Journal of Psychiatry*, *149*, 9–21. https://doi.org/10.1176/ajp.149.1.9.

Karno, M., Hough, R. L., Burnam, M. A., Escobar, J. I., Timbers, D. M., Santana, F., & Boyd, J. H. (1987). Lifetime prevalence of specific psychiatric disorders among Mexican Americans and Non-Hispanic Whites in Los Angeles. *Archives of General Psychiatry*, *44*(8), 695. https://doi.org/10.1001/archpsyc.1987.01800200021004.

Kavanagh, D. J. (1992). Recent developments in expressed emotion and schizophrenia. *British Journal of Psychiatry*, *160*(05), 601–620. https://doi.org/10.1192/bjp.160.5.601.

Kim, E. Y., & Miklowitz, D. J. (2004). Expressed emotion as a predictor of outcome among bipolar patients undergoing family therapy. *Journal of Affective Disorders*, *82*(3), 343–352. https://doi.org/10.1016/j.jad.2004.02.004.

Koenig, H. G., Larson, D. B., & Weaver, A. J. (1998). Research on religion and serious mental illness. *New Directions for Mental Health Services*, *1998*(80), 81–95. https://doi.org/10.1002/yd.23319988010.

Kopelowicz, A., Zarate, R., Gonzalez, V., Lopez, S. R., Ortega, P., Obregon, N., & Mintz, J. (2002). Evaluation of expressed emotion in schizophrenia: A comparison of Caucasians and Mexican-Americans. *Schizophrenia Research*, *55*(1–2), 179–186. https://doi.org/10.1016/S0920-9964(01)00193-1.

Kopelowicz, A., Zarate, R., Wallace, C. J., Liberman, R. P., Lopez, S. R., & Mintz, J. (2012). The ability of multifamily groups to improve treatment adherence in Mexican Americans with schizophrenia. *Archives of General Psychiatry*, *69*(3), 265. https://doi.org/10.1001/archgenpsychiatry.2011.135.

Kymalainen, J. A., Weisman, A. G., Rosales, G. A., & Armesto, J. C. (2006). Ethnicity, expressed emotion, and communication deviance in family members of patients with schizophrenia. *The Journal of Nervous and Mental Disease*, *194*(6), 391–396. https://doi.org/10.1097/01.nmd.0000221171.42027.5a.

Lambert, M., Conus, P., Cotton, S., Robinson, J., Mcgorry, P. D., & Schimmelmann, B. G. (2010). Prevalence, predictors, and consequences of long-term refusal of antipsychotic treatment in first-episode psychosis. *Journal of Clinical Psychopharmacology*, *30*(5), 565–572. https://doi.org/10.1097/jcp.0b013e3181f058a0.

Leff, J., Wig, N. N., Menon, D. K., Bedi, H., Kuipers, L., Ghosh, A., … Jablensky, A. (1987). Expressed emotion and schizophrenia in North India. *British Journal of Psychiatry*, *151*(02), 166–173. https://doi.org/10.1192/bjp.151.2.166.

Lefley, H. P. (1990). Culture and chronic mental illness. *Psychiatric Services*, *41*(3), 277–286. https://doi.org/10.1176/ps.41.3.277.

Liberman, R. P., Wallace, C. J., Blackwell, G., Kopelowicz, A., Vaccaro, J. V., & Mintz, J. (1998). Skills training versus psychosocial occupational therapy for persons with persistent schizophrenia. *American Journal of Psychiatry*, *155*(8), 1087–1091. https://doi.org/10.1176/ajp.155.8.1087.

Lieberman, J. A., Stroup, T. S., McEvoy, J. P., Swartz, M. S., Rosenheck, R. A., Perkins, D. O., … Severe, J. (2005). Effectiveness of antipsychotic drugs in patients with chronic schizophrenia. *New England Journal of Medicine*, *353*(12), 1209–1223. https://doi.org/10.1056/NEJMx100054.

Lincoln, E. C., & Mamiya, L. H. (1990). *The black church in the African American experience*. Durham, NC: Duke University Press.

Lindenmayer, J. P. (2000). Treatment refractory schizophrenia. *Psychiatric Quarterly*, *71*(4), 373–384. https://doi.org/10.1023/A:1004640408501.

Ling, S., Zhao, C., Yang, W., Wang, R., Jin, Z., Ma, T., et al. (1999). Efficacy of family interventions on schizophrenics in remission in community: Result of one year follow-up study. *Chinese Mental Health Journal*, *13*(6), 325–327.

López, S. R., Nelson Hipke, K., Polo, A. J., Jenkins, J. H., Karno, M., Vaughn, C., & Snyder, K. S. (2004). Ethnicity, expressed emotion, attributions, and course of schizophrenia: Family warmth matters. *Journal of Abnormal Psychology*, *113*(3), 428–439. https://doi.org/10.1037/0021-843X.113.3.428.

López, S. R., Nelson, K. A., Snyder, K. S., & Mintz, J. (1999). Attributions and affective reactions of family members and course of schizophrenia. *Journal of Abnormal Psychology*, *108*(2), 307–314. https://doi.org/10.1037/0021-843X.108.2.307.

Lyne, J., O'Donoghue, B., Roche, E., Renwick, L., Cannon, M., & Clarke, M. (2018). Negative symptoms of psychosis: A life course approach and implications for prevention and treatment. *Early Intervention in Psychiatry*, *12*(4), 561–571. https://doi.org/10.1111/eip.12501.

Mairs, H., Lovell, K., & Keeley, P. (2013). Psychosocial interventions for the negative symptoms of psychosis. *Mental Health Practice*, *16*(5), 19–23. https://doi.org/10.7748/mhp2013.02.16.5.19.s9539.

Marín, G., & Marín, B. (1991). *Research with Hispanic populations*. Thousand Oaks, CA: Sage Publications. Retrieved from http://psycnet.apa.org/record/1991-97676-000.

Martínez, J., & García, J. (2002). Family intervention programme in schizophrenia: Two-year follow-up of the Andalusia study. *Psychology in Spain*, *6*(1), 56–60. Retrieved from http://www.psychologyinspain.com/content/full/2002/6004.pdf.

Marwaha, S., & Johnson, S. (2004). Schizophrenia and employment: A review. *Social Psychiatry and Psychiatric Epidemiology*, *39*(5), 337–349. https://doi.org/10.1007/s00127-004-0762-4.

Maura, J., & Weisman de Mamani, A. (2018). The feasibility of a culturally informed group therapy for patients with schizophrenia and their family members. *Psychotherapy*, *55*, 27–38. https://doi.org/10.1037/pst0000109.

McCullough, M. E., Hoyt, W. T., Larson, D. B., Koenig, H. G., & Thoresen, C. (2000). Religious involvement and mortality: A meta-analytic review. *Health Psychology*, *19*(3), 211–222. https://doi.org/10.1037/0278-6133.19.3.211.

McDonald, H. P., Garg, A. X., & Haynes, R. B. (2002). Interventions to enhance patient adherence to medication prescriptions. *JAMA*, *288*(22), 2868. https://doi.org/10.1001/jama.288.22.2868.

McDonell, M. G., Short, R. A., Hazel, N. A., Berry, C. M., & Dyck, D. G. (2006). Multiple-family group treatment of outpatients with schizophrenia: Impact on service utilization. *Family Process*, *45*(3), 359–373. https://doi.org/10.1111/j.1545-5300.2006.00176.x.

McFarlane, W. (2002). *Multifamily groups in the treatment of severe psychiatric disorders*. New York, NY: Guilford Press.

McFarlane, W. (2016). Family interventions for schizophrenia and the psychoses: A review. *Family Process*, *55*(3), 460–482. https://doi.org/10.1111/famp.12235.

McFarlane, W., Dushay, R., Stastny, P., Deakins, S., & Link, B. (1996). A comparison of two levels of family-aided assertive-community treatment. *Psychiatric Services*, *47*(7), 744–750. https://doi.org/10.1176/ps.47.7.744.

McFarlane, W. R., Dixon, L., Lukens, E., & Lucksted, A. (2003). Family psychoeducation and schizophrenia: A review of the literature. *Journal of Marital and Family Therapy*, *29*(2), 223–245. https://doi.org/10.1111/j.1752-0606.2003.tb01202.x.

McFarlane, W. R., Lukens, E., Link, B., Dushay, R., Deakins, S. A., Newmark, M., … Toran, J. (1995). Multiple-family groups and psychoeducation in the treatment of schizophrenia. *Archives of General Psychiatry*, *52*(8), 679. https://doi.org/10.1001/archpsyc.1995.03950200069016.

Medina-Pradas, C., Navarro, J. B., Lopez, S. R., Grau, A., & Obiols, J. E. (2011). Dyadic view of expressed emotion, stress, and eating disorder psychopathology. *Appetite*, *57*(3), 743–748. https://doi.org/10.1016/j.appet.2011.08.016.

Miklowitz, D., & Goldstein, M. J. (1997). Bipolar disorder: A family-focused treatment approach. *American Journal of Psychiatry*, *157*, https://doi.org/10.1176/appi.ajp.157.4.657-a.

Miklowitz, D. J., & Chung, B. (2016). Family-focused therapy for bipolar disorder: Reflections on 30 years of research. *Family Process*, *55*(3), 483–499. https://doi.org/10.1111/famp.12237.

Miklowitz, D. J., & Goldstein, M. J. (1990). Behavioral family treatment for patients with bipolar affective disorder. *Behavior Modification*, *14*(4), 457–489. https://doi.org/10.1177/01454455900144005.

Mino, Y., Tanaka, S., Inoue, S., Tsuda, T., Babazono, A., & Aoyama, H. (1995). Expressed emotion components in families of schizophrenic patients in Japan. *International Journal of Mental Health*, *24*(2), 38–49. https://doi.org/10.1080/00207411.1995.11449311.

Mohr, S., Borras, L., Nolan, J., Gillieron, C., Brandt, P., Eytan, A., … Huguelet, P. (2012). Spirituality and religion in outpatients with schizophrenia: A multi-site comparative study of Switzerland, Canada, and the United States. *International Journal of Psychiatry in Medicine*, *44*(1), 29–52. https://doi.org/10.2190/PM.44.1.c.

Moline, R. A., Singh, S., Morris, A., & Meltzer, H. Y. (1985). Family expressed emotion and relapse in schizophrenia in 24 urban American patients. *American Journal of Psychiatry*, *142*(9), 1078–1081. https://doi.org/10.1176/ajp.142.9.1078.

Moreira-Almeida, A., Lotufo Neto, F., & Koenig, H. G. (2006). Religiousness and mental health: A review. *Brazilian Journal of Psychiatry*, *28*(3), 242–250. https://doi.org/10.1590/S1516-44462006005000006.

Morgan, V. A., Waterreus, A., Jablensky, A., Mackinnon, A., McGrath, J. J., Carr, V., … Saw, S. (2012). People living with psychotic illness in 2010: The second Australian national survey of psychosis. *Australian and New Zealand Journal of Psychiatry*, *46*, 735–752. https://doi.org/10.1177/0004867412449877.

Mueser, K., & Glynn, S. (1999). *Behavioral family therapy for psychiatric disorders* (2nd ed.). Oakland, CA: New Harbinger.

Murphy, H., & Raman, A. (1971). The chronicity of schizophrenia in indigenous tropical peoples: Results of a twelve-year follow-up survey in Mauritius. *The British Journal of Psychiatry*, *118*(546), 489–497. https://doi.org/10.1192/bjp.118.546.489.

Nasrallah, H. A., & Lasser, R. (2006). Improving patient outcomes in schizophrenia: Achieving remission. *Journal of Psychopharmacology*, *20*(6 Suppl), 57–61. https://doi.org/10.1177/1359786806071248.

Ng, R. M. K., Mui, J., Cheung, H. K., & Leung, S. P. (2001). Expressed emotion and relapse of schizophrenia in Hong Kong. *Hong Kong Journal of Psychiatry*, *11*(1), 4–12.

Oswald, P., Souery, D., Kasper, S., Lecrubier, Y., Montgomery, S., Wyckaert, S., … Mendlewicz, J. (2017). Current issues in bipolar disorder: A critical review. *European Neuropsychopharmacology*, *17*, 687–695. https://doi.org/10.1016/j.euroneuro.2007.03.006.

Oyserman, D., Coon, H. M., & Kemmelmeier, M. (2002). Rethinking individualism and collectivism: Evaluation of theoretical assumptions and meta-analyses. *Psychological Bulletin*, *128*(1), 3–72. https://doi.org/10.1037/0033-2909.128.1.3.

Ozerdem, A., Oguz, M., Miklowitz, D., & Cimilli, C. (2009). Family focused treatment for patients with bipolar disorder in Turkey: A case series. *Family Process*, *48*(3), 417–428. https://doi.org/10.1111/j.1545-5300.2009.01292.x.

Pargament, K. (1997). *The psychology of religion and coping: Theory, research, practice*. New York, NY: Guilford Press.

Pitschel-Walz, G., Leucht, S., Bäuml, J., Kissling, W., & Engel, R. R. (2001). The effect of family interventions on relapse and rehospitalization in schizophrenia—A meta-analysis. *Schizophrenia Bulletin*, *27*(1), 73–92. https://doi.org/10.1093/oxfordjournals.schbul.a006861.

Potash, J. B., Willour, V. L., Chiu, Y. F., Simpson, S. G., MacKinnon, D. F., Pearlson, G. D., … McInnis, M. G. (2001). The familial aggregation of psychotic symptoms in bipolar disorder pedigrees. *American Journal of Psychiatry*, *158*(8), 1258–1264. https://doi.org/10.1176/appi.ajp.158.8.1258.

Ran, M.-S., Leff, J., Hou, Z.-J., Xiang, M.-Z., & Ran, M.-S. (2003). The characteristics of expressed emotion among relatives of patients with schizophrenia in Chengdu, China. *Culture, Medicine and Psychiatry*, *27*(1), 95–106. https://doi.org/10.1023/A:1023636021995.

Resnick, S. G., Rosenheck, R. A., Dixon, L., & Lehman, A. F. (2005). Correlates of family contact with the mental health system: Allocation of a scarce resource. *Mental Health Services Research*, *7*(2), 113–121. https://doi.org/10.1007/s11020-005-3782-0.

Rogan, R. G., & Hammer, M. R. (1998). An exploratory study of message affect behavior. *Journal of Language and Social Psychology*, *17*(4), 449–464. https://doi.org/10.1177/0261927X980174002.

Rosenfarb, I. S., Bellack, A. S., & Aziz, N. (2006). Family interactions and the course of schizophrenia in African American and White patients. *Journal of Abnormal Psychology*, *115*(1), 112–120. https://doi.org/10.1037/0021-843X.115.1.112.

Ruhrmann, S., Schultze-Lutter, F., & Klosterkötter, J. (2010). Probably at-risk, but certainly ill: Advocating the introduction of a psychosis spectrum disorder in DSM-V. *Schizophrenia Research*, *120*(1–3), 23–37. https://doi.org/10.1016/j.schres.2010.03.015.

Rund, B. R., Moe, L., Sollien, T., Fjell, A., Borchgrevink, T., Hallert, M., & Næss, P.o. (1994). The Psychosis Project: Outcome and cost-effectiveness of a psychoeducational treatment programme for schizophrenic adolescents. *Acta Psychiatrica Scandinavica*, *89*(3), 211–218. https://doi.org/10.1111/j.1600-0447.1994.tb08094.x.

Saravanan, B., Jacob, K. S., Johnson, S., Prince, M., Bhugra, D., & David, A. S. (2007). Assessing insight in schizophrenia: East meets West. *British Journal of Psychiatry*, *190*(03), 243–247. https://doi.org/10.1192/bjp.bp.106.029363.

Sartorius, N., Shapiro, R., & Jablensky, A. (1974). The international pilot study of schizophrenia. *Schizophrenia Bulletin*, *1*(11), 21–34. https://doi.org/10.1093/schbul/1.11.21.

Satcher, D. (2001). *Mental health: Culture, race, and ethnicity—A supplement to mental health: A report of the surgeon general*.

Shiraishi, N., & Reilly, J. (2018). Positive and negative impacts of schizophrenia on family caregivers: A systematic review and qualitative meta-summary. *Social Psychiatry and Psychiatric Epidemiology*, 1–14. https://doi.org/10.1007/s00127-018-1617-8.

Siddle, R., Haddock, G., Tarrier, N., & Faragher, E. B. (2002). Religious delusions in patients admitted to hospital with schizophrenia. *Social Psychiatry and Psychiatric Epidemiology*, *37*(3), 130–138. https://doi.org/10.1007/s001270200005.

Silverstein, S. M., Moghaddam, B., & Wykes, T. (2013). *Schizophrenia: Evolution and synthesis*. Cambridge, MA: MIT Press.

Snowden, L. R. (2007). Explaining mental health treatment disparities: Ethnic and cultural differences in family involvement. *Culture, Medicine and Psychiatry*, *31*(3), 389–402. https://doi.org/10.1007/s11013-007-9057-z.

Sperry, L. (2001). *Spirituality in clinical practice: Incorporating the spiritual dimension in psychotherapy and counseling*. Ann Arbor, MI: Sheridan Books.

Stander, V., Piercy, F. P., Mackinnon, D., & Helmeke, K. (1994). Spirituality, religion and family therapy: Competing or complementary worlds? *The American Journal of Family Therapy*, *22*(1), 27–41. https://doi.org/10.1080/01926189408251295.

Susser, E., Moore, R., & Link, B. (1993). Risk factors for homelessness. *Epidemiologic Reviews*, *15*(2), 546–556.

Suro, G., & Weisman de Mamani, A. G. (2013). Burden, interdependence, ethnicity, and mental health in caregivers of patients with schizophrenia. *Family Process*, *52*(2), 299–311. https://doi.org/10.1111/famp.12002.

Tompson, M., Goldstein, M., Lebell, M., Mintz, L., Marder, S., & Mintz, J. (1995). Schizophrenic patients' perceptions of their relatives' attitudes. *Psychiatry Research*, *57*, 155–167. https://doi.org/10.1016/0165-1781(95)02598-Q.

Torrey, E. (1995). *Surviving schizophrenia: A manual for families, consumers, and providers* (3rd ed.). New York, NY: Harper Perennial. Retrieved from http://psycnet.apa.org/record/1995-97138-000.

Tseng, W., & Streltzer, J. (2001). *Culture and psychotherapy: A guide to clinical practice*. Arlington, VA: American Psychiatric Publishing, Inc.

United Nations-Habitat (UN-Habitat), & United Nations Human Settlements Programme. (2005). *The state of the world's cities 2004/2005. Globalization and urban culture*. Nairobi: UN-HABITAT.

United Nations–Habitat (UN-Habitat), & United Nations Human Settlements Programme. (2011). *State of the world's cities 2010/2011: Bridging the urban divide*. Nairobi: UN-HABITAT.

Valenstein, M., Blow, F. C., Copeland, L. A., McCarthy, J. F., Zeber, J. E., Gillon, L., … Stavenger, T. (2004). Poor antipsychotic adherence among patients with schizophrenia: Medication and patient factors. *Schizophrenia Bulletin*, *30*(2), 255–264. https://doi.org/10.1093/oxfordjournals.schbul.a007076.

Vaughn, C. E., Snyder, K. S., Jones, S., Freeman, W. B., & Falloon, I. R. H. (1984). Family factors in schizophrenic relapse. *Archives of General Psychiatry*, *41*(12), 1169. https://doi.org/10.1001/archpsyc.1984.01790230055009.

Ventura, J., Lukoff, D., Nuechterlein, K. H., Liberman, R. P., Green, M., & Shaner, A. (1993). Brief Psychiatric Rating Scale (BPRS), Expanded Version 4.0: Scales, anchor points, and administration manual. *International Journal of Methods in Psychiatric Research*, *3*, 227–243.

Waters, F., Blom, J. D., Jardri, R., Hugdahl, K., & Sommer, I. E. C. (2018). Auditory hallucinations, not necessarily a hallmark of psychotic disorder. *Psychological Medicine*, *48*(4), 529–536. https://doi.org/10.1017/S0033291717002203.

Wedig, M. M., & Nock, M. K. (2007). Parental expressed emotion and adolescent self-injury. *Journal of the American Academy of Child and Adolescent Psychiatry*, *46*(9), 1171–1178. https://doi.org/10.1097/chi.0b013e3180ca9aaf.

Weisman, A., López, S. R., Karno, M., & Jenkins, J. (1993). An attributional analysis of expressed emotion in Mexican-American families with schizophrenia. *Journal of Abnormal Psychology*, *102*(4), 601–606. https://doi.org/10.1037/0021-843X.102.4.601.

Weisman, A., Nuechterlein, K., Goldstein, M., & Snyder, K. (1998). Expressed emotion, attributions, and schizophrenia symptom dimensions. *Journal of Abnormal Psychology, 107*(2), 355–359. https://doi.org/10.1037/0021-843X.107.2.355.

Weisman, A., Rosales, G., Kymalainen, J., & Armesto, J. (2005). Ethnicity, family cohesion, religiosity and general emotional distress in patients with schizophrenia and their relatives. *Journal of Nervous and Mental Disease, 193*(6), 359–368. https://doi.org/10.1097/01.nmd.0000165087.20440.d1.

Weisman, A., Weintraub, M. J., Gurak, G., & Maura, J. (2014). A randomized clinical trial to test the efficacy of a family-focused, culturally informed therapy for schizophrenia. *Journal of Family Psychology*, https://doi.org/10.1037/fam0000021.

Weisman, A. G., & López, S. R. (1997). An attributional analysis of emotional reactions to schizophrenia in Mexican and Anglo-American cultures. *Journal of Applied Social Psychology, 27*(3), 223–244. https://doi.org/10.1111/j.1559-1816.1997.tb00630.x.

Weisman de Mamani, A., Dunham, R., Aldebot, S., Tuchman, N., & Wasserman, S. (2009). Family-focused psychoeducational programs for minorities with serious mental illness. In S. Loue & M. Sajatovic (Eds.), *Determinants of minority mental health and wellness*. (pp. 1–18). New York, NY: Springer New York. https://doi.org/10.1007/978-0-387-75659-2_12.

Weisman de Mamani, A. G., Kymalainen, J. A., Rosales, G. A., & Armesto, J. C. (2007). Expressed emotion and interdependence in White and Latino/Hispanic family members of patients with schizophrenia. *Psychiatry Research, 151*(1–2), 107–113. https://doi.org/10.1016/j.psychres.2006.11.007.

Weisman de Mamani, A. G., & Suro, G. (2016). The effect of a culturally informed therapy on self-conscious emotions and burden in caregivers of patients with schizophrenia: A randomized clinical trial. *Psychotherapy, 53*(1), 57–67. https://doi.org/10.1037/pst0000038.

Weisman de Mamani, A. G., Tuchman, N., & Duarte, E. A. (2010). Incorporating religion/spirituality into treatment for serious mental illness. *Cognitive and Behavioral Practice, 17*(4), 348–357. https://doi.org/10.1016/J.CBPRA.2009.05.003.

Wing, J. K. (1978). The social context of schizophrenia. *American Journal of Psychiatry, 135*(11), 1333–1339.

World Health Organization. (1979). *Schizophrenia: An international follow-up study* Retrieved from http://apps.who.int/iris/handle/10665/41598.

World Health Organization Division of Mental Health. (1994). *The ICD-10 classification of mental and behavioural disorders: Conversion tables between ICD-8, ICD-9 and ICD-10*. [WHO/MNH? 9]. Geneva: World Health Organization. Retrieved from http://apps.who.int/iris/bitstream/handle/10665/62577/WHO_MNH_92.16.Rev.1.pdf?sequence=1&isAllowed=y.

Zygmunt, A., Olfson, M., Boyer, C. A., & Mechanic, D. (2002). Interventions to improve medication adherence in schizophrenia. *American Journal of Psychiatry, 159*(10), 1653–1664. https://doi.org/10.1176/appi.ajp.159.10.165.

PART 5

Conclusion

CHAPTER 21

Human families: Universalities, cultural variations, and family therapy practice implications

W. Kim Halford
School of Psychology, University of Queensland, St. Lucia, QLD, Australia

Families and culture: Research and practice

The current Chapter analyzes the universal human aspects of family, the important cultural variations in families, and the implications of those universalities and cultural variations for practice aimed at enhancing family functioning. I begin by examining families through an evolutionary lens to identify what is universal about the forms and functions of human families. The next section of the Chapter describes the role of culture as a complement to evolution in guiding human adaptation, and the central role human families have in the transmission of culture. The third section of the Chapter explores how culture and family interact. This includes analysis of changing patterns of family formation and dissolution, and examines how culture moderates these patterns. Cultural variations in selected aspects of family functioning are analyzed, and how culture moderates the effects of important influences on family functioning. The last section of the Chapter synthesizes what is known about interventions to promote family functioning across diverse cultures.

Evolution and human families

Evolution has shaped humans and human families to have some universal defining features. Humans have behaviors of complexity unmatched by any other species; we have much more complex language and problem-solving abilities, we invent complex tools and technologies, and have social relationships through which we convey an unparalleled array of knowledge within and across generations (Suddendorf, 2013; von Hippel, 2018). In large part,

Cross-Cultural Family Research and Practice
https://doi.org/10.1016/B978-0-12-815493-9.00021-1

these characteristics reflect the synergy of three interrelated evolutionary adaptations. First, we have a large brain that provides the biological basis (the "hardware") for intelligence and its associated cognitive and social abilities (Isler & van Schaik, 2012a). Second, humans have an extended developmental period (the "software") that allows the brain and behavior to develop after birth, which aids behavioral flexibility and adaptation to diverse and changing environments (Lagenhof & Komdeur, 2018). Third, we have committed pair-bonding between parents (Fletcher, Simpson, Campbell, & Overall, 2015), and family-focused lives (Hrdy, 2016), which provide the context in which cooperative care of young humans promotes their survival during their extended developmental period.

The notably big brain of humans has, relative to nonhuman primates, a disproportionately large expanse of certain aspects of the cerebral cortex (van Essen, 2018). These enlarged parts of the cerebral cortex are associated with complex cognitive abilities like language, tool use, complex reasoning, and sophisticated social behaviors (van Essen, 2018). Human large brains help make us smart but also provide certain challenges to survival. First, a large brain requires a large head, which complicates childbirth, and there seems to be an upper limit to how big human brains can be at birth (Grabowski, 2013). The human brain is also energy-intensive, with about 20% of total energy intake derived from food used to run the brain even though the brain represents just 2.3% of body mass, which necessitates a reliable, energy-intensive supply of food (Isler & van Schaik, 2012a).

Contemporary theories of human evolution propose that there were concurrent adaptations across the last 5–6 million years that allowed an expanding brain size (Isler & van Schaik, 2012b). Humans evolved an extended developmental period and high level, intelligence-based social cooperation (Isler & van Schaik, 2012b; Kaplan, Hill, Lancaster, & Hurtado, 2000; von Hippel, 2018). The extended developmental period allows the brain to expand and develop after birth, which attenuated the problem of head size for birth (Isler & van Schaik, 2012b). It also means that human brain development occurs when young humans are interacting with the environment, allowing the environment to influence brain development (Lagenhof & Komdeur, 2018).

The long postbirth neurodevelopment means human infants behaviors can be shaped to adapt to the environment in which they are developing, but it also means they are dependent upon caregivers for survival for an extended period. Humans manage this survival challenge, and addressing the high-energy demands of the human brain, through social cooperation in care of offspring and food gathering (Hrdy, 2016). From an evolutionary perspective, the closest relatives have the most investment in infants surviving, and social cooperation in raising young focuses most strongly upon family members (Antfolk, Karlsson, Söderlund, & Szala, 2017; Hrdy, 2016).

Romantic attachment between parents is an evolved mechanism that facilitates parents committing to each other and provides a family relationship context for cooperation in raising offspring (Buss & Schmitt, 2019; Fletcher et al., 2015). Parental efforts in childcare are usually supported by other relatives, such as grandmothers, aunts, sisters, grandfathers, uncles, and brothers, who are termed alloparents (Chapter 2). Community members other than family can contribute in important ways to childcare (Keller, 2017), but parents and close relatives make the largest contributions (Antfolk et al., 2017). Thus, couple relationships—and the families they create—are an evolutionary co-adaptation to an expanding brain and an extended developmental period. Families facilitate survival of offspring, and families are central to the cluster of characteristics that define being human.

An important concept in understanding human families from an evolutionary perspective is parent-offspring conflict (Trivers, 1974). In essence, Trivers proposes that offspring seek maximum duration of parental investment, and seek parental investment in them individually rather than sibs. In contrast, parents seek to limit the duration of investment in offspring and limit investment in any individual offspring, to allow the parent to mate and produce more offspring. This parent-offspring conflict is argued to be a key reason why families are not just relationships of high cohesion promoting offspring well-being, but also contain tensions around investment.

Behavioral plasticity and families

Human evolution has selected for complex, flexible behavioral repertoires that allow for adaptation to a wide range of circumstances (Lagenhof & Komdeur, 2018), and family form and functions show this same plasticity. A noteworthy example described by Bjorklund et al. (Chapter 2) is the effect of early environment on the life cycle of the human family. In harsh environments where immediate survival is threatened, humans adopt a fast life trajectory characterized by more rapid aging, early onset of puberty, frequent sex with multiple partners, young age of reproduction, having more children, investing less effort in caring for each child, and greater risk-taking (Ellis et al., 2012). Presumably, this fast life trajectory optimizes the chance of producing at least some surviving offspring when the survival of parents and offspring is at high risk. In contrast, in resource-rich environments where survival is under less immediate threat, humans adopt a slow life trajectory characterized by slower aging, later development of puberty, later age of starting sexual activity, commitment to a long-term pair bond, high investment in each child, and avoidance of high levels of risk-taking. In a stable resource-rich environment, being highly selective in who one mates with, producing healthy children, and investing heavily in those children, produces optimal evolutionary fitness in offspring.

Buss and Schmitt (2019) review a large body of evidence showing that humans have a diverse range of mating strategies that have developed through evolution, and the environment shapes the occurrence of one strategy versus another. Two examples are the long-term pair bonding within a slow life cycle versus the fast life cycle casual mating strategy. As another example, in the regions of the world with high economic inequalities and traditional gender roles, the Middle East and North Africa, a small proportion of men have access to the vast majority of resources (Memon, Jamil, & Khan, 2019). In those regions, rich men typically legally can have multiple wives, whereas in most countries of the world polygamy is illegal (Chapter 5). Cultural beliefs are often used to explain polygyny, but an evolutionary perspective suggests these resource-rich men are getting reproductive access to multiple female partners because that is the only way those women can gain access to scarce resources to raise children.

Sex differences

Sex differences in many behaviors, including cognitive performance, personality, many social behaviors, and psychological well-being, are close to zero and of such small magnitude as to be trivial (Hyde, 2014). From an evolutionary perspective, high levels of similarity between the sexes seem unsurprising; many of the survival challenges that shaped human evolution are similar for men and women, such as not being killed by predators; getting adequate food, water, and shelter to survive; and cooperating with other humans effectively to achieve these goals. However, there are some important, large magnitude sex differences in certain behaviors of high reproductive relevance involving sex and parenting (Archer, 2019; Euler, 2020; Petersen & Hyde, 2010). These sex differences are potentially important in understanding some universals of human families.

There are some shared and some divergent evolutionary challenges for men and women in their reproductive mating strategies, which have been described within the framework of sexual strategies theory (SST) (Buss & Schmitt, 2019). SST is based on the premise that both sexes share the need to select a mate who is likely to produce offspring high in evolutionary fitness and to out-compete rivals for desirable mates. However, the minimum required parental investment to produce an offspring is much greater for women than men. (After conception, women must bear the child through 9 months of pregnancy, whereas a man only has to participate in conception.) Due to this asymmetry of minimum parental investment, SST predicts men are more open to short term mating than women, and there is a large effect size sex difference in the predicted direction across all cultures where it has been studied (Gray, Garcia, & Gesselman, 2019; Lippa, 2009). In addition, mothers on average expend more time and effort in parenting than fathers, which also is attributed to the higher evolutionary investment mothers have in each child (Buss & Schmitt, 2019). That said, human fathers exert substantially more parenting time and effort than fathers do our close, polygamous primate relatives (Gray, 2013).

Sexual strategies theory (SST) postulates that both human males and females seek evidence of relationship commitment, but the evolutionary challenges for men and women in committed relationships vary somewhat (Buss & Schmitt, 2019). For men, because the fetus develops inside the woman, there is some degree of paternal uncertainty, and this prompts males to look for signs of sexual fidelity as an attempt to ensure their efforts in raising offspring are devoted to their genetic children. In contrast, women are certain of their genetic relationship to their child, but seek signs of male commitment to the relationship as a means to ensure the male continues to invest in care and protection of the offspring.

An important influence on modern SST is Hamilton's (1964) notion of inclusive fitness, which proposes that organisms behave in ways that enable their genes to pass onto successive generations. That goal can involve direct reproduction, or assisting the reproduction and raising of offspring by relatives. For example, sibs, aunts and uncles, and grandparents enhance their inclusive fitness by assisting in the raising of sibs, nieces and nephews, and grandchildren, respectively (Buss & Schmitt, 2019).

Evolution and families in the 21st century

A key concept in evolution is that genes are selected that enhance survival and reproduction within the particular environment in which an organism lives, which is referred to as the environment of adaptation. For most of human evolution, humans lived in small clan groups in a hunter-gatherer existence (Diamond, 2014), and hence much of human behavioral adaptations—including those related to family functioning—arose in response to selection pressures in that environment. However, in 2020 most humans live in cities, towns or villages, and we lead lives very different from those of nomadic hunter-gatherers (Satherwaite, McGranahan, & Tacoli, 2010).

Homo sapiens diverged from other humanoid ancestors between 250,000 and 350,000 years ago (Mounier & Lahr, 2019; Schlebusch et al., 2017), but the development of agriculture only began about 10,000 years ago (Bocquet-Appel, 2011). Agriculture allowed humans to generate sufficient food in one place to allow settlements to develop (Bocquet-Appel, 2011). The 10,000 years from the start of the agricultural revolution is a short time in for large-scale genetic adaptation to occur, and there is substantial evidence that human mating behavior and family formation is still influenced by adaptations that occurred across millions of years in the hunter-gatherer environments of prehumans and humans (Buss & Schmitt, 2019; Mounier & Lahr, 2019). For example, the well-established sex difference in males' greater desire for casual sex is reflected in modern online environments. Men are more likely than women to watch pornography (Petersen & Hyde, 2010), to approve of and send sexually explicit texts of words and/or images ("sexts") to strangers (Burén & Lunde, 2018; Rappleyea, Taylor, & Fang, 2014), and to view positively sexts from opposite sex people that indicate sexual availability (McMahon, 2018).

As Bjorklund et al. (Chapter 2) states, evolutionary influenced human behaviors are not intrinsically desirable or "natural," the behaviors simply reflect what was selected in the environment of adaptation. For example, there is clear evidence that men are much more likely to engage in sexual violence than women across all cultures in which it has been studied (Chapter 8). From an evolutionary perspective, a male raping a female could be seen as adaptive in the narrow sense that it can reproduce the male's genes, but that does not provide any moral justification of the behavior, or suggest the behavior cannot be changed. It is well documented that rates of sexual violence against women vary enormously across countries, with much lower rates when women have more economic power (Chapter 8), when divorce is easier to access (Chapter 3), and when there is legal recognition of women's rights (Chapter 5). Thus, rates of sexual violence against women need not be an inevitable result of evolution, but rather cultural change can greatly reduce sexual violence.

It is important to reemphasize the point made earlier that human evolution has selected for considerable behavioral flexibility that can accommodate to rapidly changing environments. One striking example is the increasingly wide availability of the oral contraceptive pill since 1960, which has given women a cheap and effective means to control their fertility (Tyrer, 1999). This easy access to contraception changed numerous aspects of family relationships in many societies, not just by reducing fertility, but also by contributing to the restructuring of gender roles within relationships, liberating women to expand their education and workforce participation, and thereby giving women much greater economic power (Bernstein & Jones, 2019). In Western countries, the current generation of young adults report greater acceptance and engagement in casual sex and premarital sex relative to previous generations (Petersen & Hyde, 2010; Sassler, 2010), which likely reflect the very different consequences of sex once effective fertility control is available.

In summary, evolution has shaped human sexuality and family life to produce some universalities, such as the pervasiveness of human long-term pair bonding, sex differences in some mating behaviors, and cooperation within extended families in child-rearing. At the same time, parent-child conflict is an evolutionary influence that underpins tension within families. Family can be seen as an evolutionary co-adaptation to the expanding brain, and the long developmental period, that collectively help define what it is to be human. At the same time, evolution has selected for behavioral flexibility that allows adaptation to variable environments that are changing rapidly, such as changes in urbanization, technology, and access to contraception.

Culture and families

Culture and evolution

Heinrich (2016) makes the intriguing point that humans are puzzling in their survival abilities. A lone human left in the wilderness often fails to meet basic survival needs like getting food, water, and shelter. Yet 60,000 to 80,000 years ago groups of humans that evolved in Africa began exploring the world and were able to address survival challenges in diverse environments ranging from the frozen tundra of Siberia, through the remote islands of the Pacific, to the central deserts of Australia. Evolution through local genetic adaptations operate too slowly to explain much of this adaptability, and there is much evidence that the collective knowledge of groups of humans was (and is) crucial to human adaptability (Heinrich, 2016; Kashima, Bain, & Perfors, 2019). In other words, particular human groups learned to find shelter in the frozen tundra, to navigate and hunt for food successfully in the ocean, and find fresh water in the deserts of Australia. Each cultural group passed on to successive generations an ever more complex accumulation of cultural knowledge that enhanced survival within one of those different environments.

Kashima et al. (2019, p. 500) conceptualized culture as "the set of socially transmittable information in the population, which can influence cognition, affect and behavior." Human evolution has given us a brain that is well suited to the development of highly variable and complex cultures. For example, the unusually high level of brain growth that occurs postbirth relative to other species produces a brain receptive to learning opportunities available in that particular environment (Sherwood & Gomez-Robles, 2017). The human brain also is particularly well adapted to high fidelity imitation of others' behavior, which allows a cumulative culture to develop (Nielsen, 2018). Finally, humans are very social and interact in ways that develop and transmit cultural innovation (Muthukrishna & Henrich, 2016).

The development of human culture has altered the environment in which the highly flexible human brain develops, which has led to further changes in both brain structure and cognition (Colage & d'Errico, 2018). For example, Colage and d'Errico (2018) explain that the fusiform gyrus area of our brain evolved (we think) for accurate facial recognition, which is crucial to our cooperative breeding. The fusiform gyrus area has been repurposed through cultural experience of reading to enhance the functions of reading and writing. Moreover, the use of the fusiform gyrus area for literacy requires connections between it and other areas of the brain, like connections to the superior-mid-temporal and inferior parietal areas involved in spoken language, that allow readers to say what they have read. Thus, cultural change (e.g., early exposure to the written word) has led to repurposing of many existing brain structures, and the development of new neural connections, which are made possible by the long postbirth neurodevelopmental period in humans (Sherwood & Gomez-Robles, 2017).

The complexity of social interaction within human cultures creates the context for more cultural innovation. Many sources attribute innovation to individual activity; the geniuses of Johanes Gutenberg, Sir Isaac Newton, Charles Darwin, and Marie Curie, are credited in many sources as the individual inventor or discoverer of the printing press, calculus, evolution, and radioactivity, respectively. However, innovation flourishes not so much through the isolated effort of geniuses but rather through collective efforts that build upon and recombine existing ideas. For example, we know that Gutenberg adapted the existing ideas of woodblock printing, and use of moveable type, which people in Eastern Asia had used for hundreds of years, to produce the first printing press (Tsuen-Hsuin & Needham, 1985). Marie Curie was inspired and guided in her work by earlier work of Roentgen on X-rays, and Becquerel's work on the radiation emanating from uranium (Nelson, 2014). Often innovations occur independently to different people, likely because they are drawing on a shared cultural milieu that creates the context for invention. For example, Newton developed calculus concurrently and independently of Gottfried Leibniz (Hall, 1980), and Charles Darwin and Alfred Wallace independently proposed a very similar theory of evolution (Slotten, 2004).

All of the above examples are isolated anecdotes, but archeological evidence allows us more systematically to define where and when major human innovations occurred. The use of fire about 1 million years ago (Berna et al., 2012), blade production, and use of grindstones occurred between 250,000 and 500,000 years ago (Colage & d'Errico, 2018). These innovations did not develop in just one place, as would be expected if an individual genius created these ideas and then communicated the idea to different groups, rather each of these technologies developed independently in different places (Colage & d'Errico, 2018). Other innovations like pyrotechnology (use of fire to create things like metal tools and pottery glazes), formal bone tools, and fishing developed later, about 70,000–100,000 years ago, and also developed concurrently in multiple places (Colage & d'Errico, 2018). These findings suggest certain social contexts enhanced creativity, and that culture was cumulative building new ideas by recombining or adapting existing ideas. For example, control of fire allowed later use of pyrotechnology, and more recently the growth of information and communication technology has supercharged access to information and artificial intelligence (Lee, 2018).

Muthukrishna and Henrich (2016) describe how, as social groups get larger and more interconnected, the rate of cultural innovation accelerates because larger groups provide a larger collective pool of talent and ideas. In other words, cultural complexity and innovation grow with group size (Muthukrishna, Shulman, Vasilescu, & Henrich, 2014; Derex, Beugin, Godelle, & Raymond, 2013). Across the last 500,000 plus years human group sizes increased, and rates of cultural innovation accelerated (Colage & d'Errico, 2018).

In summary, a human culture is the collective, socially transmitted knowledge and skills of a group of humans. Human culture has become ever more complex and enabled humans to thrive in a diverse range of environments. As culture incorporated more innovation, the complexity of the culture fueled further innovation at an accelerating rate, as well as leading to changes in brain function and structure. Thus, genes and culture have coevolved, and the ever more complex cultures within which humans live is now a defining aspect of what it means to be a modern human.

Families, culture, and evolution

The most basic human social structure is the family. Young people usually first get exposed to cultural learning through their parents, and this often then extends to a range of alloparents (Chapters 2 and 18). Historically, the pair bonding of parents led different extended families to integrate into larger groups or clans, and outgroup pairings with other clans then led related clans to former larger clan groupings (Johnson & Earle, 2000). Larger groups developed more cultural innovation, provided its members with access to a larger collection of knowledge and skills, and larger groups displaced smaller groups or incorporated them (Boyd & Richerson, 2009). A critical innovation was agriculture, which provided more reliable food sources and allowed ever larger and complex human cultures to develop (Johnson & Earle, 2000), and which led eventually to cities, large ethnic groupings, and nation states (Flannery, 2012). In summary, the original building block for the development of human large group size and cultural complexity was the family.[a]

Like genes, different cultural traits can be more or less adaptive in different environments. For example, a high prevalence of hunting skills is essential in hunter-gatherer environments, but of limited importance in agricultural or industrialized environments. Literacy skills were nonexistent in hunter-gatherer societies, but are essential in modern countries. Cultural differences enhance human adaptation to diverse and changing environments (Creanza, Kolodny, & Feldman, 2017).

As noted earlier, cumulative culture relies in part on humans' capacity for high fidelity copying of others' behavior, but that capacity also is associated with a tendency to copy aspects of behavior not causally related to achieving survival-related outcomes (Nielsen, 2018). For example, children imitating how to open a box will imitate task-irrelevant steps (e.g., a twirl of the hands deliberately included by an experimenter) and continue to

[a]As humans develop toward market-driven economies there seems to be some weakening of kinship ties, although the family still remains central to most people's lives (Colleran, 2020).

incorporate those irrelevant steps when they open the box.[b] This means culture often embodies certain fixedness attributable to copying and incorporates rituals that have little adaptive significance. At the same time, it means that rituals whose significance might not be understood, but which do have adaptive significance are maintained, such as important steps in food preparation. Cultures that incorporate too many unhelpful rituals presumably are displaced by cultures not so encumbered (Boyd & Richerson, 2009). However, all human cultures seem to include rituals that have little survival significance.

As noted previously, human group sizes have tended to grow across time. Cues like voice and facial recognition serve well to recognize others as family members or members of small clan groups of 20 or 30 people but become inadequate for categorization of others as a group member when human group sizes exceed a few hundred people (Diamond, 2014). People who are categorized as in-group members are evaluated more positively and interacted with more than out-group members (Ellemers & Haslam, 2012), a process known as social identification. The adaptive significance of social identification is that, by favoring in-group members, we make it more likely both our genome and our culture pass on to subsequent generations. Moreover, outgroups were often a threat as they might compete for access to resources, and in hunter-gatherers societies it is quite common for out-group members to be killed or driven away (Diamond, 2014). Social identification often relies on the use of behaviors and rituals to categorize people as in-group and out-group members (Tajfel, 1979). Thus, seemingly survival irrelevant behaviors can assume significance as a social indicator of a shared group membership.

One striking example of ritualistic behavior associated with group identification is religion. Researchers argue that religions develop as a set of rituals and beliefs shared by a group of people that promote social solidarity (Norenzayan et al., 2016). More specifically, the development of moralizing backed with supernatural agents, and credible religious rituals, is associated with a sense of belonging, and historic analysis shows this was associated with high fertility and effective competition with other groups (Norenzayan et al., 2016). In addition, religious identification can promote tight adherence to cultural behaviors that can have adaptive significance in environments with high survival threat (Gelfand et al., 2011).

[b]Interestingly, our closest primate relatives, like chimpanzees do not do this ritualistic copying (Nielsen, 2018).

In summary, culture began as the accumulation of learning of adaptive, survival-related behaviors within family and clan groups. The reliance on high fidelity copying to transmit culture adventitiously incorporated survival irrelevant rituals. As human group size increased, cultural rituals assumed additional significance in signaling group identification. Belonging to a large group, in itself, has survival significance as large groups tend to incorporate or displace small groups. In some important ways, cultural evolution parallels and complements genetic evolution. Both cultural and genetic evolution have selection pressures so that the best-adapted cultures and genes survive in particular environments. However, cultural evolution is usually much quicker than genetic evolution; cultural variations develop rapidly through social cooperation that drives human innovation, while genetic variations arise slowly through random mutations. Moreover, while cultural evolution selects groups of individuals who share a complex array of knowledge, skills, and rituals; genetic evolution selects individual genotypes.

Culture and family formation and dissolution

Family formation

As argued earlier, the long-term committed couple relationship is the foundation of family formation, and for most of the human recorded history marriage was the bedrock of family life (Coontz, 2005). Marriage has been the cultural institution that most determined who could have sex together, and who was responsible for the care of children that arose from sex (Therborn, 2014). Marriage also determined who has access to the resources of extended family and tied together extended kinship groups in alliances (Coontz, 2005; Kamal, Hassan, Alam, & Ying, 2015). At times of resource shortage families arranged who would be married, primarily on economic grounds of mutual benefit to the extended families (Coontz, 2005). However, when resources are less tight, most notably in the last 200 years in high-income Western countries, the couple themselves determine if they form a relationship based on romantic attraction (Coontz, 2005).

Therborn (2014) notes that some theorists argue that the trend to romantic love as a determinant of forming couples is part of a broader cultural modernization that involves economic development, greater individual freedom, and possibly a reduced centrality of families in people's lives. Some global trends in family formation seem consistent with that view. First, across most countries, rates of marriage have seen some decline, mean age at marriage has increased (Wang & Schofer, 2018), and rates of couple cohabitation have increased (Chapter 4). Second, traditional family formation practices such as child marriage and enforcement of arranged marriages have declined in most countries (Ghimire, Axinn, Yabiku, & Thornton, 2006; Nguyen & Wodon, 2015; Twamley, 2013). In fact, almost all countries have signed international conventions prohibiting child marriage and guaranteeing human rights not to have to enter into arranged marriages (Chapter 5).[c] Third, there are cross-national trends toward later age at first marriage, and increased rates of remarriage after divorce (United Nations Department of Economic and Social Affairs, Population Division, 2020). Consistent with the cultural modernization hypothesis, all of these trends are much more marked in high-income than in low-income countries (United Nations Department of Economic and Social Affairs, Population Division, 2020), and in countries exposed to a modern trans-national culture that emphasizes individualism (Wang & Schofer, 2018).

In contrast, there is much other evidence that seems inconsistent with a trend for global convergence on a common modern family formation process. First, rates of marriage vary enormously between countries, ranging from a crude rate of around 10 per 1000 population in Fiji and Palestine to lows of one to two in Qatar and Peru, and the variability between countries in marriage rates is increasing across time (Chapter 3). Second, rates of unmarried couple cohabitation also vary greatly across countries (Chapter 4). Moreover, as Lesthaeghe documents in Chapter 4, the form and functions of cohabitation vary greatly between countries. For example, in some countries cohabitation predominantly is a transitional arrangement, in that cohabiting couples tend to either marry or separate within 4–5 years (e.g., Australia and United Kingdom), while in other countries cohabitation predominantly is an alternative to marriage involving long-term cohabitation often spanning decades (e.g., Austria, France, Norway). In some countries cohabitation is more common among the better educated, and seems to be correlated with liberal views on individual freedoms (e.g., China, Japan, and the Netherlands), while in other countries cohabitation is more common among the socially disadvantaged (e.g., much of Latin America).

A third example of divergence across countries in family formation is that in the 20th century no country had legal same-sex marriage, but in 2001 Norway became the first country ever to legalize same-sex marriage. At the time of writing, same-sex marriage is legal in 24 countries, plus

[c]Despite signing international accords prohibiting forced or child marriage, a substantial minority of countries still do not have legislation and/or legal enforcement to put into effect these guarantees (Chapter 5).

the United Kingdom and Mexico allow same-sex marriages in some jurisdictions, and a further 11 countries have civil recognition of same-sex relationships (World Population Review, 2020). This is a remarkable change in just 20 years. However, same-sex marriage is not legally recognized in more than 150 countries, any same-sex relationship is illegal in 72 countries (Equaldex, 2020), and punishable by death in eight countries (Goldberg, 2016). Rights for same-sex couples also is highly variable with respect to rights to have children, in 44 countries it is legal for same-sex couples to adopt a child, but not in any other countries (Equaldex, 2020).

There is not just diversity of family formation between countries, there is also substantial—and sometimes increasing—diversity of family formation within some countries. For example, relative to most other countries, the United States has particularly high marriage and divorce rates (Chapter 3). About 40% of current marriages in the United States involve at least one partner who was previously married, in about one-third of marriages at least one partner has a child from a prior relationship to form a stepfamily, and there are also high rates of postdivorce single-parent families (Lin, Brown, & Cupka, 2017; Livingstone, 2014). Stepfamilies have distinctive structures, often having two biological parents and at least one stepparent who can be involved in parenting, and can have complex boundaries when seeking to define the extended family (Livingstone, 2014). Thus, in the United States type of families and how they are formed are becoming more diverse with the rise of stepfamilies and single-parent families.

A second example of intra-country heterogeneity in family formation is in Malaysia. Separate marriage laws exist for Malays, who make up about 70% of the population, and non-Malays (Department of Statistics Malaysia, 2016). A Malay is defined as "a person who professes the religion of Islam, habitually speaks the Malay language, conforms to Malay custom" (Malaysian Government, 2010 Federal Constitution article 160). The marriage laws for Malays are based on Sharia law, (e.g., allow polygamy). In contrast, the marriage laws for non-Malays are quite different and based on civil law (e.g., do not permit polygamy). In these examples from the United States and Malaysia, there is great intra-country diversity in patterns of family formation, and no sign of convergence to any shared pattern within a country let alone between countries.

Family dissolution

Rates of divorce globally increased from the middle of the 20th century (Chapter 3). The extent of increase in divorce rates correlates with exposure of people within countries to ideals about individualism, gender equality, and modernist ideas about rights to end marriage (Wang & Schofer, 2018). Again, these trends have led to suggestions of a worldwide convergence on a global pattern of increasing rates of family dissolution (Wang & Schofer, 2018). Consistent with that proposition, in the last 50 years most high-income countries have made divorce easier to obtain, notably moving to a "no fault" form of divorce where the only necessary grounds for divorce is that at least one spouse wishes to end the marriage (Chapter 5).

The high threshold to initiate divorce in earlier eras reflected social attitudes that now have changed, at least in many high-income countries. In the 1960s, the majority of adults in the United States and Australia endorsed the view that divorce should only occur when the marriage had severe problems (Australian Institute of Family Studies, 1997; Thornton & Young-DeMarco, 2001). However, by the late 1990s the majority of adults in the United States (Thornton & Young-DeMarco, 2001), most countries of Western Europe (de Graaf & Kalmijn, 2006), and Australia (Australian Institute of Family Studies, 1997), endorsed the view that divorce is acceptable if the partners no longer wish to remain together. Both cross-sectional (Kalmijn & Poortman, 2006) and longitudinal studies (Amato & Rogers, 1997) show that severe marital problems like violence or adultery now predict the occurrence of only a minority of divorces in high-income countries. In other words, the threshold for marital dissolution has declined in high-income countries. However, in many low-income countries divorce remains hard to obtain and rare.

Rates of divorce greatly between countries ranges from a low of three divorces per year per 1000 married population in Peru to a high of 43 per 1000 in the Maldives (Chapter 3). Even between countries of seeming cultural similarity, rates of divorce vary greatly. For example, within Latin America Mexico and Peru have low divorce rates (both <8 per year per 1000 married population), whereas Cuba and Puerto Rica have high divorces rates (both >20/1000). There is also a considerable difference between countries in gender equality with respect to divorce. While the majority of countries allow men and women to seek divorce on the same grounds, 24% of countries require women to seek the permission of a religious court, their husband or other authority to seek divorce, and in 35% of countries women have less right than men to remarry after divorce (Chapter 5).

As noted earlier, in many countries long-term cohabitation has become a widely practiced prelude or alternative to marriage (Di Giulio, Impicciatore, & Sironi, 2019). There is not the same tradition of collecting rates of dissolution of cohabiting couple unions as there is of collecting divorce statistics, but in high-income countries where data is available rates of separation are notably higher in cohabiting couples than married couples (Hewitt & Baxter, 2015). The high rates of family dissolution among cohabiting relative to married couples adds further to the variability in patterns of family dissolution (Chapter 4).

Conclusions

There is some convergence in acceptance of certain minimum standards in family formation and dissolution, which seem to reflect acknowledgement of basic human rights. Specifically, there is a movement to prohibit child marriage and forced arranged marriage and to allow more access to divorce and separation. At the same time, there is great diversity in patterns of family formation, as exemplified by the heterogeneity in rates of marriage between countries, the highly variable legal status of same-sex relationships and parenting, the highly variable national rates of cohabitation as a prelude or alternative to marriage, and variable rates of single-parent families and stepfamilies. There are trends toward gender equality in family formation and dissolution, but the extent of such equality varies greatly from country to country, and in some cases varies greatly between cultural groups within countries.

Social and technological changes seem likely to contribute to more diversity in family formation and dissolution patterns. For example, online dating has become increasingly common and acceptable as a means to start couple relationships, most notably in high-income countries (Finkel, Eastwick, Karney, Reis, & Sprecher, 2012; Smith & Anderson, 2015), and in the United States online has displaced friends as the main way heterosexual couples meet (Rosenfeld, Thomas, & Hausen, 2019). One effect of online relationships is that they greatly increase exposure to potential mates from different backgrounds beyond individuals' immediate social networks (Sanri & Goodwin, 2014). When the pool of potential partners expands, it is well established that people become more selective about mates, defer relationship commitment (Choi & Tienda, 2019), and aspire to have higher functioning committed couple relationships (Finkel, Hui, Carswell, & Larson, 2014). Thus, while in high-income countries there seem to be ever-increasing aspirations for committed couple relationships, and increasing rates of separation when relationships fall short of these aspirations, in low-income countries marriage remains a rarely dissolved socioeconomic alliance of family and kinship groups essential to survival.

Culture and family functioning

Law and families

There are multiple lenses through which we examine how culture is associated with family functioning. One such lens is to examine the laws in different countries related to individual rights and responsibilities within families. Gautier (Chapter 5) describes major differences across 189 countries laws on the rights within marriage of men and women. Most countries accord gender equality in basic civil rights, but wives require husband's permission in about 10% of countries to do basic things like seek paid employment, get a passport, or travel outside the home. In more than 20% of countries, women have less rights than men to inherit the marital assets after spousal death, or make decisions about purchase or sale of the family home; and daughters have less rights than sons to inherit assets from their parents (Chapter 5). Specific laws against family violence, which most often protect women and children, do not exist with respect to intimate partner violence in 24% of countries, or marital rape in 58% of countries (Chapter 5). A total of 58 countries (31% of those assessed) by 2019 had prohibited the use of corporal punishment of children in the home (Global Initiative to End All Corporal Punishment of Children, 2020).

The United Nations Convention on the Rights of the Child (Office of the High Commissioner on Human Rights of the United Nations, 1989) specifies in its 54 articles a range of minimum rights accorded to every child. For example, the convention specifies every child who is capable for forming his or her own views shall have the right to express those views freely in all matters affecting the child (Article 12), and that both parents (or legal guardians) have responsibilities for the upbringing and development of their child (Article 18). All but two countries that were member states of the UN in 1989 ratified this convention, the United States and Somalia did not—and still have not—ratified the convention.

As Gautier (Chapter 5) notes, the written law as it relates to family functioning is only one influence on how families function. For example, in countries that prohibit child corporal punishment, many parents report viewing corporal punishment as necessary to raise children appropriately, and that they use corporal punishment (Chapter 18). In countries that prohibit interpartner violence, there are somewhat variable interpretations of what constitutes physical violence, and there are highly variable interpretations of what constitutes emotional abuse (Chapter 8). Moreover, even in countries with laws prohibiting intimate partner violence, there is a clear association between high rates of interpartner violence by men with patriarchal family cultural practices (Hayes & Boyd, 2017).

Couple and family relationship standards

A second lens through which to examine culture and family functioning is examining couple and family relationship standards, which are the beliefs people hold about what behaviors are associated with having a healthy couple and family relationships. Across a diverse range of cultures, a strong emotional bond between the spouses is seen as very important, which is exemplified by expressions of caring, affection, and emotional intimacy (Chapter 17). The strong endorsement of Couple Bond standards across cultures likely reflects the evolutionary imperative of promoting a strong couple bond to enhance offspring survival. At the same time, there are striking cultural differences in couple relationship standards about the importance of relationships with extended family, traditional gender roles, and maintaining family harmony (Chapter 17). There are also substantial cultural variations in beliefs about obligations of children to parents, siblings, and extended family members, such as showing respect to older family members and parents (Lansford et al., 2016).

Endorsement of the standards about the importance of relations with extended family and traditional gender roles covary with broader cultural values and are higher among cultures that prioritize collective actions to enhance survival rather than individual self-expression (Chapter 17). Survival values understandably are more emphasized in low-income than high-income countries (Dobewell & Strack, 2014; Kashima et al., 2019). Moreover, not behaving according to such standards is evaluated more negatively in those cultures in which such behavior is normative (Chapter 18).

The importance of religion in family relationships unsurprisingly is endorsed most strongly in countries with traditional-religious rather than rational-secular values (Chapter 17). Traditional values emphasize the importance of God and religion, and that development of obedience and conformity to social rule is important (sometimes called "tight cultures"); whereas secular-rational values prioritize the importance of science and rational thought, developing independence and self-determination in life ("loose cultures") (Gelfand et al., 2011; Inglehart & Baker, 2000). Tight cultures and high religiosity covary (Gelfand et al., 2011), and it might be that religion is a cultural adaptation that promotes adherence to tight social norms. For example, many religions have a strong prohibition of sex outside marriage, and strong disincentives to divorce, which has particular adaptive significance in harsh environments with strong survival pressures. Thus, family responsibility and importance of religion relationship standards seem to reflect the survival challenges, past and present, that the culture developed to address.

High endorsement of couple bond standards is associated with high couple relationship satisfaction across a diverse range of cultures (Halford et al., 2018a; Hiew, Halford, van de Vijver, & Liu, 2015a). This again supports the suggestion made earlier of universal adaptiveness of the couple bond. In contrast, it is an agreement between partners on the endorsement of Family Responsibility standards, rather than absolute strength of endorsement, that predicts high couple satisfaction (Halford et al., 2018a; Hiew et al., 2015a). That suggests, it is largely culture that defines acceptable behaviors in these domains (Hiew et al., 2015a).

Couple communication

The specific ways in which couples communicate differs between cultures (Chapter 7; Hiew, Halford, van de Vijver, & Liu, 2015b). Even within culturally similar, Western individualistic cultures like Germany and Australia, Germans tend to be more direct, particularly about expressing negative feelings, than Australians (Halford, Hahlweg, & Dunne, 1990), and these differences in normative rates of particular communication behaviors can create misunderstanding in cross-cultural couples (Hiew et al., 2015b). Communication patterns within couples often reflect broader cultural values. For example, in a comparison of rates of observed couple communication in Pakistani and US couples, there was an interaction of culture with gender, such that US wives were more negative than Pakistan wives, but Pakistani husbands were more negative than US husbands (Rehman & Holtzworth-Munroe, 2007). The authors suggest that the patriarchal power structure in Pakistan means Pakistani men but not women directly assert their criticisms of their spouse. Consistent with this interpretation, self-reported communication of Pakistani wives and husbands also suggest husbands are much more critical of their wives than vice versa, and wives are more likely to withdraw and go quiet than husbands (Rehman & Holtzworth-Munroe, 2006).

While there are different normative rates of communication behaviors across cultures, the association of communication behaviors with satisfaction within a culture seems to be similar across cultures. For men and women, high rates of positive communication, and low rates of negative communication, are associated with high relationship satisfaction across diverse cultures (Chapter 7). Moreover, the same associations are evident in cross-cultural couples (Hiew et al., 2015b).

Parenting

There are some universal functions of parenting: to ensure offspring's physical health and safety, cognitive stimulation, and behavioral socialization, and the provision of affection (Chapters 18 and 19). However, the parenting behaviors that achieve these central care-giving functions can be highly variable across environments. For example, in hunter-gatherer or subsistence farmer contexts where an unprotected child can be in imminent physical danger, infants are often strapped to their mother's body for much of the day, and parents respond quickly to signals of infant distress (LeVine et al., 1996). In contrast, in a high-income country where both parents are in paid employment, parents being skilled at arranging good childcare might serve a similar function of ensuring infant safety.

Parenting behaviors often reflect the importance attached to offspring learning particular skills (Chapter 18). For example, in Argentina, Belgium, Israel, Italy, and the United States, mothers differ in the emphasis they place on caregiving in nurturing, physical, social, and language domains (Bornstein, Putnick, Cote, Haynes, & Suwalsky, 2015). Differences in maternal behavior relate to infants' behaviors in corresponding domains so that infants whose mothers encouraged physical development were more physically developed, whereas infants whose mothers emphasized language were more proficient at language.

A key moderator of the effects of parenting is cultural normativeness, which refers to the extent of use, and acceptability of, particular parenting behaviors within a culture (Chapter 18). One noteworthy example is the use of physical punishment, which is associated with poorer child adjustment in all cultures where it has been studied, but the effects are most negative in cultures where physical punishment is rare and socially unacceptable (Lansford et al., 2005). Similarly, parenting behaviors that encourage responsibility toward extended family, parental monitoring of child location and activity, and child academic study, all have more positive effects on children's behavior in achieving those outcomes in cultures where those behaviors are socially normative (Lansford et al., 2018).

The social normativeness of parenting behaviors, and the child behaviors they seek to promote, usually reflect broader cultural values (Fung, Wong, & Park, 2018). For example, within the United States cultural groups drawn from more hierarchical cultural traditions (e.g., Chinese Americans) value parental control over children' lives more strongly than groups drawn from more egalitarian cultural traditions (e.g., European ancestry Americans) (Fung et al., 2018).

Economics of family functioning

Economists conceptualize the formation of a couple relationship as an investment in which the partners pool their resources, and allow some specialization in what they do, to maximize their collective benefit (Chapter 6). There is considerable evidence to support the premise that marriage usually accrues advantages to the spouses, and their children, in high-income countries (Waite & Gallagher, 2000) and in low-income countries (Lancaster, Maitra, & Ray, 2006). For example, in a high-income country like Canada, couples spend about one-third less per head on heating, lighting, food, housing, and transport to achieve the same standard of living as singles (Browning, Chiappori, & Lewbel, 2013). Economic efficiency has profound effects on long-term accumulation of wealth, which is survival critical in low-income countries (Lancaster et al., 2006).

The benefits to individuals within a family depend on how the resources of the family are spent (Chapter 6). A spouse's power to determine family expenditure is higher when their share of total resources brought to the relationship is high, their income earning in the relationship is high, and their education is high (Chapter 6). In particular, when spouses are similar in education and income-earning, family resource allocations are likely to be more equitable across all family members including adults and children.

Contextual factors like culture, the laws governing marriage and divorce, and local gender ratios of potential alternative mates, influence spousal economic power. Culture and laws about these power arrangements in low-income countries, which tend to be patriarchal, often disadvantage women and children (Chapter 6). For example, in Malawi men alone can own assets, and men tend to get a disproportionately large share of family expenditure relative to women and children (Dunbar, Lewbel, & Pendakur, 2013). Legal or social policy initiatives sometimes can attenuate these inequities. For example, in Mexico the government paid money directly to women in low-income households, and found it increased their power to determine household expenditure (Rubalcava, Teruel, & Thomas, 2002). Women used their power to allocate more family expenditure to food, clothing, and education for the family's children.

Family abuse and violence

No analysis of culture and family functioning would be complete without consideration of a dark side of families: abuse and violence. Globally more than half (58% in 2017) of intentional murders of women are perpetrated by their intimate partner or other family members,[d] being in the home is the most dangerous place for women, and rates of murder of women are strongly related to lack of gender equality in law and culture (United Nations Office on Drugs and Crime, 2019).

There are considerable cultural variations in what is socially normative behavior with respect to abuse and neglect. For example, married women may only go out from the family home with their husband's permission in some countries of Northern Africa, parts of South Asia, and the Middle East. In contrast, many classify a man living in a Western country attempting to control where his wife goes as abusive and as using coercive control. However, as Parsons et al. (Chapter 8) describe, considerable progress has been made in developing research-based, widely accepted criteria for defining what constitutes abuse and violence within families. These criteria provide what they describe as bright line definitions of harmful behaviors.

Conclusions

There are some universals of family functioning, and some cross-cultural variations in the role of partners in a couple relationship, their parenting, and relationships to extended family. Widely endorsed international conventions commit governments to promote certain minimum standards of family functioning, (e.g., the right of each spouse to decide if they have sex with the other; that interpartner violence is unacceptable; and that both fathers and mothers have shared responsibilities toward their children). Legal systems in many countries lag in enforcing the stated international conventions, but there does seem to be progressing across time in making minimum standards of family functioning more widely practiced.

There seems almost universal acceptance of the importance of an emotionally intimate, caring bond as desirable in couple relationships, but there is much cross-cultural diversity in couple relationship standards about the importance of relations with extended family, the maintenance of relation-

[d]About 5% of all intentional homicides of men are perpetrated by an intimate partner or other family members. However, as men are about 4 times more likely than women to be victims of intentional homicide by anyone, men constitute about 20% of intentional homicides by an intimate partner or other family members (United Nationals Commission of Crime and Drugs, 2019).

ship harmony, gender roles, and the role of religion in couple relationships. Gender roles that disempower women are present to varying degrees in most countries, but the disparities in gender rights are extreme in some countries and associated with high rates of male-to-female violence and abuse. Couples' communication show some universals, with high rates of positive communication and low rates of negative communication being associated with relationship satisfaction across diverse cultures. However, there are important cultural differences in the rates of both positive and negative communication, and the strength of their association with relationship satisfaction.

Parenting universally has a range of key functions (e.g., physical safety, cognitive development). In different contexts, these different key functions are differently prioritized, and rates of parenting behaviors reflect the functional priorities. Cross-culturally, different forms of parenting behaviors can achieve similar functions, and the effects of parenting behaviors on child behavior often are moderated by the social normativeness of those parenting behaviors. Economic analyses also point to a universal in family functioning that being in a couple relationship is—on average—economically more efficient than singlehood. However, the relative benefits spouses and any children derive from the family depend upon the power the spouses bring to their relationship.

Culture and influences on family functioning

Social disadvantage, social policy, and the family

There is a reciprocal influence between family functioning and family economic prosperity. Socioeconomic disadvantage is well established as a risk factor for family dysfunction, as reflected in high rates of violence and child neglect (Chapter 8), ineffective parenting, and family breakdown (Hancock, Christensen, & Zubrick, 2018; Johnson, 2012). Conversely, family breakdown is associated with exacerbating economic disadvantage (Hancock et al., 2018; Johnson, 2012).

Economists classify socioeconomic disadvantage into absolute poverty and relative poverty (World Bank, 2015). Absolute poverty is having insufficient income to meet essential needs like food and shelter, and economists operationalize extreme absolute poverty as a person living on less than US $1.90 per day. About 10% of the world's population live in absolute poverty, and—as might be expected—absolute poverty is much more common in low-income than high-income countries, although fortunately, the levels of absolute poverty in most low-income countries are in decline (World Bank, 2015).

Relative poverty is having income so low relative to the rest of the society that it prevents access to resources very widely available within that society, like access to electricity, telecommunication, healthy food, or decent housing (World Bank, 2015). Relative poverty is defined as earning less than half of the median national income (often called the "poverty line"), rates of relative poverty has been increasing in most high-income countries of the world (Organization for Economic Cooperation and Development, 2018), and as relative poverty increases social problems like family breakdown increase (Pikkety, 2014).

Inequality is highly variable between countries and often is operationalized as the percentage of households that fall below the poverty line. Stanley (Chapter 12) suggests inequality in high-income countries reflects broader cultural values, which often create a dialectic between an individualistic focus that emphasizes rewards for individual effort versus a collectivistic focus that emphasizes social inclusion of all in the benefits of society. In the individualistic approach, governments favor low taxes, providing minimal social welfare that imposes stringent requirements for access, and emphasize individuals' responsibility to provide for their own and their family's welfare. The collectivistic approach favors higher taxes, which underwrite certain minimum standards of service to all individuals and families. There are many ways in which these social policies impact on families, such as the costs of education, access to health care, availability of child support payments for low-income families, provisions for maternity and other parental leave from paid employment, and the affordability of childcare.

Stanley (Chapter 12) provides an in-depth case study of social policy influencing urban development in the large Australian city of Melbourne, noting how areas of the city that are socially disadvantaged have poorer infrastructure than the rest of the city, which impact upon family life. For example, housing that is cheaper is often located on the periphery of cities, those locations lacks public transport, which necessitates long commutes on crowded roads to get to work, which reduces the time available to spend with families. Stanley (Chapter 12) notes how social policies framed around individual responsibility can contribute to ineffective social planning of urban development, and how the difficulties to access welfare payments can undermine family functioning.

Work-family life interface

Work-family balance reflects how work and family life reciprocally influence each other, both positively and negatively (Chapter 10). Culture moderates numerous aspects of the work-family interface, such as cultural norms about who will care for children (and who will pay for that care), social policies in place to support families, and work cultures. In low-income countries, the boundaries between family and work-life often blur, parents and extended family most often provide childcare, and in some places, there is a substantial risk that children lack sufficient stimulation to optimize development for a modern world (Neuman & Okeng'o, 2019). High-income countries vary enormously in parental leave provisions ranging from nationally mandated legal minimums of 12 months or more of maternity leave, plus parental leave for either parent (e.g., Finland, Iceland) to no mandated requirements (e.g., United States) (Chapter 10). Similarly, access to affordable and high quality childcare varies greatly between countries. When parental leave and childcare is hard to access, this often results in selective disadvantage for the poor (Chapter 10).

Workplace flexibility (workload, work schedule, and work location flexibility) substantially reduces the negative impacts of work on families and enhances workplace productivity (Hill & Holmes, 2018). However, access to flexibility is highly variable across workplaces, and may be easier to implement in some work settings (e.g., offices) than others (e.g., agriculture).

Health

There is a reciprocal influence between family functioning and the physical and mental health of family members (Chapter 11; Robles, Slatcher, Trombello, & McGinn, 2014). On the one hand, when a family member has serious health problems, this can undermine the quality of relationships between family members, particularly those families that lack effective coping skills to manage the challenges of poor health (Sher & Halford, 2017). For example, living with a family member who has a psychosis is associated with high rates of psychological and physical health problems in other family members, although families' coping skills and support attenuate these effects (Chapter 20).

On the other hand, family functioning affects the physical and mental health of family members (Sher & Halford, 2017). Specifically, family conflict can have direct effects on health through the physiological changes that accompany chronic conflict, and indirect effects by compromising family members' health-related behaviors (e.g., not exercising, drinking too excess) (Robles, Slatcher, Trombello, & McGinn, 2014). Moreover, families affect a range of health-related behaviors (e.g., healthy eating, exercise, stress management), with spouses influencing each other, and parents influencing children and vice versa (Chapter 11).

Culture moderates the association between health and family functioning (Chapter 20). For example, cultural beliefs about the cause of health problems influence whether family members blame the ill person for their health problems, which in turn moderates the extent of support offered to the ill person by family members (Chapter 20). As the second example, culture can influence family members' health-related behaviors. Around the world, women tend to be less physically active than men, but this gender difference is most marked among Arab countries, where culturally based gender separation often precludes female participation in exercise (World Health Organization, 2019).

Catastrophes

War and natural disasters have a huge impact on families, particularly catastrophes that involve large-scale death and injury, and where individuals feel personally threatened (Chapter 15). Catastrophes often destroy a community's resources, take away access to safe territory, and separate family members. There are some cross-cultural consistencies in how humans react to the trauma of catastrophes, with increased anxiety, hypervigilance, sleep disturbance, and avoidance of trauma-related stimuli being common (Chapter 15). At the same time, there are also considerable differences in how different cultures interpret, and believe they should respond to, the experience of the trauma (Chapter 15). For example, mental health professionals in Western countries often interpret trauma symptoms as a maladaptive individual response that should be the focus of treatment. In contrast, in some African countries trauma symptoms are addressed by the mass practice of disassociation as a healing ritual, often based on beliefs that the ritual drives away malevolent forces (Chapter 15).

One challenge faced by families is the reintegration of individuals affected by war or other catastrophes into the family (Bakhurst, Lowe, McGuire, Halford, & Markman, 2017). Often trauma-related symptoms (e.g., hyperarousal, emotional numbing), directly affect how the traumatized person interacts with family members (Bakhurst et al., 2017). Culture moderates how these symptoms affect families. For example, de Jong (Chapter 15) describes how families struggle with reintegration of child soldiers into the family, and how support programs to ease such reintegration need to accommodate cultural beliefs about the cause of trauma-related symptoms.

Migration

Rates of international migration have increased markedly in the first two decades of the 21st century relative to the 20th century (International

Organization for Migration, 2020; United Nations Department of Economic and Social Affairs, 2019). It is estimated that there are about 280 million international migrants in the world (about 3.5% of the world's population), with about 27 million being refugees fleeing war or oppression (International Organization for Migration, 2020). Two-third of international migrants move to high-income countries, and only 5% to low-income countries (International Organization for Migration, 2020).

Most migrants leave countries that have quite different cultures to those in the countries to which they migrate (International Organization for Migration, 2020). Migrant families face multiple challenges in an unfamiliar context including; establishing careers, finding housing, getting education for children, and developing social networks; resolving identification with their culture of origin and with the cultures of their new country of residence; and managing relationships with family left behind (Chapter 13). Within migrant families, individual family members often differ in how they acculturate (Halford et al., 2018b). For example, couples coming to countries with greater gender equality than in their country of origin often differ in how readily each spouse identifies with that gender equality (Halford et al., 2018a, 2018b). Parents used to a culture that includes tight parental control over children's activities often find their children resist such control if that control is not normative in the culture of the new country of residence (Chapter 2). However, individual migrants who marry a member of the majority culture often show rapid and successful acculturation to their new country of residence (Kaya, Halford, Hiew, Sheffield, & van de Vijver, 2019).

How migrants adjust to their new country of residence depends upon the culture of the new country of residence. Acculturation seems to be easier for migrants in countries that have high rates of migration from diverse source countries, rather than destination countries with a large majority drawn from a single culture (van de Vijver, Berry, & Celenk, 2016), and spouses who develop a bicultural identity have higher couple relationship adjustment than other couples (Kaya et al., 2019). Government policy and community acceptance of migrants influence migrant adjustment. There has been some strong antiimmigrant sentiment in some destination countries, notably the United States, Australia, and many countries in Europe (Chapter 14), and these are associated with discrimination against migrants.

Culture and family interventions

Interventions to promote positive family functioning can take many forms. For example, political action can develop international covenants committing governments to guarantee certain standards in family support and functioning, as exemplified in the United Nations (2015) Sustainable Development Goals for 2030 that includes as goal 5 "Achieve gender equality and empower all women and girls." National, state, or local government social and economic policies can enhance family functioning, as exemplified by the many national and state/provincial governments who make financial support payments to poor families that attempt to make sure all children have access to necessities (e.g., clean water, enough food to eat, a safe place to live, and access to education). Other examples include police and courts developing policies to enhance effective responses to family violence, private developers and local authorities collaborating in urban planning to develop family-friendly towns, and employers developing flexible, family-friendly work practices.

Important as all types of family initiatives are, as editors of the current book my late colleague Fons van de Vijver and I focused the practice section of the current book on family psychological interventions. That focus reflects that we saw a large gap in research-based information on how culture moderates the effectiveness of psychological interventions for families. That focus also reflects our expertise as psychologists.

Development of couple and family therapies

Couple and family therapies (CFT) developed beginning in the 1940s and 1950s in high income, Western countries, predominantly in the United States (Chapter 16). The emergence of CFT reflected clinical observations and research that family functioning, and individual family member's health and well-being, reciprocally influenced each other (Chapter 16). Subsequently, research has found that psychological interventions that addressed the family can be very effective, for some presenting problems more effective than just working with individuals (Halford & Pepping, 2019). That family focus is entirely consistent with the arguments made in this Chapter about the centrality of family to the human experience.

Practitioners (and researchers) in non-Western countries began practicing CFT by reading publications, watching videos, and attending training workshops of the pioneers who developed CFT (Chapter 16). In different parts of the world, practitioners modified CFT models and practices to try to make them appropriate to the local culture (Roberts et al., 2014). In many collectivistic cultures, the family system focus of CFT better fitted local cultural norms than individual psychotherapies, but even in those contexts, there were challenges in developing culturally appropriate CFT (Chapter 16).

Cultural variations in presenting problems

Across different cultures, there are variations in how socially acceptable it is to seek psychological therapy in general and CFT in particular. Cultural gatekeepers (e.g., elders, religious leaders) can greatly influence in determining if families will seek help (Chapters 15 and 19; Yu, Roberts, Wong, & Shen, 2011). Perceptions of the cultural knowledge and appropriateness of the professionals offering the service influence acceptability of seeking help. For example, gay and lesbian people report perceptions that professionals do not understand gay and lesbian family issues is a major barrier to seeking CFT (Pepping, Lyons, Halford, Cronon, & Pachankis, 2017).

Once individuals or families do present for help, there is a substantial rate of premature discontinuation of therapy (Swift & Greenberg, 2012). Experienced CFT therapists believe that the predictors of discontinuation vary cross-culturally (Yu et al., 2011). Family therapist working in collectivistic cultures attach importance to engaging family members as a routine part of therapy than those working in individualistic cultures, and also perceive family members as more influential in determining if therapy continues (Yu et al., 2011).

There are some consistencies and some cultural variations in presenting problems for those seeking CFT (Dillon et al., 2015; Miller, Wang, Sandel, & Cho, 2002; Su et al., 2015). The consistencies seem to relate to family functioning relevant to evolutionary fitness. For example, across diverse cultures, couple conflict is often about finances, sharing of roles, communication of affection, sex, and parenting, which is argued to reflect their relevance to sustaining the couple bond and raising offspring (Dillon et al., 2015). However, the frequency of nominating these problems still does vary somewhat cross-culturally. For example, relative to American couples, Taiwanese Chinese couples more frequently report problems in raising children as an issue of couple conflict (Su et al., 2015).

Sometimes variations in rates of presenting problems reflect broader cultural values about how to present problems. In many Eastern cultures, there is a need to avoid loss of face, which might lead to underreporting of problems (Chapter 16), which can be exacerbated by social norms of not disclosing family problems to strangers (Chapter 15). In addition, different behaviors can be valued differently across cultures. For example, self-esteem in children tends to be highly valued in the United States, and CFT therapists frame some parenting interventions around promoting children's self-esteem through praise of achievement and avoidance of criticism (Miller et al., 2002). In contrast, in China child achievement and respect for elders is prioritized, praise sometimes is seen by parents as undermining child striving, and appropriate criticism is seen to enhance striving to achieve (Miller et al., 2002).

Culture also moderates beliefs about the causes of client problems. Beliefs about the causality of psychological problems can emphasize different domains: intra-human causes, as exemplified in psychodynamic individual therapies; interhuman causes, as exemplified in CFT; and suprahuman causes, often based on religious/spiritual beliefs (Chapter 15). CFT therapists often need to incorporate attention to these differing levels of explanation in ensuring culturally appropriate interventions. For example, among religious people, their spouse's religiosity is associated strongly with relationship satisfaction (Perry, 2016), and numerous CFT therapists describe incorporating attention to religiosity into their CFT practice (e.g., Chapter 20; Stanley, Trathen, McCain, & Bryan, 2014). The challenge in such adaptations is retaining the crucial elements of an evidence-based CFT, while adapting to address culturally important goals and beliefs.

Evidence-based approaches to couple and family therapy and culture

There is a substantial evidence base for the effectiveness of a range of CFT therapies within Western cultural contexts. Some of this evidence is reviewed in other chapters of this book (e.g., Chapters 15, 17, 19, and 20). In working with couples, assessment with feedback and skill-based couple relationship education programs enhance relationship satisfaction for many couples (Halford, Pepping, & Petch, 2018). There are well-replicated findings that cognitive behavioral and emotion-focused couple therapies are effective in reducing relationship distress (Halford & Pepping, 2019; Chapter 17). Moreover, enhancing the couple relationship can improve child adjustment (Halford, Rhoades, & Morris, 2018). In addition, variants of cognitive behavioral couple therapy are effective in treatment of a range of individual psychological disorders (Fischer & Baucom, 2018). Family psychoeduation and coping training can reduce conflict and stress within families experiencing a major psychological disorder in a family member (Chapter 20). Furthermore, appropriately adapted couple therapy can assist couples and families with severe physical health problems (Baucom, Porter, Kirby, & Hudepohl, 2012).

Cognitive behavioral approaches to parent training have a well-replicated effect in enhancing child adjustment (Chapter 19; Sanders & Mazzucchelli, 2018), and improving health-related behaviors (Chapter 18). Family therapy is effective in treating adolescents with multiple problems including substance abuse, delinquency, and behavioral conduct problems (van der Pol et al., 2017), eating disorders (Jewell, Blessitt, Stewart, Simic, & Eisler, 2016), the effects of child abuse, and affective disorders (Carr, 2019).

A major limitation of the research on CFT is that the vast majority of studies have evaluated CFT in the Western countries of North America, Western Europe, or Australia (Chapter 16). This reflects that Western countries do the vast majority of all published psychology research, with the United States being by far the largest contributor to psychological research (Arnett, 2008; O'Gorman, Shum, Halford, & Ogilvie, 2012). Across multiple areas of CFT it is unclear if we have a robust general approach to helping human families, or if CFT is only effective within a range of Western cultures.

One area of CFT, parenting programs for young children, has been extensively evaluated cross-culturally. Provided some modest adaptations are made to address cultural considerations, effectiveness is robust cross-culturally (Gardner, Montgomery, & Knerr, 2016). One noteworthy example is the Triple P parenting program, which has been deployed and evaluated in 39 countries, and in 22 languages; spanning low, middle, and high-income countries; and across countries with great cultural diversity (Chapter 19).

Adaptations of couple and family therapy for diverse cultures

In offering CFT across diverse cultures it is important to address who might best be placed to offer services. This involves attention to the local cultural expectations of who should provide what sorts of services, as well as who is available to deliver services. Partnering with cultural leaders to promote accessing services, and or partnering with culturally appropriate agencies to conjointly offer services can be effective (Chapters 15 and 18). For example, in Jordan, local imams were enlisted to encourage fathers to attend parent training programs, and as imams were seen by men as leaders with respected views in the community, this boosted father participation (Al-Hassan & Lansford, 2011). As another example, partnering with Indigenous run organizations to offer services enhances rates of presentation for mental health services by Indigenous Australians (Isaacs, Pyett, Oakley-Browne, Gruis, & Waples-Crowe, 2010).

Middle- and low-income countries often lack mental health professionals who can provide CFT (Chapters 15 and 19). In fact, often there are shortages of professionals in the most socially disadvantaged areas where need is often greatest, particularly in low-income countries

(Parra-Cardona et al., 2018). There are lots of examples where CFT programs have been provided by nonspecialists and proved effective (Healy, Kaiser, & Puffer, 2018). For example, community health workers were trained to provide a psychological treatment program for young children in Uganda, which proved effective (Mutamba et al., 2018), and paraprofessional community members have been trained to provide CFT effectively to traumatized refugees in range of settings (Slobodin & de Jong, 2015).

Most CFT evidence-based programs have been developed in Western countries, and modifying program content to be culturally appropriate is crucial. The *Collaborative Participation Adaptation Model* (CPAM) described by Turner, Sanders, Keown, and Shepherd (2018) explicate a process for consultation and partnership with members of cultural groups to ensure the Triple P parenting program content is culturally relevant. This can involve structured meetings, focus groups, or surveys, to get feedback and guide modification of the existing program. Numerous other writers describe similar processes of community consultation to adapt program content (e.g., Chapters 15 and 18). Sometimes the scope of modifications can be very minor, with only adjustments of phrasing of explanations and examples to make them salient to the culture. Other times, there might be substantial modifications to address cultural needs, such as engaging cultural leaders in the offering of the program, or relating program principles to key cultural ideas. For example, Beach, Fincham, Hurt, McNair, and Stanley (2008) incorporated religious concepts, and the use of prayer, into couple therapy and relationship education for cultural groups where religion is very important. As a second example, Weisman de Mamani et al. (Chapter 20) describe celebrating extended family closeness as part of a CFT approach to supporting Latino families with a member who has a psychotic disorder.

A number of evidence-based CFT programs use a self-regulation focus, meaning that participants consider multiple options in how to address a family problem, and the participants themselves select the solution they wish to implement (e.g., Chapter 19; Halford, Moore, Wilson, Farrugia, & Dyer, 2004). Such a self-regulation focus allows the clients to select what changes feel culturally appropriate for them to implement. This approach has proved helpful in adapting evidence-based CFT programs like the Triple P parenting program (Chapter 19) and Couple CARE couple relationship program (Halford & Pepping, 2019) to be effective in diverse cultural environments.

Conclusions

There are some evolutionarily derived near universals in both the form and function of families, including long-term commitment of romantic partners as a couple, and the commitment to the cooperative raising of children. There is an ongoing movement to establish in law universals of what is desirable in the family establishment and dissolution (e.g., no underage marriage, freedom to marry and divorce), and in family functions (e.g., no violence, protection of children, and gender equality). However, these universalist aspirations are legislated and promoted with differential effectiveness across cultures. There are also considerable cultural variations in gender roles, particularly the rights of women and children, the role of extended family in parenting and the couple relationship, and the role of religion and religious authorities. Some cultural variations are responses to the environment of adaptation, such as the emphasis on physical skills in hunter-gatherer contexts. Other cultural variations represent effects of stickiness of cultures, and their adaptation environment history.

Couple and Family Therapy developed in the Western high-income countries, primarily in the United States. Some evidence-based programs have been adapted for more diverse cultural contexts, primarily by adapting the rationale for programs to be acceptable within local cultures, partnering with culturally appropriate partners in the delivery of programs, and the use of self-regulation frameworks to give families control over exactly how they implement change. The shortage of well-trained family therapists in many parts of the world is a major limitation on access to services.

References

Al-Hassan, S. M., & Lansford, J. E. (2011). Evaluation of the better parenting program in Jordan. *Early Child Development and Care, 181*, 587–598. https://doi.org/10.1080/03004431003654925.

Amato, P. R., & Rogers, S. J. (1997). A longitudinal study of marital problems and subsequent divorce. *Journal of Marriage and Family, 59*, 612–624.

Antfolk, J., Karlsson, L. C., Söderlund, J., & Szala, A. (2017). Willingness to invest in children and psychological kinship estimates and emotional closeness. *Evolutionary Psychology*, (April-June), 1–10. https://doi.org/10.1177/1474704917705730.

Archer, J. (2019). The reality and evolutionary significance of human psychological sex differences. *Biological Reviews, 94*, 1381–1415. http://dx.doi.morg/ 10.111/brv.12507.

Arnett, J. J. (2008). The neglected 95%: Why American psychology needs to become less American. *American Psychologist, 63*(7), 602–614. https://doi.org/10.1037/0003-066X.63.7.602.

Australian Institute of Family Studies. (1997, April 3). *The courier mail mail* (p. 4).

Bakhurst, M., Lowe, B., McGuire, A., Halford, W. K., & Markman, H. J. (2017). Relationship education for military couples. *Family Process, 56*, 302–316. https://doi.org/10.1111/famp.12211.

Baucom, D. H., Porter, L. S., Kirby, J. S., & Hudepohl, J. (2012). Couple-based interventions for medical problems. *Behavior Therapy, 43*(1), 61–76. https://doi.org/10.1016/j.beth.2011.01.008.

Beach, S. R. H., Fincham, F. D., Hurt, T. R., McNair, L. M., & Stanley, S. M. (2008). Prayer and marital interaction: A conceptual framework. *Journal of Social and Clinical Psychology, 27*(7), 641–669. https://doi.org/10.1521/jscp.2008.27.7.641.

Berna, F., Goldberg, P., Horwitz, L. K., Brink, J., Holt, S., Bamford, M., & Chazan, M. (2012). Microstratigraphic evidence of in situ fire in the Acheulian Strata of Wonderwerk Cave, Northern Cape Province, South Africa. *Proceedings of the National Academy of Sciences, 109*, E1215–E1220. https://doi.org/10.1073/pnas.1117620109.

Bernstein, A., & Jones, K. M. (2019). *The economic effects of contraceptive access: A review of evidence.* Washington DC: Center on the Economics of Reproductive Health. https://iwpr.org/wp-content/uploads/2019/09/B381_Contraception-Access_Final.pdf.

Bocquet-Appel, J.-P. (2011). When the world's population took off: The springboard of the Neolithic demographic transition. *Science, 33*(6042), 560–561. https://doi.org/10.1126/science.1208880.

Bornstein, M. H., Putnick, D. L., Cote, L. R., Haynes, O. M., & Suwalsky, J. T. D. (2015). Mother-infant contingent vocalizations in 11 countries. *Psychological Science, 26*, 1272–1284. https://doi.org/10.1177/0956797615586796.

Bornstein, M. H., Putnick, D. L., Lansford, J. E., Deater-Deckard, K., & Bradley, R. H. (2015). A developmental analysis of caregiving modalities across infancy in 38 low- and middle-income countries. *Child Development, 86*, 1571–1587. https://doi.org/10.1111/cdev.12402.

Boyd, R., & Richerson, P. J. (2009). Culture and the evolution of human cooperation. *Philosophical Transactions of the Royal Society, 364*, 3281–3288. https://doi.org/10.1098/rstb.2009.0134.

Browning, M., Chiappori, P. A., & Lewbel, A. (2013). Estimating consumption economies of scale, adult equivalence scales, and household bargaining power. *The Review of Economic Studies, 80*, 1267–1303. https://doi.org/10.1093/restud/rdt019.

Burén, J., & Lunde, C. (2018). Sexting among adolescents: A nuanced and gendered online challenge for young people. *Computers in Human Behavior, 85*, 210–217. https://doi.org/10.1016/j.chb.2018.02.003.

Buss, D. M., & Schmitt, D. P. (2019). Mate preferences and their behavioral manifestations. *Annual Review of Psychology, 70*, 77–110. https://doi.org/10.1146/annurev-psych-010418-103408.

Carr, A. (2019). Family therapy and systemic interventions for child-focused problems: The current evidence base. *Journal of Family Therapy, 41*, 153–213. http://dx.doi.org/10.111/1467-6427.122226.

Choi, K. H., & Tienda, M. (2019). Marriage-market constraints and mate selection behavior: Racial, ethnic and gender differences in intermarraige. *Journal of Marriage and the Family, 79*(2), 301–317. http://dx.doi.org/ 10.111/jomf.12346.

Colage, I., & d'Errico, F. (2018). Culture: The driving force of human cognition. *Topics in Cognitive Science, 2018*, 1–19. http://dx.doi.org/ 10.111/tops.12372.

Colleran, H. (2020). Market integration reduces kin density on women's ego networks in rural Poland. *Nature Communications, 11*, 266. https://doi.org/10.1038/s41467-019-14158-2.

Coontz, S. (2005). *Marriage, a history: From obedience to intimacy, or how love conquered marriage.* New York: Viking Press.

Creanza, N., Kolodny, O., & Feldman, M. W. (2017). Cultural evolutionary theory: How culture evolves and why it matters. *PNAS, 114*, 7782–7789. https://doi.org/10.1073/pnas.1620732114.

de Graaf, P. M., & Kalmijn, M. (2006). Divorce motives in a period of rising divorce: Evidence from a Dutch life-history survey. *Journal of Family Issues*, *27*, 483–505.

Department of Statistics Malaysia. (2016). *Current population estimates, Malaysia* (pp. 2014–2016). https://www.statistics.gov.my/index.php?r=column/cthemeByCat&cat=155&bul_id=OWlxdEVoYlJCS0hUZzJyRUcvZEYxZz09&menu_id=L0pheU43NWJwRWVSZklWdzQ4TlhUUT09.

Derex, M., Beugin, M.-P., Godelle, B., & Raymond, M. (2013). Experimental evidence for the influence of group size on cultural complexity. *Nature*, *503*, 389–391.

Di Giulio, P., Impicciatore, R., & Sironi, M. (2019). The changing pattern of cohabitation: A sequence analysis approach. *Demographic Research*, *40*, 1211–1248. https://doi.org/10.4054/DemRes.2019.40.42.

Diamond, J. (2014). *The world until yesterday: What can we learn from traditional societies?*. New York: Allen Lane.

Dillon, L. M., Nowak, N., Weisfeld, G. E., Weisfeld, C. C., Shattuk, K. S., Imamaglu, O. E., … Shen, J. (2015). Sources of marital conflict in five cultures. *Evolutionary Psychology*, *13*(1), 1–15. https://doi.org/10.1177/147470491501300101.

Dobewell, H., & Strack, M. (2014). Relationship of Inglehart's and Schwartz's value dimensions revisited. *International Journal of Psychology*, *49*(4), 240–248. https://doi.org/10.1002/ijop.12004.

Dunbar, G. R., Lewbel, A., & Pendakur, K. (2013). Children's resources in collective households: Identification, estimation, and an application to child poverty in Malawi. *American Economic Review*, *103*, 438–471. https://doi.org/10.1257/aer.103.1.438.

Ellemers, N., & Haslam, S. A. (2012). Social identity theory. In P. A. M. Van Lange, A. W. Kruglanski, & E. T. Higgins (Eds.), *Handbook of theories of social psychology: Volume 2* (pp. 379–398). London: Sage. https://doi.org/10.4135/9781446249222.n45.

Ellis, B. J., Del Giudice, M., Dishion, T. J., Figueredo, A. J., Gray, P., Griskevicus, V., … Wilson, D. S. (2012). The evolutionary basis of risky adolescent behaviour: Implications for science, policy, and practice. *Developmental Psychology*, *48*(3), 598–623. https://doi.org/10.1037/a0026220.

Equaldex. (2020). *Status of lesbian gay bisexual and transexual laws by country*. https://www.equaldex.com/.

Euler, H. A. (2020). Gender differences in attachment from an evolutionary perspective: What is good for the goose may not be good for the gander. *Attachment & Human Development*, *22*(1), 4–8. https://doi.org/10.1080/14616734.2019.1589055.

Finkel, E. J., Eastwick, P., Karney, B. R., Reis, H. T., & Sprecher, S. (2012). Online dating: A critical analysis from the perspective of psychological science. *Psychological Science in the Public Interest*, *13*(1), 3–66. https://doi.org/10.1077/1529100612436522.

Finkel, E. J., Hui, C. M., Carswell, K. L., & Larson, G. M. (2014). The suffocation of marriage: Climbing Mount Maslow without enough oxygen. *Psychological Inquiry*, *25*(1), 1–41. https://doi.org/10.1080/1047840X.2014.863723.

Fischer, M. S., & Baucom, D. H. (2018). Couple-based interventions for relationship distress and psychopathology. In J. N. Butcher, J. Hooley, & P. C. Kendall (Eds.), *Psychopathology: Understanding, assessing and treating adult mental disorders* (pp. 661–686). Washington, DC: American Psychological Association.

Flannery, K. (2012). *The creation of inequality: How our prehistoric ancestors set the stage for Monarchy, Slavery, and Empire*. Cambridge, MA: Harvard University Press.

Fletcher, G. O., Simpson, J. O., Campbell, L., & Overall, N. C. (2015). Pair-bonding, romantic love, and evolution: The curious case of *Homo sapiens*. *Perspectives on Psychological Science*, *10*, 20–36. https://doi.org/10.1177/1745691614561683.

Fung, J., Wong, M. S., & Park, H. (2018). Cultural backgrond and religious beliefs. In M. R. Sanders & A. Morawska (Eds.), *Handbook of parenting and child development across the lifespan* (pp. 469–494). New York: Springer.

Gardner, F., Montgomery, P., & Knerr, W. (2016). Transporting evidence-based parenting programs for child problem behavior (age 3-10) between countries: Systematic review and meta-analysis. *Journal of Clinical Child and Adolescent Psychology, 45*(6), 749–762. https://doi.org/10.1080/15374416.2015.1015134.

Gelfand, M. J., Raver, J. L., Nishii, L., Leslie, L. M., Lun, J., Lim, B. C., ... Yamaguchi, S. (2011). Differences between tight and loose cultures: A 33-nation study. *Science, 332*, 1100–1104. http://dx.doi.org/ science.1197754.

Ghimire, D. J., Axinn, W. G., Yabiku, S. T., & Thornton, A. (2006). Social change, premarital nonfamily experience, and spouse shoice in an arranged marriage society. *American Journal of Sociology, 111*(4), 1181–1218. https://doi.org/10.1086/498468.

Global Initiative to End All Corporal Punishment of Children. (2020). *Global report 2019: Progress toward ending corporal punishment of children*. http://endcorporalpunishment.org/wp-content/uploads/global/Global-report-2019.pdf.

Goldberg, A. (Ed.), (2016). 1-3. *The SAGE encyclopedia of LGBTQ studies*. Thousand Oaks, CA: SAGE Publications Ltd. https://doi.org/10.4135/9781483371283.

Grabowski, M. W. (2013). Hominin obstetrics and the evolution of constraints. *Evolutionary Biology, 40*, 57–75. https://doi.org/10.1007/s11692-012-9174-7.

Gray, P. B. (2013). Father Darwin: Effects of children on men, viewed form an evolutionary perspective. *Fathering: A Journal of Theory, Research, & Practice About Men as Fathers, 12*(2), 121–142.

Gray, P. B., Garcia, J. R., & Gesselman, A. N. (2019). Age-related patterns in sexual behaviors and attitudes among single U.S. Adults: An evolutionary approach. *Evolutionary Behavioral Sciences, 13*(2), 111–126. https://doi.org/10.1037/ebs0000126.

Halford, W. K., Hahlweg, K., & Dunne, M. (1990). Cross-cultural consistency of marital communication associated with marital distress. *Journal of Marriage and the Family, 52*, 487–500.

Halford, W. K., Leung, P. W. L., Hyng-Chueng, C., Chau-Wan, L., Hiew, D., & Van de Vijver, F. J. R. (2018a). Relationship standards and relationship satisfaction in Chinese, Western, and Intercultural couples living in Australia and Hong Kong. *Couple and Family Psychology: Research and Practice*, 7, 127–142. https://doi.org/10.1037/cfp0000104.

Halford, W. K., Leung, P. W. L., Hyng-Chueng, C., Chau-Wan, L., Hiew, D., & Van de Vijver, F. J. R. (2018b). Couple relationship standards and migration: Comparing Hong Kong Chinese with Australian Chinese. *Family Process, 57*(4), 996–1011. http://dx.doi.org/10.111/famp.12337.

Halford, W. K., Moore, E., Wilson, K. L., Farrugia, C., & Dyer, C. (2004). Benefits of a flexible delivery relationship education: An evaluation of the Couple CARE program. *Family Relations, 53*, 469–476. https://doi.org/10.1111/j.0197-6664.2004.00055.x.

Halford, W. K., & Pepping, C. A. (2019). What every therapist needs to know about couple therapy. *Behavior Change, 36*(3), 121–142. https://doi.org/10.1017/bec.2019.12.

Halford, W. K., Pepping, C., & Petch, J. (2018). Promoting healthy relationships. In A. Vangelisti & D. Pearlman (Eds.), *Cambridge handbook of personal relationships* (pp. 511–522). Cambridge, UK: Cambridge University Press.

Halford, W. K., Rhoades, G. L., & Morris, M. (2018). Effects of the parents' relationship on children. In A. Morawska & M. R. Sanders (Eds.), *Handbook of parenting and child development across the lifespan* (pp. 97–120). New York: Springer.

Hall, A. R. (1980). *Philosophers at war: The quarrel between Newton and Gottfried Leibniz*. Cambridge: Cambridge University Press.

Hamilton, W. (1964). The genetical evolution of social behaviour I. *Journal of Theoretical Biology*, 7(1), 1–16. https://doi.org/10.1016/0022-5193(64)90038-4.

Hancock, K. J., Christensen, D., & Zubrick, S. (2018). Development and assessment of cumulative risk measures of family environments and parental investments in a longitudinal study of Australian children. *Social Indicators Research, 137*, 665–694. https://doi.org/10.1007/s11205-017-1607-3.

Hayes, B. E., & Boyd, K. A. (2017). Influence of individual-and national-level factors on attitudes toward intimate partner violence. *Sociological Perspectives*, *60*, 685–701. https://doi.org/10.1177/0731121416662028.

Healy, E. A., Kaiser, B. N., & Puffer, E. S. (2018). Family-based youth mental health interventions delivered by nonspecialist providers in low- and middle-income countries: A systematic review. *Families, Systems & Health*, *36*(2), 182–197. https://doi.org/10.1037/fsh0000334.

Heinrich, J. (2016). *The secret of our success: How culture is driving human evolution, domesticating our species, and making us smarter*. Princeton, NJ, United States: Princeton University Press.

Hewitt, B., & Baxter, J. (2015). Relationship dissolution. In G. Heard & D. Arunachalam (Eds.), *Family formation in 21st century Australia* (pp. 77–99). Dordrecht: Springer. https://doi.org/10.1007/978-94-017-9279-0_5.

Hiew, D. N., Halford, W. K., van de Vijver, F. J. R., & Liu, S. (2015a). Relationship standards and satisfaction in Chinese, Western, and intercultural Chinese-Western Couples in Australia. *Journal of Cross-Cultural Psychology*, *46*(5), 684–701. https://doi.org/10.1177/0022022115579936.

Hiew, D. N., Halford, W. K., van de Vijver, F. J. R., & Liu, S. (2015b). Communication and relationship satisfaction in Chinese, Western, and intercultural Chinese-Western couples. *Journal of Family Psychology*, *30*(20), 193–202. https://doi.org/10.1037/fam0000144.

Hill, E. J., & Holmes, E. K. (2018). Families and workplaces. In B. H. Fiese, M. Celano, K. Deater-Deckard, E. N. Jouriles, & M. A. Whisman (Eds.), *Applications and broad impact of family psychology*: Volume 2. American Psychological Association handbook of contemporary family psychology (pp. 379–395). Washington, DC: American Psychological Association.

Hrdy, S. B. (2016). Of marmosets, men, and the transformative power of babies. In W. R. Trevanthan & K. R. Rosenberg (Eds.), *Costly and cute: Helpless infants and human evolution* (pp. 177–203). Albuquerque, NM: University of New Mexico Press.

Hyde, J. S. (2014). Gender similarities and differences. *Annual Review of Psychology*, *65*, 373–398. https://doi.org/10.1146/annurev-psych.010213-115057.

Inglehart, R., & Baker, W. E. (2000). Modernization, cultural change, and the persistence of traditional values. *American Sociological Review*, *65*(1), 19–51. https://doi.org/10.2307/2657288.

International Organization for Migration. (2020). *World migration report 2020*. Geneva, Switzerland: International Organization for Migration. https://publications.iom.int/books/world-migration-report-2020?language=en.

Isaacs, A. N., Pyett, P., Oakley-Browne, M. A., Gruis, H., & Waples-Crowe, P. (2010). Barriers and facilitators to the utilization of adult mental health services by Australia's Indigenous people: Seeking a way forward. *International Journal of Mental Health Nursing*, *19*(2), 75–82. https://doi.org/10.1111/j.1447-0349.2009.00647.x.

Isler, K., & van Schaik, C. P. (2012a). Allomaternal care, life history, and brain size evolution in mammals. *Journal of Human Evolution*, *63*, 52–63.

Isler, K., & van Schaik, C. P. (2012b). How our ancestors broke through the gray ceiling: Comparative evidence for cooperative breeding in early Homo. *Current Anthropology*, *53*, S453–S465. https://doi.org/10.1086/667623.

Jewell, T., Blessitt, E., Stewart, C., Simic, M., & Eisler, I. (2016). Family therapy for child and adolescent eating disorder: A critical review. *Family Process*, *55*(3), 577–594. https://doi.org/10.1111/famp.12242.

Johnson, M. D. (2012). Healthy marriage initiatives: On the need for empiricism in policy implementation. *American Psychologist*, *67*(4), 296–308. https://doi.org/10.1037/a0022743.

Johnson, A. W., & Earle, T. K. (2000). *The evolution of human societies: From foraging group to Agrarian State*. Stanford, CA: Stanford University Press.

Kalmijn, M., & Poortman, A. (2006). His or her divorce? The gendered nature of divorce and its determinants. *European Sociological Review*, *22*, 201–214.

Kamal, S. M. M., Hassan, C. H., Alam, G. M., & Ying, Y. (2015). Child marriage in Bangladesh: Trends and determinants. *J. Biosocial Sci.*, *47*(1), 120–139. https://doi.org/10.10107/S0021932013000746.

Kaplan, H., Hill, K., Lancaster, A., & Hurtado, A. M. (2000). A theory of human life history evolution: Diet, intelligence, and longevity. *Evolutionary Anthropology*, *9*, 156–185.

Kashima, Y., Bain, P. G., & Perfors, A. (2019). The psychology of cultural dynamics: What is it, what do we know, and what is yet to be known? *Annual Review of Psychology*, *70*, 499–529. https://doi.org/10.1146/annurev-psych-010418-103112.

Kaya, M., Halford, W. K., Hiew, D. N., Sheffield, J., & van de Vijver, F. J. R. (2019). Ethnic identification and relationship satisfaction in Chinese, Western and intercultural Chinese-Western couples. *Couple and Family Psychology*, *8*(3), 121–136. https://doi.org/10.1037/cfp0000120.

Keller, H. (2017). Culture and development: A systematic relationship. *Perspectives on Psychological Science*, *12*(5), 833–840. https://doi.org/10.1177/1745691617704097.

Lagenhof, M. R., & Komdeur, J. (2018). Why and how early-life environment affects development of coping behaviors. *Behavioral Ecology and Sociobiology*, *72*(34), https://doi.org/10.1007/s00265-018-2452-3.

Lancaster, G., Maitra, P., & Ray, R. (2006). Endogenous intra-household balance of power and its impact on expenditure patterns: Evidence from India. *Economica*, *73*, 435–460. https://doi.org/10.1111/j.1468-0335.2006.00502.x.

Lansford, J. E., Chang, L., Dodge, K. A., Malone, P. S., Oburu, P., Palmérus, K., … Quinn, N. (2005). Physical discipline and children's adjustment: Cultural normativeness as a moderator. *Child Development*, *76*, 1234–1246. https://doi.org/10.1111/j.1467-8624.2005.00847.x.

Lansford, J. E., Godwin, J., Alampay, L. P., Uribe Tirado, L. M., Zelli, A., Al-Hassan, S. M., … Tapanya, S. (2016). Mothers', fathers', and children's perceptions of parents' expectations about children's family obligations in nine countries. *International Journal of Psychology*, *51*, 366–374. https://doi.org/10.1002/ijop.12185.

Lansford, J. E., Godwin, J., Zelli, A., Al-Hassan, S. M., Bacchini, D., Bombi, A. S., … Uribe Tirado, L. M. (2018). Longitudinal associations between parenting and youth adjustment in twelve cultural groups: Cultural normativeness of parenting as a moderator. *Developmental Psychology*, *54*, 362–377. https://doi.org/10.1037/dev0000416.

Lee, K.-F. (2018). *AI superpowers: China, Silicon Valley and the New World Order*. Boston: Houghton Mifflin Harcourt.

LeVine, R. A., LeVine, S., Dixon, S., Richman, A., Leiderman, P. H., Keefer, C. H., & Brazelton, T. B. (1996). *Child care and culture: Lessons from Africa*. New York, NY: Cambridge University Press.

Lin, I.-F., Brown, S. L., & Cupka, C. J. (2017). A national portrait of stepfamilies in later life. *Journals of Gerontology: Social Sciences*, *73*(6), 1043–1054. https://doi.org/10.1093/geronb/gbx150.

Lippa, R. A. (2009). Sex differences in sex drive, sociosexuality, and height across 53 nations: Testing evolutionary and social structural theories. *Archives of Sexual Behavior*, *38*(5), 631–651.

Livingstone, G. (2014). Four in ten couples are saying "I do" again: Growing numbers of adults have remarried. Pew Social Center Report. http://www.pewsocial trends.org/2014/11/14/four-in-ten-couples are saying-I-do again/#.

Malaysian Government. (2010). *Federal constitution*. http://www.agc.gov.my/agcportal/uploads/files/Publications/FC/Federal%20Consti%20(BI%20text).pdf.

McMahon, H. G. (2018). *An investigation of sexting across the course of relationships: From dating, to couple relationships and revenge porn.* Doctoral Thesis. Brisbane Australia: University of Queensland. https://doi.org/10.1426/uql.2019.126.

Memon, N. S., Jamil, S., & Khan, S. (2019). Global inequalities: An overview. *International Journal of Academic Research Business and Social Sciences*, *9*(5), 214–231. https://doi.org/10.6007/IJARBSS/v9-i5/5851.

Miller, P. J., Wang, S., Sandel, T., & Cho, G. E. (2002). Self-esteem as folk theory: A comparison of European American and Taiwanese mothers' beliefs. *Parenting: Science and Practice*, *2*, 209–239. https://doi.org/10.1207/S15327922PAR0203_02.

Mounier, A., & Lahr, M. (2019). Deciphering African late middle Pleistocene hominin diversity and the origin of our species. *Nature Communications*, *10*(1), 3406. https://doi.org/10.1038/s41467-019-11213-w.

Mutamba, B. B., Kane, J. C., de Jong, J.T.V.M., Okello, J., Musisi, S., & Kohrt, B. A. (2018). Psychological treatments delivered by community health workers in low-resource government health systems: Effectiveness of group interpersonal psychotherapy for caregivers of children affected by nodding syndrome in Uganda. *Psychological Medicine*, *48*(15), 2573–2583. https://doi.org/10.1017/S0033291718000193.

Muthukrishna, M., & Henrich, J. (2016). Innovation in the collective brain. *Philosophical Transactions of the Royal Society, B: Biological Sciences*, *371*(1690), 2015.0192. https://doi.org/10.1098/rstb.2015.0192.

Muthukrishna, M., Shulman, B. W., Vasilescu, V., & Henrich, J. (2014). Sociality influences cultural complexity. *Proceedings of the Royal Society B: Biological Sciences*, *281*(1774), 20132511. https://doi.org/10.1098/rspb.2013.2511.

Nelson, C. (2014). *The age of radiance: The epic rise and dramatic fall of the atomic era*. New York: Simon & Schuster.

Neuman, M. J., & Okeng'o, L. (2019). Early childhood policies in low and middle-income countries. *Early Years*, *39*(3), 223–228. https://doi.org/10.1080/09575146.2019.1636571.

Nguyen, M. C., & Wodon, Q. (2015). Global and regional trends in child marriage. *The Review of Faith & International Affairs*, *13*(3), 6–11. https://doi.org/10.1080/15570274.2015.1075756.

Nielsen, M. (2018). The social glue of cumulative culture and ritual behavior. *Child Development Perspectives*, *12*(4), 264–268. https://doi.org/10.1111.cdep.12297.

Norenzayan, A., Shariff, A. F., Gervais, W. M., Willard, A. K., McNamara, R. A., Slingerland, E., & Henrich, J. (2016). The cultural evolution of prosocial religions. *Behavioral and Brain Sciences*, https://doi.org/10.1017/S0140525X14001356.

O'Gorman, J., Shum, D., Halford, W. K., & Ogilvie, J. (2012). The state of psychological science: World trends in national research output and impact. *International Perspectives in Psychology*, *1*(4), 268–283. https://doi.org/10.1037/a0030520.

Office of the High Commissioner on Human Rights of the United Nations. (1989). *Convention on the rights of the child*. https://www.ohchr.org/en/professionalinterest/pages/crc.aspx.

Organization for Economic Cooperation and Development. (OECD). (2018). *An overview of growing income inequalities in OECD countries: Main findings*. http://www.oecd.org/social/inequality-and-poverty.htm.

Parra-Cardona, R., Leijten, P., Lachman, J. M., Mejía, A., Baumann, A. A., Amador Buenabad, N. G., … Domenech Rodríguez, M. M. (2018). Strengthening a culture of prevention in low- and middle-income countries: Balancing scientific expectations and contextual realities. *Prevention Science*, 1–11. https://doi.org/10.1007/s11121-018-0935-0.

Pepping, C. A., Lyons, A., Halford, W. K., Cronon, T. J., & Pachankis, J. E. (2017). Couple interventions for same sex couples: A consumer survey. *Couple and Family Psychology: Research and Practice*, *6*, 258–273. https://doi.org/10.1037/cfp0000092.

Perry, S. L. (2016). Perceived spousal religiosity and marital quality across racial and ethnic groups. *Family Relations*, *65*(2), 327–341. https://doi.org/10.1111/fare.12192.

Petersen, J. L., & Hyde, J. S. (2010). A meta-analytic review of research on gender differences in sexuality, 1993-2007. *Psychological Bulletin*, *136*(1), 21–38. https://doi.org/10.1037/a0017504.

Pikkety, T. (2014). *Capital in the twenty-first century*. Cambride, Massachusetts, USA: Harvard University Press.

Rappleyea, D. L., Taylor, A. C., & Fang, X. (2014). Gender differences and communication technology use among emerging adults in the initiation of dating relationships. *Marriage & Family Review*, *50*(3), 269–284. https://doi.org/10.1080/01494929.2013.879552.

Rehman, U. S., & Holtzworth-Munroe, A. (2006). A cross-cultural analysis of the demand-withdraw marital interaction: Observing couples from a developing country. *Journal of Consulting and Clinical Psychology*, *74*(4), 755–766. https://doi.org/10.1037/0022-006X.74.4.755.

Rehman, U. S., & Holtzworth-Munroe, A. (2007). A cross-cultural examination of the relation of marital communication behavior to marital satisfaction. *Journal of Family Psychology*, *21*(4), 759–763. https://doi.org/10.1037/0893-3200.21.4.759.

Roberts, J., Abu-Baker, K., Fernandez, C. D., Garcia, N. C., Fredman, G., Kayma, H., ... Vega, R. Z. (2014). Up close: Family therapy challenges and innovations around the world. *Family Process*, *53*, 544–576. https://doi.org/10.1111/famp.12093.

Robles, T. F., Slatcher, R. B., Trombello, J. M., & McGinn, M. M. (2014). Marital quality and health: A meta-analytic review. *Psychological Bulletin*, *140*(1), 140–187. https://doi.org/10.1037/a0031859.

Rosenfeld, M. J., Thomas, R. J., & Hausen, S. (2019). Disintermediating your friends: How online dating in the United States displaces other ways of meeting. *PNAS*, *116*(36), 17753–117758. https://doi.org/10.1073/pnas.1908630116.

Rubalcava, L., Teruel, G., & Thomas, D. (2002). Welfare design, women's empowerment and income pooling. In *Documento de Trabajo Numero 244. CIDE (Centro de Investigatión y Docencia Económicas)*. http://aleph.academica.mx/jspui/bitstream/56789/3833/1/DOCT2065350.PDF.

Sanders, M. R., & Mazzucchelli, T. G. (2018). How parenting influences the lives of children. In M. R. Sanders & T. G. Mazzucchelli (Eds.), *The power of positive parenting: Transforming the lives of children, parents, and communities using the triple P system* (pp. 5–31). New York: Oxford University Press.

Sanri, C., & Goodwin, R. (2014). The influence of globalization and technological development on intimate relationships. In C. R. Agnew (Ed.), *Social influences on romantic relationships* (pp. 11–32). Cambridge: Cambridge University Press.

Sassler, S. (2010). Partnering across the life course: Sex, relationships, and mate selection. *Journal of Marriage and Family*, *72*(3), 557–575. https://doi.org/10.1111/j.1741-3737.2010.00718.x.

Satherwaite, D., McGranahan, G., & Tacoli, C. (2010). Urbanization and its implications for food and farming. *Philosophical Transactions of the Royal Society of London B Biological Sciences*, *265*(1554), 2809–2820. https://doi.org/10.1098/rstb.2010.0136.

Schlebusch, C. M., Malström, H., Günther, T., Sjödin, P., Couthino, A., Edlund, H., ... Jakobsson, M. (2017). Southern African ancient genomes estimate modern human divergence to 350,000 to 260,000 years ago. *Science*, *358*(6363), 652–655. https://doi.org/10.1126/science.aao6266.

Sher, T. G., & Halford, W. K. (2017). Couples and health. In J. Fitgerald (Ed.), *Foundations for couples therapy: Research for the real world* (pp. 176–185). Oxford: Taylor and Frances.

Sherwood, C. C., & Gomez-Robles, A. (2017). Brain plasticity and human evolution. *Annual Review of Anthropology*, *46*, 399–419. https://doi.org/10.1146/annurev-anthro-102215-100009.

Slobodin, O., & de Jong, J.T.V.M. (2015). Family interventions in traumatized immigrants and refugees: A systematic review. *Transcultural Psychiatry*, *52*(6), 723–742. https://doi.org/10.1177/1363461515588855.

Slotten, R. A. (2004). *The heretic in Darwin's court: The life of Alfred Russel Wallace*. New York: Columbia University Press.

Smith, A., & Anderson, M. (2015). 5 Facts about online dating. Pew Research Center. https://www.pewresearch.org/fact-tank/2020/02/06/10-facts-about-americans-and-online-dating/.

Stanley, S. M., Trathen, D., McCain, S., & Bryan, M. (2014). *A lasting promise* (2nd ed.). New York: Jossey-Bass.

Su, L. P., Miller, R. B., Canlas, J. M., Tsui-Chan, L., Hsiao, Y.-L., & Willoughby, B. J. (2015). A cross-cultural study of perceived marital problems in Taiwan and the United States. *Contemporary Family Therapy*, *37*, 165–175. https://doi.org/10.1007/s10591-015-9337-7.

Suddendorf, T. (2013). *The gap: The science of what separates us from other animals*. New York: Basic Books.

Swift, J. K., & Greenberg, R. P. (2012). Premature discontinuation in adult psychotherapy: A meta-analysis. *Journal of Consulting and Clinical Psychology*, *80*(4), 547–559. https://doi.org/10.1037/a0028226.

Tajfel, H. (1979). Individuals and groups in social psychology. *British Journal of Social & Clinical Psychology*, *18*, 183–190. https://doi.org/10.1111/j.2044-8260.1979.tb00324.x.

Therborn, G. (2014). Family systems of the world: Are they converging? In J. Treas, J. Scott, & M. Richards (Eds.), *The Wiley Blackwell Companion to the sociology of families* (pp. 3–19). Chichester, UK: John Wiley & Sons, Ltd.

Thornton, A., & Young-DeMarco, L. (2001). Four decades of trends in attitudes toward family issues in the United States: The 1960s through the 1990s. *Journal of Marriage and the Family*, *63*(4), 1009–1038. http://dx.doi.org/ 10.111/j.1741-3737/2001.01009.x.

Trivers, R. L. (1974). Parent-offspring conflict. *American Zoologist*, *14*, 249–264. https://doi.org/10.1093/icb/14.1.249.

Tsuen-Hsuin, T., & Needham, J. (1985). *Science and civilisation in China 5 Part 1*. Cambridge: Cambridge University Press.

Turner, K. M. T., Sanders, M. R., Keown, L. J., & Shepherd, M. (2018). A collaborative participation adaptation model. In M. R. Sanders & T. G. Mazzucchelli (Eds.), *The power of positive parenting: transforming the lives of children, parents, and communities using the triple P system* (pp. 310–320). New York: Oxford University Press.

Twamley, K. (2013). The globalisation of love? Examining narratives of intimacy and marriage among middle-class Gujarati Indians in the UK and India. *Families, Relationships and Societies*, *2*(2), 267–283. https://doi.org/10.1332/204674313X664923.

Tyrer, L. (1999). Introduction of the pill and its impact. *Contraception*, *59*(1 Supplement), 11S–16S. https://doi.org/10.1016/s0010-7824(98)00131-0.

United Nations. (2015). *2030 agenda for sustainable development*. https://sustainabledevelopment.un.org/post2015/transformingourworld.

United Nations Department of Economic and Social Affairs. (2019). *Population facts. 2019/4*. https://www.un.org/en/development/desa/population/migration/publications/populationfacts/docs/MigrationStock2019_PopFacts_2019-04.pdf.

United Nations Department of Economic and Social Affairs, Population Division. (2020). *World marriage patterns 2020*. https://www.un.org/en/development/desa/population/publications/pdf/family/marriage-patterns.pdf.

United Nations Office on Drugs and Crime. (2019). *Global study on homicide*. https://www.unodc.org/unodc/en/data-and-analysis/global-study-on-homicide.html.

van de Vijver, F. J. R., Berry, J. W., & Celenk, O. (2016). Assessment of acculturation. In D. L. Sam & J. W. Berry (Eds.), *The Cambridge handbook of acculturation psychology* (pp. 293–314). Cambridge, United Kingdom: Cambridge University Press.

van der Pol, T. M., Hoeve, M., Noom, M. J., Stams, G.J.J.M., Doreleijers, T. A. H., van Domburgh, M., & Vermeiren, R.R.J.M. (2017). Research review: The effectiveness of multidimensional family therapy in treating adolescents with multiple behaviour problems: A meta-analysis. *Journal of Child Psychology and Psychiatry, and Allied Disciplines*, *58*(5), 532–545. https://doi.org/10.1111/jcpp.12685.

van Essen, D. C. (2018). Scaling of human brain size: Higher cognitive regions are preferentially expanded in individuals with large brains. *Science*, *360*(6394), 1184–1185. https://doi.org/10.1126/science.aat8948.

von Hippel, W. (2018). *The social leap: How and why humans connect.* Melbourne, Australia: Scribe Publications.

Waite, L. J., & Gallagher, M. (2000). *The case for marriage: Why married people are happier, healthier, and better off financially*. New York: Doubleday.

Wang, C.-T.L., & Schofer, E. (2018). Coming out of the penumbras: World culture and cross-national variation in divorce rates. *Social Forces, 97*(2), 675–704. https://doi.org/10.1093/sf/soy070.

World Bank. (2015). *Poverty*. https://www.worldbank.org/en/topic/poverty/overview.

World Health Organization. (2019). *Global action plan on physical activity 2018–2030: More active people for a healthier world*. Geneva: World Health Organization Press. https://www.who.int/ncds/prevention/physical-activity/global-action-plan-2018-2030/en/.

World Population Review. (2020). *Countries where same sex marriage is legal*. http://worldpopulationreview.com/countries/countries-where-gay-marriage-is-legal/.

Yu, J., Roberts, M., Wong, M., & Shen, Y. (2011). Acceptability of behavioral family therapy among caregivers in China. *Journal of Child and Family Studies, 20*(3), 272–278. https://doi.org/10.1007/s10826-010-9388-1.

Index

Note: Page numbers followed by *f* indicate figures and *t* indicate tables.

F

G

Q

R

T

U

Made in the USA
Monee, IL
13 September 2026

62977008R20422